Introduction to Fuel Cells

San Ping Jiang · Qingfeng Li

Introduction to Fuel Cells

Electrochemistry and Materials

Springer

San Ping Jiang
WA School of Mines: Minerals, Energy and Chemical Engineering
Curtin University
Perth, WA, Australia

Qingfeng Li
Department of Energy Conversion and Storage
Technical University of Denmark
Lyngby, Denmark

ISBN 978-981-10-7625-1 ISBN 978-981-10-7626-8 (eBook)
https://doi.org/10.1007/978-981-10-7626-8

This Springer imprint is published by the registered company Springer Nature Singapore Pte Ltd.
The registered company address is: 152 Beach Road, #21-01/04 Gateway East, Singapore 189721, Singapore

Preface

Fuel cells were initially developed for the space mission programs in 1960s and have today well demonstrated for automobiles, portable electronics, and large-scale power generation plants. The development has appreciably endorsed by the ever-growing concern of fossil fuel depleting and global warming, and in the last years by the potential synergies with renewable energies and the future hydrogen economy.

There are quite a number of fuel cell books available in the market. A majority of the books are technically orientated, aimed at updating the detailed progress from material science, kinetic mechanisms, characterization methodologies to technological integration of membrane-electrode assemblies, stacks, fuel processors, and system demonstration. These types of books are interdisciplinary including thematic monographs, symposium proceedings, and a variety of comprehensive handbooks. They are written by fuel cell experts and scientists and suitable for the reader of researchers, professional engineers, and other fuel cell workers in the field.

As the interest in fuel cells diffuses well beyond the scientific and technical community, the fuel cell topics is of daily discussion in news media and popular teaching courses in university campuses. The fundamental principles of fuel cells are covered by several well-written textbooks. This type of books is introducing thermodynamics, basic electrochemistry, functional materials, components, performance, and applications. They are suitable for the reader of varied engineering backgrounds to learn what fuel cells are, how fuel cells work, and why they offer the high efficiency, zero emission, and potential sustainability.

There is a need for fuel cell textbooks to fill up the gap in between. One, for example, outlines the field at a fundamental thermodynamics, electrochemistry, and material level that can be understood by those who are new in the field with a general engineering background, while the content is advanced enough to provide essential knowledge of technical and material insights for those who are doing or wish to do research in the field. This is what the authors of the present book have attempted to do.

Both authors of the book have been doing research and teaching fuel cell courses for more than thirty years. A difficulty encountered in teaching is the lack of examples and problems in fuel cells, which are essential to bridge between the basic knowledge and skill to handle practical problems for students and engineers. It is therefore attempted to include illustrative and practical examples and problems through the chapters of this book. From numerous examples and referecenes, readers can also find detailed information in the experimental design and techniques commonly used in the studies of fuel cells.

The book is organized in four parts. Part I of the book is a brief introduction of fundamentals, though in a practical way, of fuel cells including thermodynamics, electrochemistry, and fueling scenarios. The other three parts of the book are devoted to fuel cell technologies. Of types of the technologies, emphasis is placed on polymer electrolyte membrane fuel cells (Part II) and solid oxide fuel cells (Part III), each covering principle and materials, reaction, characterization, microstructure, and fabrication. Many examples of procedures and protocols are from our own lab practice while well selected information from literature is also provided. The last part of the book presents the rest types of fuel cells, i.e., alkaline fuel cells, phosphoric acid fuel cells, molten carbonate fuel cells, and the emerging type of protonic ceramic fuel cells, microbial fuel cells, and biofuel cells.

The book would not be finished without help and contributions from past and present colleagues, students, research fellows, and collaborators of both authors who would like to take this opportunity to thank:

- Prof. Chen Kongfa and Dr. Ai Na of Fuzhou University; Dr. Zhang Lan of Nanyang Technological University; Prof. Cheng Yi of Central South University; Prof. Zhao Ling of China University of Geosciences; Prof. He Tianmin of Jilin University; Dr. Zhang Jin and Prof. Lu Shanfu of Beihang University; Dr. He Shuai of University of St. Andrew; and Dr. Liu Yu, Dr. Zhao Shiyong, Dr. Zhang Xiao, Ms. Sun Yi and Ms. Zhang Xiaoran of Curtin University for providing raw materials and numerous drawings
- Prof. Li Jian and Prof. Pu Jian of Huazhong University of Science and Technology; Prof. Shao Zongping of Curtin University; and Prof. John Zhu of Queensland University for valuable comments and suggestions on the SOFC chapters
- Prof. Gordon Parkinson of Curtin University for painstakingly reading and editing of the book chapters
- Prof. Jens Oluf Jensen for sharing a lecturing course (hydrogen energy and fuel cells) in last 18 year at DTU from which one of the authors has received many inspiration in writing this book
- Dr. Lars N. Cleemann, Dr. Erik Christensen, and Dr. David Aili for sharing an experimental course (hydrogen and fuel cell chemistry) at DTU for the last 15 years. The teaching materials have been the basis for some of the chapter content
- Dr. David Aili and Dr. Yang Hu for reading and commenting book chapters and providing graphs.

These acknowledgements would not be complete without thanking our family members. SPJ would like to take this opportunity to thank his wife, Choo, his life companion for 30 years, for taking care of all other aspects of his life while he was involved in writing this book. Prof. Jiang would also like to thank his three beautiful children, Yin, Weiping, and Danhua, for the understanding, patience, and support. QL would like to thank his family, wife Tianqing and son Mike, for their understanding and constant supportive presence through the time spent on writing this book. Without their encouragement and support, this book would not be possible.

Last but not least, the publication of the book would not be possible without the encouragement and help from the editorial staff at Springer and special thanks go to June Tong for the initiation of the project, Umamagesh A P, Sridevi Purushothaman, Sunny Guo, and Nobuko Hirota for the constant support and conductive environment for pursuing this type of project.

Perth, Australia San Ping Jiang
Lyngby, Denmark Qingfeng Li

Contents

Part I Fundamentals

1 Introduction 3
- 1.1 Fuel Cells in the Hydrogen Chain 3
- 1.2 A Brief History of Fuel Cells 5
- 1.3 Types, Construction, and Components of Fuel Cells 8
  - 1.3.1 Fuel Cell Classification 8
  - 1.3.2 Construction and Components of Fuel Cells 9
  - 1.3.3 Brief Summary of Each Type of Fuel Cells 12
- 1.4 Fuel Cells Versus Batteries 16
- 1.5 Unitized Regenerative Fuel Cells and Reversible Fuel Cells 17
- 1.6 Applications and Prospect of Fuel Cell Technologies 19
- 1.7 Summary 23
- 1.8 Questions 24
- 1.9 General Readings 25
- References 25

2 Fuel Cell Thermodynamics 27
- 2.1 Internal Energy, Heat, Work, and Entropy 27
- 2.2 Enthalpy and Gibbs Free Energy 31
  - 2.2.1 Definition of Enthalpy and Gibbs Free Energy 31
  - 2.2.2 Gibbs Free Energy from First and Second Thermodynamics Laws 32
  - 2.2.3 Enthalpy of Formation 33
  - 2.2.4 Effect of Temperature on the Change in Enthalpy and Entropy 36
  - 2.2.5 Effect of Temperature on Gibbs Free Energy 40
  - 2.2.6 Gibbs Free Energy and Electrical Work 41
  - 2.2.7 Thermodynamic Reversible Potential and Thermoneutral Potential 43

2.2.8 Effect of Temperature and Pressure on Reversible Potential 46
2.3 Nernst Equation 48
2.4 Comparison of Work Done by a Heat Engine and Fuel Cell 51
2.4.1 Work Done in a Heat Engine and Carnot Efficiency 52
2.4.2 Fuel Cell Efficiency, Stoichiometry Number, Fuel and Oxygen Utilization Efficiency 54
2.4.3 Other Attributes of ICE and Fuel Cells 60
2.5 Thermodynamics and Efficiency of Water Electrolysis 60
2.5.1 Basic Thermodynamics of Water Electrolysis 61
2.5.2 Thermal Balance of Electrolysis Cells 62
2.5.3 Energy Efficiency of Water Electrolysis 63
2.5.4 High-Temperature Electrolysis 64
2.6 Summary 65
2.7 Questions 66
2.8 General Readings 68

3 Fuel Cell Electrochemistry 69
3.1 Electrochemistry at the Interface 69
3.1.1 Origin of Double Layer Structure 70
3.1.2 Standard Hydrogen Reference Electrodes 72
3.1.3 Other Reference Electrodes 75
3.1.4 Electrode Potential and Standard Reduction Potential 77
3.1.5 Concentration Cells 79
3.2 Activation and Activation Polarization Loss 81
3.2.1 Why Charge Transfer at the Interface 81
3.2.2 Butler–Volmer Equation 83
3.2.3 Tafel Plot 86
3.2.4 Effect of Exchange Current Density 88
3.2.5 Effect of Tafel Slope 91
3.3 Ohmic Polarization Loss 92
3.4 Mass Transport and Concentration Polarization in Steady State 95
3.5 I-V Characteristics 99
3.5.1 Cell Voltage and Various Types of Voltage Losses 99
3.5.2 Open-Circuit Voltage Loss 101
3.5.3 I-V Characteristics and Materials Issues 104
3.6 Basic Electrochemical Techniques 106
3.6.1 Rotating Disk and Rotating Ring Disk Electrodes 106
3.6.2 Linear Sweeping and Cyclic Voltammetry Techniques 112
3.6.3 Electrochemical Impedance Spectroscopy (EIS) 114
3.7 Summary 119

3.8 Questions ... 121
3.9 General Readings ... 122
References ... 122

4 Fuels for Fuel Cells ... 123
4.1 Introduction ... 123
4.2 Hydrogen Production ... 126
4.2.1 Steam Reforming ... 128
4.2.2 Water Electrolysis ... 130
4.3 Hydrogen Storage ... 141
4.3.1 Physical Storage of Hydrogen ... 141
4.3.2 Reversible and Irreversible Hydrides ... 142
4.4 Methanol, Ethanol, Formic Acid, and Other Liquid Fuels ... 144
4.4.1 Electrochemical Oxidation of Alcohols ... 145
4.4.2 Effect of pH and Temperature on Alcohol Oxidation Reactions ... 153
4.4.3 Energy Density, Environment Effect, and By-Products of Various Fuels ... 157
4.5 Ammonia, Urea, and Hydrazine ... 159
4.5.1 Ammonia Oxidation Reaction ... 159
4.5.2 Hydrazine Oxidation Reaction ... 162
4.5.3 Toxicity Issues ... 164
4.6 Natural Gas, Hydrocarbons, and Biomass-Derived Synthetic Fuels ... 164
4.7 Summary ... 166
4.8 Questions ... 167
References ... 168

Part II Polymer Electrolyte Membrane Fuel Cells

5 Polymer Electrolyte Membrane Fuel Cells: Principles and Materials ... 173
5.1 Introduction ... 173
5.2 Proton-Conducting Electrolytes ... 174
5.2.1 Electrolytes for Low-Temperature Fuel Cells ... 174
5.2.2 Proton Conduction Mechanisms and Materials ... 177
5.3 Proton Exchange Membranes ... 181
5.3.1 Poly(Perfluorosulfonic Acid) Membranes ... 182
5.3.2 Short Side-Chain PFSA Membranes ... 190
5.3.3 PFSA Composite Membranes ... 191
5.3.4 Alternative Membranes ... 195
5.4 Electrocatalysts ... 196
5.4.1 Anode Catalysts ... 197
5.4.2 Cathode Catalysts ... 198
5.4.3 Carbon Supports ... 207
5.5 Electrodes and Membrane Electrode Assembly ... 214

5.5.1 Gas Diffusion Layer 214
5.5.2 Microporous Layers 218
5.5.3 Catalyst Layer 219
5.6 Bipolar Plates and Seals 222
5.7 Summary 223
5.8 Questions 224
5.9 General Readings 226
References 227

6 Polymer Electrolyte Membrane Fuel Cells: Fabrication and Characterization 229
6.1 Proton Exchange Membranes 229
6.1.1 PFSA Membrane Fabrication 229
6.1.2 Ion-Exchange Capacity and Equivalent Weight 231
6.1.3 Water Uptake and Swelling 232
6.1.4 Proton Conductivity 235
6.1.5 Electroosmotic Drag of Water 238
6.1.6 Water and Methanol Diffusion 241
6.1.7 Solubility and Diffusion of Oxygen 242
6.1.8 Permeability of Hydrogen 244
6.1.9 Methanol Crossover—Liquid and Vapor 248
6.1.10 Fenton Test 252
6.1.11 Mechanical Properties 254
6.1.12 Thermal Stability Test 256
6.2 Catalyst Synthesis 257
6.2.1 Impregnation–Reduction Method 257
6.2.2 Colloidal Adsorption Method 259
6.2.3 Self-assembly Method on CNT and Graphene Supports 261
6.3 Ex Situ Characterization of Catalysts 262
6.3.1 Catalyst Ink and Thin Catalyst Film 263
6.3.2 Three-Electrode Cell Configuration 263
6.3.3 Preconditioning and ORR Activity Measurement 265
6.3.4 Determination of the ECSA 266
6.3.5 Analysis of the Polarization Curves 268
6.3.6 Stability Test 271
6.3.7 Half-Cell Characterization 272
6.4 Fuel Cell Test for in Situ Characterization 275
6.4.1 Cell Assembling and Pretest Conditioning 275
6.4.2 Determination of ECSA 276
6.4.3 Membrane Durability Test 277
6.4.4 Catalyst Stability Test 279
6.5 Summary 285
6.6 Questions 286
6.7 General Readings 287
References 288

7 Polymer Electrolyte Membrane Fuel Cells: Performance and Operation 291
7.1 Performance and Analysis 291
7.1.1 Analysis of i-V Curves 293
7.1.2 Ohmic Losses 294
7.1.3 ORR Kinetic Parameters 295
7.1.4 Mass Transport Losses 298
7.2 Fueling Options and CO Poisoning 299
7.2.1 CO Adsorption on Pt 300
7.2.2 CO Poisoning Effect 301
7.2.3 Relative Activity and Temperature Effect 302
7.2.4 CO Poisoning Mitigation and Bifunctional Mechanism 302
7.2.5 Catalysts for Methanol Electro-Oxidation 304
7.2.6 Other Effects of CO 304
7.3 Water Management 305
7.3.1 Water Balance and Transport 305
7.3.2 Saturated Water Vapor Pressure and Relative Humidity 306
7.3.3 Self-Humidification 307
7.3.4 External Humidification 309
7.3.5 Modified and Thinner Membranes 310
7.4 Thermal Management 311
7.4.1 Heat Generation 311
7.4.2 Cooling Duty and Radiator Area 312
7.4.3 Air Cooling Versus Water Cooling 313
7.4.4 Operating Points and Stack Calculation 314
7.5 Durability Issues and Mitigation 314
7.5.1 Status and Target of Durability 314
7.6 Summary 321
7.7 Questions 322
References 323

8 High-Temperature Polymer Electrolyte Membrane Fuel Cells 325
8.1 Why and How High Temperatures 325
8.2 Polybenzimidazoles and Their Interactions with Phosphoric Acids 326
8.2.1 Synthesis and Molecule Weight Determination 326
8.2.2 Membrane Casting from DMAc Solutions and Acid Doping 328
8.2.3 Direct Cast Membranes 329
8.2.4 Membrane Characterization 330
8.2.5 PBI Variants 332
8.2.6 Cross-linking, Blends, and Thermal Curing 332
8.2.7 PBI Composite Membranes 336

8.3 Catalysts and Electrodes 336
8.3.1 Catalysts 336
8.3.2 Gas Diffusion Electrodes 341
8.3.3 Membrane Electrode Assemblies 341
8.4 Fuel Cell Performance and Operation 344
8.4.1 Performance and Pt Loading 344
8.4.2 Impact of Fuel Impurities and Water 345
8.5 Alternative High-Temperature Membranes 346
8.5.1 Non-PBI Based High-Temperature Membranes 346
8.5.2 Inorganic High-Temperature Membranes 347
8.6 Durability and Commercialization 349
8.7 Summary 351
8.8 Questions 351
8.9 General Readings 352
References 352

Part III Solid Oxide Fuel Cells

9 Solid Oxide Fuel Cells: Principles and Materials 357
9.1 Introduction 357
9.1.1 Operation Principles 357
9.1.2 Charge Transport in Oxide Materials 360
9.2 Electrolyte 369
9.2.1 Zirconium Oxides 372
9.2.2 Cerium Oxides 375
9.2.3 Lanthanum Strontium Magnesium Gallate Perovskites 377
9.2.4 Apatite Oxides 379
9.2.5 Bismuth Oxides 380
9.3 Anode 381
9.3.1 Ni-Based Cermet Anode 383
9.3.2 Cu/CeO_2/YSZ-Based Anode 387
9.3.3 Ceramic Oxide Anode 387
9.4 Cathode 389
9.4.1 Lanthanum Strontium Manganite 390
9.4.2 Lanthanum Strontium Cobalt Ferrite 395
9.4.3 Other Perovskites, Double Perovskites, and Ruddlesden-Popper Structured Oxides 396
9.4.4 Composite Cathode 400
9.5 Interconnect and Sealant 403
9.5.1 Interconnect 403
9.5.2 Sealants 410
9.6 Cell Structures and Stack Design 412
9.6.1 Cell Structures 412
9.6.2 Stack Design 413

9.7 Variations of SOFCs 414
9.7.1 Single-Chamber Solid Oxide Fuel Cells 415
9.7.2 Metal-Supported Solid Oxide Fuel Cells 416
9.7.3 Direct Carbon Fuel Cells 417
9.8 Summary 418
9.9 Questions 420
References 422

10 Solid Oxide Fuel Cells: Reactions 425
10.1 Surface Segregation of Oxide Electrodes 425
10.1.1 Surface Segregation Under Open Circuit 426
10.1.2 Surface Segregation Under Polarization 428
10.1.3 A Note 435
10.2 Reactions at the Cathode 436
10.2.1 Activation Process 436
10.2.2 Oxygen Reduction Reaction 439
10.2.3 Effect of Oxygen Vacancies 444
10.3 Reactions at the Anode 446
10.3.1 Hydrogen Oxidation Reaction 446
10.3.2 Hydrocarbon Fuel Oxidation Reaction 454
10.3.3 Redox Reaction 457
10.4 Interface and Interface Reactions 458
10.4.1 LSM and YSZ (Doped Ceria) Systems 458
10.4.2 Interface and Interaction Between LSCF and YSZ 462
10.4.3 LSCF/GDC/YSZ Interface 466
10.4.4 Other Interfaces 466
10.5 Reaction with Contaminants 467
10.5.1 Reactions Between Cathode and Contaminants 468
10.5.2 Reaction Between Anode and Contaminants 478
10.5.3 Contaminant-Tolerant Electrodes 483
10.6 Summary 487
10.7 Questions 490
References 492

11 Solid Oxide Fuel Cells: Techniques and Characterization 497
11.1 Electrode Arrangement and Test Stations 497
11.1.1 Test Station Design and Arrangement 497
11.1.2 Electrode Arrangements 501
11.1.3 Comparative Experimental Approach 504
11.2 Cell Configurations and Performance Scalability 505
11.2.1 Cell Configurations 505
11.2.2 Symmetric Cell 506
11.2.3 Polarization Performance Analysis 509
11.3 Equipotential Line and Constriction Effect 517
11.3.1 Equipotential Line 517

11.3.2 Special Voltage Probe and Resistance Distribution ... 518
11.3.3 Relationship Between Validity of Reference Electrode and Electrolyte Thickness ... 525
11.3.4 Constriction Effect ... 527
11.4 Equivalent Circuit Analysis in Electrochemical Impedance Spectroscopy ... 529
11.4.1 Equivalent Circuit for Capacitance Impedance ... 531
11.4.2 Equivalent Circuit for Inductance Impedance ... 533
11.4.3 Physical Significance of Equivalent Circuit Elements ... 535
11.4.4 Impedance Measurement in SOFCs ... 537
11.5 Galvanostatic Current Interruption Technique ... 539
11.5.1 Comparison of EIS and GCI Techniques ... 541
11.6 Conductivity Measurement of SOFC Components and Materials ... 543
11.6.1 Two-Probe and Four-Probe Methods ... 543
11.6.2 Conductivity Measurement of Porous Electrode Coating ... 545
11.6.3 Contact Resistance Between Electrode and Current Collector ... 547
11.6.4 Ionic Conductivity of Electrolyte by EIS ... 548
11.7 Other Properties ... 550
11.7.1 Oxygen Surface Diffusion and Exchange Coefficient Measurements ... 550
11.7.2 Porosity and Density Measurement ... 552
11.7.3 Focus Ion Beam and Scanning Transmission Electron Microscopy (FIB-STEM) ... 553
11.8 Summary ... 557
11.9 Questions ... 558
References ... 559

12 Solid Oxide Fuel Cells: Fabrication and Microstructure ... 561
12.1 Introduction ... 561
12.2 Powder Synthesis Methods ... 562
12.2.1 Solid-State Reaction and Physical Mixing Methods ... 562
12.2.2 Solution Combustion Method ... 563
12.2.3 Co-precipitation Method ... 566
12.2.4 Pechini and Polymeric Complexing Method ... 568
12.2.5 Sol–Gel Method ... 570
12.2.6 Gel-Casting Method ... 571
12.3 Fabrication Techniques for Electrolyte and Electrode Coatings ... 574
12.3.1 Die Pressing Method ... 575

12.3.2 Tape-Casting Method ... 576
12.3.3 Screen-Printing Method ... 579
12.3.4 Chemical Vapor Deposition and Atomic Layer Deposition ... 580
12.3.5 Electrochemical Vapor Deposition ... 580
12.3.6 Magnetron Sputtering Techniques ... 581
12.3.7 Pulsed Laser Deposition ... 584
12.3.8 Plasma Spray Deposition ... 584
12.3.9 Slurry-Coating, Spin-Coating and Dip-Coating Methods ... 586
12.3.10 Electrophoretic Deposition ... 589
12.3.11 Spray Pyrolysis and Flame-Assisted Vapor Deposition ... 591
12.4 High-Temperature Sintering Process ... 592
12.4.1 Anode ... 593
12.4.2 Cathode ... 595
12.4.3 Sintering Profile ... 597
12.4.4 High-Temperature Sintering Aids and Solid-State Reactive Sintering ... 599
12.5 Direct Assembly and Polarization-Induced Interface ... 600
12.6 Nano-Structured Electrodes ... 603
12.6.1 Wet-Infiltration Techniques ... 603
12.6.2 Decoration and Polarization-Induced Phase Migration ... 608
12.6.3 Nanosized Catalysts by in Situ Exsolution ... 611
12.6.4 Microstructure and Microstructure Stability ... 612
12.7 Cell Fabrication ... 614
12.7.1 Planar Cells ... 614
12.7.2 Tubular Cells ... 615
12.8 Summary ... 616
12.9 Questions ... 618
References ... 619

Part IV Other Fuel Cells

13 Alkaline Fuel Cells ... 623
13.1 Introduction ... 623
13.2 Alkaline Fuel Cells ... 624
13.2.1 Electrolyte for AFC ... 624
13.2.2 Catalysts ... 625
13.2.3 Gas Diffusion Electrodes for Liquid Electrolytes ... 627
13.2.4 Electrolyte Configuration ... 632
13.2.5 Stack Design ... 636
13.2.6 Carbonization and CO_2 Scrubbing ... 636
13.2.7 Performance and Durability ... 638

13.3 Anion Exchange Membrane Fuel Cells ... 639
13.3.1 AEM Electrolyte ... 640
13.3.2 Catalyst Development ... 643
13.3.3 Performance ... 644
13.4 Summary ... 646
13.5 Questions ... 647
References ... 648

14 Phosphoric Acid Fuel Cells ... 649
14.1 Introduction ... 649
14.2 Phosphoric Acid ... 649
14.2.1 Acid Chemistry ... 649
14.2.2 Conductivity and Mechanism ... 651
14.2.3 Vapor Pressure and Thermal Stability ... 653
14.3 Cell Components ... 654
14.3.1 Catalysts ... 654
14.3.2 Gas Diffusion Electrodes ... 660
14.3.3 Matrix and Acid Reservoir ... 661
14.3.4 Bipolar or Separate Plates ... 663
14.3.5 Seals and Coolers ... 663
14.4 Performance and Cell Management ... 664
14.4.1 Performance and Lifetime ... 664
14.4.2 Acid Evaporation and Mitigation ... 667
14.4.3 Migration of Acid and Management ... 667
14.5 Summary ... 668
14.6 Questions ... 669
References ... 670

15 Molten Carbonate Fuel Cells ... 673
15.1 Introduction ... 673
15.2 Electrolytes ... 676
15.2.1 Alkali Metal Carbonates ... 676
15.2.2 Acid—Base Chemistry ... 677
15.2.3 Properties of Molten Carbonate Systems ... 678
15.2.4 Comparison of Li/K and Li/Na Systems ... 679
15.3 Cathode Reactions and Materials ... 680
15.3.1 Oxygen Reduction Reaction in Molten Carbonates ... 680
15.3.2 Cathode Materials ... 681
15.3.3 NiO Dissolution and Precipitation ... 681
15.3.4 Electrolyte Additives and Alternative Materials ... 682
15.4 Anode Reactions and Materials ... 683
15.5 Matrix and Immobilization of Molten Carbonate Electrolyte ... 685
15.5.1 Matrix Materials ... 685
15.5.2 Immobilization and Distribution of Electrolyte ... 685
15.5.3 Matrix Stability and Fabrication ... 686

15.6 Construction Materials 688
15.7 Performance and Durability 689
15.8 Summary 691
15.9 Questions 692
References 692

16 Protonic Ceramic Oxide Fuel Cells, Microbial Fuel Cells, and Biofuel Cells 695
16.1 Protonic Ceramic Oxide Fuel Cells 695
16.1.1 Operation Principle of Protonic Ceramic Oxide Fuel Cells 696
16.1.2 Proton-Conducting Oxides 698
16.1.3 Electrode Materials for PCFCs 701
16.1.4 Other Applications of Proton-Conducting Oxides 703
16.1.5 Proton Conductivity Measurement 705
16.2 Microbial Fuel Cells 708
16.2.1 Microbial Fuel Cell Configuration 708
16.2.2 Catalysis in Microbial Fuel Cell Anode 710
16.2.3 Catalysis in Microbial Fuel Cell Cathode 712
16.2.4 Electron Transfer Mechanism 714
16.2.5 Challenges in MFCs 716
16.3 Biofuel Cells 716
16.4 Summary 718
16.5 Questions 719
References 719

About the Authors

San Ping Jiang is John Curtin Distinguished Professor at WA School of Mines: Minerals, Energy and Chemical Engineering, Deputy Director of Fuels and Energy Technology Institute, Curtin University and Adjunct Professor of the University of Sunshine Coast, Australia. He obtained his BEng from South China University of Technology, Guangzhou in 1982 and Ph.D. from The City University, London in 1988. He has over 30 years academic and industry R&D experience and has held positions at Essex University in UK, Nanyang Technological University in Singapore, and CSIRO Materials and Manufacturing Division and Ceramic Fuel Cells Ltd in Australia. His research interests encompass fuel cells, water splitting, supercapacitors, solid oxide electrolyzers, CO_2 electrolysis, photocatalysis, electrocatalysis, solid-state ionics, and high-temperature solid-state electrochemistry. Professor Jiang has made significant contributions to the fundamental knowledge and technological development of electrochemical energy conversion and storage areas and is one of the most cited researchers in fuel cells. Professor Jiang was recognized as a Most Cited Researcher (*MCR*) in Energy Science and Engineering by Shanghai Jia Tong University's ARWU/Elsevier in 2016 and a Highly Cited Researcher (*HCR*) in Cross-Fields by Clarivate Analytics/Web of Science in 2018.

Qingfeng Li is a full professor at Department of Energy Conversion and Storage, Technical University of Denmark. His research areas include proton-conducting electrolytes, electrocatalysts, and the related technologies particularly fuel cells and electrolyzers. He received his Ph.D. in electrochemistry from Northeastern University, China, in 1990 and was awarded Doctor Degree of Technices at DTU in 2006. As a research fellow, he started in the middle of 1990s the research on high-temperature polymer electrolyte membrane fuel cells at DTU. He has participated/coordinated more than 20 EU and Nordic research projects within the fuel cell area and leading the 4M Centre devoted to fundamental research on mechanisms, materials, manufacturing, and management of high-temperature polymer electrolyte membrane fuel cells, funded by the Danish Council for Strategic Research and later Innovation Fund Denmark. Professor Li has been involved in teaching a lecturing course (hydrogen energy and fuel cells) since 2004 and an experimental course (hydrogen and fuel cell chemistry) since 2008 at DTU.

Abbreviations

AAS	Atomic absorption spectroscopy
AB-PBI	Poly(2,5-benzimidazole)
ADL	Acid doping level
AEC	Alkaline electrolysis cell
AEM	Anion exchange membrane
AEMFC	Anion exchange membrane fuel cell
AFC	Alkaline fuel cell
AFL	Anode function layer
ALD	Atomic layer deposition
AM	Acrylamide ($C_2H_3CONH_2$)
AOR	Alcohol oxidation reaction
APS	Ammonium persulfate or atmosphere plasma spray
APU	Auxiliary power unit
ASR	Area-specific resistance
AST	Accelerated stress test
ATR	Autothermal reforming
BCC	Body centered cubic
BCV	Battery electrical vehicle
BFC	Biofuel cell
BoL	Beginning of life
BoP	Balance of plant
BPP	Bipolar plate
BSCF	Barium strontium cobalt ferrite
CA	Citric acid ($C_6H_8O_7$)
CCCL	Cathode current collection layer
CCL	Cathode contact layer
CCM	Catalyst-coated membrane
CE	Counter electrode
CFL	Cathode functional layer
CGH_2	Compressed gaseous hydrogen
CHP	Cogeneration of heat and power
CL	Catalyst layer

CN	Coordination number
CNT	Carbon nanotube
CPE	Constant-phase element
CV	Cyclic voltammetry
CVD	Chemical vapor deposition
DAFC	Direct alcohol fuel cell
DCD	Dicyandiamide ($C_2H_4N_4$)
DCFC	Direct carbon fuel cell
DEFC	Direct ethanol fuel cell
DFAFC	Direct formic acid fuel cell
dFAOR	Direct formic acid oxidation reaction
DFT	Density functional theory
DHE	Dynamic hydrogen electrode
DHzFC	Direct hydrazine fuel cell
DMAc	N,N-dimethylacetamide
DMF	N,N-dimethylformamide
DMFC	Direct methanol fuel cell
DUFC	Direct urea fuel cell
EC	Electrolysis cell
ECM	Equivalent circuit mode
ECR	Electrical conductivity relaxation
ECSA	Electrochemical surface area or electrochemically active surface area
EDTA	Ethylenediaminetetraacetic acid
EG	Ethylene glycol
EIS	Electrochemical impedance spectroscopy
EMF	Electromotive force
EoL	End of life
EOR	Ethanol oxidation reaction
EPD	Electrophoretic deposition
ESB	Erbium-stabilized bismuth
EVD	Electrochemical vapor deposition
EW	Equivalent weight
FA	Formic acid
FAOR	Formic acid oxidation reaction
FC	Fuel cell
FCC	Face centered cubic
FCV	Fuel cell vehicle
FER	Fluorine ion emission rate
GC	Glassy carbon
GCE	Glassy carbon electrode
GDC	Gadolinium-doped ceria
GDE	Gas diffusion electrode
GDL	Gas diffusion layer
GNP	Glycine–nitrate process

HER	Hydrogen evolution reaction
HHV	High heat value
HOR	Hydrogen oxidation reaction
HP	Hemin porcine ($C_{34}H_{32}ClFeN_4O_4$)
HPA	Heteropolyacids
HPW or PWA	Phosphotungstic acid ($H_3PW_{12}O_{40}$)
HR-TEM	High-resolution transmission electron microscopy
HT-PEMFC	High-temperature polymer electrolyte membrane fuel cell
H_{UPD}	Underpotential hydrogen deposition
HzOR	Hydrazine oxidation reaction
ICE	Internal combustion engine
ICP-AES	Inductive coupled plasma-atomic emission spectrometry
IEC	Ion exchange capacity
IPA	Isopropanol or isopropyl alcohol
IPAOR	Isopropanol or isopropyl alcohol oxidation reaction
IT-SOFC	Intermediate temperature solid oxide fuel cell
K-L equation	Koutecky–Levich equation
LCCr	Lanthanum calcium chromite
LH_2	Liquid hydrogen
LHV	Low heat value
LSC	Lanthanum strontium cobaltite or long side chain in polymer
LSCF	Lanthanum strontium cobalt ferrite
LSCM	Lanthanum strontium chromium manganite
LSCr	Lanthanum strontium chromite
LSF	Lanthanum strontium ferrite
LSGM	Lanthanum strontium magnesium gallate
LSM	Lanthanum strontium manganite
LSV	Linear scan voltammetry
LT-PEMFC	Low-temperature polymer electrolyte membrane fuel cell
MA	Mass activity (of catalysts)
MBAM	*N,N″*-methylenebis-acrylamide
MCFC	Molten carbonate fuel cell
MD	Machine (extrusion) direction
MEA	Membrane-electrode assembly
MET	Mediated electron transfer
MFC	Microbial fuel cell
MIEC	Mixed ionic and electronic conductivity
MOR	Methanol oxidation reaction
MPL	Microporous layer
MSAA	Mass specific active area
MSE	Mass specific energy density
MS-SOFC	Metal-supported solid oxide fuel cell
NEXAFS	Near-edge X-ray absorption fine structure
NG	Natural gas
NMP	N-methyl-2-pyrrolidone

NMR	Nuclear magnetic resonance spectroscopy
NP	Nano-particle
NPMC	Non-precious metal catalyst
OCP	Open-circuit potential
OCV	Open-circuit voltage
OER	Oxygen evolution reaction
ORR	Oxygen reduction reaction
P2X	Power-to-x(chemicals)
PA	Phosphoric acid, H_3PO_4
PA/PBI	Phosphoric acid/polybenzimidazole
PAFC	Phosphoric acid fuel cell
PBI	Polybenzimidazole
PDDA	Poly(diallyldimethylammonium chloride)
PEEK	Poly(ether ether ketone)
PEFC	Protonic ceramic fuel cell
PEI	Poly(ethyleneimine)
PEM	Polymer electrolyte membrane
PEMEC	Polymer electrolyte membrane electrolysis cell or proton exchange membrane electrolysis cell
PEMFC	Polymer electrolyte membrane fuel cell or proton exchange membrane fuel cell
PES	Polyethersulfone
PEWE	Polymer electrolyte water electrolyzer
PFSA	Perfluorosulfonic acid
PGM	Precious group metals
PLD	Pulsed laser deposition
PMG	Precious metal group
POx	Partial oxidation
PPA	Polyphosphoric acid
PPD	Peak power density
PSS	Polysodium-p-styrenesulfonate
PSSA	Poly(styrenesulfonic acid)
PSU	Poly(ether sulfone)
PTFE	Polytetrafluoroethylene
PTL	Porous transport layer
PVDF	Poly(vinylidene fluoride)
PVP	Poly(vinyl pyrrolidone)
PWA	Phosphotungstic acid
QA	Quaternary ammonium
RDE	Rotating disk electrode
RDS	Rate determining step
RE	Reference electrode
RF	Roughness factor
RFC	Regenerative fuel cell
RH	Relative humidity (%)

RHC	Relative humidity cycle
RHE	Reversible hydrogen electrode
RP	Ruddlesden–Popper structure
RRDE	Rotating ring disk electrode
SA	(area) specific activity
SAC	Single-atom catalysts
SCE	Saturated calomel electrode
SC-SOFC	Single-chamber solid oxide fuel cell
ScYZ	Scandia-stabilized zirconia
SDC	Samnium-doped ceria
SEM	Scanning electron microscopy
SHE	Standard hydrogen electrode
SMR	Steam methane reforming
SOC	Solid oxide cell
SOEC	Solid oxide electrolysis cell
SOFC	Solid oxide fuel cell
SR	Steam reforming
SSC	Short side chain
SSR	Solid-state reaction
STP	Standard temperature and pressure
TCO	Triple conducting oxide
TD	Transverse direction
TEC	Thermal expansion coefficient
TEM	Transmission electron microscopy
TEMED	*N*,*N*,*N*′*N*′-tetra methyl-ethylene diamide
TEOS	Tetraethoxysilane
TFE	Tetrafluoroethylene
TF-RDE	Thin-film rotating disk electrode
T_g	Glass transition temperature
TGA	Thermogravimetric analysis
THF	Tetrahydrofuran
TOF	Turnover frequency
TPB	Three-phase boundary
URFC	Unitized regenerative fuel cell
VP	Voltage probe
VSE	Volume-specific energy density
WE	Working electrode
WFR	Water formation reaction
WGS	Water–gas shift reaction
XAFS	X-ray absorption fine structure
XANES	X-ray absorption near-edge structure
XRD	X-ray diffraction
YSB	Yttria-stabilized bismuth
YSZ	Yttria-stabilized zirconia

Physical Constants and Conversion

Speed of light	c	3×10^8 m/s
Avogadro's number	N_A	6.02×10^{23} atoms/mol
Boltzmann constant	k	1.38×10^{-23} J/K
		8.61×10^{-5} eV/K
Electron charge	q	1.60×10^{-19} C
Electron mass	m_e	9.11×10^{-31} kg
Proton mass	m_p	1.67×10^{-27} kg
Faraday's constant	F	96485.34 C/mol
Gas constant	R	8.3145 J/mol·K
		82.06 atm·cm^3/mol·K
		83.145 bar·cm^3/mol·K
		1.987 cal/mol·K
		1.987 btu·1b/mol·K
		10.740 Pa·ft^3 1b/mol·K
Planck's constant	$\hbar$	6.626×10^{-34} J·s
		4.136×10^{-15} eV·s
Force		1 N = 1 kg/m·s^2
		1 N = 10^5dyn
Distance		1 km = 0.622 mile
		1 m = 3.28 ft
Energy		1 J = 6.241506×10^{18} eV
		1 J = 9.478134×10^{-4} Btu
		1 J = 2.777778×10^{-7} kWh
		1 calorie = 4.184 J
Pressure		1 Pa = 1 N/m^2
		1 bar = 100 kPa
		1 atm = 1.01325 bar

(continued)

(continued)

		1 atm = 0.1013 MPa
		1 atm = 14.6959 psi
		1 atm = 760 mmHg
Power		1 W = 1 J/s = 1 Nm/s
		1 W = 1.34×10^{-3} horsepower
		1 W = 3.415 Btu/h
Volume		1 m^3 = 1000 L
		1 L = 0.265 gal
		1 L = 3.53×10^{-2} ft^3
Mass		1 kg = 2.20 1b

Part I
Fundamentals

Chapter 1
Introduction

1.1 Fuel Cells in the Hydrogen Chain

Our modern civilization is founded on the consumption of a vast quantity of energy primarily from fossil fuels, three major types of which are coal, oil, and natural gas. The current energy technology converts the chemical energy of fuels first to thermal energy by combustion, to mechanical energy by thermal engines and most often to electricity by means of generators as schematically shown in Fig. 1.1. It is recognized by S. Carnot in 1824 that there is an upper limit in the efficiency of a thermal engine. For modern power plants, the efficiency is normally below 40%. For small engines, e.g., vehicle motors, the efficiency is as low as 10–20%. Another critical concern is that combustion of fossil fuels is the major source of greenhouse gases, i.e., CO_2, more than 71% of all air emissions.

A fuel cell is an electrochemical cell, which can directly and continuously convert the chemical energy of fuels such as hydrogen, methanol, ethanol, or natural gas to electrical energy. The principle of fuel cells is illustrated in the lower left corner of Fig. 1.1 where a proton-conducting electrolyte is used. It is essential that no collision of the fuel (e.g., hydrogen) at the anodic side and oxygen at the cathodic side occurs owing to the separating electrolyte, otherwise a chemical combustion of hydrogen in oxygen will take place. Hydrogen is catalytically oxidized at the anode to give protons (H^+) and electrons. The electrolyte is unique in a way that it allows only the passage of ions, e.g., H^+ but not the electrons. This electrolyte forces the electrons to flow away from the anode via the external circuit toward the cathode. The protons pass through the electrolyte to the cathode, where oxygen reacts with the incoming electrons from the external circuit and protons from the electrolyte to produce water. In this way a flow of electrons (electricity) is directly generated. Due to this electrochemical nature of the energy conversion process, fuel cells inherently have a higher efficiency than those of conventional energy conversion technologies with much less greenhouse gas emission.

Replacement of fossil fuels by renewable energy resources is the ultimate solution. Renewable energies come either directly or indirectly from the sun. Solar energy can

S. P. Jiang and Q. Li, *Introduction to Fuel Cells*,
https://doi.org/10.1007/978-981-10-7626-8_1

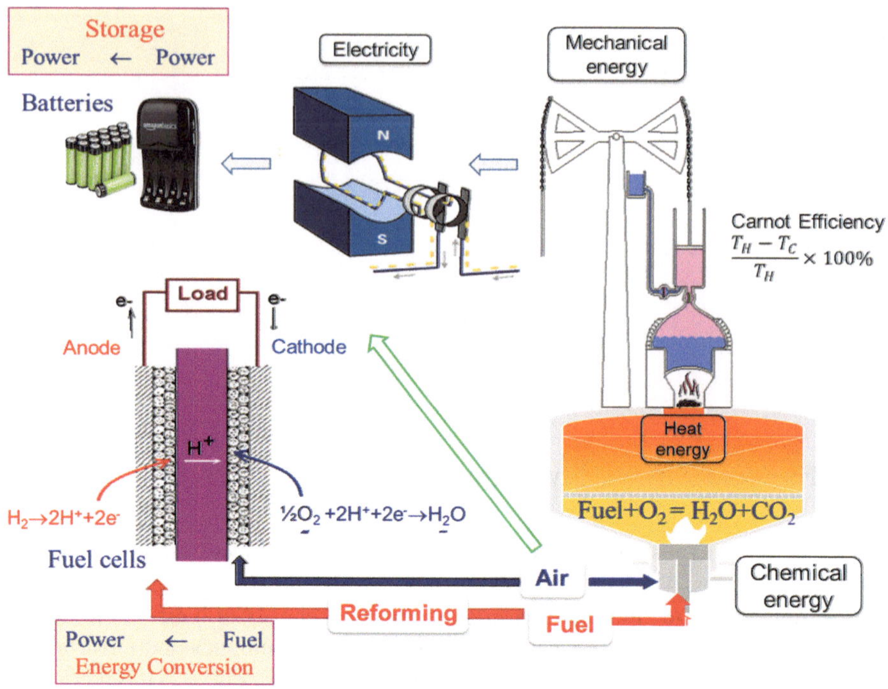

Fig. 1.1 Routes of energy conversion for fossil fuels

be used directly for heating and generating electricity. Heat from the sun also drives the winds, and the wind energy can be captured by windmill. The blow of wind and heat of the sun cause water to evaporate and therefore rain or snow, which are the source of hydroelectric power.

Electricity from these renewable sources is of intermittent nature. A key issue for any kind of renewable energies is the effective energy storage at different time scales. Water electrolysis, among others, offers a practical way of storing energy from renewable sources in the form of generated hydrogen. The generated hydrogen is the energy carrier, which can be stored on an energy scale matching the grid and for flexible time periods. Hydrogen and its combination with the electrochemical energy conversion technology—the fuel cell—have been recognized as a reliable, secure, and clean energy technology in association with renewable energy sources. For mobile applications, hydrogen can be further stored physically, for example, as compressed gas, liquid, or metal hydrides. An alternative way of hydrogen storage is via synthesis of liquid biofuels, e.g., ammonia, methanol and ethanol.

Along with the rain and snow, on the other hand, the sunlight causes plants to grow. The organic matter making up those plants is biomass, based on which biofuels have already emerged as an alternative to gasoline and conventional diesel in transportation sectors, with an exponential increase in the consumption in recent years. From readily available energy crops such as sugar, starch, and edible feedstock, biofuels are produced including biodiesel, biomethanol, and bioethanol, among many others.

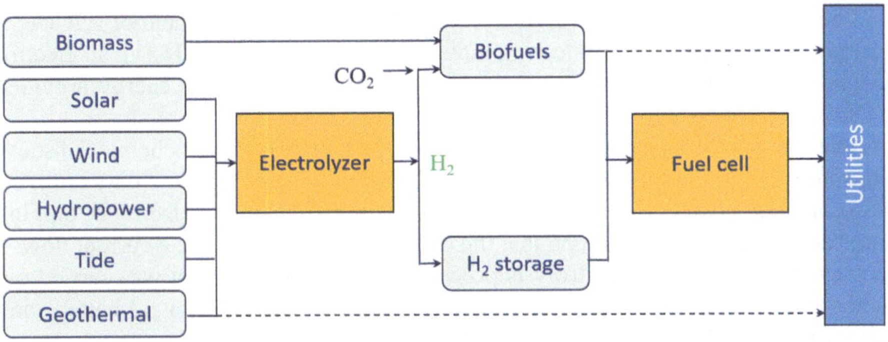

Fig. 1.2 Renewable power chain with fuel cell and electrolyzer as the central linking technologies

These biofuels can be electrochemically converted into power, i.e., either by direct oxidation or via reforming in fuel cells. In the future renewable energy systems, fuel cells and water electrolyzers are the central linking technologies on the hydrogen chain, as shown in Fig. 1.2.

In brief, fuel cells are a technology that can be fitted into the fossil fuel scenario offering advantages such as lower emissions and higher efficiency. They play also a central role in the future renewable energy scenario where hydrogen via water electrolysis is the energy carrier. As an energy conversion device, fuel cells have also an advantage of being modular and can therefore be built in a wide range of power scales from a few hundred watts up to kilo- and megawatt. Thus, fuel cell-based power units can be constructed in a decentralized manner, i.e., at locations close to users and hence with reduced the cost of power transmission. As an electrochemical device, the efficiency of a fuel cell is not dependent on the size of the power unit nor on the load level of the power output. A fuel cell of a rated power of, e.g., 50 kW, will have a higher efficiency when it operates at a low load of less than 50 kW. This is to the contrary of a thermal engine!

1.2 A Brief History of Fuel Cells

The fuel cell is an old energy conversion technology invented slightly later than the steam engine and internal combustion engines (ICEs) [1]. The schematic thermal engine shown in Fig. 1.1 was an invention by T. Newcomen in 1712, and an improved version with nearly doubled efficiency was the Watt engine from 1770s. That was the time when the phenomenon of electricity was discovered by the Galvani's frog leg experiment (1780). The first prototype of a galvanic cell, i.e., the first battery was the voltaic pile invented by A. Volta in 1799.

The voltaic pile was an invention that has led to the establishment of science of electricity and electrical engineering including the Faraday disk (1831), an electromagnetic generator that can generate electricity from mechanical energy provided by steam or gas turbine.

The voltaic pile had also enabled a long and rapid series of electrochemical discoveries including the water electrolysis in 1800. Sir William Robert Grove was in 1838 also conducting water electrolysis experiment with two platinum electrodes dipping into "acidified" water and found that the evolved gases could drive a current flow—the reverse of the water electrolysis process. A similar experiment was carried out at about the same time by C. F. Schönbein. The device was called a gas battery, and the year is regarded as the invention of fuel cells.

The idea of the gas battery was not made any use in the first half century until L. Mond and C. Langer in 1888 patented an idea of trapping liquid electrolyte in a porous matrix onto which platinum black was applied. The device was fueled by coal gas and delivered a useful power output (6.5 mA/cm^2 @ 0.73 V). The word *fuel cell* was also coined by Mond and Langer. The practical design of immobilized liquid electrolyte in a porous matrix is still used today.

Professor F. W. Ostwald, the founder of physical chemistry, provided in 1893 the early theoretical understanding of fuel cells in terms of the electrode, electrolyte, oxidation, and reduction, as well as anions and cations. This has inspired the early year research on direct oxidation of various fuels including the carbon fuel cells by W. W. Jacques in 1890s.

These early year efforts primarily resulted in little technological progress apparently due to the lack of knowledge of electrochemistry, electrode kinetics, and heterogeneous catalysis. During this period the development of ICEs, on the other hand, was so fast that further research into the fuel cell technology was abandoned.

During the first decades of the nineteenth century, a name to be mentioned is Professor Baur of the Swiss Federal Institute of Technology, Zurich, who devoted his lifetime on development of high-temperature fuel cells including solid oxide fuel cell (SOFC) and molten carbonate fuel cell (MCFC). F. T. Bacon had since the late 1930s made efforts to develop fuel cells using alkaline electrolyte, i.e., alkaline fuel cell (AFC), and nickel electrodes. The effort was continued until 1959 when a 5–6 kW stack was successfully demonstrated, which attracted great scientific and technological attention. As a result, Pratt & Whitney licensed Bacon's work for the Apollo spacecraft fuel cells. Also in 1960s was the first prototype MCFC developed by Dutch scientists Broers and Ketelaar, the first phosphoric acid fuel cells (PAFCs) reported by Elmore and Tanner, and the first polymer electrolyte membrane fuel cell (PEMFC) invented by Grubb and Niedrach at General Electric.

A marvelous feature of fuel cells is their energy capacity limited by the fuel tank, which was attractive for the space mission programs of US National Aeronautics and Space Administration (NASA) agency in 1960s. The Gemini program used the early PEMFCs, and the Apollo program employed the AFC power units. The renewed research interest in fuel cells in 1990s is largely related to the well-known pubic events such as oil and energy crises in 1970s and 80s and debate and signing of Kyoto Protocol in 1997. These historical events of the fuel cell development are listed in Fig. 1.3.

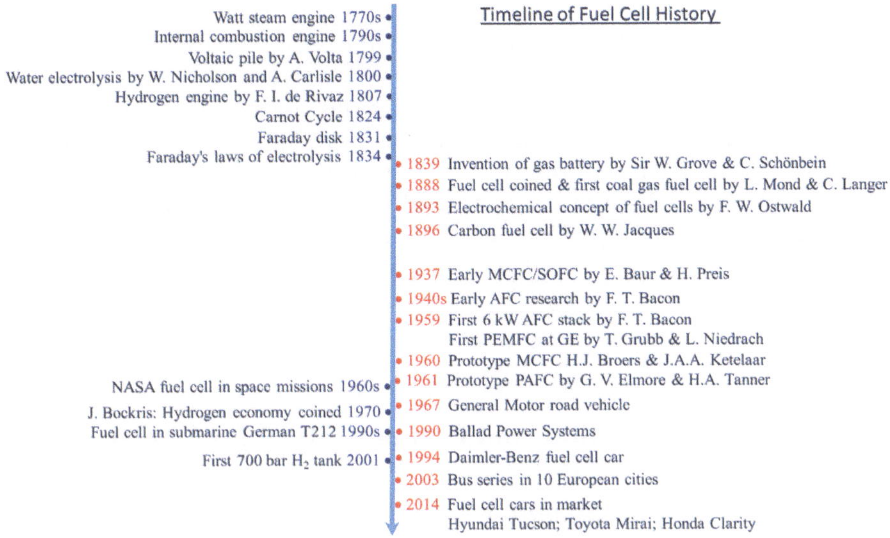

Fig. 1.3 Timeline of fuel cell development

During the last 30–40 years, fuel cells have received enormous attention worldwide as alternative and clean electrical energy conversion technologies because of their potentials for power generation in portable, transport, and stationary applications. The ever-growing public awareness of the finite reserves of fossil fuels, greenhouse gas emissions, and resultant global warming have been the major driving force of the fuel cell development. In the future, fuel cells will play a key role in increasing the share of the renewable energies in the global energy systems, particularly the electricity generation which has reached 28% in 2020.

First Fuel Cell

There are some disagreements on who actually discovered the first working fuel cell. The discovery of fuel cell phenomena, i.e., the generation of electricity from an electrochemical cell can be dated back to the beginning of nineteenth century and ascribed to the English chemist Humphry Davy in 1802. In his primitive cell he used a carbon anode and aqueous nitric acid as reactant at the cathode, a weak electrical signal was observed. In 1839, the Swiss scientist Christian Friedrich Schönbein first asserted the possibility of a fuel cell that combined hydrogen with oxygen. However, we can see in the letter that Schönbein wrote to Michael Faraday on April 4, 1840, as follows: "...*Our friend Grove wrote me the other day communicating to me some very interesting results of his later researchers with which you are no doubt acquainted. What do you think about the fact that the transfer or oxidation of particles in the voltaic arc is definite for a definite current? Important as such fact is I cannot yet understand it, that is to say, I cannot conceive that by the simple oxidation of the positive electrode the current can be conducted in the same way as by electrolysis, both actions being so very different from each other ...*". This shows that Grove was probably the first to introduce and demonstrate the hydrogen fuel cell. In the experiment conducted in 1838, Grove used hydrogen as fuel and oxidized it with oxygen in a voltameter with two platinum electrodes in a solution of sulphuric acid and observed a constant current flowing between the two electrodes. The finding was called "*gas battery*" and was published in Phil. Mag., vol. 14, p.129 in 1839.

1.3 Types, Construction, and Components of Fuel Cells

1.3.1 Fuel Cell Classification

Types of fuel cells can be classified according to different categories based on temperatures, electrolytes, or fuels, as shown in Fig. 1.4.

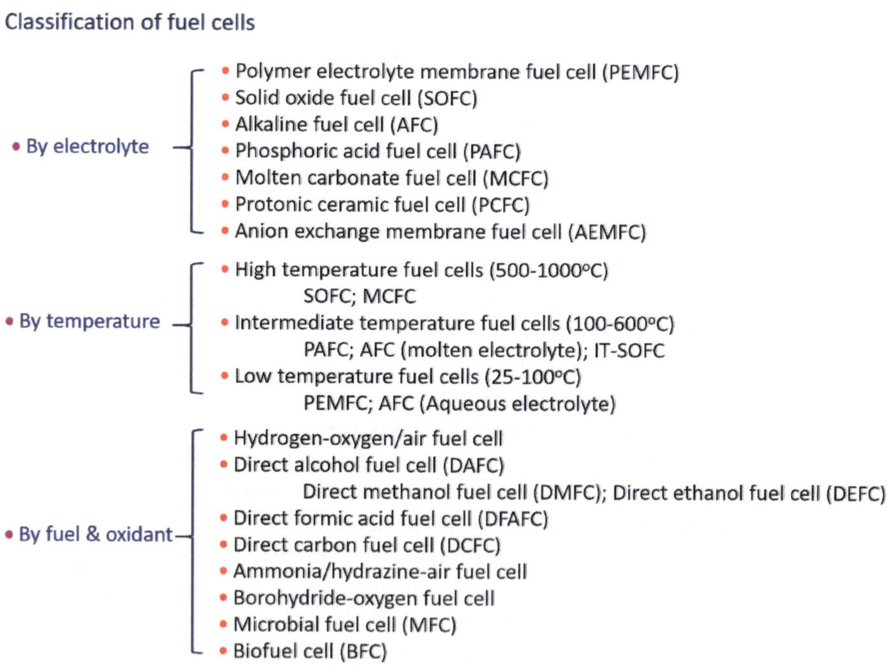

Fig. 1.4 Examples of fuel cell classification

The popular classification of fuel cells is based on the electrolyte materials employed and is also used in this book. Based on the nature of the electrolytes, fuel cells can be categorized into *proton exchange or polymer electrolyte membrane fuel cells* (PEMFCs), *solid oxide fuel cells* (SOFCs), *alkaline fuel cells* (AFCs), *anion exchange membrane fuel cells* (AEMFCs), *phosphoric acid fuel cells* (PAFCs) and *molten carbonate fuel cells* (MCFCs). The electrolytes in fuel cells must have a good conductor of $O^{=}$ or H^{+} or other charge carriers containing $O^{=}$ or H^{+} such as OH^{-}, H_3O^{+} or $CO_3^{=}$ with negligible electronic conductivity. The choice of electrolytes largely determines the operating temperatures, electrode or half-cell reactions, materials, and the features of the construction and operation of fuel cells. Fuel cells based on aqueous electrolytes like AFCs and PAFCs are limited to temperatures of up to 200 °C because of the high vapor pressure of electrolytes. Solid oxides and/or molten salt-based electrolyte can be operated to temperatures of 600 °C or higher.

In addition to the clarification based on the electrolyte types, fuel cells can also be clarified according to the fuels and oxidants used, for example, fuel cells using alcohols such as methanol, ethanol, propanol, etc. without reforming are called *direct alcohol fuel cells* (DAFCs), which can be further classified into *direct methanol fuel cells* (DMFCs), *direct ethanol fuel cells* (DEFCs), *direct formic acid fuel cells* (DFAFCs), etc. SOFCs directly using solid carbon fuels are called *direct carbon fuel cells* (DCFCs). Another example is *microbial fuel cells* (MFCs) or *biofuel cells* (BFCs) using the biochemical substance as the fuel: A biochemical substance is decomposed by means of an enzyme or bacteria in solutions to produce hydrogen.

1.3.2 Construction and Components of Fuel Cells

A single fuel cell has three basic functional components: ionic conductive electrolyte, anodic electrode, and cathodic electrode. These three components are assembled together and often referred to as the membrane-electrode assembly (MEA), as shown in Fig. 1.5.

The electrolyte has several functions. It serves as a barrier to molecule permeation and hence separates the fuel, e.g., hydrogen and the oxidant, typically oxygen. It is essential that the electrolyte prevents the direct chemical combustion of the fuel. It is an ion conductor, letting ions migrate across it.

By means of the electrolyte, the overall cell reaction is split into two half-cell reactions. In the case of oxide ionic conducting electrolytes, one has:

$$\text{Cathode:}\quad \frac{1}{2}O_2 + 2e^- \rightarrow O^{=} \tag{1.1}$$

$$\text{Anode:}\quad H_2 + O^{=} \rightarrow H_2O + 2e^- \tag{1.2}$$

$$\text{Overall:}\quad H_2 + \frac{1}{2}O_2 \rightarrow H_2O \tag{1.3}$$

Similarly for cells based on proton-conducting electrolytes:

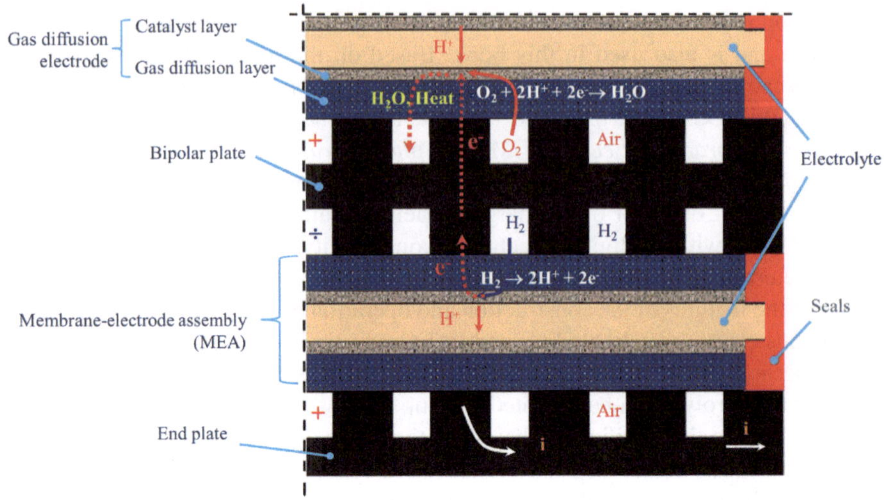

Fig. 1.5 Schematic of a PEMFC construction

$$\text{Anode:}\quad H_2 \rightarrow 2H^+ + 2e^- \tag{1.4}$$

$$\text{Cathode:}\quad \frac{1}{2}O_2 + 2H^+ + 2e^- \rightarrow H_2O \tag{1.5}$$

$$\text{Overall:}\quad H_2 + \frac{1}{2}O_2 \rightarrow H_2O \tag{1.6}$$

More half-cell reactions for other types of electrolytes are shown in Fig. 1.6. For hydrogen–oxygen fuel cells, the conducting ions are preferably the product of either hydrogen oxidation (H^+) or the oxygen reduction ($O^=$). Other conducting species containing either H^+ or $O^=$ such as OH^-, H_3O^+ or $CO_3^=$ are also possible but need to manage the balance of the carrying atoms or molecules. Figure 1.6 schematically shows the configurations of five types of fuel cells. With cation conducting electrolytes water is formed on the cathode (air) side while for anion conducting electrolytes it is on the anode (H_2) side. In the case of $CO_3^=$ conducting electrolytes CO_2 is needed at the cathode and formed at the anode. For electrolytes conducting OH^-, water should be supplied on the cathode and extra water is formed at the anode. In the case of H_3O^+ conducting electrolytes, on the other hand, water should be added to the fuel stream and extra water is formed on the cathode side. The red dashed lines indicate the issues of CO_2 and H_2O management during the fuel cell operation.

At the same time the electrolyte should be mechanically and chemically stable. In the case of liquids, the electrolytes are often immobilized in an inert matrix for easy handling and cell assembling.

As also seen from Fig. 1.5 the electrode is typically made of a porous substrate onto which catalysts are applied. The porous substrate is also called a gas diffusion

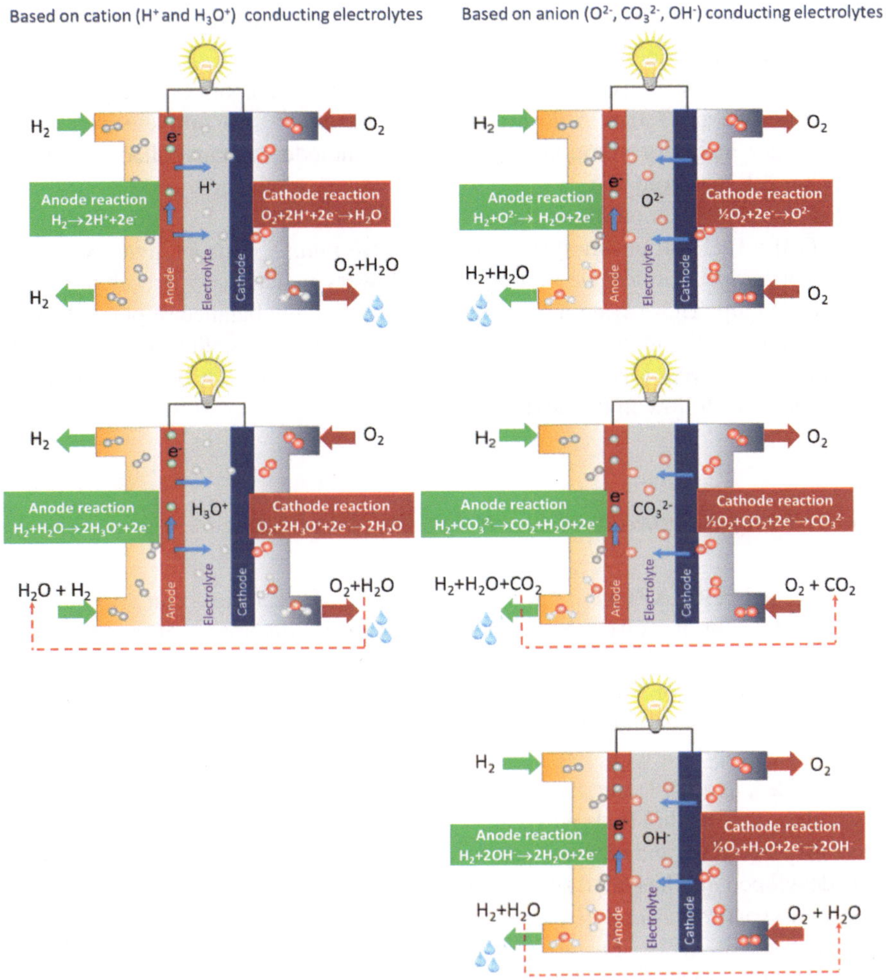

Fig. 1.6 Schematic representation of configurations and operations of fuel cells based on cation (left) and anion (right) conducting electrolytes

layer (GDL) and such an electrode called a gas diffusion electrode (GDE). The GDL should be an electronic conductor allowing for a series connection of cells, which is probably the simplest configuration of stacking many cells. The porous structure of the GDL is important as both reactants (H_2 and O_2) and products (H_2O) need to transport to/from the catalyst active sites to the gas channels.

When multiple cells are stacked on top of each other, an interconnect material is needed. In the configuration shown in Fig. 1.5, this interconnect provides gas channels for both hydrogen and oxygen and serves as cathode (positive) on one side and as anode (negative) on the other. It is therefore often referred to as bipolar plates. This material should be electronically and thermally conductive, dense for gas impermeability, mechanically strong, and most importantly corrosion resistant toward the highly reducing and oxidative atmospheres on each side.

In addition, plastic gaskets for low-temperature fuel cells or ceramic glass fibers for high-temperature fuel cells are used for gas tight seal and electrical insulation. Other auxiliary components of a stack include current collector, cooling plate/channel, manifold, etc.

A stack, often in module, is the power generating unit of a system. Except the stack itself a fuel cell power system includes also pumps, sensors, heat exchanger, compressors, gas humidifier, recirculation blowers or catalytic burner for anode exhaust cleanup. These system components are called the balance of plant (BoP).

Table 1.1 lists the main different types of fuel cells, as classified according to the electrolyte employed, together with their main characteristics. A brief discussion of these items will follow in the next section.

1.3.3 Brief Summary of Each Type of Fuel Cells

Alkaline fuel cells (AFCs) use concentrated KOH as electrolyte and depending on the concentration of KOH, AFC can operate at high temperature of ca. 200 °C for 85% KOH and low temperatures of 65–100 °C for less concentrated KOH. AFC can be classified as immobilized electrolyte type using porous matrix (e.g., asbestos) for space application and circulated electrolyte type for terrestrial application. Due to the alkaline electrolyte, a wide range of electrocatalysts can be used, including Ni, Ag, metal oxides, and noble metals. CO is a poison gas and CO_2 either from air or from the oxidation of alcohol fuels will react with KOH to form K_2CO_3 precipitates, which alters the electrolyte and blocks electrode pores. AFC was one of the first modern fuel cells developed and has been successfully employed as on-board electric power for the Apollo space missions. However, its terrestrial application has been challenged by its high sensitivity to CO and CO_2. As a result, the need for high purity of hydrogen and CO_2 scrubbers has a significant impact on the size and cost of the system.

Replace the aqueous alkaline electrolyte with a solid alkaline anion exchange membrane (AEM) provides a paradigm shift with the combined advantages of both PEMFCs and traditional AFCs. In AEMFCs, AEM is a polymer electrolyte that conducts anions, e.g., OH^-. When the positively charged (cationic) groups are bounded to a polymer backbone, the CO_2 poisoning is significantly mitigated as no free carbonates are precipitated, though the carbonation still decreases the anion conductivity. A challenge in the development of AEMs is its long-term stability [2].

Polymer electrolyte membrane fuel cells (PEMFCs) use polymeric materials as the electrolyte. The most common PEMs are based on perfluorosulfonic acid (PFSA) such as Nafion®. The materials exhibit high ionic conductivities above 0.1 S/cm; however, under highly hydrated conditions, which limits the operating temperatures at around 80 °C. PEMFCs are the most mature technologies for the development of *fuel cell vehicles* (FCVs). In PEMFCs, platinum (Pt) is the state-of-the-art electrocatalyst for both *oxygen reduction reaction* (ORR) at the cathode and *hydrogen*

Table 1.1 Classification and characteristics of fuel cells

	AFC	PEMFC	PAFC	MCFC	SOFC
Temperature/°C	40–80/200–260	40–80	180–210	600–650	600–1000
Electrolyte (in matrix)	Aqueous/molten KOH (Asbestos/polymer/ceramic)	Perfluorosulfonic acid	H_3PO_4 (soaked in SiC)	Li–K(Na)CO_3 (soaked in $LiAlO_2$)	Y_2O_3–ZrO_2
Charge carrier	OH^-	H^+ (H_3O^+)	H^+	$CO_3^=$	$O^=$
Anode	Ni, metal oxides	Pt/C	Pt/C	Ni alloy	Ni/YSZ
Cathode	Pt/C/NiO	Pt/C	Pt/C	Lithiated NiO	LSM
Interconnect/bipolar plate	Metal	Graphite/metal/composite	Carbon composites	Stainless steel	Metal/ceramic
Primary fuel	H_2	H_2 (+CO_2)	H_2 (+CO_2 + 1%CO)	H_2/CO	H_2/CO/CH_4
Oxidant	O_2 (CO_2 scrubbed air)	O_2/air	Air	Air + CO_2	Air
Efficiency	45–60%	40–60%	40–45%	45–60%	60–65%
Power/size	1–100 kW	10 W–1 MW	200–400 kW	0.3–3 MW (300 kW module)	1 kW–2 MW

(continued)

Table 1.1 (continued)

	AFC	PEMFC	PAFC	MCFC	SOFC
Operating feature	+ Fast cathode kinetics + Cheap electrolyte + Low-cost catalysts & stack materials – Sensitive to CO_2 in both fuel and air – H_2 circulation – Complex systems	+ Solid construction + High power density + Low temperature + Quick start-up + Fast load response – Noble metal catalysts – CO intolerance – Water management – Intensive cooling – Low waste heat value	+ CO tolerance (External CH_4 reforming) + No water management + Easy cooling + Good heat value (CHP) – Poor ORR kinetics – Noble metal catalysts – Slow start-up – Low power density	+ High efficiency + Fuel flexibility (Internal CH_4 reforming) + Non-noble metal catalysts + Cheap stack materials + Good heat value (CHP) – Large Ohmic resistance – Low current/power density – HT corrosion – Slow start-up	+ High efficiency + Solid construction + Fuel flexibility (Internal CH_4 reforming) + Non-noble metal catalysts + Good heat value (CHP) – Thermal mismatch – Poor thermal cyclability – Slow start
Applications	• Military • Space	• Backup power • Portable power • Transport • Distributed generation	• Distributed generation	• Electric utility • Distributed generation	• Auxiliary power • Electric utility • Distributed generation
Notes	• Active development of anion exchange membrane	• Active development of HT-PEMFC			• Active developement of IT-SOFC

Note Plus sign " + " indicates advantage and negative sign " – " indicates disadvantage

oxidation reaction (HOR) at the anode, imposing issues of the scarce resource and high cost, though the metals can be recycled at the end of their service life. Significant reduction in the precious metal usage has been achieved, and further efforts are being made in optimizing the composition (by alloying) and nano-morphologies of electrocatalysts as well as use of structured carbon supports. In addition, development of non-precious metal (NPM) catalysts as replacement of platinum has made considerable progress.

In terms of system construction and operation, PEMFC faces challenges such as water management, effective cooling at very high current densities. Sensitivity to the fuel impurities, e.g., CO is also an issue. Alternative technologies that can operate at temperatures above 100 °C are under active development. The most successful one is the phosphoric acid doped polybenzimidazole (PA/PBI) membrane fuel cells with operating temperature at 160—170 °C.

Phosphoric acid fuel cells (PAFCs) use 100% H_3PO_4 as electrolyte and typically operates at temperatures of 150–220 °C. Silicon carbide is stable in hot phosphoric acid and is a common matrix used to retain the acid. The high operating temperature makes PAFCs less sensitive to CO poisoning than that of PEMFC and AFC and can tolerate about 1% CO. Thermal management is relatively simple, and the waste heat from PAFCs can in principle be used in the industrial cogeneration applications. PAFCs have been developed primarily for stationary applications, and the first commercially available unit and also best known is the ONSI PC-25 systems developed by UTC fuel cells with 200 kW electricity power and 220 kW heat. However, the highly corrosive nature of phosphoric acid electrolyte has a significant impact and demand on the electrocatalysts and stack component requirements.

Molten carbonate fuel cells (MEFCs) use a mixture of alkali metal carbonates as the electrolyte, which is retained in a ceramic matrix of $LiAlO_3$. MCFC operates at 600–650 °C which makes it possible to use non-precious Pt metal and metal oxides such as Ni and NiO as electrode catalysts. The high operating temperature also offers other advantages such as direct use of CO as fuel and CH_4 via internal reforming, simplifying the ballance of plant (BoP). Better utilization of waste heat is also an option with improved overall efficiency. Stainless steel can be used as the construction materials. The main challenge of MCFCs is the highly corrosive nature of molten carbonate electrolyte and further improvement of the lifetime.

Solid oxide fuel cells (SOFCs) are an all solid devices and do not require precious metal electrocatalysts due to the high operation temperatures (600–1000 °C) [3]. The high operation temperature also makes it possible to use a wide range of fuels by internal reforming of natural gas, hydrocarbons, and alcohols without the need for extra fuel processing and purifying units, a distinct advantage over other fuel cell types. Fast electrode kinetics allows SOFC to operate at high energy efficiency, also with high quality of waste heat. For this reason, SOFCs are particularly suitable for stationary and distributed power supply applications, ranging from kW to MW. A typical SOFC consists of a porous Ni anode supported on yttria-stabilized zirconia (YSZ), a dense oxide-conducting YSZ solid electrolyte, and a porous lanthanum strontium manganite (LSM) oxide cathode. Traditional SOFCs operate at high temperatures (900–1000 °C) because of the relatively low oxide conductivity and high activation energy of oxide electrolytes such as YSZ.

The key challenges in SOFC technologies are the thermal stability or thermal cycle capability, stack durability and system controllability, which are interrelated. In all solid-state devices like SOFCs, any thermal mismatch between components would lead to thermal instability, but a good system control and management can significantly minimize the thermal stress and subsequently reduce the degradation and prevent system failure during the start-up and shutdown cycles and in the case of emergencies. The stack durability depends strongly on the activity and stability of electrode materials. However, a major challenge associated with a SOFC system over a 5-year lifetime is the gradual degradation and deactivation of the electrodes by contaminants which can be either in the fuel and air streams or from the volatile species of cell components, such as metallic interconnect, sealant, and manifold.

Lowering of the operating temperature to an intermediate temperature range of 500–800 °C, the so-called intermediate temperature SOFCs or IT-SOFCs brings substantial technical and economic benefits. The cost of an SOFC system can be substantially reduced by using less costly metallic alloys as interconnect and compliant temperature gaskets. Furthermore, as the operation temperature is reduced, system reliability and durability of SOFCs increase which may open the possibility of using SOFCs for a wide variety of applications including residential and automotive applications.

1.4 Fuel Cells Versus Batteries

Electrochemically, the basic principles of a fuel cell are the same as a battery, which is an indispensable part of many activities of our everyday life. The most important difference is that a fuel cell can generate power continuously by converting the chemical energy from an external fuel storage or supply while a battery stores energy, i.e., from power to power.

In the case of batteries, all chemicals are stored inside the device which has therefore a finite capacity. For fuel cells, electrodes are made of inconsumable catalyst while the reactants, fuel to the anode and oxidant to the cathode, are stored outside of the device and supplied when they are needed. Oxygen is often supplied as air so that it does not need to store. At the same time the electrode reaction product, water when hydrogen is used and CO_2 as well when hydrocarbons are used as fuels, are released from the device. The requirement of fuel and oxidant supply dictates that the BoP is complicated and should be taken into account as an integrated part of the power unit when considering the mass and volume power density of a fuel cell system.

A fuel cell can operate continuously as long as the fuel is supplied, like a gasoline or diesel engine while a battery operates discontinuously—when the chemical energy has been converted to electricity, the battery must be thrown away if it is a primary battery or recharged if it is a secondary or rechargeable battery. In other words, the capacity of a fuel cell is limited by the size of the fuel tank while the capacity of a battery is limited by the size of the battery itself. To double the power capacity of a battery, one needs to double the mass and volume of a battery. For fuel cells,

Table 1.2 Similarities and differences between fuel cell and battery technologies

Fuel cells	Batteries
Similar electrochemistry principles	Similar electrochemistry principles
Energy conversion—fuel to electricity	Energy storage—electricity to electricity
Useful unity—BoP and system	Useful unity—cell
Infrastructure—fuel supply	Infrastructure—electricity supply
Technology—R&D	Technology—mature
Road to commercialization—still very long	Road to commercialization—commercially available

however, one needs only to double the size of the fuel tank, not the fuel cell stack. This makes fuel cell a favorable power technology for large-scale applications, for example, heavy duty vehicles or ferries.

In addition, the recharging of batteries is often a time-consuming process, involving conversion of electrical energy to chemical energy—fast recharging often compromises the lifetime of batteries. Recharging a battery-based electric car needs several hours from empty-to-full and at least half an hour for addition of 150 km range. Refueling a fuel tank for fuel cells, on the other hand, is a mechanical process and can be done within 3–5 min.

Another remark to add is that, as a power-to-power storage device, the battery-based electrical cars are beneficial for reducing the emissions if the power is from a renewable source otherwise batteries are only shifting the greenhouse gas emission. These discussions are summarized in Table 1.2.

1.5 Unitized Regenerative Fuel Cells and Reversible Fuel Cells

It is desirable that a power system can operate reversibly, i.e., combining water electrolysis cell (EC) and fuel cell, the so-called regenerative fuel cell (RFC) system. In association of renewable energy sources from solar and wind power the RFC system is an attractive approach for the energy storage and regeneration via H_2 the energy carrier:

$$H_2O + \text{electricity} \xrightarrow{\text{electrolysis mode}} H_2 + \frac{1}{2}O_2 \tag{1.7}$$

The produced hydrogen during the peak power period can be stored and converted back to electricity when needed during the low power period:

$$H_2 + \frac{1}{2}O_2 \xrightarrow{\text{fuel cell mode}} H_2O + \text{electricity} \tag{1.8}$$

The RFC can be an integrated system consisting of discrete EC and FC modules that operate alternatively during the charging (EC for power storage) and discharging (FC for power generation). Another system is the unitized regenerative fuel cell (URFC), which is a reversibly bifunctional device that can work in either EC or FC mode. An obvious advantage of URFCs is the combination of the fuel cell and electrolysis into one device with dual functions of power and fuel (in most cases, hydrogen) generation. Technical features of URFCs include reduced weight, increased power density and improved reliability and simplified and cost-effective system design and management.

As required by the reversible operation, the electrodes used in URFCs must be bifunctional and this actually puts a very high standard on the electrochemical activity and stability of electrode materials. The bifunctionality is dependent on the types of electrolytes used which are based on proton exchange membranes (PEM), anion exchange membranes (AEM), and solid oxides.

For PEM-based URFCs, Pt catalyst is an excellent bifunctional hydrogen catalyst with high activity for the H_2 oxidation reaction (HOR) and H_2 evolution reaction (HER). For the oxygen electrode, the bifunctional catalyst is a combination of Pt catalyst for O_2 reduction reaction (ORR) with the best O_2 evolution reaction (OER) catalyst, i.e., Ir/Ru oxides. For AEM-based URFCs the alkaline electrolyte allows for uses of Pt or Ni alloys as the hydrogen electrode catalysts and noble/non-noble metal oxides as the oxygen electrode catalysts.

The challenges facing the PEM- and AEM-based URFCs is the low round-trip efficiency (around 40–50%) due to the sluggish electrode kinetics and high cost due to the use of precious metal catalysts and low stability due to the increased complexity in the water and heat management [4]. The technologies are primarily limited to aerospace and military defense fields.

High-temperature solid oxide-based URFCs or more often called reversible *solid oxide cells* or SOCs in the SOFC communities offer a practical potential in civil areas in power grid, renewable solar and wind energy distribution, backup power supply. The electrode materials and structure in solid oxide electrolysis cell (SOEC) can be essentially the same as in SOFC. Thus, SOC can operate reversibly between fuel cell mode and electrolysis mode. Most interestingly, reversible operation can in fact increase the stability of the solid oxide cells particularly at the oxygen electrode [5, 6]. Figure 1.7 shows the operation principles of a SOC. Using renewable electricity generated from solar and wind power (e.g., during the peak period, the day time), SOCs can store the electricity in H_2 by electrolysis and generate electricity using the stored hydrogen when needed during the low period (e.g., the night time) without carbon penalty. Such combination of renewable energy storage with electricity regeneration is a unique feature of high-temperature solid oxide cells and is attributed to the regeneration of the electrode/electrolyte interface induced by the polarization [7]. This high-temperature process can reach a round-trip efficiency of 60–80%.

In addition to the H_2-H_2O-electricity cycles, fuel cells can also be used for the production of hydrocarbon or alcohol fuels. In the electrolysis mode of high-temperature SOC, captured CO_2 from flue gas or atmosphere and water can be

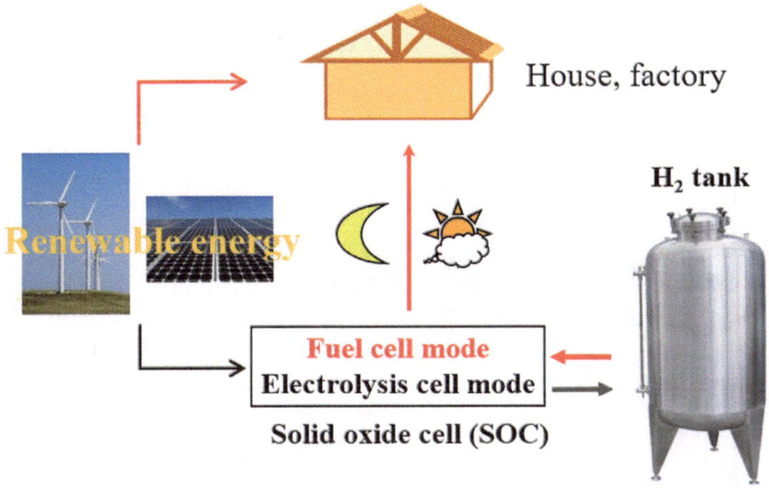

Fig. 1.7 Operation principle of SOC in SOEC operation mode to store the renewable energy in H_2 and SOFC operation mode to generate electricity using the stored H_2

co-electrolyzed to form syngas ($CO + H_2$), which then can be converted to CH_4 or liquid hydrocarbon fuels through the Fischer–Tropsch reaction or chemical engineering process. The process is called power-to-*x* (P2X), where *x* represents a variety of value-adding chemicals such as hydrogen, methane, methanol, ethanol, etc. These chemicals can be used as industrial feedstock or as fuels to be converted back to power whenever and wherever it is required.

1.6 Applications and Prospect of Fuel Cell Technologies

Fuel cells can be used in a wide range of applications meeting energy demand: stationary and portable electricity generation, land and marine transportation, communications and portable electronic devices. The range of fuel cell applications and the size of the potential markets for fuel cell-based energy devices are enormous. Figure 1.8 shows applications and the trend of fuel cells in general [8].

Worldwide, the transport sector accounts for about 27% of the fuel consumption (primarily fossil fuels) and is responsible for ca. 25% of the total carbon dioxide emissions [9]. Fossil fuels are unsustainable. Thus, a number of alternative, clean, and sustainable energy solutions are under active development, including *battery electric vehicles* (BEVs) and fuel cell vehicles (FCVs). PEMFCs have been the technology of choice for transportation because of their low-temperature operation and rapid start-up/shutdown capability. The commercialization of hydrogen-powered FCV are however slow. This is largely due to lack of the infrastructure for pure H_2 production, storage, transportation and distribution [10]. Since H_2 cannot be found

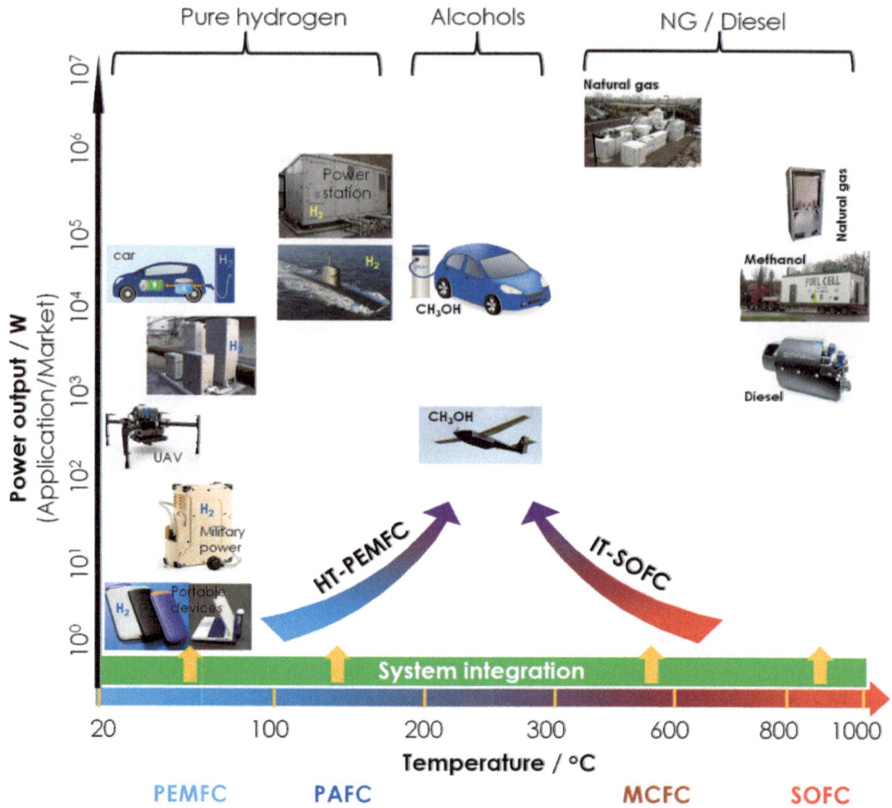

Fig. 1.8 Application types and power ranges of main fuel cells as a function of operation temperatures. The arrows show the R&D trend in the development of HT-PEMFCs and IT-SOFCs. Reproduced from Ref. [8] under Creative Commons CC BY license

free in nature, it has to be extracted from different energy sources. Currently, the majority of H_2 (~80–90%) is made from fossil fuels by stream reforming of natural gas. The storage, transportation, and distribution of pure H_2 are all very expensive. Safety of H_2 filling stations, particularly in populated cities, is also of concern for hydrogen-fuel-powered FCVs or *h*-FCVs.

Liquid alcohol fuels like methanol for FCVs have a higher volume-specific energy density (17.28 MJ/L) compared to that of H_2 (1.9 MJ/L@20 MPa). In addition, methanol fuels are less expensive and can be produced from renewable energy sources, e.g., agricultural wastes and other biomass by thermochemical processes, and can also be obtained from feedstock such as coal by the gasification and steam reforming processes. Methanol can be stored and distributed easily via the existing network of petrol filling stations, and in the context of the transport sector, the transfer from pure hydrogen fuel to methanol fuel will substantially accelerate the wide and practical development and application of methanol-fuel-powered FCVs, or *m*-FCVs. Scenarios of hydrogen and methanol-fueled FCVs are schematically shown in Fig. 1.9.

However, there are significant and technological hurdles facing the development of PEMFCs based on renewable alcohol fuels, such as methanol. The activity of the state-of-the-art Pt-based electrocatalysts for methanol oxidation reactions is very low at ambient temperatures. Increasing the operating temperature of PEMFCs can significantly enhance the activity of the electrocatalysts, but the key challenges are the relatively low proton conductivity and thermal stability of polymer electrolyte membranes at elevated temperatures. Development of new and alternative PEMs with high proton conductivity and stability at elevated temperatures is an active research area in PEMFCs [11].

One of the significantly growing application areas for SOFC systems in the 1–5 kW size is the residential combined heat and power (CHP) systems operating on natural gas. For example, the units with power output of up to 2 kW and 60% electrical efficiency have been demonstrated to power an average home. Residential CHP

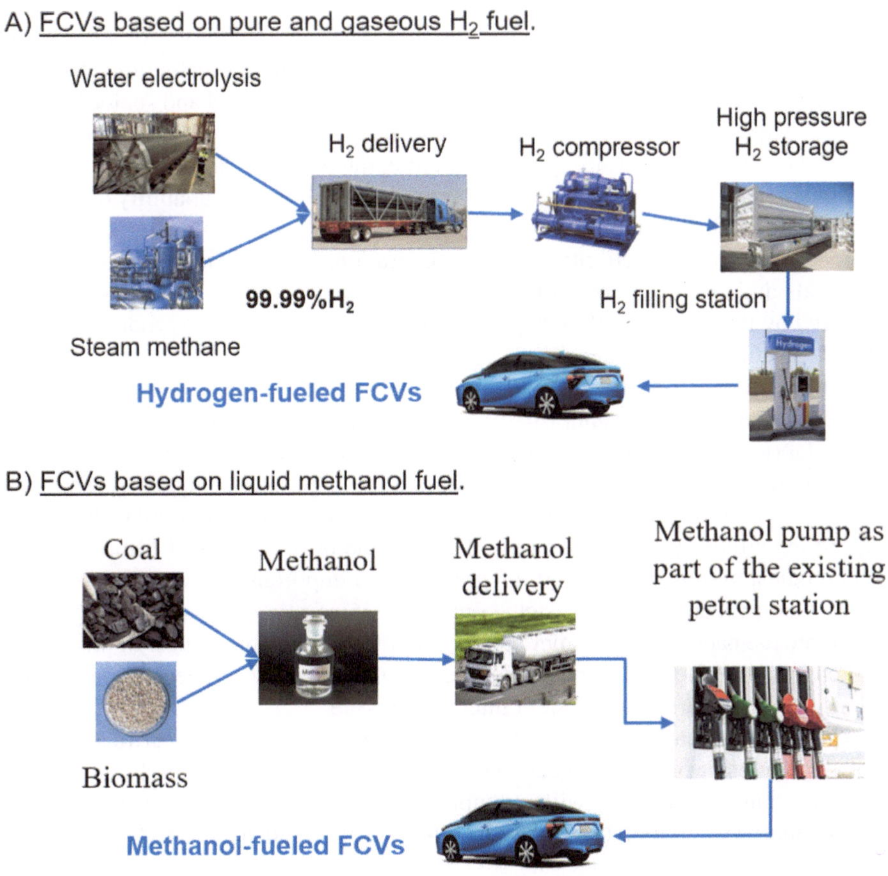

Fig. 1.9 Fuel supply chain for hydrogen and methanol-fueled fuel cell vehicles

systems using SOFCs could be the first commercial product. By lowering the operation temperature, faster start-up and shutdown time, and more rugged construction by means of compliant seals and metallic interconnects, IT-SOFCs can be a viable technology for mobile applications, e.g., as auxiliary power units (APUs) and traction power in heavy duty vehicles.

The power density requirement for portable power sources is ever on the increase, and power consumption is forecast to pose long-term technical challenges for the portable electronic industries, which are working to find ways to extend the running time of mobile devices such as portable computers and mobile phones. The power range is 2–5 W for mobile phone and 15–30 W for laptop computers. In fact, the performance, mass and volumes, and lifetimes of power supplies presently limit most applications of microelectromechanical systems technology. Miniaturized fuel cells or microfuel cells are particularly very attractive for powering these portable electronic devices. The most apparent benefit over batteries is longer service time between recharges, given the considerably greater energy density of liquid fuels than that of state-of-the-art batteries.

There is a significant and growing interest in the fuel cell technology for various applications including submarine power, unmanned aerial and underwater vehicles, base camp power and personal power supplier. The integration of biofuels such as biomethanol and bioethanol is a promising and forthcoming technology for electricity and heat cogeneration along with profound environmental and socioeconomic benefits. High-temperature fuel cells create many synergies, e.g., recovery of the waste heat in external thermal cycles. Such integrated systems are well suited for the distributed power generation for remote areas due to the wide availability of biomass and flexible modulability of the fuel cell technology.

In addition to the electricity and heat generation, fuel cells can also be used for chemicals and energy cogeneration. In cogeneration mode, the fuel cells are characterized by their current efficiency, selectivity with respect to the product obtained and/or current and power densities.

A key difference of fuel cell technologies from other electrochemical energy devices is that fuel cells cannot be used without the system integration. Fuel cells can be fabricated in micrometer and even nano-scale, but the system required for a workable fuel cell would be orders of magnitudes larger and complex. As shown in Fig. 1.8, the commercial viability and successful application of fuel cell technologies require ingenious system integration. Thus, in the development of future fuel cell technologies, system control and integration is as important as the fuel cells themselves. Compact and reliable system control and integration play an enabling role to accelerate the pace of the commercialization of fuel cell technologies. In this aspect, challenges in the fuel cell technologies are not only scientific and technological ones but also an engineering and system integration issue.

Among many challenges in fuel cell technologies, one of the most critical issues is probably the electrolyte. A fundamental breakthrough in AEM or high-temperature proton exchange membranes will undoubtedly open new opportunities in the development of non-precious metal catalysts and integrated fuel cells/on-board fuel reformers for the practical employment of fuel cell technologies in transportation, communication, and portable devices. In SOFCs, the main benefit of the development of new or alternative electrolytes with high ionic conductivity and structure stability at intermediate and low temperatures is the substantially increased device durability and direct utilization of active and mixed ionic and electronic conducting cobaltite-based perovskite cathodes, but this requires innovation in materials and substantially enhanced ionic conductivity, chemical and structural stability under realistic fuel cell operation conditions.

Despite its long history, the fuel cell is a new player in energy conversion and storage in comparison with the mutual and well established ICE and battery technologies. The barriers facing the commercial viability of fuel cell technologies are both technical and marketing. The technical issues of fuel cells include total cost of materials and manufacturing of fuel cell stack, system durability, dynamic operation environment, and fuel infrastructure while the market barriers are related to the environment impact, government policies, regulations, and interface with the renewable power market. This is a complicated challenge, requiring considerable R&D efforts, collaborations and investment not only from government policy-makers, research institutes and academies but also from the industrial developers and private investors. With continuing and sustainable efforts, we believe there is a bright future for the truly commercially viable and competitive fuel cell technologies for the energy conversion and power source applications.

1.7 Summary

In this chapter, we have introduced and discussed the following topics:

- Fuel cell is an electrochemical device to directly convert the chemical energy of fuels such as hydrogen, methanol, and natural gas to electricity. Due to the nature of electrochemical process, fuel cell is inherently more efficient than that of internal combustion engine (ICE). In fuel cells, the most common fuel is hydrogen and air is the most common oxidant.
- Fuel cell was invented almost at the same time as that for the ICE but found little practical use in the first century due to the lack of understanding of electrochemistry. The fuel cell development in early years was primarily stimulated by applications in the space programs. The more recent interests in fuel cells have been driven mainly by the need of high energy efficiency and capacity, non-interrupted power sources and increased concerns in the reduction in the greenhouse gas emission and the renewable energy conversion and storage. The major barriers for the fuel cell commercialization are the high cost and lack of infrastructure of supplying hydrogen fuel.
- Fuel cell is a family consisting of different types which can operate from room temperatures to as high as 1000 °C. Five main types are: alkaline fuel cell (AFC), phosphoric acid fuel cell (PAFC), polymer electrolyte membrane fuel

cell (PEMFC), molten carbonate fuel cell (MCFC), and solid oxide fuel cell (SOFC). In fuel cells, electrolyte materials are the most critical component, which determines the features of fuel cell configurations, materials, constructions, and operation.

- Unlike batteries, the electrodes of fuel cells are not consumed during the discharge—they merely act as reaction sites for the reactants (fuel and oxidant) which are stored externally to the cell. The requirement of fuel and oxidant supply dictates that the system control is an integrated and important part of the fuel cell, very different from batteries.
- Not all fuel cells can be reversibly operated. The key requirement for the unitized regenerative, regenerative, or reversible fuel cell operation is the dual functionality of both anode and cathode.
- Different fuel cells have both pros and cons for the specific applications. Applications of fuel cells as a power source for transportation are most attractive from zero emission operation when the hydrogen fuel is produced from renewable energy sources. For hydrogen fuel-based FCVs, a major challenge is the generation, storage, transportation, and distribution of pure hydrogen. The alternative is to replace the gaseous hydrogen fuel with liquid fuels such as methanol.

1.8 Questions

1.1 Explain the following terms and materials:

Gas battery, fuel cell, internal combustion engine, and battery
Oxide ion conducting electrolytes and proton conducting electrolytes
Cation conducting electrolytes and anion conducting electrolytes
Reversible fuel cell
Power-to-x
Fuel cell vehicles.

1.2 Discuss the advantages and disadvantages of various fuel cells listed in Table 1.1. Give specific examples of fuel cell applications.

1.3 Discuss the prospective and implications of fuel cell vehicles (FCVs) and battery electric vehicles (BEVs) as the main transportation means in the society.

1.4 Can fuel cell-based devices be as reliable as ICEs? (Clue: consider the differences in the interfaces in fuel cells and ICEs).

1.5 Francois Isaac de Rivaz invented the first hydrogen-powered internal combustion engine (ICE) with an electric ignition in 1807, and it took ca. 100 years for Ford Motor Company to produce first affordable automobile Model T car in 1908. The first fuel cell was demonstrated in 1838 and about 180 years late, fuel cell cars are beginning to enter the market. Explain the reasons behind the very different technological development paths of ICE and fuel cells.

1.6 Hydrogen, instead of fuelling fuel cells, can also be used as a fuel for internal combustion engine. What are the issues of the hydrogen internal combustion engine vehicles compared to hydrogen fuel cell vehicles?

1.7 Discuss the fuel production, transportation, storage, and distribution of hydrogen and methanol fuels for FCVs. What are the technical and market barriers for the commercialization of hydrogen- and methanol-fueled FCVs?

1.8 Discuss the issues and challenges facing UR-PEMFCs. Why water management is more complicated in UR-PEMFCs in comparison with PEMFCs?

1.9 General Readings

1. Jiang SP, Wang X (2011) Fuel cells: advances and challenges. In: Kharton VV (ed) Solid state electrochemistry II. Wiley-VCH, Weinheim, Germany, pp 179–264
2. Winter M, Brodd RJ (2004) What are batteries, fuel cells, and supercapacitors? Chem Rev 104:4245–4269
3. Dicks AL, Rand DAJ (2018) Fuel cell systems explained, 3rd edn. Wiley, New York.
4. O'Hayre R, Cha SW, Colella W, Prinz FB (2016) Fuel cell fundamentals. Wiley, New York.
5. Minh NQ, Takahashi T (1995) Science and technology of ceramic fuel cells. Elsevier, Amsterdam

References

1. Andujar JM, Segura F (2009) Fuel cells: history and updating. A walk along two centuries. Renew Sust Energ Rev 13(9):2309–2322
2. Varcoe JR, Atanassov P, Dekel DR, Herring AM, Hickner MA, Kohl PA, Kucernak AR, Mustain WE, Nijmeijer K, Scott K, Xu TW, Zhuang L (2014) Anion-exchange membranes in electrochemical energy systems. Energy Environ Sci 7(10):3135–3191
3. Mori T, Wepf R, Jiang SP (2020) Future prospects for the design of 'state-of-the-art' solid oxide fuel cells. J. Phys. Energy 2(3):031001
4. Wang YF, Leung DYC, Xuan J, Wang HZ (2016) A review on unitized regenerative fuel cell technologies, part-A: Unitized regenerative proton exchange membrane fuel cells. Renew Sust Energ Rev 65:961–977
5. Graves C, Ebbesen SD, Jensen SH, Simonsen SB, Mogensen MB (2015) Eliminating degradation in solid oxide electrochemical cells by reversible operation. Nat Mater 14(2):239–244
6. Chen KF, Liu SS, Ai N, Koyama M, Jiang SP (2015) Why solid oxide cells can be reversibly operated in solid oxide electrolysis cell and fuel cell modes? Phys Chem Chem Phys 17(46):31308–31315
7. Jiang SP (2015) Thermally and electrochemically induced electrode/electrolyte interfaces in solid oxide fuel cells: an AFM and EIS study. J Electrochem Soc 162(10):F1119–F1128
8. Wang SY, Jiang SP (2017) Prospects of fuel cell technologies. Natl Sci Rev 4(2):163–166

9. Ajanovic A, Haas R (2019) Economic and environmental prospects for battery electric- and fuel cell vehicles: a review. Fuel Cells 19:515–529
10. Apostolou D, Xydis G (2019) A literature review on hydrogen refuelling stations and infrastructure. Current status and future prospects. Renew Sust Energ Rev 113:109292
11. Zhang J, Aili D, Lu S, Li Q, Jiang SP (2020) Advancement toward polymer electrolyte membrane fuel cells at elevated temperatures. Research 2020:9089405.

Chapter 2
Fuel Cell Thermodynamics

Before the discussion of other aspects such as electrochemistry, materials, and types of fuel cells, it is useful to briefly recap the basic thermodynamics. Thermodynamics is the study of the transformation of energy from one form to another, i.e., the study of energetics. As fuel cells are energy conversion devices, thermodynamics is important for the fundamental understanding of the conversion process of chemical energy into electrical energy. As in any systems, thermodynamics can tell how much work, e.g., electricity in electrochemical systems, can theoretically be done from the totally available energy or whether a reaction is energetically spontaneous or impossible. In this chapter, we will only introduce the basic thermodynamics which is essential for fuel cell reactions and the treatment of thermodynamics is also simplified.

2.1 Internal Energy, Heat, Work, and Entropy

In the context of thermodynamics, we can define a chemical reaction or electrochemical reaction as a system, see Fig. 2.1. The *internal energy* (U) of the system depends on the inherent properties of the materials in the system. Internal energy is the energy associated with microscopic movement and interaction between particles on the atomic and molecular scales. In fuel cells, the internal energy can be quantified by the intrinsic energy of fuels and oxidants. While the absolute value of the internal energy cannot be measured, it is the change in internal energy (ΔU) of the system between the final and initial state that is of practical significance.

For a chemical reaction where reactants A and B are consumed and C and D are produced as the products, there must be energy involved in the reaction, i.e., a change in the internal energy of the system. Based on the *first law of thermodynamics*, energy cannot be consumed or created from nothing. Thus, any change in the internal energy

S. P. Jiang and Q. Li, *Introduction to Fuel Cells*,
https://doi.org/10.1007/978-981-10-7626-8_2

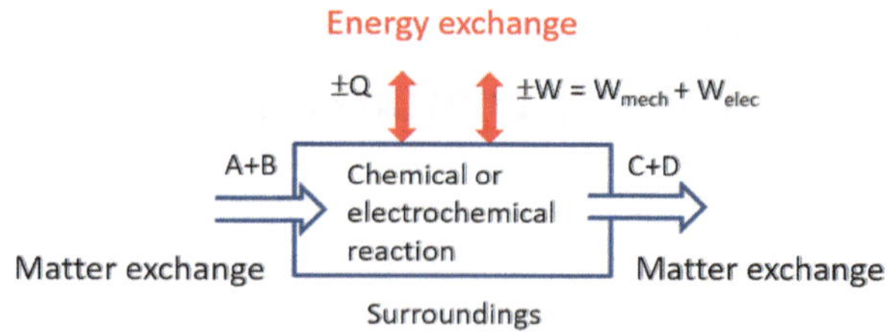

Fig. 2.1 Materials and energy balance in a chemical or electrochemical reaction

of a system, e.g., a reaction, must be fully accounted for by energy transfer between the system and the surroundings.

In thermodynamics a *system* is defined as part of the universe to study. The rest of the universe is called the *surroundings*. Between the system and its surroundings there might be exchanges of matter or energy. The energy exchange can be performed by doing work or by releasing or taking heat. According to the types of these exchanges, systems are classified as (1) *open system* that has both matter and energy exchanges with the surroundings; (2) *closed system* that has no matter but only energy exchange with the surroundings; (3) *isolated system* that has neither matter nor energy exchange with the surroundings. For study of a chemical or electrochemical reaction, an open system is defined as illustrated in Fig. 2.1.

For a classic example of thermodynamics dealing with expansion work done by heating an ideal gas, there are two forms of energy considered, *heat* (Q) and *work* (W). Heat (Q) is energy transferred between the system and surroundings because of a temperature difference. The work transferred between the system and its surroundings can be in form of expansion work as in the following examples or of electricity in fuel cells. From a system point of view, the work is defined as positive when done by the surroundings to the system. The same definition applies to the heat, i.e., Q is positive if it is transferred from the surroundings to the system.

The *first law of thermodynamics* of conservation of energy states that energy can be neither created nor destroyed, so

$$\Delta(\text{Energy of system}) + \Delta(\text{Energy of surrounding}) = 0 \tag{2.1}$$

$$\Delta(\text{Energy of system}) = -\Delta(\text{Energy of surrounding}) \tag{2.2}$$

Energy has the unit of N·m or joules (J). If the energy is measured over unit mass, kg, or mole of materials, the energy divided by the unit mass is termed specific energy with unit of J/kg or J/mol. For the purpose of simplicity, the thermodynamic parameters in the following sections are all based on the unit mass and thus the word "specific" is omitted. For a *closed system* with no matter exchange with the surrounding, the internal energy change, ΔU, can be given as follows:

$$\Delta U = \mathrm{Q} - \mathrm{W} \tag{2.3}$$

Here, to make the expression simple, we adopt Δ rather than d to represent the incremental changes in internal energy. This implies that the change in internal energy of the system (ΔU) is equal to the heat transferred to the system (Q) minus the work done by the system (W).

For a system that exchanges only the expansion work with its surroundings during a thermodynamic process at constant pressure:

$$\mathrm{W} = P\Delta V \tag{2.4}$$

where ΔV is the volume change of the system under a constant external pressure, P. The thermal energy added to the system (i.e., Q is positive) is accomplished by the increase of the internal energy plus the work done by the system. Thus, Eq. (2.3) can be expressed as

$$\Delta U = \mathrm{Q} - P\Delta V \tag{2.5}$$

The negative sign of work in the above equation is because the system does work on the surroundings. This shows that the energy released by the system is either dissipated as heat or transformed onto mechanical energy without specifying how it is distributed between these two forms.

In an electrochemical device like fuel cells the work is in form of electrical work. From an efficiency point of view, it would be ideal if the energy output of the reaction is transformed completely into useful work like mechanical or electrical work ($Q = 0$). Or even better if the system absorbs heat from its surroundings, so more work could be done.

The first law of thermodynamics does not tell us how much heat can be transferred to work. To know the limits of the fraction of heat that can be converted to work, we need the *second law of thermodynamics.*

Reversible and irreversible processes

Let us make a distinction between the reversible and irreversible processes in thermodynamics. A process is called reversible if the initial state of the system can be restored with no observable effects on the system or the surroundings. Considering a gas expanding into a vacuum chamber, the resistance to the gas is zero. To restore the system back to its original state before the expanding (i.e., vacuum chamber without the gas), the gas must be compressed. Work will be done to the gas. This will result in an energy increase in the system and has an observable effect on the system and surroundings. Such a process is irreversible. The work is irreversible as there is a finite pressure difference between the system (i.e., the gas) and the surroundings in which the movement is taking place. However, if the forces at the boundary of the system and surroundings are nearly balanced such that the resisting force at the boundary differs from the force inside the system by only an infinitesimal amount, the work is reversible.

State and path functions

Some thermodynamic properties are only a function of the initial and final state, called the state functions, e.g., the internal energy. Some thermodynamic properties depend on the path from the initial to the final state, called the path function, e.g., the work and heat in Eq. (2.3). The distinction of a reversible process is of importance in thermodynamics because it is a unique path between the initial and final state (with constant internal energy) where the maximum work can be achieved with the minimum heat loss.

How much heat can be converted into work depends on the quality of heat, which in turn depends on the temperature. The higher the temperature, the higher value of the heat. A measure of the heat quality can be defined as the ratio of the amount of the heat to its temperature. This ratio is, in thermodynamics, called *entropy* (S):

$$S = \frac{Q}{T} \tag{2.6}$$

As the heat is associated with the thermal movement of atoms or molecules of a substance and temperature is a measure of the speed of the random thermal movement, the entropy is also related to the *degree of disorder* of the system. As a reference point, the entropy of a perfect crystalline substance at absolute temperature of 0 K is zero. The absolute value of entropy of a substance or a system is however impossible to measure. For practical use, the most important is the change of the entropy, ΔS, between the final and initial states of the system. Here again we emphasize that entropy is a state function depending only on the initial and final state of the system. The heat energy (Q), as well as the work done, is a path function depending on the path of change from the initial to the final state. The way to make the heat energy a state function is to assume that the process takes place in a reversible manner so that the maximum amount of work and minimum amount of heat are involved. This is to say that, the change of entropy can be defined based on the *reversible heat* (Q_{rev}):

$$\Delta S = \frac{Q_{\text{rev}}}{T} \tag{2.7}$$

Here Q_{rev} becomes a state function. Entropy has unit of energy divided by temperature, J/K. For any irreversible processes the amount of heat involved must be larger and so is the entropy change. In other words, the increase in entropy accounts for the irreversibility of a process.

2.2 Enthalpy and Gibbs Free Energy

2.2.1 Definition of Enthalpy and Gibbs Free Energy

Equation (2.5) can be rearranged as:

$$\mathrm{Q} = \Delta U + P\Delta V \tag{2.8}$$

At constant pressure ($\Delta P = 0$), thus $P\Delta V = \Delta(PV)$. We can write:

$$\mathrm{Q} = \Delta(U + PV) \tag{2.9}$$

The combination, $U + PV$, occurs frequently in thermodynamics and is conveniently defined as *enthalpy* (H):

$$H = U + PV \tag{2.10}$$

Like internal energy, enthalpy is also a function of state. Its absolute value is of little importance and the most significant is its incremental change:

$$\Delta H = \Delta U + \Delta(PV) = \Delta U + P\Delta V + V\Delta P \tag{2.11}$$

At constant pressure, $\Delta P = 0$, ΔH is simply $\Delta U + P\Delta V$, and is therefore equal to the heat, Q added to the system.

$$\Delta H = \Delta U + P\Delta V = \Delta U + \mathrm{W} = \mathrm{Q} \tag{2.12}$$

For this reason, ΔH is also called the heat content. With the definition of the enthalpy and entropy, we can derive the *useful work* that can be done on a system under conditions of constant temperature and pressure as the difference between the enthalpy and irreversible heat loss due to the entropy of the system, *TS*.

$$G = H - TS \tag{2.13}$$

G represents the useful work that can be done and is best known as the *Gibbs free energy*. The free energy change of a chemical reaction is a measure of the maximum network obtainable from the reaction and can be written at constant temperature as a function of the enthalpy and entropy change of the reaction ($\Delta T = 0$):

$$\Delta G = \Delta H - T\Delta S \tag{2.14}$$

It can be seen from the above discussion that if in a chemical reaction, the number of moles of gaseous products and reactants are equal, the entropy change of such a reaction is nearly zero. This is because the main contribution to the entropy change in a reaction is a change in the number of ways of configuring a system or translational entropies, which is zero for a reaction with no net change in the number of molecules in the gas phase.

In the expression for Gibbs free energy and enthalpy, *P*, *V*, and *T* are easily measurable parameters while *U*, *H*, or *S* could not be measured and only their changes are used. Thus, the solution to this problem is to define a reference (standard) state based on which the values at any other states are calculated. By convention, the enthalpy of all elements under *standard temperature and pressure* (STP) conditions, i.e., 298.15 K and 1 atm (1 atm = 101.325 kPa) is taken as zero. For example, at 298.15 K and 1 atm, the enthalpies of H_2 and O_2 are zero but those of monatomic oxygen and monatomic hydrogen are not, as O and H are not in equilibrium at 298.15 K and 1 atm. As in the case of enthalpies, by convention, at standard conditions of 298.15 K and 1 atm, the Gibbs free energy of all elements is taken as zero and Gibbs free energy of a substance at the standard conditions is called *standard Gibbs free energy*, ΔG^o.

G represents the useful work or the minimum energy required to create or initiate the system. On the other hand, *G* is the maximum work that one could ever get back out of the system. Based on Gibbs free energy the following thermodynamic criteria are established:

- $\Delta G < 0$: a process can take place spontaneously and release an energy with maximum value of ΔG (energetically favorable).
- $\Delta G = 0$: the system is at equilibrium.
- $\Delta G > 0$: a process cannot take place spontaneously and require an energy supply of at least ΔG (energetically unfavorable).

2.2.2 Gibbs Free Energy from First and Second Thermodynamics Laws

The relationship between Gibbs free energy, enthalpy, and entropy can be derived from the first and second thermodynamics law. As the internal energy of the system is independent of path, its change from state 1 to state 2 can be represented by both the reversible and the irreversible paths:

$$\Delta U = Q_{rev} + W_{rev} = Q + W \tag{2.15}$$

Rearranging Eq. (2.15), we have

$$Q_{rev} = Q - W_{rev} + W = Q - (W_{rev} - W) \tag{2.16}$$

The term $(W_{rev} - W)$ is the difference between the work done in the reversible process and the work done in an actual and irreversible process and is a measure of the irreversibility involved in the transfer of energy between the system and surroundings.

Combining Eqs. (2.7) and (2.16) yields an expression for the *change in entropy* of the system:

$$\Delta S = \frac{Q}{T} - \frac{W_{rev} - W}{T} \tag{2.17}$$

Multiplying both sides of the above equation by T:

$$T\Delta S = Q - (W_{rev} - W) \tag{2.18}$$

Combining Eqs. (2.18) and (2.15) yields:

$$\Delta U = T\Delta S + W + W_{rev} - W = T\Delta S + W_{rev} \tag{2.19}$$

The reversible work, W_{rev}, represents all forms of the work. For a process involving only expansion work, we have:

$$\Delta U = T\Delta S - P\Delta V \tag{2.20}$$

This equation relates the ΔU of a system to changes in entropy and volume. Noting that $\Delta H = \Delta U + P\Delta V + V\Delta P$, we obtain:

$$\Delta H = T\Delta S + V\Delta P \tag{2.21}$$

Let us consider the thermodynamics of the chemical reactions at constant pressure. In this case, the system can freely expand or contract against constant pressure. Then energy required to transfer to create the system, i.e., to make the reaction take place, is the reversible work plus the PV term. If we define this as the "*useful*" work, W^*_{rev}, we have:

$$W^*_{rev} = W_{rev} + P\Delta V = \Delta U - T\Delta S + PV = \Delta(H - TS) \tag{2.22}$$

Then we have:

$$\Delta(H - TS) = \Delta H - T\Delta S = \Delta G \tag{2.23}$$

Equation (2.23) is the definition of the *Gibbs free energy*, which is also called *useful work*. Therefore, Gibbs free energy is also defined as useful work.

2.2.3 Enthalpy of Formation

By the convention as discussed so far, we have the assumptions that elements have zero enthalpy in their reference or STP states. Based on this condition, we can obtain

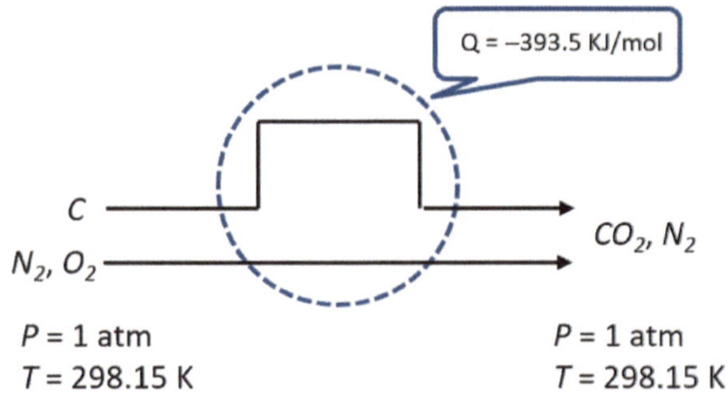

Fig. 2.2 Combustion reaction of carbon

the value of enthalpy of the compounds at the same reference state. To illustrate this we consider the combustion of charcoal in air. In this reaction, charcoal or carbon reacts with oxygen in air at STP conditions (298.15 K and 1 atm) to form carbon dioxide at the same temperature (see Fig. 2.2).The combustion reaction of charcoal is:

$$C + O_2 \rightarrow CO_2 \tag{2.24}$$

Assuming the system is in steady state, i.e., the system does not change with time even though substances may enter or leave the system, we have:

$$\sum_{n_i} H_i n_i - \sum_{n_o} H_o n_o + Q - W = \Delta U = 0 \tag{2.25}$$

where n_{i} and n_{o} are the coefficients of the substance entering and leaving the system. In general, the enthalpy change for a chemical reaction at a specified temperature can be expressed as the sum of the enthalpies of formation of the products minus the sum of the enthalpies of formation of the reactants at the same temperature. Because no work is done ($W = 0$) one has:

$$Q = \sum_{n_o} H_o n_o - \sum_{n_i} H_i n_i \tag{2.26}$$

As nitrogen does not participate in the reaction, we only need to consider the combustion reaction between carbon and O_2. Thus,

$$Q = H_{\mathrm{CO_2}} n_{\mathrm{CO_2}} - H_{\mathrm{C}} n_{\mathrm{C}} - H_{\mathrm{O_2}} n_{\mathrm{O_2}} = \Delta H \tag{2.27}$$

We can measure the value of the heat of the reaction quite easily. For the combustion reaction of charcoal, the heat produced is 393.5 kJ/mole to burn one mole of carbon with one mole of oxygen in the system, i.e., there would be heat of 393.5 kJ/mole transferred from the system to the surroundings. As the enthalpy of elements, C and O_2, is zero at STP conditions, we have:

$$\Delta H_{CO_2} = H_{CO_2} n_{CO_2} = -393.5\,\text{kJ/mol} \tag{2.28}$$

We define this value as the *enthalpy of formation* of CO_2, $\Delta H_f^0(CO_2) = -393.5$ kJ/mol where the superscript "o" refers to standard conditions, i.e., at 298 K and 1 atm. This value is negative as the heat is produced from the reaction and released from the system to the surrounding. The process is called *exothermic*. If the enthalpy change is positive, then heat must be added to the process to make it happen, and this is called *endothermic*. ΔH_f is also called *heat of formation* or *heat of combustion* for the combustion reactions. If both reactants and products are at the STP conditions, the enthalpy of formation is called the *standard enthalpy of formation*, ΔH_f^0.

The enthalpy change for a chemical reaction can be calculated from the heats of formation of the compounds and elements involved in the reaction.

Example 2.1 Calculate the enthalpy change for the oxidation or combustion of methane at standard conditions. The methane combustion reaction is:

$$CH_4 + 2O_2 \rightarrow CO_2 + 2H_2O_{(g)} \tag{2.29}$$

The subscript *g* attached to water indicates that the product of water is in the gaseous state.

Solution:

The above reaction can be written as the sum of three reactions:

$$C + O_2 \rightarrow CO_2 \qquad \Delta H = \Delta H_f^0(CO_2) = -393.5\,\text{kJ/mol} \tag{2.30}$$

$$2H_2 + O_2 \rightarrow 2H_2O_{(g)} \qquad \Delta H = \Delta H_f^0\left(H_2O_{(g)}\right) = -241.8\,\text{kJ/mol} \tag{2.31}$$

$$C + 2H_2 \rightarrow CH_4 \qquad \Delta H = \Delta H_f^0(CH_4) = -74.8\,\text{kJ/mol} \tag{2.32}$$

Thus, the ΔH_f^0 of the reaction at 298.15 K in terms of the enthalpy change of reactions (2.30–2.32) can be written as follows (note that enthalpy change of element O_2 is zero):

$$\begin{aligned}\Delta H_{f,rxn} &= \Delta H_f^0(CO_2) + 2\Delta H_f^0\left(H_2O_{(g)}\right) - \Delta H_f^0(CH_4)\\ &= -393.5 + 2(-241.8) - (-74.8) = -802.3\ \text{kJ/mol}\end{aligned} \tag{2.33}$$

The enthalpy change for the methane oxidation reaction with oxygen is also called the enthalpy of combustion, or heat of combustion of methane and is –802.3 kJ/mol. The above calculated value is also called *low heat value* or *LHV of combustion* for the gaseous water as product.

Let us turn to the most simple combustion reaction of hydrogen:

$$H_2 + \frac{1}{2}O_2 \rightarrow H_2O_{(g)} \tag{2.34}$$

Combination of 2 hydrogen atoms and 1 oxygen atom to form 1 water molecule is an exothermic reaction. If H_2 reacts directly with O_2 in air, only heat is produced because electrons and ions collide and their energies are randomized. The product water can exist in liquid or gaseous state. Thus, depending on the state of water product, all fuels containing hydrogen will have two different heats of combustion according to the two enthalpies of water formation for gaseous and liquid water, respectively:

$$\Delta H_f^0\left(H_2O_{(g)}\right) = -241.8\,\text{kJ/mol} \quad \text{(LHV)} \tag{2.35}$$

$$\Delta H_f^0\left(H_2O_{(l)}\right) = -285.9\,\text{kJ/mol} \quad \text{(HHV)} \tag{2.36}$$

The subscript "g" and "l" indicate the gaseous and liquid states of the product water, respectively. The difference of −44.1 kJ/mol between the enthalpies of formation of liquid and gaseous water is the *latent heat of condensation*, ΔH_{con}, of water. Clearly, $\Delta H_{\text{con}} = -\Delta H_{\text{vap}}$, where ΔH_{vap} is the *latent heat of vaporization.* For the gaseous water as product, the enthalpy change is called *low heat value* (LHV) of combustion. When the water produced is liquid, the heat released for liquefying the gaseous water is included and the enthalpy change is called *high heat value* (HHV).

2.2.4 *Effect of Temperature on the Change in Enthalpy and Entropy*

If we want to calculate enthalpy changes for chemical reactions at temperatures other than the reference temperature ($T_o = 298.15$ K), we will need to have the temperature dependence of the enthalpies of the substance involved, i.e., the *heat capacity* at constant pressure, C_P, which is the amount of thermal energy required to change the temperature of a substance at a constant pressure. Thus, the variation of enthalpy with temperature can be expressed by

$$\Delta H_f = \Delta H_f^0 + \int_{T_o}^{T} C_P(T)\mathrm{d}T \tag{2.37}$$

where ΔH_f is the enthalpy of formation of a substance at temperature T and $C_P(T)$ is the constant-pressure heat capacity which is also a function of temperature. The C_P has the unit of J/mol·K. When the heat capacity of the substance does not change much with the temperature, the above equation becomes:

$$\Delta H_{\mathrm{f}} = \Delta H_{\mathrm{f}}^{0} + C_{\mathrm{P}}(T - 298) \tag{2.38}$$

The dependence of the heat capacity on temperature can be expressed using a polynomial expression. Usually three terms in a polynomial are sufficient to describe the heat capacity with sufficient accuracy:

$$C_P = a + bT + cT^2 \tag{2.39}$$

Then:

$$\begin{aligned}\Delta H_{\mathrm{f}} &= \Delta H_{\mathrm{f}}^{0} + \int_{T_o}^{T} (a + bT + cT^2)\mathrm{d}T \\ &= \Delta H_{\mathrm{f}}^{0} + a(T - 298) + \frac{b}{2}\left(T^2 - 298^2\right) + \frac{c}{3}\left(T^3 - 298^3\right)\end{aligned} \tag{2.40}$$

The coefficients a, b and c for common substances or compounds can be found in physical chemistry handbooks.

Because enthalpy is a state function, its change for a chemical reaction ($\Delta H_{f,\,\mathrm{rxn}}$) at temperatures other than 298 K can be calculated by:

$$\Delta H_{\mathrm{f,rxn}} = \Delta H_{\mathrm{f}}^{0} + \int_{298}^{T} \left[\sum_{\mathrm{Products}} n_{\mathrm{P}} C_{\mathrm{P}} - \sum_{\mathrm{Reactants}} n_{\mathrm{R}} C_{\mathrm{R}} \right] \mathrm{d}T \tag{2.41}$$

$$\Delta H_{\mathrm{f,rxn}} = \Delta H_{\mathrm{f}}^{0} + \Delta a(T - 298) + \frac{\Delta b}{2}(T - 298) + \frac{\Delta c}{3}(T^3 - 298^3) \tag{2.42}$$

where Δa, Δb, and Δc are the differences between the coefficients, a, b, and c, respectively, for products and reactants, and n_P and n_R are the coefficient of products and reactants, respectively. As illustrated in Example 2.2, the effect of temperature (i.e., the heat capacity) on ΔH_f of gases is minor.

Table 2.1 Parameters used in calculating C_p in J/mol·K for water formation

Chemical species	a	b	c
H_2	28.91404	−0.00084	2.01×10^{-6}
O_2	25.84512	0.012987	-3.9×10^{-6}
$H_2O_{(g)}$	30.62644	0.009621	1.18×10^{-6}

Example 2.2 The standard enthalpy of formation of water vapor is −241.8 kJ/mol at 298 K. What is the enthalpy of formation of water vapor

when both reactant and product are at 800 °C?

$$H_2 + \frac{1}{2}O_2 \rightarrow H_2O \quad (2.43)$$

To calculate the enthalpy of formation of water at 800 °C, we need to know the specific heats at constant pressure of H_2, O_2 and $H_2O_{(g)}$.

Solution:

From Table 2.1, we have:

$$\Delta a = a_{H_2O} - a_{H_2} - \frac{1}{2}a_{O_2}$$
$$= 30.62644 - 28.91404 - \frac{1}{2}25.84512 = -11.21016 \quad (2.44)$$

$$\Delta b = b_{H_2O} - b_{H_2} - \frac{1}{2}b_{O_2} = 0.009621 + 0.00084 - \frac{1}{2} \times 0.012987 = 0.0039675 \quad (2.45)$$

$$\Delta c = c_{H_2O} - c_{H_2} - \frac{1}{2}c_{O_2} = 1.1810^{-6} - 2.0110^{-6} + \frac{1}{2} \times 3.9 \times 10^{-6} = 3.07 \times 10^{-6} \quad (2.46)$$

At 800 °C, $T = 1073$ K,

$$\Delta H_{1073} = \Delta H_{298} + \Delta a(1073 - 298) + \frac{\Delta b}{2}(1073 - 298) + \frac{\Delta c}{3}(1073^3 - 298^3)$$
$$= -241.8 \times 10^3 + (-11.21016) \times 775 + 0.00198375 \times 775 + 1.02333 \times 10^{-6} \times (1182448833)$$
$$= -249276.3 \text{ J/mol} = -249.3 \text{ kJ/mol} \quad (2.47)$$

The enthalpy of formation of water vapor at 800 °C is −249.3 kJ/mol, slightly different to the value at 25 °C (−241.8 kJ/mol). This indicates that the effect of temperature on ΔH_f for the water vapor formation is quite small.

An approximate answer can be obtained from Eq. (2.38) using constant value of the heat capacity, C_P. By taking only the first term of Eq. (2.39), $C_P = \Delta a$, then,

$$\Delta H_{1073} = \Delta H_{298} + \Delta a(1073 - 298) = -241.8 \times 10^3 - 8687.874 = -250.5 \text{ kJ/mol} \quad (2.48)$$

This is very close to the value calculated using Eq. (2.40).

The entropy change for a chemical reaction can be calculated in a similar manner. The entropy of a substance varies with temperature and can be described as a function of the heat capacity:

$$\Delta S_{\mathrm{T}} = \Delta S^{0} + \int_{T_o}^{T} \frac{C_{\mathrm{P}}(T)}{T} \mathrm{d}T = \Delta S^{0} + a \ln \frac{T}{T_o} + b(T - T_o) + \frac{c}{2}\left(T^2 - T_o^2\right) \tag{2.49}$$

If the heat capacity of the substance is constant and T_{o} is 298 K, the above equation becomes:

$$\Delta S_{\mathrm{T}} = \Delta S^{0} + C_{\mathrm{P}} \ln \frac{T}{298} \tag{2.50}$$

The entropy change for a chemical reaction is simply the sum of the standard entropies of the products minus the sum of the standard entropies of the reactants, similar to the enthalpy of formation of the reactions, Eq. (2.41). The change in entropy at any temperatures for a chemical reaction ($\Delta S_{T,\mathrm{rxn}}$) can thus be calculated in a similar way to that of enthalpy change:

$$\Delta S_{\mathrm{T,rxn}} = \Delta S^{0} + \int_{298}^{T} \left(\sum_{\text{Products}} \frac{n_{\mathrm{P}} C_{\mathrm{P}}}{T} - \sum_{\text{Reactants}} \frac{n_{\mathrm{R}} C_{\mathrm{R}}}{T} \right) \mathrm{d}T \tag{2.51}$$

$$\Delta S_{\mathrm{T,rxn}} = \Delta S^{0} + \Delta a \ln\left(\frac{T}{298}\right) + \Delta b(T - 298) + \Delta c\left(\frac{T^2 - 298^2}{2}\right) \tag{2.52}$$

Similarly, the effect of temperature on the heat capacity of gases is small and ΔS^{o} can be assumed to be independent of temperature, see Example 2.3.

Table 2.2 Thermodynamic parameters of water formation reaction

Chemical species	$\Delta H_{\mathrm{f}}^{\mathrm{o}}$, kJ/mol	ΔS^{o}, kJ/mol·K	C_{P}, kJ/mol·K at 298 K
O_2	0	0.205	0.02891
H_2	0	0.1306	0.02884
H_2O(g)	– 241.8	0.1887	0.03359
H_2O(l)	– 285.9	0.070	0.07537

Example 2.3 Calculate the standard entropy change for the formation of gaseous water at 298 K and 1073 K (800 °C).

Solution:

When water is formed as gas at 298 K, we have:

$$\Delta S^{0} = S^{0}_{\mathrm{H_2O}} - S^{0}_{\mathrm{H_2}} - 0.5 \times S^{0}_{\mathrm{O_2}} = 0.1887 - 0.1306 - 0.5 \times 0.205 = -0.0444\,\mathrm{kJ/mol \cdot K} \tag{2.53}$$

This should be compared with the standard entropy change of −0.1631 kJ/mol K for the reaction with liquid water formation in Example 2.4. At 1073 K also with gaseous water formation, from Eq. (2.49), we have:

$$\Delta S_{1073} = \Delta S^0 + \Delta a \ln\left(\frac{1073}{298}\right) + \Delta b(1073 - 298) + \Delta c\left(\frac{1073^2 - 298^2}{2}\right)$$
$$= -0.0444 \times 10^3 - 11.21016 \ln\left(\frac{1073}{298}\right) + 0.0039675 \times 775$$
$$+3.07 \times 10^{-6} \times 531,262.5 = -0.038\,\text{kJ/mol} \cdot \text{K} \tag{2.54}$$

The standard entropy change for the formation of gaseous water at 298 K and 1073 K is −0.0444 kJ/mol·K and −0.038 kJ/mol·K, respectively. The effect of temperature on ΔS is minor.

2.2.5 *Effect of Temperature on Gibbs Free Energy*

Since we already have the relation of H and S as a function of temperature, we can calculate the change in Gibbs free energy at other temperatures. At a constant temperature, we have

$$\Delta G = \Delta H - T\Delta S \tag{2.55}$$

If the enthalpy and entropy for a reaction are known, the ΔG of the reaction ($\Delta G_{\text{T,rxn}}$) can also be obtained at any other temperature:

$$\Delta G_{\text{T,rxn}} = \Delta H_{\text{T,rxn}} - T\Delta S_{\text{T,rxn}} \tag{2.56}$$

As we can see from above examples, the effect of temperature on enthalpy of formation and entropy of substances is minor. Thus, assuming that the enthalpy and entropy of reaction is independent of temperature, we can approximate above the equation as:

$$\Delta G_{\text{T,rxn}} = \Delta H^0_{\text{rxn}} - T\Delta S^0_{\text{rxn}} \tag{2.57}$$

Similarly to the enthalpy and entropy of a reaction, the Gibbs free energy of reaction can also be obtained from the difference of the sum of the Gibbs free energies of products minus the sum of the Gibbs free energies of reactants.

$$\Delta G_{\text{rxn}} = \sum_{\text{Products}} n_{\text{P}} G_{\text{P}} - \sum_{\text{Reactants}} n_{\text{R}} G_{\text{R}} \tag{2.58}$$

Example 2.4 Calculate the Gibbs free energy of liquid water formation at 25 °C and for gaseous water formation at 800 °C and 1 atm.

Solution:

At 25 °C (298 K, standard conditions), from Table 2.2, H_2O is in liquid state and we have:

$$\Delta H^0_{\mathrm{rxn}} = \Delta H^0_{\mathrm{f}}(\mathrm{H_2O})_{(\mathrm{l})} - \left[\Delta H^0_{\mathrm{f}}(\mathrm{H_2}) + \frac{1}{2}\Delta H^0_{\mathrm{f}}(\mathrm{O_2})\right]$$
$$= -285.83 - (0 + 0) = -285.83\,\mathrm{kJ/mol} \tag{2.59}$$

$$\Delta S^o_{\mathrm{rxn}} = \Delta S^0_{\mathrm{f}}(\mathrm{H_2O})_{(\mathrm{l})} - \left[\Delta S^0_{\mathrm{f}}(\mathrm{H_2}) + \frac{1}{2}\Delta S^0_{\mathrm{f}}(\mathrm{O_2})\right]$$
$$= 0.070 - \left(0.1306 + \frac{1}{2} \times 0.205\right) = -0.1631\,\mathrm{kJ/mol \cdot K} \tag{2.60}$$

$$\Delta G^0_{\mathrm{rxn}} = \Delta H^0_{\mathrm{rxn}} - T\Delta S^0_{\mathrm{rxn}} = -285.83 - 298 \times (-0.1631) = -237.23\,\mathrm{kJ/mol} \tag{2.61}$$

At 800 °C (1073 K), H_2O is a gas and we have:

$$\Delta H^0_{\mathrm{rxn}}(\mathrm{H_2O})_{(\mathrm{g})} = -249.3\,\mathrm{kJ/mol}; \tag{2.62}$$

$$\Delta S^0_{\mathrm{rxn}} = \Delta S^0_{\mathrm{f}}(\mathrm{H_2O})_{(\mathrm{g})} - \left[\Delta S^0_{\mathrm{f}}(\mathrm{H_2}) + \frac{1}{2}\Delta S^0_{\mathrm{f}}(\mathrm{O_2})\right]$$
$$= 0.1887 - \left(0.1306 + \frac{1}{2} \times 0.205\right) = -0.0444\,\mathrm{kJ/mol \cdot K} \tag{2.63}$$

$$\Delta G_{\mathrm{rxn}} = \Delta H_{\mathrm{rxn}} - T\Delta S_{\mathrm{rxn}} = -249.3 - 1073 \times (-0.0444) = -201.66\,\mathrm{kJ/mol} \tag{2.64}$$

The Gibbs free energy for water formation at 25 °C and 800 °C is −237.23 kJ/mol and −201.66 kJ/mol, respectively.

2.2.6 *Gibbs Free Energy and Electrical Work*

In classic thermodynamics the form of work is always exemplified as the expansion (pressure–volume) work as discussed above. In electrochemical systems, the electrical work should be considered. Let us start with the most basic reaction in a H_2–O_2

fuel cell in acidic electrolyte. In this fuel cell, the overall cell reaction takes place in two half-cell reactions: hydrogen loses electrons and forms hydrogen cations while oxygen, at the same time, gains electrons and protons to form water. The half-cell reactions can be written as:

$$H_2 \rightarrow 2H^+ + 2e^- \tag{2.65}$$

$$\frac{1}{2}O_2 + 2e^- + 2H^+ \rightarrow H_2O \tag{2.66}$$

The two half-cell reactions are separated by an electrolyte which is an ionic (here proton) conductor but a barrier for electrons and molecules of hydrogen and oxygen. As a result, a flow of electrons is forced through an external circuit, and therefore, an electric work is obtained. An example of the electrolyte material that conducts protons is called *proton exchange membranes* (PEM).

The reaction for the breakdown of hydrogen into hydrogen ions is the *hydrogen oxidation reaction* and the electrode for oxidation is called an *anode*. Similarly, we call the process of oxygen molecules taking electrons to form oxygen ions or water the *oxygen reduction reaction* and the electrode for reduction is the *cathode*. Combination of Eqs. (2.65) and (2.66) gives an overall cell reaction in the same form as the hydrogen combustion reaction, Eq. (2.34).

$$H_2 + \frac{1}{2}O_2 \rightarrow H_2O \tag{2.67}$$

The electrons flowing through the external circuit doing electrical work, W_{elec}. The process can also be treated as a thermodynamic system, as shown in Fig. 2.3. However, not all of the energy associated with the fuel cell reactions will be released

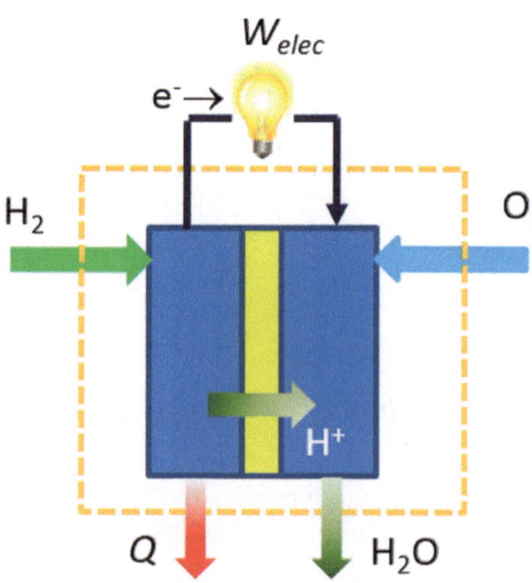

Fig. 2.3 Fuel cell as a thermodynamic system with exchange of both matter and energy with the surrounding

in the form of electricity. It is important to know how much of the heat of formation of water can be converted to electricity and why some of the energy must take the form of heat (Q).

To establish the relationship between the electrical work and Gibbs free energy, we need to define the electrical work. As the electrical work is done by moving a charge q, measured in Coulombs, through an electrical potential difference E, measured in volts, between two positions or electrodes:

$$W_{\text{elec}} = Eq \tag{2.68}$$

Here, the charge is carried by electrons, we have:

$$q = nF \tag{2.69}$$

where n is number of moles of electrons involved in the reaction and F is total charge of one mole electrons, the so-called Faraday's constant (96,485 C/mol e). Let us go back to Eq. (2.7) and to include the electrical work term in the equation:

$$\Delta U = T\Delta S - W = T\Delta S - (W_{\text{elec}} + P\Delta V) \tag{2.70}$$

With $\Delta H = \Delta U + P\Delta V + V\Delta P$, we have

$$\Delta G = \Delta H - T\Delta S = \Delta U + P\Delta V + V\Delta P - T\Delta S = -W_{\text{elec}} + V\Delta P \tag{2.71}$$

At constant pressure, then

$$\Delta G = -W_{\text{elec}} \tag{2.72}$$

The maximum electrical work that can be obtained from a fuel cell reaction at a constant temperature and pressure is related to the Gibbs free energy.

2.2.7 *Thermodynamic Reversible Potential and Thermoneutral Potential*

Combining Eqs. (2.68), (2.69), and (2.72), we have

$$\Delta G = -nFE \tag{2.73}$$

This is the most fundamental relationship between the electrical potential of the reaction and the change in Gibbs free energy between the reactants and products. The value of ΔG sets the magnitude of the reversible voltage for an electrochemical reaction. This fundamental equation also shows that the term of thermodynamics

(G) and term of electrochemistry (E) are linked by the quantity, nF, expressed as electrical charge, q. If each of the reactants and products in an electrochemical cell is in its standard state, the electrochemical potential observed is also called the *standard potential* of the cell.

$$\Delta G^0 = -nFE^0 \tag{2.74}$$

Faraday's constant

Each mole of electrons contains $N = 6.022 \times 10^{23}$ electrons (the *Avogadro's number*). The electrical charge of an electron is 1.602×10^{-19} Coulombs. Thus, the charge for each mole electrons is $F = qN = 1.602 \times 10^{-19} \times 6.022 \times 10^{23} = 96{,}485$ C/mol. This constant is called the Faraday constant and often used to convert the flow rate (mol/s) of fuel cell reactants to the generated current (A = C/s).

The total energy of a cell reaction can be thought of consisting of two parts: an entropy free part, called Gibbs free energy, that can be completely converted to electricity and a part that must be handled as heat.

$$\Delta H = \Delta G + T\Delta S \tag{2.75}$$

In most cases, $\Delta G < \Delta H$. Thus, the energy from the reaction exceeds the electric energy delivered to an external load even in an ideal reversible fuel cell. The excess energy, ($T\Delta S$), must appear as heat. Let us consider the change in entropies, ΔS, involved in the reaction. Each substance has a certain entropy which depends on its state. In the H_2–O_2 reaction at 298 K, when 1 mol of gaseous water is formed, 1 mol of H_2 and 0.5 mol of O_2 are consumed and so the corresponding entropy change for the reaction is −0.0444 kJ/mol·K (see Example 2.3). In a closed system, the entropy of any spontaneous reaction cannot decrease; at best, under reversible conditions, its change is zero. Thus, the maximum amount of the entropy heat must be dealt with as heat at 298 K is:

$$Q = T\Delta S = 298.15 \times (-0.0444) = -13.24\,\text{kJ/mol} \tag{2.76}$$

The sign of the heat is negative as the system emits the heat to its surroundings, as shown in Fig. 2.3. The emitted heat increases with the temperature. Thus, a portion of the chemical energy or enthalpy of the chemical reaction which can be converted to electrical energy in a fuel cell is in the case of gaseous water formation at STP is:

$$\Delta G^0_{\text{LHV}} = \Delta H_{\text{LHV}} - T\Delta S = -241.8 - (-13.24) = -228.56\,\text{kJ/mol} \tag{2.77}$$

If the product water is liquid, $\Delta G^0_{HHV} = -237.2\,\text{kJ/mol}$ at standard conditions. Thus, at standard conditions of 1 atm and 298.15 K, the maximum or theoretical voltage of the reaction, E^0, is:

$$E^0_{LHV} = \frac{\Delta G^0_{LHV}}{-nF} = \frac{228{,}560\,\text{J/mol}}{2 \times 96{,}485\,\text{A} \cdot \text{s/mol}} = +1.184\,\text{V} \quad \text{for}\, H_2O_{(g)} \tag{2.78}$$

$$E^0_{HHV} = \frac{\Delta G^0_{HHV}}{-nF} = \frac{237{,}200\,\text{J/mol}}{2 \times 96{,}485\,\text{A} \cdot \text{s/mol}} = +1.229\,\text{V} \quad \text{for} \quad H_2O_{(l)} \tag{2.79}$$

Thermodynamics dictates the highest reversible voltage which could be achieved in an electrochemical cell. However, the actual voltage of an electrochemical cell will be lower than the thermodynamic values due to various irreversible processes. This will be discussed in late chapters in detail.

As shown above, the total thermodynamic energy available from a fuel cell reaction is the enthalpy change (ΔH), only part of which can be used for doing work (ΔG) and the rest has to be given away in the form of heat ($T\Delta S$). However, from the total energy point of view, if a fuel cell is hypothetically assumed to operate at a voltage where no heat is released, i.e., the total enthalpy energy is completely converted to the electrical work, we would have

$$\Delta H = -nFE_{TN} \tag{2.80}$$

where E_{TN} is the hypothetical voltage from the total enthalpy change, the so-called *thermoneutral voltage.* At this voltage, no net heat flux to and from the fuel cells occurs. Using the high heat value at room temperature, a hydrogen–oxygen fuel cell has a thermoneutral voltage under STP conditions:

$$E^0_{TN} = \frac{\Delta H^0_{HHV}}{-nF} = \frac{285.9 \times 1000}{2 \times 96{,}485} = 1.482\,\text{V} \tag{2.81}$$

If low heat value of water is used, the thermoneutral voltage under STP conditions for a hydrogen–oxygen fuel cell is

$$E^0_{TN} = \frac{\Delta H^0_{LHV}}{-nF} = \frac{241.8 \times 1000}{2 \times 96{,}485} = 1.253\,\text{V} \tag{2.82}$$

2.2.8 *Effect of Temperature and Pressure on Reversible Potential*

To derive the temperature dependence of reversible potential, we need to go back to the definition of Gibbs free energy by taking into account the effect of temperature

$$\Delta G = \Delta H - \Delta(TS) = \Delta H - T\Delta S - S\Delta T \tag{2.83}$$

With equations: $\Delta U = T\Delta S - P\Delta V$ and $\Delta H = \Delta U + P\Delta V + V\Delta P$, then

$$\Delta G = -S\Delta T + V\Delta P \tag{2.84}$$

At the constant pressure, and with differentiation of the above expression, then

$$\left(\frac{\mathrm{d}\Delta G}{\mathrm{d}T}\right)_P = -S \tag{2.85}$$

Using the equation of Gibbs free energy with reversible potential (2.73), the above equation can be written as:

$$\left(\frac{\mathrm{d}E}{\mathrm{d}T}\right)_P = \frac{S}{nF} \tag{2.86}$$

If E_T is the reversible potential at an arbitrary temperature T, at the constant pressure, and assuming that entropy is independent of temperature, we will have:

$$E_{\mathrm{T}} = E^0 + \frac{S}{nF}(T - T_{\mathrm{o}}) \tag{2.87}$$

This indicates that if S of the reaction is positive, then E_{T} will increase with temperature. If S is negative, E_{T} will decrease with temperature.

Considering the H_2–O_2 fuel cell, the entropy of the reaction is negative, $\Delta S^0_{\mathrm{rxn}} = -44.4$ J/mol · K, then

$$E_{\mathrm{T}} = E^0 + \frac{-44.4}{2 \times 96{,}485}(T - T_o) = E^0 - 2.3 \times 10^{-4}(T - T_o) \tag{2.88}$$

Thus, for every 100° increase in temperature, there will be 23 mV decrease in cell voltage. The entropy changes for some fuel cell related reactions will be further discussed in Example 2.6.

The reversible cell voltage or open circuit voltage for other selected fuel cells can be calculated and examples are shown in Fig. 2.4. Table 2.3 shows thermodynamic data of typical fuels at STP.

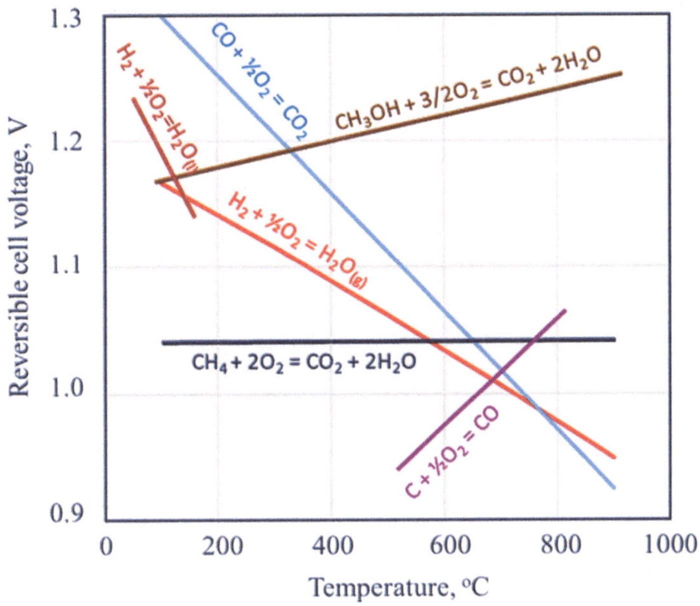

Fig. 2.4 Reversible cell voltage of selected fuel cell reactions as a function of temperature

Table 2.3 Thermodynamic data for typical fuels at 1 atm and 298.15 K

Chemical species	ΔH_f^0, kJ/mol	ΔS^0, kJ/mol·K	ΔG_f^0, kJ/mol
O_2 (g)	0	0.2051	0
H_2 (g)	0	0.1306	0
Water, H_2O (g)	−241.8	0.1887	−228.6
Water, H_2O (l)	−285.9	0.0699	−237.2
Methane, CH_4 (g)	−74.9	0.1862	−50.8
Methanol, CH_3OH (l)	−238.7	0.1268	−166.3
Ethanol, C_2H_5OH (l)	−277.7	0.1607	−174.8
Carbon monoxide, CO (g)	−111.6	0.1977	−137.4
Carbon dioxide, CO_2 (g)	−393.8	0.2136	−394.6
Ammonia, NH_3 (g)	−46.05	0.1923	−16.7

At constant temperature, Eq. (2.84) becomes

$$\Delta G = V \Delta P \tag{2.89}$$

Or in its deferential form (here both ΔG and ΔV are taken as variable quantities of the reaction)

$$\left(\frac{\mathrm{d}\Delta G}{\mathrm{d}P}\right)_T = \Delta V \tag{2.90}$$

After integration and with $\Delta G = -nFE$, we have:

$$\left(\frac{\mathrm{d}E}{\mathrm{d}P}\right)_T = -\frac{\Delta V}{nF} \tag{2.91}$$

This indicates that the variation of the reversible potential with pressure is related to the volume change of the reaction. The volume change of the reaction is typically related to the gaseous species components. For an ideal gas,

$$PV = n_g \mathrm{RT} \tag{2.92}$$

where n_g is the mole number of gases and R the gas constant, 8.314 J/mol·K. The pressure is proportional to both gas concentration, n_g/V, and temperature. This is the gas law applicable to an ideal gas. Then we have

$$\left(\frac{\mathrm{d}E}{\mathrm{d}P}\right)_T = -\frac{\Delta n_g \mathrm{RT}}{nFP} \tag{2.93}$$

where n is the mole number electrons and Δn_g is the change in the mole number of gases from the reactants to the products of the reaction, i.e., the difference between the mole number of product gases (n_p) and the mole number of reactant gases (n_r), $\Delta n_g = n_p - n_r$. By integration of above equation, we have:

$$E = E^0 - \frac{\Delta n_g \mathrm{RT}}{nF} \ln \frac{P}{P_o} \tag{2.94}$$

In general, the effect of pressure on the reversible potential, in the logarithmic term, is small.

2.3 Nernst Equation

We now consider a reversible chemical reaction in general form:

$$\alpha A + \beta B \rightleftharpoons \gamma C + \delta D \tag{2.95}$$

Each chemical species in the reaction has a chemical potential, μ_i, which is defined as:

$$\mu_i^\alpha = \left(\frac{\partial G}{\partial n_i}\right)_{T,P,n_{j\neq i}} \tag{2.96}$$

and

$$\mu_i = \mu_i^0 + \text{RT}\ln a_i \tag{2.97}$$

In the above equations, μ_i^α is the chemical potential of species in phase α, the μ_i^0 is the chemical potential of species at STP conditions, and a_i is the activity of the species. The term $\left(\frac{\partial G}{\partial n_i}\right)_{T,P,n_{j\neq i}}$ defines how much the Gibbs free energy of the reaction changes for an infinitesimal increase in the quantity of the species while T, P, and the quantities of all other species in the system are constant. The chemical potential is a thermodynamic quantity explicitly related to the number of moles of species and its impact on the system. By combination of Eqs. (2.96) and (2.97), we have:

$$\text{d}G = \sum_i \mu_i \text{d}n_i = \sum\nolimits_i (\mu_i^0 + \text{RTln}a_i)\text{d}n_i \tag{2.98}$$

For reaction (2.95), ΔG can be written as

$$\Delta G = \gamma G_C + \delta G_D - (\alpha G_A + \beta G_B) \tag{2.99}$$

Integrating Eq. (2.98) and substituting Eq. (2.99), then:

$$\Delta G = \Delta G^0 + \text{RTln}\left[\frac{(C)^\gamma (D)^\delta}{(A)^\alpha (B)^\beta}\right] \tag{2.100}$$

Introducing Eq. (2.73), we have

$$E = E^0 - \frac{\text{RT}}{nF}\ln\left[\frac{(C)^\gamma (D)^\delta}{(A)^\alpha (B)^\beta}\right] = E^0 - \frac{\text{RT}}{nF}\ln\frac{\prod a_{i,\,\text{products}}^{v_i}}{\prod a_{i,\,\text{reactants}}^{v_i}} \tag{2.101}$$

This is known as the *Nernst equation.* In a more general form, a_i is the activity of species i and v_i is the stoichiometric coefficient of species i. Thus, a_i is a measure of concentration of the species in the system. For reactions involving only gaseous species (e.g., the cases for high-temperature SOFCs), $a_i = p_i/p_o$ in which p_i is the partial pressure and for an ideal gas, p_o is the standard pressure (1 atm), the Nernst equation can be written as:

$$E = E^0 - \frac{RT}{nF}\ln\left[\frac{(p_C)^{\gamma}(p_D)^{\delta}}{(p_A)^{\alpha}(p_B)^{\beta}}\right] \tag{2.102}$$

In fuel cells, the gaseous species involved in the reaction are generally treated as an ideal gases. In the case of solutions, a_i is the molar concentration of species, $c_{i.}$ For bulk and pure components, a_i is equal to unity. For the H_2–O_2 fuel cell reaction, the Nernst equation can be written as:

$$E = E^0 - \frac{RT}{2F}\ln\frac{a_{H_2O}}{a_{H_2} \times a_{O_2}^{\frac{1}{2}}} \tag{2.103}$$

If the fuel cell is operating at temperatures below 100 °C, the activity of water can be set to 1, $a_{H_2O} = 1$, because product water exists as liquid. At standard pressure of 1 atm for H_2 at the fuel side and pure O_2 on the air side of the fuel cell, the logarithmic function of Eq. (2.103) is zero. Thus, $E = E^o$. This is the standard electrode potential when the products and reactants are in their standard conditions. In general, the activities of hydrogen and oxygen can be represented by their dimensionless partial pressure, then

$$E = E^0 - \frac{RT}{2F}\ln\frac{1}{p_{H_2} \times p_{O_2}^{\frac{1}{2}}} \tag{2.104}$$

If air is used instead of pure oxygen, the reversible potential of a hydrogen–air fuel cell at 298.15 K can be calculated as follows:

$$E = 1.229 - \frac{8.314\left(\frac{J}{mol\cdot K}\right) \times 298.15}{2 \times 96{,}485\left(\frac{C}{mol}\right)}\ln\frac{1}{1 \times 0.21^{\frac{1}{2}}} = 1.219\,V \tag{2.105}$$

The reduction in the reversible potential is 10 mV when the oxidant is switched from pure oxygen to air. However, in reality, the water partial pressure cannot be ignored. This is especially true when both air and hydrogen are humidified and therefore affects the Nernst potential (See Example 2.5).

For the reactions occurred at elevated temperatures, the effect of temperature needs to be taken into account in the Nernst equation:

$$E = E_T - \frac{RT}{nF}\ln\frac{\prod a_{i,\,products}^{\upsilon_i}}{\prod a_{i,\,reactants}^{\upsilon_i}} \tag{2.106}$$

By using E_T from Eq. (2.87), we have the change of the reversible potential with the temperature:

$$E = E^0 + \frac{\Delta S_{rxn}^0}{nF}(T - T_o) - \frac{RT}{nF}\ln\frac{\prod a_{i,\,products}^{\upsilon_i}}{\prod a_{i,\,reactants}^{\upsilon_i}} \tag{2.107}$$

If the activities of reactants and/or products are presented by partial pressure, Eq. (2.107) can be used for calculation of the reversible potential of the reaction at any given temperatures and pressures.

Example 2.5 Calculate the reversible voltage of a hydrogen-oxygen fuel cell.

(a) A PEMFC at 80 °C with fully humidified H_2 and air under ambient total pressure;
(b) A SOFC at 800 °C with H_2 gas that is bubbling through a room temperature water bath and dry air is used as oxidant.

Solution:

(a) The water vapor pressure at 80 °C is $p_{\mathrm{H_2O}} = 0.47$ atm. In fully humidified gases the partial pressure of hydrogen and oxygen is hence

$$p_{\mathrm{H_2}} = 1 - 0.47 = 0.53 \text{ atm}$$

$$p_{\mathrm{O_2}} = 0.21 \times (1 - 0.47) = 0.11 \text{ atm}$$

Using Eqs. (2.88) and (2.103), one has

$$E = (1.229 - 0.00023 \times (80 - 25)) + \frac{8.314 \times 353}{2 \times 96485} \ln\left(\frac{0.53 \times 0.11^{1/2}}{1}\right) = 1.139 \text{ V} \tag{2.108}$$

(b) In SOFC H_2 gases that are bubbling through a room temperature water bath is assumed to be saturated with water vapor, which is $p_{\mathrm{H_2O}} = 0.03$ bar. As water is in gaseous state, the activity of water (a_{H_2O}) is 0.03 and H_2 fuel is diluted, $p_{\mathrm{H_2}} = 0.97$ atm, and $p_{\mathrm{O_2}} = 0.21$. Take $E^0 = 1.184$ V.

$$E = (1.184 - 0.00023 \times (800 - 25)) + \frac{8.314 \times 1073}{2 \times 96485} \ln\left(\frac{0.97 \times 0.21^{1/2}}{0.03}\right) = 1.131 \text{ V} \tag{2.109}$$

2.4 Comparison of Work Done by a Heat Engine and Fuel Cell

It is very often that when we talk about the efficiency of fuel cells, we use the efficiency of a heat engine, e.g., an internal combustion engine (ICE) for comparison.

The reason is also obvious as ICE is the most advanced and currently used energy conversion technology. Thus it is necessary to discuss the efficiency of ICE and fuel cells from the view point of thermodynamics.

For energy conversion devices, the efficiency is the most important parameter. We can define the efficiency (ε) of the conversion process as the ratio of the amount of work done or useful energy to the amount of total input energy.

$$\varepsilon = \frac{\text{useful energy}}{\text{total energy}} = \frac{\text{useful energy}}{\Delta H} \tag{2.110}$$

The efficiency derived from the reversible process is called the *thermodynamic efficiency* or ideal efficiency, while the one measured from a working energy conversion device is the practical or *real efficiency*. In real fuel cells, the real efficiency is generally lower than the ideal efficiency because of a number of losses and irreversible processes that occur in the fuel cells. There are substantial research efforts and activities around the world to reduce the losses in fuel cells and thus to increase the real efficiency.

2.4.1 Work Done in a Heat Engine and Carnot Efficiency

A heat engine is a machine which can convert heat to mechanical work, as shown in Fig. 2.5. The engine operates between the high and low temperature heat sources with temperatures T_H and T_L. The machine absorbs heat from the high-temperature heat source T_H and discharges the waste heat to the low temperature reservoir at T_L, while the machine is doing work. The input is an amount of heat, Q_{in}, W is the useful output energy, and Q_{out} is an amount of heat that is rejected.

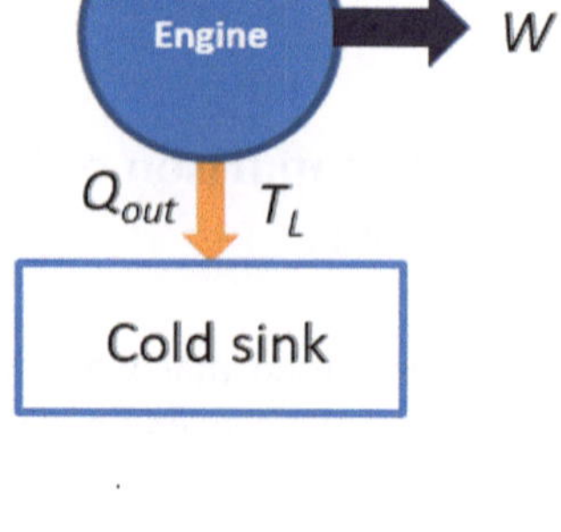

Fig. 2.5 Heat engine (machine) operating between a hot source, T_H and a cold sink, T_L

We consider an ideal situation that the operation of the machine is reversible; thus, the maximum amount of work can be done, $W = W_{rev}$. The efficiency of such an engine can be expressed as the output energy, W, divided by the total quantity of heat input that is absorbed from the high-temperature reservoir (Q_{in}):

$$\varepsilon = \frac{\text{net work done by heat engine}}{\text{heat absorbed by heat engine}} = -\frac{W}{Q_{in}} \tag{2.111}$$

For such a system, we have

$$Q_{in} + Q_{out} + W = 0 \tag{2.112}$$

As this is a reversible process ($\Delta S = 0$), then

$$\sum \frac{Q}{T} = \frac{Q_{in}}{T_H} + \frac{Q_{out}}{T_L} = 0 \tag{2.113}$$

$$Q_{out} = -\frac{T_L}{T_H} Q_{in} \tag{2.114}$$

Combining with Eq. (2.112)

$$-W = \frac{Q_{in}}{T_H}(T_H - T_L) \tag{2.115}$$

$$\varepsilon_{Carnot} = -\frac{W}{Q_{in}} = \frac{T_H - T_L}{T_H} = 1 - \frac{T_L}{T_H} \tag{2.116}$$

in which T is absolute temperature in Kelvin. This equation is called the *Carnot law* and ε_{Carnot} is the *Carnot efficiency*, which depends only on the temperatures between which the engine works. As it is impossible to have a low temperature reservoir at absolute zero temperature, there will be always some heat discharged to the low temperature reservoir. In another word, the efficiency of a heat engine cannot be 100% because T_L cannot be zero. Thus, in a heat engine that operates between a cylinder temperatures of 700 °C and an exhaust temperature of 50 °C, the maximum efficiency of conversion can be calculated,

$$\varepsilon_{Carnot} = 1 - \frac{T_L}{T_H} = 1 - \frac{50 + 273.15}{700 + 273.15} = 67\% \tag{2.117}$$

If the exhaust temperature is 200 °C, the Carnot efficiency becomes 51%. This is shown in Fig. 2.6. These numbers are the maximum efficiency of heat engines. The practical values depend on the scale of the engines. For megawatt power plants it can be as high as over 40% with an average efficiency of 33–35%. For small engines like gasoline motors in automobiles, the efficiency is less than 20%.

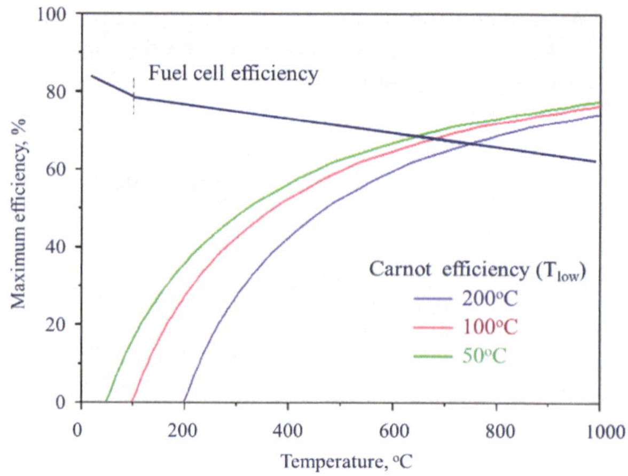

Fig. 2.6 A comparison of the theoretical efficiencies of a heat engine and a fuel cell

It is important to note that the limitation imposed by the Carnot efficiency applies only to the conversion of heat into an equivalent amount of work, not to other conversion processes. For example, there is no such a limit on the conversion efficiency of electrical work into heat by passing the current through a resistor. There is no such a limit either for an electrochemical process like batteries and fuel cells.

2.4.2 Fuel Cell Efficiency, Stoichiometry Number, Fuel and Oxygen Utilization Efficiency

2.4.2.1 Energy Efficiency and Power

The energy conversion efficiency of a fuel cell is the ratio of the useful work generated to the enthalpy change of the chemical reaction involved. Consequently, the maximum useful work is the maximum electrical energy a fuel cell can produce from chemical energy of the fuel. Thermodynamically the maximum amount of energy of the reaction available to be converted to electrical work is the Gibbs free energy. Thus, the thermodynamic efficiency of a fuel cell is given by:

$$\varepsilon_{\text{th}} = \frac{\Delta G}{\Delta H} = 1 - \frac{T\Delta S}{\Delta H} \tag{2.118}$$

Thus, the thermodynamic efficiency of an H_2–O_2 fuel cell based on HHV of water at 298.15 K is:

$$\varepsilon_{\text{th(HHV)}} = \Delta G/\Delta H_{\text{(HHV)}} = 237.2/285.9 = 83\% \tag{2.119}$$

If the produced water is in the gaseous state, the corresponding thermodynamic efficiency is:

$$\varepsilon_{th(LHV)} = \Delta G/\Delta H_{(LHV)} = 228.6/241.8 = 94.5\% \tag{2.120}$$

As seen from Eq. (2.118), the efficiency losses in energy conversion are due to the entropy term, $T\Delta S$. In general, solids have a more regular structure than liquids while gases have a much higher extent of disorder. As a simple principle, any reactions involving a larger mole number of gases in products than in reactants, i.e., an increase in disorder of the system, have a positive entropy change of the reaction.

Example 2.6 Estimate the entropy change of the following reactions by looking at the mole number of gases in the reactant side and product side. Comment on the thermodynamic efficiency if the reactions are arranged in an electrochemical cell, e.g., fuel cell.

$$H_{2(g)} + \frac{1}{2}O_{2(g)} = H_2O_{(g)} \tag{2.121}$$

$$H_{2(g)} + \frac{1}{2}O_{2(g)} = H_2O_{(l)} \tag{2.122}$$

$$C_{(s)} + O_{2(g)} = CO_{2(g)} \tag{2.123}$$

$$C_{(s)} + \frac{1}{2}O_{2(g)} = CO_{(g)} \tag{2.124}$$

Solution:

(a) There are 1½ moles of gases in reactants and 1 mol of gases in products. The mole number of gases decreases ($\Delta n_g = -0.5$) through the reaction. The reaction entropy change of the reaction must be negative ($\Delta S^0 = -44.4$ J/mol) and the thermodynamic efficiency is less than 100% (94.5%).

(b) There are 1½ moles of gas reactants but no gases in products ($\Delta n_g = -1.5$). The reaction entropy change should be more negative ($\Delta S^0 = -163.2$ J/mol). And the thermodynamic efficiency is even lower (83%).

(c) This is a reaction in a direct carbon fuel cell using solid carbon as fuel. There is 1 mol of oxygen in reactants and 1 mol of carbon dioxide in products. The mole number of gases as well as the reaction entropy remains basically unchanged, *i.e.*, $\Delta S^0 \approx 0$. The thermodynamic efficiency is essentially 100%.

(d) There is ½ mole of gases in reactants and 1 mol of gases in products. The mole number of gases increases ($\Delta n_g = +0.5$), so the reaction entropy change must be positive ($\Delta S^0 = 89.5$ J/mol). The thermodynamic efficiency should be higher than 100 (actually 124%!). This high efficiency does not violate the thermodynamic laws as it is accounting for the conversion of heat into work. The thermodynamic efficiency of 124% means that this electrochemical device can convert more (than the reaction enthalpy ΔH) heat into electrical work if extra heat is externally supplied.

Note: The temperature coefficient of the reversible voltage also depends on the entropy change. For those reactions with a negative entropy changes, the temperature coefficient is negative, and for those reactions with a positive entropy changes, the temperature coefficient is positive (see Fig. 2.4).

Figure 2.6 compares the thermodynamic efficiency of a H_2–O_2 fuel cell and a heat engine. The fuel cell efficiency decreases with temperature. In contrast, for heat engine, the efficiency increases with the increase in temperature so long as the temperature of cold sink, T_L, is kept constant. The change in the slope of the fuel cell efficiency at 100 °C is due to the change of water product of the fuel cell reaction from liquid to gas.

In practice, the actual voltage of a fuel cell is less than the predicted thermodynamic voltage due to irreversible losses when the current is drawn from the cell. The voltage of the cell in operation, E_{cell}, would be lower than the thermodynamic cell voltage E^0 under STP conditions. The real energy conversion efficiency for a fuel cell, ε_{cell} can be obtained by the cell voltage ratio:

$$\varepsilon_{cell} = \frac{E_{cell}}{E_{TN}} \tag{2.125}$$

And the power (P), expressed in unit of watts, delivered by a fuel cell is:

$$P = z \times I \times E_{cell} \quad \text{(W)} \tag{2.126}$$

where z is the number of cells in stack and I is the total current (A). From the number of cells, the stack voltage can be obtained by the product of $z \times E_{cell}$. For $z = 1$, the power obtained is for a single cell. If current density is used in the unit of A/cm^2, the power obtained will be power density, W/cm^2 based the geometric area of the active electrodes. Power is the rate at which energy (En) is made available. Conversely, energy, expressed in Watt-hours, is

$$En = P \times t \quad (\text{W} \cdot \text{h}) \tag{2.127}$$

It is noted from Eq. (2.125) that the fuel cell efficiency is based on the total available energy (ΔH) or the thermoneutral voltage ($E_{TN} = 1.48$ V based on HHV of water at 298.15 K). A fuel cell operating at 0.7 V is converting 47% of the total energy into electric power and 53% released as heat. If the same cell operates at 0.9 V, then the electrical efficiency is ~61% and the remaining energy (39%) will be in the form of heat.

2.4.2.2 Stoichiometry Number and Fuel Utilization Efficiency

The power generation of a fuel cell depends on the cell voltage and current. While the cell voltage is linked to the energy conversion efficiency the current in unit of C/s is linked to the reactant supply, governed by Faraday's law. The current is related to the fuel, e.g., hydrogen supply as:

$$I = n \times F \times J_{\mathrm{f}}^{0} \times U_{\mathrm{f}} \tag{2.128}$$

where I is the total current, n is the number of electrons involved in the fuel oxidation reaction, J_{f}^{0} is the input fuel flow in mol/s, and U_{f} is the fuel utilization efficiency. U_{f} is defined as the ratio of the used fuel flow (J_{f}) to the input fuel flow at a given current, the reciprocal of which is the stoichiometry number λ.

$$U_{\mathrm{f}} = \frac{1}{\lambda} = \frac{J_{\mathrm{f}}}{J_{\mathrm{f}}^{0}} \tag{2.129}$$

Similarly, we can define the oxygen utilization efficiency, U_{O_2}, as the ratio of used oxygen flux to the input oxygen flux at a given current:

$$U_{O_2} = \frac{J_{O_2}}{J_{O_2}^{0}} \tag{2.130}$$

Using equation (2.128), we can calculate the flow rate required for a fuel cell stack of rated power P (W) consisting of z single cells operating at an average single cell voltage (E_{Cell}, V) and a current of I (A). The oxygen consumption rate, in mol/s, is a linear function of current and therefore

$$\text{Oxygen consumption rate} = \frac{zI}{nF} = \frac{P}{4 \times F \times E_{\mathrm{cell}}} \quad \text{mol/s} \tag{2.131}$$

where $n = 4$ is the molar number of electrons for one mole of O_2. The oxygen consumption rate can also be expressed per cell:

$$\text{Oxygen consumption rate} = \frac{I}{nF} = \frac{I}{4F} \quad \text{mol/s} \cdot \text{cell} \tag{2.132}$$

When air instead of oxygen is supplied, the oxygen content of 21 mol% O_2 should be used

$$\text{Air consumption rate} = \frac{I}{0.21 \times 4 \times F} = \frac{I}{0.84 \times F} \quad \text{mol/s} \cdot \text{cell} \tag{2.133}$$

Similarly, for hydrogen consumption, we have:

$$\text{Hydrogen consumption rate} = \frac{I}{2 \times F} \quad \text{mol/s} \cdot \text{cell} \tag{2.134}$$

In real fuel cells, oxygen and hydrogen gas flow rates are always higher than the theoretical values. The ratio of theoretical oxygen or hydrogen flow rate to the actual flow rate is defined as stoichiometry number.

$$\lambda = \frac{\text{actural flow rate}}{\text{theoretical flow rate}} = \frac{\text{actual rate}}{\text{consumption rate}} \tag{2.135}$$

Thus, the stoichiometry number λ is the actual rate provided to a fuel cell relative to the rate at which it is consumed. For $\lambda < 1$, a fuel cell is in a starvation state which limits the fuel cell current to a value that corresponds to the point of $\lambda \approx 1$ where a limiting current will be observed with seriously reduced cell voltage. To avoid the starvation and additional performance loss in fuel cells, typically $\lambda = 1.2$–1.5 for hydrogen and $\lambda = 1.5$–3.0 for air are used. With the stoichiometry number, the flow rate can be expressed as

$$\text{Oxygen flow rate} = \lambda \frac{I}{4 \times F} \quad \text{mol/s} \cdot \text{cell} \tag{2.136}$$

$$\text{Air flow rate} = \frac{\lambda}{0.21} \frac{I}{4 \times F} \quad \text{mol/s} \cdot \text{cell} \tag{2.137}$$

$$\text{Hydrogen flow rate} = \lambda \frac{I}{2 \times F} \quad \text{mol/s} \cdot \text{cell} \tag{2.138}$$

Example 2.7 A 5 kW PEMFC stack operating at an average single cell voltage of 0.65 V. What is the flow rate (in L/min) of hydrogen at $\lambda_{H_2} = 1.2$ and air at $\lambda_{Air} = 2.0$ needed under standard conditions?

Solution:

The hydrogen flow rate

$$\text{Hydrogen flow rate} = \frac{5000}{2 \times 96485 \times 0.65} \times 1.2 = 0.048\ \text{mol}\ H_2/\text{s} \tag{2.139}$$

Under standard conditions (1 atm and 298 K) assuming an ideal gas, this corresponds to a volume flow rate

$$\text{Hydrogen flow rate} = 0.048\ \frac{\text{mol H}_2}{\text{s}} \times 0.082 \frac{\text{atm} \cdot \text{L}}{\text{mol} \cdot \text{K}} \times \frac{298\ \text{K}}{1\ \text{atm}} \times \frac{60\ \text{s}}{\text{min}}$$
$$= 70.4\ \text{L/min} \tag{2.140}$$

And

$$\text{Air flow rate} = \frac{5000}{0.84 \times 96485 \times 0.65} \times 2 = 0.190\ \text{mol/s} \tag{2.141}$$

$$\text{Air flow rate} = 0.19\ \frac{\text{mol air}}{\text{s}} \times 0.082 \frac{\text{atm} \cdot \text{L}}{\text{mol} \cdot \text{K}} \times \frac{298\ \text{K}}{1\ \text{atm}} \times \frac{60\ \text{s}}{\text{min}} = 278.6\ \text{L/min} \tag{2.142}$$

Some researchers suggest to include the fuel utilization efficiency in calculation of the fuel cell energy efficiency by assuming that the excess fuel is wasted. It should be noted that the excess fuel only passes through the cell without being converted and in principle can be recovered and circulated. The fuel utilization is an operational issue, not the intrinsic characteristics of fuel cell operation.

From a system point of view, as hydrogen is not readily available and does not exist in nature. The efficiency of producing hydrogen fuel needs to be taken into account in calculating the overall energy efficiency in H_2 fuel-based fuel cells. If hydrogen is generated by electrolysis using power from traditional coal or natural gas power plants, the overall or life cycle energy efficiency of the process would be much lower, according to:

$$\text{Fossil fuel} \xrightarrow{\text{Generation}} \text{Electricity} \xrightarrow{\text{Electrolysis}} \text{Hydrogen} \xrightarrow{\text{Fuel cells}} \text{Electricity} \tag{2.143}$$

If the generation efficiency of electricity generation is 40%, the electrolysis efficiency of hydrogen is 80%, and fuel cell efficiency to generate electricity is 57%, then the overall energy conversion efficiency is ~20% through the entire renewable power chain. The overall efficiency of the energy conversion process is very much dependent on the efficiency of the individual steps involved. However, the energy efficiency is not the sole criterion on which to judge fuel cells.

It should be noted that in case mass flow is needed one can use the molar mass (W_M) of 28.97 g/mol for air and 2.02 g/mol for hydrogen, as illustrated in Example 2.8.

Example 2.8 Calculate the hydrogen and oxygen mass flow rates for generation of 1 A in a H_2–O_2 cell assuming the fuel utilization of 90% and oxygen utilization of 50%.

Solution:

For the complete electrochemical oxidation of H_2:

$$H_2 + \frac{1}{2}O_2 \rightarrow H_2O \tag{2.144}$$

$n = 2$, $F = 96{,}485$ C/mol, the H_2 mass flow rate is:

$$J_{H_2}^0 = \lambda_{H_2}\frac{I}{n \times F}W_{M,H_2} = \frac{1}{0.8} \times \frac{1}{2 \times 96485} \times 2.02 = 1.3 \times 10^{-5} \quad \text{g/s} \tag{2.145}$$

For O_2 $n = 2$, the O_2 mass flow rate is:

$$J_{O_2}^0 = \lambda_{O_2}\frac{I}{4 \times F}W_{M,O_2} = \frac{1}{0.5} \times \frac{1}{4 \times 96485} \times 28.97 = 1.5 \times 10^{-5} \quad \text{g/s} \tag{2.146}$$

2.4.3 *Other Attributes of ICE and Fuel Cells*

In addition to the remarkable differences in the operation principles (i.e., thermodynamics and energy forms), the materials used in a heat engine and fuel cell are very different. The heat engine consists of a fixed cylinder and a moving piston. The base of a reciprocating internal combustion engine is the engine block, which is typically made of cast iron or aluminum. Commonly used materials for IC engine pistons are cast iron, cast steel, forged steel, cast aluminum alloys, and forged aluminum alloy. Based on the mechanical grinding mechanism of the metallic components used in the IC engines with experimental verification, the overhaul life of a heat engine can be estimated.

The materials used in fuel cells vary from metallic and oxide electrocatalysts, polymeric and ceramic electrolytes to graphite and alloy construction materials. The structure of these materials also requires specific micro- and nano-architectures to maximize the functionalities in catalysis and conduction of ions, electrons, reactants and products of fuel cell reactions. Although fuel cells are free of moving parts during operation, compressors, pumps and other components are always needed in the system balance-of-the-plants. In addition, the nature of the nano-structured materials and the electrochemical processes at the interface would make the life or durability of a fuel cell less predictive as compared to that of a heat engine.

2.5 Thermodynamics and Efficiency of Water Electrolysis

In this section, we will briefly introduce the thermodynamics and efficiency of water electrolyzers, another key technology to produce hydrogen from renewable energy sources. As it is a reverse process of a fuel cell, much of the thermodynamics discussed so far is equally applicable to water electrolysis.

2.5.1 Basic Thermodynamics of Water Electrolysis

Water electrolysis is the reverse reaction of hydrogen and oxygen combustion reaction and is consisted of oxygen evolution reaction (OER) at anode and hydrogen evolution reaction (HER) at cathode. The overall water electrolysis can be written as follows:

$$H_2O_{(l)} \rightarrow H_2 + \frac{1}{2}O_2 \tag{2.147}$$

Thus, the total energy demand to split water into its constitutes of hydrogen and oxygen is the opposite of the enthalpy of water formation, ΔH. Under constant temperature and pressure, this is the sum of the electrical energy demand, ΔG, and thermal energy demand, Q,

$$\Delta H = \Delta G + Q = \Delta G + T\Delta S \tag{2.148}$$

The entropy change associated with the splitting of water can be obtained according to reaction (2.147) and Table 2.1 under STP:

$$\begin{aligned} \Delta S &= S_{H_2} + 0.5 \times S_{O_2} - S_{H_2O} = 0.1306 + 0.5 \times 0.2051 - 0.0699 \\ &= 0.1631\,\text{kJ/mol} \cdot \text{K} \end{aligned} \tag{2.149}$$

And the thermal energy demand is:

$$Q = T\Delta S = 298 \times 0.1631 = 48.7\,\text{kJ/mol} \tag{2.150}$$

Thus, the total energy demand is:

$$\Delta H = \Delta G + Q = 237.2 + 48.7 = +285.9\,\text{kJ/mol} \tag{2.151}$$

Thermodynamically, the minimum or reversible cell voltage for the onset of water electrolysis under STP conditions is defined as (note: the number of electrons in the case of water electrolysis is 2 and we use V instead of E to distinguish the electrolysis cells to fuel cells):

$$V^0 = -\frac{\Delta G^0}{nF} = \frac{237.2 \times 1000}{2 \times 96485} = 1.23\,\text{V} \tag{2.152}$$

Therefore, water electrolysis can only take place when the cell voltage is higher than this reversible voltage, i.e.

$$V_{\mathrm{cell}} > V^0 \tag{2.153}$$

As shown early, the effect of temperature on the enthalpy and entropy changes is minor. Assuming that ΔH and ΔS are constant over a temperature range, the thermodynamic cell voltage as a function of temperature can be simply written as, using thermodynamic data at 298 K:

$$V_{\mathrm{T}}^0 = \frac{\Delta G}{nF} = \frac{\Delta H - T\Delta S}{nF} = \frac{285.9 \times 1000 - 163.3 \times T}{2 \times 96485} = 1.482 - 8.46 \times 10^{-4} T \tag{2.154}$$

According to the Nernst equation, we can write the thermodynamic cell voltage as a function of temperature and partial pressure of reactant (H_2O) and products (H_2 and O_2):

$$V_{\mathrm{T,P}}^0 = V_{\mathrm{T}}^0 + \frac{\mathrm{RT}}{nF} \ln \frac{p_{\mathrm{H_2}} \times p_{\mathrm{O_2}}^{1/2}}{p_{\mathrm{H_2O}}} = 1.482 - 8.46 \times 10^{-4} T + 4.3 \times 10^{-5} T \times \ln \frac{p_{\mathrm{H_2}} \times p_{\mathrm{O_2}}^{1/2}}{p_{\mathrm{H_2O}}} \tag{2.155}$$

In real water electrolysis cells, the gases produced may be saturated with water vapor.

2.5.2 *Thermal Balance of Electrolysis Cells*

In electrolysis cells, the thermal balance is defined as the difference between the reversible heat and irreversible heat of the reaction. The reversible heat is associated with the entropy change of the reaction,

$$Q_{\mathrm{rev}} = T\Delta S = \Delta H_{\mathrm{rev}} - \Delta G_{\mathrm{rev}} = nF(V_{\mathrm{TN}} - V^0) \tag{2.156}$$

where V_{TN} is the thermoneutral voltage calculated from the enthalpy change. This is the heat demand if an electrolysis cell operates at the reversible voltage, i.e., with no overpotentials. Any overpotentials above the reversible cell voltage, often called the polarization losses, will generate heat. This heat is called the irreversible heat and associated with the internal energy dissipation due to the polarization losses under electrolysis current:

$$\eta_{\mathrm{loss}} = \eta_{\mathrm{a}} + \eta_{\mathrm{c}} + \eta_{\mathrm{ohm}} \tag{2.157}$$

where η_a, η_c, and η_{ohm} are the anode, cathode, and ohmic overpotentials of the electrolysis process. The significance of polarization losses associated with electrochemical reactions will be discussed in details in Chap. 3. The actual cell voltage of electrolyzer under operation conditions is thus composed of:

$$V_{cell} = V^0 + (\eta_a + \eta_c + \eta_{ohm}) = V^0 + \eta_{loss} \quad (2.158)$$

where V_{cell} is the operating cell voltage of electrolysis. As the polarization losses (η_{loss}) are not zero under electrolysis operation conditions, V_{cell} is greater than V^0. The polarization losses will be dissipated as irreversible heat,

$$Q_{irrev} = nF\eta_{loss} = nF\left(V_{cell} - V^0\right) \quad (2.159)$$

Using Eqs. (2.156) and (2.159), we have the balance between the heat demand and heat generation in an electrolysis cell:

$$Q_{cell} = Q_{rev} - Q_{irrev} = nF(V_{TN} - V^0) - nF(V_{cell} - V^0) = nF(V_{TN} - V_{cell}) \quad (2.160)$$

This equation shows the total heat exchanged between the surroundings and the electrolysis cell depends on the operating cell voltage. For example, under STP conditions, if the cell voltage is below thermoneutral voltage (e.g., 1.482 V), Q_{cell} is positive, then the electrolyzer will need a supply of heat and operates endothermically. Conversely, if $V_{cell} > V_{TN}$, Q_{cell} is negative, indicating that excess heat is produced by the electrolysis cell and dissipated to the surroundings. The production of heat causes energy inefficiency. Thus, it is desirable to operate the cell voltage as close as possible to the thermoneutral voltage where $Q_{cell} = 0$, i.e., the cell operates in a thermal neutral mode without exchange of heat between the electrolysis cell and surroundings.

2.5.3 Energy Efficiency of Water Electrolysis

With the supply of electrical work ($W_{elec} = \Delta G$) and heat ($\Delta H - \Delta G$), the product gases carrying the total energy of ΔH are produced by an electrolyzer. From a thermodynamic point of view, only the electrical work (V_{cell}) is accounted as the energy input. Thus, the energy efficiency of an electrolyzer is defined as the ratio of ΔH to the total electrical energy input:

$$\varepsilon_{cell} = \frac{\Delta H}{W_{elec}} = \frac{V_{TN}}{V_{cell}} \quad (2.161)$$

The water electrolysis can take place at a voltage slightly higher than the reversible voltage V^0 but lower than the thermoneutral voltage V_{TN}. Under this condition, the

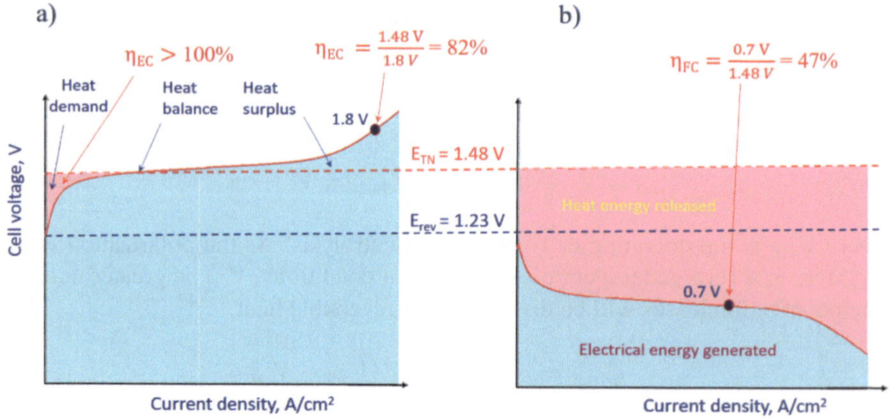

Fig. 2.7 Illustration of energy efficiency terms of **a** water electrolysis and **b** fuel cell

electrical work larger than the minimum amount of ΔG is provided, of course, a certain amount of heat has to be supplied to meet the heat demand. The definition of the energy efficiency takes only the electrical work as input but ignores the necessary heat supply. It is therefore that an energy efficiency of higher than 100% is possible when a cell operates at $V_{cell} < V_{TN}$. This is illustrated in Fig. 2.7. At the point of $V_{cell} = V_{TN}$, the energy efficiency of the electrolysis cell is 100%, meaning that the energy input in form of electric work is equal to the total enthalpy contained in the product hydrogen. In the figure the fuel cell efficiency is also presented as a comparison.

2.5.4 High-Temperature Electrolysis

In Fig. 2.7 the thermoneutral and reversible voltages are values for room temperature. At elevated temperatures, the enthalpy change for the reaction, ΔH, increases slightly while the thermal energy demand ($T\Delta S$) is increased and the electrical energy demand (ΔG) is reduced with the increase of operation temperature of electrolysis, see Fig. 2.8. This indicates that the high-temperature water electrolysis such as solid oxide electrolysis cells is more effective (i.e., less electrical energy need) than low temperature cells if the heat energy can be supplied by external sources such as waste heat or steam from nuclear power station or from solar thermal heater. The heat can also be provided by the Joule heat generated from ohmic resistance, activation, and transportation polarization losses.

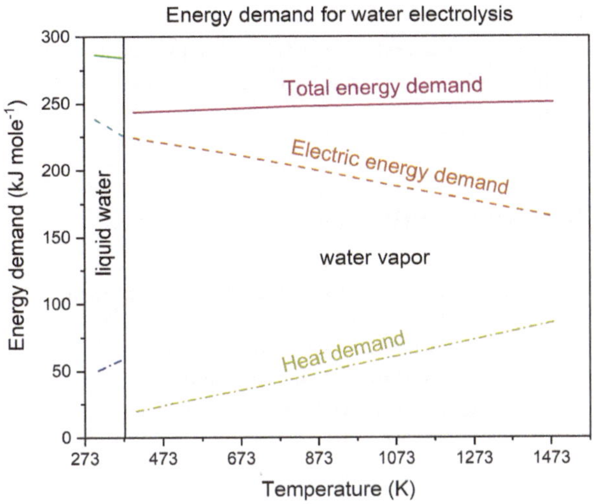

Fig. 2.8 Total energy demand, electric energy demand, and heat demand as function of temperature for water electrolysis

2.6 Summary

In this chapter, we have introduced and discussed the following topics:

- Electrochemical reactions in fuel cells are treated in the context of thermodynamics, i.e. the conversion of the chemical energy of the fuel to heat (Q) and work (W) during operation of fuel cells where the work is electrical.
- The total energy available from an electrochemical reaction, ΔH, consists of two parts, the Gibbs free energy, ΔG, representing the fraction that can be used for doing work, and the entropy part ($T\Delta S$), represnting the fraction that has to be handle as heat. ΔG is also the most important parameter for the transformation of the system, energetically spontaneous or impossible.
- By assuming that elements have zero enthalpy at STP states, one can obtain the enthalpy change of substances. If the reaction is at STP states, the enthalpy of formation is called the standard enthalpy formation, ΔH_f^0. If the enthalpy change of a reaction is negative, heat is produced from the reaction and the process is called exothermic; otherwise, the process is called endothermic.
- $\Delta G = -nFE$ is the most fundamental relationship between the electrical potential of the reaction and the change in Gibbs free energy, i.e., the link between thermodynamics and electrochemistry. Thus, the electrochemical potential of the cell can be obtained from the change of the Gibbs free energy.
- Nernst equation is a fundamental equation in electrochemistry, relating the electrical potential of the reaction to the activity of reactants and products.
- The thermodynamic efficiency of a fuel cell is the ratio of the maximum useful work generated to the enthalpy change of the chemical reaction involved, $\Delta G/\Delta H$. The thermodynamic efficiency can be less or greater than unity, depending on the entropy change of the reaction. In a real fuel cell, any losses in the conversion of the useful energy to electrical energy will appear as heat.

- The thermodynamics of water electrolysis is briefly introduced. Water electrolysis is an essentially reversible process of the $H_2 - O_2$ fuel cell. Thus, the thermodynamics discussed for fuel cells are equally applicable to electrolysis processes.

2.7 Questions

2.1 Explain the following terms and materials:

First law of thermodynamics
Second law of thermodynamics
Reversible and irreversible processes
Enthalpy, entropy, and internal energy
State and path functions
Gibbs free energy and electrical work
Nernst equation
Thermoneutral potential
Carnot efficiency
Stoichiometry number
Energy demand and heat demand.

2.2 The standard entropy of formation of water vapor is 1.887×10^{-1} kJ/mol·K at 298 K. What is the entropy of formation of water vapor when both reactant and product are at 800 °C? (Answer: 0.179 kJ/mol·K)

2.3 In direct methanol fuel cells, methanol is used as a fuel. The methanol oxidation reaction occurs as follows:

$$CH_3OH_{(l)} + \frac{3}{2}O_{2\,(g)} \rightarrow CO_{2\,(g)} + 2H_2O_{(l)}$$

Calculate the ΔH^0_{rxn} and ΔS^0_{rxn} for the methanol oxidation (or combustion) reaction.
(Answer: $\Delta H^0_{rxn} = -726.67\,\text{kJ/mol}$ and $\Delta S^0_{rxn} = -81.0\,\text{J/mol}$.

2.4 Water gas shift (WGS) reaction is an important reaction in hydrogen generation by steam reforming of e.g. natural gas. In the presence of water, carbon monoixde is converted into carbon dioxide with additional hydrogen produced. Determine the maximum temperature above which WGS will not occur spontaneously. (Answer: 706 °C)

$$\text{WGS}: \quad CO + H_2O_{(g)} \rightleftharpoons CO_2 + H_2$$

2.5 Calculate the entropy change when 1 mol of water is evaporated at its boiling temperature under 1 atm pressure? (Answer: 109 J/K)

2.6 Calculate the standard Gibbs formation energy and reversible potential for the following electrochemical reactions under STP conditions.

(a) $C + \frac{1}{2}O_2 \rightarrow CO$
(b) $C + O_2 \rightarrow CO_2$
(c) $CO + \frac{1}{2}O_2 \rightarrow CO_2$

(Answer: (a) −137.4 kJ/mol and 0.71 V; (b) −394.6 kJ/mol and 1.02 V; (c) −257.11 kJ/mol and 1.33 V)

2.7 Calculate the standard enthalpy of the combustion reaction of propene, C_3H_6. For this problem, we have following standard enthalpy values for hydrogenation of propene
$CH_2CHCH_{3(g)} + H_{2(g)} \rightarrow CH_3CH_2CH_{3(g)}\ \Delta H^0_{rxn} = -124\,\text{kJ/mol}$
Oxidation reaction of propane
$CH_2CH_2CH_{3(g)} + 5O_{2(g)} \rightarrow 3CO_2 + 4H_2O_{(l)}\ \ \Delta H^0_{rxn} = -2220\,\text{kJ/mol}$
(Answer: −2058 kJ/mol)

2.8 Assume a 100 cm^2 fuel cell is operating, under typical conditions of 1 atm pressure and 80 °C, at 0.7 V and generating 0.6 A/cm^2 of current. What is the excess heat generated by this fuel cell? The ideal or thermodynamic voltage of the cell is 1.16 V @ 80 °C.
(Answer: 47.4 W (J/s))

2.9 Derive the efficiency of a Carnot cycle using the temperature-entropy diagram.

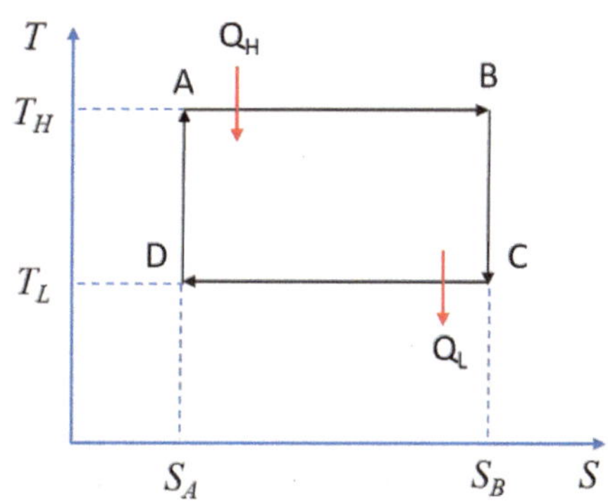

In the diagram, S_B is the entropy at T_L and S_A is the entropy at T_H.

2.10 Calculate the reversible cell voltage of a H_2−O_2 fuel cell on 3 atm pure hydrogen and 4 atm air at 298.15 K. (Answer: 1.242 V)

2.11 The following reaction may be made to occur in fuel cells at 298 K:

$$1)\ \ CH_{4(g)} + 2O_{2(g)} \rightleftharpoons CO_{2(g)} + 2H_2O_{(l)}$$

$$\Delta H^0_{298} = -890.4\,\text{kJ/mol};\ \Delta G^0_{298} = -818.0\,\text{kJ/mol}$$

$$2)\ \ C_2H_{6(g)} + \frac{7}{2}O_{2(g)} \rightleftharpoons 2CO_{2(g)} + 3H_2O_{(l)}$$

$$\Delta H^0_{298} = -1560\,\text{kJ/mol};\ \Delta G^0_{298} = -1467.5\,\text{kJ/mol}$$

$$3)\ \ C_3H_{8(g)} + 5O_{2(g)} \rightleftharpoons 3CO_{2(g)} + 5H_2O_{(l)}$$

$$\Delta H^0_{298} = -2220\,\text{kJ/mol};\ \Delta G^0_{298} = -2108.0\,\text{kJ/mol}$$

$$4)\ \ CH_3OH_{(l)} + \frac{3}{2}O_{2(g)} \rightleftharpoons CO_{2(g)} + 2H_2O_{(l)}$$

$$\Delta H^0_{298} = -764.0\,\text{kJ/mol};\ \Delta G^0_{298} = -706.9\,\text{kJ/mol}$$

For each reaction, calculate (a) the number of electrons transferred in the cell reaction; (b) the reversible emf or voltage of the cell at 298 K; and (c) the maximum efficiency of the cell. (Answer: 1) 8, 1.059, 91.9%; 2) 14, 1.086, 94.1%; 3) 20, 1.092, 94.9%; 4) 6, 1.221, 92.5%)

2.12 Does the entropy increase or decrease for the following reactions? How will the open circuit voltage change with temperature?

(a) $H_{2(g)} + \frac{1}{2}O_{2(g)} = H_2O_{(g)}$
(b) $H_{2(g)} + \frac{1}{2}O_{2(g)} = H_2O_{(l)}$
(c) $C_{(s)} + \frac{1}{2}O_{2(g)} = CO_{(g)}$
(d) $CO_{(g)} + \frac{1}{2}O_{2(g)} = CO_{2(g)}$
(e) $CH_{4(g)} + 2O_{2(g)} = CO_2 + 2H_2O_{(g)}$
(f) $CH_3OH_{(g)} + \frac{3}{2}O_{2(g)} = CO_2 + H_2O_{(g)}$

2.8 General Readings

1. Rogone DV (1995) Thermodynamics of materials. Wiley, New York

Chapter 3
Fuel Cell Electrochemistry

3.1 Electrochemistry at the Interface

Electrochemistry is a branch of science studying the relationship between electrical energy and chemical reactions. These reactions involve electric charge transfer between an electronically conductive phase and an ionically conductive phase. The boundary of these two phases is the electrochemical interface. A typical electrochemical interface consists of a solid conductive electrode and an ionic conducting electrolyte. The electronically conducting electrodes used for fuel cell include metals, oxides, and their composites. The electrolyte is conventionally an aqueous solution of acids (e.g., H_2SO_4 or H_3PO_4) or bases (e.g., KOH or NaOH) or salts (e.g., NaCl) that are dissociated into charge-carrying ions. The electrolyte can also be non-aqueous solutions, for example, ionic liquids or molten salts (e.g., molten carbonates). The charge carriers can also be dissolved into a solid medium such as polymers or ceramics, the so-called solid electrolytes. All these types of electrolytes are in fact used in fuel cells.

As long as an electrode is in contact with an electrolyte, an electric potential is set up across the two phases due to the transfer of charged species across the interface, specific adsorption of ions at the interface or orientation of permanent or individual dipoles. The magnitude of the potential is a function of both the composition and nature of the two phases. In connection to fuel cells the most common types of interfaces include:

- Metal (metal oxide)/aqueous electrolyte, e.g., in alkaline fuel cell (AFC) and phosphoric acid fuel cell (PAFC)
- Metal/solid polymeric membrane, e.g., in proton exchange membrane fuel cell (PEMFC) and anion-exchange membrane fuel cell (AEMFC)
- Metal (metal oxide)/molten carbonate electrolyte, e.g., in molten carbonate fuel cell (MCFC)
- Metal (metal oxide)/solid oxide electrolyte, e.g., in solid oxide fuel cell (SOFC).

S. P. Jiang and Q. Li, *Introduction to Fuel Cells*,
https://doi.org/10.1007/978-981-10-7626-8_3

An interface sounds two dimensional but is in fact more than a flat surface. The solid surface is always very rough or more often porous. The electrolyte phase involves orientation of mobile ions which is also extended to a thickness of several mono atomic layers. In general the regions in both electrode and electrolyte phases near their interface are considered of three dimensional nature, across which a *double layer* is formed, which has electrical, compositional, and structural characteristics. Such characteristics are directly related and relevant to the electrochemical reactions that occur in fuel cells. To understand why it is so, we will start with a brief description of the evolution of theoretical aspects of the structure of double layers between a metallic conductor and surrounding aqueous electrolyte solution.

3.1.1 Origin of Double Layer Structure

In an acidic aqueous solution containing, e.g., hydrogen ions or protons H^+ the ions are always in random movement. When an inert metal (e.g., platinum) wire, an electronic conductor with free mobile electrons, is dipped in a solution, the hydrogen ions constantly hit the metal surface. Some of the ions, by a good chance, may take an electron from the surface, forming a chemically adsorbed atom. Some of the chemisorbed atoms, on one hand, vibrate so as to escape from the surface by breaking the chemical bond and returning the electron to the metal. Some, on the other hand, combine with the neighboring adsorbed atoms to form a hydrogen molecule H_2. When a gaseous hydrogen is steadily supplied by bubbling around the platinum wire, the following equilibrium is established at the interface between the metal electrode and the aqueous electrolyte solution, depending on the concentration of the hydrogen ion and the pressure of the gaseous hydrogen:

$$2\mathrm{H}^+ + 2\mathrm{e}^- \rightleftharpoons 2\mathrm{H}_{\mathrm{ads}} \rightleftharpoons \mathrm{H}_2 \qquad K = \frac{P_{\mathrm{H}_2}}{a_{\mathrm{H}^+}^2} \tag{3.1}$$

The involvement of electrons in Eq. (3.1) indicates that a good electronic conductor (Pt wire) is necessary in the equilibrium. It is also important in this connection that the platinum electrode is assumed to be chemically inert, functioning only as a lead of electrons. This assumption is, however, not always true when other reactant gases, e.g., oxygen is involved. As a result of the equilibrium, electrons are building up on the platinum surface. The excess electrons are distributed in a uniform layer, making the platinum surface negatively charged. The charged surface is, of course, neutralized by redistribution of the positively charged hydrogen ions in the surrounding solution. In this case, we have electrons in the metal and positive ions in solution, as shown in Fig. 3.1a.

It was von Helmholtz who, first in 1850s, assumes that the interaction between the charged metal surface and surrounding ions is physically electrostatic, i.e., the charged electrode repels ions of the like charge while attracts those of the opposite charger. In the case discussed above, the attracted ions approach the electrode surface and form a layer of positive charge near the electrode surface. As a result an electrified

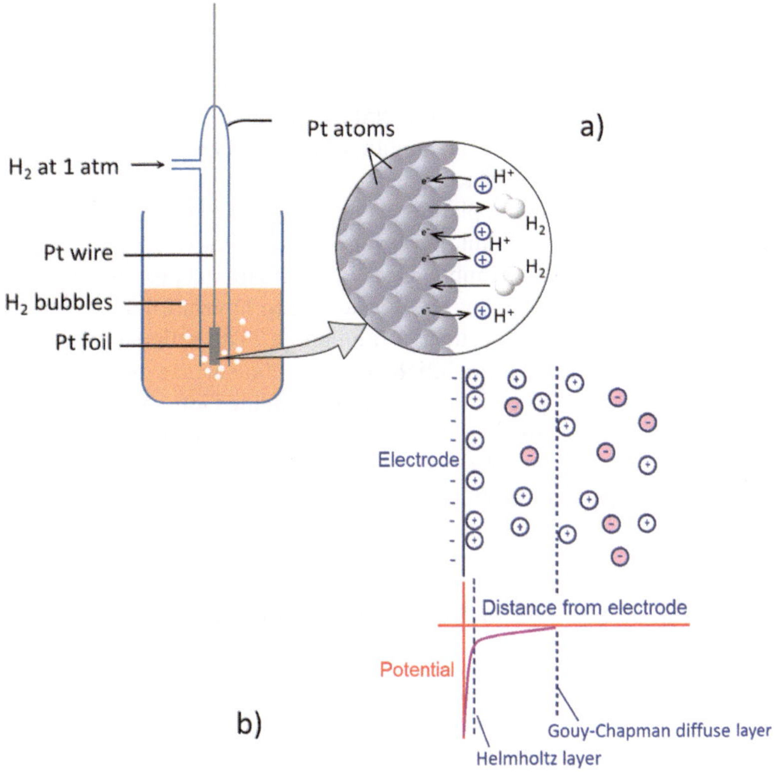

Fig. 3.1 **a)** Construction of a hydrogen electrode with a platinum foil dipped in an aqueous acidic solution onto the surface of which gaseous hydrogen is bubbled around. To the right is a view of interactions between hydrogen molecules and platinum surface atoms. **b)** A simplified illustration of the Stern double layer on a negatively charged solid surface with a Helmholtz layer and a Gouy–Chapman diffusion layer

interface is established with two layers of opposite charges, called the electrical double layer. Between the opposite charges is a linear potential gradient, analogous to an electrical capacitor consisting of two charged plates separated by a dielectric layer. As also rationalized by von Helmholtz the separation of the double layer is in fact of distance of the ion radius and therefore called the Helmholtz layer. As the potential drop across the interface is linear, the capacitance, C, of the double layer as in the case of a parallel plate condenser:

$$C = \frac{\varepsilon}{4\pi d} \tag{3.2}$$

where ε is the dielectric constant of the medium between the two layers of opposite charges and d is the distance between them.

Considering the fact that ions in solution are in the random Brownian motion, the electrostatic interaction of the mobile ions and the stationary electrode surface should exist in an extended layer outside the Helmholtz layer. It was proposed by Gouy in 1910 and Chapman in 1913 that ions in the electric double layer are subjected to electrical and thermal fields, and thus there will be charge distribution of ions as a function of distance away from the metal surface. The extended layer is called the Gouy–Chapman diffusion layer, within which the electric potential decreases exponentially away from the surface.

It is more realistic that, at the platinum surface, some ions strongly adhere to the electrode surface forming a Helmholtz layer and some move around in the Gouy–Chapman diffusion layer. This combination is a hybrid model proposed by Otto Stern in 1924. In this hybrid model, ions are considered to have a finite size and located at a finite distance from the electrode. The charge distribution in the electrolyte consists of two components: immobilized close to the surface as in the Helmholtz model and distributed and spread out in solution as in the Gouy–Chapman model, as shown in Fig. 3.1b. There are other double layer structures and potential differences across interface due to the specific adsorption generally associated with aqueous electrochemical systems and reactions.

3.1.2 Standard Hydrogen Reference Electrodes

Now we consider an oxygen electrode by immersing a Pt foil in an acid solution, e.g., 1 M HCl with oxygen bubbling around at pressure of 1 bar. Similar to the hydrogen electrode discussed, oxygen molecules adsorb on the Pt surface and interact with H^+ ions in the solution. Eventually a double layer as well as the following equilibrium is established at the interface:

$$\frac{1}{2}O_2 + 2H^+ + 2e^- = H_2O \tag{3.3}$$

This charge separation establishes a potential difference across the interface. Now the question is, can we measure this electrode potential with a voltmeter? Of course, not, because we only have one terminal to connect to the oxygen electrode while the other terminal is hanging loose; see Fig. 3.2a. The absolute value of this potential is difficult, and most often, impossible to measure. This is analogous to many other physical chemical properties such as enthalpy, entropy, and Gibbs free energy—the absolute values of them are by no means to know. Only the changes of them from a reference point or the standard state are of practical importance.

In physical chemistry, a reference point is defined as the standard state of a material, the choice of which in principle is arbitrary but a set of conventional conditions are recommended for general use. For an ideal gas, it is the hypothetical pressure of 1 bar. For liquids and solids it is the pure substance while for a solute in solutions it is a concentration of 1 M or more precisely an activity of 1. Strictly speaking, temperature is not part of the definition of a standard state but the thermodynamic quantities are most often collected at 298.15 K (25 °C).

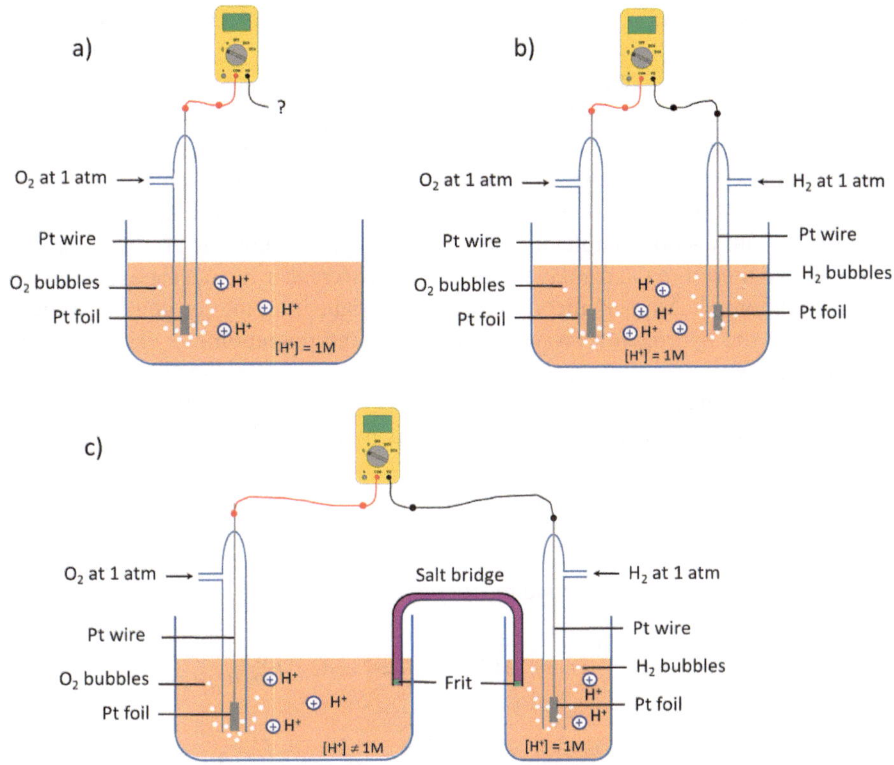

Fig. 3.2 **a)** An oxygen electrode in 1 M acid solution; **b)** measurement of the oxygen electrode potential with help of a hydrogen reference electrode; and **c)** measurement of the oxygen electrode potential with a standard hydrogen electrode using a salt bridge

In electrochemistry the *standard hydrogen electrode* (SHE) is defined as the basis of a universal scale of electrode potentials. The construction of an SHE is essentially the same as that shown in Fig. 3.1a. For equilibrium (3.1) the Nernst equation can be written as

$$E = E^0 - \frac{RT}{2F}\ln\frac{P_{H_2}}{a_{H^+}^2} \tag{3.4}$$

Under standard conditions, i.e., the hydrogen pressure of 1 bar and the H^+ concentration of 1 M $\left(a_{H^+} = 1\right)$, the logarithm term is zero, $E = E^0$. The standard hydrogen electrode potential, E^0, is arbitrarily defined as zero at any temperatures.

The reason for using the hydrogen electrode as a reference is that the forward and backward reactions (3.1) are very fast on a Pt electrode, which is often made of a platinum foil coated with platinum black to increase the effective surface area. Such an electrode may allow passage of a small current with little change in the potential across the interface and therefore exhibit a constant and stable potential. Such an electrode is also called an *ideally non-polarizable electrode* and suitable to serve as a reference electrode.

As long as the acid solution of the oxygen electrode is the same as the standard hydrogen electrode, i.e., 1 M H^+, and the two electrodes can easily be immersed into the solution (precaution should be taken to prevent any mixing of the two gases). A potential difference can be measured by a voltmeter (Fig. 3.2b). This is in fact a poor example to illustrate the potential measurement with hydrogen reference electrode because platinum is not completely inert when oxygen is bubbling around in an acid solution, and therefore the displayed potential is different from that for a real oxygen electrode potential. Anyway theoretically when connecting the oxygen electrode as positive and the hydrogen reference as negative, the potentiometer should show a value of 1.23 V.

In electrochemistry such a cell construction consisting of an SHE (H^+/H_2) and an O_2/H_2O half-cell is called a galvanic cell and is notated as:

$$\mathrm{Pt}\left|\mathrm{H}_{2(\mathrm{g},\ 1\mathrm{bar})}\right|\mathrm{H}^+_{(\mathrm{aq},1\mathrm{M})}\left\|\mathrm{H}^+_{(\mathrm{aq},1\mathrm{M})}\right|\mathrm{O}_{2(\mathrm{g},\ 1\mathrm{bar})}|\mathrm{Pt} \tag{3.5}$$

which corresponds to the electrode reactions:

$$\text{Oxygen electrode:}\quad \frac{1}{2}\mathrm{O}_2 + 2\mathrm{H}^+ + 2\mathrm{e}^- \rightleftharpoons \mathrm{H}_2\mathrm{O} \quad E^0_{\mathrm{O}_2,\mathrm{H}^+/\mathrm{H}_2\mathrm{O}} = ? \tag{3.6}$$

$$\text{Hydrogen electrode:}\quad 2\mathrm{H}^+ + 2\mathrm{e}^- \rightleftharpoons \mathrm{H}_2 \qquad E^0_{\mathrm{H}^+/\mathrm{H}_2} = 0 \tag{3.7}$$

The measured potential difference under standard conditions is

$$E^0 = E^0_{\mathrm{O}_2/\mathrm{H}_2\mathrm{O}} - E^0_{\mathrm{H}^+/\mathrm{H}_2} = +1.23\,\mathrm{V} \tag{3.8}$$

Thus, one has $E^0_{\mathrm{O}_2/\mathrm{H}_2\mathrm{O}} = +1.23\,\mathrm{V}$. This value is therefore the potential difference between the oxygen electrode and the SHE.

When the acid solution has a concentration different from that of the standard hydrogen electrode, the measurement can be done in two ways. One way is to isolate the hydrogen reference electrode from the oxygen electrode compartment to avoid the solution contamination while the ionic connection is achieved by a salt bridge to maintain the electrical neutrality (Fig. 3.2c). A salt bridge is a glass tube filled with a saturated solution of salts. The ends of the tube are often plugged with a porous ceramic frit which allows for ionic connection but prevents the excessive diffusion.

A more practical method is to immerse a hydrogen electrode in the actual acid solution whatever its concentration is. A reliable and reproducible potential can be obtained which however deviates from that of the SHE, such a reference electrode is a reversible hydrogen electrode (RHE). Under a hydrogen pressure of 1 bar, the RHE potential can be expressed as a function of pH:

$$2H^+ + 2e^- \rightleftharpoons H_2 \qquad E_{H^+/H_2} = E^0_{H^+/H_2} - \frac{2.303\,RT}{F}pH \tag{3.9}$$

The constant $2.303\,RT/F$ at room temperature (25 °C) is 0.059 V, a deviation of the hydrogen electrode potential when the H^+ concentration changes by 10 times or pH change by 1. Thus, the Nernstian potential scale of a RHE depends on the H^+ concentration or pH of the actual solution. The potential dependence of −0.059 V/pH tells that a hydrogen electrode potential decreases from zero at $pH = 0$ to −0.83 V at $pH = 14$. This in fact is the reduction potential of the hydrogen evolution reaction in the alkaline electrolyte (see Chap. 4). Thus, at room temperature (25 °C), we have the RHE potential as:

$$E_{RHE} = E_{SHE} + 0.059 \times pH \tag{3.10}$$

In practice the studied reactions involving the hydrogen ions such as the oxygen reduction reaction (ORR) in the same electrolyte shifts also with pH by exactly the same value:

$$\frac{1}{2}O_2 + 2H^+ + 2e^- \rightleftharpoons H_2O \quad E_{O_2/H_2O} = E^0_{O_2/H_2O} - \frac{2.303\,RT}{F}pH \tag{3.11}$$

As a result, the potential difference between a studied reaction, e.g., ORR and the RHE, remains the same in an electrolyte of any pH. This behavior of the RHE makes it an ideal reference electrode for practical uses. Instead of platinum, a palladium hydrogen electrode is also a popular reference electrode, which has most characteristics of a reversible hydrogen electrode while palladium has a better capability to absorb molecular hydrogen. Both reversible reference electrodes are commercially available.

In some studies on fuel cells, for example, electrochemical measurements of the hydrogen permeability, the overall current is relatively small and the hydrogen electrode is sometimes used as a pseudo-reversible hydrogen reference electrode. In cells where hydrogen is not available, e.g., in direct methanol fuel cell (DMFC), a *dynamic hydrogen electrode* (DHE) is often used. It consists of two platinum wires between which a very small and constant current of 10–50 μA is applied by means of a battery (See Fig. 3.3). Under this condition, the Pt wire as the negative electrode is covered by the deposited hydrogen, resulting in a stable reference potential, which may deviate from the RHE potential by a few mV. The advantages of the DHE include no need for a hydrogen supply and minimized contamination.

3.1.3 Other Reference Electrodes

Hydrogen reference electrodes have the advantages of accuracy and reliability but in practice it is not always convenient to use due to the need to supply high-purity hydrogen. It is thus a common practice in fuel cell research to use secondary reference

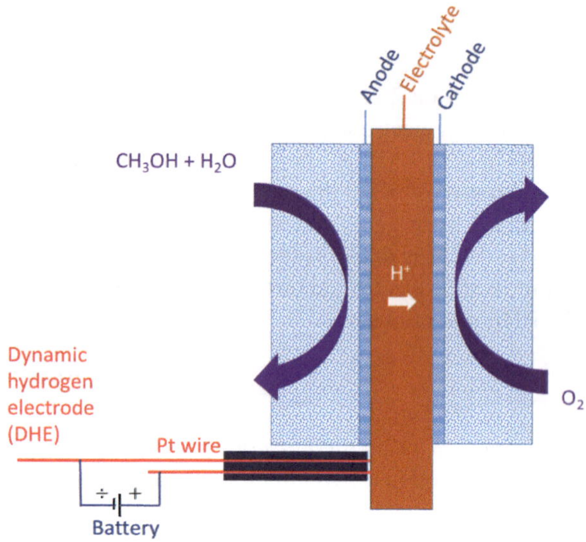

Fig. 3.3 Schematic of a dynamic hydrogen electrode in a direct methanol fuel cell

electrodes whose potentials have been accurately determined with respect to that of a hydrogen reference electrode. These reference electrodes have some common characteristics: metal in contact with one of its sparingly soluble salts, placed in a solution containing a strongly ionized salt with a common anion.

The most commonly used is Calomel reference electrodes which is based on the reaction between elemental mercury and mercury(I) chloride in an aqueous solution of potassium chloride:

$$Hg_2Cl_2 + 2e^- \rightleftharpoons 2Hg + 2Cl^- \quad E^0_{Hg_2Cl_2/Hg,Cl^-} = +0.27\,V \tag{3.12}$$

The electrode reaction is in combination with the precipitation reaction

$$Hg_2^{2+} + 2Cl^- \rightleftharpoons Hg_2Cl_{2(s)} \tag{3.13}$$

The solubility product for reaction (3.13) is

$$K_{sp} = a_{Hg_2^{2+}} \times a^2_{Cl^-} = 1.18 \times 10^{-18} \tag{3.14}$$

showing that at high concentration of chloride ions, the mercury ion concentration to be handled is very low. The potential of this reference electrode can be expressed as

$$E_{Hg_2Cl_2/Hg,Cl^-} = E^0_{Hg_2Cl_2/Hg,Cl^-} - \frac{RT}{F}\ln a_{Cl^-} \tag{3.15}$$

Table 3.1 Reference electrodes commonly used in fuel cells

Reference electrode	Potential, V (vs. SHE)
Calomel electrode (0.1 M KCl)	+0.336
Calomel electrode (1.0 M KCl)	+0.283
Calomel electrode (saturated KCl, SCE)	+0.244
Ag/AgCl (saturated KCl)	+0.199
Hg/Hg_2SO_4 (saturated K_2SO_4)	+0.615V
Hg/HgO (20 wt% KOH)	+0.098V

It is seen that the electrode potential is a function of the activity of the chloride ions. The saturated KCl solution is most often used, and the electrode is called the *saturated calomel electrode* (SCE). At 25 °C, the potential of SCE against SHE is 0.244 V under standard conditions. In the case of electrolyte solution with pH $\neq$ 0, Eq. (3.10) needs to be used and the measured potential can be converted to that referred to RHE.

$$E_{\text{RHE}} = E_{\text{SHE}} + 0.059 \times \text{pH} = E_{\text{SCE}} + 0.244 + 0.059 \times \text{pH} \tag{3.16}$$

Calomel electrodes with other KCl concentrations and other widely used reference electrodes are listed in Table 3.1 with their potentials relative to SHE under standard conditions.

3.1.4 Electrode Potential and Standard Reduction Potential

With the help of hydrogen reference or other reference electrodes the potential of a single electrode can be measured. The *electric potential* is the energy per unit charge that is moving through an electric field, having a unit of volt (1 V = 1 J/C). It varies with temperature, concentration, and pressure of the reactants and products of the electrode reaction, as expressed by Nernst equation. Thus, for any redox reactions, we have:

$$\text{Ox} + \text{ne}^- \rightleftharpoons \text{Re} \qquad E = E^o - \frac{\text{RT}}{nF}\ln\frac{[\text{Re}]}{[\text{Ox}]} \tag{3.17}$$

Here E^0 is the standard electrode potential under standard conditions, i.e., the concentration of oxidant and reductant, Re and Ox, is unity. The activity of metallic electrode is taken as 1. For example, a zinc bar is immersed in a solution containing 1 M Zn^{2+} (1 M $ZnSO_4$) and a reaction occurs between metallic Zn bar and surrounding solution. This is an electrochemical reaction as it releases two electrons and charge separation occurs at the interface: negatively charge electrons on the Zn bar solid side and excess of positively charged zinc ions on the solution side. This charge separation

Table 3.2 Standard reduction potentials of selected reactions

Cathode (reduction) half reaction	Standard potential E^o (V)
$H_2O_2 + 2H^+ + 2e^- = 2H_2O$	+1.78
$O_2 + 4H^+ + 4e^- = 2H_2O$	+1.23
$HO_2^- + H_2O + 2e^- = 3OH^-$	+0.88
$Hg^{2+} + 2e^- = Hg$	+0.80
$O_2 + 2H^+ + 2e^- = H_2O_2$	+0.68
$O_2 + 2H_2O + 4e^- = 4OH^-$	+0.40
$Cu^{2+} + 2e^- = Cu$	+0.34
$AgCl + e^- = Ag + Cl^-$	+0.22
$2H^+ + 2e^- = H_2$	0.00
$Co^{2+} + 2e^- = Co$	−0.28
$Fe^{2+} + 2e^- = Fe$	−0.44
$Zn^{2+} + 2e^- = Zn$	−0.76
$2H_2O + 2e^- = H_2 + 2OH^-$	−0.83
$Al^{3+} + 3e^- = Al$	−1.66
$Na^+ + e = Na$	−2.71
$Li^+ + e^- = Li$	−3.05

establishes a potential difference across the interface. Thus, the electrode potential across the Zn electrode/electrolyte (1 M $ZnSO_4$) interface can be measured against the SHE, which is −0.76 V.

$$Zn \rightleftharpoons Zn^{2+} + 2e^- \qquad E^o_{Zn^{2+}/Zn} = -0.76\,V \tag{3.18}$$

The $E^0_{Zn^{2+}/Zn}$ value measured against SHE is called the standard electrode potential of the Zn^{2+}/Zn couple. Table 3.2 lists the values of standard electrode potentials of selected electrode reactions.

It should be noted that the standard electrode potential is corresponding to the reduction reaction (receiving electrons) and also called the *standard reduction potential*. The sign of the electrode potential is a relative indication of the ability of the electrode to take or give electrons relatively to the hydrogen electrode. For example, the zinc electrode has a negative potential, indicating that the zinc electrode (i.e., zinc metal) is easy to give electrons relatively to the hydrogen electrode. The hydrogen electrode, when coupled with the zink electrode, becomes positive, indicating that hydrogen ions tend to receive electrons forming hydrogen gas at the hydrogen electrode half-cell. A higher (positive) value of the electrode potential means a greater tendency for the reduction reaction to occur while a lower (negative) one means a greater tendency for the oxidation reaction to occur.

When assembling a galvanic cell or voltaic cell to derive electrical energy from a spontaneous redox reaction, e.g., fuel cells or batteries, the electrode of higher potential is used as the cathode where reduction takes place and the electrode of lower potential is used as the anode where the oxidation reaction takes place. The cell voltage between the cathode potential ($E_{cathode}$) and anode potential (E_{anode}) is the maximum voltage the cell can deliver, also called *equilibrium or reversible cell voltage*, denoted as E_{rev} in this book:

$$E_{rev} = E_{cathode} - E_{anode} \tag{3.19}$$

This voltage is related to the Gibbs free energy changes of the corresponding cell reaction

$$\Delta G = -nFE_{rev} \tag{3.20}$$

When $E_{rev} > 0$, then the cell reaction is spontaneous.

Example 3.1 Calculate the reversible voltage of a hydrogen–oxygen fuel cell operating at 60 °C with H_2 and air fully humidified.

Solution:

Refer to Chap. 2, we have the Nernst equation as a function of temperatures for an H_2–O_2 fuel cell:

$$\begin{aligned} E &= E_T - \frac{RT}{2F}\ln\left(\frac{a_{H_2O}}{p_{H_2}\cdot p_{O_2}^{1/2}}\right) \\ &= E^0 - 2.3\times 10^{-4}(T-298) - \frac{RT}{2F}\ln\left(\frac{a_{H_2O}}{p_{H_2}\cdot p_{O_2}^{1/2}}\right) \end{aligned} \tag{3.21}$$

At 60 °C water has a saturated vapor pressure of 0.2 atm (i.e., $a_{H_2O} = 1$) while the partial pressure of hydrogen is $p_{H_2} = 1 \times 0.8 = 0.8$ atm and the partial pressure of oxygen in fully humidified air is $P_{O_2} = 0.21 \times 0.8 = 0.17$ atm.

$$\begin{aligned} E &= 1.229 - 2.3\times 10^{-4} \times (333-298) + \frac{8.314\times 333}{2\times 96485} \\ &\ln\left(\frac{0.8\times 0.17^{1/2}}{1}\right) = +1.205\text{ V} \end{aligned} \tag{3.22}$$

3.1.5 Concentration Cells

The case we described above is for the electrode reaction under standard conditions. However, as the concentration of oxidant species such as Cu^{2+} changes due to the respective oxidation and reduction reactions, the cell potential changes according to

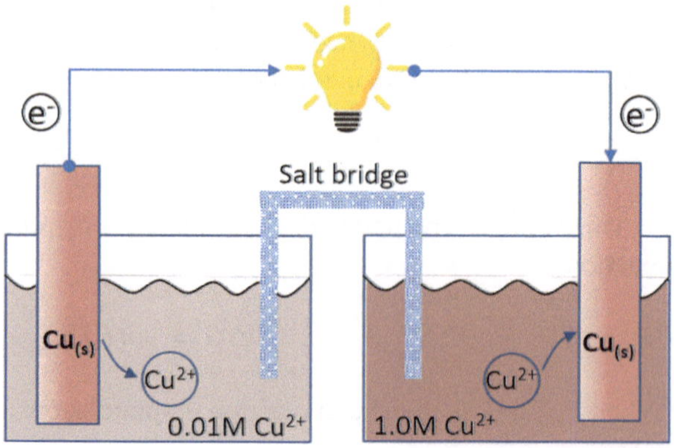

Fig. 3.4 Concentration cell with copper electrode and $CuSO_4$ solution

the Nernst equation. Thus, for cells made of two half-cells with the same electrode but differing in the concentration of the common component, such cells are called the concentration cells in electrochemistry. Figure 3.4 shows an example of concentration cell with the same copper electrode but different concentration of $CuSO_4$.

For this cell, based on the Nernst equation, the electrode potential for the copper in the compartment of 0.01 M $CuSO_4$ solution at 298 K is

$$E_{0.01\,\mathrm{M\,Cu^{2+}/Cu}} = E^0 - \frac{\mathrm{RT}}{nF}\ln\frac{a_{\mathrm{Cu}}}{a_{\mathrm{Cu^{2+}}}} = 0.34 - \frac{0.0591}{2}\log\frac{1}{0.01} = +0.28\,\mathrm{V} \quad (3.23)$$

The electrode potential at the Cu/$CuSO_4$ half-cell with 0.01 M $CuSO_4$ solution is 0.28 V, lower than 0.34 V for the other half-cell with 1.0 M $CuSO_4$ solution. Thus, according to the Eq. (3.20), for a spontaneous reaction, the electrode reaction at the Cu electrode in the Cu/$CuSO_4$ half-cell with 0.01 M $CuSO_4$ solution is oxidation reaction and the Cu electrode is anode. Similarly the electrode reaction at the Cu electrode in the Cu/$CuSO_4$ half-cell with 1.0 M $CuSO_4$ solution is reduction reaction and the Cu electrode is cathode. The oxidation half-cell reaction or *anodic reaction* can be written as:

$$\mathrm{Cu_{(s)}} \rightarrow \mathrm{Cu^{2+}(aq.\ 0.01M)} + 2\mathrm{e^-} \qquad E_{0.01\,\mathrm{MCu^{2+}/Cu}} = +0.28\,\mathrm{V} \quad (3.24)$$

And reduction half-cell reaction or *cathodic reaction* is

$$\mathrm{Cu^{2+}(aq.\ 1.0\,M)} + 2\mathrm{e^-} \rightarrow \mathrm{Cu_{(s)}} \qquad E_{1.0\,\mathrm{MCu^{2+}/Cu}} = +0.34\,\mathrm{V} \quad (3.25)$$

The overall cell reaction is

$$\mathrm{Cu^{2+}(aq.\ 1.0\,M)} \rightarrow \mathrm{Cu^{2+}(aq.\ 0.01\,M)} \quad (3.26)$$

Thus, we have the cell reversible potential, E_{rev}, as

$$E_{\mathrm{rev}} = E_{\mathrm{cathode}} - E_{\mathrm{anode}} = 0.34 - 0.28 = +0.06\,\mathrm{V} \quad (3.27)$$

Cu will be dissolved in the compartment with 0.01 M $CuSO_4$, leading to the increase in the Cu^{2+} ion concentration, while in the compartment with 1.0 M $CuSO_4$, the Cu will be deposited to the Cu electrode and the Cu^{2+} ion concentration will decrease. The reaction will stop when the concentration in both compartment is equal.

One example is the concentration cell for high-temperature SOFC. The galvanic cell reaction (3.5) in SOFC can be written as the combination of hydrogen from anode and oxygen from cathode to form water:

$$H_{2(\text{anode})} + \frac{1}{2}O_{2(\text{cathode})} = H_2O_{(\text{anode})} \tag{3.28}$$

At high temperatures the reactants and water products are all in gas phase in the case of SOFCs and the equilibrium potential of the cell depends on the partial pressure of oxygen at the cathode and anode, which is in turn determined by the thermodynamics of the gaseous species at the electrode compartments. From the Nernst equation, we have:

$$E_{\text{rev}} = \frac{\text{RT}}{2nF}\ln\frac{p_{O_{2(c)}}}{p_{O_{2(a)}}} \tag{3.29}$$

The number of electrons involved in Eq. (3.29) is two. This is actually a concentration cell for high-temperature SOFC in H_2–O_2. Similar concentration cell based on the partial pressure of hydrogen also occurs for the protonic ceramic fuel cells (see Chap. 16). This is a convenient equation to calculate the cell potential or voltage from known partial pressure of oxygen at both sides or vice versa. However, there is a fundamental difference between the concentration cells in aqueous electrochemical cells and SOFC. In aqueous electrochemical cells, the reactant species are consumed and produced at the cathode and anode, respectively, and salt bridge is required to maintain the electrical neutrality of the cells. In the case of SOFC, the reactant species (e.g., oxygen molecules) transfer through the electrolyte from cathode to anode and electrical neutrality is maintained by the electrochemical reactions at the anode. For detailed explanation and deriving above equation, please see Chap. 9.

3.2 Activation and Activation Polarization Loss

3.2.1 Why Charge Transfer at the Interface

We consider the hydrogen electrode again, as shown in Fig. 3.1. An equilibrium is established at the platinum–solution interface involving electron transfer between gaseous hydrogen and hydrogen ions or protons involving electron transfer. The electron exchange at the interface occurs essentially due to the fact that there is a difference in the energy of electrons in the solid metal electrode and the active species in the electrolyte solution. According to the band theory, the electron energy in a metal is measured by the *Fermi level*. Thus, electron transfer generally occurs between the highest available energy level in the electrode (the *Fermi level*) and the lowest energy orbitals of the electrochemically active species in solution. In the case of the hydrogen electrode one has the electron transfer process:

$$H^+ + e^-(Pt) \rightleftharpoons (Pt)H_{ads} \tag{3.30}$$

If the energy of electrons in the platinum electrode, as measured by the Fermi level, E_F, is close to the energy of electrons that are associated with protons in the solution, the transfer of electrons in either direction may occur and an equilibrium will be established. The potential measured under this conditions is the reversible or equilibrium potential, E_{rev} while no net current can be measured. Remember the potential is a measure of electron energy. Application of an external potential will disturb the equilibrium. If a relatively more negative (more cathodic) potential is applied than the equilibrium potential, transfer of electrons occurs from the metal electrode to the protons in solution, forming adsorbed hydrogen atoms on the platinum surface; i.e., a reduction reaction takes place and a net cathodic current is obtained. In a similar way, if a potential is relatively more positive (more anodic) than the equilibrium potential, transfer of electrons occurs in the opposite direction. A net anodic current can thus be measured. Figure 3.5a illustrates this process of charge transfer for a hydrogen electrode at the interface.

The reaction process can be described graphically using the Gibbs free energy profile for the forward (reduction) and backward (oxidation) reaction. For the cathodic process to take place, the first step is the chemical absorption of hydrogen atoms on platinum $(Pt)H_{ads}$. The surface absorption of hydrogen atoms reduces its free energy, which however increases with the distance from the electrode surface; see Fig. 3.5b. For the cathodic reaction to take place, $(Pt)H_{ads}$ needs to be in the activated state (point A marked in the diagram) in order to be converted to H^+. In the case of reactive species H^+, its free energy increases as it approaches the electrode surface due to the electrostatic repulsive forces between the charged ions and the electrode surface. Thus energy is required to bring H^+ ions toward the electrode surface for the reaction to occur. Here the reaction profile involves the transition state with activation energy barrier, $\Delta G^{\#}$; see Fig. 3.5c. Because the presence of the activation energy barrier impedes the reaction, the reaction rate is finite. The same is also true for the backward reaction.

Fermi Level and Fermi Energy

Fermi level is a measure of the energy of the most mobile electrons within a solid. It is an important parameter in determining the electrical and thermal properties of solids, while Fermi energy is the Fermi level at absolute zero (0 K) and is a constant for each solid.

The Fermi energy is only defined at absolute zero, while the Fermi level can be defined for any temperature.

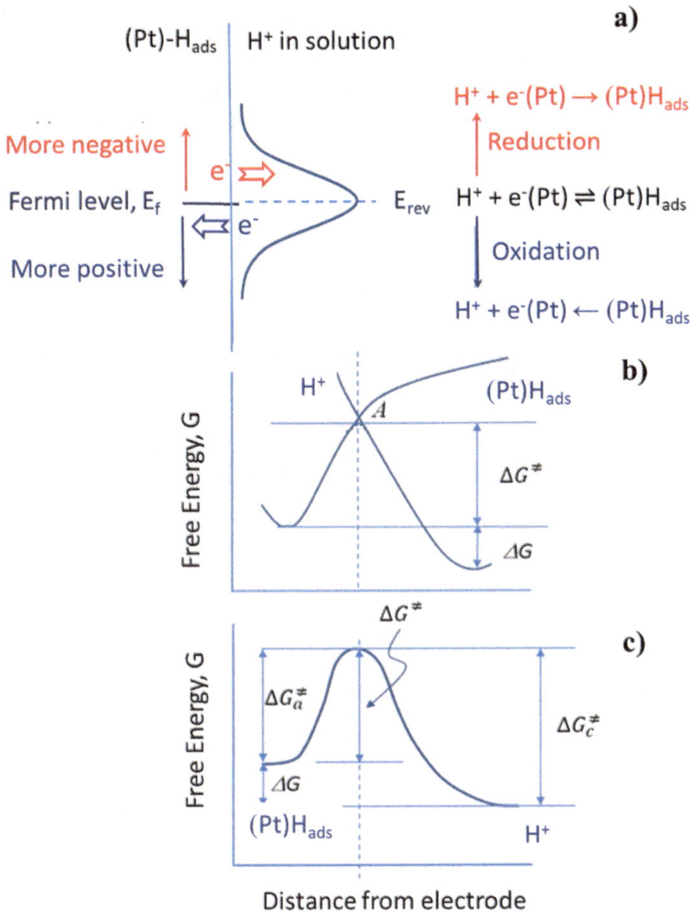

Fig. 3.5 **a)** Illustration of dependence of a hydrogen electrode reaction on the Fermi level of electrons; **b)** Gibbs free energy profile for the reaction species, H^+ and (Pt)H_{ads}; and **c)** Gibbs free energy profile of the electrode reaction (3.1). Point A indicates the activation energy barrier for the reaction. Other symbols are explained in the text

3.2.2 *Butler–Volmer Equation*

Let us consider a general redox reaction;

$$\mathrm{Ox} + \mathrm{ne}^- \rightleftharpoons \mathrm{Re} \tag{3.31}$$

In this redox reaction, k_c is the reaction rate constant for the cathodic or reduction reaction (i.e., the forward reaction) and k_a is the reaction rate constant for anodic or oxidation reaction (i.e., the backward reaction). For the reaction to occur, either in the forward direction or backward direction, the active species (i.e., Ox and Re) must have sufficient energy to go over an energy barrier to reach other side. The probability to make them over this energy barrier or *activation energy* is governed by the Arrhenius law. Thus, the rates of the process may be expressed in terms of the Arrhenius equation for anodic and cathodic reactions, respectively:

$$v_a = k_a[\mathrm{Re}]\mathrm{e}^{-\frac{\Delta G_a^{\neq}}{\mathrm{RT}}} \tag{3.32}$$

$$v_c = k_c[\mathrm{Ox}]\mathrm{e}^{-\frac{\Delta G_c^{\neq}}{\mathrm{RT}}} \tag{3.33}$$

where $\Delta G_c^{\neq}$ and $\Delta G_a^{\neq}$ are the activation energy for the cathodic and anodic direction reactions, respectively. When the reaction reaches equilibrium, $v_a = v_c$, we have

$$k_a[\mathrm{Re}]\exp\left[-\frac{\Delta G_a^{\neq}}{\mathrm{RT}}\right] = k_c[\mathrm{Ox}]\exp\left[-\frac{\Delta G_c^{\neq}}{\mathrm{RT}}\right] \tag{3.34}$$

For the overall reaction $\Delta G = \Delta G_c^{\neq} - \Delta G_a^{\neq}$. In electrochemistry the free energy change is often translated into the electrode potential by $\Delta G = -\mathrm{nFE}$. We have

$$\exp\left[-\frac{\Delta G}{\mathrm{RT}}\right] = \frac{k_c}{k_a} \times \frac{[\mathrm{Ox}]}{[\mathrm{Re}]} = \exp\left[\frac{nEF}{\mathrm{RT}}\right] \tag{3.35}$$

Therefore, the electrode potential can be expressed as

$$E = \frac{\mathrm{RT}}{nF}\ln\left(\frac{k_c}{k_a}\right) - \frac{\mathrm{RT}}{nF}\ln\frac{[\mathrm{Re}]}{[\mathrm{Ox}]} = E^0 - \frac{\mathrm{RT}}{nF}\ln\frac{[\mathrm{Re}]}{[\mathrm{Ox}]} \tag{3.36}$$

This equation derived kinetically is in the same form as Nernst equation that is previously derived thermodynamically. It is important to note that E^o in Eq. (3.36) is a function of the rate constants of the two processes occurring at the electrode.

At equilibrium or reversible potential, E_{rev}, no perceptible reaction occurs. The transition state can only be formed by application of an extra energy or potential. The deviation of the electrode potential from its reversible value is called *overpotential*, η, which is the driving force of the reaction proceeding in either cathodic or anodic direction depending on the negative or positive potential change applied. Thus, to effectively promote the cathodic direction of the redox reaction, the magnitude of the Gibbs free energy change of the reaction becomes

$$\Delta G_c^{\neq} = \left(\Delta G_c^0\right)^{\neq} + \alpha nF\eta \tag{3.37}$$

$$\Delta G_a^{\neq} = \left(\Delta G_a^0\right)^{\neq} - (1-\alpha)nF\eta \tag{3.38}$$

Here the overpotential (η) has two functions: to accelerate the cathodic reaction with a fraction α and at the same time to inhibit the anodic reaction with the remaining fraction $(1-\alpha)$. Fraction α is known as the *transfer coefficient*.

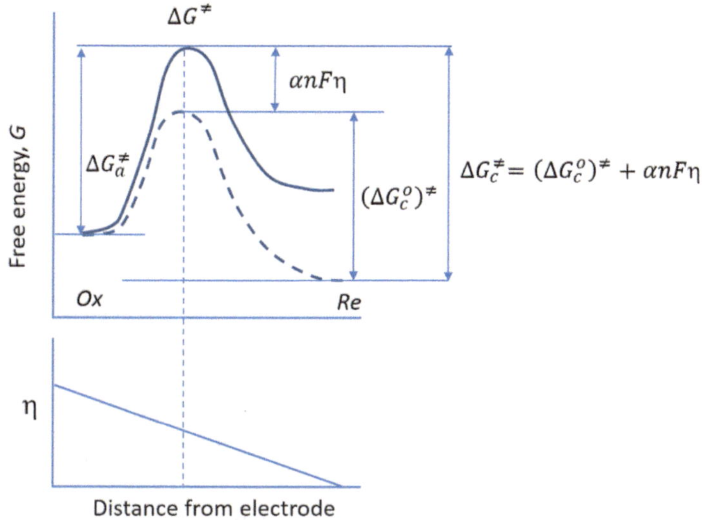

Fig. 3.6 Electrode reaction free energy profile for an irreversible electrode reaction, showing the formation of a transition state by imposing of a cathodic overpotential, η, and the distribution of η across the region between the electrode and the outer Helmholtz plane

Figure 3.6 shows schematically the effect of the overpotential η on the transition state and the region between the electrode surface and the reaction layer as a function of distance from the electrode surface. When approaching the transition or activated state, the active species will experience the increased influence by η. However, only a fraction of η, α, is required to produce the transition state which occurs at an intermediate distance. In this case, with new activation energies, the rates of cathodic and anodic processes become:

$$v_c' = k_c[\text{Ox}]\exp\left[-\frac{\left(\Delta G_c^0\right)^{\neq} + \alpha nF\eta}{\text{RT}}\right] \tag{3.39}$$

$$v_a' = k_a[\text{Re}]\exp\left[-\frac{\left(\Delta G_a^0\right)^{\neq} - (1-\alpha)nF\eta}{\text{RT}}\right] \tag{3.40}$$

In this case, the rates for the forward and backward reactions are not equal, a net cathodic or anodic reaction occurs. This will produce a net current, and thus the rates can be expressed in the form of cathodic and anodic current densities as follows:

$$v_c' = \frac{i_c}{nF} \tag{3.41}$$

$$v_a' = \frac{i_a}{nF} \tag{3.42}$$

in which

$$i_c = nFk_c[\mathrm{Ox}]\exp\left[-\frac{(\Delta G_c^0)^{\neq}}{\mathrm{RT}}\right] \times \exp\left[-\frac{\alpha nF\eta}{\mathrm{RT}}\right] = i^0\exp\left[-\frac{\alpha nF\eta}{\mathrm{RT}}\right] \quad (3.43)$$

and

$$i_a = nFk_a[\mathrm{Re}]\exp\left[-\frac{(\Delta G_a^0)^{\neq}}{\mathrm{RT}}\right] \times \exp\left[\frac{(1-\alpha)nF\eta}{\mathrm{RT}}\right] = i^0\exp\left[\frac{(1-\alpha)nF\eta}{\mathrm{RT}}\right] \quad (3.44)$$

i^0 is the exchange current density and will be discussed below. For the proposed cathodic reaction, a net cathodic current density is given by

$$i = i_c - i_a = i^0\left\{\exp\left[-\frac{\alpha nF\eta}{\mathrm{RT}}\right] - \exp\left[\frac{(1-\alpha)nF\eta}{\mathrm{RT}}\right]\right\} \quad (3.45)$$

This is the well-known *Butler–Volmer equation.* Strictly speaking this expression only holds for processes involving a single electron. When electrode reactions involve more than one electron, it is necessary to replace $(1 - \alpha)$ by β since $\alpha + \beta \neq 1$. Therefore, we have a more general Butler–Volmer equation

$$i = i^0\left\{\exp\left[-\frac{\alpha nF\eta}{\mathrm{RT}}\right] - \exp\left[\frac{\beta nF\eta}{\mathrm{RT}}\right]\right\} \quad (3.46)$$

3.2.3 *Tafel Plot*

The Butler–Volmer equation is now discussed in two extreme conditions. If the overpotential is small, the overpotential terms in the Butler–Volmer equation can be expanded and all terms except the first two can be neglected. The equation is then reduced to

$$i = i^0\frac{nF\eta}{\mathrm{RT}} \quad (3.47)$$

This indicates that current and overpotential are linearly related for a small value of overpotentials at regions very close to the equilibrium and are independent of fraction α. Thus, theoretically, the exchange current density, i^0 can be obtained from the measurement of i versus η at low values of η. In practice, i^0 is generally obtained by extrapolation from relatively high value region of η, as shown in Fig. 3.7 where the overpotential against $\log(i_a)$ and $\log(i_c)$ is plotted assuming $\alpha = 0.5$ and $i^0 = 5 \times 10^{-7}$ A/cm^2.

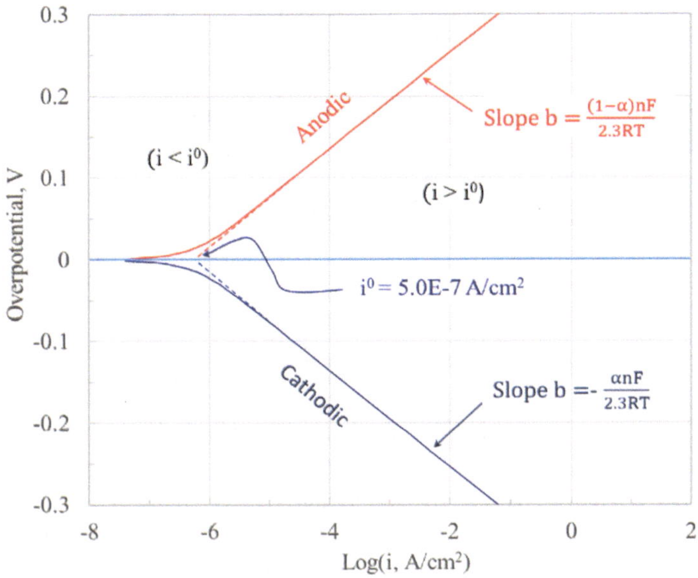

Fig. 3.7 Logarithmic plot of Eq. (3.45) assuming $\alpha = 0.5$ and $i^0 = 5 \times 10^{-7}$ A/cm^2

The ratio η/i has dimensions of resistance and is often called the charge transfer resistance, R_{ct}:

$$R_{\mathrm{ct}} = \frac{\eta}{i} = \frac{\mathrm{RT}}{nFi^0} \tag{3.48}$$

This parameter serves as a convenient and useful indicator of kinetic facility of the reaction.

The second conditions are for large η values corresponding to current densities significantly larger than the exchange current density. For a large overpotential to the cathodic reaction, the second exponential term in the Butler–Volmer equation becomes negligible. The forward reaction (i.e., cathodic reaction) is dominant, and the dependence of the cathodic current density on overpotential is then given by:

$$\ln i = \ln i^0 - \frac{\alpha nF\eta_c}{\mathrm{RT}} \tag{3.49}$$

Thus, the cathodic overpotential, η_c, can be expressed by

$$\eta_c = -\frac{2.3\mathrm{RT}}{\alpha nF}\log\left(\frac{i}{i^0}\right) = \frac{2.3\mathrm{RT}}{\alpha nF}\log i^0 - \frac{2.3\mathrm{RT}}{\alpha nF}\log i \tag{3.50}$$

Similarly for a large overpotential to the backward reaction, i.e., the anodic process, only the second term in the Butler–Volmer equation is significant. Then, we have

$$\eta_a = -\frac{2.3\text{RT}}{nF}\log\left(\frac{i}{i^0}\right) = \frac{2.3\text{RT}}{\beta nF}\log i^0 - \frac{2.3\text{RT}}{\beta nF}\log i \tag{3.51}$$

Both equations have the form identical to an empirical equation proposed by a Swiss Chemist Julius Tafel in 1905 as:

$$\eta = a + b\log i \tag{3.52}$$

in which

$$a = -b\log i^0 \text{ and } b = \frac{2.3\text{RT}}{\alpha nF} \text{ or } \frac{2.3\text{RT}}{\beta nF} \tag{3.53}$$

This is known as *Tafel equation* and b is called the *Tafel slope.* Tafel equation is a very important equation in the electrode kinetics. As seen from Fig. 3.7 in the region where current densities are significantly larger than the exchange current density, either anodic oxidation or cathodic reduction is dominating, and a relationship between η and $\log(i_a)$ or $\log(i_c)$ is linear, from which the Tafel slope can be obtained.

3.2.4 Effect of Exchange Current Density

The exchange current density is an important parameter of the electrochemical reactions. For fast electrode reactions, the i^0 values are high while low i^0 values imply slow reactions. As seen from Eqs. (3.50) and (3.51), it is not the absolute current density but the ratio of the current density to the exchange current density (i/i^0) that determines the polarization of the electrode.

For equilibrium reactions in a hydrogen–oxygen fuel cell, the reversible potential is 1.2 V at 60 °C (See Example 3.1). Assuming a Tafel slope of 60 mV/dec the calculated Tafel plot is shown in Fig. 3.8 with varied exchange current densities. When the net current density is in the vicinity of the exchange current density, electrode reactions in both directions are significant and the polarization curves are bent off. At an overpotential higher than 100 mV, corresponding to a current density *ca.* 17 times larger than the exchange current density (see discussion below), the reaction in either reduction or oxidation direction is dominating and a clear Tafel plot is obtained. Extrapolating of the linear part of the Tafel plot intercepting on the line $\eta = 0$ (i.e., the reversible potential) gives an estimation of the exchange current density from $\log i^0$.

During operation of a fuel cell, the current density is in a range of up to A/cm^2, much larger than the exchange current density of the cathodic reaction. The electrode potential is often plotted in form of i–E, instead of the semi logarithm curves. Such curves are shown in Fig. 3.9 also with a variation of exchange current densities. Here the overpotential is defined as the difference of the reversible potential (1.2 V) and the

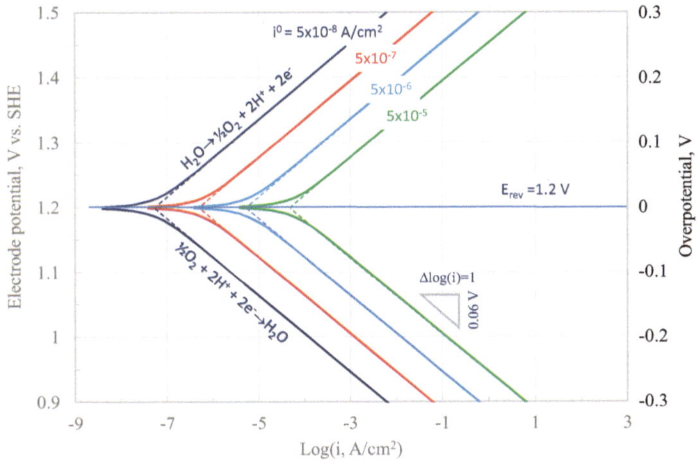

Fig. 3.8 Calculated Tafel plots for oxygen reduction/evolution reactions with varied exchange current densities and a Tafel slope of 60 mV/dec

measured electrode potential of the *i-E* curves. At a fixed Tafel slope of 60 mV/dec, increasing the exchange current density by 10 times leads to the *i-E* curve moving up by 60 mV; i.e., the potential loss due to the ORR is reduced by 60 mV.

The exchange current density of an electrode reaction is intrinsic on the surface of a catalyst in contact with an electrolyte. For example, for smooth platinum electrodes in concentrated phosphoric acid, e.g., 85–100% H_3PO_4, a strong poisoning of the catalyst surface by the acid anions takes place. As a result, a low exchange current density in the range of 10^{-10}–10^{-11} A/cm^2 is reported. For dilute acid electrolytes, e.g., $HClO_4$, which exhibits little adsorption characteristics as the polyperfluorinated sulfonic acid ionomer does for proton exchange membrane fuel cells, the exchange current density is at least an order of magnitude higher (See Table 3.3). It should be remarked that these exchange current density numbers are based on the true platinum surface area, not the geometric area of a fuel cell electrode (see Example 3.2). In fuel cells, a practical way to increase the electrode performance is by the development of porous gas diffusion electrodes made of Pt nanoparticles uniformly dispersed, which have an active surface area of 10^2–10^3 times larger than the geometric area of the electrode.

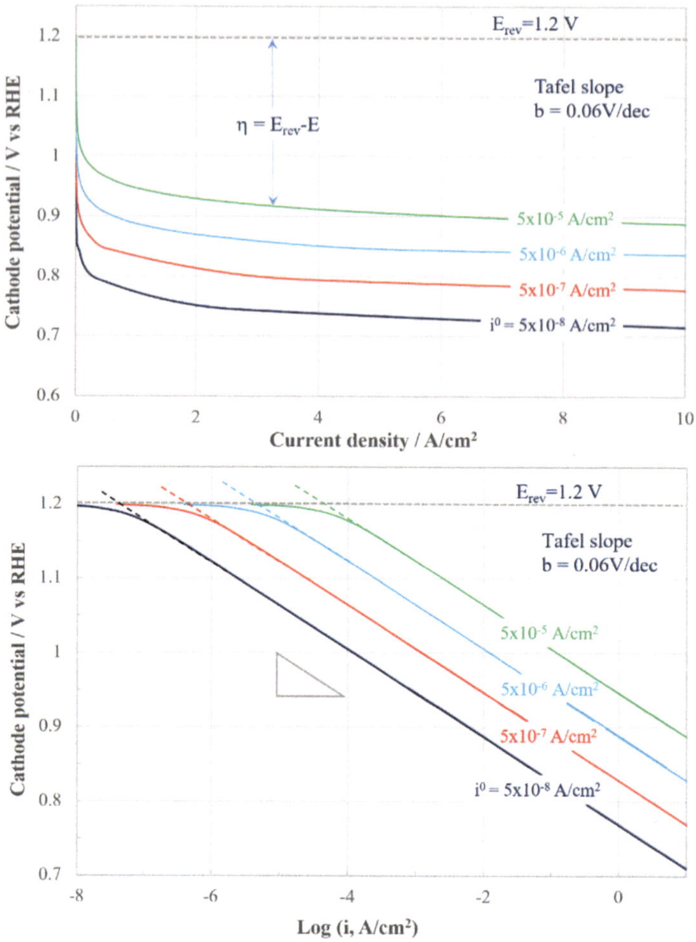

Fig. 3.9 Calculated polarization plots in form of *i-E* (upper) and *E*-log*i* (bottom) curves for the oxygen reduction reaction using varied exchange current densities as indicated in the figure. The Tafel slope is 60 mV/dec

Table 3.3 Typical exchange current densities for hydrogen oxidation reaction (HOR) and oxygen reduction reaction (ORR) on smooth metal electrodes in acidic electrolyte at 300 K and 1 atm

Electrode reaction	Catalyst	Exchange current density (A/cm^2)
HOR	Pt	10^{-3}
	Pd	10^{-4}
	Ni	10^{-5}
ORR	Pt	10^{-10}–10^{-9}
	Pd	10^{-10}
	Rh	10^{-11}

Example 3.2 A PEMFC operates with ambient pressure oxygen at 65 °C. The cathode is made of carbon-supported platinum nanoparticles at a loading on 0.4 mg $Pt/cm^2{}_{Geo}$ (based on the geometric electrode area—the current density of a fuel cell electrode should always be understood on this basis unless otherwise specified). By voltammetric measurements, the electrochemical surface area (ECSA) of platinum is found to be 500 $cm^2/cm^2{}_{Geo}$.

1. Calculate apparent exchange current density using $i^0 = 10^{-9}$ A/cm² based on smooth platinum metal electrode.
2. Calculate the mass specific surface area of platinum catalysts.

Solution:

1. The exchange current density listed in Table 3.3 is based on the real metal surface area which corresponds to the ECSA of a fuel cell electrode. The exchange current density based on the geometric electrode area is then

$$i^0_{geo} = 10^{-9}\frac{A}{cm^2_{ECSA}} \times 500\frac{cm^2_{ECSA}}{cm^2_{Geo}} = 5.0 \times 10^{-7}\frac{A}{cm^2_{Geo}} \tag{3.54}$$

2. The mass specific active area (MSAA) of platinum catalysts is

$$\begin{aligned}\text{MSAA} &= \frac{\text{total active surface area}(\text{cm}^2)}{\text{total Pt mass}(\text{mgPt})}\\ &= \frac{500\,\text{cm}^2_{\text{ECSA}}/\text{cm}^2_{\text{Geo}}}{0.4\,\text{mgPt}/cm^2_{\text{Geo}}} = 1250\,\text{cm}^2_{\text{ECSA}}/\text{mgPt} = 250\,\text{M}^2/\text{g}_{\text{Pt}}\end{aligned} \tag{3.55}$$

3.2.5 *Effect of Tafel Slope*

It is seen from Eq. (3.53) that the Tafel slope is a function of the electron number involved in the electrode reaction and the charge transfer coefficient. In a reaction involving only one step with one electron transfer, the Tafel slope will be determined by the symmetry factor, which is usually 0.5. This gives a Tafel slope of 0.120 V/dec. In a more complex reaction involving several steps and several electron transfers, the Tafel slope is then determined by the rate-determining step. In brief, any change in the Tafel slope indicates a change in the mechanism of the electrode reaction. The physical meaning of the Tafel slope is the overpotential loss when the operating current density is increased by 10 times. Changes in the Tafel slope, due to changes in the charge transfer coefficient, have effects on the *i-E* and logarithmic plots, as shown in Fig. 3.10 when the exchange current density is assumed to be a constant of $i^0 = 5 \times 10^{-7}$ A/cm².

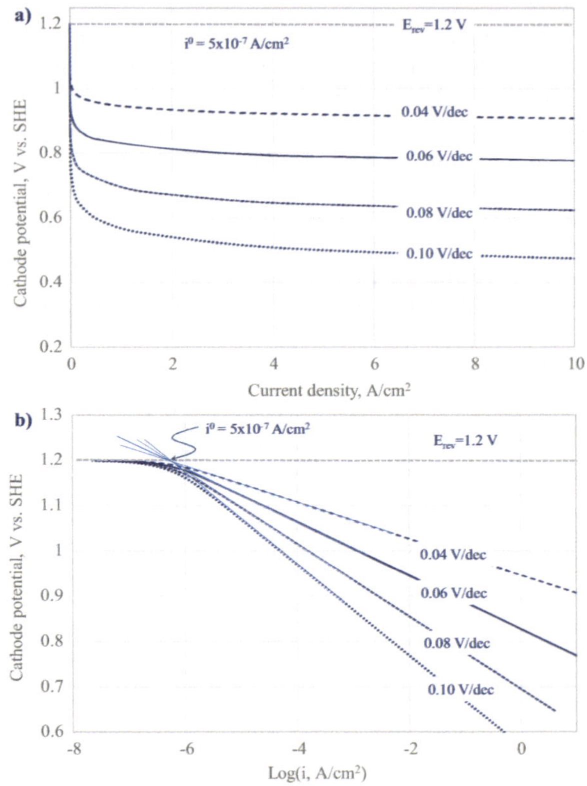

Fig. 3.10 Effect of changes in the Tafel slope on **a)** the *i-E* and **b)** logarithmic plots calculated by assuming a constant exchange current density of $i^0 = 5 \times 10^{-7}$ A/cm^2

3.3 Ohmic Polarization Loss

Another type of the voltage loss is known as the ohmic polarization loss. Both electronic conducting components, such as bipolar plates, gas diffusion and catalyst layers, and the ionic conducting electrolyte contribute to the cell voltage loss of a fuel cell upon the passage of current (see Table 3.4). These losses obey Ohm's law which is

$$\eta_{\text{ohm}} = i \sum R_i \tag{3.56}$$

where R_i is the sum of ohmic resistance of the above-mentioned fuel cell components. For a conductor of a uniform cross section, for example, in form of a pellet, its resistance is proportional to its resistivity (ρ) and length or thickness (L) and inversely proportional to its cross-sectional area (A):

$$R = \rho \frac{L}{A} \tag{3.57}$$

Table 3.4 Conductivity of some electronic and ionic conducting materials and possible ohmic losses in fuel cells

Materials		Electrical conductivity (S/cm)	Thickness in FC use (cm)	ASR (Ω·cm)	Ohmic loss at 1.0 A/cm^2 (mV)
Electronic conductors	Silver	6.3×10^7 (25 °C)	–	–	–
	Platinum	9.4×10^6 (25 °C)	–	–	–
	Stainless steel	1.5×10^6 (25 °C)	–	–	–
	Graphite (in-plane)	2.5×10^5 (25 °C)	–	–	–
	Graphite (through plane)	3.3×10^2 (25 °C)	0.5	0.0015	1.5
	$La_{1-x}Sr_xMnO_3$	250 (800 °C)	0.002	8×10^{-6}	0.008
	Composite carbon	150 (25 °C)	0.5	0.003	3
	Carbon cloth	5	0.03	0.006	6
	Carbon paper	10	0.02	0.002	2
Ionic conductors	100% H_3PO_4	0.60 (190 °C)	0.10	0.17	170
	27% KOH	0.55 (80 °C)	0.10	0.18	180
	Nafion 117	0.18 (80 °C, 100%RH)	0.02	0.11	110
	Molten $Li_2CO_3-Na_2CO_3$	0.85 (600 °C)	0.08	0.09	90
	8 mol%$Y_2O_3-ZrO_2$	0.05 (800 °C)	0.001	0.02	20

In fuel cells and other electrochemical devises, the *area specific resistance* (ASR in Ω cm^2) is of more engineering importance, which also can be expressed as the product of the thickness and the resistivity, $\rho \times L$.

The reciprocal of the resistivity is called conductivity, which has a unit of S/cm.

$$\sigma = \frac{1}{\rho} = \frac{1}{R}\frac{L}{A} \tag{3.58}$$

The conductivity of electronic conductors, e.g., metals, decreases as the temperature increases due to vibration of the crystal lattice that the electrons travel through. At very low temperatures some materials lose their electric resistance, a phenomenon known as superconductivity. For semiconductors, however, electrons can be promoted to the conduction energy band by thermal activation and both the promoted electrons and the left holes behind them are mobile. As a result, the conductivity of semiconductors increases with temperature.

Electrolytes of fuel cells are ionic conductors. The conductivity is achieved by traveling of charge-carrying ions, which is a thermally activated process and can be expressed as an exponent function of temperature:

$$\sigma = \sigma^0 \exp\left(-\frac{E_a}{RT}\right) \tag{3.59}$$

where σ^0 is the pre-exponential factor and E_a is the activation energy of conductivity. For the thermally activated conductivity via hopping mechanism process, the conductivity is also often represented as:

$$\sigma = \frac{\sigma^0}{T} \exp\left(-\frac{E_a}{RT}\right) \tag{3.60}$$

The activation energy, E_a, can be obtained from the slope of the ln σ versus $1/T$ plot or ln σT versus $1/T$ plot. Table 3.4 compiles conductivity data of some electronic and ionic conducting materials.

The conductivity is the character of a conductor. For electrochemical engineers, the most important is ASR, which can be expressed as the ratio of the thickness and the conductivity (L/σ). The physical meaning of the ASR can be understood as the voltage loss (V) of the electrolyte at a unit current density (A/cm^2). The expression of the thickness to conductivity ratio indicates that a high conductivity (σ) and a small electrolyte thickness (L) give a lower cell ohmic polarization loss.

Example 3.3 A PEMFC fuel cell of 20 cm^2 active area using a Nafion 117 membrane. Under operation at 80 °C with fully humidified hydrogen and air, the cell resistance is measured to be 10 mΩ. (1) calculate the area specific resistance, (2) the membrane conductivity, and (3) the voltage loss at a current density of 1.5 A/cm^2.

Solution:

1. $$\text{ASR} = \text{resistance}(\Omega) \times \text{area}(\text{cm}^2) = 0.010 \times 20 = 0.20\ \Omega \cdot \text{cm}^2 \tag{3.61}$$

2. Nafion 117 membranes have a dry thickness of 175 μm. When it is fully hydrated, the thickness of the swollen membrane is around 210 μm at room temperature. At the fuel cell operating temperature 80 °C, the real thickness is unknown. In practice the room temperature thickness is used in calculation of the conductivity from the measured resistance.

$$\text{Conductivity} = \frac{\text{thickness}}{\text{ASR}} = \frac{0.021\ \text{cm}}{0.20\ \Omega\ \text{cm}^2} = 0.105\ \text{S/cm} \tag{3.62}$$

3. At current density of 1.5 A/cm^2 the voltage loss is

$$\eta_{\text{loss}} = \text{ASR} \times \text{current density} = 0.20\ \Omega\ \text{cm}^2 \times 1.5\ \text{A/cm}^2 = 0.30\ \text{V} \tag{3.63}$$

3.4 Mass Transport and Concentration Polarization in Steady State

Operation of a fuel cell needs efficient supply of reactants and removal of products, i.e., the phenomenon of mass transport. Hydrogen and oxygen in air enter the fuel cell channels and are reacting within the front catalyst layer at the catalyst–membrane interface where the reaction products are formed. Any depletion of the reactants and/or accumulation of products within the catalyst layer will cause losses in the fuel cell performance, which is called the concentration polarization or concentration overpotential, η_{conc}.

Figure 3.11 illustrates the structure of a PEMFC and the concentration gradient of oxygen through the gas channel with a bulk concentration and diffusion layer reaching the front catalyst surface layer with a lower concentration. According to Fick's first law of diffusion, the flux rate (mol/cm^2) is proportional to the concentration gradient,

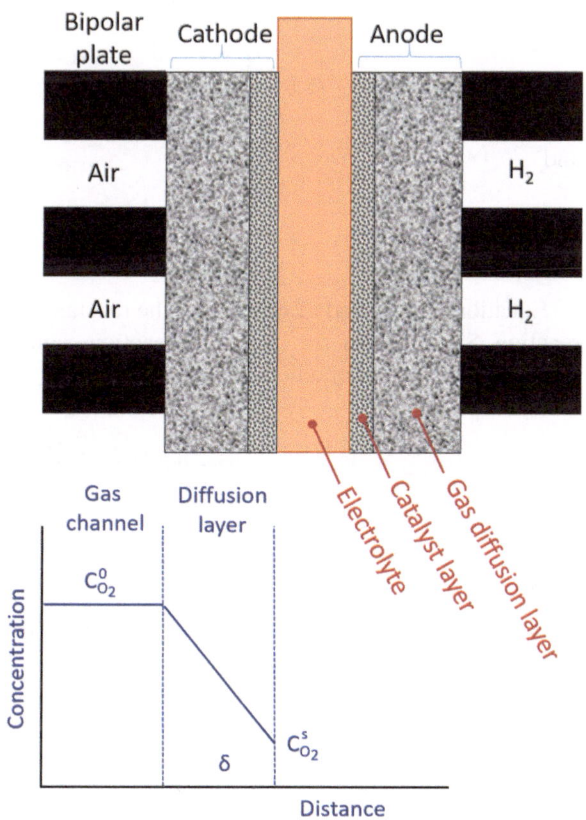

Fig. 3.11 Schematic representation of fuel cell construction and concentration profile of oxygen through the gas channel and diffusion layer

$$J = D\frac{\Delta c}{\delta} = D\frac{\left(c^0_{O_2} - c^s_{O_2}\right)}{\delta} \tag{3.64}$$

where $c^0_{O_2}$ and $c^s_{O_2}$ are the bulk and surface concentrations of oxygen, in mol/cm^3, respectively. δ is the thickness of the diffusion layer and D is the diffusion coefficient.

The reactant flux is consumed at the catalyst surface, generating a current density

$$i = nFD\frac{c^0_{O_2} - c^s_{O_2}}{\delta} \tag{3.65}$$

When the surface concentration $c^s_{O_2} = 0$, the reaction reaches the maximum current, the cathodic limiting current

$$i_{(L,c)} = nFD\frac{c^0_{O_2}}{\delta} \tag{3.66}$$

Combining Eqs. (3.65) and (3.66), one has

$$c^0_{O_2} = c^s_{O_2} - \frac{i\delta}{nFD} = \frac{i_{L,c}\delta}{nFD} - \frac{i\delta}{nFD} \tag{3.67}$$

and

$$c^s_{O_2} = c^0_{O_2}\left(1 - \frac{i}{i_{L,c}}\right) \tag{3.68}$$

Equations (3.65) and (3.68) are for the reactants of the reaction, i.e., the cathodic reaction. Similarly for the backward reaction, same equations can be obtained for the products, i.e., the anodic reaction.

$$i = nFD\frac{c^s_{O_2} - c^0_{O_2}}{\delta} \tag{3.69}$$

$$c^s_{O_2} = c^0_{O_2}\left(1 + \frac{i}{i_{L,a}}\right) \tag{3.70}$$

Here, the limiting current is for the backward reaction and is taken as positive value.

A more general Butler–Volmer equation can be obtained by multiplying the reaction rates, v_a and v_c with concentration terms, c^s_{Ox}/c^0_{Ox} and c^s_{Re}/c^0_{Re}, in the forms similar to Eqs. (3.68) and (3.70). Similar to Eq. (3.46), we have

$$i = i^0\left\{\left(1 - \frac{i}{i_{L,Ox}}\right)\exp\left[\frac{\alpha nF\eta}{RT}\right] - \left(1 + \frac{i}{i_{L,Re}}\right)\exp\left[-\frac{\beta nF\eta}{RT}\right]\right\} \tag{3.71}$$

This equation includes the activation polarization and concentration polarization contributions in the whole η ranges, and the magnitude of activation and concentration polarizations are explicitly presented by exchange current density and limiting current density, i^0 and i_L, respectively. The equation is applicable to reversible as well as irreversible electrode reactions. We can discuss the equation in terms of i^0/i_{L} ratios where $i_L = i_{\mathrm{L,c}} = -i_{\mathrm{L,a}}$ (for convenience and consistence, we use $i_{\mathrm{L,c}}$ for $i_{\mathrm{L,Ox}}$ and $i_{\mathrm{L,a}}$ for $i_{\mathrm{L,Re}}$).

For $i^0/i_{\mathrm{L}} \gg 1$, which corresponds to a very large exchange current density or a very facile kinetics and reversible behavior of the reaction, Equation (3.71) becomes

$$\eta = \frac{\mathrm{RT}}{nF}\left[\ln\left(1 + \frac{i}{i_{\mathrm{L,a}}}\right) - \ln\left(1 - \frac{i}{i_{\mathrm{L,c}}}\right)\right] = \eta_{\mathrm{conc}} \tag{3.72}$$

In this case, the reaction is dominated by the concentration polarization and the reaction overpotential is completely caused by the concentration overpotential. This is typical reversible electrode reaction. If $i^0/i_{\mathrm{L}} \ll 1$, the reaction is irreversible. For $\eta \gg \frac{\mathrm{RT}}{\alpha nF}$, the backward reaction can be ignored and thus we have

$$i = i^0\left(1 - \frac{i}{i_{\mathrm{L,c}}}\right)\exp\left[\frac{\alpha nF\eta}{\mathrm{RT}}\right] \tag{3.73}$$

By re-arranging above equation, we have

$$\eta = \frac{\mathrm{RT}}{\alpha nF}\ln\frac{i}{i^0} + \frac{\mathrm{RT}}{\alpha nF}\ln\left(\frac{i_{\mathrm{L,c}}}{i_{\mathrm{L,c}} - i}\right) \tag{3.74}$$

This equation consists of two components, i.e., the activation overpotential and concentration overpotential, η_{act} and η_{conc}, respectively.

$$\eta_{\mathrm{act}} = \frac{\mathrm{RT}}{\alpha nF}\ln\frac{i}{i^0} \tag{3.75}$$

$$\eta_{\mathrm{conc}} = \frac{\mathrm{RT}}{\alpha nF}\ln\left(\frac{i_{\mathrm{L,c}}}{i_{\mathrm{L,c}} - i}\right) \tag{3.76}$$

For small overpotentials, $\eta \ll \frac{\mathrm{RT}}{\alpha nF}$, the exponential terms in Eq. (3.71) can be expanded and a linearized relation can be used. By re-arranging the equation, we have

$$i = i^0\left(-\frac{i}{i_{\mathrm{L,c}}} - \frac{i}{i_{\mathrm{L,a}}} + \frac{nF\eta}{\mathrm{RT}}\right) \tag{3.77}$$

Or

$$\frac{\eta}{i} = \frac{RT}{nF}\left(\frac{1}{i^0} + \frac{1}{i_{L,c}} + \frac{1}{i_{L,a}}\right) \tag{3.78}$$

This equation shows that the polarization curves are linear at low η values in the vicinities of equilibrium. The slope of the linear region can be seen as the polarization resistance, consisting of three resistances, activation polarization resistance, R_{ct}, cathodic and anodic concentration polarization resistances, $R_{c,conc}$ and $R_{a,conc}$ in series.

$$\frac{\eta}{i} = R_p = R_{ct} + R_{c,conc} + R_{a,conc} \tag{3.79}$$

Here we have situation that for reversible electrode reactions, exchange current density is much greater than the limiting current densities, $R_{ct} \ll R_{c,conc} + R_{a,conc}$, and the polarization, even near the equilibrium, is dominated by concentration polarization. On the other hand, if the exchange current density is much smaller than the limiting current densities, the overpotential near the equilibrium is due to the activation polarization related to charge transfer. This is related to the cases where the electrode reactions are irreversible. Then, from Eq. (3.78) by ignoring the concentration overpotential terms, we can have:

$$i^0 = \frac{RT}{nF} \times \frac{i}{\eta} = \frac{RT}{nF} \times \frac{1}{R_{ct}} \tag{3.80}$$

This information is important to judge the significance of the polarization resistance obtained from the slopes of linear regions of the overpotential curves.

In real fuel cells the limiting current depends on the porosity and thickness of both gas diffusion layer and catalyst layer as well as hydrophobic characteristics with respect to the water management for low-temperature fuel cells. For PEMFCs the limiting current density can be in the range of 1–5 A/cm^2. The value can be much lower for fuel cells with liquid electrolytes, e.g., AFC and PAFC. In high-temperature SOFCs, reactants as well as products are in gaseous phases and the concentration polarization is closely related to the gas distribution and flow field design of interconnector and manifolds. The practical limiting current density of fuel cells is in general poorly defined and difficult to measure because, in the current density range approaching the limit, other effects are also in a critical play though the contribution of the mass transport term is dominating. Figure 3.12 shows typical concentration polarization loss curve. In this case, i_L is assumed to be 2.5 A/cm^2.

It is seen from the figure that, as the current density approaches the limiting value a significant mass transport loss occurs. At typically operational current density of, e.g., 1 A/cm^2, the contribution of the mass transport loss is about 50 mV, but the concentration polarization losses increase substantially with the increase in current.

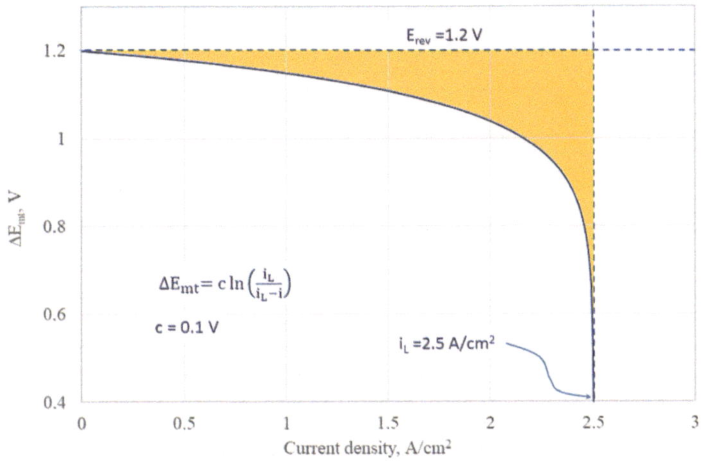

Fig. 3.12 Typical activation and concentration polarization losses for a PEMFC at 25 °C with a reversible cell voltage of 1.2 V in H_2 and air. The cathode has a limiting current density of 2.5 A/cm^2 and constant, $c = RT/\alpha nF = 0.1$ V

3.5 I-V Characteristics

3.5.1 Cell Voltage and Various Types of Voltage Losses

As discussed in Chap. 2, the total thermodynamic energy available from a fuel cell reaction is the enthalpy change (ΔH), only part of which can be used for doing electrical work (ΔG) and the rest has to be dissipated in the form of heat ($T\Delta S$). Based on the Gibbd free energy change ΔG the reversible cell voltage can be calculated, corresponding to the maximum electrical work. At STP conditions, this reversible cell voltage at room temperature is $E_{rev} = 1.23$ V. It should be noted that at this reversible cell voltage, a fuel cell is converting 100% of ΔG into the electrical work but there is still heat generation corresponding to the entropy term $T\Delta S = \Delta H - \Delta G$. It should be the total energy (ΔH) that is the basis for calculation of the energy efficiency of a fuel cell. To illustrate this a hypothetical cell voltage, the so-called thermoneutral voltage, E_{TN}, is defined based on the total enthalpy

$$E_{TN} = \frac{\Delta H^o_{HHV}}{nF} = \frac{285.9 \times 1000}{2 \times 96485} = 1.48\,\text{V} \tag{3.81}$$

The actual cell voltage measured at open circuit, E_{OCV}, in a fuel cell is often lower than E_{rev}. The difference between the reversible voltage and OCV is the OCV loss which is caused by the fuel and oxidant crossover, electrode irreversibility and internal current leaking due to the electronic conductivity of the electrolyte under open circuit. When a current is drawn, further voltage losses occur. The operating cell voltage can be expressed as below and illustrated in Fig. 3.13:

$$E_{cell} = E_{rev} - \sum E_{loss} \tag{3.82}$$

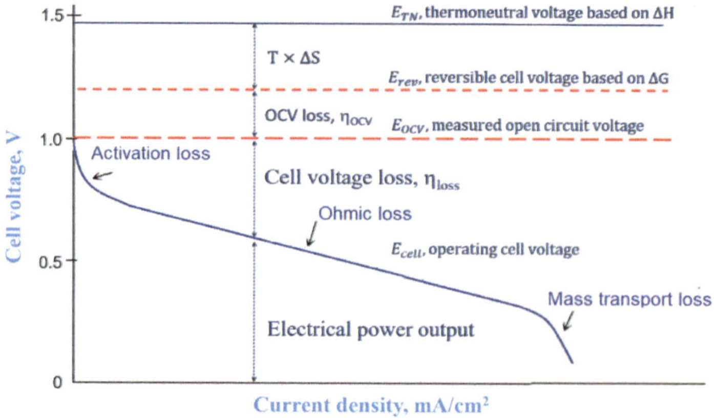

Fig. 3.13 Cell voltages and current–voltage curve characteristics of a typical PEMFC

Example 3.4 A fuel cell stack operating with hydrogen and air at 60 °C delivers an electrical power of 2 kW at an average single cell voltage of 0.6 V. The reversible cell voltage is 1.2 V. For design of the system cooling, estimate how much heat the stack is generating.

Solution:

It seems that the stack is operating at a voltage efficiency of 0.6/1.2 = 50%. When it delivers 2 kW electricity, 2 kW heat should be generated. This answer is however incorrect. This 2 kW heat is the inefficient part from ΔG. There is another term of heat generation—$T\Delta S$. The voltage efficiency should be based on the total enthalpy change (ΔH), from which the thermoneutral voltage of $V_{TN} = 1.48$ V is calculated; i.e., the system electrical efficiency is 0.6/1.48 = 40%. That is to say, about 60% of the total energy is released as heat and 40% as the electrical power. Thus, the heat release should be 2 kW × 0.6/0.4 = 3 kW. The cooling duty of the stack is 3 kW!

When a fuel cell delivers a power output by a passage of current, additional voltage losses occur including activation polarization losses at both cathode and anode, the ohmic losses of electrolyte and electrodes and the loss due to the mass transport. The cell working voltage by including these losses is expressed as follows:

$$E_{\text{cell}} = E_{\text{OCV}} - \left(\eta_{\text{cathode}} + \eta_{\text{anode}} + i\sum R_i + \eta_{\text{mt}}\right) \tag{3.83}$$

where η_{mt} represents mass transport related polarization losses or concentration overpotential. In practical use of Eq. (3.83) the cathode potential deviates significantly from the reversible potential particularly for low-temperature fuel cells due to the slow electrode kinetics, as discussed above. Thus, the cathode polarization loss is the major contributor to the overall cell voltage loss.

In the case of hydrogen as fuel, the anode overpotential (η_{anode}) for hydrogen oxidation reaction (HOR) with Pt-based catalysts and acidic electrolytes, as a result of fast kinetics, i.e., a high exchange current density, is small and a linear function of the current density. It is either negligible or practically included in the ohmic resistance term, which is a sum of all ohmic contributions including resistances of the electrolyte, electrode substrate, bipolar plates, and leading wires. Of these resistances are the electrolyte accounting for the major part.

In the case of high-temperature SOFCs, the ohmic resistance of solid electrolyte is considered to be the major component of the overall polarization losses. In fuel cells using fuels other than H_2, the anode overpotential can be significant and cannot be ignored. Good examples are fuel cells using methanol or ethanol as fuels, and in these cases, the electrooxidation reaction of methanol and ethanol becomes the limiting rate of the cell performance.

3.5.2 Open-Circuit Voltage Loss

For an idle cell without drawing any current, the cell voltage experimentally measured between the two electrode terminals, i.e., the OCV, is often lower than the reversible cell voltage. For PEMFC, for example, the typical open-circuit voltage (OCV) value is around 1.0 V, i.e., about 200 mV lower than the reversible cell voltage. There are several reasons for the voltage loss at OCV including the fuel and oxidant crossover, internal leakage current and lack of equilibrium of the electrode particularly the oxygen electrode.

3.5.2.1 Fuel Crossover

First of all the electrolyte is not perfectly impermeable for reactant gases, oxygen, hydrogen, or other types of gaseous or liquid fuels. Any permeation of fuels, e.g., hydrogen or methanol from the anode, may reach the cathode side and react with oxygen. In the same way, permeation of oxygen may reach the anode side and react with the fuel. This is a chemical oxidation, the transfer of electrons from a fuel to the oxidant is not contributing to the cell current. In other words, the hydrogen or oxygen permeation through the electrolyte corresponds to the leakage of an internal current—a current that is not externally measureable between the two electrodes. In the case of low-temperature PEMFCs, the hydrogen permeation rate corresponds to an internal leaking current density of a few mA/cm^2. The voltage loss due to this internal leaking current depends on the electrode kinetics, i.e., the voltage–current relationship of the oxygen cathode.

Compared to the exchange current density of, e.g., 5×10^{-7} A/cm^2 for the ORR at the cathode (see Example 3.2), this internal leaking current density is about 3–4 orders of magnitude larger. Using a Tafel slope of 0.06 V/dec, we can estimate the consequent voltage loss due to the H_2 leaking current density of e.g. $i_{\text{leak}} =$ 1.6 mA/cm^2

$$\eta_{\text{leak}} = -b\log\left(\frac{i_{\text{leak}}}{i^0}\right) = -0.06\,\frac{V}{\text{dec}}\log\left(\frac{1.6\times10^{-3}}{5\times10^{-7}}\right) = 0.21\ \text{V} \tag{3.84}$$

This open-circuit voltage loss due to the hydrogen crossover is quite significant. In practice the contribution of the hydrogen crossover current can be included in the cathode overpotential by using the following equation:

$$\eta_{\text{cathode}} = b\ \log\left(\frac{i + i_x}{i^0}\right) \tag{3.85}$$

The presence of a leaking current density can be understood as a shift of the i-V curve along the x-axis, which causes losses according to the polarization profile of the opposite electrodes, as schematically shown in Fig. 3.14. In the case of methanol-based fuel cells, the crossover of fuels through the polymer electrolyte membrane is 2–3 orders of magnitude higher than that of hydrogen. An addition effect of the methanol crossover is the poisoning of the Pt-based electrocatalysts for the oxygen reduction reaction at the cathode. Therefore, methanol permeability is a critical issue to achieve high fuel cell performance of direct methanol fuel cells.

There is also permeation of oxygen from the cathode to the anode, which also causes a voltage loss on the anode by chemical reaction with hydrogen. The hydrogen electrode has a much faster kinetics, with an exchange current density of the same order of magnitude as the leaking current density. As a result the anode potential loss is negligible as schematically shown in Fig. 3.14.

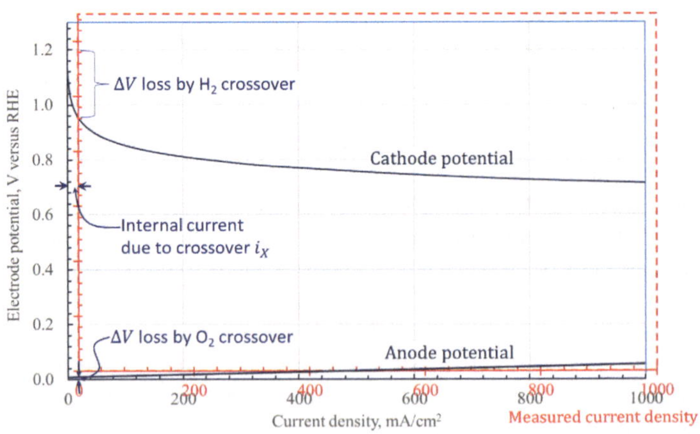

Fig. 3.14 Schematic representation of the internal current (i_x) due to hydrogen and oxygen crossover and resultant potential loss at the cathode and anode

3.5.2.2 Electrode Irreversibility

Another major origin of the OCV loss is due to the electrode irreversibility. A platinum electrode is assumed inert, serving merely as an electronic conductor for delivering or collecting electrons. When oxygen is bubbled around, the measured potential is assumed to correspond to the oxygen reduction potential. This is, however, not the case. Platinum is not inert in acidic media in the presence of oxygen. In the early years of discovery of fuel cells around 1839, William Grove used platinum gauzes immersed in an acid solution with bubbled hydrogen and oxygen and noticed this discrepancy between the observed and theoretical values of the reversible oxygen potentials. Formation of a layer of oxides on the platinum surface has long been proposed, and the measured open-circuit potential is a result of varied compositions of poorly defined mixture oxides consisting of PtO_2 and PtO or its hydrated form $Pt(OH)_2$ or $PtO{\cdot}H_2O$, which have reversible potentials of 1.1 and 0.98 V, respectively, according to Watanabe et al. [1]:

$$PtO_2 + 2H^+ + 2e^- \rightleftharpoons Pt(OH)_2 \qquad E^0 = +1.1\,V \tag{3.86}$$

$$Pt(OH)_2 + 2H^+ + 2e^- \rightleftharpoons Pt + 2H_2O \qquad E^0 = +0.98\,V \tag{3.87}$$

As soon as the oxide covered platinum electrode is immersed in the acid solution the oxide layer is broken down, and a mixed potential is established in a range from 0.9 to 1.05 V. In the treatment of electrochemistry this kind of OCV loss is often included as part of the kinetic or overpotential loss of ORR though this loss occurs at the zero current.

3.5.2.3 Electronic Conductivity of Electrolytes

Some electrolytes have a fraction of the electronic conductivity which causes an internal short circuit. This is the case for the solid oxide electrolyte materials in high-temperature SOFCs. In SOFCs, the OCV depends strongly on the oxygen ionic and electronic conductivity properties of the electrolytes. The relative importance of the ionic and electronic (including electron and hole conductivity) of the materials can be measured by an ionic transfer number, t_i,

$$t_i = \frac{\sigma_i}{\sigma_i + \sigma_e' + \sigma_h^\bullet} \tag{3.88}$$

where σ_i, σ_e', and $\sigma_h^\bullet$ are the ionic, *n*-type (electron), and *p*-type (hole) conductivities, respectively. The ionic transfer number is a critical property of the electrolytes used in SOFCs. The electronic conductivity of the electrolyte causes a leaking current through the electrolyte; i.e., an internal short circuit will lead to a decrease in the fuel cell OCV. For example, the doped ceria is a high oxygen ion conducting electrolyte under oxidizing conditions; however, it exhibits a significant electronic conductivity under reducing conditions at the fuel electrode side. Thus, the measured OCV is generally 200–300 mV lower than the reversible voltage due to the significant leaking current. As a comparison, the yttrium stabilized zirconia (YSZ) electrolyte has a negligible electronic conductivity under either oxidizing or reducing conditions and its ionic transfer number is close to one. Therefore, in the case of the YSZ electrolyte-based SOFCs, the OCV is generally close to that calculated from Nernst equation.

The leakage current reduces the current or Faraday efficiency of fuel cells, which can be expressed as

$$\varepsilon_i(\%) = \frac{i}{i + i_{\text{leak}}} \times 100 \tag{3.89}$$

At very low current densities where the leak current is significant, the fuel cell energy efficiency is dominated by the Faraday efficiency while at high current densities, it is dominated by the voltage efficiency, as shown below:

$$\varepsilon_{\text{FC}}(\%) = \frac{E_{\text{cell}}}{E_{\text{TN}}} \times \frac{i}{i + i_{\text{leak}}} \times 100 \tag{3.90}$$

3.5.3 *I-V* Characteristics and Materials Issues

The polarization curves can be characterized in different forms depending on the dominant polarization losses. Figure 3.15 shows examples of various types of *i-V* curves with particular dominant polarization losses. All the *i-V* curves pass through a performance point of, e.g., 0.7 V @ 0.8 A/cm^2 but the shapes of *i-V* curves show different characteristics and hence material issues.

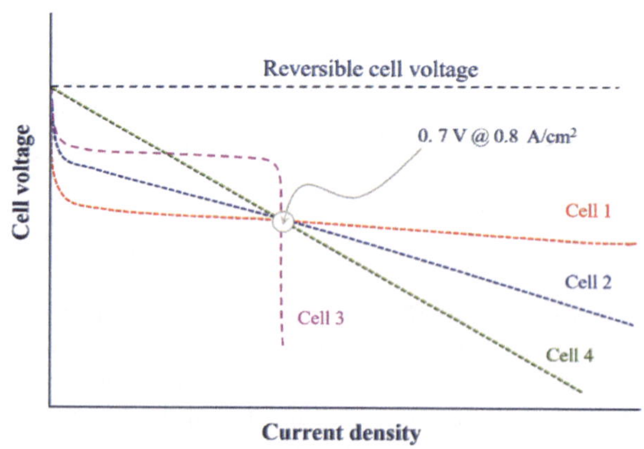

Fig. 3.15 Illustration of fuel cell performances due to various mechanisms of voltage losses

Cell 1 shows a large activation loss in the very small current density range but very small ohmic loss. It likely has poor catalysts, electrodes, or the electrode–electrolyte contacting interface which are dominating the activation loss. On the other hand, the electrolyte must have a high conductivity or/and a small thickness.

Cell 2 shows a smaller activation loss indicating good catalysts and electrodes; however, in the intermediate to large current density range the ohmic loss is significant, suggesting an electrolyte layer of low ionic conductivity or/and a big thickness.

Cell 3 has the smallest activation loss and ohmic resistance, apparently showing very good catalysts/electrodes as well as the electrolyte. It suffers from poor mass transport, likely because of dense or low porosity catalyst layers and/or gas diffusion layers. A fuel or oxidant starvation or electrode flooding is also possible.

Cell 4 shows almost pure ohmic resistance loss. Such an ohmic polarization dominated behavior can be found in high-temperature SOFCs where the electrode kinetics is fast. This is largely due to the high resistivity of solid electrolyte materials. Therefore, for SOFCs operating at reduced temperatures, e.g., 600–800 °C, the polarization losses due to the resistance of the electrolyte become dominant. In order to reduce the ohmic resistance of SOFCs, a thin electrolyte cell design is generally used. In this case, the electrolyte thickness can be as small as 10 μm and is supported on a thick electrode substrate; in most cases the anode is used as the substrate. Such a cell design is called an electrode-supported SOFC.

In reality, the polarization curves of practical fuel cells are far more complicated and can be mixed types of different polarization losses. Figure 3.16 shows typical

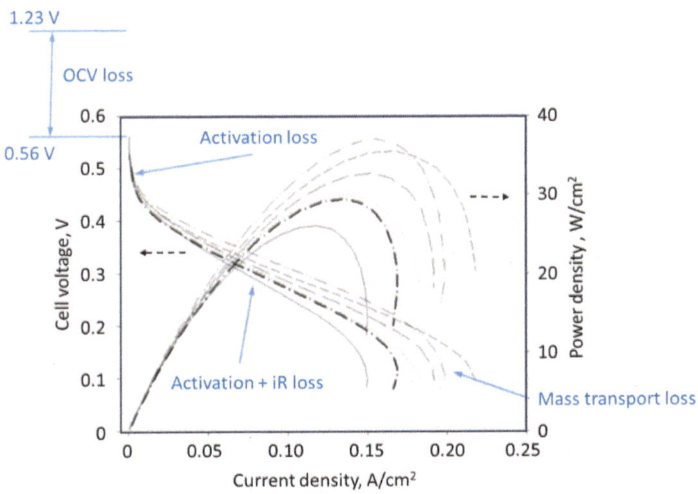

Fig. 3.16 Polarization curves and power output of a direct methanol fuel cell (DMFC) using pristine Nafion 1135 (solid line) and its composites with surface modification by self-assembly of various polyelectrolytes, measured at 30 °C. Operating conditions: 2 M CH_3OH, 1 bar; cathode, oxygen, 1.7 bara. Modified from Ref. [2] with permission from John Wiley & Sons Inc., Copyright 2006

polarization curves of fuel cells operating with 2 M CH_3OH aqueous solution as the fuel and O_2 as the oxidant measured at 30 °C. The differences in the cell power performance are related to the self-assembled polyelectrolyte layers, which are beneficial to reduce the methanol crossover and thus increase the cell performance [2]. The OCV loss is mainly related to the methanol crossover and mixed potential at the cathode. The rapid initial drop in the polarization current is largely due to the activation loss related to the poor activity of Pt/C-based electrocatalysts for the methanol oxidation reaction as well as the relatively low activity of Pt/C for ORR. The linear region of the polarization curves is primarily dependent on the ohmic (iR) loss. At current density of 0.15–0.20 A/cm^2, the polarization curves show typical mass transport loss characteristics. However, the mass transport loss in this case is in fact related to the electrocatalytic activity of Pt/C catalysts on the cathode side for ORR. This is indicated by the fact that the surface modified cells shows a higher mass transport limiting current due to the reduced methanol crossover. This in turn increases the activity of the Pt/C cathode for ORR, and thus the limiting current densities for the methanol fuel cell.

The real cells can behave in a very complicated way!

3.6 Basic Electrochemical Techniques

In studies of basic electrochemistry and materials evaluation, a range of electrochemical techniques is used including rotating disk and rotating ring disk electrodes, cyclic voltammetry and electrochemical impedance spectroscopy, which will be briefly introduced in this section. For other electrochemical techniques the readers are advised to consult basic electrochemistry textbooks.

3.6.1 Rotating Disk and Rotating Ring Disk Electrodes

3.6.1.1 Rotating Disk Electrode

The rotating disk electrode (RDE) and rotating ring disk electrode (RRDE) are the most common techniques used in the study of the electrochemical reactions and catalyst behaviors related to fuel cells. The surface area of a disk electrode is well-defined, so the electrochemical activity of the electrocatalysts can be conveniently obtained as mass or surface area specific activity. In RDE, an active ion or molecule is being conveyed to the electrode surface in a steady-state laminar flow. Benjamin Levich developed the general theory describing mass transport of such laminar flow of material from the bulk solution to the electrode surface at a RDE in 1950s. While the bulk solution far away from the electrode remains well-stirred by the convection induced by rotation, the portion of the solution in the vicinity to the electrode surface tends to rotate with the electrode. Thus, if one views the solution from the position of the rotating electrode surface, the solution near the electrode surface appears relatively stagnant. This relatively stagnant layer is known as the hydrodynamic boundary layer, and its thickness δ_H can be approximated by,

$$\delta_H = 3.6\left(\frac{v}{\omega}\right)^{1/2} \tag{3.91}$$

where v is kinematic viscosity of the electrolyte solution and ω is the angular rotation rate ($\omega = 2\pi f/60$, where f is the rotation rate in revolutions per minute, rpm). In an aqueous solution at a moderate rotation rate (~1000 rpm), the stagnant layer is approximately 300–400 μm thick.

Net movement of active species to the electrode surface can be described mathematically by applying general convection–diffusion concepts from fluid dynamics. Mass transport of the species from the bulk solution into the stagnant layer occurs by convection due to the stirring generated by the rotating electrode. After the species enters the stagnant layer and moves closer to the electrode surface, convection becomes less important than diffusion. The final movement of the active species to the electrode surface is dominated by diffusion across a very thin layer of solution immediately adjacent to the electrode known as the diffusion layer. The diffusion layer is much thinner than the hydrodynamic layer, and the diffusion layer thickness δ_F can be calculated as follows.

$$\delta_F = 1.61 D_F^{1/3} v^{1/6} \omega^{-1/2} \tag{3.92}$$

where D_F is the diffusion coefficient of the species and typically in the range of 10^{-5} cm^2/s in an aqueous solution. The diffusion layer is approximately 20 times thinner than the stagnant layer, $\delta_F = 0.05\ \delta_H$. Figure 3.17 shows the schematic diagram of typical RDE and RRDE. Both disk and ring are working electrodes and can be independently controlled.

Considering the case where the reaction reaches the limiting current (i_L) with the oxidized or reduced form of an active species of interest in the electrochemical cell, such electrochemical and hydrodynamic processes taking place at RDE can be described by Koutecky–Levich (K–L) equation:

$$\frac{1}{i} = \frac{1}{i_L} + \frac{1}{i_K} = \frac{1}{0.620\text{nFA} D_a^{2/3} v^{-1/6} \omega^{1/2} c_a} + \frac{1}{nFkc_a} \tag{3.93}$$

where A is the electrode area, n is the number of electrons involved in the reaction, D_a and c_a are the diffusion coefficient and concentration of active species of the reaction, and k is the electron transfer rate constant. n and k can be obtained from the slope and intercept of K–L equation, respectively. The current equals to kinetic current, $i = i_K$ when $\omega^{1/2} \rightarrow \infty$.

In the case of an oxidation reaction, the cathodic limiting current, $i_{L,c}$, is related to the diffusion coefficient and concentration of the oxidized form of the reaction species (D_O and c_O) in the solution. For a reduction reaction, the anodic limiting

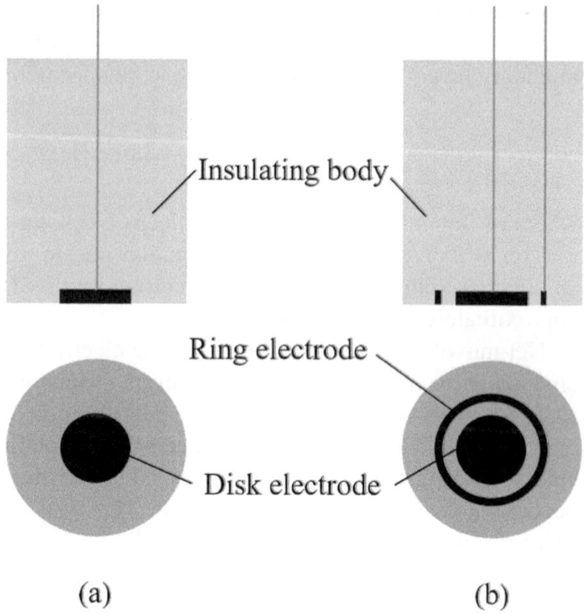

Fig. 3.17 Schematic diagram of typical **a)** RDE and **b)** RRDE

current, $i_{L,a}$, is related to the diffusion coefficient and concentration of the reduced form of the reaction species (D_R and c_R) in the solution. The limiting current density is generally written as:

$$i_{\mathrm{L}} = 0.620\mathrm{nFA}D_a^{2/3}\upsilon^{-1/6}\omega^{1/2}c_a \tag{3.94}$$

This is also called the *Levich equation* applying to the totally mass transfer-limited conditions at the RDE. The equation indicates that if the electrochemical reaction is a simple and reversible half reaction with no complications associated with sluggish kinetics or coupled chemical reaction, the limiting current measured will increase linearly with the square root of the rotation rate with a slope of $0.620\,\mathrm{nFA}D_{\mathrm{a}}^{2/3}\nu^{-1/6}c_{\mathrm{a}}$, and the line intercepts the vertical axis at zero. Such plots of limiting currents against the square root of the rotation rate are called Levich plots.

An alternate method of presenting the data is based on a re-arrangement of the Levich equation in terms of the reciprocal current.

$$\frac{1}{i_{\mathrm{L}}} = \left(\frac{1}{0.620nFAD_a^{2/3}\upsilon^{-1/6}c_a}\right)\omega^{-1/2} \tag{3.95}$$

A plot of the reciprocal current versus the reciprocal square root of the angular rotation rate is called a K–L plot. Again, for a simple and reversible half reaction with no complications, the plot is a straight line with the intercepts at the vertical axis equal to zero.

In the K–L plots, if the line intercepts the vertical axis above zero, this is a strong indication that the rate of a half reaction is limited by a combination of mass transport and sluggish kinetics. In this case, however, it is often possible to use a RDE to elucidate both the mass transport parameters (such as the diffusion coefficient) and the kinetic parameters (such as the standard rate constant, k^0) from a properly designed Levich study. A full treatment of this kind of analysis is beyond the scope of this chapter, but a general description of how to extract kinetic information from a set of RDE data can be found in reference electrochemistry books.

3.6.1.2 Rotating Ring Disk Electrode

The RRDE is a RDE with a concentric ring electrode around the rotating disk electrode. RRDE is a particularly powerful tool to study the reaction of intermediates which can be formed on the disk electrode and monitored by the ring electrode. RRDE is characterized by two parameters, the collection efficiency and the transit time and can be obtained from the RRDE manufacturer. The collection efficiency is the fraction of the reaction intermediates from the disk electrode which subsequently passes the ring electrode. A typical RRDE has a collection efficiency between 20 and 30%. The transit time is the time required for the intermediates produced at the disk electrode to travel across the gap between the disk and the ring electrode. This depends on both the gap distance and rotation rate.

RRDE is particularly useful in the study of mechanism and kinetics of the oxygen reduction reaction (ORR). ORR can proceed with two-electron in acid solution:

$$O_2 + 2H^+ + 2e^- \rightleftharpoons H_2O_2 \qquad E^0 = +0.682\,\mathrm{V} \tag{3.96}$$

or four-electron process:

$$O_2 + 4H^+ + 4e^- \rightleftharpoons 2H_2O \qquad E^0 = +1.229\,\mathrm{V} \tag{3.97}$$

or in alkaline solution

$$\frac{1}{2}O_2 + H_2O + 2e^- \rightleftharpoons 2OH^- \qquad E^0 = +0.401\,\mathrm{V} \tag{3.98}$$

The hydrogen peroxide can be further reduced by the following reation:

$$H_2O_2 + 2H^+ + 2e^- \rightleftharpoons 2H_2O \quad E^0 = +1.776\,\mathrm{V} \tag{3.99}$$

In the case of an unbuffered solution, an equilibrium exists between H_2O_2 and HO_2^- ions:

$$HO_2^- + H^+ \rightleftharpoons H_2O_2 \tag{3.100}$$

A suitable potential window between 1.1 and 1.5 V *versus* RHE can be used as the set potential of the Pt ring electrode of RRDE for the collection and analysis of the formation of intermediates, hydrogen peroxide, produced at the disk electrode. The fraction of current due to peroxide ions, HO_2^- recorded by the ring electrode, $x_{H_2O_2}$, and electron transfer number, *n*, for the ORR on elecrocatalysts can be obtained by the measurement of the disk and ring current.

$$x_{H_2O_2} = 2 \times \frac{i_R/\delta}{\left(\frac{i_R}{\delta} + i_D\right)} \tag{3.101}$$

$$n = 4 \times \frac{i_D}{\left(\frac{i_R}{\delta} + i_D\right)} \tag{3.102}$$

where i_D is the disk current, i_R is the ring current, and δ is the collection efficiency of RRDE, which is a constant usually supplied by the manufactures of the RRDE. Figure 3.18 shows a typical ORR polarization curve of electrocatalysts (e.g., Pt/C) corrected on a disk electrode and the corresponding oxidation current collected by

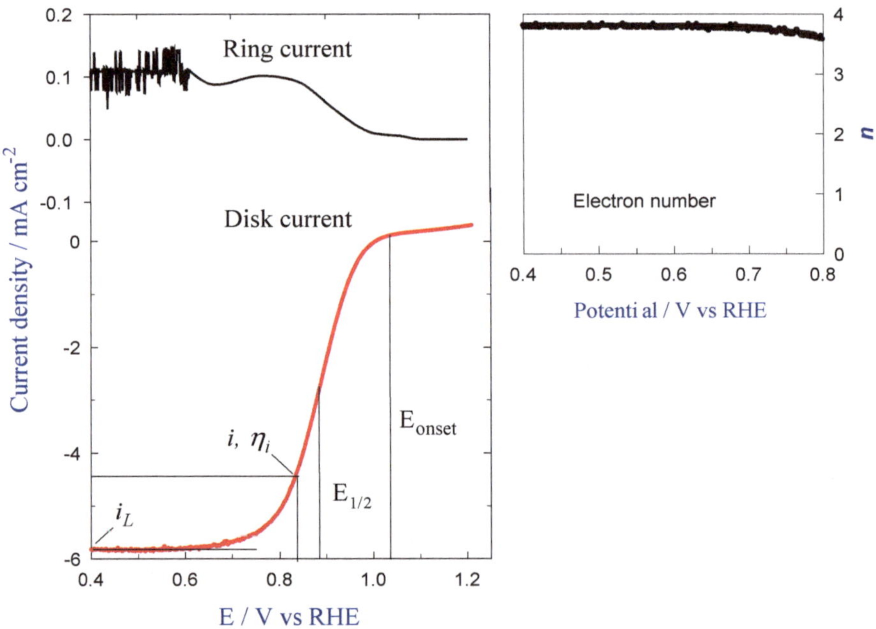

Fig. 3.18 A typical ORR polarization curve of the Pt/C catalyst corrected on disk and anodic ring current for the oxidation of peroxide intermediates of the reaction at a chosen set potential. The RRDE consists of a 5.6 mm outer diameter glassy carbon disk electrode and a Pt ring (inner diameter 6.25 mm, and outer diameter 7.92 mm) with a collection efficiency of 0.26. The catalysts loading was 25.6 μg/cm^2. The ORR curve was obtained at a scan rate of 10 mV/s in O_2-saturated 0.1 M $HClO_4$ solution by a linear scan voltammetric technique using RRDE with rotating rate of 1600 rpm

ring electrode at a chosen set potential. From the ORR polarization curve, some performance indicators can be obtained including:

- Onset potential (E_{onset}) — the starting potential of the reaction;
- Half-wave potential ($E_{1/2}$) — the potential at which the current is equal to half of the current density between initial and the limiting current density. In the case of difficulty in defining the half-wave potential, the potential obtained at a current density of 3 mA/cm^2 can be used as $E_{1/2}$;
- Limiting cathodic or anodic current density (i_L) — the current at mass transfer limiting conditions;
- Overpotential at a specific current density (η_i).

Based on the ring and disk currents measured, the H_2O_2 yield was calculated as 8–10% for ORR on Pt/C. The electron transfer number for the ORR on Pt/C electrocatalysts is 3.8–3.9 in the potential range of 0.2–0.75 V [3]. This indicates that the ORR on Pt/C proceeds through an 4-electron transfer process with negligible H_2O_2 yield, a well-established process. The electron transfer number for ORR can also be obtained by K–L plots, based on Eq. (3.95). Figure 3.19 shows the activities of Fe single atom catalysts supported on nitrogen doped graphene (SAFe@NG) for ORR. The linear sweep voltammetry (LSV) curves were measured in O_2-saturated 0.1 M KOH solution using RDE. Similar to that on Pt/C catalysts, performance parameters for ORR on SAFe@NG in 0.1 M KOH can also be obtained, e.g., E_{onset} = 0.96 V and $E_{1/2}$ = 0.876 V. The SAFe@NG catalyst shows a better activity for ORR as compared to that measured on the state-of-the-art Pt/C. The K–L plots obtained from the LSV curves reveal good linearity, indicating the first-order reaction kinetics for ORR with a potential-independent electron transfer rate. Based on the K–L equation, the electron transfer number can be calculated from the slopes of

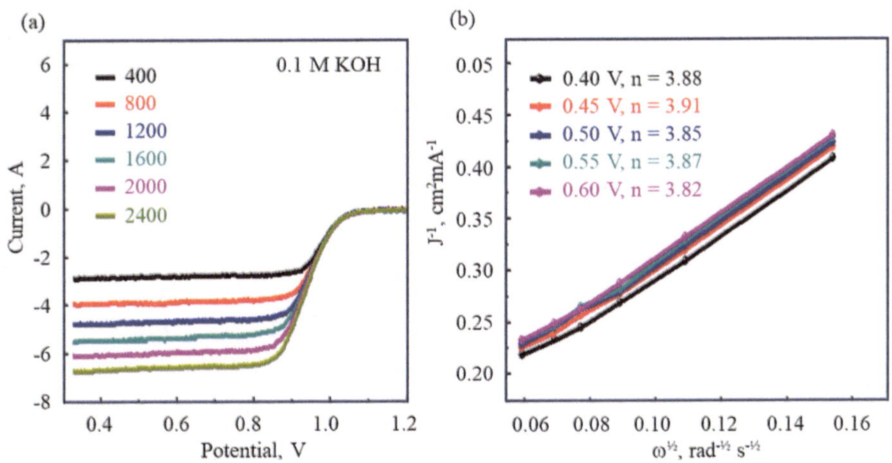

Fig. 3.19 Oxygen reduction reaction performance. **a)** Linear scan voltammetric (LSV) curves in 0.1 M KOH and **b)** K–L plots of Fe single atom catalysts supported on graphene (SAFe@NG), measured at different rotating rates in O_2-saturated 0.1 M KOH solution. The SAFe@NG loading was 0.3 mg/cm^2

the curves, which is 3.93 (Fig. 3.19b), indicating an essential 4-electron process of ORR on the SAFe@NG electrocatalyst. In this case, the concentration of the bulk concentration of O_2 in the electrolyte, c_O, is 1.26×10^{-6} mol/cm^3, the diffusion coefficient of O_2 in electrolyte, D_O, is 1.93×10^{-5} cm^2/s and the kinematic viscosity of the electrolyte, ν, is 1.009×10^{-2} cm^2/s.

3.6.2 *Linear Sweeping and Cyclic Voltammetry Techniques*

The term voltammetry in electrochemistry in general refers to a technique where the electrode potential is varied at a constant rate while the current is measured. The most common forms of voltammetry involves linear sweeping or *linear scan voltammetry* (LSV) and *cyclic voltammetry* (CV). In LSV, the electrode potential is swept from an initial value to a final value at a constant rate, while in CV, the scan is reversed and repeated between a certain potential window. LSV and CV are powerful and popular electrochemical techniques commonly used to investigate the reduction and oxidation reactions of ionic or molecular species. CV is also invaluable to study electron-initiated chemical reactions such as in catalysis.

Fuel cells represent a complex situation which involves electronic and ionic conduction in electrode and electrolyte, multiple reactions at the solid/liquid interface in the case of liquid electrolyte-based fuel cells, or the solid/solid interface in the case of solid electrolyte fuel cells. Theoretically, there are many complex equations in the interpretation of the LSV and CV data, but in practice, we probably would just need to collect some specific values from LSV and CV curves that are useful for the understanding of the electrode reactions. To start the collection of CV curves, we need to determine the starting potential (E_1), terminal or switching potential (E_2) and the scan rate (v). If the scan direction is starting positively from point *A* (see Fig. 3.20), this is referred as the anodic scan. The recorded current becomes positive and increases with the increase in potential, reaching point *B* where the peak

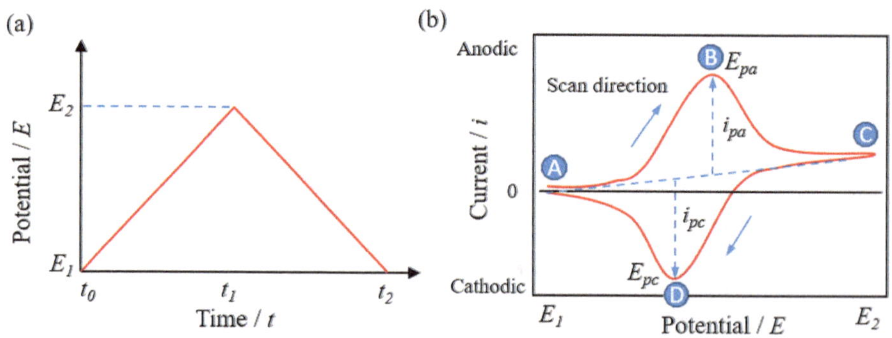

Fig. 3.20 Cyclic voltammetry. **a)** Potential profile and **b)** voltammogram of a simple and reversible reaction $O + ne^- \rightleftharpoons R$

anodic potential and current (E_{pa} and i_{pa}) are observed. At this point, the oxidation rate is the highest due to the fact that the concentration gradient in the close vicinity of the electrode is able to feed the oxidation process with reactants. At the same time, the volume of solution at the electrode surface containing the product, called the diffusion layer, continues to grow. This slows down mass transport of reactant to the electrode. Thus, upon scanning to more positive potential the supply of reactants becomes diffusion limited and the current decreases and then reaches a constant value at point *C*, i.e., switching potential, E_2. Then the scan direction is reversed and the potential is scanned in the negative (cathodic) direction back to E_1. The concentration of reactant species at the electrode surface is depleted, but the concentration of product species at the electrode surface increases, satisfying the Nernst equation. The product species present at the electrode surface is reduced back to reactant species, reaching the peak cathodic potential and current (E_{pc} and i_{pc}) at point *D*. Figure 3.20 shows typical potential profile and current responses *versus* potential curves for a simple electrode reaction.

In addition, there are onset potentials at which the anodic or cathodic reaction starts, $E_{onset,a}$ and $E_{onset,c}$. At points where the concentrations of *O* and *R* at the electrode surface are equal, following the Nernst equation, this corresponds to a half-wave potential, $E_{1/2}$. The difference between the anodic and cathodic peak potentials is called peak-to-peak separation, $\Delta E_p = E_{pa} - E_{pc}$. The recorded data are generally plotted in *x*-axis with the applied potential and in *y*-axis with the resulting current passed, referred to cyclic voltammograms or voltammograms. Significant information about the electrochemical reactions occurring at the electrode and the chemical transport kinetics can be found from the voltammogram produced.

Experimentally, CV includes the parameters of the beginning and ending potential of the scan (i.e., E_1 and E_2), the scan rate in mV/s, the direction of scan and the number of scans. The difference between E_1 and E_2 is also called electrochemical window within which the interested electrode reaction occurs. The scan rate is important. A faster scan rate leads to a decrease in the thickness of diffusion layer, resulting in a higher current. At the same time the capacitive current of charging/discharging the double layer at the electrode/electrolyte interface is also larger at a high scan rate.

Figure 3.21 shows a practical example of using CV in the study of electrooxidation of ethanol on Pd nanoparticle catalysts in 1.0 M KOH + 1.0 M ethanol solution. Pd nanoparticles were supported on carbon nanotubes (CNTs) of varied numbers of inner walls [4]. Ethanol is contained in the alkaline electrolyte. In this case the oxidation of ethanol takes place during both anodic and cathodic scans. The voltammograms are characterized by well-separated anodic peaks in forward and reversed scans. The magnitude of the anodic current density in the forward scan is directly related to the amount of ethanol oxidized at Pd catalysts. The electrocatalytic activity of Pd catalysts strongly depends on the types of CNT supports, and better results were observed on Pd nanoparticles supported on CNTs-2 (triple-walled CNTs) and CNTs-3 (CNTs with average seven walls) than those achieved on CNTs-1 (single-walled CNTs) and CNTs-4 (multi-walled CNTs). The significantly better performance of Pd catalysts supported on CNTs with specific numbers of inner tubes has been attributed to the fast charge transfer process via electron tunneling between the outer wall and inner tubes of CNTs for the ethanol oxidation reaction [4].

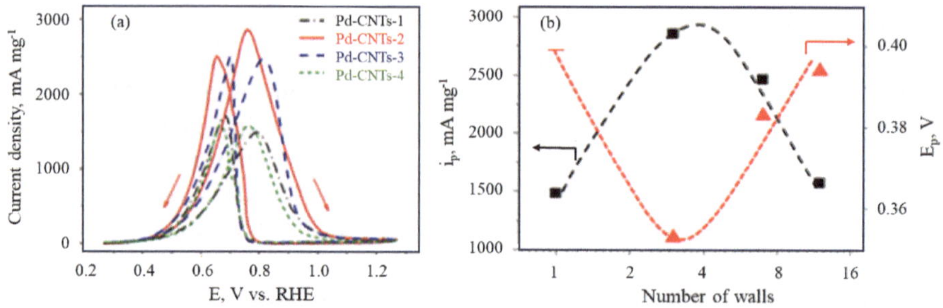

Fig. 3.21 **a)** Cyclic voltammograms of Pd-CNTs catalysts for ethanol oxidation in 1.0 M KOH + 1.0 ethanol at scan rate of 50 mV/s and **b)** The corresponding plots of the forward peak current density and onset potential of Pd-CNTs catalysts as a function of the wall number of CNTs. Pt plate and Hg/HgO (1.0 M KOH) (0.92 V vs. RHE) electrode were used as the counter and reference electrodes, respectively. Blank cyclic voltammetry tests were conducted in 1.0 M KOH solution. Pd catalyst loading was 0.05 mg/cm^2 on the glassy carbon electrode substrate. Modified from Ref. [4] with permission from Wiley-VCH, Copyright 2015

The appearance of the anodic peaks in the forward and reverse scans rather than the anodic peak in the forward scan and cathodic peak in the reverse scan as shown in Fig. 3.20b is typically associated with the incomplete oxidation reactions and irreversible oxidation/reduction reactions. Such characteristics of CV is most often observed in the case of alcohol or liquid fuel oxidation reactions due to the fact that total or complete oxidation of alcohols such as methanol, ethanol and formic acid seldom occurs. Additional examples of such CV behavior can be seen in Chap. 4. CV is also commonly used for the measurement of electrochemical surface area (ECSA) of electrocatalysts such as Pt, PtRu, and Pd NPs (see Chap. 6).

3.6.3 Electrochemical Impedance Spectroscopy (EIS)

Electrochemical impedance spectroscopy (EIS) is a powerful tool for characterization of a wide range of electrochemical systems and determining the contribution of individual electrode or electrolyte processes. The technique has been in use in aqueous electrochemistry as well in a range of ionic conductors which include polymers, ceramics, and glasses. Normally, various processes from electrode reactions at the electrode/electrolyte interface (diffusion, adsorption/dissociation, charge transfer) and those associated with mass transport within the solid electrolyte grains and across grain boundaries all have a different time constant and can be separated in the frequency domain. Combined with microstructure and surface characterization techniques such as XRD, SEM, and TEM, the EIS provides an invaluable tool for the electrochemical characterization of fuel cells.

3.6.3.1 Basic Theory

The EIS technique in general involves the measurement of current through an electrolyte cell when a sinusoidal voltage of low amplitude is applied. The sine wave is used in impedance measurements because for a sinusoidal signal (irrespective of its amplitude or frequency), the inputs and outputs have the same form and the angular frequency. Also for a linear system, the magnitude of the response is directly related to the electrical stimulus for any given frequency. Hence, in a linear system, the applied potential is given by:

$$E(t) = \Delta E \ \exp(j\omega t) \tag{3.103}$$

The current output of the system is also sinusoidal and has the same angular frequency ω, but differences in amplitude and phase from the voltage signal may occur depending on the elements in the circuit and can be represented by:

$$I(t) = \Delta I \exp(j\omega t + \phi) \tag{3.104}$$

where ϕ is the phase angle, ΔE is the amplitude of the voltage and ΔI the amplitude of the current signal.

Since Ohm's law holds true in the time or the frequency domain, the impedance of a circuit consisting of resistors, capacitors, and/or inductors is the ratio of the voltage signal divided by the current flowing through the circuit. The impedance of the circuit (Z(ω)) at any frequency ω can be represented in both polar and Cartesian form and has both the magnitude Z and the phase angle ϕ.

$$Z(\omega) = \frac{E(t)}{I(t)} = \frac{\Delta E}{\Delta I} \exp(-j\phi) \tag{3.105}$$

Equation (3.105) can also be expressed as:

$$Z(\omega) = \frac{E(t)}{I(t)} = Z\cos\phi - jZ\sin\phi = Z_{\mathrm{R}} - jZ_{\mathrm{I}} \tag{3.106}$$

where j is a complex number with value of $\sqrt{-1}$. Z_R and Z_I are real and imaginary parts of the impedance. The relationships between various terms are as follows:

$$\emptyset = \tan^{-1}\left(\frac{Z_{\mathrm{I}}}{Z_{\mathrm{R}}}\right) \tag{3.107}$$

$$Z = \sqrt{(Z_{\mathrm{R}}^2 + Z_{\mathrm{I}}^2)} \tag{3.108}$$

For a pure resistor (R), capacitor (C), and an inductor (L), the impedance is given by the following equations, respectively:

$$\text{Resistance:} \quad Z = R + 0j \tag{3.109}$$

$$\text{Capacitance:} \quad Z = 0 - j/\omega C \tag{3.110}$$

$$\text{Inductance:} \quad Z = 0 + j\omega L \tag{3.111}$$

The impedance of a capacitor or an inductor has the same dimensions as the resistor; however, it differs from the impedance of a resistor in that it is dependent on the frequency of the AC signal.

3.6.3.2 Equivalent Circuit Analysis

The fundamental laws governing the relationship between the charge and potential and the properties of linear systems are similar in passing from electronic to ionic materials. Therefore as a first approximation, it is reasonable to assume that a direct connection exists between the behavior of a real electrochemical system and that of an electrical circuit consisting of discrete components (resistors, capacitors, inductors). Thus, various processes in an electrochemical system can be approached by an electrical equivalent circuit which would depict the way in which various conducting elements in the electrochemical cell are connected. The equivalent circuit approach in data analysis is most popular in the study of solid ionic materials like SOFCs as detailed microscopic models of all possible electrode and electrolyte processes are generally not available or are extremely complex.

In the simplest representation, a solid electrolyte cell can be considered to consist of a series network of several resistors and capacitors in parallel with each sub-circuit representing a different electrode or an electrolyte process. It can be shown that for each capacitor, *C*, in parallel with a resistor, *R*, the response in the complex plane is a semicircle with its center on the real or *x*-axis, as shown in Fig. 3.22. The high-frequency intercept of the semicircle on the real axis gives the ohmic resistance of the cell, R_O, and the difference of low- and high-frequency intercepts of the semicircle

Fig. 3.22 .

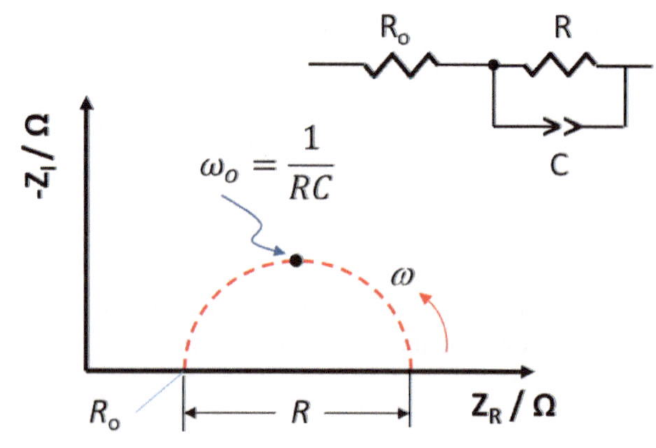

on the real axis gives the value of the resistor. The value of the capacitor and the time constant of the circuit can be calculated from the frequency (f_o) at the top of the semicircle ($\omega_o = 2\pi f_o = 1/\tau = 1/RC$). If each sub-circuit (or an electrochemical process represented by it), denoted by RQ where Q can be capacitance C or *constant-phase element* (CPE), has a different time constant associated with it then the response of the equivalent circuit in the complex impedance plane, will consist of several semicircles each corresponding to a sub-circuit.

Figure 3.23 gives an example of the application of equivalent circuit in the study of ORR on an $La_{0.8}Sr_{0.2}MnO_3$ (LSM) electrode in SOFC. The LSM electrode coating was applied to an YSZ electrolyte pellet and sintered at 1150 °C with a Pt counter electrode and a ring-shaped reference electrode on the opposite side of the YSZ pellet. Equivalent circuit with three RQ sub-circuits is used. In electrochemical cells, the ohmic resistance of the electrolyte and electrode is represented as the resistance, R_0 in series with the circuit and can be obtained at the high-frequency intercept. The inductance in solid electrolyte cells can occur at high or low frequencies. The fitting between the experimental data and fitted one is excellent, indicating the ORR on the porous LSM electrode is controlled by at least three electrode steps. If the time constants associated with each process are not at least two orders of magnitude different, a considerable overlap of the individual frequency domains over which each process relaxes will occur, leading to overlapping arcs in the complex impedance plane. However, the impedance arcs for individual electrode processes

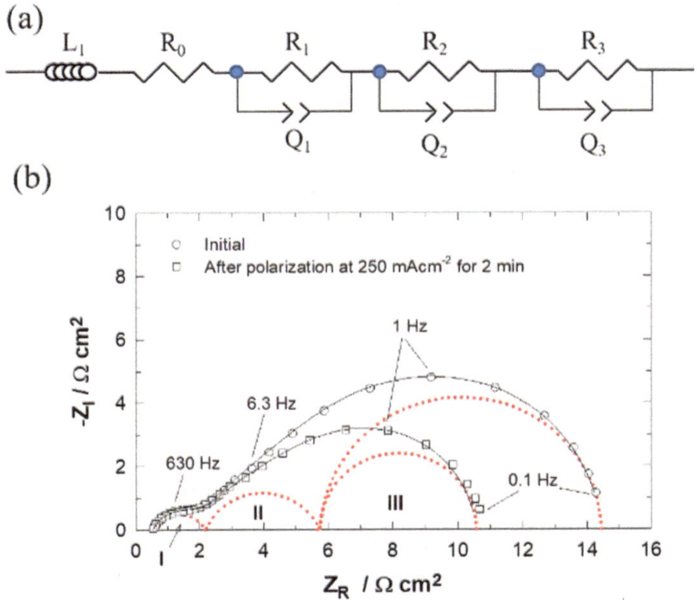

Fig. 3.23 **a)** Equivalent circuit representing ORR at a $La_{0.8}Sr_{0.2}MnO_3$ electrode of SOFCs and **b)** impedance responses of a $La_{0.8}Sr_{0.2}MnO_3$ electrode before and after polarization treatment at 250 mA/cm^2 for 2 min. EIS was measured in the frequency range of 0.1 Hz–100 kHz and signal amplitude of 10 mV at 800 °C in air. The symbols are experimental data, and lines are fitted data from the equivalent circuit

can be separated by variation and manipulation of testing conditions. The change of the electrode process associated with the low-frequency arc (step III) with the polarization treatment at 250 mA/cm^2 for 2 min is due to the activation effect of the cathodic polarization on the dissociation and diffusion of oxygen for ORR on the LSM electrode [5].

For a simple electrode involving a charge transfer process, the current is determined by both charge transfer (faradaic) and the electrical double layer charging (non-faradaic) processes. The electrode/electrolyte interface can be represented by a simple equivalent circuit of a charge transfer or polarization resistance, R_p in parallel with a double layer capacitance C_{dl} as shown in sub-circuit I (R_1 and Q_1). The effective combined impedance of the circuit can be written as:

$$Z = R_0 + \frac{1}{R_p} + \frac{1}{\frac{-J}{\omega C_{dl}}} = R_0 + \frac{R_p}{1 + \omega^2 C_{dl}^2 R_p^2} - j\frac{\omega^2 C_{dl} R_p^2}{1 + \omega^2 C_{dl}^2 R_p^2} \tag{3.112}$$

This is also called Randles cell impedance. The Randles equation can be analyzed in two frequency limits. If $\omega = 0$, then the imaginary term is zero and the impedance is reduced to $R_0 + R_p$. As $\omega \rightarrow \propto$, imaginary terms of the equation again approaches zero and the impedance of the system is equal to the ohmic resistance of the electrolyte, R_0. The double-layer capacitance can be determined from the apex frequency ($\omega_o = 1/\tau = 1/R_p C_{dl}$).

At very low frequencies, and especially for low concentrations of electroactive species, the diffusion polarization may be observed. This can be approximated by a Warburg impedance relationship given by:

$$W = \sigma\omega^{-0.5} - j\sigma\omega^{-0.5} \tag{3.113}$$

where σ is Warburg coefficient which is related to the diffusion coefficients for the charge-carrying species. The Warburg impedance is originally proposed for the diffusion process of neutral mobile species at the metal electrode/electrolyte interface and is often observed in aqueous electrolyte systems. For a circuit with Warburg impedance in series with the charge transfer resistance, the impedance is given by:

$$Z_R = R_{el} + \frac{R_p + \alpha\omega^{-0.5}}{(C_{dl}\alpha\omega^{0.5} + 1)^2 + \omega^2 C_{dl}^2 (R_p + \alpha\omega^{-0.5})^2} \tag{3.114}$$

$$Z_I = \frac{\omega C_{dl}(R_p + \alpha\omega^{-0.5})^2 + \sigma^2 C_{dl} + \sigma\omega^{-0.5}}{(C_{dl}\alpha\omega^{0.5} + 1)^2 + \omega^2 C_{dl}^2 (R_p + \alpha\omega^{-0.5})^2} \tag{3.115}$$

The simulated behavior of the circuit in the impedance planes is typically characterized by a straight line on the low-frequency end of the electrode arc intersecting the real impedance axis at an angle of 45°. Depending upon the value of the Warburg coefficient, considerable overlap can occur with the charge transfer process.

The impedance of solid electrolyte systems in reality cannot be well-approximated by the impedance of an equivalent circuit involving a limited or finite number of ideal resistors and capacitors, which give rise to perfect semicircular arcs with discrete time constants for each electrochemical process. The commonly observed behavior of a solid electrolyte system in the impedance plane is rarely in the form of semicircular arcs with their origin on the real impedance axis, but instead consists of depressed arcs. These arcs can often be skewed on the low or the high-frequency end of the impedance spectrum. This type of behavior often indicates heterogeneity in the system and distribution of time constants and can be simulated by including distributed or CPE in the equivalent circuit. The empirical fractional-power-law for CPE is given by:

$$Z_{\mathrm{CPE}} = A(j\omega)^{-n} \tag{3.116}$$

where A is independent of ω and $1 > n > 0.5$. The introduction of CPE into the equivalent circuit is an empirical approach in order to get acceptable fitting between the experimental data and equivalent circuit. However, the physical origin of CPE has not been fully established and is poorly understood.

It should be emphasized that the equivalent circuit approach to the impedance response of the electrochemical system is seldom unique. It is common to fit the same impedance responses with two or more different equivalent circuits. An important consideration with this approach is that an equivalent circuit involving three or more circuit elements can often be rearranged in various ways and still produce similar impedance responses [6]. To verify the validity of the selected circuit model, the electrochemical system should also be studied under wide experimental conditions (e.g., temperature, partial pressure of reactant gas, concentration of electroactive species and physical and chemical nature of electrode and electrolyte materials) to obtain reaction parameters and kinetic information to predict responses. The verification of equivalent circuit approaches will be discussed in more details in Chap. 11.

3.7 Summary

In this chapter, we have introduced and discussed the following topics:

- Electrolyte is the core component of any electrochemical systems. Fuel cells are constructed with various forms of electrolyte materials including aqueous solutions in AFC and PEFC, molten salts in MCFC and solid polymers in PEMFC and solid oxides in SOFC. All electrochemical reactions take place at the electrode/electrolyte interface, which is described by the electrical double layer theory. The theory is illustrated with a solid metal electrode in aqueous electrolytes and is also essential for the fundamental understanding of solid electrolyte electrochemistry and principles.

- The potential cross the electrode/electrolyte interface is of thermodynamic nature, the absolute value of which is by no means to measure. A standard hydrogen electrode (SHE) consisting of a Pt wire immersed in 1M H^+ solution with bubbling H_2 under 1 atm pressure is defined as the zero point reference. For practical convenience, various reference electrodes are used including SCE, Ag/AgCl and Hg/HgO electrodes.
- Standard reduction potentials for electrochemical pairs are compiled and available, the change of which with the temperature, concentration, and pressure of the reactants and products of the reactions is expressed by the Nernst equation.
- At equilibrium, electron transfer occurs at the electrode/electrolyte interface but with equal rates at both directions, as expressed by the exchange current density i^0, though no net current can be measured. For the reaction to proceed in cathodic or anodic direction, an external potential has to be applied to disturb the equilibrium. Such deviation of the electrode potential from its equilibrium value is called overpotential, the driving force for the reaction.
- The relationship between the current and overpotential is expressed by the Butler–Volmer equation, the most basic equation in electrochemistry. The equation can be simplified in small and large overpotentials, η. At low value of η, the exchange current density, i^0 can be obtained from the measurement of i *versus* η. Under conditions of large overpotentials, the equation is reduced to the Tafel equation, from which, the Tafel slope can be obtained. Tafel slope is an important parameter related to the electron number involved in the electrode reaction and the charge transfer coefficient. As Tafel slope is generally determined by the rate-determining step, any change in the Tafel slope implies the change in the reaction mechanism.
- Overpotential can also be caused by the insufficient mass transfer either of the reactants to the reactive sites or of the removal of the products from the reactive sites. This is called mass transfer or concentration overpotential. The electric resistance of the electrode and electrolyte to the electron and ion flow causes the overpotential, i.e., ohmic overpotential. The electrode irreversibility and internal leakage current through the electrolyte also cause the loss in the electrode potential and are generally represented by the loss in the open-circuit voltage.
- The most common electrochemical techniques used in fuel cell electrocatalysts and reactions are rotating disk electrode (RDE) and rotating ring disk electrode (RRDE) via linear sweeping and cyclic voltammetry. Reaction parameters such as the mass transfer coefficient, electron transfer number, etc. can be obtained from plots of the reciprocal current versus the reciprocal square root of the angular rotation rate, the K-L plots.
- In solid electrolyte cells such as PEMFC and SOFC, the use of electrochemical techniques is limited due to the restriction and location of reference electrode within the solid electrolyte. In PEMFC and SOFC, the most common electrochemical technique is the electrochemical impedance spectroscopy (EIS).The EIS data is most commonly interpreted by the equivalent circuit analysis with the introduction of CPE to take into account of the heterogeneity in the system.

3.8 Questions

3.1 Explain the following terms and materials:

Electrochemical interface and double layer
Standard hydrogen electrode (SHE) and saturated calomel electrode (SCE)
Anode and anodic reaction
Cathode and cathodic reaction
Concentration cell
Butler–Volmer equation and cell potential
Exchange current density and reaction rate
Tafel equation and Tafel slope
Polarization losses and area specific resistance
Onset potential, half-wave potential, limiting current, and kinetic current
Rotating disk electrode, rotating disk ring electrode, K-L plots and electron transfer number
Equivalent circuit analysis and constant-phase element

3.2 Derive the electrode potential relation for the following oxygen reduction reaction. Assuming $P_{O_2} = 1$ atm at room temperature, what is the electrode potential for the oxygen reduction reaction at pH = 0 and pH = 14. (Answer: 1.23 V at pH = 0 and 0.40 V at pH = 14)

$$O_2 + 4H^+ + 4e^- = 2H_2O$$

3.3 In a concentration cell with the same copper electrode but different concentration of $CuSO_4$, as shown in Fig. 3.4. Find out what is the concentration of $CuSO_4$ solution when the reaction stops? (Answer: equal concentration)

3.4 In the case of iron corrosion, one of the important reactions is

$$Fe + 2HCl + \frac{1}{2}O_2 = FeCl_2 + H_2O$$

If the activity of Fe^{2+} and H^+ in the solution is unity, calculate (a) the cell potential and the spontaneous direction of this reaction and b) at what activity of Fe^{2+} iron corrosion will stop in 1 M HCl solution.
(Answer: (a) 1.669 V and the reaction runs to the right and (b) 2.5×10^{56})

3.5 Consider a fuel cell with a I-V characteristics of $E_{cell} = 0.92 - 50 \times 10^{-2} \times I$ (in which I is in A/cm^2 and E_{cell} is in V) operated under standard temperature and pressure (298.15 K and 1 atm). The product water is removed from the cell in vapor form. Calculate (a) open-circuit voltage, (b) the maximum power density, (c) under the maximum power output, what is the amount of heat generated, and (d) efficiency of the cell.
(Answer: (a) 0.92 V, (b) 0.423 V, (c) 0.73 W/cm^3, (d) 37%)

3.6 Continue on problem 3.4. For this cell under operation at maximum power output, what is the Joule heat, thermodynamic heat and heat losses due to the OCV losses (i.e., internal leakage current losses).
(Answer: 0.423 W/cm^3, 0.063 W/cm^3, and 0.243 W/cm^3)

3.7 For a H_2–O_2 fuel cell stack, the polarization curve can be presented as $E_{cell} = 0.8 - 1 \times 10^{-4} \times I$ (in which I is in A and E_{cell} is in V) under standard temperature and pressure conditions. Assuming 100% current efficiency, calculate a) hydrogen consumption rate in mg/s when the power output is 1 kW, and b) what is the heat generated by the stack cell? Liquid water is produced.
(Answer: (a) 144.5 mg/s and (b) 1296 W)

3.9 General Readings

1. Bard AJ, Faulkner LR (1980) Electrochemical methods—fundamentals and applications. Wiley, New York

References

1. Watanabe N, Devanathan MAV (1964) J Electrochem Soc 111:615
2. Jiang SP, Liu ZC, Tian ZQ (2006) Layer-by-layer self-assembly of composite polyelectrolyte-nafion membranes for direct methanol fuel cells. Adv Mater 18(8):1068–1072
3. Cheng Y, He S, Lu SF, Veder JP, Johannessen B, Thomsen L, Saunders M, Becker T, De Marco R, Li QF, Yang SZ, Jiang SP (2019) Iron single atoms on graphene as nonprecious metal catalysts for high-temperature polymer electrolyte membrane fuel cells. Adv Sci 6(10):e1802066
4. Zhang J, Lu SF, Xiang Y, Shen PK, Liu J, Jiang SP (2015) Carbon-nanotubes-supported Pd nanoparticles for alcohol oxidations in fuel cells: effect of number of nanotube walls on activity. Chemsuschem 8(17):2956–2966
5. Jiang SP, Love JG (2001) Origin of the initial polarization behavior of Sr-doped $LaMnO_3$ for O_2 reduction in solid oxide fuel cells. Solid State Ionics 138(3–4):183–190
6. Jiang SP, Love JG, Ramprakash Y (2002) Electrode behaviour at (La, Sr)MnO_3/Y_2O_3-ZrO_2 interface by electrochemical impedance spectroscopy. J Power Sour 110(1):201–208

Chapter 4
Fuels for Fuel Cells

4.1 Introduction

As an energy conversion device, fuel cells generate the electrical power from chemical fuels. Different types of fuel cells operate at temperatures from subfreezing up to 1000 °C and cover a wide power spectrum of applications from a few watts to megawatts. For all these applications, hydrogen (H_2) is the most popular fuel. First of all, the H_2 oxidation and evolution are electrochemically reversible on the platinum electrode in acidic media. This can be translated into a statement that H_2 has the fastest kinetics and therefore the highest conversion efficiency of all types of fuels.

Hydrogen is abundant in resources, however, not in form of pure gas but combinations with other elements. As an energy carrier, hydrogen must be first produced. The ultimate hydrogen must be produced by water electrolysis using power from renewable energies. In this way, hydrogen is a completely carbon-free green fuel allowing for operation of fuel cells with zero emission of greenhouse gases.

Various fuels have been explored for use in fuel cells. Table 4.1 lists some selected gaseous and liquid fuels and their properties for comparison.

The hydrocarbon fuels such as natural gas and petroleum (gasoline and diesel) are typical sources of hydrogen via reforming. Methanol and ethanol are potentially renewable and can be used either by direct oxidation or as hydrogen-rich reformat. Thus, obtained hydrogen-rich gases contain carbon dioxide, carbon monoxide, and a small quantity of inert gases such as nitrogen and water vapor. Traces of sulfur may also be present in the reformat. The impact of these gaseous constitutes on various fuel cells is summarized in Table 4.2. For all types of fuel cells, the most sensitive gas constitute is sulfur. Carbon monoxide is another poison for low-temperature fuel cells. For high-temperature MCFC and SOFC, CO can be used as fuel.

In term of the energy content, the mass and volume specific energy densities of these fuels are compared in Table 4.1. These numbers are based on both high heat value (HHV) and low heat value (LHV) of the enthalpy changes (ΔH^0, kJ/mol) of combustion with the product water formed as liquid and vapor, respectively. The difference between these two values is the energy of water condensation. From the

S. P. Jiang and Q. Li, *Introduction to Fuel Cells*,
https://doi.org/10.1007/978-981-10-7626-8_4

Table 4.1 Selected properties of hydrogen and other fuels

	Hydrogen (H_2)	Methane (CH_4)	Ammonia (NH_3)	Methanol (CH_3OH)	Ethanol (C_2H_5OH)	Gasoline[a] (C_4–C_{12})	Diesel[a] (C_4–C_{12})
Molar mass (g/mol)	2.016	16.04	17.03	32.04	46.07	60–150	150 ~ 250
Freezing point (°C)	−259.2	−182.5	−77.7	−98.8	−114.1	−55 to −70	−8 to −12
Boiling point (°C)	−252.8	−161.5	−33.4	64.7	78.3	<210	<350
Mass specific energy density (MJ/kg):							
HHV	142.8	55.5	22.5	22.7	29.7	ca. 45	ca. 46
LHV	120	50.1	18.8	19.8	26.8	ca. 42	ca. 43
Density (kg/L):							
Liquid	77	425	674	786	789	720–775	820–840
Gas under ambient pressure	0.00008	0.00066	0.00073	–	–	–	–
Storage pressure (Bar)	700	250	1	1	1	1	1
Volume energy density (HHV, MJ/L):							
Liquid	10.1	25.2	11.6	17.8	21.2	ca. 33	ca. 35
Gas under ambient pressure	0.011	0.036	–	–	–	–	–
Compressed gas under storage pressure	5.6[b]	9.0[b]	–	–	–	–	–

[a]Gasoline and diesel are mixtures of hydrocarbons with varied compositions; [b]Under 700 bar for H_2 and 250 bar for CH_4.

Table 4.2 Gaseous constituents in the reformate and their impacts on various fuel cells

Gas species	PEMFC	AFC	PAFC	MCFC	SOFC
H_2	Fuel	Fuel	Fuel	Fuel	Fuel
CO_2	Diluent	Poison	Diluent	Re-circulated	Diluent
CO	Poison (<20 ppm)	Poison	Poison (<2%)	Fuel	Fuel
CH_4	Diluent	Unknown	Diluent	Diluent[a]	Diluent[a]
Sulfur	Poison	Poison	< 20 ppm H_2S < 50 ppm(H_2S + COS)	< 10 ppm H_2S in fuel < 1 ppm SO_2 in air	< 5–10 ppm H_2S
NH_3	Poison	Unknown	Poison	Unknown	Fuel < 5000 ppm
Halogens (HCl)	Unknown	Unknown	Poison (<4 ppm)	Poison (<1 ppm)	Poison (<1 ppm)

[a]As a fuel for internal reforming; COS—carbonyl sulfide

ΔH^0 value, the *mass specific energy density* (MSE) is calculated using the molar mass (W_M) and the *volume specific energy density* (VSE) using the molar volume (V_M) of the fuel:

$$MSE = \frac{\Delta H^0,\ \text{kJ/mol}}{W_M,\ \text{g/mol}}\quad \text{MJ/kg} \tag{4.1}$$

$$VSE = \frac{\Delta H^0,\ \text{kJ/mol}}{V_M,\ \text{cm}^3/\text{mol}} = MSE\left(\frac{\text{MJ}}{\text{kg}}\right) \times d\left(\frac{\text{kg}}{\text{L}}\right)\quad \text{MJ/L} \tag{4.2}$$

Here d is the density of the fuel (kg/L). Example 4.1 illustrates the calculation of MSE and VSE for hydrazine.

Hydrogen as the lightest element has the highest mass energy density (141 MJ/kg) of all listed fuels. In term of volume specific energy density, however, hydrogen has the lowest value as either liquid or gas. In high-pressure tanks of 700 bar, the technology for fuel cell cars, the compressed hydrogen has a volume specific energy density of 5.6 MJ/L.

Example 4.1 Calculate mass and volume energy density of hydrazine.

Hydrazine was used as rocket fuel starting in World War II under name B-Stoff (hydrazine hydrate) and in a mixture with methanol (M-Stoff) and water called C-Stoff. It is also used as a monopropellant for spacecraft, where the hydrazine, with the help of, e.g., iridium metal catalyst, is decomposed into ammonia, hydrogen, and nitrogen:

$$3N_2H_4 \rightarrow 4NH_3 + N_2 \tag{4.3}$$

$$N_2H_4 \rightarrow N_2 + 2H_2 \tag{4.4}$$

$$4NH_3 + N_2H_4 \rightarrow 3N_2 + 8H_2 \tag{4.5}$$

These reactions are extremely exothermic (the catalytic chamber can reach 800 °C within a few milliseconds) and produce a large volume of hot gas from a small volume of liquid hydrazine. Hydrazine has also been suggested as fuel for alkaline fuel cells:

$$\text{Overall reaction}: \quad N_2H_4 + O_2 = N_2 + 2H_2O \tag{4.6}$$

For the overall reaction above, the standard change of enthalpy and free energy are $\Delta H^0{}_{HHV} = -622$ kJ/mol and $\Delta G^0 = -601$ kJ/mol, respectively. Hydrazine can be used as a liquid in its monohydrate form ($N_2H_4{\cdot}H_2O$). The liquid contains 64% hydrazine and has a density of 0.94 kg/L. Calculate the gravimetric (in MJ/kg) and volumetric (in MJ/L) energy density of the $N_2H_4{\cdot}H_2O$ as a liquid fuel.

Solution:

Using the molar mass of hydrazine 0.032 kg/mol, the MSE of hydrazine can be calculated from the enthalpy change (ΔH^0_{HHV}) for the overall cell reaction

$$\text{MSE} = 622\frac{\text{kJ}}{\text{mol}} \times \frac{1}{0.032\,\text{kg/mol}} = 19,438\,\text{kJ/mol} = 19.4\,\text{MJ/kg} \tag{4.7}$$

1 kg N_2H_4 H_2O contains 0.64 kg N_2H_4, the MSE of $N_2H_4{\cdot}H_2O$ is therefore 12.4 MJ/kg. And the VSE is

$$\text{VSE} = 12.4\frac{\text{MJ}}{\text{kg}} \times 0.94\frac{\text{kg}}{\text{L}} = 11.7\,\text{MJ/L} \tag{4.8}$$

4.2 Hydrogen Production

A main issue in the use of hydrogen is that it does not occur as a free element on the earth. It is not a new form of primary energy but an energy carrier. Therefore, H_2 must be produced from hydrogen-containing resources such as water and hydrocarbons with input of external energy. There are numerous processes for the production of hydrogen, including:

- Hydrocarbon reforming (steam reforming, partial oxidation, or autothermal reaction)
- Electrolysis (low- and high-temperature electrolysis)
- Thermolytic (thermal decomposition, thermolysis, thermochemical cycles)
- Photosynthesis (artificial photocatalysis and photo-electrochemcial catalysis)
- Biological.

Currently, up to 96% of hydrogen used in industry is produced from fossil fuels, including steam reforming of natural gas, cracking of hydrocarbons as well as coal gasification processes, but all these methods consume precious fossil resources, thermal and electrical energy and produce a large amount of CO_2. Thus, hydrogen produced from fossil fuels is not environmentally friendly nor sustainable. In addition, the hydrogen reformat also contains contaminants of CO and H_2S, which must be removed. Pt-based electrocatalysts used in PEMFCs are not tolerant to these impurities and can be poisoned by very low (ppm) levels of CO and H_2S. AFCs are susceptible to carbonation of the electrolyte by CO_2 from the hydrogen reformat as well as the ambient air, the latter containing up to 410 ppm CO_2. Preferential oxidation of CO is an effective method for cleanup of trace CO from the reformate stream.

A more sustainable and clean method is to extract hydrogen from water. Water electrolysis is considered to be one of the potential options to store renewable energy sources such as solar and wind power by using hydrogen as an energy carrier in the framework of energy transition. The production of hydrogen via electrolysis is attractive as the share of renewable energies is steadily increasing.

Figure 4.1 shows the hydrogen supply chain through the reforming of fossil (natural gas, petroleum, coal) fuels and biomass as well as water electrolysis. Natural

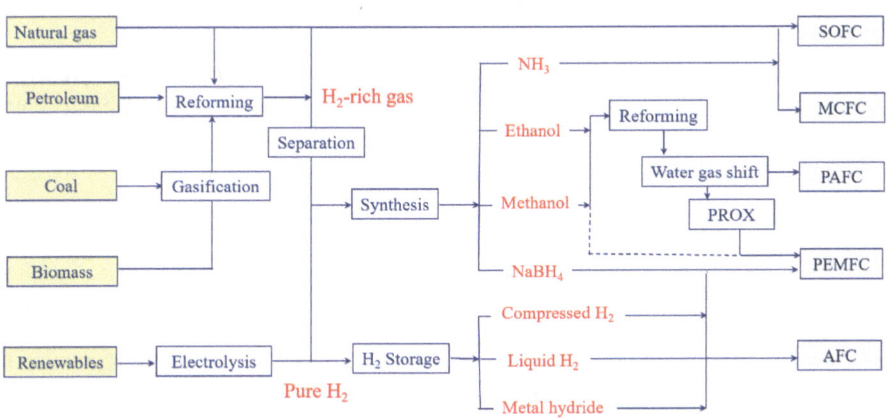

Fig. 4.1 Hydrogen supply chain for fuel cells

gas can be used as fuel for SOFC and MCFC via internal reforming while petroleum, coal, and biomass need to be proceeded by external reforming. Pure hydrogen from water electrolysis can be stored in form of compressed gas, liquid, or in metal hydrides and used in PEMFC and AFC. Alternatively hydrogen can be converted into chemicals by synthesis of, e.g., ammonia, methanol, ethanol, or hydrides. These hydrogen carriers can be directly used in fuel cells or via reforming. Only the hydrogen produced from water and renewable energy such as solar and wind power is considered to be truly green and sustainable. The key issues in the green hydrogen production is the integration of the water electrolyzers with the renewable energies which are of the intermittent nature and therefore demand a dynamic mode of operation of large-scale water electrolyzers.

4.2.1 Steam Reforming

Steam reforming (SR) of fossil fuels like natural gas is the most common and cheapest way to produce hydrogen on a large scale. In steam reforming, the fuel reacts with water that adds hydrogen to the fuel but does not introduce nitrogen into the reformate. This is very different from partial oxidation or autothermal reaction with air as oxidant, which introduces nitrogen and dilutes the hydrogen fuel. Steam reforming of a hydrocarbon can proceed according to:

$$C_nH_m + nH_2O \rightarrow nCO + \left(\frac{m + 2n}{2}\right)H_2 \tag{4.9}$$

The produced CO and H_2 mixture is also called *syngas*. In the case of natural gas, methane first reacts with steam to produce carbon monoxide and hydrogen.

$$CH_4 + H_2O \rightarrow CO + 3H_2 \quad \Delta H_f^0 = +206\ \text{kJ/mol} \tag{4.10}$$

This reaction is strongly endothermic and requires heat input. The above reaction occurs at high temperatures (700–1100 °C) in the presence of metal-based catalysts (e.g., Ni). Additional hydrogen can be produced by the water–gas shift (WGS) reaction at lower temperatures:

$$CO + H_2O \rightarrow CO_2 + H_2 \quad \Delta H_f^0 = +41\ \text{kJ/mol} \tag{4.11}$$

The WGS is also a mildly endothermic reaction, essential to adjust the H/C ratio of syngas. The shift reaction is used to minimize CO, a poison on the ppm level for catalysts of low-temperature fuel cells. It is necessary and a common practice to include extra CO removal units in the fuel cell systems. High-temperature SOFCs and MCFCs do not have this problem. The energy efficiency of hydrogen production from natural gas reforming is approximately 65–85%, which is illustrated in Example 4.2.

Example 4.2 Calculation of energy efficiency for the natural gas reforming.

Solution:

Both steam reforming and water–gas shift reactions are endothermic, requiring in total 247 kJ/mol-CH_4. The high heat value for combustion of methane and hydrogen is 894 and 286 kJ/mol, respectively, i.e.,

$$H_2 + {}^1/_2O_2 \rightarrow H_2O_{(l)} \quad \Delta H^0_{HHV} = -286\ \text{kJ/mol} \tag{4.12}$$

$$CH_4 + O_2 \rightarrow CO_2 + 2H_2O_{(l)} \quad \Delta H^0_{HHV} = -894\ \text{kJ/mol} \tag{4.13}$$

From the basic stoichiometry of the steam reforming reaction and WGS reactions:

$$1\ \text{mol}\ CH_4 \rightarrow 4\ \text{mol}\ H_2 \quad \varepsilon = \frac{4 \times 286}{894} = 129\% \tag{4.14}$$

The molar ratio of hydrogen to methane is 4, giving an energy efficiency of 129%. This indicates that the product of 4 mol H_2 contains 29% more energy than the reactant of 1 mol CH_4. This should, of course, be balanced by the reaction heat supply by burning extra (29%) methane. As a result, the real reforming stoichiometry is:

$$\text{Molar ratio of } H_2 \text{ to } CH_4 = \frac{4}{1 + 0.29} = 3.1 \tag{4.15}$$

That is to say that the real molar ratio of hydrogen to methane is 3.1 instead of 4. The energy efficiency including the excess of methane is therefore:

$$\varepsilon = \frac{4 \times 286}{1.29 \times 894} = 99.2\% \tag{4.16}$$

A set of data for an industrial steam methane reforming (SMR) process is as follows (From John Jechura https://inside.mines.edu/~jjechura/EnergyTech/07_Hydrogen_from_SMR.pdf):

	Basic SMR process	SMR with heat recovery
Energy input		
CH_4 inlet:	89.4 MJ/hr	89.4 MJ/hr
Steam boiler, Reformer, Purification, Compressor, Pump etc	70.5 MJ/hr	51.6 MJ/hr
Energy output		
Product (H_2) out:	108.6 MJ/hr	108.6 MJ/hr

$$\text{Overall efficiency } \varepsilon = \frac{108.6}{89.4 + 70.5} = 68\%; \text{ Overall efficiency with heat recovery}$$

$$\varepsilon = \frac{108.6}{89.4 + 51.6} = 77\% \tag{4.17}$$

In addition to the methane feeding (89.4 MJ/h), energy input also includes steam boiler, maintaining temperature of reformer, compressor, pumps as well as hydrogen purification (by absorbing CO_2 using amine or membrane process), all together being 70.5 MJ/h for the basic process without heat recovery. The heat recovery saves the energy and reduces this part of the energy input to 51.6 MJ/h. Based on the total output in form of hydrogen, the overall energy efficiency without and with heat recovery is calculated to be 68 and 77%, respectively.

4.2.2 Water Electrolysis

Water electrolysis is the process of electrically splitting water into oxygen and hydrogen. The overall water electrolysis can be described by the following equation:

$$H_2O \xrightarrow{\text{Electricity}} H_2 + \frac{1}{2}O_2 \tag{4.18}$$

The overall process is composed of the hydrogen evolution reaction (HER) on the cathode and the oxygen evolution reaction (OER) on the anode of the electrolyzer. Water electrolysis is the reverse process of an H_2-O_2 fuel cell. Water electrolysis produces hydrogen of high purity which is particularly relevant and important to fuel cell applications as platinum electrocatalysts used in low-temperature fuel cells are sensitive to the poisoning of fuel impurities e.g. CO from hydrocarbon reforming.

Due to the high energy demand, the environmental benefits of using electrolysis depend on the methods and electricity source to split the water. Hydrogen production from water electrolysis derived by renewable energy is sustainable and provides an environmentally friendly pathway to contribute toward meeting the constantly growing demand for energy supply and storage. For example, conversion of excess solar power into chemical energy of hydrogen by water electrolysis can be used to store surplus solar energy during peak generation periods. During low generation periods (e.g., the night), the H_2 fuel can then be used to efficiently regenerate electricity via fuel cells. As a general energy carrier based on renewable energies H_2 can also be used as heat supply and industrial chemicals. Figure 4.2 shows a schematic of

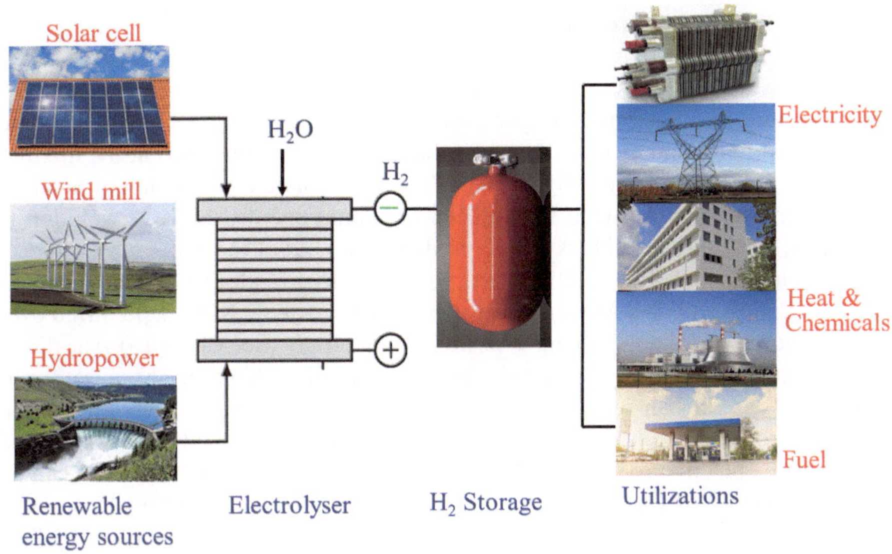

Fig. 4.2 Schematic representation of the energy cycle using water electrolysis to store excess solar electrical energy and fuel cells to provide the electricity during the low peak period

an energy cycle using water electrolysis as the central link of the renewable energy chain where fuel cells are also a key energy conversion technology. As can be seen from the figure, an effective H_2 storage facility is an essential process.

From a technological point of view, water can be electrolyzed by using low-temperature electrolyzers such as alkaline electrolysis cells (AECs), PEM electrolysis cells (PEMECs), and high-temperature solid oxide electrolysis cells (SOECs). Major industrial electrolyzers are based on alkaline electrolytes, i.e., aqueous solutions of potassium hydroxide. Alkaline electrolyzers are available on MW scales suitable to operate with a fairly constant power supply with well-demonstrated long lifetime. To fit renewable sources of the intermittent nature, PEM electrolyzers have a much better dynamic performance in terms of startup/shutdown and load cycling. Both alkaline and PEM electrolyzers operate in the low temperature range and therefore greatly constrained by the sluggish OER kinetics.

As the splitting of water is an endothermic reaction, electric demand and decomposition voltage decrease with the increase of temperature. Therefore, it is beneficial to elevate the temperature of water electrolysis process. Overpotentials and ohmic voltage drops are also decreased considerably at high operating temperatures. High-temperature solid oxide electrolysis cells (SOECs) can reach an energy efficiency of ~90% [1]. Compared with alkaline and PEM electrolyzers, SOECs are still in the early stage of development. However, it is a promising technology for large-scale hydrogen production and attracts wide research interests.

4.2.2.1 Alkaline Electrolysis Cells

Alkaline electrolysis cells (AECs) are a mature technology and have been used for over a century in industrial hydrogen production. In AECs, anode and cathode are immersed in the liquid electrolyte, and a diaphragm is used to prevent the mixture of evolved H_2 and O_2 gases. Figure 4.3a shows schematic construction of an alkaline electrolysis cell. Through both electrode compartments, the KOH solution is circulated, carrying the bubbles of two product gases out of the cell.

In commercial alkaline electrolyzers, potassium hydroxide (KOH) with concentration of 25–30 wt% (~6 M) is generally used as electrolyte. KOH is preferred over sodium hydroxide (NaOH) owing to its high conductivity (See Chap. 13). The direct contact with air must be avoided during the electrolyte handling because atmospheric CO_2 in air can react with hydroxide to form solid carbonates. Hydrogen produced from KOH electrolyzers contains fine droplets of KOH and water, and steps must be taken to remove it. After drying, high-purity H_2 (99.5–99.9%) can be obtained. Current AECs operate at a cell voltage of 1.8–2.2 V at current densities of 100–300 mA/cm^2 and temperature of 70–90 °C. The power consumption of AECs depends on the cell scale, or in other words, on the hydrogen generation rate. At a nominal production rate above 10 Nm^3 H_2 per hour, the energy consumption is around 4.5 kWh/Nm^3-H_2, corresponding to a value of 50 kWh/kg-H_2. The high heat value of hydrogen is 142 MJ/kg, which is equal to 39.4 kWh/kg-H_2. This energy consumption for an alkaline electrolyzer is therefore corresponding to an energy efficiency of 79% based on the HHV. On top of this, a pressurized electrolyzers need

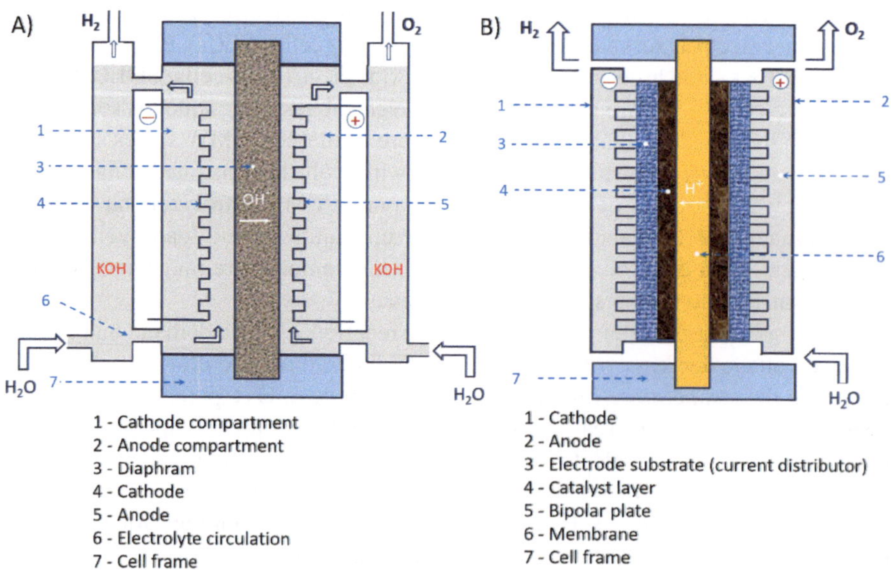

Fig. 4.3 Schematic of **a** alkaline and **b** proton exchange membrane electrolyzer cell constructions

Table 4.3 Specifications and operational parameters of alkaline and PEM electrolyzers. Data were taken from Ref. [2] with permission from Elsevier, Copyright 2013

	Alkaline electrolyzers	PEM electrolyzers
Cell temperature (°C)	60–80	60–80
Cell pressure (bar)	< 30	< 30
Current density (A/cm^2)	0.2–0.4	0.6–2.0
Cell voltage (V)	1.8–2.4	1.8–2.2
Power density (W/cm^2)	< 1	< 4.4
Voltage efficiency (%, HHV)	62–82	67–82
Stack specific energy consumption (kW h N/m^3)	4.2–5.9	4.2–5.6
System specific energy consumption (kW h N/m^3)	4.5–7.0	4.5–7.5
Lower partial load change (%)	20–40	0–10
Cell area (m^2)	> 4	> 0.03
Stack/system H_2 production rate (Nm^3/h)	< 760	< 10
Stack lifetime (h)	< 90,000	< 20,000
System lifetime (yr)	20–30	10–20
Degradation rate (μV/h)	< 3	< 14

a slightly higher power demand. A set of specifications and operational parameters of AECs are listed in Table 4.3.

The characteristics of AEC are summarized as follows where plus signs (+) indicate advantages and minus signs (−) indicate disadvantages:

- Traditional and mutual technology with a long history of commercialization (+)
- Non-noble electrode and construction materials and hence low cost (+)
- Long-term stability (+)
- Large scale in MW size (+)
- Operational pressures of up to 1–32 bar (+)
- Low current density and hence bulky construction (−)
- Gas crossover and hence hydrogen purity issues (−)
- Low partial load range (operational at 20–40% full load) and poor dynamics of operation (start-up/shutdown and load cycling) (−)
- Circulation of highly corrosive electrolyte (−).

Electrochemically an AEC involves the anodic oxygen evolution reaction (OER) and cathodic hydrogen evolution reaction (HER), as given by following equations.

For cathodic HER:

$$2H_2O + 2e^- \rightarrow 2OH^- + H_2 \quad E^0 = -0.828\,V \tag{4.19}$$

For anodic OER:

$$2OH^- \rightarrow \frac{1}{2}O_2 + H_2O + 2e^- \quad E^0 = +0.401\,V \tag{4.20}$$

The overall electrolysis reaction:

$$H_2O \rightarrow H_2 + \frac{1}{2}O_2 \quad E^0 = -1.229\,V \tag{4.21}$$

Based on Eq. (2.73), the change of Gibbs free energy of reaction (4.21) $\Delta G > 0$, indicating that the electrolysis process cannot take place spontaneously and requires an electrical energy supply. The low operating current density and sensitivity to differential pressures are the main limitation of the alkaline electrolyzer technology. The major material issues related to alkaline electrolyzers are the development of advanced diaphragms and electrocatalysts.

The most common cathode and anode materials are Ni and Ni-based oxides or compounds due to their reasonably high activity, low price, and super stability in alkaline electrolytes. Significant efforts have been made in the development of new electrocatalysts. In comparison with HER, the OER requires a high activation or overpotential (η) to overcome the four-electron sluggish kinetics processes, significantly limiting the efficiency of the technology [3]. Many different earth-abundant transition-metal-based materials have been reported to be active electrocatalysts in alkaline media. For example, metal oxides, hydroxides, phosphates, borates, and perovskites have been investigated for OER and HER. The kinetics of OER are related to the chemisorption and dissociation steps of OH^- and the intermediates (OH^*, OOH^*) on the surface of the catalysts, while the HER is kinetically dominated by the equilibrium coverage between the decomposition of trapped H_2O and the subsequent chemisorption of the dissociated intermediates (OH^- and H^*). Therefore, the challenge for the development of effective and/or bifunctional electrocatalysts for the efficient overall water electrolysis is the stringent requirement of the catalysts which can accelerate the chemisorption, dissociation, or the transformation of intermediates for both OER and HER.

One strategy is to develop electrocatalysts with heterointerfaces, for example, heterostructured $Ni(OH)_2/NiS_2$, $Ni(OH)_2/Ni_3S_2$, and Co_9S_8/Ni_3S_2 nanotubes or nano-rods [4, 5]. Defects in the surfaces or interfaces of the electrocatalysts induce active sites. The Ni_3S_2 usually occurs as the mineral heazlewoodite and can be easily synthesized. On the other hand, $Ni(OH)_2$ has the undercoordinated metal sites on the surface with unfilled *d* orbitals that possesses superior chemisorption of OH^- and oxygen-containing intermediates for water oxidation. In this case, $Ni(OH)_2/Ni_3S_2$ heterostructures are directly grown on Ni foam via a two-step hydrothermal treatments, forming $Ni(OH)_2/Ni_3S_2$/NF electrode (see Fig. 4.4a) [5]. The $Ni(OH)_2/Ni_3S_2$/NF is characterized by both needle-shaped and tree-like particles and flower-like nano-sheets (Fig. 4.4b) and defect-rich interface (Fig. 4.4c). The Ni foam substrates as supports of the electrocatalysts have advantages of high electronic conductivity, strong mechanical properties, large macropores, open space, and high flexibility. The as-synthesized $Ni(OH)_2/Ni_3S_2$/NF is highly active and durable for overall water electrolysis by combining the advantages of electrocatalytic activity of defect-activated and synergistic effects of heterostructured $Ni(OH)_2/Ni_3S_2$ interface and robustness of Ni foam. An alkaline electrolysis cell assembled by the $Ni(OH)_2/Ni_3S_2$/NF bifunctional electrodes, reached 10 and 100 mA/cm^2 at a

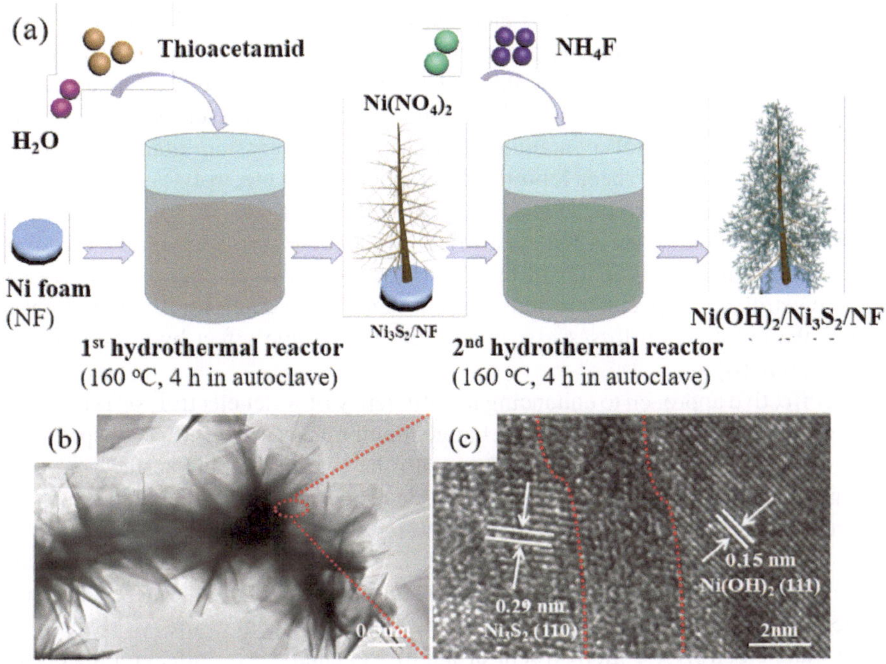

Fig. 4.4 **a)** Synthesis process of heterostructured $Ni(OH)_2/Ni_3S_2$ electrocatalysts grown on nickel form, forming $Ni(OH)_2/Ni_3S_2$/NF electrode and **b, c)** HRTEM micrograph of $Ni(OH)_2/Ni_3S_2$/NF electrodes and the heterointerface region between $Ni(OH)_2$ and Ni_3S_2 on NF (indicated by dotted line). Modified from Ref. [5] with permission from American Chemical Society, Copyright 2019

low overall cell voltages of 1.53 and 1.80 V, respectively. The cell with bifunctional $Ni(OH)_2/Ni_3S_2$/NF electrodes shows also a high stability under alkaline water splitting conditions at 100 mA/cm^2 for over hundred hours [5].

There are other strategies to increase the electrocatalytic activity of transition metal catalysts. Introducing surface defects such as oxygen vacancies on the surface of electrocatalysts is also effective to enhance the electrocatalytic activity for water electrolysis. Oxygen vacancy defects can be induced by plasma treatment, $NaBH_4$ reduction or by intercalation and removal processes [5]. Oxygen vacancy defects enhance the electronic conductivity, surface wettability, and H_2O adsorption/activation ability of the electrocatalysts.

The diaphragm is a key component of the cell. Traditionally, the diaphragm was made of asbestos in a thickness of a few millimeters. Under industrial working conditions in KOH aqueous solutions around 80 °C, this asbestos material is carcinogenic and its use is today forbidden in more than 60 countries. Alternative materials have to be ion (hydroxide) conductive and stable in a caustic environment at elevated temperatures (80 °C) and pressures. Microporous polymer matrix, often reinforced with glass fibers, is typically used, e.g., polysulfone or polyphenylene sulfide. Commercial diaphragm materials include Zirfon®, which is typically composed of a polysulfone matrix containing up to 85% ZrO_2 nano-powders in a thickness of around 500 μm. This material is porous, having a porosity of about 60% with the pore size in the submicron range. When filled by the KOH electrolyte, it exhibits an area-specific resistance of 0.1 Ωcm^2 at 80 °C and can tolerate a bubble pressure as high as 4 bar. Other composite materials containing nickel oxide or potassium titanate supported on nickel mesh have also been reported. In practical liquid electrolyte cells, hydrogen and oxygen gases are produced at active sites of the electrode surface and are supersaturated to form bubbles in the electrolyte. The bubble formation on the electrode surface and bubble dispersion in the electrolyte increase energy consumption of water electrolysis, especially at high current densities. Therefore, it is important to reduce the polarization losses associated with the bubble effect by microstructure optimization [6].

An effective approach to enhancing the efficiency of water electrolysis is to design zero-gap electrolyzers with anion exchange membranes (AEMs) (See Chap. 13). In such cells, the electrodes are in contact with the two membrane surfaces so as to minimize the voltage (IR) drop between the electrodes. The electrodes can be applied directly onto the membrane surfaces to form a membrane electrode assembly (MEA), similar to that in the fuel cell technology. The challenge is the performance and chemical stability of AEMs in hot alkaline solutions, though rapid progress has been made in recent years in polymer chemistry and enhancement of the AEM stability.

In water electrolysis, microstructural stability of catalysts is critical for the performance and efficiency of the devices. The catalytic particle agglomeration and detachment are generally related to weak catalyst support (e.g., nanostructured electrocatalysts) interaction and corrosion of the supporting materials in alkaline solutions. However, catalyst deterioration may depend on the surface states of the catalysts. In HER, the reductive environment could convert the catalyst surface of transition metals such as Pt, Ru, and Ir to a metallic state, while other metals could be dissolved as metal hydroxide ions in alkaline media. In the case of OER, the oxidative conditions could transfer the catalyst surface into MOH, MO, and MOOH, which could disintegrate into metal oxide complexes in alkaline media. These are the issues which need to be considered in the development of active and durable electrocatalysts for water electrolysis technologies.

4.2.2.2 Proton Exchange Membrane Electrolysis Cells

Proton exchange membrane electrolysis cell (PEMEC) represents a system that incorporates a solid proton-conducting membrane as electrolyte. The membrane serves the purpose of a proton conductor and a gas separator. The perfluorosulfonic acid type, e.g., Nafion® membrane, is the most common membrane electrolyte for PEMECs. Unlike AECs, there is no requirement for further purification of hydrogen in PEM electrolysis. Porous catalysts layers are directly attached on either side of the membrane, i.e., in a zero-gap configuration. Due to the low ionic resistance, PEM electrolysis can achieve high currents of 1–2 A/cm^2 at a cell voltage of 1.8–1.9 V, which corresponds to an energy efficiency of 78–82%. A set of specifications and operational parameters are listed in Table 4.3. PEMEC technologies are characterized by the following where plus signs (+) indicate advantages and minus signs (−) indicate disadvantages:

- Wide operational range of current densities (+)
- High voltage efficiency (+)
- Good partial load range (operational at 0–10% full load) (+)
- Good dynamic operation (start-up/shutdown) and rapid system response to load changes (+)
- Compact system design a possibly high operational pressure (+)
- High hydrogen gas purity (+)
- Noble metal catalysts and high cost of other components (−)
- Highly corrosive environment (−)
- Small-to-medium scales from kW to tens of kW (−)
- Still under development and long-term stability to be demonstrated (−).

Electrode reactions occurring during PEM electrolysis can be written as:
For cathodic HER:

$$2H^+ + 2e^- \rightarrow H_2 \quad E^0 = 0.0\ \text{V} \tag{4.22}$$

For anodic OER:

$$H_2O \rightarrow \frac{1}{2}O_2 + 2H^+ + 2e^- \quad E^0 = +1.229\ \text{V} \tag{4.23}$$

The overall water electrolysis reaction is the same as that occurring in alkaline solution, Eq. (4.21). Figure 4.3b shows the construction of PEM electrolysis cell. PEM water electrolyzers typically employ precious group metals (PGM), e.g., Pt or Pt/Pd alloys as the HER catalyst at the cathode. The state-of-the-art electrocatalyst for OER is IrO_2 or IrO_2/RuO_2 mixture oxides with typical mass loadings in the range of up to 2 mg/cm^2 [7]. Iridium as a PGM belongs to the scarcest elements on the earth with an annual global supply of only 7 tons.

The HER in water electrolysis involves the adsorbed hydrogen intermediate, H_{ad}, regardless the acidic or alkaline solutions. Therefore, the H_{ad} adsorption energy, which describes the bond strength between H and metal-active center, and the Gibbs free energy of H_{ad} are important for the electrocatalytic activity of the catalysts for HER. In general, the HER activity increases as the Gibbs free energy of H_{ad} decreases. Based on the density functional theory (DFT) calculations, among transitional metal catalysts, Pt-based catalysts show the highest activity for HER [8]. However, for cost-effective and stable electrocatalysts in acid media, there are only a few materials meeting the criterion, e.g., binary metallic ceramics such as transition metal phosphides and sulfides. The limited resource of PGM based catalysts is a significant hurdle to the large scale (e.g., MW) installation and implementation of PEMEC technologies for hydrogen production. Due to the use of expensive proton exchange membranes and PGM catalysts, material costs are high for PEMECs as compared to AECs. Therefore, significant effort is required to develop more efficient and cost-effective electrocatalysts [2].

4.2.2.3 Solid Oxide Electrolysis Cells

Water electrolysis can be accomplished via a high-temperature process in a solid oxide electrolysis cell (SOEC). The high-temperature operation of SOECs can significantly reduce the electrical energy demand as compared to low-temperature alkaline and PEM electrolyzers. SOECs can, in fact, operate at the thermoneutral point, i.e., at an energy efficiency of 100% (See Chap. 2). Higher than 100% energy efficiency is, in principle, also possible when an SOEC is integrated with additional high-temperature energy sources such as nuclear reactors, combustors, or solar power.

SOECs are essentially developed based on the existing SOFC materials [1, 9]. The most common electrolyte is the oxide ion conducting yttria-stabilized zirconia (YSZ). Sr-doped $LaMnO_3$ (LSM) or mixed ionic and electronic conductor $La_{0.6}Sr_{0.4}Co_{0.2}Fe_{0.8}O_{3-\delta}$ (LSCF) perovskite oxides are used as the oxygen anode, while Ni-based cermets are commonly used as the hydrogen cathode.

In the case of oxide ion conducting electrolytes, the cell reactions are written as follow:

For cathodic HER:

$$H_2O \rightarrow H_2 + O^{=} + 2e^- \tag{4.24}$$

For anodic OER:

$$O^{=} + 2e^- \rightarrow \frac{1}{2}O_2 \tag{4.25}$$

It is noted that the solid oxide electrolyte can also be proton-conducting oxides such as $BaCe_{0.5}Zr_{0.3}Y_{0.2}O_{3-\delta}$ and $SrZr_{0.9}Yb_{0.1}O_{3-\delta}$. In this case, the electrode reactions of water electrolysis are the same as that of PEM electrolysis processes of Eqs. (4.22) and (4.23). The doped ceria-based electrolyte is not suitable for SOECs due to its high electronic conductivity, as it leads to current leaking and loss of electrical efficiency.

Electrochemical activity of Ni-based cermet hydrogen electrodes is dependent on the Ni content, operating temperature, water concentration, and dc bias (overpotential) for the HER. Figure 4.5 shows an example of impedance spectroscopy curves of HER and hydrogen oxidation reaction (HOR) on Ni/YSZ cermet electrodes, measured at 800 °C in 50% H_2O/50% H_2 as a function of dc bias [10]. Similar to HOR on the cermet anodes under SOFC operation conditions, the impedance behavior for HER on the Ni/YSZ electrode is characterized by two separable impedance arcs at high and low frequencies, indicating that HER is controlled by at least two rate limiting processes (for detailed analysis of impedance and equivalent circuit in SOFCs, refer to Chap. 11). However, very different from HOR under the SOFC

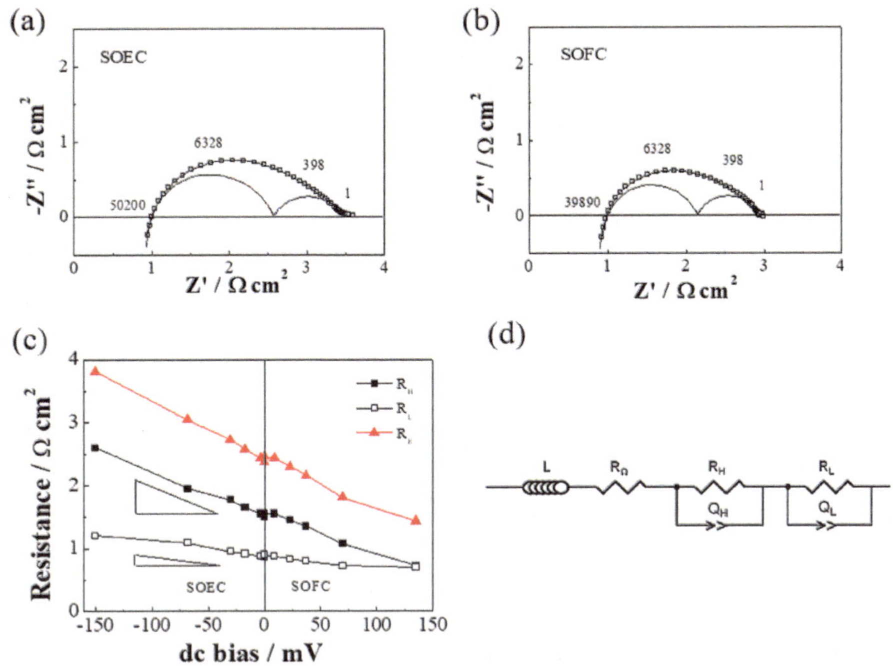

Fig. 4.5 Impedance curves for the reaction on a Ni/YSZ (76/24%) hydrogen electrode measured under **a** SOEC mode at dc bias of −30 mV and **b** SOFC mode at dc bias of 37 mV. Symbols are experimental data, and lines are the fitted results. Plots of R_E (=R_H+R_L), R_H, and R_L as a function of dc bias measured at 800 °C under 50%H_2O/50%H_2 are given in (**c**). The equivalent circuit for fitting is shown in (**d**). R_H: solid square, R_L: empty square, R_E: solid triangle. Reproduced from Ref. [10] under Creative Commons CC BY license

operation mode, for HER, both polarization resistances associated with high- and low-frequency arcs, R_H and R_L increases with the applied dc bias. This implies that the reaction mechanism and kinetics of HER would be very different from that of HOR. The significant increase in R_H with the dc bias indicates the increased reaction barrier of the electrode process under the polarization, most likely due to insufficient reactant supply prior to the charge transfer steps. The slow supply of reactants for HER is most likely related to the slow adsorption and dissociation of the water molecules on the surface of Ni/YSZ cermet electrodes.

$$H_2O_{(g)} + O - S_m \xrightarrow{\text{slow}} H_2O_{ad} - O - S_m \tag{4.26}$$

Here O-S_m represents oxygen-covered Ni or YSZ surface site or active metal surface site adjacent to an adsorbed oxygen, O_{ad}, for the dissociation of H_2O or H_2. In comparison, the reverse desorption of water molecules is a fast process. This is due to the fact that gaseous H_2O molecules are thermodynamically more stable than the one adsorbed on Ni/YSZ surface.

$$H_2O_{ad} - O - S_m \xrightarrow{\text{fast}} H_2O_{(g)} + O - S_m \quad (4.27)$$

This is followed by the diffusion of adsorbed water molecules to the three-phase boundary (TPB) region, and the H_2O_{ad} at TPB is directly dissociated into H_2 and O or OH species by charge transfer steps:

$$H_2O_{ad} - O - S_m \rightarrow H_2O_{ad,TPB} \quad (4.28)$$

$$H_2O_{ad,TPB} + 2e^- + V_O^{\bullet\bullet} \rightarrow H_2 + O_{YSZ}^x \quad (4.29)$$

or

$$H_2O_{ad,TPB} \rightarrow OH_{YSZ}^- + H_{Ni}^+ \quad (4.30)$$

$$2OH_{YSZ}^- + 2e^- + 2V_O^{\bullet\bullet} \rightarrow 2O_{YSZ}^x + 2H_{Ni}^+ \quad (4.31)$$

$$H_{Ni}^+ + H_{Ni}^+ \rightarrow H_2 \quad (4.32)$$

The electrode process associated with the low-frequency arc could be the water adsorption and diffusion on oxygen-covered Ni or YSZ surface, i.e., reactions (4.27) and (4.28), while the one at high-frequency arc is related to the charge transfer steps of reaction (4.29) or (4.31).

SOECs generally experience a relatively higher performance degradation in comparison with their counterpart operated under SOFC operation conditions, which is the most critical issue in the development of reliable and high performance SOECs for water electrolysis [11]. The main degradation phenomena as reported in the literatures can be summarized as following:

- YSZ electrolyte degradation at the interface close to the oxygen and hydrogen electrodes, characterized by grain boundary widening/coarsening, intergranular fracturing, void formation;
- Ni agglomeration, dense Ni-YSZ layer within the Ni-YSZ-based hydrogen electrode;
- Delamination, disintegration of LSM particles at the interface region, and phase change and $SrZrO_3$ formation at the GDC porous barrier layer/YSZ interface in the case of LSCF oxygen electrodes;
- Deposition and poisoning of contaminants such as chromium from Fe–Cr metallic alloys and boron from borosilicate glass sealant on oxygen electrodes, and carbon deposition and delamination of hydrogen electrodes in the case of CO_2 electrolysis.

Delamination at the oxygen electrode/electrolyte interface is fundamentally related to the ability of electrolyte and oxygen electrode to release oxygen. Thus, improving the ionic conductivity of the electrolyte and in particular oxygen electrode can in principle lead to the stabilization of the electrode/electrolyte interface and reduce the tendency of the delamination. Use of nanostructured and/or MIEC-abased oxygen electrodes have been shown to substantially enhance the stability of SOECs.

Hydrogen can also be produced from photocatalytic water splitting, solar thermochemical process, biological synthesis, and from many types of biomass such as agricultural and animal waste using pyrolysis and gasification processes. All these processes are carbon–neutral, but the technologies are not mature and require further development for production of low-cost hydrogen.

4.3 Hydrogen Storage

Hydrogen has a high mass but low volume specific energy density. A key issue for the vision of a future hydrogen-based energy system is therefore an effective way of hydrogen storage, and it is in particular true for the onboard hydrogen storage in automobiles. A number of techniques have been suggested including.

Physical storage:

- Compressed gaseous hydrogen (CGH_2)
- Liquid hydrogen (LH_2)
- Adsorption on porous materials.

Metal Hydrides:

- Metal hydrides
- Amine, amides, and imides
- Hydrogen carbons and alcohols.

4.3.1 Physical Storage of Hydrogen

From physical to chemical storage the release of hydrogen and its energy demand is gradually increasing. The physical storage technologies, in particular CGH_2 and LH_2, are most mature. The prototypes of fuel cell cars typically use CGH_2. Compression is straightforward to increase the density of the gas and its energy content. To achieve vehicles with a driving range of 500–600 km, 4–5 kg H_2 is needed to store on board. Due to the limited space of vehicles, a working pressure ranging from 350–700 bar is a common practice. At pressures greater than 70 MPa, materials for the hydrogen tank are not readily available and the deviation from the ideal gas behavior and therefore the energy demand become larger. It has been estimated that compression to 700 bar consumes about 15% of the LHV of the store hydrogen. From a material point of view, cylinders using high-strength carbon fiber composites without or with metal liners have been developed to reach the pressure of 700 bar at which the mass storage density of 5.7 wt% H_2 and volumetric energy density up to 4.90 MJ/L have been reported.

Hydrogen can be liquefied by cooling to its boiling point at 20 K. Maintaining hydrogen at such a low temperature is a challenge of the cryogenic hydrogen storage, as a result, liquid hydrogen requires well-insulated and expensive cryogenic storage vessels to prevent boil-off at temperatures below to 20 K. Liquefaction is both time and energy consuming. The average heat capacity of hydrogen for the cooling interval is about 28.5 J/mol K, and the heat of condensation at 20 K is 892 J/mol. The liquefaction is carried out via a series of processes, and the energy consumption in a modern plant ranges up to 30–40% of the hydrogen LHV. The LH_2 does not seem a practical approach of hydrogen storage for fuel cell vehicles but can be a future option for, for example, commercial aircrafts where a high volume storage with high density is required and boil-off is less of a concern.

Physisorption of hydrogen on porous materials has also been proposed as a storage method at near ambient temperatures and under safe pressures. These porous materials include carbon, boron nitride, silica, alumina, and metal organic framework as well as polymer. Structured materials, e.g., carbon nanotubes and carbon nanofibers, have been the subject of intensive research with very much varied hydrogen-storage capacities in literature. Lower temperatures and higher gas pressures favor the adsorption. At -196 °C (liquid N_2) under high pressures, as high as 5 wt%, H_2 can be adsorbed on rigid porous carbons and even higher for porous metal organic framework materials. The amount of hydrogen adsorbed at ambient temperatures is, however, too small to consider for any practical uses.

4.3.2 *Reversible and Irreversible Hydrides*

Hydrogen forms solid metal hydrides with some metals and alloys. For the purpose of hydrogen storage, the hydrogen hydride should have high hydrogen capacity per unit mass and unit volume. In term of the number of atoms per unit volume, metal hydrides have higher hydrogen-storage densities (e.g., MgH_2) than ambient pressure hydrogen gas. Hydrogen storage in metal hydrides is hence a volume efficient method for onboard vehicle applications.

A hydride can be an interstitial compound where the hydrogen atoms sit in an interstitial hole in a host metal lattice. A hydride can also be chemical compound where chemical bonds are fully formed between hydrogen and the host metal. The interstitial hydrides are generally reversible to adsorb and release hydrogen under moderate pressure and temperature

$$M + \frac{x}{2}H_2 \rightleftharpoons MH_x + \text{heat} \tag{4.33}$$

Heat must be removed during the adsorption of hydrogen but has to be added to release hydrogen. Large number of metal hydrides are available, including magnesium, titanium, zirconium, yttrium, lanthanum, palladium, etc. However, current metal hydrides have relatively low mass storage for hydrogen. This ranges from 1 to 2 wt% for materials such as $CaNi_5H_4$ at room temperature and 3.6 wt% for magnesium-based materials such as Mg_2NiH_4 which has a hydrogen desorption temperature above 300 °C. The most studied interstitial hydrides include MgH_2, Mg_2NiH_4, and $NaAlH_4$, and some properties are listed in Table 4.4. The challenges in metal hydride materials are the low gravimetric density and high-temperature operation required to release hydrogen.

Certain chemicals contain significant amount of hydrogen that can be recovered and thus can be considered as prospective candidates for hydrogen storage. Examples are liquid ammonia (NH_3), hydrazine hydrate ($N_2H_4{\cdot}H_2O$), and ammonia borate (NH_3BH_3). Among them, NH_3BH_3 is well known with a very high hydrogen content, containing 19.6 wt% of H_2 and 150 g H_2/L by weight and volume, respectively. This compound is thermally stable in the solid state at ambient temperatures, and upon heating, hydrogen is gradually released in a three-step reaction:

$$\text{Step 1}: \quad NH_3BH_{3(s)} \xrightarrow{383K} NH_2BH_{2(s)} + H_{2(g)} \tag{4.34}$$

$$\text{Step 2}: \quad NH_2BH_{2(s)} \xrightarrow{423K} NHBH_{(s)} + H_{2(g)} \tag{4.35}$$

$$\text{Step 3}: \quad NHBH_{(s)} \xrightarrow{1673K} NB_{(s)} + H_{2(g)} \tag{4.36}$$

Apparently, releasing the last equivalent of H_2 in Step 3 requires rather high temperature, and thus is not considered practical for hydrogen-storage applications. In practice, their use may present safety and operational difficulties such as slow kinetics of dehydrogenation and formation of other volatile by-products, especially when the hydrogen is intended for use in vehicles.

Hydrogen can also be stored chemically in the form of so-called complex chemical hydrides such as alanates $NaAlH_4$ and $NaAlH_6$ and sodium borohydride ($NaBH_4$). Thermal decomposition of $NaAlH_4$ to discharge hydrogen takes place at temperatures

Table 4.4 Reversible capacity, heat, and temperature of desorption of some selected interstitial hydrides. Data were taken from Ref. [12] with permission from Elsevier, Copyright 2007

Metal hydride	Reversible capacity ($wt\%H_2$)	Heat of desorption (kJ/mol H_2)	Desorption temperature at 1 bar (°C)
Interstitial hydrides	1–2	ca. 30 (ca. 12.4% LHV)	Room temperature
MgH_2	7.6	74.5 (30.8% LHV)	300
Mg_2NiH_4	3.6	64.5 (26.7% LHV)	255
$NaAlH_4$ (1st step)	3.7	37 (15.3% LHV)	150
Na_3AlH_6	1.9	47 (19.4% LHV)	110

up to 180 °C, while $NaBH_4$ is thermally stable up to 400 °C. Chemically $NaBH_4$ is stable in an alkaline aqueous solution. However, $NaBH_4$ reacts with water with help of catalysts to release hydrogen:

$$NaBH_4 + 2H_2O \rightarrow NaBO_2 + 4H_2 \quad \Delta H = -212\ \mathrm{kJ/mol} \tag{4.37}$$

The hydrogen-storage capacity of this hydride is 21.2 wt%, but its practical capacity is much lower due to the need of water. The above reaction is exothermic with the enthalpy of −53 kJ/mol H_2 (22% of LHV).

These chemical compounds of hydrides are irreversible in a sense that hydrogen, after being released, cannot be recharged and must be regenerated through other chemical pathways. This means that a much higher energy demand is needed, similar to other hydrogen carriers such as hydrocarbons, alcohols, and other liquid fuels, as to be discussed in the following sections.

4.4 Methanol, Ethanol, Formic Acid, and Other Liquid Fuels

The direct use of liquid fuels in fuel cells is of significant importance due to the potentially higher energy density and higher thermodynamic efficiencies. Liquid fuels such as methanol and ethanol have several advantages with respect to hydrogen. Compared to hydrogen, the energy density of liquid fuels is significantly higher. For example, the energy density of methanol and ethanol is 17.84 and 23.27 MJ/L, respectively, much higher than that of hydrogen under 200 bar which is 1.9 MJ/L. Liquid fuels have more flexible resources and are easier in transportation and storage [13]. In addition, liquid alcohol fuels are less expensive, low toxicity, and can be produced from renewable energy sources. Ethanol can be produced by fermentation of sugars, while methanol can be produced from wood, agriculture wastes, and other biomass by thermochemical processes [14].

Fuel cells based on liquid alcohol fuels are commonly called direct alcohol fuel cells (DAFCs). The challenge in the practical and wide application of DAFCs is the very low electrochemical activity of catalysts for oxidation of liquid alcohols such as methanol and ethanol. There have been significant research efforts in the development of non-precious metal catalysts such as Ni and Fe for the electrochemical oxidation of alcohols. Nevertheless, their activities are generally too low for practical uses in DAFCs. The most effective and state-of-the art electrocatalysts in DAFCs are precious group metal (PGM)-based including Pt, Ru, Pd, and Au [15]. Alloys of PGM electrocatalysts such as PtRu and PtSn show a high stability, fast kinetics, and high tolerance toward the CO poisoning. High catalyst loading is needed to produce the reasonable power output of DAFCs, leading to the high cost of the technology. Very different from the HOR, the electrochemical oxidation reaction of alcohol fuels involves tranfer of multiple electrons. Thus environmental effects of intermediate products of the liquid fuel oxidation reactions should also be taken into account in design and development of DAFC systems [16]. Due to the flammability, toxicity, and also the high crossover, liquid fuels are often used in solutions with water as a natural solvent.

4.4.1 Electrochemical Oxidation of Alcohols

4.4.1.1 Methanol Oxidation Reaction

Methanol is the simplest alcohol with chemical formula CH_3OH (often abbreviated MeOH). It is a light, colorless, volatile, and flammable liquid. The fuel cell using methanol fuel is also called direct methanol fuel cells (DMFCs). The complete methanol oxidation reaction (MOR) in acidic and alkaline electrolytes is a six-electron process which proceeds as follows:

$$\text{Acid:} \quad CH_3OH + H_2O \rightarrow CO_2 + 6H^+ + 6e^- \quad E^0 = +0.02\text{ V} \tag{4.38}$$

$$\text{Alkaline:} \quad CH_3OH + 6OH^- \rightarrow CO_2 + 5H_2O + 6e^- \quad E^0 = -0.81\text{ V} \tag{4.39}$$

It should be noted that the presence of CO_2 in the anode reaction product prevents the use of alkaline electrolye since the chemical reaction between CO_2 and hydroxide leads to precipitation of solid metal carboonates. The recently developed anion echange membrane is an exception as the organic cationic sites are fixed in the polymer backbones (See Chap. 13). The MOR involving tranfer of six electrons always takes place via multiple steps with the formation of intermediates, either through a direct pathway or an indirect pathway (i.e., CO pathway). In MOR, the adsorption of MeOH molecules on the catalyst surface is the first step, followed by the dehydrogenation with cleavage of the C–H bond and/or dehydrogenation with abstract of acidic hydrogen and formation of adsorbed methoxyl CH_3O [17]. For MOR occurring through the indirect pathway, the cleavage of C–H bond produces hydroxymethyl, $(CH_2O)_{ads}$ and ultimately, CO_{ads}, which poisons the Pt-based electrocatalysts. In the case of the direct pathway, the cleavage of the O–H bond leads to the CH_3O_{ads} formation and the dehydrogenation of CH_3O_{ads} produces formaldehyde which can be dissolved in solution or be further oxidized. Both CO and format are stable intermediates of MOR on Pt-based electrocatalysts. The formation of stable reaction intermediates greatly affects the efficiency and kinetics of MOR. The power performance of DAFCs based on methanol is also significantly affected by the methanol crossover from anode to cathode, which reduces the open circuit voltage of the cells and poisons the electrocatalytic activities of Pt-based catalysts for ORR.

In MOR, CO_{ads} is a stable species at low temperatures and potentials and poisons the catalyst by occupying the active sites. One of the solutions to enhance the CO-tolerance of Pt electrocatalysts is to use Pt alloys with, e.g., Ru. The PtRu alloy promotes the MOR via the bifunctional mechanism, *i.e.*, the adsorption of methanol with dehydrogenation on Pt atoms and the oxidative desorption of CO_{ads} via its reaction with OH_{ads} groups that appear on neighboring Ru atoms [18]. The adsorption of hydroxyls on the oxophilic Ru facilitates the removal of CO species adsorbed on the Pt and thus releases the occupied active sites. However, the electrochemical activities of PtRu critically depend on the size and distribution of the catalyst nanoparticles. The activity of Pt-based nanoparticles is also significantly affected by the nature of their interaction with the carbon-based supports such as carbon nanotubes (CNTs).

Figure 4.6 shows typical dependence of the electrocatalytic activities of the PtRu catalysts for MOR on distribution and size of PtRu nanoparticles, measured in 0.5 M H_2SO_4 + 1.0 M CH_3OH solution. In the figure, PtRu/1-AP-MWCNTs were synthesized on 1-AP (aminopyrene)-functionalized multiwalled CNTs (MWCNTs) by microwave-assisted polyol process, Pt/AO-MWCNTs were synthesized on acid-treated CNTs and PtRu/C was carbon black (XC-72) supported PtRu nanoparticles

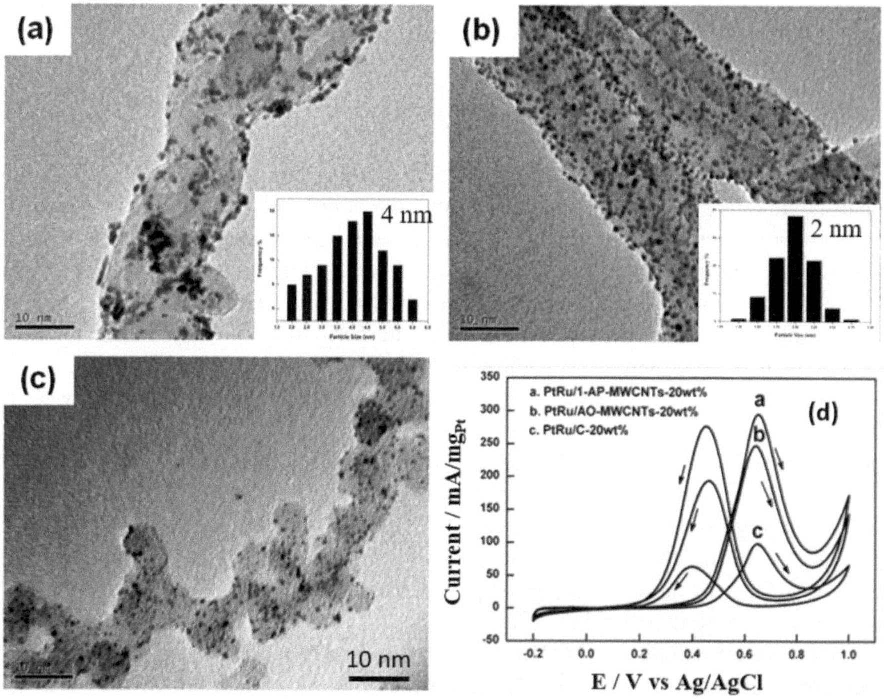

Fig. 4.6 TEM images of PtRu NPs supported on **a)** AO-MWCNTs (PtRu/AO-MWCNTs), **b)** 1-AP-MWCNTs (PtRu/1-AP-MWCNTs), and **c)** carbon black (PtRu/C). The PtRu loading of the electrocatalysts was 20wt%. Mass-normalized cyclic voltammograms of PtRu catalysts supported on 1-AP-MWCNTs, AO-MWCNTs and carbon black with the 20wt% PtRu loading in nitrogen saturated 0.5 MH_2SO_4 + 1.0 M CH_3OH at a scan rate of 50 mV/s were given in (**d**). Reproduced from Ref. [19] with permission from American Chemical Society, Copyright 2008

(NP) [19]. In the case of PtRu/AO-MWCNTs, the dispersion of PtRu NPs on CNTs is poor, characterized by the formation of a large number of aggregates and the average particle size of PtRu NPs is 3 ± 0.4 nm. The poor distribution of PtRu NPs is clearly due to the non-uniform distribution of the defects generated on CNTs by the acid oxidation treatment. In contrast, PtRu NPs deposited on the 1-AP-functionalized CNTs are uniform with no agglomeration and the average particle size is 2 ± 0.2 nm. This is also indicated by the high electrochemical surface area (ECSA) of PtRu catalysts obtained from the area of hydrogen desorption peak after correcting for the double layer charging current from the CVs measured in 0.5 MH_2SO_4 solutions. The calculated ECSA values for 20wt% PtRu/1-AP-MWCNTs (PtRu/1-AP-MWCNTs-20wt%) is 423 cm^2 per mg Pt, higher than 370.6 cm^2 per mg Pt of 20wt% PtRu/AO-MWCNTs (PtRu/AO-MWCNTs-20 wt%), and 164.9 cm^2 per mg Pt of 20wt% PtRu/C (PtRu/C-20 wt%).

The Faradic current for MOR exhibits the well-known features of methanol oxidation on Pt-based electrocatalysts. The forward oxidation current peak occurs at −0.65 V (vs. Ag/AgCl) and the backward oxidation current peak at −0.46 V. The magnitude of the forward anodic current peak is an indicator of the activity of PtRu electrocatalysts for MOR. Under identical PtRu catalyst loading, the anodic current for the reaction on PtRu/1-AP-MWCNTs and PtRu/AO-MWCNTs is significantly higher than that on PtRu/C, indicating a much higher electrocatalytic activity of CNTs supported PtRu electrocatalysts. The higher electrocatalytic activity for MOR on PtRu/1-AP-MWCNTs than that on PtRu/AO-MWCNTs again shows the importance of the distribution and dispersion of PtRu electrocatalysts on CNT supports.

4.4.1.2 Ethanol Oxidation Reaction

For liquid fuels containing C–C bonds, a complete electrochemical oxidation to CO_2 and H_2O is very difficult in low temperature fuel cells, e.g., PEMFCs. As a result, the CO_2 yield is usually very low in the case of ethanol oxidation reaction (EOR) because of the high energy barrier due to the presence of surface O and OH species [20]. The EOR in acid electrolyte solutions is dominated by reactions without the formation of CO_2, forming acetaldehyde at low potentials (<0.6 V vs. RHE) and acetic acid at high potentials (>0.8 V vs. RHE) [21].

$$C_2H_5OH \rightarrow CH_3CHO + 2H^+ + 2e^- \quad E^0 = +0.28\ \text{V} \tag{4.40}$$

$$C_2H_5OH + H_2O \rightarrow CH_3COOH + 4H^+ + 4e^- \quad E^0 = +0.06\ \text{V} \tag{4.41}$$

Acetic acid is not electrochemically active under fuel cell operation conditions, while the electrocatalytic activity of acetaldehyde is much lower than that of ethanol. The kinetics of EOR on Pt-based electrocatalysts in acid electrolytes is much slower than HOR, leading to high polarization overpotentials. Compared to methanol, ethanol shows a lower crossover through acidic membranes (e.g., Nafion) and lesser cathode poisoning. The rate-determining step (RDS) for EOR is the removal of the adsorbed ethoxi, CH_3CO_{ads}, and CO [22]. Similar to MOR, alloyed Pt-based electrocatalysts are more active for EOR as compared to monometallic electrocatalysts. For example, binary PtRu and PtSn catalysts are much more active for EOR than pristine Pt.

The electrocatalytic activity of binary PtSn catalysts is closely related to the atomic ratio of Pt:Sn. One of the effective methods to control the atomic ratio of PtSn is the deposition of binary PtSn NPs on tetrahydrofuran (THF)-functionalized CNTs. THF is an oxygen-containing heterocycle with five-membered rings. The electronegativity difference between carbon and oxygen makes the C–O bond moderately polar. Thus, THF is a polar, aprotic solvent, and has a sterically accessible oxygen atom. THF can be protonated in a chloroplantic acid solution. Thus, negatively charged $PtCl_6^{2-}$ ions and positively charged Sn^{4+} ions are attracted to the sterically accessible oxygen atoms in THF by electrostatic self-assembly process. On the other hand, the σ–π attractive force between the π bonds of CNTs and the σ bonds of cyclopentanes of THF also makes the CNTs easily dispersible in solution. After treated in H_2 at 300 °C for 2 h, PtSn NPs on CNTs with precisely controlled PtSn atomic ratios can be formed, see Fig. 4.7a [23].

PtSn/CNTs exhibit significantly enhanced catalytic activities for EOR as compared to that on Pt/CNTs. PtSn/CNTs with Pt:Sn atomic ratio of 2:1, Pt_2Sn/CNTs, shows the best electrocatalytic activity for the electrooxidation of ethanol (in this case, data at a potential of 0.5 V are used for comparison). The onset potentials for the ethanol oxidation on Pt_2Sn/CNTs catalysts are 0.15 V, significantly lower than 0.38 V measured on Pt/CNT catalysts. The enhanced catalytic activities and the reduced onset potentials of the PtSn catalysts can be explained, in part, by the relatively fast oxidation of adsorbed CO by the adjacent Sn oxide. The best electrocatalytic activity for the ethanol oxidation was observed on the Pt_2Sn/CNTs catalyst (Fig. 4.7c).

The oxidation mechanism of ethanol in acid solutions is considered to occur via two adsorbed intermediate fragments with one or two carbon atoms, $C1_{ad}$ and/or $C2_{ad}$. Despite the significant discrepancies in the reaction mechanism, it is generally in consensus that breaking C–C bond for a total oxidation to CO_2 is a major challenge in the electrocatalysts for EOR. The limited electrocatalytic activity on Pt/C or Pt/CNT catalysts indicate the high yields of partial oxidation products, CH_3CHO and CH_3COOH, on pure Pt catalysts. Alloying of Pt with Sn leads to the oxidation of ethanol at lower potentials with higher current and higher stability. This is due to the fact that Sn tends to donate electrons to Pt because of their differences in electronegativity, leading to a downshift of the *d*-band center of Pt and consequently weakening the chemisorption energy with the CO intermediates on PtSn electrocatalysts.

In alkaline electrolytes, the major product of EOR on Pt-based electrocatalysts is acetate:

$$C_2H_5OH + 5OH^- \rightarrow CH_3COO^- + 4H_2O + 4e^- \quad E^0 = -0.743\ \mathrm{V} \tag{4.42}$$

The product distribution of EOR in alkaline electrolytes also depends on the electrocatalysts.

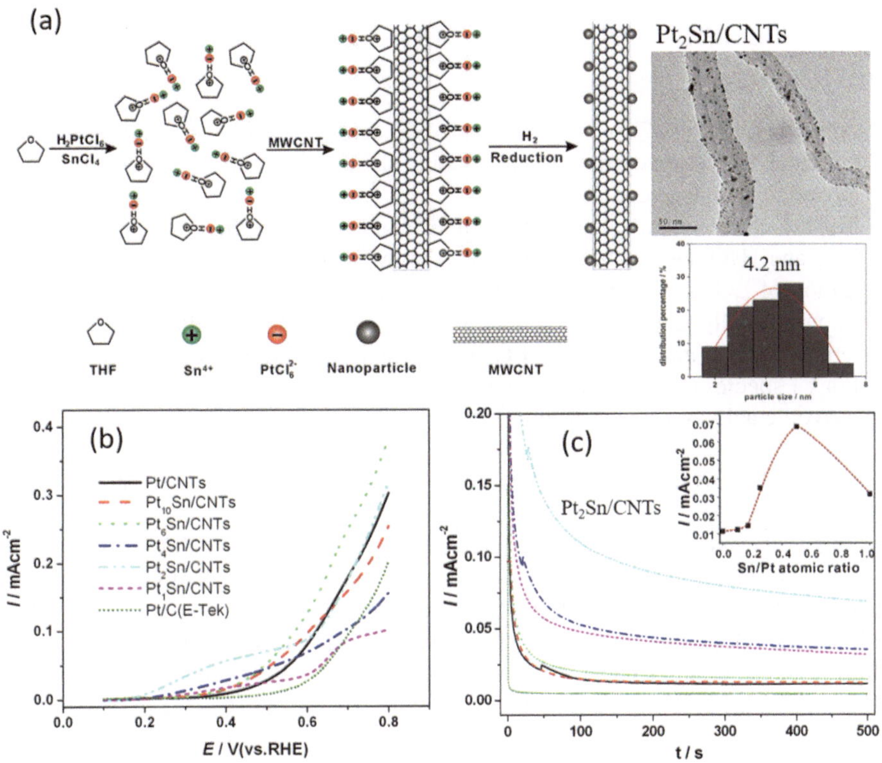

Fig. 4.7 **a)** Scheme of the synthesis of PtSn NP catalysts on THF-functionalized CNTs and the inset is the TEM image and particle size distribution of $Pt_2Sn/CNTs$ with Pt:Sn = 2:1, **b)** Linear sweep voltammetry (LSV) and **c)** chronoamperometry curves for EOR on Pt/CNTs and PtSn/CNTs with different Pt/Sn atomic ratios in a N_2-saturated $0.5MH_2SO_4 + 1.0MC_2H_5OH$ solution at room temperature. LSV was measured at a scan rate of 1 mV/s and CA was measured at 0.5 V (vs RHE). Inset in (**c**) is the plots of current measured at 0.5 V after 500 s test *versus* the Sn:Pt atomic ratios of PtSn/CNT catalysts. Modified from Ref. [23] with permission from Elsevier, Copyright 2010

4.4.1.3 Isopropanol, Ethylene Glycol, and Glycerol Oxidation Reactions

Isopropanol or isopropyl alcohol (IPA) is relatively inexpensive, has a low toxicity, and is miscible with water. Glycerol is the main product in biodiesel production by transesterification of plant oils and animal fats, and is nontoxic. As in the case of C_2 alcohols like ethanol, the total oxidation of IPA, ethylene glycol (EG), and glycerol has not been demonstrated.

Isopropanol oxidation reaction (IPAOR) can occur in both acidic and alkaline electrolytes. In both cases, acetone is the single oxidation product.

$$CH_3CHOHCH_3 \rightarrow CH_3COCH_3 + 2H^+ + 2e^- \quad E^0 = +0.13\ V \tag{4.43}$$

$$CH_3CHOHCH_3 + 2OH^- \rightarrow CH_3COCH_3 + 2H_2O + 2e^- \quad E^0 = -0.67\ V \tag{4.44}$$

At high potentials, CO_2 could be formed. In acidic electrolytes, only PGM-based electrocatalysts can be used, while in alkaline electrolytes, both PMG and non-PMG electrocatalysts have been used.

For the electrochemical oxidation of EG and glycerol, the complete oxidation to CO_2 and H_2O has not been achieved (the number of electrons for the complete oxidation of EG is 10) and mixed products are generally obtained. The oxidation reaction generally takes place via several consecutive and/or parallel steps involving different reaction intermediates of C_2 (glycolate, glyoxal, glyoxylate, and oxalate) and/or C_1 species (formate and carbonate). In alkaline solution, the oxidation of EG gives oxalic acid as a major product,

$$HOCH_2CH_2OH + 8OH^- \rightarrow (COOH)_2 + 6H_2O + 8e^- \quad E^0 = -0.69\,V \tag{4.45}$$

The oxidation reaction of glycerol in acidic electrolytes on a Pt/C catalyst gives a mixture of products with glyceraldehyde as the main one. In alkaline solution, Pd-based electrocatalysts show a higher electrocatalytic activity for glycerol oxidation reaction as compared to Pt-based electrocatalysts. However, Pd/C electrocatalysts alone have a low activity and very poor stability for the electrooxidation reaction of alcohols. Addition of oxides such as Co_3O_4, NiO, Mn_3O_4, and CeO_2 can significantly promote catalytic activity and stability of the Pd/C electrocatalysts for the alcohol oxidation reaction. A study has showed that Pd-Co_3O_4 composite with weight ratio of 2:1 supported on carbon, Pd-Co_3O_4(2:1, w:w)/C, has the highest activity for the electrooxidation of methanol, EG, and glycerol, while the most active catalyst for EOR is Pd-NiO(6:1, w:w)/C [24]. On the other hand, Pd-Mn_3O_4/C shows significantly better performance stability than other Pd-oxide composite catalysts for the alcohol electrooxidation. Pd-Co_3O_4/C electrocatalysts show a poor stability most likely due to the limited solubility of cobalt oxides in alkaline solutions.

The RDS of the electrocatalysts for AORs strongly depends on the charge transfer and the effective oxidation/removal of the adsorbed intermediates, which are related to the nature of alcohol liquid fuels. Studies show that the electrocatalytic activity of PGM-based catalysts depends on the alcohol fuels. For example, it has been reported that the electrochemical oxidation rate of different alcohols in alkaline electrolyte solutions on Pd electrocatalysts decreases in the order of *n*-propanol > isopropanol > ethanol > ethylene glycol > glycerol > methanol, while on Pt electrocatalysts, a different order is observed: isopropanol > ethanol > glycerol > ethylene glycol > *n*-propanol > methanol [25].

4.4.1.4 Formic Acid Oxidation Reaction

Formic acid with hydrogen content of 4.4 wt% is a popular fuel for PEMFCs due to easy transportation and storage of formic acid and high theoretical open circuit voltage (1.48 V). Formic acid has a lower crossover through the Nafion-based membrane and a higher kinetic activity than methanol. Lower crossover of formic acid through a membrane allows a much higher concentration to be used as the fuel (e.g., 10–15 M). However, the development of PEMFCs using formic acid fuel or direct formic acid fuel cells (DFAFCs) is retarded by the poor durability and low power output primarily due to the sluggish kinetics of formic acid oxidation reaction (FAOR). The direct electrochemical oxidation of FA can proceed via two parallel pathways, for example, with Pt-based catalysts, the "direct pathway or dehydrogenation of molecule" gives CO_2 via the $HCOO_{ads}$.

$$HCOOH \rightarrow CO_2 + 2H^+ + 2e^- \quad E^0 = -0.22\ V \tag{4.46}$$

And the "indirect pathway or dehydration of molecule" transforms the formic acid into CO_2 via the formation of intermediates CO_{ads}:

$$HCOOH \rightarrow CO_{ads} + H_2O \tag{4.47}$$

The adsorbed CO_{ads} blocks the active sites of Pt for the adsorption and oxidation of formic acid molecules. On the other hand, the two-electron process of the oxidation reaction of format in alkaline media is proceeding as follows:

$$HCOO^- + 3OH^- \rightarrow CO_3^{2-} + 2H_2O + 2e^- \quad E^0 = -1.05\ V \tag{4.48}$$

Thus, the high energy barrier for the breaking of C-H bond and the inevitable formation of CO intermediate by the indirect pathway of FA electrochemical oxidation leads to the sluggish kinetics of FAOR in DFAFC at low operating temperatures [26]. Pt-based electrocatalysts are prone to be poisoned by CO_{ads} intermediate during the dehydration step of FAOR. In contrast to Pt, dehydration of formic acid molecules does not occur on Pd. Thus, Pd shows a high efficiency due to the dominant direct pathway (dehydrogenation reaction) transformation of formic acid to H_2 and CO_2 [27]. However, the electrocatalytic activity of Pd-based catalysts is sensitive to both the shape and size of Pd nanoparticles.

Figure 4.8 compares the electrochemical stability and stability of Pd nano-wire networks on carbon (NWN/C) and Pd/C electrocatalysts for FAOR in $0.5MH_2SO_4$ + 0.5MHCOOH solution at room temperature. Chronoamperometric experiments were carried out at 0.1 V (*vs* SCE). Highly dispersed and mesoscale porous Pd NWNs catalysts were prepared by a templateless and polysodium-p-styrenesulfonate (PSS) polyelectrolyte-mediated self-assembly process [28]. The diameter of the Pd NWs is ~3 nm, similar to the size Pd NPs. Pd NWN/C and Pd NP/C show characteristic and separated anodic peaks in the forward and reverse scans, and the magnitude of the anodic peak current in the forward scan is proportional to the amount of formic

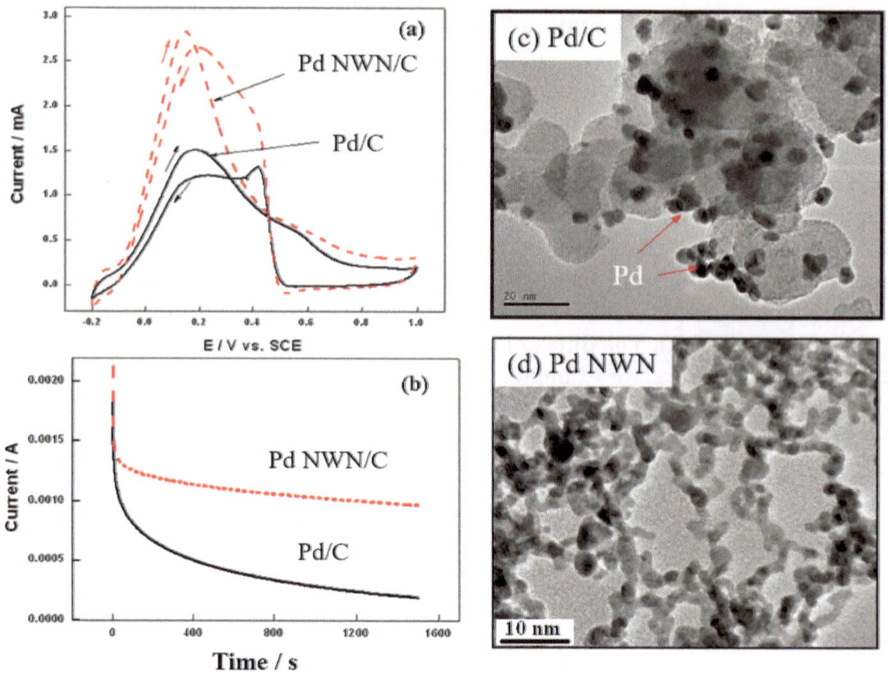

Fig. 4.8 **a)** Cyclic voltammograms measured at a scan rate of 50 mV/s, and **b)** chronoamperometric curves for formic acid oxidation reaction at 0.1 V (*vs* SCE) on Pd NWN/C and Pd NP/C electrodes in a 0.5 MH_2SO_4 + 0.5 MHCOOH solution, using glassy carbon electrode. Pd loading was 0.01 mg/cm^2. Reproduced from Ref. [28] with permission from IOP Publishing, Copyright 2008

acid oxidized at the Pd electrocatalysts, i.e., a direct measure of the electrocatalytic activity of Pd catalysts. The peak current for the FAOR is 2.82 mA on the Pd NWN/C electrode, significantly higher than 1.50 mA on the conventional Pd/C electrode. The forward peak potential for the FAOR on Pd NWN/C is 0.145 V (*vs* SCE), lower than 0.192 V (*vs* SCE) for the reaction on the Pd/C. The Pd NWN/C electrocatalysts show a high peak oxidation current and low anodic peak potential as compared to conventional Pd/C electrocatalysts, indicating a higher electrocatalytic activity and faster kinetics of Pd NWN/C electrocatalysts. This is most likely due to the high density of active sites and large surface area of nano-wire network nanostructures with high-aspect-ratio, as shown in Fig. 4.8d. The polarization current for FAOR on both Pd catalysts shows a rapid decay (Fig. 4.8b), indicating the intermediate species poisoning of the HCOOH electrooxidation via the CO pathway or the deactivation of the Pd electrocatalysts by an organic poisoning species [29]. Nevertheless, the current decay for the reaction on the Pd NWN/C electrocatalysts is much slower, as compared to that on Pd/C catalysts. With the increase of time, the difference between the current output for the reactions on Pd NWN/C and Pd/C electrocatalysts increases, indicating the excellent tolerance of Pd NWN/C electrocatalysts toward poisoning of the intermediates of the reaction. This indicates that Pd NWN structure enhances the electrochemical stability of Pd electrocatalysts for FAOR, demonstrating that the shape or morphology of nanostructures is important for the electrochemical activity of electrocatalysts.

4.4.2 Effect of pH and Temperature on Alcohol Oxidation Reactions

The important factors affecting the kinetics and mechanism of AORs are the adsorption and interaction between reactants/products and PMG-based electrocatalysts. This is critically related to the pH of the reaction medium, *i.e.*, the acidic (H^+ conductor) or alkaline (OH^- conductor) electrolyte solutions or membranes in fuel cells. The remarkable dependence of the reaction rate on pH is due to the significant effect of pH on the bonding energy of organic adsorbates with catalyst surface, changes in the structure of reacting species, and the concurrent adsorption of organic and oxygen-containing species, interaction of OH_{ads} species and OH^- ions with the alcohol-adsorbed products [30]. The passing from acidic media to basic ones significantly increases the alcohol (e.g., methanol, ethanol, EG, etc.) oxidation rate and performance of DAFCs [31]. Nevertheless, the most significant challenge in the development of alkaline alcohol fuel cells is the stability and durability of anion exchange membranes (AEMs) with high anion conductivity [32]. Another challenge of alkaline alcohol fuel cells (e.g., using methanol and ethanol fuel) is the formation of alkali metal carbonates and bicarbonates that crystallize in the electrolyte-filled pores, which blocks the porous electrode structures and reduces the ionic conductivity of the electrolyte. Similar to proton exchange membrane fuel cells, operation at elevated high temperatures can also significantly increase the power output of AEMs based fuel cells, but the difficulty facing high-temperature AEMFCs is the low chemical stability of the membranes. Table 4.5 lists the oxidation potential of selected oxidation reaction of liquid fuels at STP conditions.

Temperature is probably the most important parameter to affect the activity and performance of DAFCs. The reaction rate of AOR on Pt-based catalysts is a complex function of several parameters such as adsorption/desorption of intermediates, thermal stability of the reactants, supporting-electrolyte anions, and adsorbed oxygen forms. An example is the existence of a kinetic transition temperature for FAOR on Pt/C catalysts at elevated temperatures, as shown in Fig. 4.9 [33].

The polarization behavior (η) of FAOR as measured by pseudo-dynamic hydrogen electrode (pseudo-DHE) shows a significant dependence on temperature. At 160 °C, the current density of FAOR increases slowly at low η. With the increase of temperature, the current density for the reaction increases dramatically. This is indicated by the change of apparent electrode polarization conductivity, the reciprocal of apparent electrode polarization resistance, $\sigma_E = 1/R_E = i/\eta$, measured from the slope of the i-η curve at low currents. σ_E increases significantly with the temperature, and the

Table 4.5 Standard oxidation potentials of oxidation reactions of selected fuels and oxygen reduction reaction

Fuels	Oxidation reactions	Reversible potential /V[a]	Note
Hydrogen	$H_2 \rightarrow 2H^+ + 2e^-$	0.00	Acid
	$H_2 + 2OH^- \rightarrow 2H_2O + 2e^-$	−0.83	Alkaline
Oxygen	$½O_2 + 2H^+ + 2e^- \rightarrow H_2O$	+1.23	Acid
	$½O_2 + H_2O + 2e^- \rightarrow 2OH^-$	+0.40	Alkaline
Formic acid	$HCOOH \rightarrow CO_2 + 2H^+ + 2e^-$	−0.22	Acid
	$HCOO^- + 3OH^- \rightarrow CO_3^= + 2H_2O + 2e^-$	−1.05	Alkaline
Methanol	$CH_3OH + H_2O \rightarrow CO_2 + 6H^+ + 6e^-$	+0.02	Acid
	$CH_3OH + 6OH^- \rightarrow CO_2 + 5H_2O + 6e^-$	−0.81	Alkaline
Ethanol	$C_2H_5OH \rightarrow CH_3CHO + 2H^+ + 2e^-$	+0.28	Acid
	$C_2H_5OH + H_2O \rightarrow CH_3COOH + 4H^+ + 4e^-$	+0.06	Acid
	$C_2H_5OH + 5OH^- \rightarrow CH_3COO^- + 4H_2O + 4e^-$	−0.743	Alkaline
Isopropanol	$CH_3CHOHCH_3 \rightarrow CH_3COCH_3 + 2H^+ + 2e^-$	+0.13	Acid
	$CH_3CHOHCH_3 + 2OH^- \rightarrow CH_3COCH_3 + 2H_2O + 2e^-$	−0.67	Alkaline
Ethylene glycol	$HOCH_2CH_2OH + 8OH^- \rightarrow (COOH)_2 + 6H_2O + 8e^-$	−0.69	Alkaline
Hydrazine	$N_2H_5^+ \rightarrow N_2 + 5H^+ + 5e^-$	−0.33	Acid
	$N_2H_4 + 4OH^- \rightarrow N_2 + 4H_2O + 4e^-$	−1.16	Alkaline
Urea	$CO(NH_2)_2 + 6OH^- \rightarrow N_2 + CO_2 + 5H_2O + 6e^-$	−0.746	Alkaline
Ammonia	$2/3NH_3 + 2OH^- \rightarrow 1/3N_2 + 2H_2O + 2e^-$	−0.77	Alkaline

[a]The potential is given at pH = 0 for acid and pH = 14 for alkaline

increase in σ_E becomes slow at temperatures above 200 °C (see the inset in Fig. 4.9a). The significant change of σ_E values for the reaction implies the change of the reaction kinetics of FAOR at elevated temperatures. To study the kinetics of the reaction, the exhaust gas of the anode side of the cell was analyzed at OCV as a function of temperature (Fig. 4.9b). Both CO and H_2 were detected, indicating the thermal decomposition of formic acid in the anode chamber of the cell. The results indicate the exhaust H_2 gas only becomes dominant at elevated temperatures, reaching 13.43 ± 0.5 mL/min (i.e., 5.42×10^{-1} mmol/min) at 240 °C. The molar H_2 yield rate is

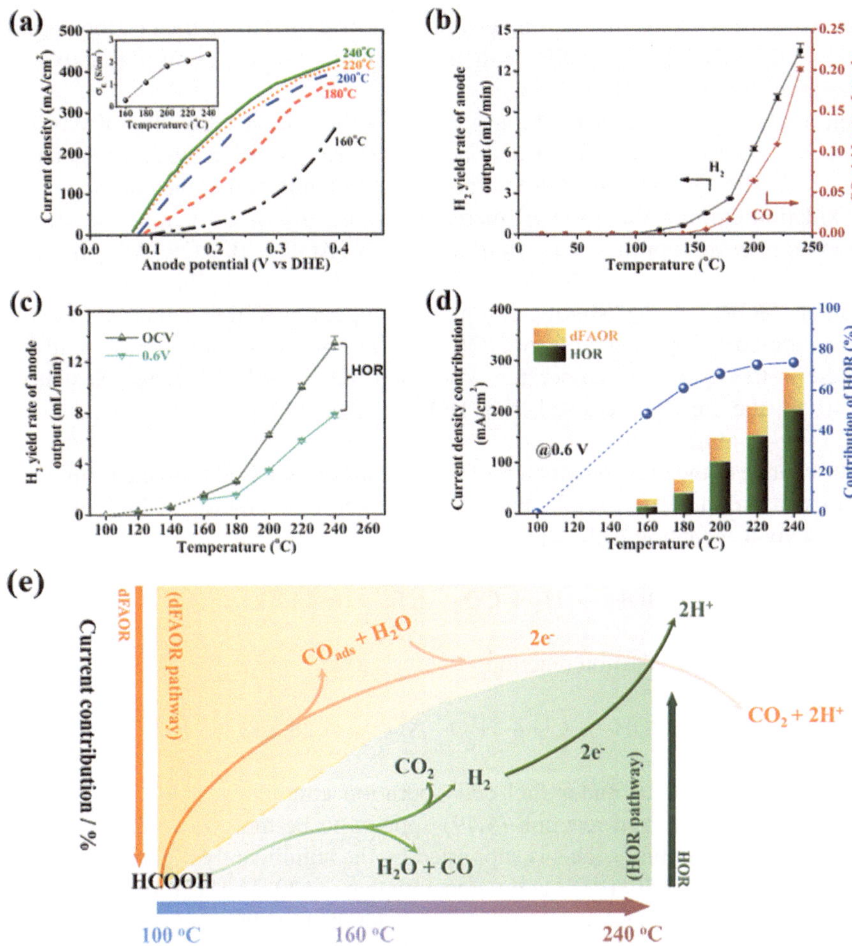

Fig. 4.9 **a)** Polarization curves of FAOR measured at different temperatures. H_2 was supplied to the cathode with flow rates of 100 mL/min as the pseudo-dynamic hydrogen electrode (pseudo-DHE) reference electrode. The inset is the plots of electrode conductivity, σ_E against temperature. **b)** H_2 and CO yield rate of exhaust gas from the anode measured at OCV. **c)** H_2 yield rate of exhaust gas from the anode measured at OCV and 0.6 V, respectively, and different temperatures. **d)** Polarization current density of H_2 oxidation reaction (HOR) and direct FAOR (dFAOR) reaction and relative contribution of HOR to the overall FAOR at a constant cell voltage of 0.6 V, measured at different temperatures. 12 mol/L FA was fed to the anode with flow rate of 1 mL/min. **e)** Scheme of gradual transformation of reaction kinetics from the dFAOR to HOR for the polarization performance of DFAFCs due to the in situ decomposition of FA. Modified from Ref. [33] with permission from Wiley–VCH, Copyright 2020

about two orders of magnitude higher than that of CO. This indicates the significant decomposition reaction of FA with dominant H_2 formation on Pt/C catalysts at temperatures above 100 °C.

The reduced H_2 content or the gap between the one at OCV and at 0.6 V is evidently consumed by the H_2 oxidation reaction (HOR) inside the cell to generate the current. Thus, with the increase of the operation temperature, the contribution of the oxidation currents due to HOR increases. Thus, the overall Faradaic current is composed of current produced by HOR and direct FAOR or dFAOR. At operational temperature below 100 °C, the decomposition of FA is negligible and dFAOR is dominant. HOR becomes significant only at temperatures higher than 100 °C. At 160 °C, the current produced from HOR is 48.7% of the overall current. With the increase in the operating temperature, the contribution of HOR increases sharply. At 240 °C, the current produced from HOR reaches 73.6% of the total current of DFAFC (Fig. 4.9d).

The change of the reaction kinetics of FAOR can be discussed from the formic acid decomposition on the Pt-based catalysts at elevated temperatures, which proceeds via the dehydrogenation pathway:

$$HCOOH \rightarrow H_2 + CO_2 \quad \Delta G^0 = -48.8\,\mathrm{kJ/mol} \tag{4.49}$$

and/or via the dehydration reaction

$$HCOOH \rightarrow CO + H_2O \quad \Delta G^0 = -28.5\ \mathrm{kJ/mol} \tag{4.50}$$

Both reactions occur under fuel cell operation conditions at Pt/C catalysts at elevated temperature, and reaction (4.49) appears to be more favorable due to its thermodynamic feasibility. This is supported by the significantly higher molar yield rate of H_2 at high temperature as compared with that of CO. The H_2 formed due to the in situ decomposition of FA will be oxidized under the DFAFC operation conditions, contributing to the overall power output of DFAFCs:

$$H_2 \rightarrow 2H^+ + 2e^- \tag{4.51}$$

Kinetically, HOR is much faster than that of dFAOR, *i.e.*, reactions (4.46) and (4.47), indicated by the increase of the HOR contributing to the Faradaic current from 48.7% at 160 °C to 73.6% at 240 °C. This will lead to a gradual transition of the reaction kinetics from dFAOR at low temperatures to HOR at elevated temperatures for the DFAFCs (Fig. 4.9e). At temperatures below 100 °C, the performance of DFAFCs is dominated by the sluggish dFAOR. With the increase in the temperature, the contribution of HOR to the overall reaction kinetics increases, due to the significant increase in the production of H_2 at elevated temperatures through the decomposition of FA. Similar significant change in the power output at elevated temperatures has also been reported for DMFCs, measured at temperatures up to 260 °C [34].

4.4.3 *Energy Density, Environment Effect, and By-Products of Various Fuels*

Ethanol can be easily produced by hydration of acetylene or can be derived by fermentation of sugar-rich raw materials and is less harmful as compared to methanol. Thus, ethanol is a renewable biofuel with fairly high energy density, 29.7 kJ/g, but electrooxidation of ethanol (C_2H_5OH, containing two carbons) is much more complex when compared to methanol oxidation. The complete oxidation of ethanol to CO_2 requires the transfer of 12 electrons, and this is almost impossible to be accomplished using the current electrocatalysts and at room temperature. The electrochemical oxidation of ethanol on PtRu catalyst, one of the most common catalysts for electrochemical oxidation of liquid alcohols, would lead to the formation of acetaldehyde as the predominant product, followed by its further oxidation to acetic acid at more positive potential, and to small extent, to CO_2 [35]. Acetaldehyde (CH_3CHO) is toxic, and electrooxidation of acetaldehyde would require a higher potential than that for the oxidation of ethanol. This indicates that it is more difficult to electrochemically oxidize or clean the by-product of the ethanol oxidation than the ethanol fuel itself.

The electrochemical oxidation of the majority liquid fuels in low-temperature fuel cells is hardly a complete reaction. In most cases, the electrooxidation of liquid fuels occurs through the formation of various intermediate or by-products. Thus, the environmental effects of the intermediates or by-products need to be taken into consideration in the design of fuel cell systems. For example, the main by-products of the MOR are formic acid and formaldehyde. Formaldehyde is highly irritant, corrosive, carcinogenic, and toxic. In the case of ethanol fuel, the final products of the EOR are acetic acid and acetaldehyde, which are also highly flammable and carcinogenic. Dimethyl ether (DME) can also be used as fuel for fuel cells as DME is the simplest ether with no C–C bonds and low toxicity as compared to methanol. However, the main by-products of direct dimethyl ether fuel cells are methanol and methyl formate, and the formation of methanol does not depend on the current but increases with increasing temperature. In the case of the oxidation of 1-propanol and 2-propanol fuels in a PEMFC, acetone is the main by-products of the reaction. Thus, treatment processes must be incorporated into fuel cell systems or be created as the by-products of the oxidation reactions of the liquid fuels cannot be directly discharged without proper cleaning treatment. Table 4.6 lists specific energy density, hazard, environmental effects, and possible main by-products of the common liquid and selected solid fuels used in fuel cells.

Direct liquid fuel cells using methanol, ethanol, propanol, formic acid, etc., suffer from an additional and more specific problem: CO species from the indirect decomposition of liquid fuels significantly poisons the Pt-based electrocatalysts. Development of advanced nanoparticle synthesis methods and supporting materials is needed to significantly enhance the electrocatalytic activity and durability of the Pt-based catalysts for liquid fuels based fuel cells.

Table 4.6 Theoretical specific energy density, hazard, environmental effects, and main by-products of the most common fuels used in fuel cells

Name	Chemical formula	n_e[a]	Specific energy density		Health effects and safety issue	Main by-product
			kWh/kg	kWh/L		
Hydrogen	H_2 (g)	$2e^-$	39.41	0.0033[b]	Highly flammable	Water
	H_2 (l)	$2e^-$	39.41	2.79	Highly flammable	Water
Methane	CH_4 (g)	$4e^-$	15.46	0.011[b]	Highly flammable	CO_2, water
Propane	$CH_3CH_2CH_3$ (g)	$20e^-$	13.98	0.028	Highly flammable	CO_2, water
Ethanol	C_2H_5OH (l)	$12e^-$	8.26	6.47	Irritant, highly flammable	Formic acid, acetaldehyde
Ethylene glycol	C_2H_6O (l)	$10e^-$	5.2	5.9	Irritant, harmful if swallowed, flammable	Oxalic acid
Glycerol	$HOCH_2CH(OH)CH_2OH$ (l)	$14e^-$	5.0	6.4	Flammable	Oxalic acid
Dimethyl ether	CH_3OCH_3 (g)	$12e^-$	8.81	0.02[b]	Highly flammable	Methanol, methyl formate
Methanol	CH_3OH (l)	$6e^-$	6.31	4.96	Toxic, highly flammable	Formic acid, formaldehyde
Formic acid	HCOOH (l)	$2e^-$	1.63	1.73	Irritant, harmful, flammable	CO
1-Propanol	$CH_3CH_2CH_2OH$ (l)	$18e^-$	9.34	7.48	Irritant, highly flammable	Acetone
2-Propanol	$CH_3CH(OH)CH_3$ (l)	$18e^-$	9.29	7.25	Irritant, highly flammable	Acetone
Ammonia	NH_3 (l)	$3e^-$	6.26	3-5[c]	Irritant, toxic	N_2, NO
Ammonia borane	$H_3N{:}BH_3$ (s)	$6e^-$	8.4			B_2O_3, $(NHBH)_3$
Urea	$CO(NH_2)_2$ (s)	$6e^-$		4.69	Irritant	N_2, CO_2
Hydrazine	N_2H_4 (l)	$4e^-$	4.19	4.27	Toxic, flammable	N_2, water
Sodium borohydride	$NaBH_4$ (s)	$8e^-$	9.3	10[d]		

(continued)

Table 4.6 (continued)

Name	Chemical formula	n_e[a]	Specific energy density		Health effects and safety issue	Main by-product
			kWh/kg	kWh/L		
Typical gasoline (l)		–	13.15	9.42	Highly flammable	

[a]The number of electrons (e^-) per fuel molecule involved in fuel oxidation, [b]Gas under 101.3 kPa and room temperature, [c]Depending on the conditions, [d]For a solution of 500 g $NaBH_4$ in 1 L solution at 25 °C. 1 kWh = 3.597 MJ

4.5 Ammonia, Urea, and Hydrazine

4.5.1 Ammonia Oxidation Reaction

Ammonia (NH_3) and its derivatives e.g. urea, $CO(NH_2)_2$, are the most important fertilizer used in agriculture with world production exceeding 200 million tons per annum. NH_3 is traditionally produced by the Haber–Bosch process by using N_2 from the atmosphere and H_2 from natural gas (CH_4) at high pressures and temperatures (>200 atm, 400–500 °C), which is highly energy intensive with high CO_2 emissions, accounting for ~1.6% of total global emissions [36]. However, NH_3 has a higher boiling temperature of −33.34 °C under atmospheric pressure as compared to H_2 (−251.95 °C), so it can be easily stored and transported as a liquid. Consequently, NH_3 with 17.6 wt% H_2 and a high volume energy density (13.6 GJ/m^3) with easy transportation characteristics is a potential energy carrier for hydrogen conversion and electricity generation at the point of use, if required, via a fuel cell, but without the carbon penalty. Thus, ammonia in solid or liquid form is also considered as an emerging renewable energy storage and is a carbon-free chemical energy carrier for hydrogen. Other ammonia-related chemicals such as metal amine salts, ammonium carbonate, and urea can also be used for hydrogen storage with no toxicity and volatility issues.

Ammonia can also be synthesized by electrochemical (EC) and/or photo-electrochemical (PEC) routes, using renewable and CO_2-free electricity such as solar and wind power. The development of N_2-to-NH_3 conversion technology based on renewable energy is both economically viable and technologically important. The N_2-to-NH_3 fixation reaction is thermodynamically feasible.

$$N_{2(g)} + 3H_{2(g)} \rightarrow 2NH_{3(g)} \quad \Delta G^0_{298K} = -32.8\,\text{kJ/mol} \tag{4.52}$$

However, this reaction cannot take place spontaneously under ambient conditions. There are many factors hindering the cleavage and hydrogenation of dinitrogen in nature, primarily including the thermodynamically strong cleavage energy (410 kJ/mol) of the first bond in N_2, the large energy gap between the HOMO (the

σ_g 2p bonding orbital) and LUMO (the π_g^* 2p anti-bonding orbital), and the low proton affinity [37]. EC and/or PEC production of ammonia in H_2O-based systems is both thermodynamically and kinetically challenging. Thermodynamically, N_2 gas is very unreactive because the atoms are held together by strong triple bonds and splitting of the strong N≡N bond requires a high activation energy. This leads to low Faraday efficiency of 10–40% and an extremely low ammonia formation yield rate in the magnitude of $\mu g/cm^2 \cdot h$ for N_2-to-NH_3 conversion under ambient conditions. Innovation in the photo-electrochemical and electrochemical catalysts is essential to enhance the efficiency of the N_2-to-NH_3 fixation and to increase the ammonia yield [38]. Another potential platform for the production of the ammonia fuel is high-temperature solid oxide electrolysis.

The present approach of using ammonia in a fuel cell is to decompose or crack it into H_2 and N_2 at high temperatures and then using the evolved H_2 in conventional low-temperature fuel cells. In principle, ammonia can be used as fuel directly in low-temperature alkaline fuel cells or anion exchange membrane fuel cells. Based on the standard convention, the electrochemical oxidation reaction of ammonia occurs as one-electron oxidation reaction with simultaneous oxygen reduction reaction at the cathode [39]:

$$\frac{2}{3}NH_3 + 2OH^- \rightarrow \frac{1}{3}N_2 + 2H_2O + 2e^- \quad E^0 = -0.77\,V \tag{4.53}$$

$$\frac{1}{2}O_2 + H_2O + 2e^- \rightarrow 2OH^- \quad E^0 = +0.40\,V \tag{4.54}$$

The overall cell reaction for a direct ammonia fuel cell is

$$\frac{2}{3}NH_3 + \frac{1}{2}O_2 \rightarrow \frac{1}{3}N_2 + H_2O \quad E^0 = +1.17\,V \tag{4.55}$$

The cell voltage of direct ammonia fuel cells compares favorably with the theoretical cell voltage of 1.23 V of a standard H_2-O_2 fuel cell. It should be remarked that traces of ammonia poison the acidic proton exchange membrane electorlyte via the acid-base reactions and hence only can be used as fuel for alkaline fuel cells.However, the power output of ammonia fuel cells is very low due to the low catalytic activity of the electrocatalysts for ammonia oxidation reaction at low temperatures. The ammonia decomposition temperature is close to the operating conditions of solid oxide fuel cells (SOFCs). Therefore, the integration of these two devices is beneficial in terms of efficient heat and energy managements and will lead to the development of simplified generation systems. In oxide conducting YSZ electrolyte-based SOFCs, oxidation of ammonia may lead to the generation of small amounts of toxic NO:

$$2NH_3 + 5O^{=} \rightarrow 2NO + 3H_2O + 10e^- \tag{4.56}$$

$$2NH_3 + 3NO \rightarrow \frac{5}{2}N_2 + 3H_2O \tag{4.57}$$

Ammonia oxidation is limited by reaction (4.56) due to the slow supply of oxygen ions from the electrolyte, and this leads to the formation of NO at the SOFC anode. However, the formation of NO can be suppressed by the optimization of electrocatalysts of the anode or by using a protonic ceramic fuel cells (PCFCs) based on proton-conducting oxide electrolyte, such as doped $BaCeO_3$ and $BaZrO_3$. Under this circumstance, the anode reaction is as following.

$$2NH_3 \rightarrow N_2 + 6H^+ + 6e^- \tag{4.58}$$

Protons produced will transfer through proton-conducting electrolyte to the cathode and react with oxygen, producing water. Water produced at the cathode will not dilute the fuel. However, at high operation temperatures, the thermal decomposition of ammonia is facile, producing H_2 and N_2 before reaching the electrode/electrolyte interface. The rate of decomposition depends on the temperature. At temperatures of 600 °C, complete decomposition of ammonia is possible. The challenges of ammonia-fed SOFCs and PCFCs are the performance durability and activity of anode for the decomposition of ammonia and hydrogen oxidation reaction.

Urea and $CO(NH_2)_2$ belong to the member of ammonia-related hydrogen-storage materials for low-temperature fuel cells. Urea is a white solid at room temperature and is not toxic, with melting temperature of 133–135 °C, but highly soluble in water (1079 g/L at 20 °C). Urea is synthesized from ammonia and carbon dioxide and is widely used in fertilizers as a source of nitrogen and is an important raw material for the chemical industry.

Urea has a high energy density of 16.9 MJ/L and can be used as fuel in low-temperature fuel cells [40]. In *direct urea fuel cells* (DUFCs), AEM is generally used to avoid the possible poisoning of cation exchange membrane. In fuel cells, urea is oxidized at the anode to N_2, releasing six electrons:

$$CO(NH_2)_2 + 6OH^- \rightarrow N_2 + CO_2 + 5H_2O + 6e^- \quad E^0 = -0.746\,\text{V} \tag{4.59}$$

In practical DUFCs, the open circuit voltage is in the range of 0.2–0.6 V, much lower than the theoretical equilibrium cell voltage, 1.146 V. The most common catalysts for the electrooxidation of urea is Ni-based alloy or nanostructured materials and the power output of the DUFC is limited by the sluggish reaction kinetics of the direct electrochemical oxidation of urea. At the present, urea is mainly produced from NH_3 and CO_2, and thus, the cost of urea is largely dependent on the cost of NH_3. Another source of urea is urine, which contains approximately 22 g/L of urea. One of the potential application of urea-based fuel cells is its use for wastewater treatment particularly in urine treatment to remove N and to generate electricity simultaneously.

4.5.2 Hydrazine Oxidation Reaction

Non-carbon-containing hydrazine hydrate ($N_2H_4{\cdot}H_2O$) is a weak base with a pK_a of 8.1, reactive compound, toxic, and will irritate eyes and skin. Similar to ammonia, hydrazine can also be directly used as fuel in fuel cells, or direct hydrazine fuel cells (DHzFCs). DHzFCs are promising power devices because of easy storage and transportation of liquid hydrazine fuel, high energy density (4.27 kWh/L), high theoretical cell voltage (1.56 V), absent of CO_2 gas emission and CO poisoning. Hydrazine can be electrochemically oxidized in acidic and neutral media:

$$N_2H_5^+ \rightarrow N_2 + 5H^+ + 5e^- \quad E^0 = -0.33\,\text{V} \tag{4.60}$$

or in alkaline media:

$$N_2H_4 + 4OH^- \rightarrow N_2 + 4H_2O + 4e^- \quad E^0 = -1.16\,\text{V} \tag{4.61}$$

The overall cell reaction generates only nitrogen and water, with a high reversible cell potential of 1.56 V in both acid and alkaline solutions.

$$N_2H_4 + O_2 \rightarrow N_2 + 2H_2O \quad E^0 = +1.56\text{ V} \tag{4.62}$$

N_2 is the only by product formed for a purely electrochemical oxidation reaction of hydrazine. However, hydrazine can also undergo catalytic chemical decomposition reaction on catalyst surface, forming H_2 and N_2:

$$N_2H_4 \rightarrow N_2 + 2H_2 \tag{4.63}$$

$$3N_2H_4 \rightarrow N_2 + 4NH_3 \tag{4.64}$$

Chemical decomposition to form H_2 reduces the efficiency and mixed potentials, i.e., reduced OCV, of fuel cells. In fuel cells, electrochemical oxidation reaction of hydrazine is plausible under both acid and alkaline conditions and is sensitive to the nature of the metal used as anode and to its surface structure. In acid solution, hydrazine oxidation involves the initial reversible adsorption of hydrazinium ions ($N_2H_5^+$) followed by successive deprotonation steps to the final product N_2 [41], while in alkaline media hydrazine oxidation reaction primarily proceeds through a stepwise electrochemical dehydrogenation reaction without the activation of the N–H bond[42].

The focus of the DHzFCs research is mainly on the alkaline hydrazine fuel cells due to the fact that alkaline anion exchange membrane electrolytes enable the use of non-precious group metal (NPGM) catalysts (see Chap. 13). For the hydrazine oxidation reaction (HzOR), noble metals including Pt, Pd, and Ag nanoparticles are used as electrocatalysts. However, Pt preferentially catalyzes the hydrazine into H_2 and N_2 in alkaline solutions via a chemical decomposition process. In alkaline solutions, Pt is less active for HzOR as compared to Ag, Ni, and Co, and this opens opportunities for the development of NPGM electrocatalysts for fuel cells. Transition metal catalysts including Ni and their alloys are the most active for HzOR and can efficiently oxidize hydrazine to nitrogen and water in alkaline solution by a four-electron step process. Nevertheless, Ni-based electrocatalysts suffer from oxidation after exposed to air (*i.e.*, surface passivation) and agglomeration during HzOR in alkaline conditions, leading to the deterioration of shelf life and electrocatalytic activity of catalysts over time.

Modification and doping with non-metal elements including phosphorous, nitrogen, boron, etc., and optimization of nanostructures have been shown to improve the performance of Ni-based electrocatalysts [43]. The doping of phosphorous on metal catalysts significantly improves their electrocatalytic activity by altering bond distance, coordination number, and electronic structure of metal catalysts. This would lead to favorable adsorption and desorption of reactants for reactions such as water splitting and HzOR. The surface electronic states of crystalline phosphide can be effectively influenced by a core metal via core–shell structure to tune interaction between phosphide and reaction intermediates to enhance electrocatalytic activity [44]. Moreover, the amorphous material increases electron and ion conductivity by providing large specific surface area and pathways for ion diffusion. Thus,

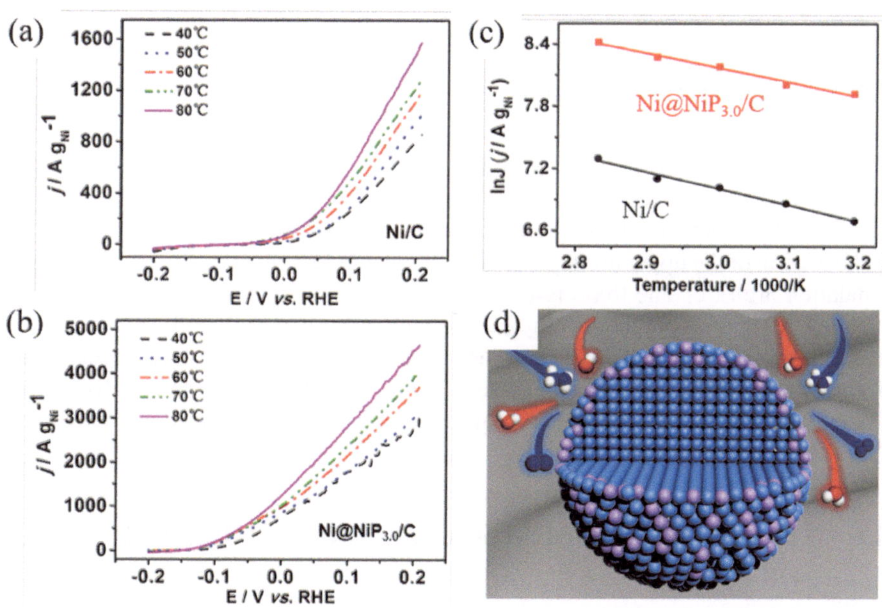

Fig. 4.10 LSV curves of **a)** Ni/C and **b)** Ni@NiP catalyst in 1.0 M KOH and 1.0 M hydrazine solution with a scan rate of 10 mV/s. **c)** Arrhenius plots of Ni/C and Ni@NiP$_{3.0}$/C at 0.2 V. The scheme of Ni@NiP catalyst is shown in (**d**). Reproduced from Ref. [43] with permission from American Chemical Society, Copyright 2019

heterostructured core–shell catalysts are desirable for efficient hydrazine oxidation reactions.

Figure 4.10 shows typical electrocatalytic activities of Ni/C and Ni@$NiP_{3.0}$/C for HzOR at different temperatures [43]. Ni@$NiP_{3.0}$/C is a Ni crystalline core@Ni phosphide amorphous shell heterostructured catalyst supported by active carbon with P:Ni molar ratio of 3:100. The NiP shell effectively protects the Ni core from passivation in air. The intrinsic effect between the crystalline Ni core and amorphous NiP shell increases the hydrazine adsorption energy and thus the activity for HzOR. The core–shell structured Ni@NiP/C catalysts demonstrate significantly enhanced electrocatalytic performance for HzOR as compared to Ni NPs supported on carbon (Ni/C). With increasing the operating temperature of HzOR, the current density for HzOR increases while the onset potential decreases. More specifically, the current density of Ni@$NiP_{3.0}$/C at 0.2 V increases from 3049 A/g_{Ni} at 40 °C to 4653 A/g_{Ni} at 80 °C (Fig. 4.10c), which is 2.54 and 1.84 times higher than that of the Ni/C at the same test condition, respectively. The onset potential decreases from −0.102 V at 40 °C to −0.146 V at 80 °C. According to the Arrhenius equation, the activation energy of Ni@$NiP_{3.0}$/C is 11.6 kJ/mol, lower than 13.3 kJ/mol for the reaction on Ni/C catalyst. Tafel slopes of Ni@NiP/C catalyst for HzOR are ~50 mV/dec, lower than ~60 mV/dec for the reaction on Ni/C catalysts. The results indicate that the formation of NiP shell significantly increase the kinetics of HzOR.

4.5.3 Toxicity Issues

Ammonia, urea, and hydrazine are toxic. Ammonia usually stored in liquid as ammonium hydroxide or gaseous form. Ammonia vapor is colorless but has a distinct and pungent odor. Contact with ammonia can cause eye damage, skin burn, lachrymation, and irritation of the upper airway. However, fatal incidents to humans due to ammonia inhalation are rare. The toxic issues of urea is similar to ammonia. Hydrazine has a higher toxicity. Contact with hydrazine can cause irritation and caustic burns if exposed to skin. Hydrazine is classified as a carcinogen. Therefore, safety issues need to be carefully considered in direct ammonia and hydrazine fuel cells because of the high toxicity and flammability in particular of N_2H_4 and its derivatives.

4.6 Natural Gas, Hydrocarbons, and Biomass-Derived Synthetic Fuels

Given the lack of a supply infrastructure and difficulties in storing hydrogen, the ability of fuel cells to operate at high efficiency on hydrocarbon fuels is considered a major advantage of high-temperature fuel cells such as MCFCs and SOFCs.

Natural gas (primary methane) is seen as an ideal fuel due to its abundance, existing distribution infrastructure, and low cost.

The steam methane reforming (SMR) is an established process used on an industrial scale for the production of hydrogen, as discussed in Sect. 4.2.1. Methane can also be directly or indirectly used as fuel in high-temperature MCFCs and SOFCs. As sulfur generally exists in natural gas, desulfurization needs to be incorporated into natural gas reforming processes as sulfur in the fuels generally poisons the electrochemical activity of anodes used in SOFCs.

Diesel, a commonly used and widely available fuel in industry and automobile markets, contains hydrogen and carbon that can be used for fuel cells. The density of petroleum diesel is about 0.832 kg/L and contains ~75% saturated hydrocarbons and ~25% aromatic hydrocarbons having a boiling point in the range of 180–360 °C. The diesel reacts with air or steam or both at temperatures over 800 °C on catalysts and converts to reformate that can be used for high-temperature fuel cells such as MCFCs and SOFCs. There are several methods to reform diesel, including SR, partial oxidation (POx), and autothermal reforming (ATR). Different to SR, POx is an exothermic reaction that requires only air. The advantages of POx is the rapid start-up, fast reaction speed, and compact reactor size, but suffers the catalyst agglomeration and carbon deposition, resulting in the process's degradation. ATR is a combination of SR and POx, slightly exothermic and would be an appropriate method to reform diesel. However, as ATR requires air and steam, the final product of diesel reformate has a relatively high amount of steam and nitrogen and thus low fuel concentration. For example, a typical diesel reformate would contain 15% H_2 and 3% CO with 40% N_2 and 32% steam. The low fuel concentration leads to high oxygen partial pressure and low open circuit voltage, reducing the efficiency of a fuel cell system. The high steam content at the anode chamber of SOFCs could also lead to the oxidation of Ni/YSZ cermet anodes, leading to the degradation of cell performance.

Renewable fuels derived from biomass (particularly from non-food biomass) are attractive for fuel cells as the electricity generation from biomass-based renewable fuels through fuel cells is not only highly efficient but it is also carbon neutral. The typical product gas of the biomass gasification consists of H_2, CO, CH_4, CO_2, H_2O, N_2 as well as some impurities such as sulfur, chlorine, and alkali metals. The composition of biogas is dependent on the type of biomass and production processes. Table 4.7 shows the typical composition of biogas derived from Australia mallee wood and wheat straw by high-temperature gasification [45, 46]. The gas products of the biomass were reformed by passing through a catalytic reactor using char catalysts.

The fuel from the biomass gasifier can also contain considerable amounts of tars depending on the type of gasifier used. Tar is a complex of aromatics and can be represented by a mixture of toluene, naphthalene, phenol, and pyrene. Thus, certain pre-reforming and cleaning of biomass gasification fuels are needed for fuel cell applications. Alkali metals are highly corrosive toward materials associated with syngas processing and utilization, including ceramic particulate filters and metallic components. The removal of alkali contaminants, consequently, is an important step in the cleaning of coal and biomass-derived syngas.

Table 4.7 Composition of gasification product gas (N_2 free) before (A1 and A2) and after (B1 and B2) being reformed by the char catalyst[a]

Components	Mallee wood		Wheat straw	
	A1	B1	A2	B2
H_2	50.62	54.26	49.18	53.31
CO	17.05	18.54	16.83	17.58
CO_2	23.99	19.61	25.47	21.00
CH_4	8.34	7.59	8.52	8.12

[a]The sulfur and chlorine contents of the gasification product gases were not measured, and N_2 content in the biogas was in the range of 50–64%

4.7 Summary

In this chapter, we have introduced and discussed the following topics:

- Hydrogen is the best fuel for all types of fuel cells. In addition, there are other fuels which can be used in fuel cells either by direct oxidation or through steam reforming or other processes to release hydrogen. This includes gaseous fuels such as natural gas and liquid fuels such as methanol and ethanol. Energy density of fuels is defined as mass and volume specific energy density. It is important to specify the energy density values based on high or low heat value enthalpy.
- Hydrogen can be produced by various routes from hydrocarbon and alcohol reforming, water electrolysis, chemical looping, to photosynthesis. Hydrogen production by steam reforming of fossil fuels is not sustainable. By definition, only the hydrogen produced by electrochemical, photo-electrochemical, or photocatalytic processes from water and renewable energy sources like solar and wind power is green and sustainable. The dominant technology for the green hydrogen production is the water electrolysis using renewable electricity. Nevertheless, the intermittent nature of the renewable energy supply has significant implications in the applicability and installation of large-scale water electrolysis technologies.
- There are three water electrolysis technologies for the hydrogen production ranging from low-temperature AEC and PEMEC to high-temperature SOEC. Among them, AEC is the most mutual technology and has been commercially available for over a century. The technology is, however, inherently bulky and less dynamic in operation. PEMEC has the significant advantages of being compact, dynamic operation, and high efficiency, but its application in large-scale is hindered by the extremely scare supply and high cost of iridium-based electrocatalysts. High-temperature SOEC has the potential for large-scale hydrogen production, but the challenge is the long-term durability and thermal cyclability.
- Effective hydrogen storage is a significant challenge in the hydrogen-based energy society. The most viable technology for hydrogen storage for the vehicle application is the compression tank at pressures of up to 700 bar.

- From storage and transportation points of view, the development of liquid fuel-based fuel cells is a clearly attractive option. However, breaking C-H or O–H bonds is electrochemically challenging. The most critical issues in the increase of power output of liquid fuels such as methanol, ethanol, formic acid, and hydrazine are the development of effective electrocatalysts and the increase in the operating temperature of fuel cells.
- Among various liquid fuels, ammonia (NH_3) is of special interest as it is carbon-free and contains 17.6 wt% H_2. It has a boiling temperature of -33.34 °C and can be liquefied under mild pressures. Ammonia can be used as a direct fuel for AFCs and SOFCs, however, not for PEMFC. Nevertheless, alternative synthesis of ammonia through renewable energy under ambient conditions proves very challenging.
- The zero-emission mode of fuel cell operation can be achieved based on the pure hydrogen fuel. Other fuels of hydrocarbons or alcohols are sustainable only when they are synthesis from the captured CO_2 and renewable power. Other issues such as the toxicity and possible emissions of the products of direct electrochemical oxidation of hydrocarbon or alcohol fuels in fuel cells need to be taken into account in selection of fuels.

4.8 Questions

4.1. Explain the following terms and materials:

Mass specific energy density and volume specific energy density
Electrolysis and zero-gap configuration
AEC, PEMEC and SOEC
Heterostructured catalysts
Steam reforming and water–gas shift reaction
Reversible capacity and heat of adsorption
Liquid fuel and alcohol oxidation reaction
Catalytic activity of alloy catalysts and atomic ratio
Direct pathway and indirect pathway
Nano-wire network structured Pd catalysts
Core-shell structured Ni@NiP/C catalysts

4.2. Water contains 11.1 wt% H_2, while ammonia contains 14.7 wt% H_2. Why ammonia not water is considered as a hydrogen carrier?

4.3. Given the energy efficiency of a water electrolysis is 61.5%, calculate the energy consumption in kWh for the production of 1 cubic meter of hydrogen at 1 atm and 273 K. If the efficiency is 100%, what is the energy consumption? (Answer: 4.79 kWh/m^3 H_2 and 2.94 kWh/m^3 H_2)

4.4. In PEM electrolysis cells, the state-of-the-art electrocatalyst for water oxidation or oxygen evolution reaction is IrO_2 or IrO_2/RuO_2 mixture oxides.

Discuss the prospects of the development of alternative electrocatalysts for iridium metal oxides based electrocatalysts for PEMEC.

4.5. Discuss the advantages, disadvantages, and application areas of alkaline water electrolyzer, PEM electrolyzer, and solid oxide electrolyzer.

4.6. Calculate the volume of 1 kg hydrogen at ambient atmosphere and 80 MPa at 20 °C. (Answers: 11,931 Lt at ambient atmosphere and 15.1 Lt under 80 MPa.)

4.7. Referred to Fig. 4.9 and related literatures, discuss the kinetic transition process of FAOR on Pt/C catalysts as a function of temperatures. If the electrocatalyst is Pd/C, would similar kinetic transition of FAOR occur?

4.8. Discuss the challenges and possible solutions for the promotion of the efficiency and yield of ammonia synthesized via electrochemical and/or photo-electrochemical processes.

4.9. Discuss the potential challenges in the development of direct ammonia fuel cells.

4.10. Discuss the environmental effects of DMFC, DEFC, and DHzFC.

References

1. Hauch A, Ebbesen SD, Jensen SH, Mogensen M (2008) Highly efficient high temperature electrolysis. J Mater Chem 18(20):2331–2340
2. Carmo M, Fritz DL, Merge J, Stolten D (2013) A comprehensive review on PEM water electrolysis. Int J Hydrogen Energy 38(12):4901–4934
3. Cheng Y, Jiang SP (2015) Advances in electrocatalysts for oxygen evolution reaction of water electrolysis -from metal oxides to carbon nanotubes. Prog Nat Sci Mater Int 25(6):545–553
4. Li JW, Xu PM, Zhou RF, Li RC, Qiu LJ, Jiang SP, Yuan DS (2019) CO_9S_8-Ni_3S_2 heterointerfaced nanotubes on Ni foam as highly efficient and flexible bifunctional electrodes for water splitting. Electrochim Acta 299:152–162
5. Li JW, Jiang LJ, He S, Wei LC, Zhou RF, Zhang JM, Yuan DS, Jiang SP (2019) Heterostructured $Ni(OH)_2/Ni_3S_2$ supported on Ni foam as highly efficient and durable bifunctional electrodes for overall water electrolysis. Energy Fuels 33(11):12052–12062
6. Wang MY, Wang Z, Gong XZ, Guo ZC (2014) The intensification technologies to water electrolysis for hydrogen production—a review. Renew Sustain Energy Rev 29:573–588
7. Danilovic N, Subbaraman R, Chang KC, Chang SH, Kang YJJ, Snyder J, Paulikas AP, Strmcnik D, Kim YT, Myers D, Stamenkovic VR, Markovic NM (2014) Activity-stability trends for the oxygen evolution reaction on monometallic oxides in acidic environments. J Phys Chem Lett 5(14):2474–2478
8. Norskov JK, Bligaard T, Logadottir A, Kitchin JR, Chen JG, Pandelov S, Norskov JK (2005) Trends in the exchange current for hydrogen evolution. J Electrochem Soc 152(3):J23–J26
9. Laguna-Bercero MA (2012) Recent advances in high temperature electrolysis using solid oxide fuel cells: a review. J Power Sources 203:4–16
10. Pan WP, Chen KF, Ai N, Lu Z, Jiang SP (2016) Mechanism and kinetics of Ni-Y_2O_3-ZrO_2 hydrogen electrode for water electrolysis reactions in solid oxide electrolysis cells. J Electrochem Soc 163(2):F106–F114
11. Chen K, Jiang SP (2016) Review-materials degradation of solid oxide electrolysis cells. J Electrochem Soc 163(11):F3070–F3083

12. Jensen JO, Vestbo AP, Li Q, Bjerrum NJ (2007) The energy efficiency of onboard hydrogen storage. J Alloy Compd 446:723–728
13. Soloveichik GL (2014) Liquid fuel cells. Beilstein J Nanotechnol 5:1399–1418
14. Nichols RJ (2003) The methanol story: a sustainable fuel for the future. J Sci Ind Res 62(1–2):97–105
15. Zhang J, Lu SF, Xiang Y, Jiang SP (2020) Intrinsic effect of carbon supports on the activity and stability of precious metal based catalysts for electrocatalytic alcohol oxidation in fuel cells: a review. Chemsuschem 13(10):2484–2502
16. Demirci UB (2007) Direct liquid-feed fuel cells: thermodynamic and environmental concerns. J Power Sources 169(2):239–246
17. Cao D, Lu GQ, Wieckowski A, Wasileski SA, Neurock M (2005) Mechanisms of methanol decomposition on platinum: a combined experimental and ab initio approach. J Phys Chem B 109(23):11622–11633
18. Wang K, Gasteiger HA, Markovic NM, Ross PN (1996) On the reaction pathway for methanol and carbon monoxide electrooxidation on Pt-Sn alloy versus Pt-Ru alloy surfaces. Electrochim Acta 41(16):2587–2593
19. Wang SY, Wang X, Jiang SP (2008) PtRu nanoparticles supported on 1-aminopyrene-functionalized multiwalled carbon nanotubes and their electrocatalytic activity for methanol oxidation. Langmuir 24(18):10505–10512
20. Kavanagh R, Cao XM, Lin WF, Hardacre C, Hu P (2012) Origin of low CO_2 selectivity on platinum in the direct ethanol fuel cell. Angew Chem Int Ed 51(7):1572–1575
21. Hitmi H, Belgsir EM, Leger JM, Lamy C, Lezna RO (1994) A kinetic, analysis of the electrooxidation of ethanol at a platinum-electrode in acid-medium. Electrochim Acta 39(3):407–415
22. Liang ZX, Zhao TS, Xu JB, Zhu LD (2009) Mechanism study of the ethanol oxidation reaction on palladium in alkaline media. Electrochim Acta 54(8):2203–2208
23. Wang DL, Lu SF, Jiang SP (2010) Tetrahydrofuran-functionalized multi-walled carbon nanotubes as effective support for Pt and PtSn electrocatalysts of fuel cells. Electrochim Acta 55(8):2964–2971
24. Xu CW, Tian ZQ, Shen PK, Jiang SP (2008) Oxide (CeO_2, NiO, Co_3O_4 and Mn_3O_4)-promoted Pd/C electrocatalysts for alcohol electrooxidation in alkaline media. Electrochim Acta 53(5):2610–2618
25. Xie S-W, Chen S, Liu Z-Q, Xu C-W (2011) Int J Electrochem Sci 6:882–888
26. Neurock M, Janik M, Wieckowski A (2009) A first principles comparison of the mechanism and site requirements for the electrocatalytic oxidation of methanol and formic acid over Pt. Faraday Discuss 140:363–378
27. Larsen R, Ha S, Zakzeski J, Masel RI (2006) Unusually active palladium-based catalysts for the electrooxidation of formic acid. J Power Sources 157(1):78–84
28. Wang SY, Wang X, Jiang SP (2008) Controllable self-assembly of Pd nanowire networks as highly active electrocatalysts for direct formic acid fuel cells. Nanotechnology 19(45)
29. Zhu YM, Khan Z, Masel RI (2005) The behavior of palladium catalysts in direct formic acid fuel cells. J Power Sources 139(1–2):15–20
30. Tarasevich MR, Korchagin OV (2013) Electrocatalysis and pH (a review). Russ J Electrochem 49(7):600–618
31. Yu EH, Krewer U, Scott K (2010) Principles and materials aspects of direct alkaline alcohol fuel cells. Energies 3(8):1499–1528
32. Varcoe JR, Atanassov P, Dekel DR, Herring AM, Hickner MA, Kohl PA, Kucernak AR, Mustain WE, Nijmeijer K, Scott K, Xu TW, Zhuang L (2014) Anion-exchange membranes in electrochemical energy systems. Energy Environ Sci 7(10):3135–3191
33. Yan WR, Xiang Y, Zhang J, Lu SF, Jiang SP (2020) Substantially enhanced power output and durability of direct formic acid fuel cells at elevated temperatures. Adv Sustain Syst 4(7):2000065
34. Cheng Y, Zhang J, Lu SF, Jiang SP (2020) Significantly enhanced performance of direct methanol fuel cells at elevated temperatures. J Power Sources 450:227620

35. Wang H, Jusys Z, Behm RJ (2004) Ethanol electrooxidation on a carbon-supported Pt catalyst: reaction kinetics and product yields. J Phys Chem B 108(50):19413–19424
36. Wang L, Xia MK, Wang H, Huang KF, Qian CX, Maravelias CT, Ozin GA (2018) Greening ammonia toward the solar ammonia refinery. Joule 2(6):1055–1074
37. Jia HP, Quadrelli EA (2014) Mechanistic aspects of dinitrogen cleavage and hydrogenation to produce ammonia in catalysis and organometallic chemistry: relevance of metal hydride bonds and dihydrogen. Chem Soc Rev 43(2):547–564
38. Zheng JY, Lyu YH, Qiao M, Wang RL, Zhou YY, Li H, Chen C, Li YF, Zhou HJ, Jiang SP, Wang SY (2019) Photoelectrochemical synthesis of ammonia on the aerophilic-hydrophilic heterostructure with 37.8% efficiency. Chem 5(3):617–633
39. Rees NV, Compton RG (2011) Carbon-free energy: a review of ammonia- and hydrazine-based electrochemical fuel cells. Energy Environ Sci 4(4):1255–1260
40. Xu W, Wu ZC, Tao SW (2016) Urea-based fuel cells and electrocatalysts for urea oxidation. Energy Technol 4(11):1329–1337
41. Alvarez-Ruiz B, Gomez R, Orts JM, Feliu JM (2002) Role of the metal and surface structure in the electro-oxidation of hydrazine in acidic media. J Electrochem Soc 149(3):D35–D45
42. Aldous L, Compton RG (2011) The mechanism of hydrazine electro-oxidation revealed by platinum microelectrodes: role of residual oxides. Phys Chem Chem Phys 13(12):5279–5287
43. Zhang J, Cao XY, Guo M, Wang HN, Saunders M, Xiang Y, Jiang SP, Lu SF (2019) Unique Ni crystalline core/Ni phosphide amorphous shell heterostructured electrocatalyst for hydrazine oxidation reaction of fuel cells. ACS Appl Mater Interfaces 11(21):19048–19055
44. Li X, Liu W, Zhang M, Zhong Y, Weng Z, Mi Y, Zhou Y, Li M, Cha JJ, Tang Z, Jiang H, Li X, Wang H (2017) Strong metal-phosphide interactions in core-shell geometry for enhanced electrocatalysis. Nano Lett 17(3):2057–2063
45. Dong L, Asadullah M, Zhang S, Wang XS, Wu HW, Li CZ (2013) An advanced biomass gasification technology with integrated catalytic hot gas cleaning Part I. Technology and initial experimental results in a lab-scale facility. Fuel 108:409–416
46. Chen KF, Zhang L, Ai N, Zhang S, Song Y, Song YC, Yi Q, Li CZ, Jiang SP (2016) Feasibility of direct utilization of biomass gasification product gas fuels in tubular solid oxide fuel cells for on-site electricity generation. Energy Fuels 30(3):1849–1857

Part II
Polymer Electrolyte Membrane Fuel Cells

Chapter 5
Polymer Electrolyte Membrane Fuel Cells: Principles and Materials

5.1 Introduction

The very first fuel cell invented by W. R. Grove in 1838 was based on using sulfuric acid as electrolyte. About 100 years later when F. T. Bacon made the first demonstration fuel cell he used alkaline electrolyte which was then adapted for the first space application. These acid and alkaline electrolytes have advantages of high conductivity and low cost; however, as highly corrosive liquids they need to be circulated or immobilized in a matrix as in phosphoric acid fuel cell (PAFC), alkaline fuel cell (AFC), and molten carbonate fuel cell (MCFC).

As a comparison, a solid polymer electrolyte membrane possesses mechanical strength and flexibility and therefore is simple and possible to construct all-solid cells with water as the only liquid in the system. Also as the cell separator material a polymer electrolyte exhibits an excellent resistance to the crossover of gaseous reactants. In addition, functional sites responsible for ion transport are chemically bonded to the polymer itself, issues such as sealing, interconnecting, assembling, and handling are less complex for PEMFC than for other types of fuel cells using liquid electrolytes. The PEMFC has achieved very high power densities. While other fuel cell technologies operate at current densities of a few hundred mA/cm^2, a PEMFC can operate at above 1 A/cm^2. Other desirable attributes include fast start-up capability and rapid response to load changes. As a result, the PEMFC is particularly suited for vehicular propulsion besides for stationary power applications of scale less than 100 kW.

From material points of view, three types of low-temperature fuel cells, PAFC, AFC, and PEMFC are based on electrolytes that are conducting proton-containing species. In the following sections, general aspects of these proton conducting materials are first discussed. The rest of this chapter is then devoted to other key materials of PEMFC.

S. P. Jiang and Q. Li, *Introduction to Fuel Cells*,
https://doi.org/10.1007/978-981-10-7626-8_5

5.2 Proton-Conducting Electrolytes

5.2.1 Electrolytes for Low-Temperature Fuel Cells

Polymer electrolyte membranes (PEMs) usually consist of a polymer network onto which functional groups for ion exchange are attached. The functional groups are typically sulfonic acid, and therefore, the membranes are proton conductive. For a fuel cell to function, the membrane must prevent passage of electrons and permeation or crossover of reactants. At the same time, the membrane provides mechanical support for construction of the membrane electrode assembly (MEA). Finally, the membrane must be resistant to the extreme reducing and oxidative environments at each side. These essential requirements for polymer electrolyte membrane materials of PEMFC are listed in Table 5.1.

Electrolytes can conduct different kinds of ions. For fuel cells operating with hydrogen and oxygen, the first criterion for selection of electrolyte materials is that the conducting ions through the electrolyte should be relevant to the electrode reactions of either hydrogen or oxygen, preferably H^+ from the hydrogen oxidation or $O^=$ from the oxygen reduction. Other transporting ions that are carrying either H^+ or $O^=$, for example, OH^-, H_3O^+ or $CO_3^=$ are also possible; however, the balance of these charge carrying species must be managed in configuration and operation of the fuel cell.

For an overall reaction of a hydrogen–oxygen fuel cell:

$$H_2 + \frac{1}{2}O_2 = H_2O \tag{5.1}$$

Table 5.1 Requirements and motivations of PEM

Requirements	Motivation
• High conductivity	• To minimize the ohmic voltage losses
• Good electronic insulation	• To avoid short circuit of the cell
• Little reactant permeability	• To improve fuel utilization and open circuit voltage
• Mechanical strength and flexibility	• To withstand thermal/load/humidity cycling
• Chemical inertness and stability	• To minimize corrosion and degradation
• Low cost and high availability	• To promote commercialization
• Compatibility with electrodes	• To get good interfacial contact

When the charge carrier is protons (H^+) transporting from anode to cathode, the anode oxidation product of hydrogen must be bare protons. At the cathode side, the protons migrated from the anode must be consumed in order to avoid any buildup of the ion concentration at the electrode surface. This is illustrated in Scheme 5.1, where red boxes indicate the matching of the electrode reaction with the transporting species of the electrolyte. Formation and consumption of protons at two electrodes are straightforward and such a fuel cell may have a very simple configuration, i.e., by supply of hydrogen on the anode side and oxygen or air at the cathode side. A proton-conducting electrolyte means also a chemically acidic environment, which can be translated into challenges for stabilities of catalysts and construction materials.

It is, however, very seldom that a proton-conducting electrolyte exhibits conductivity of bare protons. Concentrated phosphoric acid is almost of the type. Proton exchange membranes, as the name hints, have capability of exchanging protons. The conducting species are, however, protons solvated by water molecules, i.e., the hydrated form of protons, e.g., H_3O^+. In this case, the anode oxidation of hydrogen should produce the hydrated protons, instead of bare protons H^+. A fuel cell based on H_3O^+ conducting electrolytes will only work when hydrogen together with water is supplied to the anode. The cathode reaction must therefore consume the hydrated protons and generate more water, as shown in Scheme 5.2.

It is seen from the scheme that use of an electrolyte conducting hydrated protons, instead of simple protons, makes the configuration of a fuel cell complicated in terms of supplying water on the anode side and removing water from the cathode side. In fact, each proton normally carries more than one water molecule, as to be discussed

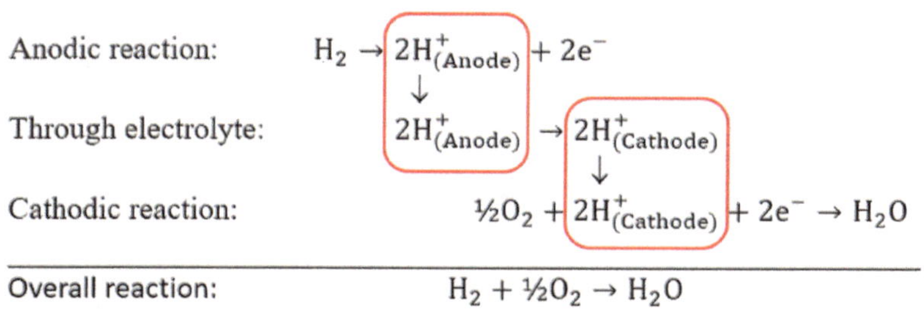

Scheme 5.1 Electrode reactions for fuel cells based on bare proton-conducting electrolytes

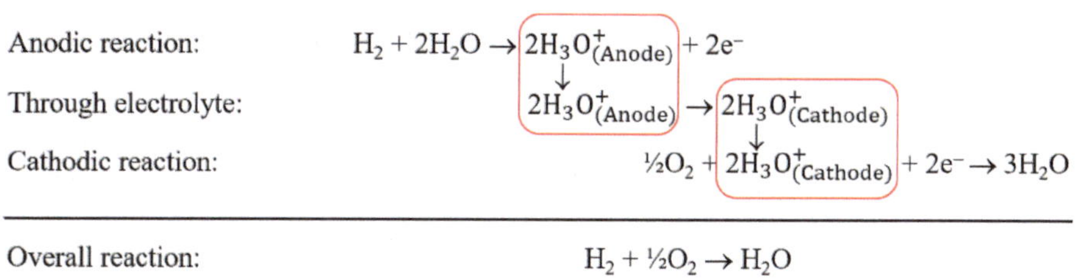

Scheme 5.2 Electrode reactions for fuel cells based on hydrated proton-conducting electrolytes

in the electroosmotic water drag in Chap. 6. The involvement of water in fuel cells limits the operational temperatures below its boiling point, which in turn dictates the problems of, e.g., poisoning of fuel impurities.

When the electrolyte is an anion conductor, e.g., OH^-, $O^=$ or $CO_3^=$, these negative charge carriers transport from cathode to anode. This type of fuel cells has a feature that the by-product water is generated at the anode side. When the anion species is hydroxide (OH^-), the cathode reaction of oxygen reduction must generate the hydroxide ions — which needs a supply of water together with the oxygen. At the anode side, the hydrogen oxidation should involve consumption of the hydroxide (see Scheme 5.3). There is apparently an issue of water management when water is consumed at the cathode and produced in a doubled amount at the anode.

Figure 5.1 summarizes the above discussion on the matching of electrolyte charge carriers and fuel cell electrode reactions. The bare H^+ conductivity of the electrolyte gives the simplest electrode reactions in terms of water management, which is nearly the case for PAFC. Conduction of H_3O^+ instead of H^+ needs to adapt the electrode reactions in order to manage the balance of the charge carrier, which means involvement of parasitic transport of water molecules and therefore need of humidification of the fuel (and air as well). This is the case for PEMFC.

In the case of AFCs where the conduction of OH^- anions is involved, water management is also an issue; i.e., addition of water is required to the air stream. Management of water is critical for low-temperature AFC when water is generated as a liquid, which leads to changes of the electrolyte concentration, volume and distribution within the fuel components.

Three types of electrolytes involving the proton-conducting species are used for the development of low-temperature fuel cells, potassium hydroxide, phosphoric acid, and proton exchange membranes. Characteristics of these types of fuel cells are briefly listed in Table 5.2.

Cathodic reaction: $\frac{1}{2}O_2 + H_2O + 2e^- \rightarrow 2OH^-_{(Cathode)}$

Through electrolyte: $2OH^-_{(Cathode)} \rightarrow 2OH^-_{(Anode)}$

Anodic reaction: $H_2 + 2OH^-_{(Anode)} \rightarrow 2H_2O + 2e^-$

Overall reaction: $H_2 + \frac{1}{2}O_2 \rightarrow H_2O$

Scheme 5.3 Electrode reactions for fuel cells based on hydroxide conducting electrolytes

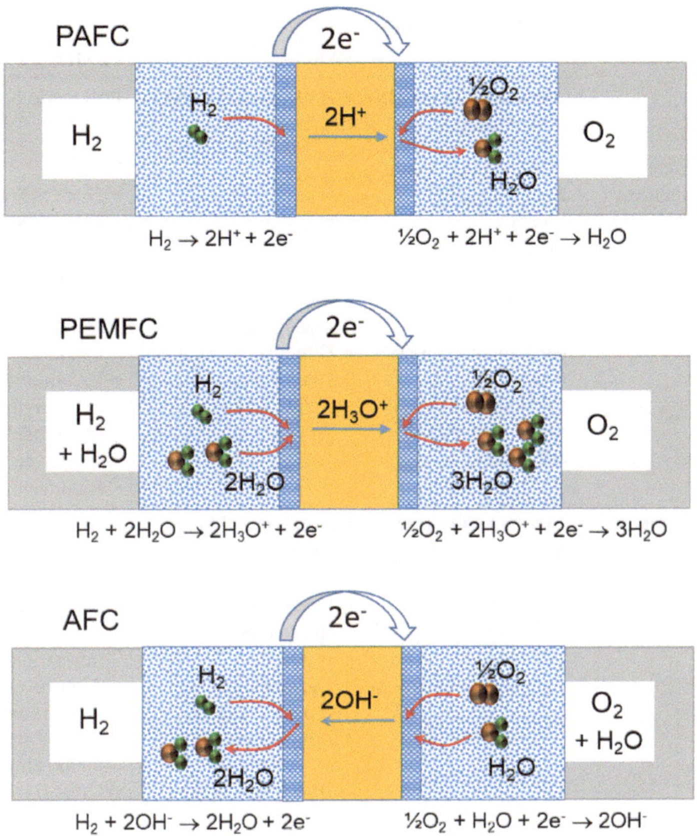

Fig. 5.1 Summary of three types of low-temperature fuel cells

5.2.2 Proton Conduction Mechanisms and Materials

5.2.2.1 Proton Chemistry

The hydrogen atom is unique in a way that it possesses only one electron. When the electron is lost, it has no electron shell to shield the nucleus. It is therefore that this chemical ion (H^+) is called a proton in physics. It is the only ion that has a femtometer dimension, 0.84 fm (1 fm = 10^{-15} m or 10^{-6} nm), about 1/64,000 of the radius of a hydrogen atom, much smaller than any other ions. Proton has following specific characteristics:

- No electron shell of its own;
- Extremely small in size;
- Strong interaction with neighbors;

Table 5.2 Characteristics of three low-temperature fuel cells

	AFC	PAFC	PEMFC
Electrolyte	KOH soaked in matrix or circulating	H_3PO_4 soaked in SiC matrix	Polymeric bound -SO_3H
Temperature	70–250 °C	180–210 °C	60–90 °C
Major Advantages	• Excellent ORR kinetics • Wide catalyst selection (Ni, Ag, and oxides) • Excellent performance on H_2 and O_2 • Cheap construction materials (Nickel and plastics)	• Use of reformate hydrogen (CO_2 rejecting, CO tolerant) • Easy water management • Effective cooling and heat recovery • Value of heat recovery	• Fast ORR kinetics • High current or power density • Minimal corrosion for construction materials (carbon or metals) • Essentially solid electrolyte with resistance to gas crossover • Rapid start-up
Major disadvantages	• CO- and CO_2-free hydrogen (high purity H_2) • Air scrubbing • Bulky system	• Noble metal-based catalysts • Slow ORR kinetics • Highly corrosive (limited construction materials e.g. graphite separator plates) • Low current and power density • Bulky and heavy systems • External fuel processing (e.g. natural gas reforming) and necessary WGS	• Noble metal-based catalysts • Strong CO poisoning • Nearly pure H_2 and associated infrastructure issue • A narrow temperature range • Critical water management • Critical cooling at high loads • Little value of heat recovery
Applications	• On-board power for space applications • Limited terrestrial uses	• Stationary uses	• Electricity utility • Automobile • Portable

- Has extremely strong attraction to electrons—forming covalent bonds by sharing electron pairs and hydrogen bonds.

A proton, with a positive charge and no steric restriction, may approach very close to a neighboring atom or ion, which always is surrounded by a negatively charged electron shell. As a result, this extremely reactive cation exists only in some very special cases such as in plasma, solar wind or in a synchrotron ring. In condensed phases, protons interact strongly with the environment forming associated entities. In aqueous media, the proton interacts with other molecules or ions forming multinuclear species such as OH^-, H_3O^+, NH_4^+, $H_4PO_4^+$, and $H_2PO_4^-$. A special type

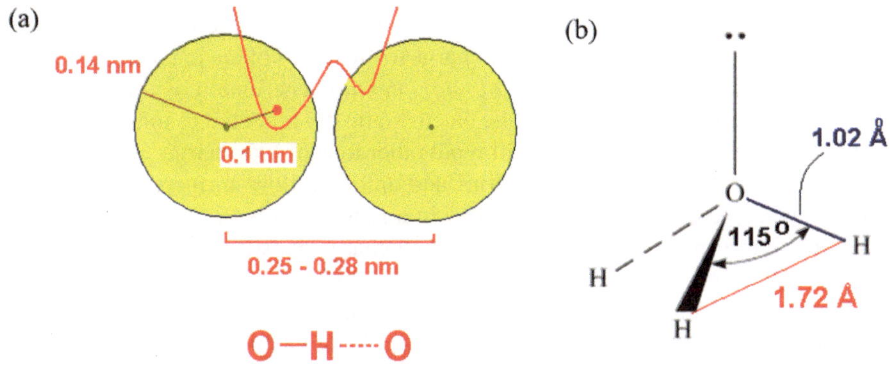

Fig. 5.2 **a)** Formation of hydrogen–oxygen bonds and **b)** molecular structure of a H_2O molecule

Table 5.3 List of various hydrogen bond and hydrogen bond strength

Hydrogen bond	Hydrogen bond strength (kJ/mol)
F–H···:F	155
O–H···:N	29
O–H···:O	21
N–H···:N	13
N–H···:O	8
HO–H···:OH_3^+	18
HO–H···:OH_2	7.9

of attraction between a hydrogen atom and a very electronegative atom with a lone pair of electrons such as N, O, or F atoms is called the *hydrogen bond*, as shown in Fig. 5.2. The hydrogen bond is essentially an electrostatic attraction with some covalent nature, having a bond strength ranging from 4 to 50 kJ per mole except the very strong covalent bond F—H···F (see Table 5.3). In form of an extended network in aqueous media, the hydrogen bond is characterized as being variable, temporary, easy broken and therefore allowing for structure diffusion of protons through the forming-breaking dynamics.

5.2.2.2 Proton Conduction Mechanisms

The unique chemistry of protons suggests two possible mechanisms of proton conduction: the proton transfer with a carrying species (*the vehicle mechanism*) or through a hydrogen bond chain by jumping from one site to another (*the Grotthuss mechanism*). The *vehicle mechanism* involves migration of protons associated with a molecular or ionic carrier (vehicle), while the unloaded carrier species diffuse in the opposite direction. The transport of protons with the help of carrier molecules does not require or take advantage of an infinite hydrogen bond network and therefore follows Stokes law. The mobility of an ion in a viscous solution under an electric field depends on the ratio of the carried charge to the radius of the ion, i.e., the surface charge density of the ion, also called *ionic potential.* The ionic potential describes how strongly or weakly the ion will be electrostatically attracted to ions of opposite charge and to what extent the ion will repel other ions of like charge.

As shown in Fig. 5.3, typical cations and anions follow an increasing trend of mobility with increasing the ionic radius, in a range within 10^{-7} $m^2/V{\cdot}s$. The bare proton has a too small size and hence too high charge density to have any mobility. When a proton is associated with an oxide or water molecule, the formed OH^- or H_3O^+ ions have a similar size as F^- or K^+, respectively; however, they have unusually high mobility. The difference is even more significant in solid phases, for example, in ice, the proton mobility is about 10 million times larger than that of Li^+ (see Table 5.4). This clearly indicates that protons conduct in a radically different mechanism from other ions.

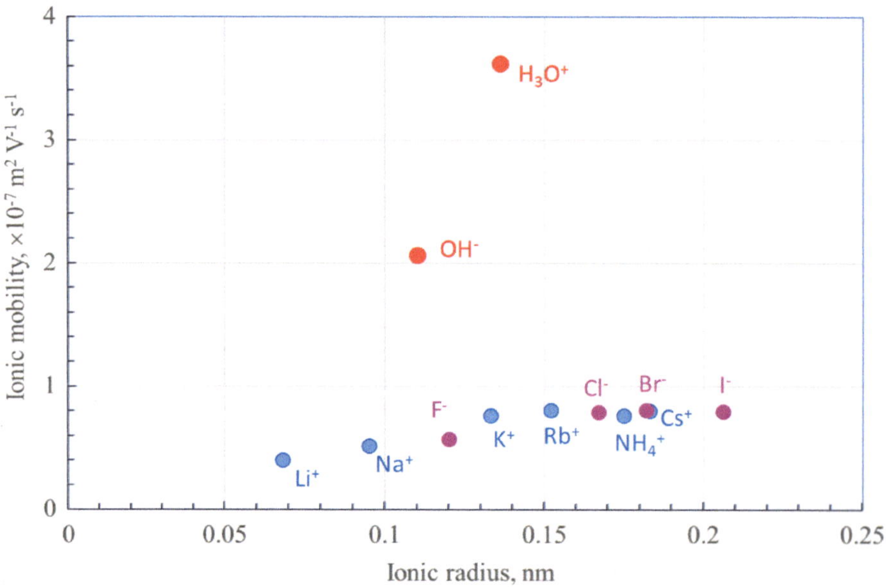

Fig. 5.3 Mobility of some monovalent cations and anions as a function of their ion radii in dilute aqueous solution at room temperature

Table 5.4 Mobility of alkali cations and protons in liquid and solid phases

	Charge carriers	Ionic mobility ($m^2/V{\cdot}s$)
Liquid	K^+ in aqueous solution	7.5×10^{-8}
	Hydrated proton (H_3O^+) in aqueous solution	3.6×10^{-7}
Solid	Li^+ cation in ice (e.g., Li^+)	$\ll 10^{-15}$
	Protons in ice	*ca.* 10^{-7}~10^{-6}

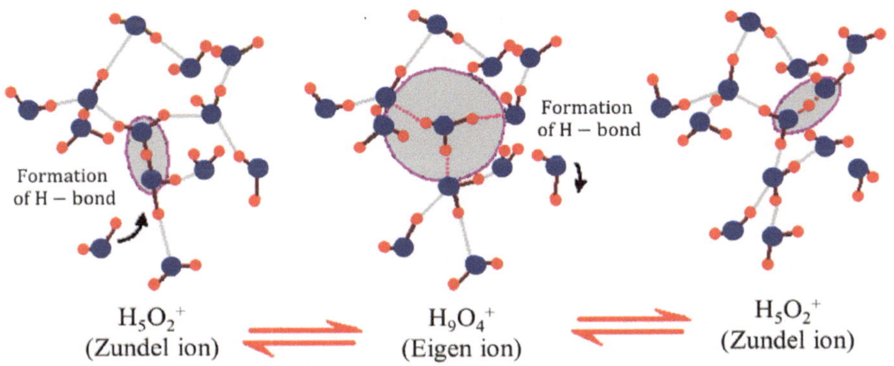

Fig. 5.4 Hydrogen bond clusters via Zundel ($H_5O_2^+$) or Eigen ($H_9O_4^+$) ions. Redrawn from Ref. [1] with permission from Elsevier, Copyright 2000

The mechanism of the proton hopping was first proposed about 200 years ago by T. Grotthuss. A proton or proton defect can jump via formation and breakage of hydrogen bonds with neighboring water molecules through an extensive network. The hydrogen bonds have an intermediate strength, typically between values of van der Waals interactions and covalent chemical bonds. The energetics for formation and breaking of hydrogen bonds can be achieved, resulting in a time-varying distribution of hydrogen bonds among different donor–acceptor pairs. In fact, protons are changing the hydrogen bonding partners within a timescale of 200 femtoseconds though any individual hydrogen atom for 85–90% of the time is involved in hydrogen bonding with one or another oxygen atom.

It should be understood that the hopping mechanism, by breaking an O–H bond in one molecule and forming the same type of bond with another, is a structure diffusion rather than transfer of any individual protons. In order to migrate, the ions must be associated with hydrogen-bonded clusters from H_3O^+ to the more likely Zundel ($H_5O_2^+$) or eigen ($H_9O_4^+$) ions; see Fig. 5.4.

In contrast to the proton or hydrated protons, the hydroxide anions (OH^-) are essentially a water molecule losing a proton. The transport of hydroxide ions through the hydrogen bond chain is achieved by diffusion of the "proton hole," as shown in Fig. 5.5, and is also often associated with water molecules.

5.3 Proton Exchange Membranes

Polymer electrolyte membranes usually consist of a polymer network onto which functional groups for ion exchange are attached. The functional groups are typically sulfonic acid, and therefore, the membranes are proton conducting.

The PEMFCs were invented by Thomas Grubb and Leonard Niedrach at General Electric (GE) in the early 1960s [2]. The first membrane was prepared by sulfonation of poly(phenolformaldehyde), the commercial resin synthesized from phenol

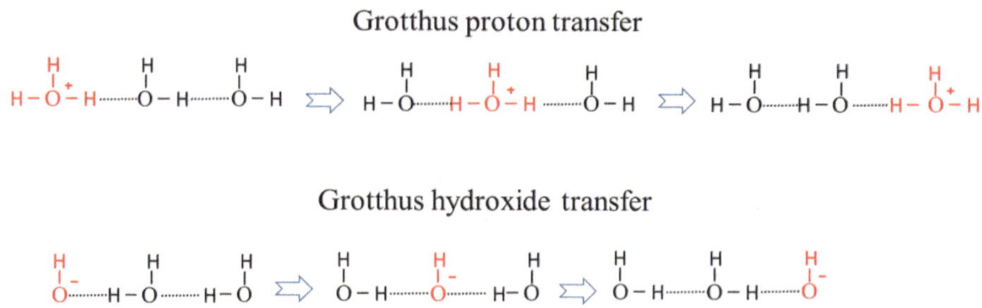

Fig. 5.5 Grotthuss mechanism of proton and hydroxide transfer

(C_6H_5OH) with formaldehyde (HCHO). The obtained poly(phenolformaldehyde sulfonic acid) was however brittle in the dry state and susceptible to hydrolysis. The membrane that exhibited sufficiently promising characteristics was made of sulfonated polystyrene (poly(styrenesulfonic acid), PSSA), showing a lifetime of about 200 h at 60 °C. Further crosslinking the PSSA was made into an inert fluorocarbon matrix. Such mechanically reinforced membranes exhibited acceptable strength in both the wet state and the dry state. The PSSA-based composite membranes were used in several Gemini space missions during the 1960s. This type of composite membranes was, however, not sufficiently durable in the strongly oxidative environment of fuel cells. In the following years, however, PEMFCs lost to AFC in the competition for an auxiliary power source in Apollo and space shuttle programs. Poly(perfluorosulfonic acid) (PFSA) membranes, the first class of so-called ionomers initially developed for the chloro-alkali industry, were adopted in fuel cells for powering the biosatellite mission in later 1960s and led to the renaissance of the PEMFCs from the late 1980s [3].

In the 1980s, Dow Chemical produced a perfluorinated ionomer of a similar structure to Nafion but with a shorter pendant side chain carrying the functional sulfonic acid groups, called short side-chain (SSC) ionomer. Ballard Power Systems showed improvements in fuel cell performance using the SSC ionomer. After Dow Chemical abandoned the industrial production, Solvay developed a new synthetic route for production of this SSC ionomer under a trademark of Aquivion®. Another types of the SSC perfluorosulfonic ionomer are produced by 3M. There are more commercial products of perfluorosulfonic ionomers in the market; however, with polymer structures that are not fully described, for example, those from Asahi Glass or Asahi Kasei.

5.3.1 *Poly(Perfluorosulfonic Acid) Membranes*

In the chemical structures of the above-mentioned poly(phenolformaldehyde sulfonic acid and poly(styrenesulfonic acid) polymers, ionic groups that are covalently bonded

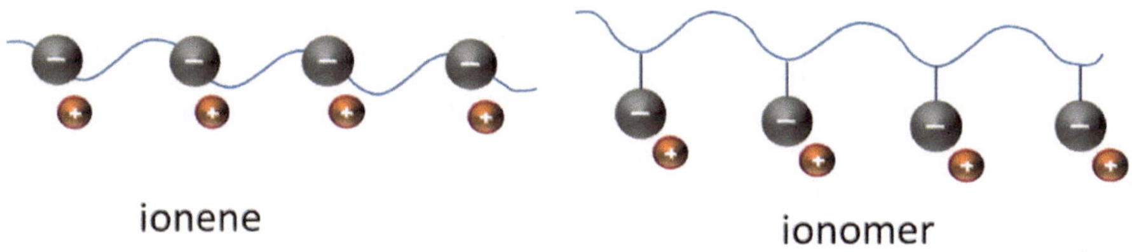

Fig. 5.6 Schematic structure of ionene and ionomer attached to polymer backbones

to the polymer backbone as pendant group moieties are called ionomers. Polyperfluorinated sulfonic acid membranes are also ionomers. Another type of polymers with ionic groups being part of the actual polymer backbones is called ionenes. An example of ionenes is polybenzimidazoles, which after acid doping contain protonated imidazole ring in the backbones; see Fig. 5.6. A difference between the two types of ionic polymers is that ionomer may display a micro-phase separation between the non-polar backbone and the ionic polar domain, the former providing the mechanical stability and the latter the ionic conductivity.

Figure 5.7 shows examples of these ionomers commercialized by DuPont, 3M, and Aquivion. The structures vary slightly in the length of the backbones but mainly in the length of the sulfonic acid terminated side chains. Nafion® is called the long side chain (LSC) while the 3M and Aquivion the short side-chain membranes. In this section detailed discussion is given to properties of Nafion® membrane, followed by brief discussions of the other types.

The nomenclature of Nafion® and Aquivion® membranes contains the molecular weight and thickness. For example, Nafion® 117 represents an EW (equivalent weight) 1100 g/mol and a thickness of 7 mil (1 mil equals to one-thousandth of an

Fig. 5.7 Chemical structures of poly(perfluorosulfonic acid) membranes

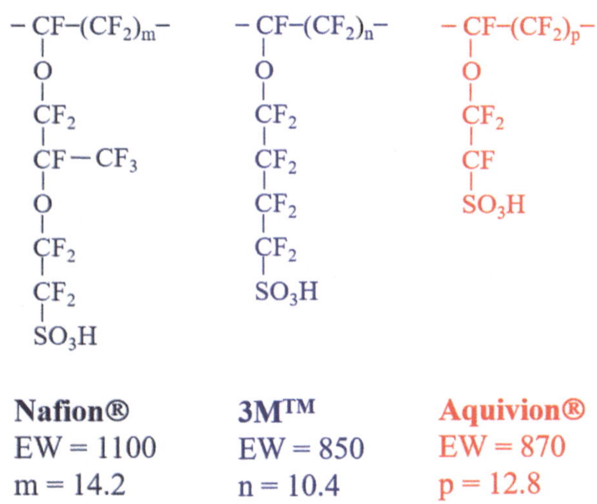

inch or 25 μm). Aquivion® E98-05S represents an EW 980 g/mol and a thickness of 50 μm while the S-suffix indicates a chemically stabilized polymer. Dimensions of both membranes are based on conditions at 23 °C and 50% relative humidity (RH) and change with the water contents. Table 5.5 summarizes properties of typical Nafion and Aquivion membranes.

Table 5.5 Some properties of Nafion and Aquivion type membranes measured at 23 °C and 50% RH after conditioning by water soaking at 100 °C

Membrane type	Nafion® 115	Nafion® 117	Aquivion® E98-05S	Aquivion® E98-09S
Equivalent weight (g/mol)	1100		980	
Thickness (μm)	127	183	50	90
Density (g/cm^3)	1.98		1.93	
Specific weight (g/m^2)	250	360	97	174
Tensile modulus at 23 °C and 50%RH (MPa), Conditioning: 23 °C and 50% RH	249		270	
Tensile modulus at 23 °C and 50% RH (MPa), Conditioning: Water soaked at 100 °C	64		–	
Max. tensile strength (MD/TD, MPa), Conditioning: 23 °C and 50% RH	43/32		40/30 (break)	
Max. tensile strength (MD/TD, MPa), Conditioning: Water soaked at 100 °C	25/24			
Elongation at break (MD/TD, %), Conditioning: 23 °C and 50% RH	225/310		150/200	
Elongation at break (MD/TD, %), Conditioning: Water soaked at 100 °C	180/240		–	
Conductivity (S/cm) 80 °C and 100%HR Uptake (wt%)	0.12		0.16	
From dry to 23 °C and 50%RH	5		–	
From dry to water soaked at 100 °C	38		<25	

5.3.1.1 Morphology and Hydration

The poly(perfluorosulfonic acid)s are copolymers synthesized from tetrafluoroethylene (TFE) and a perfluorinated monomer containing the sulfonic acid functionality. The structure of the latter monomer can vary, which determines the length of the sulfonic acid terminated side chains. The ratio between the two co-monomers determines the equivalent weight of the resulting polymer. The polymers are melt processed into membranes, followed by conversion of the membranes into the salt form by hydrolysis in strong base and finally to the acid form by conversion in nitric acid.

The water uptake behavior and ionic clustering structure of the perfluorinated ionomer membranes is well-documented. The general understanding is that the hydrophobic perfluorinated backbones aggregate in a region with certain crystallinity, which provides the structural integrity to the membrane. The hydrophilic acidic groups accommodate the absorbing water and swelling, which solvates the protons and enables the proton conductivity. Between the ion clusters are likely cylindrical channels, allowing for hydrodynamic permeability of charged as well as uncharged species. On a microscopic scale, this is schematically illustrated in Fig. 5.8. The scale bar to the left indicates, from top to bottom, the increase in hydrophobicity and therefore water content with accompanying transition from crystalline, amorphous to ionic phases.

Water uptake and swelling of perfluorinated ionomer membranes are essential to achieve proton conductivity. The water uptake of a membrane is the weight percent of absorbed water by the dry polymer. The water uptake is often normalized and expressed as hydration number, λ, defined as the number of water molecules per

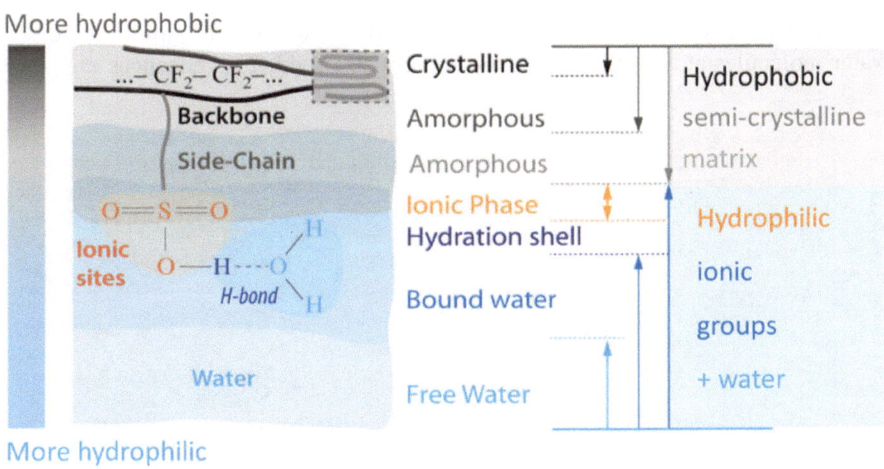

Fig. 5.8 Schematic illustration of hydrated PFSA structure with separation of the hydrophobic backbone and hydrophilic pendant ionic domain as well as bound hydration water. Modified from Ref. [4] with permission from American Chemical Society, Copyright 2017

sulfonic acid group.

$$\lambda = \frac{\text{molar number of } H_2O}{\text{molar number of} - SO_3H} = \frac{\text{weight of } H_2O\ (g)/18\ g/mol}{\text{weight of dry ionomer } (g)/EW\ g/mol} \quad (5.2)$$

The hydration number λ primarily depends on the humidity, temperature, and swelling behavior of the membrane. The latter is a function of the particular chemical structure, sulfonic acid counter cation, thermal history as well as the ion exchange capacity (IEC) or EW of the polymer. For Nafion 117 membranes, for example, $\lambda = 21–22$ after equilibrium in liquid water at room temperature, i.e., around each sulfonic acid group, there are about 21 water molecules absorbed. This corresponds to a water uptake of 34–36 wt% on basis of the dry polymer weight. This water uptake level approximately results in 15–20% increase in the dry thickness; e.g., the typical wet thickness of Nafion 117 membranes is about 200–210 μm, as shown in Fig. 5.9.

From a vapor phase, the water uptake is strongly dependent on the atmospheric humidity. By exposing membrane samples above LiCl solutions of known water activity at room temperature, the water uptake under isopiestic conditions increases with the activity of water. As seen from the figure, over the low water activity range up to 0.75–0.80, there is a slow increase in the water uptake of the membranes. This can be attributed to the strong interaction of the first water molecules with the sulfonic acid sites of the membranes. These water molecules are tightly bonded to the polymer. It has been reported that proton transfer from the sulfonic acid to water forming the H_3O^+ occurs when the very first water molecule is absorbed. During drying of a PFSA membrane, the very last water molecule is found to remain associated with the sulfonic acid groups even at moderate temperatures and under vacuum conditions.

At atmospheric water activities above 0.80, the water content rapidly increases. Water molecules at a level of $\lambda > 6$ are weakly absorbed. This enhances the phase

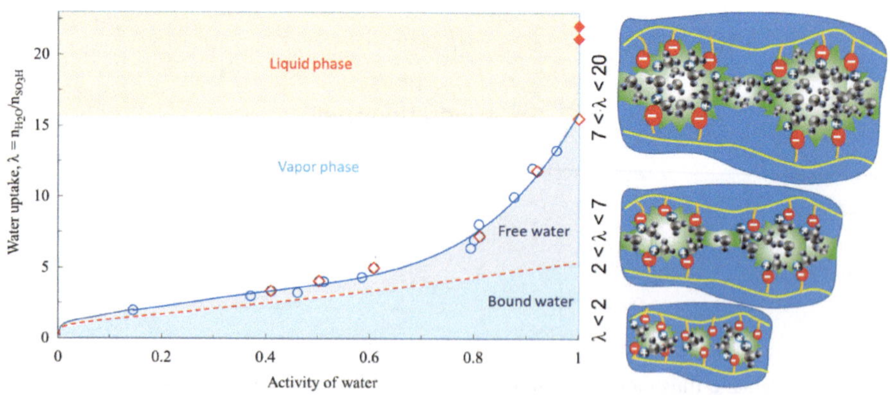

Fig. 5.9 Water uptake of Nafion 117 (in the H^+ form) membranes as a function of water activity and illustration of cluster and network morphologies at varied hydration regions. Open circles are data from Zawodzinski et al. [4] and rest from Li et al. [5]

separation with formation of multiple solvation shells around each sulfonic acid site. Eventually a percolated network of ionic domains is formed, leading to a more bulk-like water region where protons can be conducted over longer length scales.

As the atmospheric humidity approaching to unity, the water uptake reaches a hydration number of about $\lambda = 14$. It is interesting to note that this water uptake is much lower than the number of $\lambda = 21$–22 from the liquid phase. This phenomenon, connected to the skin effect of the membrane, is called Schröders paradox as it does not follow the expectation from thermodynamic equilibrium between liquid and vapor phases of water. In addition, the two types of water uptake show different temperature dependences. While the water uptake from the liquid phase increases with temperature, the water uptake from the vapor phase decreases, reaching about 10 water molecules per sulfonic acid site at 100% RH and 80 °C.

5.3.1.2 Proton Conductivity and Electroosmotic Drag of Water

Early measurements of the ionic conductivity were often made by submerging membrane samples into a liquid supporting electrolyte, e.g., in an aqueous NaCl solution. This ensures the water saturation but also causes sorption of electrolyte species onto the membrane. For PEMFC the membrane should be exposed to an environment of specified water activity, free from any supporting electrolyte.

Typical PFSA membranes in the H^+ form, e.g., Nafion, exhibit an ionic conductivity of about 0.05–0.06 S/cm at room temperature, which is increased to above 0.1 S/cm at about 80 °C and 100% RH. When the temperature approaches the boiling point of water, the proton conductivity declines as water evaporates. The strong dependence of the conductivity on the RH or the water content in the membrane hints a challenge in controlling the water vapor pressure during conductivity measurements. The water uptake also results in dimensional changes of the membrane, which makes it difficult to have a correct definition of the dimensions of the membrane for calculation of the conductivity. The reported proton conductivities in the literature vary to a good deal, and the same is the dependence of proton conductivity on temperatures. The latter can be translated into the activation energy of conductivity, which has been reported in a range of 10–20 kJ/mol, depending very much on the humidity [6].

Example 5.1 What is the minimum requirement for the membrane conductivity for fuel cell applications?

Solution:

This is an often asked question for the development of new electrolyte membranes; however, the question itself is dependent on the cell operation and performance. For a fuel cell operating at 80 °C the reversible cell voltage is about 1.2 V. At a practically operational current density of, say, 1 A/cm^2 the cell voltage should be above 0.6 V, which corresponds to an energy efficiency of higher than ca. 40%. At this current density, the cathode overpotential is about 0.4 V and the anode overpotential is about 0.03 V. In order to achieve a practical cell voltage above 0.6 V, then the maximum ohmic loss of the electrolyte membrane will be

$$1.2 - 0.4 - 0.03 - \Delta V_{\text{iR}} > 0.6\ \text{V} \tag{5.3}$$

$$\Delta V_{\text{iR}} < 0.17\,\text{V} \tag{5.4}$$

This ohmic loss can be translated into the area specific resistance of the electrolyte at the operating current density of 1 A/cm^2:

$$\text{ASR} < \frac{0.17\ \text{V}}{1\ \text{A/cm}^2} = 0.17\ \Omega\text{cm}^2 \tag{5.5}$$

The area specific resistance of the electrolyte membrane is a ratio of the membrane thickness to its conductivity:

$$\text{ASR} = \frac{\text{thickness (cm)}}{\text{conductivity (S/cm)}} < 0.17\ \Omega\text{cm}^2 \tag{5.6}$$

To compromise the gas permeability and mechanical strength, the practical membrane thickness is assumed to be 50 μm (0.005 cm). The minimum conductivity of the membrane is therefore

$$\sigma > \frac{0.005\ \text{cm}}{0.17\ \Omega\ \text{cm}^2} = 0.03\ \text{S/cm} \tag{5.7}$$

For any practical use the membrane should have a conductivity higher than 3×10^{-2} S/cm.

It is the water molecules that solvate the protons from the sulfonic acid groups, forming hydrated complexes such as hydronium cations (H_3O^+), Zundel cations ($H_5O_2^+$), or eigen cations ($H_9O_4^+$). These proton-containing charge carriers are mobile and involved in the charge transport, i.e., via the vehicle mechanism. As discussed in Sect. 5.2.2, this means that water migrates through the membrane during fuel cell operation. The water movement due to the proton transport is called the *electroosmotic drag* and therefore dependent on the current passing through the membrane. An electroosmotic drag coefficient of water is defined as the molar ratio of water moved with each proton in the absence of a concentration gradient — any concentration gradient drives additional movement of water by diffusion. For liquid water equilibrated Nafion membranes at room temperature, a coefficient of 2.5–3.0 H_2O/H^+ has been reported. For water vapor equilibrated membranes with low water content,

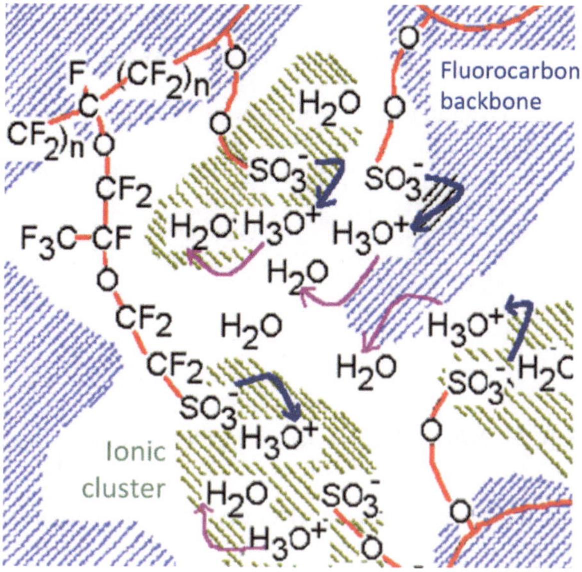

Fig. 5.10 Schematic representation of electroosmotic drag of water through membrane

however, the coefficient is between 1.0 and 1.4 H_2O/H^+, varying with the water vapor activity.

Any water concentration gradient, developed from the humidification of reactant gases, electroosmotic drag or electrochemical formation at the cathode, is a driving force of water diffusion; see Fig. 5.10. Water diffusion is the primary process to balance the electroosmotic drag process. It also determines the dynamic response of the membrane and fuel cells to changes of, e.g., current density, humidification, or temperature. In addition, the water diffusion can also be a major mechanism of the limiting current density of a fuel cell. The water diffusion coefficient for Nafion membranes reported in literature falls in a range of 10^{-6}–10^{-5} cm^2/s depending on the water content.

Example 5.2 Membrane cost calculation.

DuPont has once announced that the future Nation® membrane prices could be as low as $10 per kilowatt assuming an annual production of 150,000 mid-size vehicles. For a fuel cell stack operating at a single cell performance of 0.75 V at 1.0 A/cm^2, estimate the membrane cost of such a power system.

Solution:

Operating at 0.75 V and 1.0 A/cm^2 the single cell power density is

$$P = 0.75\,\mathrm{V} \times 1\,\mathrm{A/cm^2} = 0.75\,\mathrm{W/cm^2} \tag{5.8}$$

The total active area of the membrane used for a 1 kW stack is:

$$\text{Area} = \frac{\text{power}}{\text{power density}} = \frac{1000\,\text{W}}{0.75\,\text{W/cm}^2} = 1333\,\text{cm}^2 = 0.13\,\text{m}^2 \tag{5.9}$$

Assuming that the utilization of membranes is 80% (with edge areas for sealing) and the membrane needed is 0.16 m^2/kW. The membrane cost can then be estimated from the announced 10 US$/kW as

$$\frac{10\,\text{US\$/kW}}{0.16\,\text{m}^2/\text{kW}} = 62.5\,\text{US\$/m}^2 \tag{5.10}$$

5.3.2 *Short Side-Chain PFSA Membranes*

For a fuel cell with an electrolyte with a certain conductivity, the area specific resistance (ASR) increases with increasing distance between the electrodes. The ohmic loss of a PEM fuel cell can thus be lowered by reducing the thickness of the membrane. Another advantage of using thinner membranes is the improved water transport via electroosmotic migration and diffusion, which is an issue when operating at higher current densities. The mechanical strength should be of course improved, which can be achieved by modifying the side-chain structures of the membranes. This was the initial motivation for the development of the SSC membranes.

The SSC PFSA membranes were intensively developed by Dow Chemical in the early 1990s, which was however abandoned due to the challenging synthesis of the sulfonic acid-containing monomers. A simpler production process of the key monomer was later developed by Solvay Solexis, and the SSC PFSA membrane was commercialized under a trade name of Hyflon® and then Aquivion®.

The short side-chain PFSA materials have a high concentration of sulfonic acid groups and therefore proton conductivity. At the same time, the materials have a high degree of crystallinity within the hydrophobic regions, as indicated by the high heat values of fusion, compared with Nafion (Fig. 5.11a). In general Nafion® is completely amorphous (zero crystallinity) in the region 700–900 EW where Aquivion® still maintains a certain crystallinity. The high crystallinity gives the membrane better mechanical properties in the low EW range.

In addition the short side-chain PFSA membranes show better hydration characteristics than Nafion® type membranes. Shown in Fig. 5.11b is the water uptake in liquid water at 100 °C, which is considerably lower for the long side-chain (LSC) (Nafion®) membrane at a given EW or in other words the same water uptake of a low EW LSC ionomer can be obtained with a high EW SSC ionomer. For example, a water uptake of 35% from liquid the phase at 100 °C is typical for 1100 EW Nafion® membranes, which can be achieved with a 925 EW Aquivion® membrane.

As a result, the SSC membranes have a higher softening temperature, e.g., *ca.* 100 °C for Nafion®, *ca.* 125 °C for 3M™ and *ca.* 140 °C for Aquivion® membranes.

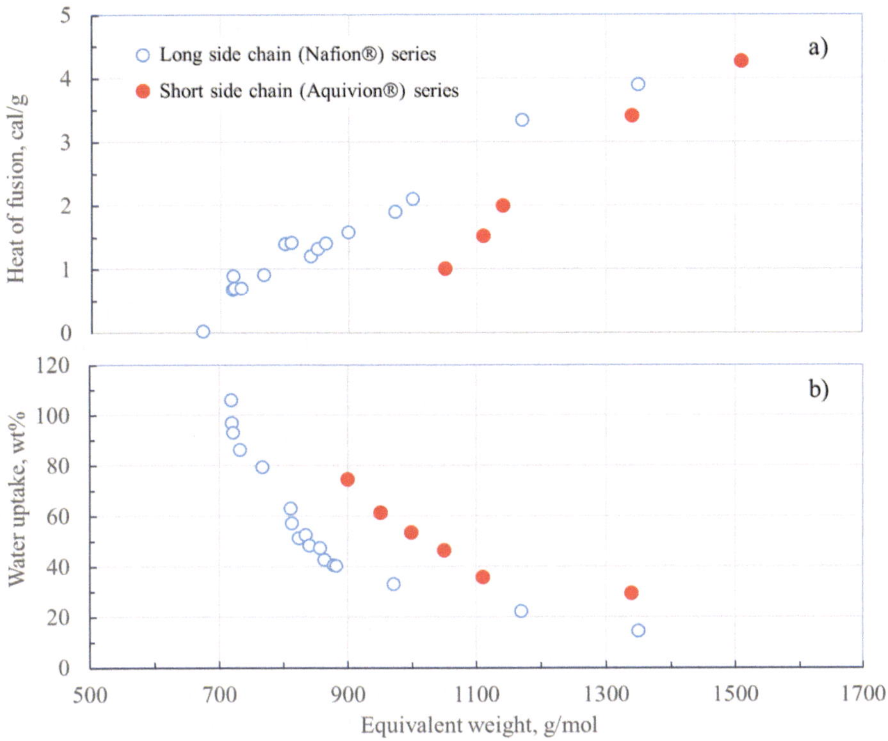

Fig. 5.11 **a)** Heat of fusion and **b)** water uptake of Aquivion® and Nafion® membranes as a function of equivalent weight. The water uptake was measured from liquid water at 100 °C. Replotted from Ref. [7] with permission from Elsevier, Copyright 2005

In addition to the improved thermal stability, the SSC membranes exhibit better hydration characteristics and therefore higher proton conductivity due to the high concentration of highly hydrophilic sulfonic acid groups. For example, the water uptake for Dow membranes with an EW of 858 g/eq. is higher than for Nafion® type membranes and at a degree of hydration λ of, say 15, the proton conductivity of the Dow membrane is 0.11 S/cm at room temperature compared with 0.06 S/cm for a Nafion® 117 membrane [8]. Aquivion® membranes show a proton conductivity of *ca.* 0.1 S/cm at 110–120 °C in the relatively low relative humidity range of < 60%.

5.3.3 PFSA Composite Membranes

A thin and robust membrane is desirable to reduce the area specific resistance. Due to the effective back diffusion of water from the cathode to the anode side through such thin membranes, water management and therefore the average conductivity can also be improved. A challenge for thin membranes is the reduced mechanical strength, especially under swelling.

5.3.3.1 Reinforced Composite Membranes

In the early 1980s, reinforced PFSA membranes were first developed for chlor-alkali electrolysis cells where PTFE woven fabrics were embedded in composite membranes typically of 200-300 μm thickness. In 1990s, W.L. Gore & Associates commercialized composite membranes (Gore-Select®) using thin PTFE porous sheets impregnated with PFSA ionomer from both sides. Reinforced membranes by dispersion of the PTFE fibrils such as Flemion® have also been prepared by extrusion. A set of data of Gore-Select® membranes with comparison with Nafion 117 are compiled in Table 5.6. The reinforced membranes are characterized by small thickness (5–30 μm) and excellent mechanical properties. They have in general lower conductivity than Nafion®. However, by taking the thickness into account, the reinforced membranes have a significantly lower ASR.

Large-scale manufacturing of PTFE supported membranes involves two cycles of dip-coating of the PTFE fibers into the ionomer bath and infrared oven drying, followed by boiling water and air drying. In laboratories, the reinforced composite membrane can be made by vacuum impregnation using Nafion ionomers and porous PTFE matrix [10]. In this process, Nafion ionomers (e.g., 5 wt.% Nafion in a mixture of water and propanol, DuPont) are first converted to the Na^+ form by slowly adding NaOH solution till pH = 6.5–7.0 and the PTFE matrix is converted to hydrophilic form by immersing the porous PTFE in sodium naphthalene solution for a short period. The treated porous PTFE matrix is mounted on a plastic frame and placed into a vacuum chamber which is then filled with the Nafion ionomer solution. The

Table 5.6 A set of characteristic data of Gore-Select® membranes compared with Nafion® 117 [9]

	Gore-Select®	Nafion®-117
Thickness (μm)	20	200
Equivalent weight (g/mol)	1100	1100
Water uptake (wt%)	32	32–34
Conductivity in 1 M H_2SO_4 at 25 °C (S/cm)	0.05	0.14
Tensile strength (MPa)		
• Machine direction (MD)	32.4	15.9
• Transverse direction (TD)	17.7	10.8
Water vapor transmission rates in 24 h (kg/m^2)	25.0	23.6
Shrinkage at dehydration (%)		
• Machine direction (MD)	2.90	10.55
• Transverse direction (TD)	2.95	11.80

introduced negative pressure will induce the Nafion to fill the PTFE micropores. The PTFE-reinforced membrane is then heat-treated at 270 °C for 2 min to remove the solvent and induce the cross-link of the Nafion ionomer. These impregnation and heat-treatment steps can be repeated to obtain a dense composite membrane. The membrane is finally treated in sulfuric acid to restore the acidic form of the membrane.

5.3.3.2 Inorganic Fillers-Based Composite Membranes

Another important type of composite membranes is based on the metal oxides or inorganic fillers or inorganic–organic hybrid nanocomposite. There are various techniques to fabricate metal oxides-based composite membranes including casting, in situ sol–gel reaction and liquid phase deposition (LPD). Among them, the casting method is the most simple one. In this method, nano-sized inorganic particles are blended uniformly with PFSA ionomers to form nanocomposite membranes. The commonly used inorganic fillers include (1) hygroscopic oxides, e.g., silica (SiO_2), titanium oxide (TiO_2), zirconia (ZrO_2), tin oxide (SnO_2), etc.; (2) hygroscopic proton conductors such as zirconium hydrogen phosphates, phosphotungstic acid, silicotungstic acid. Among these inclusion of hygroscopic oxides is reported to improve the water retention characteristics. A nano-dispersed secondary inorganic phase enhances the elastic modulus of the membrane, especially at temperatures close to the softening temperatures. This, in combination with the improved hydration characteristics, allows for a slightly elevated operating temperature. It has been reported that, in some cases, such composite systems show improved conductivity at lower water contents particularly when the inorganic fillers have functionalities adding to the total ion-exchange capacity of the membrane. Most importantly for DMFCs, incorporation of inorganic particles within the polymer matrix gives a reduced permeability of reactants, i.e., methanol.

The challenge to prepare inorganic–organic composite membranes is to achieve good dispersion of the inorganic fillers in the polymeric phase in order to increase interfacial synergies. Instead of simple mixing, in situ formation of inorganic nanoparticles by means of, e.g., sol/gel or ion-exchange techniques are preferred. Figure 5.12 gives an example of the significantly enhanced interface between the SiO_2 nanoparticles and Nafion polymer matrix by modification of the surface properties of the inorganic nanoparticles [11]. In this case, the SiO_2 nanoparticles (NPs) were formed in a mixed solution of tetraethoxysilane (TEOS) and Nation/NMP (*N*-methyl-2-pyrrolidone). In the presence of the Nafion ionomer, self-assembly occurs between the silica with positively charged $SiOH_2^+$ groups and Nafion ionomers with negatively charged sulfonated end group (SO_3^-) by electrostatic force. The Nafion assembled on SiO_2 stabilizes the SiO_2 NPs and prevent the grain growth. The self-assembled Nafion-SiO_2 NPs with particle size 2.8 nm were uniformly distributed. The Nafion/SiO_2 nanocomposite membranes with 5 wt% SiO_2 NPs were then prepared by recasting with uniformly and highly dispersed SiO_2 NPs chemically bonded to Nafion polymer matrix.

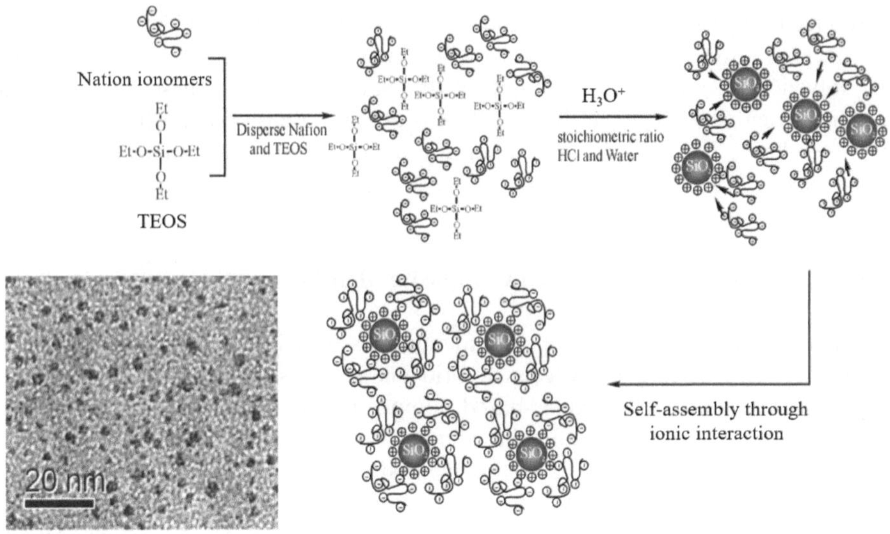

Fig. 5.12 Scheme for the synthesis of Nafion-SiO_2 nanoparticles by self-assembly method. The inset is the TEM image of the as-synthesized Nafion-SiO_2 nanoparticles. Reproduced from Ref. [11] with permission from Elsevier, Copyright 2007

5.3.3.3 Membrane Durability Test

As to be detailed in Chap. 6 the mechanical durability of the membranes can be characterized by measuring the H_2 crossover current in PEM fuel cells under an accelerated wet/dry RH cycling test [11]. In this test, the membrane is stressed by the wet/drying cycling at a constant temperature, e.g., 90 °C. H_2 gas is fed to the anode side which is used as the reference and counter electrode and N_2 is fed to the cathode side which is the working electrode. By scanning potential of the working electrode from 0 to 0.5 V *versus* the hydrogen electrode at a scan rate of, e.g., 0.5 mV/s, a constant oxidation current of the permeated H_2 can be measured, which is directly related to the structural stability of the membrane. Figure 5.13 shows a typical example of the oxidation current of H_2 crossover through a Nafion 212 membrane, a conventional Nafion/SiO_2 composite membrane prepared by a colloidal method and a self-assembled Nafion/SiO_2 composite membrane (see Fig. 5.12) [11]. For the pristine Nafion 212 membrane, the H_2 crossover current started to increase after 3400 wet/dry cycles, however, for the conventional Nafion/SiO_2 composite membrane, the H_2 crossover current increase started much earlier at ~2700 cycles. In the case of the self-assembled Nafion/SiO_2 composite membrane the H_2 crossover current starts to increase after ~5400 cycles, even better than pristine Nafion 212 membrane. This is clearly due to the improved microstructure of self-assembled Nafion/SiO_2 composite membranes as the result of the excellent interface between the Nafion-SiO_2 nanoparticles and Nafion matrix. In the case of the conventional Nafion/SiO_2 composite membrane, the microstructure after tested for 3900 cycles is characterized by the formation of large number of through and open pores with size of 6–10 μm

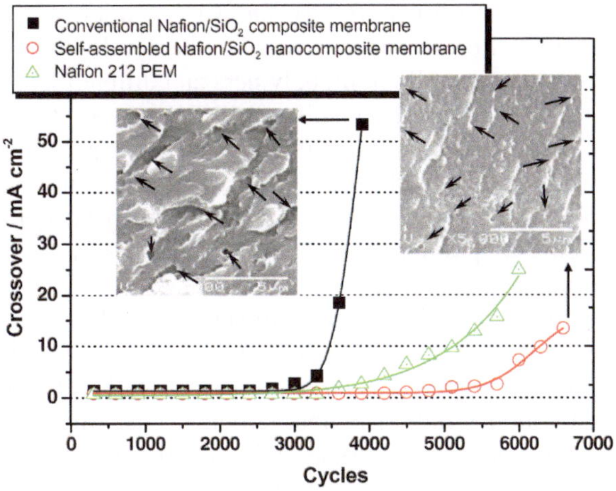

Fig. 5.13 Plots of limiting current densities of the H_2 crossovered through Nafion and Nafion/SiO_2 composite membranes under accelerated wet/dry cycling tests at 90 °C. The inset shows the cross-section SEM images of the conventional and self-assemble Nafion/SiO_2 composite membranes after the cyclic tests. In the insert, the scale bar = 5 μm. Reproduced from Ref. [11] with permission from Elsevier, Copyright 2007

(see the inset, Fig. 5.13). The phase separation indicates the poor interface between Nafion matrix and unmodified SiO_2 nanoparticles of the conventional Nafion/SiO_2 composite membranes.

5.3.4 Alternative Membranes

High price and performance loss at elevated temperature and low relative humidity are limiting the development of PEMFC systems with PFSA electrolyte. A large number of alternative membranes, mostly hydrocarbons based, have been developed and studied in a search for a lower cost and durable alternative. Various types of hydrocarbon membranes have been used in PEMFC devices for portable and vehicle applications. However, hydrocarbon membranes are still under development to achieve satisfactory performance, chemical and mechanical properties, and stability. Furthermore, the apparent cost advantages of many hydrocarbon materials may prove unconvincing when added synthesis and processing costs are considered.

Sulfonated polystyrenes were investigated in the 1960s as the first generation of polymer electrolytes for fuel cells. This type of polymers has shortcomings that the tertiary C–H bonds in the styrene chains are sensitive to oxidation by oxygen and hydrogen peroxide. Partially fluorinated polystyrenes have been investigated by sulfonation of membranes based on α, β, β-trifluoro-styrene monomers. Radiation grafting has also been extensively explored as an approach to introduce acidic functionalities on activated sites in pre-formed polymer films from PTFE, poly(tetrafluoroethylene-*co*-hexafluoropropylene) (FEP), poly(ethylene-*alt*-tetrafluoroethylene) (ETFE), poly(vinylidene fluoride) (PVDF), and others.

The superior chemical stability of perfluorinated polymers is connected to the high bond strength of the C–F bonds. Furthermore, the hydrophobic nature of the backbone structure repels nucleophiles, further protecting it from hydrolysis. Many aromatic structures are also known to exhibit excellent chemical stability. Much research effort has been devoted to the development of sulfonic acid functionalized systems, especially homopolymers and different copolymers based on poly(arylene ether)s, e.g., poly(ether sulfone) (PSU) and poly(ether ether ketone) (PEEK).

Polyimides comprise a family of aromatic polymers with good film forming properties as well as excellent thermal, chemical and mechanical stability. Sulfonated polyimides, especially the six-membered ring derivatives have been thoroughly studied as polymer electrolyte membranes with high proton conductivities and promising durability.

Polybenzimidazoles are another family of aromatic engineering plastics that can be sulfonated to give materials with sufficient proton conductivity. Most work has been done on poly[2,2′-*m*-(phenylene)-5,5′-bibenzimidazole] (PBI or *m*-PBI), which is a commercially available polymer. One way to sulfonate PBI is through electrophilic aromatic sulfonation by first doping the cast membrane with sulfuric acid or oleum followed by heat treatment at 450–600 °C. Another way to introduce the sulfonic acid groups on the phenyl moieties is by using a sulfonated monomer in the condensation polymerization reaction. Alternatively, PBI can be sulfonated through *N*-functionalization. In general, the proton conductivity of the sulfonated polybenzimidazoles is relatively low compared to other sulfonated polymers due to their amphoteric nature. Figure 5.14 shows some typical alternative sulfonated PEMs for fuel cells.

5.4 Electrocatalysts

Electrocatalysts for PEMFCs were initially based on concepts and understanding developed for PAFCs due to the same acidic nature of electrolytes. The evaluation of the catalysts is also often performed in liquid-electrolyte environments for experimental convenience. High catalytic activity is the first criterion for the catalysts. The intrinsic activity of a catalytic site, expressed by the *turnover frequency* (TOF), is defined as the number of electrons transferred by the site per second of time at a specific potential. For a particular catalyst, the practical way of optimizing the catalytic activity is to increase the TOF and the number of active sites in a unit volume or mass. In the catalyst layer or electrodes in fuel cells, another key factor is the utilization of the active sites, i.e., the fraction of the active sites that are actually used during the operation. Depending on the catalyst loading and thickness of the catalyst layer, the common goal for the fabrication of the catalyst layer or electrodes is to achieve a high catalyst utilization. Other fundamental requirements for electrocatalysts include high electrical conductivity, chemical and electrochemical stability under the oxidative or reductive atmosphere and in contact with acidic electrolytes, etc. To ensure that a fuel cell operates at a high energy efficiency, both electrode reactions need to take place close to their thermodynamic potentials. Platinum and platinum-containing catalysts thus far are the most effective catalyst materials in terms of both activity and stability.

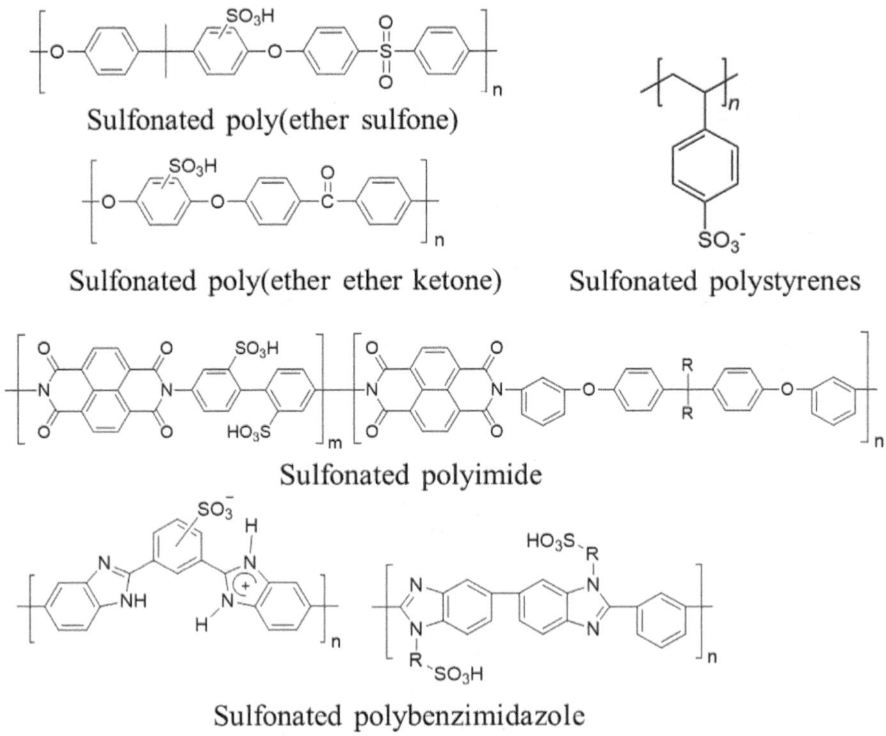

Fig. 5.14 Structures of sulfonated poly(ether sulfone), polystyrene, poly(ether ether ketone), polyimide and polybenzimidazole

5.4.1 Anode Catalysts

Hydrogen, in high purity or mixtures (e.g., reformates from hydrocarbons or oxygenated hydrocarbons), is the most common fuel for PEMFC. Oxygenated fuels such as methanol, ethanol, formaldehyde, and formic acid have also been studied, but only methanol is considered of practical significance. In acidic electrolytes, metals such as Pt and Pd are the most active and stable catalysts with an exchange current density around 10^{-3} A/cm^2. The mechanism of HOR on Pt is assumed to proceed by slow dissociative adsorption of dihydrogen followed by facile charge transfer:

$$H_2 + 2Pt \rightarrow 2Pt - H_{ads} \rightarrow 2Pt + 2H^+ + 2e^- \quad (5.11)$$

As a result of the fast kinetics, the satisfactory anode performance can be achieved with a Pt loading of less than 0.1 mg$_{Pt}$/cm^2. Gas diffusion electrodes based on carbon-supported platinum exhibit a polarization of around 20–30 mV at a current density of 1000 mA/cm^2 under fuel cell conditions.

H_2-rich reformates from steam reforming or partial oxidation of hydrocarbon fuels contain impurities of CO, CO_2 and nitrogen. Nitrogen has only the effect of dilution, while CO decreases the anode performance through poisoning Pt catalysts at the concentration of as low as 10 ppm. CO_2 itself is inert to the catalyst, but it can undergo the reverse water–gas shift reaction and hence poison the catalyst:

$$CO_2 + H_2 \rightarrow CO + H_2O \tag{5.12}$$

CO strongly binds to the Pt sites and hence reduces the number of active sites available for the H_2 adsorption and oxidation. Electrochemical oxidation of CO is thermodynamically possible through the following reaction

$$CO + H_2O \rightarrow CO_2 + 2H^+ + 2e^- \quad E^0 = -0.10\,V \tag{5.13}$$

However, a large overpotential of over 500 mV is required to let the reaction to proceed at a practically meaningful rate on pure Pt surfaces. In the typical operating potential region of a fuel cell anode (i.e., 0.0–0.1 V), CO is deemed electrochemically inert. It should be noted that CO is also an intermediate of direct oxidation reaction of alcohols such as methanol and ethanol.

Efforts to develop CO-tolerant electrocatalysts were started at General Electric in the early 1960s when addition of Ru, Rh, and Ir to Pt blacks was found to improve the tolerance toward CO. Since then, a variety of alloy catalysts such as Pt-Ru, Pt-Sn and Pt-Mo have been developed. Pt-Ru alloys with the typical atomic ratio of 1:1 are among the most promising candidates, which are able to tolerate up to 100 ppm of CO compared to *ca.* 10 ppm for pure Pt. It is proposed that the presence of alloying metals facilitates the oxidation of CO adsorbates via a bifunctional mechanism, i.e., accelerating the water dissociation reaction and/or weakening the CO adsorption. In addition, the bleeding of a small amount of air to the CO-containing anode feed stream [12] is beneficial as the bleed oxygen reacts chemically with CO, although this also consumes a certain amount of hydrogen.

5.4.2 Cathode Catalysts

5.4.2.1 Oxygen Reduction Reaction

Noble metals such as Pt, Pd, and Rh and their alloys are known to be active catalysts for the oxygen reduction reaction (ORR). Pt is the best of these metals, which, however, exhibits still about 10^4 times smaller exchange current density than that for HOR.

Complete reduction of oxygen to water is a 4-electron reaction, which may proceed via two mechanistic pathways, i.e., the dissociative (or called direct) mechanism where O_2 dissociate before it is hydrogenated and the associative (or called indirect)

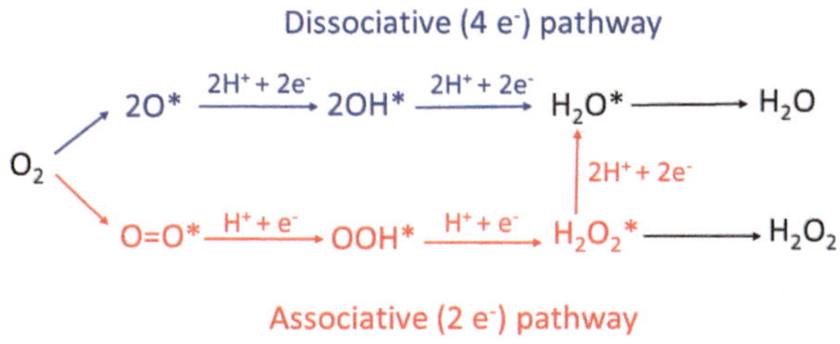

Fig. 5.15 Mechanistic pathways of oxygen reduction reaction

mechanism where O_2 does not dissociate before it is hydrogenated, as shown in Fig. 5.15.

The dissociative adsorption pathway leads to the direct formation of water

$$O_2 + 4H^+ + 4e^- \rightarrow 2H_2O \quad E^0 = +1.23\,V \tag{5.14}$$

while the associated pathway involves the formation of hydrogen peroxide

$$O_2 + 2H^+ + 2e^- \rightarrow H_2O_2 \quad E^0 = +0.695\,V \tag{5.15}$$

which is then further reduced to water:

$$H_2O_2 + 2H^+ + 2e^- \rightarrow 2H_2O \quad E^0 = +1.78\,V \tag{5.16}$$

The overall ORR process consists of the transfer of four protons and four electrons for the reduction of each oxygen molecule. On the surface of platinum-based catalysts, several intermediates such as oxygen (O^*, here the star * denotes an active site) and hydroxyl (HO^*) are involved during the reaction, and the binding energies of these intermediates determine the intrinsic activity of the surface. Besides, the reaction rate is also affected by other conditions such as the electrolyte, the double-layer charge field, the free energy of electrons, and reactants as a function of potential.

Based on theoretical calculations, Nørskov et al. [13] established a volcano plot of general metal surfaces toward the ORR, as shown in Fig. 5.16. Among them, Pt and Pd are the closest to the volcano peak. Generally speaking, for metals that bind oxygen too strongly; i.e., on the left branch of the plot, the reaction rate is limited by the removal of the product from the active site, while for metals that bind oxygen too weakly, the reaction rate is limited by the adsorption of O_2 on the active site. Based on the plot, Pt has the oxygen binding energy slightly higher than needed to be on the peak. As will be discussed in Sect. 5.4.2.4, Pt alloy catalysts have a slightly lower oxygen binding energy than pure Pt and thus should have a higher activity for ORR. This indicates that the oxygen binding energy determines the activities of the metal surface for ORR. In addition, the hydroxyl binding energy is also important for the reaction.

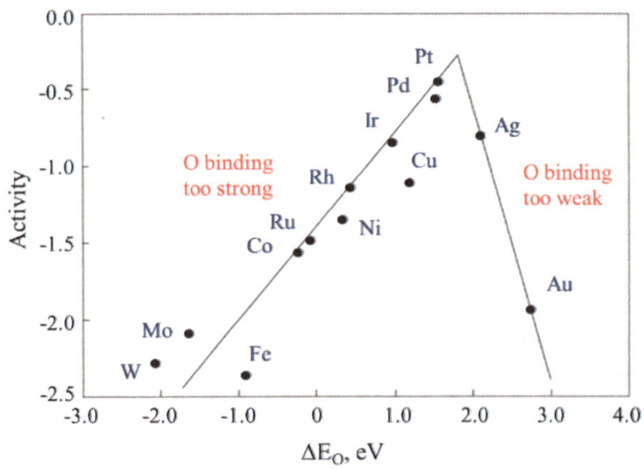

Fig. 5.16 Trends in the oxygen reduction activity plotted as a function of the oxygen binding energy. Reproduced from Ref. [13] with permission from American Chemical Society, Copyright 2004

5.4.2.2 Pt-Based Catalysts

5.4.2.2.1 Pure Pt

In the early development of PEMFC technology in 1960s, Pt and Pd blacks in the form of unsupported metal powders were used with the Pt loading of more than 10 mg/cm^2. When PEMFC technology revived in 1980s, carbon-supported Pt catalysts developed for PAFC were introduced, resulting in a decrease to one tenth of the noble metal loading, i.e., about 1 mg$_{Pt}$/cm^2. In early 1990s, high PEMFC performance has been achieved with the platinum loading of 0.3 mg/cm^2 [14].

Example 5.3 Power density, cost, and Pt usage.

A research program sets a PEMFC power target of 1.80 W/cm^2 at the cell voltage of 0.60 V and a Pt cost target of 4.2 €/kW. Assuming Pt price of 800 € per troy ounce, what is the total platinum loading required to achieve the target?

Solution:

One troy ounce platinum is 31.1 g. The platinum price is then 25.7 €/g.

Using the Pt price, the Pt cost can be converted to the Pt usage:

$$\text{Pt usage} = \frac{4.2\,€/\text{kW}}{25.7\,€/\text{g}} = 0.16\frac{\text{g}}{\text{kW}} = 0.16\,\text{mg/W} \qquad (5.17)$$

Based on the target power density, the total platinum load is

$$\text{Pt loading} = 1.8\,\frac{\text{W}}{\text{cm}^2} \times 0.16\frac{\text{mg}}{\text{W}} = 0.29\,\text{mg/cm}^2 \qquad (5.18)$$

If the anode platinum loading is 0.04 mg/cm^2, then the cathode Pt loading is 0.25 mg/cm^2.

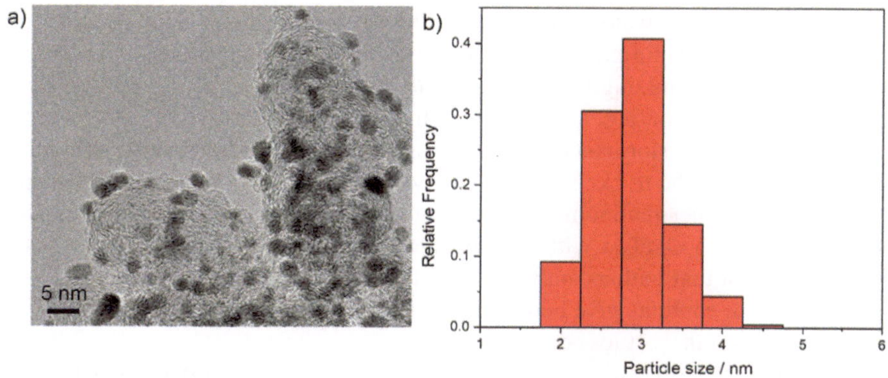

Fig. 5.17 **a)** A typical TEM image of carbon-supported Pt nanoparticles and **b)** the corresponding size distribution histogram

Extensive work has been done to establish the correlation between the size of Pt particles and their ORR activity. Smaller platinum particles have higher surface areas. Thus, one may expect that the overall ORR activity of Pt particles will increase with decreasing the size. This is true until the particle size is decreased to about 2–3 nm. Further decreasing the particle size leads to decreased activity, compromised by the decreased surface activity [15]. It can be briefly explained by the volcano plot discussed above. For a bulk Pt surface, it already binds the O_2 too strong. When the size of Pt particles decreases, it will contain more edges and defect sites on the particle surface, which bind the O_2 even stronger and thus has a lower surface activity. Apart from the particle size, other factors such as the type of anions in the electrolyte also have varied impacts on the surface of Pt particles. For example, in dilute H_2SO_4 the <111> planes are the most active, as HSO_4^- strongly adsorbs on the <100> and <110> planes. Concerning dilute $HClO_4$, which has a similar absorbing behavior to the perfluorosulfonic acid ionomers used in PEMFC, all three groups of planes exhibit the similar activity.

Experimentally the platinum particle size can be estimated from X-ray diffraction (XRD) patterns through the Scherrer equation. More accurately, it can be determined from high-resolution transmission electron microscopy (HR-TEM) images of the catalyst, as exemplified in Fig. 5.17. Besides, electrochemical methods such as underpotential hydrogen deposition (H_{upd}) and CO-stripping can be used to determine the overall electrochemical surface area (ECSA) of the Pt particles in the catalyst (see Sect. 6.3.4).

Besides containing active Pt particles, a good PEMFC catalyst must have a porous structure to allow the efficient transport of reactants and products. During the operation of PEMFC, the reactant gases, i.e., hydrogen and oxygen, need to cross the porous electrode layer to reach active Pt sites. Resistance to the transport of gases must be minimized. Water, the product of the cathode reaction and also from the humidification of reactant gases, is essential for maintaining the proton conductivity of the membrane electrolyte, but can also be a hindrance to the transport of the gases. Any condensed water in the catalyst layer may lead to the electrode flooding and block access of the gases to the active Pt sites. The water management can be improved by optimizing the hydrophilicity/hydrophobicity of the catalyst layer through the addition of, e.g., PTFE, a superhydrophobic polymer.

The loss of cell performance due to the resistance of mass transport can be minimized by reducing the thickness of the catalyst layer. This can be achieved by using catalysts with higher platinum contents. For example, at a fixed platinum loading of 0.3 mg_{Pt}/cm^2, a catalyst with 60 wt% platinum forms a catalyst layer with the thickness half of that with a catalyst of 20 wt% platinum. As a result, catalysts with higher platinum contents give better fuel cell performance at high current densities where the mass transport loss dominates.

Pt/C catalysts with varied metal contents from 10 wt % to over 60 wt % are commercially available. The sizes of platinum nanoparticles are typically in the range of 2–4 nm, and the platinum loading of PEMFC is about 0.3–0.5 mg/cm^2. With this loading, the best PEMFC may achieve a power density of 0.75 W/cm^2, e.g., the cell voltage of 0.75 V at the current density of 1 A/cm^2. Under these conditions, the metal consumption is estimated to about 0.19 g/kW, which was estimated to count for 53% MEA cost.

In addition to the cost and performance, the long-term stability is another issue for Pt-based catalysts. Toward the fuel cell lifetime targets of 5000 h for automotive and 40,000 h for stationary applications, the catalyst degradation is a major challenge. Sintering of platinum particles, dissolution of platinum, and corrosion of the carbon support are the main degradation mechanisms. Sintering of platinum particles decreases the ECSA. Small metal particles of the catalyst may dissolve in the acidic electrolyte, subsequently precipitate onto larger metal particles, and thus lead to the particle growth, which is the so-called Ostwald effect.

5.4.2.2.2 Pt Alloys

Development of carbon-supported Pt alloys as oxygen reduction catalysts was pioneered by the United Technologies Corporation (UTC) (later renamed as International Fuel Cells) in the 1980s for PAFCs. It was discovered that the enhanced specific activity was achieved with Pt-V, Pt-Co–Cr, and other binary, ternary, and quaternary Pt alloys. Using the alloy catalysts, an increase in the cell voltage by 40–60 mV was observed under practical cell voltages, which corresponds to an enhanced specific activity by a factor of 3–5, taking a Tafel slope of 90 mV/dec for PAFCs (see Example 5.4).

The knowledge was then extended to PEMFCs in the early 1990s, particularly with carbon-supported Pt_3M type of alloys (where M=Cr, Co, Ni, Fe, Mn, Ti, etc.). The observed performance gains by using these Pt alloy catalysts was around 20–40 mV, fitting well the typical Tafel slope of 60 mV/dec for PEMFC.

Table 5.7 gives a set of electrochemical data obtained by Johnson Matthey from PEMFC tests with Pt-Fe, Pt-Mn, Pt-Ti, Pt-Cr, Pt-Cu, and Pt-Ni (in atomic ration 1:1) catalysts supported on carbon [16]. These samples have been heat-treated at temperatures from 500 to 930 °C as indicated by the suffix in sample names. Carbon

Table 5.7 A set of electrochemical and kinetic data of carbon-supported 20 wt% Pt and its alloys as cathode catalysts [16]

Cathode catalysts	Cell voltage at 50 mA/cm^2_{Geo} (mV)	Current density at 0.9 V (mA/cm^2_{Geo})	Tafel slope (mV/dec)	In situ ECSA (cm^2_{Pt}/cm^2_{Geo})	Specific activity at 0.9 V ($\mu A/cm^2_{Pt}$)
Pt/C	858	12	67	98	127
Pt/C930	850	8.5	60.5	48	178
Pt-Fe/C650	892	36	58	110	327
Pt-Fe/C930	888	34	64	88	386
Pt-Mn/C930	882	27	58	68	397
Pt-Ti/C650	846	8	62	138	58
Pt-Ti/C850	864	15	66	116	129
Pt-Ti/C930	887	33	64	148	223
Pt-Cr/C650	865	13	49	41	317
Pt-Cr/C930	858	11	59	44	250
Pt-Cu/C500	857	9	61	169	53
Pt-Cu/C930	857	9	58	112	80
Pt-Ni/C650	853	10	67	133	75
Pt-Ni/C930	849	7	59	87	81

support was Vulcan XC72R. MEAs of 25 cm^2 active area were constructed with Dow XUS13204.10 membranes. Anode catalysts were 20 wt% Pt/C (Vulcan-XC72R). The catalyst loading of the anode and cathode together was 0.5 mg_{Pt}/cm^2. Fuel cell test was performed with H_2 and O_2 under pressures of 308/377 kPa and with a flow rate of 300/1500 cm^3/min. Cell temperature was 75 °C with anode inlet humidified at 75 °C and cathode inlet at 85 °C. The catalytic activities are exemplified by the cell voltage at 50 mA/cm^2 (based on geometric electrode area) and current density at 0.9 V. In the table the in situ measured platinum surface area is expressed as the ratio of ECSA with the geometric electrode area, which is often called the roughness factor. In the Table, the specific activity was calculated from the current density measured at 0.9 V divided by ECSA of the Pt-based catalysts.

Compared to pure Pt/C references, identical catalysts without and with heat treatment at 930 °C, several alloy catalysts, e.g., Pt-Ti, Pt-Mn and Pt-Fe, showed a clear improvement by 25–40 mV. The observed Tafel slopes are all around 60 mV/dec, indicating the likely same reaction mechanisms. Also included in the study is the heat treatment effect on both pure Pt and alloy catalysts. As indicated by the specific activity based on the ECSA of platinum shown in the Table, the intrinsic activity was enhanced for the heat-treated catalysts. In H_2-air fuel cells operating at practical current densities, the promising catalysts did exhibit a better performance typically by 25–40 mV.

Example 5.4 Enhancement of activity by Pt alloy catalysts.

Polarization curves typically have a Tafel slope of 90 mV/dec for PAFCs and 60 mV/dec for PEMFC. In the evaluation of Pt alloy cathode catalysts under typical operating current densities, an improvement of the cell voltage of 40–60 mV is found for PAFCs and 25–40 mV for PEMFCs. Explain the difference between these two cells.

Solution:

The cathode overpotential of a fuel cell can be expressed as

$$\eta = b \times \log\frac{i}{i^0} \tag{5.19}$$

Here b is the Tafel slope. The cell voltage difference is attributed to the change of the intrinsic activity of the catalysts expressed as the exchange current density of the catalyst based on the actual platinum surface area:

$$\eta_2 - \eta_1 = b \times \log\frac{i_2^0}{i_1^0} \tag{5.20}$$

For PAFCs $b = 90$ mV/dec, an improvement of the cell voltage by 40–60 mV

$$\eta_2 - \eta_1 = 90\frac{\text{mV}}{\text{dec}} \times \log\frac{i_2^0}{i_1^0} = 40 - 60\,\text{mV} \tag{5.21}$$

which corresponds to an exchange current density ratio of Pt alloy catalyst to pristine Pt catalyst:

$$\frac{i_2^0}{i_1^0} = \exp\left(2.303 \times \frac{40}{90}\right) \quad \text{or} \quad \exp\left(2.303 \times \frac{60}{90}\right) = 2.8 - 4.6 \tag{5.22}$$

while for PEMFC $b = 60$ mV/dec, an improvement of the cell voltage by 25–40 mV corresponds to:

$$\frac{i_2^0}{i_1^0} = \exp\left(2.303 \times \frac{25}{60}\right) \quad \text{or} \quad \exp\left(2.303 \times \frac{40}{60}\right) = 2.6 - 4.6 \tag{5.23}$$

In other words, the improvement in the cell voltage of PAFC and PEMFC corresponds to the same ca. 3–5 times improvement of the specific activity of the catalysts.

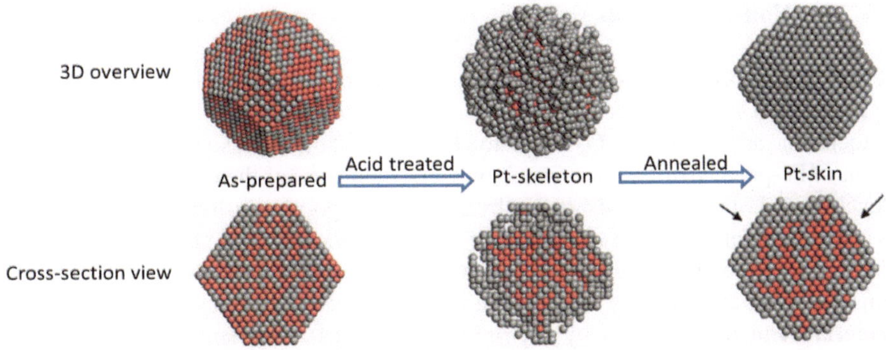

Fig. 5.18 Schematic illustration of Pt-skeleton and Pt-skin nanostructures in 3D overview (top) and cross-sectional view (bottom) of as-prepared (left), acid treated (middle), and annealed (right) PtNi alloy catalysts. Modified from Ref. [19] with permission from American Chemical Society, Copyright 2011

The observed activity enhancement by Pt alloying is suggested to be via modification of the surface electronic structure of Pt and therefore weakening the binding of the intermediates, e.g., O^* and HO^*. Among the different alloying elements, transition metals are most commonly used with a general formula of Pt_3X where X=Ni, Co, or Fe are considered the most active alloys [17]. In forms of carbon-supported nanoparticles in PEMFC, these alloy catalysts have demonstrated a twofold increase in mass activity and a fourfold increase in surface-specific activity over pure Pt catalysts. The difference in the enhancement between mass activity and surface-specific activity is due to the typically larger sizes of Pt alloy particles. Recent research has been extended to the early transition metals (Pt_3Y and Pt_3Sc) and rare earth (Re) metals in forms of Pt_5Re (Re=La, Ce, Sm, Gd, Dy, Tb, Tm, etc.) or alkaline earth metals (Pt_5Ca), though no fuel cell test results have been reported [18].

The stability of Pt alloy catalysts has always been a concern. In the nano-alloy structures, the outmost surface layer of the alloys is always composed of pure Pt atoms of 1–2 nm in thickness, due to the poor kinetic stability of transition metals against the dissolution. As shown in Fig. 5.18, the as-prepared Pt alloys are usually pretreated in acid electrolytes. Dissolution of the alloying atoms of transition metals leaves behind an atomically rough surface composed of Pt atoms with low or no lateral coordination, called the Pt-skeleton [19]. Thermal annealing the acid-treated alloys can restructure the near-surface region of the catalyst, leading to the formation of the so-called Pt-skin structure, typically being a monolayer thick. For both Pt-skin and Pt-skeleton surfaces, the effect of alloying is to modify the electronic structure of the Pt overlayer in order to weaken the binding of the surface to the ORR intermediates.

5.4.2.3 Non-Pt Catalysts

It has long been known that cytochrome-containing binuclear porphyrinato-Fe(heme)-Cu site can reduce oxygen to water and the presence of Fe-porphyrin in the cytochrome active sites for O_2 reduction has inspired the extensive work on development of Fe-porphyrin-like non-precious metal catalysts for the ORR. The early year development of non-Pt catalysts was by synthesis of porphyrin-like macromolecules. A critical concern was the stability of the materials since H_2O_2 is also involved during the oxygen reduction, which may decompose the porphyrins and release free metal ions. An innovative finding was that the stability as well as the activity of the catalyst materials can be improved by high-temperature heat treatment in an inert atmosphere [20]. The high-temperature process makes it unnecessary to synthesize the macromolecular precursors. The common practice to produce a transition metal, e.g., iron-based catalyst is by high-temperature pyrolysis of iron- and nitrogen-precursors in the presence of a carbon support. The nitrogen precursors are often porphyrin-like or other N-containing solid or gaseous compound. The carbon support is typically carbon black or carbon from thermal decomposition of the precursors. The most interesting catalyst materials are of the nitrogen (N)-coordinated iron single atoms embedded in carbon materials with general formula FeN_x/C type composites. The most likely coordination is the formation of a planar local symmetry of Fe coordination with four nitrogen atoms (Fe–N_4), because the unsaturated coordination like Fe–N_3 and Fe–N_2 would lead to the distortion of planar geometry. FeN_x/C acts as active center for ORR. However, for single atom catalysts supported on carbonaceous materials, the coordination environment, e.g., the chemical bond between the single metal atom and the surrounding coordinated atoms, e.g., carbon, nitrogen and oxygen, from the supports plays a crucial role in determining the catalytic activity for ORR. The doping-induced charge transfer from carbon atoms to their adjacent N atoms strengthens the chemisorption of O_2 and weakens the O–O bond, promoting the kinetics of ORR. However, the N in the graphitic lattice of nitrogen doped carbon can exist in several different configurations, including pyridinic nitrogen (N_P), pyrrolic nitrogen (N_{PY}), graphitic nitrogen (N_G), oxidized nitrogen (N_O), nitrile nitrogen (N_N) and primary amine nitrogen (N_A), see Fig. 5.19. Among them, N_P, N_{PY} and N_G

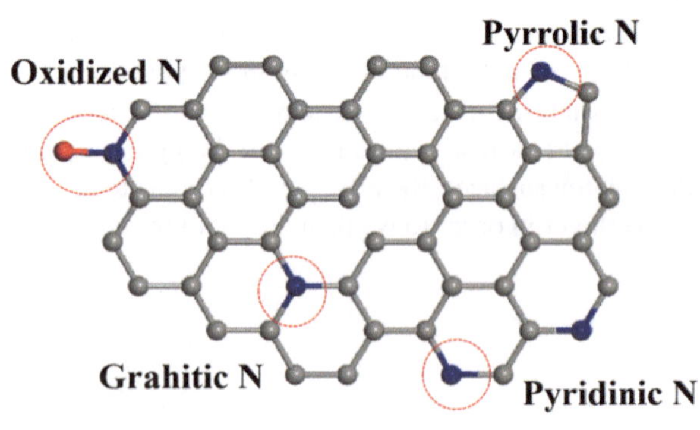

Fig. 5.19 Types of nitrogen coordination on graphitic lattice of N-doped carbon

are most influential and active for electrocatalytic activity of the coordinated metal atoms. The activity of the active sites depends on both the metal single atom and N coordination type.

Significant progress has been made since 2009 [21, 22] with breakthrough performances that match the Pt/C in the high cell voltage range, though at catalyst loadings of typically 10–15 times higher than that of the platinum based analogues. A serious consequence of the increased catalyst loading in the electrode thickness which restricts the cell performance at current densities above 200 mA/cm^2 due to mass transport.

Zelenay et al. [23] have achieved the maximum power density of 0.39 W/cm^2 at 1.0 bar partial pressure of H_2 and air with a cathode containing 4.0 mg/cm^2 of catalysts. Comparison with the performance of a cathode of 0.1 mgPt/cm^2 loading, prepared from 10 wt% Pt/C catalyst, nearly the same current densities were obtained with the iron–nitrogen–carbon (FeN_x/C) based non-precious metal catalysts (NPMC) in the kinetic region (>0.75 V) though lower in the mass transport region (see Fig. 5.20a). The poor mass transport is apparently because of the much greater thickness of the cathode layer, which was approximately four times thicker than the Pt electrode. Fuel cell performance under H_2–O_2 conditions with three different O_2 partial pressures of 0.3, 1.0 and 2.0 bar is shown in Fig. 5.20b. Much improved mass transport performance is seen at cell voltage as low as 0.2 V. The maximum power density values of 0.87 and 0.94 W/cm^2 were achieved at p_{O_2} of 1.0 and 2.0 bar, respectively. One of the significant challenges in practical applications of NPMCs is the low loading of single atom catalysts, which in turn leads to the significantly increased thickness of the catalyst layer and thus the high mass transport polarization.

5.4.3 Carbon Supports

5.4.3.1 Carbon Black

To achieve maximum activity per unit mass of precious metal catalysts, it is necessary to disperse the precious group metal (PGM) catalysts like Pt, Pd, and Au on high surface-area carbonaceous support materials in order to achieve effective dispersion of the precious metal particles and to reduce the mass transfer limitation inside the electrode structure. Carbonaceous supports such as high surface area carbons provide electronic contact throughout the catalyst layer, isolate PGM nanoparticles to maintain the high electrochemical surface area, and provide hydrophobic properties to avoid the flooding of the catalyst layer by liquid water produced. Supports with high surface areas, ranging from 4 to 25 $\times$ 10^8 m^2 per cubic meter bulk volume or a specific surface area of 200–1250 m^2/g for carbon blacks, are usually used. In addition to the high surface area, the support material must provide, at the same time, high electrical conductivity, open access for reactant gases, and supercorrosion resistance in fuel cell environment, particularly the high potential range in the cathode.

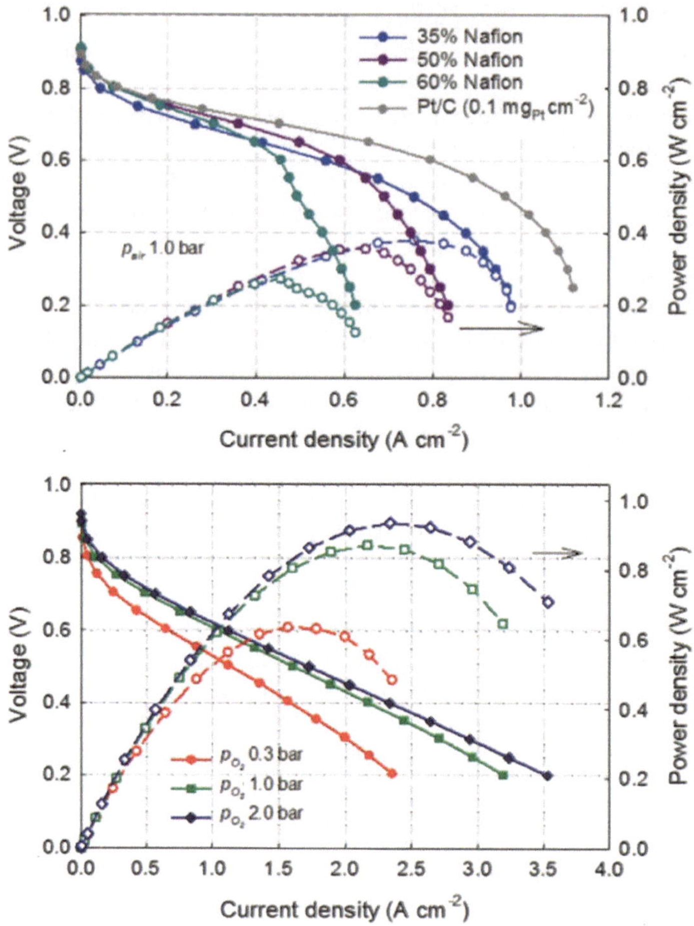

Fig. 5.20 Polarization curves of 5 cm^2 fuel cell equipped with cathode of 4.0 mg/cm^2 of NPMC, anode of 2.0 mgPt/cm^2 Pt/C and Nafion 211 membrane at 80 °C in H_2 and air at 200 ml/min under 1.0 bar partial pressure and 100% RH (upper figure) and in O_2 at 200 ml/min and 100% RH under 0.3, 1.0, and 2.0 bar partial pressures (lower figure). H_2 at 200 ml/min at 100% RH under 1.0 bar partial pressures. Reproduced from Ref. [23] with permission from American Association for the Advancement of Science, Copyright 2017

Carbon blacks are the most used support materials, which are commercially available from a number of companies, such as Cabot Corporation (Vulcan XC72R, Black Pearls 2000), Ketjen Black International, Chevron (Shawinigan) and Denka. These carbon blacks are generally manufactured by the pyrolysis of hydrocarbons in a furnace, where the starting material is burned with a limited supply of air at about 1400 °C, followed by immediate quenching with water. Some commercial furnace blacks often used for fuel cell catalysts and their typical properties are listed in Table 5.8.

Table 5.8 Common types of carbon blacks

Carbon blacks	Type	Manufacturer	Surface area (m^2/g)	Average particle size (nm)
Vulcan XC-72R	Furnace black	Cobot	250	30–50
Black Pearl 2000	Furnace black	Cobot	1470	10–15
Ketjen EC300J	Furnace black	Ketjen	800	30–40
Ketjen EC600JD	Furnace black	Ketjen	1300	30
Shavinnigan	Acetylene black	Gulf oil	70–90	40–50
Denka black	Acetylene black	Denkikagaku	58	40

The primary particles of carbon blacks consist of graphene-like pieces that agglomerate together in sizes of up to 100 nm (see Fig. 5.21). These particles are poorly defined in morphology and commonly agglomerate into clusters. The BET surface area may vary from tens to 1500 m^2/g. Another type of carbon black is manufactured by thermal decomposition of acetylene gas, whereby a high order of crystallinity and, therefore, high conductivity is achieved, although at the expense of low surface areas.

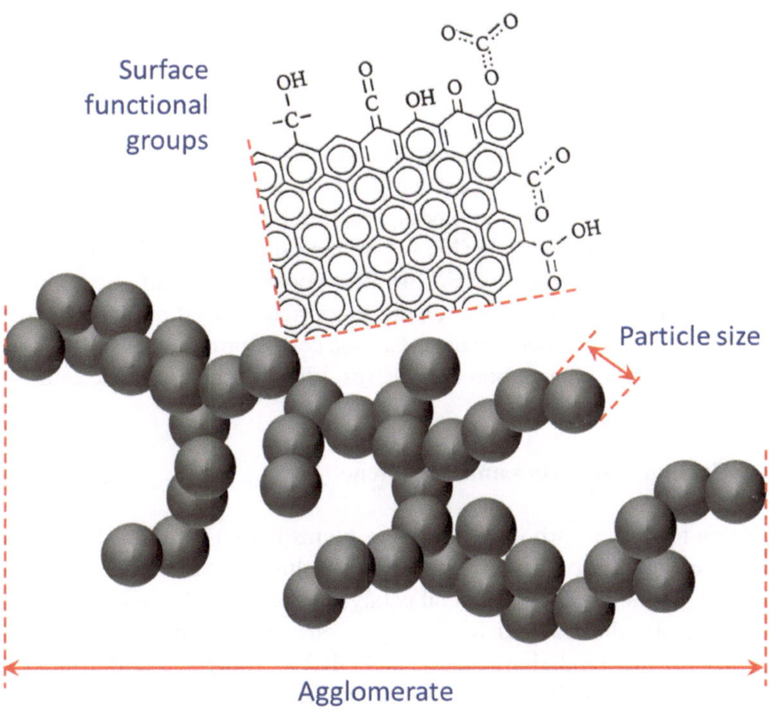

Fig. 5.21 Schematic representation of carbon black particles, agglomerates, and surface functionalities

The effect of carbon supports on the dispersion and activity of metal nanoparticles has long been a subject of study. From the synthesis point of view, carbon support can influence the catalyst dispersion because small particles are preferentially anchored to high-energy sites on the carbon surface. The density of such sites on the carbon support affects the crystallite sizes of Pt nanoparticles in the synthesized catalyst. The role of carbon blacks, of course, changes in different synthesis methods. One typical synthesis method is called the impregnation method, where the metal precursor, commonly a platinum salt, is first wet impregnated onto the carbon support, followed by drying and then reduction with hydrogen. Good metal dispersion can be achieved by this method, and it does not influence the specific surface area of the used carbon support. Another popular method is called colloidal method, where the metal nanoparticles are first reduced in the form of suspension colloidal, followed by loading the suspending nanoparticles onto carbon supports. This method is particularly useful for preparing catalysts with high metal contents. The size of Pt particles prepared by this method is found to decrease with increasing the specific surface area of the carbon support.

To tailor the number and nature of functional sites for anchoring Pt nanoparticles, carbon supports are often chemically modified or activated by a variety of oxidizing agents such as HNO_3, H_3PO_4, H_2O_2, and O_2. After the treatment, the carbon black surfaces are rich in functionalities such as carboxylic acid and many others (see Fig. 5.21). The treatment aims at destruction of basic sites and addition of acidic groups, which are believed to facilitate the metal dispersion and thus improve the catalytic activity. Besides, this treatment could also promote the development of porous structures.

One of the major issues in the use of carbon–based support materials is, however, the electrochemical corrosion of carbon. Carbon corrosion occurs under PEMFC operation conditions because the thermodynamic potential of carbon oxidation to carbon dioxide is 0.207 V (vs. RHE), while the potential of a PEMFC cathode is typically in the range of 0.6–1.2 V [24]. The corrosion of carbon support results in a loss of electronic contact between the supported catalyst particles, leading to a degradation in the fuel cell performance. Carbon corrosion can also lead to an increase in the hydrophilicity of the carbon surface, resulting in flooding in the catalyst layer.

5.4.3.2 Carbon Nanotubes and Graphene

One solution to reduce carbon corrosion problems is to use carbon-based materials with increased graphitic properties such as graphitized carbon, carbon nanotubes (CNTs), and graphene. Precious metal catalysts such as Pt, Pd, and PtRu supported on CNTs generally show a much better dispersion and higher electrocatalytic activity and stability for fuel oxidation reaction and ORR as compared to that supported on high surface area carbon. It has been shown that the crystalline Pt nanoparticles (NPs) would interact with CNTs through synergic bonding involving charge redistribution between C 2*p*-derived states and Pt 5*d* bands due to the presence of unsaturation in the graphene sheets (delocalized π orbitals) [25]. Such interaction facilitates the immobilization of metal NPs on the CNT surface, inhibiting the lateral diffusion of metal NPs under fuel cell operating conditions.

5.4.3.2.1 Surface Functionalization and Pt Deposition

However, pristine CNTs are difficult to disperse in water due to their high aspect ratios and strong van der Waals interactions, and the surface is also very inert with few active sites for homogeneously attaching metallic nanoparticles. It is a common practice to treat CNTs to exfoliate bundled CNTs and to modify the surface of CNTs to enhance the dispersion of Pt-based NPs. One of the widely used methods is the covalent modifications based on the surface oxidation treatment [26]. Various oxidants such as HNO_3, H_2SO_4, H_2O_2, and HF can be used. The strong acid oxidation treatment introduces hydrophilic groups such as hydroxyl (–OH), carboxyl(–COOH) and carbonyl (–C=O) groups to the CNT surface by the modification of the aromatic conjugate ring system of CNTs, similar to the oxidation agents treated carbon black materials, see Fig. 5.21. The oxidized CNTs can then be easily dispersed in water and the surface functionalized groups serve as the active sites for anchoring metallic NPs such as Pt and Pd in the presence of reducing agents, e.g., ethylene glycol, $NaBH_4$, etc. Pt nanoparticle catalysts on covalently functionalized CNTs exhibit enhanced electrocatalytic activity and stability as compared to Pt nanoparticles supported on high surface area carbon, e.g., Pt/C.

5.4.3.2.2 Non-covalent Functionalization

Non-covalent or physical modification treatment is a much more mild process and can preserve the intrinsic sp^2 hybrid state and intrinsic properties of CNTs [27]. A large number of molecules and compounds such as surfactants, aromatic compounds, polymers, or polyelectrolytes can be used as non-covalent functionalization agents; see Table 5.9. CNTs are functionalized by the interaction between the functionalization molecules or agents and CNTs via various interaction modes such as π-π, hydrophobic, CH-π, and charge transfer interaction [28]. Non-covalent functionalization leads to the wrap of the functionalization molecules or agents on the surface of CNTs, and the functional groups or charged sites of the functionalized molecules serve as the uniform and effective active sites for the assembly and anchoring metal precursors and NPs. The non-covalent functionalization via polyelectrolytes and surfactants is also applicable to graphenes.

Figure 5.22 gives an example of the non-covalent polyelectrolyte functionalization method for multi-walled CNTs (MWCNTs) using PDDA as supports for Pt catalysts [29]. PDDA is a water-soluble quaternary ammonium and a strong polyelectrolyte. During the functionalization process, NaCl salt is added to allow the PDDA chain to adopt a random configuration and promotes functionalization, thus leading to a uniform wrapping of PDDA on MWCNTs, PDDA-MWCNTs. The strong adsorption of the positively charged PDDA on MWCNTs is due to the π-π interaction

Table 5.9 List of most common polyelectrolytes/agents used in the non-functionalization of carbon nanotubes as support for precious group metal-based nanoparticles [27]

Polyelectrolytes/Agents	Structure
Polybenzimidazole, PBI	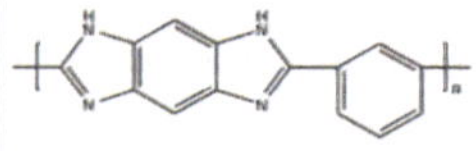
Poly (diallyldimethylammonium chloride), PDDA	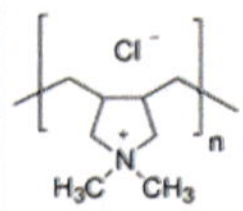
Poly(sodium 4-styrenesulfonate), PSS	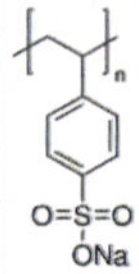
Poly(allylamine hydrochloride), PAH	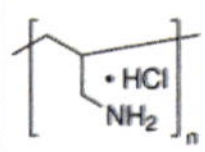
Poly(acrylic acid), PAA	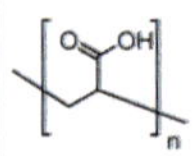
Polyaniline, PANI	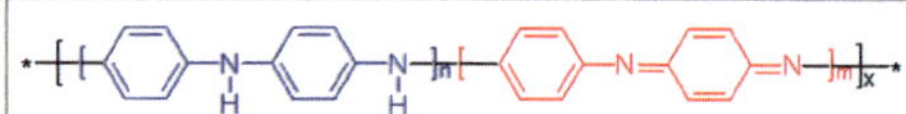
Sodium dodecylsulfate, SDS	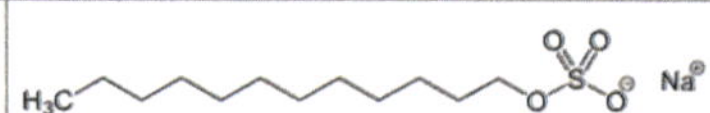
Aniline	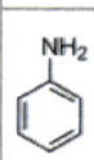
Polyvinyl pyrrolidone, PVP	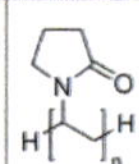
1-aminopyrene, 1-AP	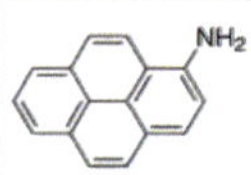
Poly(ethyleneimine), PEI	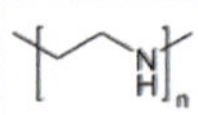
Tetrahydrofuran, THF	
Chitosan, CS	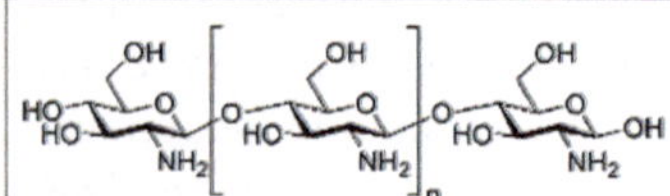

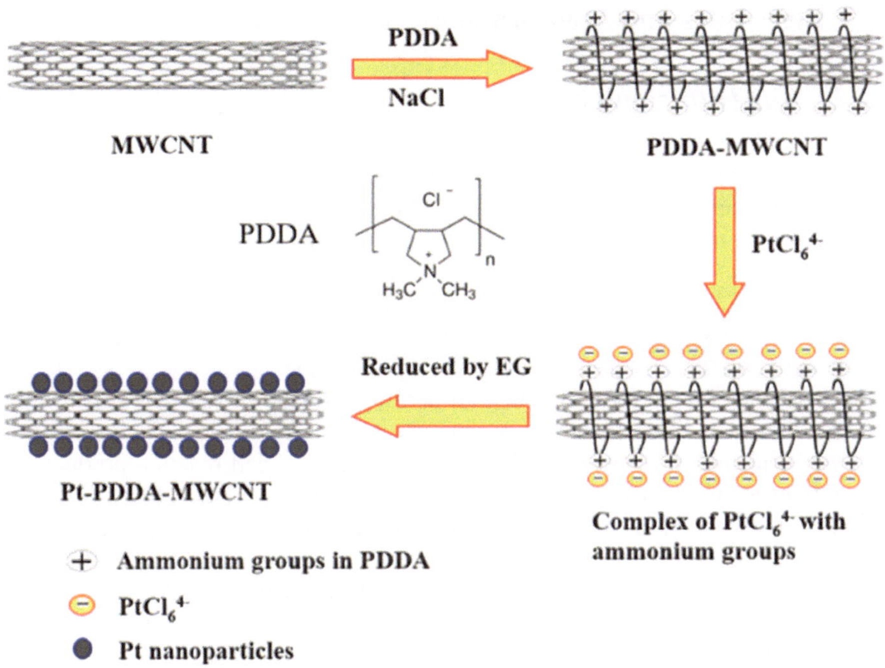

Fig. 5.22 Scheme of PDDA functionalization of MWCNTs and the in situ synthesis of Pt nanoparticles on PDDA-functionalized MWCNTs. Reproduced from Ref. [29] with permission from IOP Publishing, Copyright 2008

between PDDA and the basal plane of CNTs [30]. Pt NPs are synthesized on the PDDA-MWCNTS via the self-assembly between negatively charged Pt precursors and positively charged functional groups of PDDA and reduction in situ by EG. The assembly of Pt precursors is indicated by the negative zeta potential of Pt-PDDA-MWCNTs. Pt NPs on PDDA-functionalized CNTs show a much better uniformity as compared to that on acid-treated MWCNTs (AO-MWCNTs).

In addition to carbon-based materials, a myriad of non-carbonaceous-based catalyst support including metal oxides, perovskites, nitrides, carbides, and sulfides have been explored and show superior corrosion resistance as catalyst support. The main problem of oxide or non-carbonaceous supports is the relatively low electronic conductivity and low specific surface areas to anchor the PGM NPs.

5.5 Electrodes and Membrane Electrode Assembly

5.5.1 Gas Diffusion Layer

Gas diffusion electrodes are made of loose catalyst powders. With the help of a polymeric binder, the catalysts are coated as a thin layer on top of either a porous substrate or the membrane. The porous substrate is a *gas diffusion layer* (GDL), and the catalyst-coated GDL is also called a *gas diffusion electrode* (GDE). When the catalyst layer (CL) is applied on a membrane, it is called a *catalyst-coated membrane* (CCM). When the CCM is assembled into a cell, the porous media are needed to attach on the backside of each catalyst coating layer side.

The primary function of a GDL is to permit diffusion of reactant gases to the catalyst sites at the interface with the membrane (see Fig. 5.23). The layer plays also a critical role in removal of liquid water as the fuel cell reaction product and therefore needs to possess hydrophilic/hydrophobic characteristics. For this reason, it is sometimes called the porous transport layer (PTL), a term more popularly used for the water electrolysis fields where the water supply is in form of liquid rather than vapor. It should also be of note that the gas diffusion electrodes for PEMFC are less demanding than for AFC and PAFC where retention of liquid electrolytes between the two electrodes is vital for both the performance and lifetime.

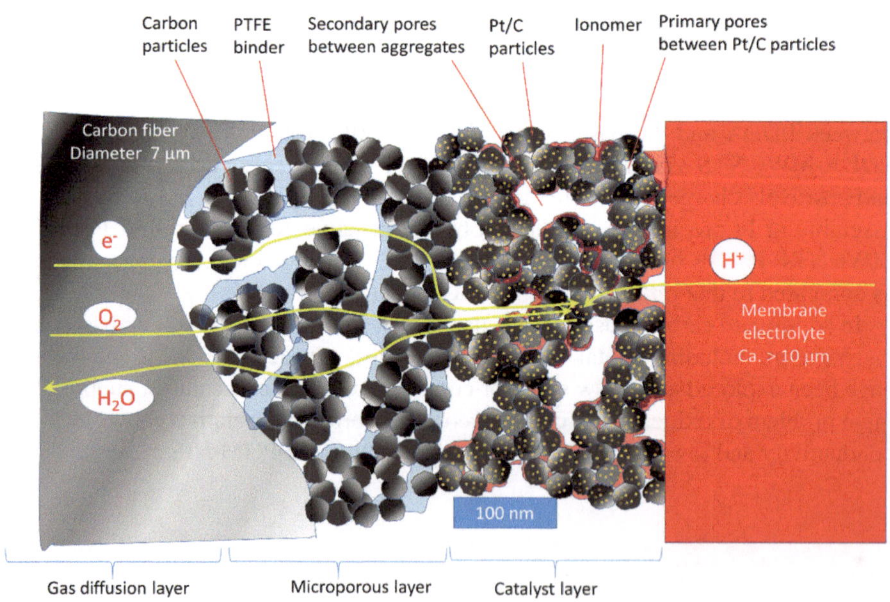

Fig. 5.23 Microstructure of the electrode-membrane interface and transport of gas, water, proton, and electron

At the same time, the layer should be an electronic conductor, allowing the passage of current generated from CL to bipolar plates, which connect the adjacent cells. A similar requirement applies for conduction of heat, which is generated within the MEA and rejected through the cooling channels located behind the bipolar plates of individual cells or several cell modules.

This porous and conductive GDL is also used to provide mechanical support to a membrane–electrode–assembly (MEA) where two GDEs are assembled with a soft polymer membrane. It is also essential that the GDL ensures a good electronic contact between the catalyst layer and bipolar plates while the layer should not be compressed into the channels blocking the gas flow or increasing the channel pressure drops. This is particularly true when a reactant pressure difference exists between the anode and cathode gas channels. This requirement for the GDL may vary according to the flow channel design of bipolar plates, for example, using mesh or foam materials without flow channels. In addition, the GDL should always be chemically and electrochemically stable and steadily available at low cost as well as low processing cost.

The most promising candidates for use as diffusion media in PEMFC are carbon fiber-based products. Carbon fibers were first commercialized in the 1950s and are useful in a variety of markets due to their high stiffness. Woven fabrics (or cloths) or nonwoven papers of carbon fibers have been commercialized for other applications and adapted to PEMFCs due to their high porosity (>70%) and good electronic conductivity. Furthermore, carbon fiber materials exhibit excellent stability and corrosion resistance along with good mechanical properties. Table 5.10 lists properties of three gas diffusion layer materials.

In terms of electronic conductivity, it should be noted that the in-plane conductivity is of more importance than the through-plane conductivity. Compared to the thickness of a GDL, which is in a range of 200–300 μm, the conducting distance of electrons generated or consumed in a catalyst site located in the middle of a gas flow channel is about 10 times longer (see Fig. 5.24).

Table 5.10 Typical materials of carbon fiber paper and cloth

Properties and units	Carbon paper (Toray TCP-H-060)	Woven carbon cloth (W0S1009)	Non-woven carbon cloth (H23C2)
Thickness (μm)	0.19	0.33	250 (at 0.025 MPa)
Area mass (g/m^2)	85	122	135
ASR (through-plane, Ωcm^2)	0.009	0.005	0.013
Bulk resistivity (through-plane, Ωcm)	0.08	–	–
Bulk resistivity (in-plane, Ωcm)	0.0055	0.009	–
Gas permeability (through-plane, Darcy)	8	55	2.3

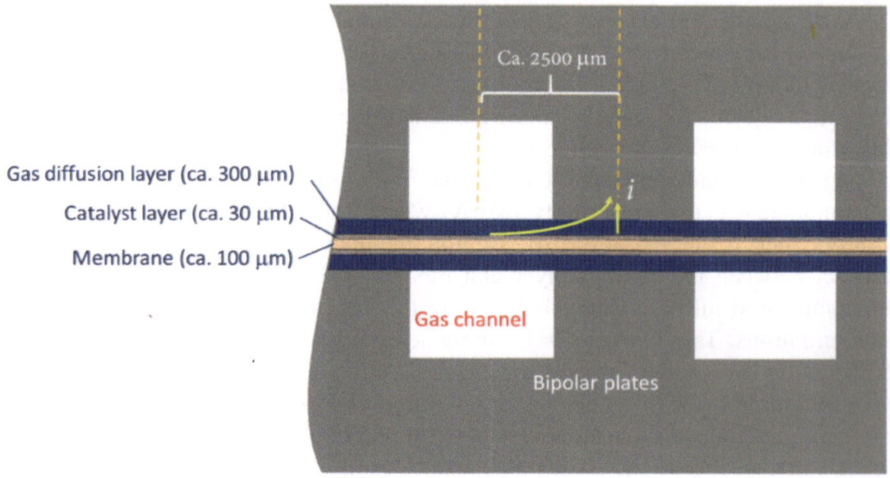

Fig. 5.24 Illustration of in-plane and through-plane conduction of electrons in through a gas diffusion layer in PEMFC

Carbon fibers can be woven into a thin layer of cloth-like materials. Carbon cloth materials are typically between 300 and 400 μm thick and have a high porosity and therefore gas permeability. It can be compressed by 30–50% when clamped into a fuel cell assembly. With this compression the electrical conductivity and gas permeability of the carbon cloth change significantly. Carbon paper materials are produced by bonding carbon fibers into a stiff paper-like sheet of a typical thickness between 150 and 250 μm. The binding is achieved by carbonization of a resin binder, which makes the carbon paper denser and stiffer.

For the purpose of facilitating the removal of cathode product of water, the GDL materials need a wet-proofing treatment to make it hydrophobic. This is achieved by using polytetrafluoroethylene (PTFE, e.g., Teflon®) in a loading range between 5 and 30%. PTFE is most commonly applied by dipping the GDL into an aqueous PTFE suspension and then baking the treated GDL in an oven at temperatures of 330–360 °C in order to remove the surfactants and sinter the PTFE particles to the GDL fibers.

In GDL, both the carbon loading and the PTFE content affect the cell performance but in a different way. Figure 5.25 shows a typical example of the effect of carbon and PTFE loadings on the cell performance, measured at 80 °C [31]. The polarization curves show distinguish polarization regions, activation, ohmic, and mass transport, dominated by three polarization losses, i.e., η_{act}, η_{ohm}, and η_{conc}. The loading of the carbon powder in the GDL significantly affects the performance of the fuel cell in all three regions. The polarization loss at activation region is most likely due to the poor support of the GDL for the catalyst layer at low carbon loadings, causing the loss of Pt catalysts in the catalyst layer. Increasing the carbon loading in the GDL can also reduce the internal resistance of the cell and the contact resistance between the interfaces of GDL/CL, hence improving the fuel cell performance. Increasing the carbon loading can also reduce the mass transport limitation due to the improvement of water

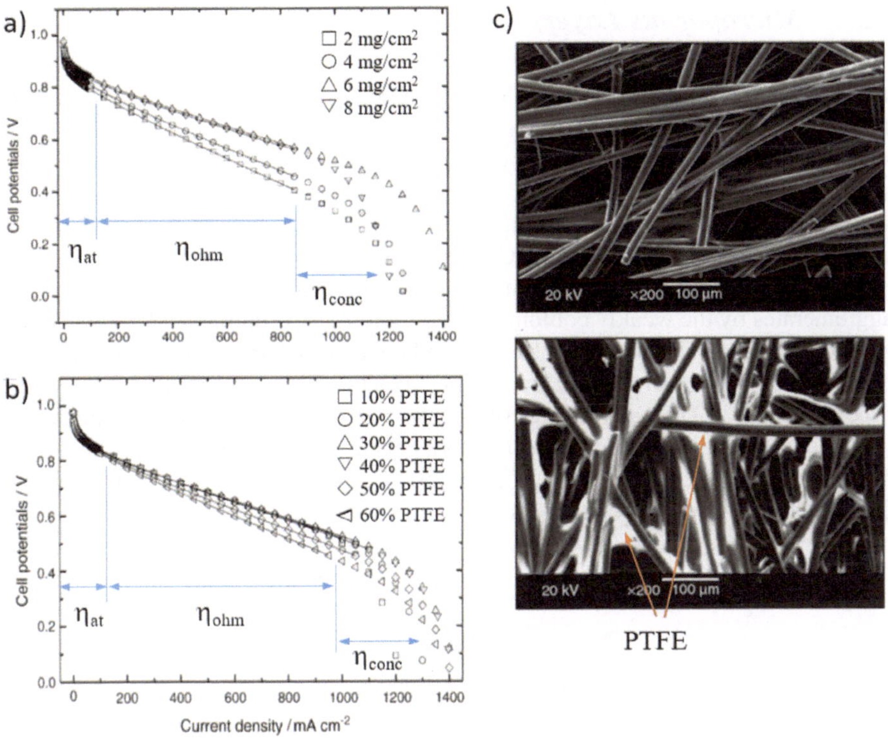

Fig. 5.25 Polarization curves of cells with **a** different carbon content and 40% PTFE in GDL, **b** different PTFE content and 6 mg/cm^2 carbon loading in GDL, measured at 80 °C in H_2/air, and **c** surface of SGL704E carbon paper before and after hydrophobic treatment (40% PTFE). Reproduced from Ref. [31] with permission from Elsevier, Copyright 2006

management. However, excessive carbon loading in GDL, e.g., 8 mg/cm^2, decreases the porosity and thus increases the concentration or mass transport polarization, η_{conc} or η_{mt}.

Different to that of carbon loading, the PTFE content of the GDL mainly affects the polarization performance of the cell in the ohmic and mass transport control polarization regions. PTFE treatment increases hydrophobic property of GDL and thus improves the water management. The mass transport polarization caused by the PTFE content comes from the resistance to gas diffusion due to the changed porosity and diffusion path in the GDL and water flooding due to the changed hydrophobicity of GDL. Best performance was obtained on cells with carbon loading of 6 mg/cm^2 and 30–40% PTFE in GDL. Excess PTFE in the GDL, e.g., 50%, will lead to the increase in the electrical contact resistance between the interface of GDL/CL.

Besides carbon fiber-based materials, finely patterned metal meshes, and porous metal foams have also explored as GDL for PEMFC. These metal-based GDL materials present significant challenges; e.g., they tend to corrode and are too hydrophilic and having a too coarse range of porosity.

5.5.2 Microporous Layers

The GDL consists of carbon fibers of a diameter around 7 μm and macropores of 10–30 μm size while the catalyst layers are made of carbon blacks of 30–50 nm size supporting metal nanoparticles of 3–5 nm. Between these two layers are a transition of a smoother intermediate layer needed, called the *microporous layer* (MPL; see Fig. 5.26). The MPL should have a minimized resistance to the gas diffusion and contact resistance between the GDL and CL. The carbon particles usually form aggregates with some kinds of bonding while the aggregates are further forming agglomerates by the weakly coulombic attraction. As a result the MPL contains two types of pores, small pores of < 10 nm size between carbon particles within aggregates and large pores of 10–100 nm size between the aggregates of the agglomerates (See Fig. 5.23). PTFE, as particles of micron-size dispersed in aqueous media and stabilized by wetting agents, forms network between the aggregates after heat treatment at the melting temperature of 330–350 °C, making the MPL hydrophobic to facilitate the removal of water.

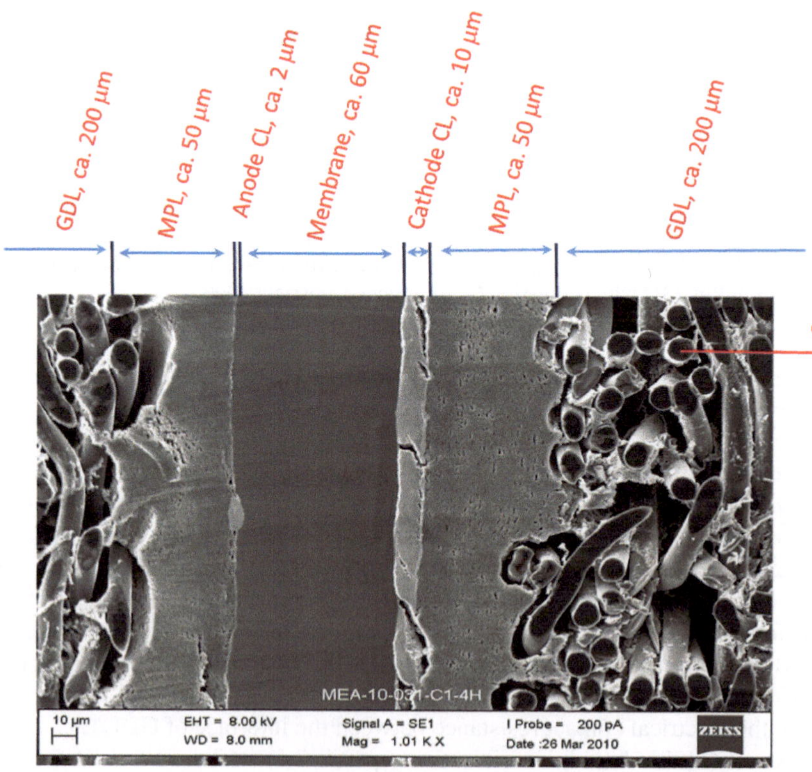

Fig. 5.26 Scanning electron microscopic image of an MEA with indicated thickness of the membrane, anode and cathode catalyst layers (CL), macro-porous layer (MPL) and gas diffusion layer (GDL). The diameter of carbon fibers is also indicated in the figure

The MPL layer is typically formed by mixing submicrometer-sized particles of graphite with a polymeric binder, usually PTFE. The mixture is applied onto one side of the GDL and heat-treated, resulting in a thin, uniform, microporous graphitic layer of 20–50 μm thick.

5.5.3 Catalyst Layer

The *catalyst layer* (CL) is the place where the electrochemical reaction occurs. For a noble metal catalyst site to be electrochemically active in the catalyst layer, the layer must be conductive for both electrons and protons, which are produced or consumed by the electrochemical reaction in the catalyst sites. At the same time the layer must be permeable to the gaseous reactants and product water. The essence of the catalytic layer is the three-phase boundaries (TPBs) where the catalyst sites are accessible for the reactant gases and connected to both the membrane and the GDL with a proton and an electron-conducting pathway.

Many recipes for preparation of catalysts ink formulations can be found in the literature. A material is first of all needed to bond the loose catalyst particles together, called a binder. Adapted from the early PAFC technology, an inert polymer, e.g., PTFE, is usually used as the binder. An additional function of this binder is to provide hydrophobicity to the catalytic layer and thus to get a balance between open pores for reactant gases and pores filled with the liquid acid in order to allow for a proton conduction pathway. For the early PEMFC applications, electrodes were pressed onto a membrane by high bonding pressures and temperatures, leading to partial penetration of the ionic conductive ionomer phase into the catalyst structure. The insufficient ionic conductivity of the catalyst layer limited the performance and, as a result, very high Pt loadings were used. It was Raistrick who first suggested to impregnate the Pt/C/PTFE catalyst layer with an ionomer solution prior to hot pressing it [32]. The ionomer impregnation enables the protonic conductivity to the catalyst sites that are not in intimate contact with the membrane. It is later a more popular practice to formulate an ink including the catalyst, ionomer, and hydrophobic agent. The PFSA ionomer is not a true solution but a dispersion of 2–2.5 nm particles in an alcohol and water mixture solvent. Such types of solvents are often used for preparation of the Pt/C catalyst ink. Aqueous inks are sometimes preferred for easier processing as well as absence of toxic organics. Typically, the Pt/C is added to water along with a surfactant or wetting agent such as isopropyl alcohol and ultrasonically blended to form a dispersion, into which an ionomer is added with further stirring. Similar to the MPL, the CL contains also two types of pores, the primary pores between Pt/C particles and secondary pore between the carbon aggregates (See Fig. 5.23). The ionomer can fill in the primary pores, is swollen with water and becomes conductive during the fuel cell operation.

5.5.3.1 Catalyst Layer Formation by Hot-Press and Catalyst-Coated Membrane

There are a variety of techniques to accomplish deposition of the ink onto the membrane or GDL. The catalyst deposition method has a strong influence on the performance of the MEA since it largely determines the structure of the catalytic layer. It may affect the micro-assembling of the catalyst powders by tuning the size of the catalyst aggregates and its spatial arrangement throughout the catalytic layer. These parameters govern the meso- and macro-porosity of the catalytic layer and thus the extent of TPBs that definitely influence on the ECSA. The most used methods include painting, spraying, coating, or printing.

There are basically two methods for the MEA fabrication: (1) application of the catalyst layer to the GDL followed by membrane addition, or (2) application of the catalyst layer to the membrane followed by GDL attachment. MEA is fabricated by hot pressing onto each side of a PEM under high pressure at relatively high temperatures (i.e., hot-pressed MEA). The other method is to first apply the catalysts onto membranes by casting or spraying a catalyst ink onto the membrane. Alternatively, the Pt ink can also be applied onto a plastic, e.g., PTFE blank and then transfer it to the membrane by hot pressing. When the plastic blank is peeled off, the thin layer of catalysts is left on the membrane, called a catalyst-coated membrane (CCM) (see Fig. 5.27).

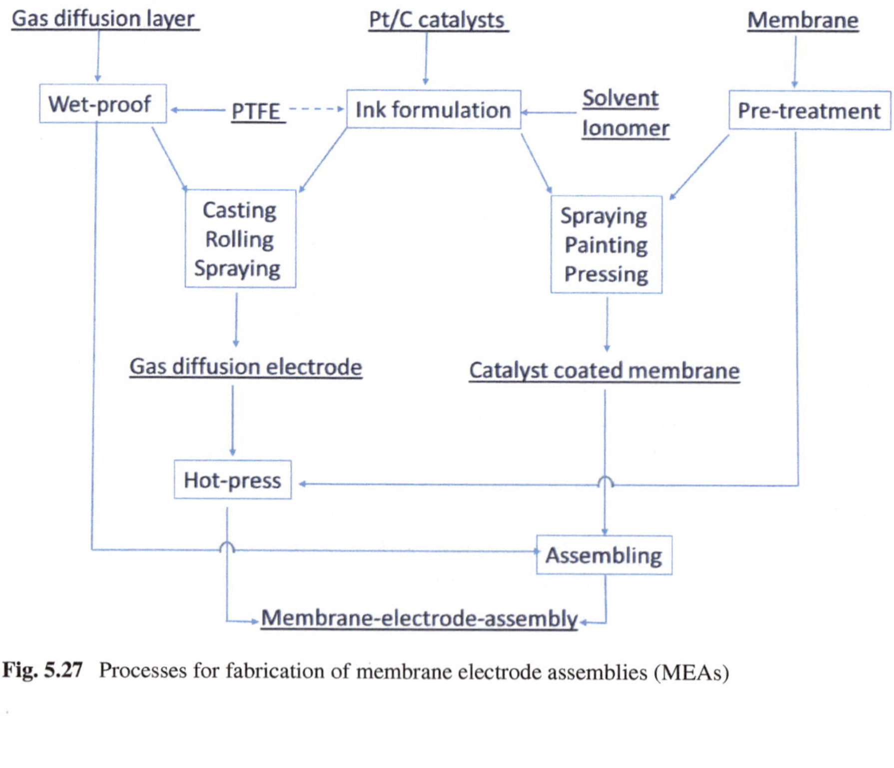

Fig. 5.27 Processes for fabrication of membrane electrode assemblies (MEAs)

The procedure and method for the preparation of the MEA play an important role in the distribution, efficiency, and electrocatalytic activity of the electrocatalysts with direct consequences to the fuel cell performance. In general, PEM fuel cells prepared using CCM method show a better performance as compared to that by the hot-pressed MEA method. The reason is due to the much more uniform Pt distribution at the interface and reduced catalyst loss to the gas diffusion media, as indicated in Fig. 5.28 [33]. In the case of MEA by the hot-press method, the Pt/C catalyst ink slurry was screen-printed onto the GDL to form the electrode catalyst layer, followed by drying at 60 °C for 10 min and at 90 °C in N_2 for 3 min. The pretreated membrane and electrode layers were bonded together by hot pressing under 10 MPa at 125 °C for 90 s. To prepare MEA by the CCM method, the same catalyst slurry was applied to PTFE thin film by spraying. After dried at 60 °C for 10 min and at 90 °C in N_2 for 3 min, the catalyst layer was then transferred onto a Nafion 112 membrane at 125 °C and 10 MPa by the decal method to form the CCM. The GDL was placed

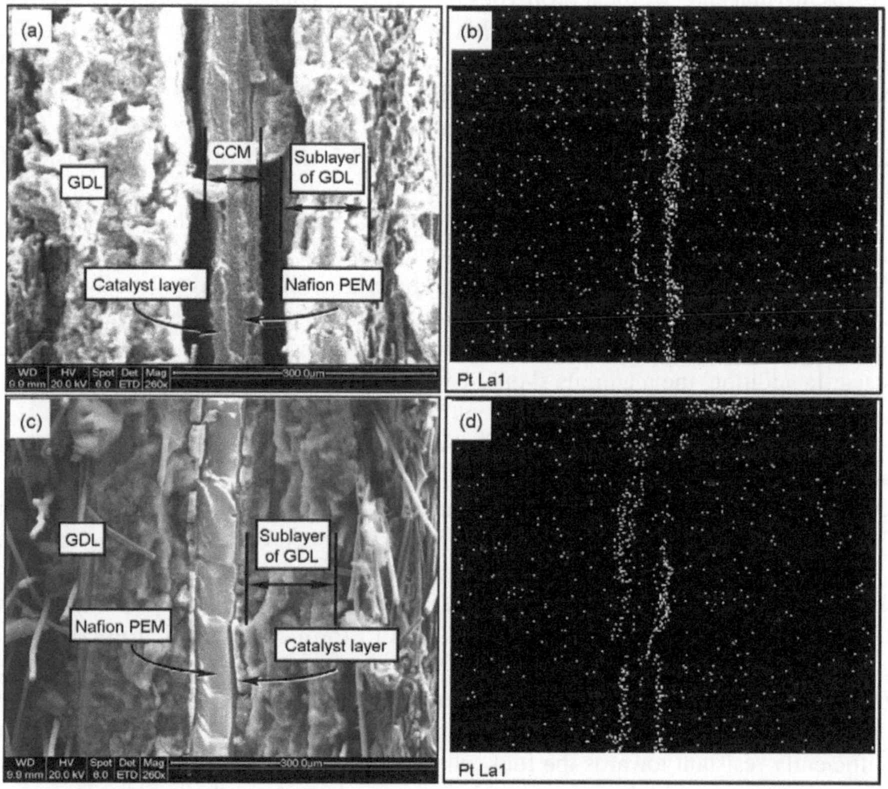

Fig. 5.28 SEM images and Pt element mapping of cross-section of MEA prepared by (**a**, **b**) CCM and (**c**, **d**) hot-press methods. Reproduced from Ref. [33] with permission from Elsevier, Copyright 2007

on the anode and cathode side of the CCM to form the MEA. The Pt loading was 0.2 mg/cm for both the anode and cathode catalyst layer in both cases.

For the MEA prepared by CCM, the Pt catalysts are uniformly distributed at the electrode/membrane interface and the catalysts have been integrated with the Nafion membrane due to the close contacts. In the case of MEA prepared by hot-press, a clear interface is visible and Pt catalyst is more scattered at the interface region, although hot-press produces a better interface between the catalyst layer and GDL. MEA prepared by CCM produces a high power output as compared to that by hot-press, indicating the importance of the catalyst layer/membrane interface.

5.6 Bipolar Plates and Seals

Bipolar plates (BPP) are a key component determining the dimension, weight as well as thermal and electrical properties of a fuel cell stack. First of all the materials should be electrically conductive, with a target conductivity of over 100 S/cm. On top of the bulk conductivity, the interfacial contact resistance between the BPP and backside of electrodes is also critical. Adding these two items together and including the thickness, the overall area specific resistance is targeted at a value below 0.01 Ωcm^2, which can be translated into a voltage loss of <15 mV at a current density of 1.5 A/cm^2.

The main function of BPP is to distribute reactant gases in the flow channels and, at the same time, separate the anode and cathode reactants between adjacent cells. These two functions require a dense material with low gas permeability that can be formed into with flow channels. Other functions of BPP are to transport the generated water and heat out of the cells and provide mechanical stability of the stack. In addition, the materials should be inexpensive, light in weight and resistant toward the highly corrosive environment of fuel cells.

Materials that satisfy these functional demands include metals, graphite, and its composites. Graphite, in both its natural and synthetic form, is a good electronic conductor with excellent chemical stability. It is, however, suffering from the high cost, low mechanical strength, and need for machining in order to form flow channels.

Metallic BPPs are the primary choice for the automobile applications as metals with high electrical and heat conductivity can be formed into very thin sheets with flow channels. Additionally, they do not tend to form cracks, which in combination with their non-porous structure offers good gas tightness. As a result metallic BPPs have advantages in volume and weight of the stack. Metal plates are, on the other hand, inherently prone to corrosion. Even high grade stainless steels are not sufficiently resistant towards the fuel cell environment involving acidic electrolyte, oxygen, steam, and high electrochemical potential on the cathode side. The most investigated metals include aluminum, stainless steel, titanium, and nickel. Considerable effort has been devoted to the development of protective coatings for steel plates, both precious metal coatings and preferentially cost-effective coatings and surface treatments.

Carbon composite BPP is often made from fillers like graphite or other electrically conducting carbons and binder polymers. Carbon fillers, most commonly based on synthetic graphite, constitute at least 80 wt% of the bipolar plate material. Carbon additives such as highly conductive carbon nanotubes are often used in order to improve the electrical conductivity. The key challenge is a good balance between the electrical conductivity, provided by the carbon component, and mechanical stability as well as gas tightness, provided by the binder polymer. Molding by injection or compression is the widely used processing technique, allowing for cost and quantity effective production.

Besides the bipolar plate, the gasket material is an important component of the fuel cell stack and tends to be heavily underestimated. Main functions of seals in a fuel cell stack are to seal and prevent leakage of anode and cathode area as well as the cooling plate. It plays also a key role in stacking by compensation of tolerances and mechanical compression of the MEA. The often used sealing materials, primarily due to their low cost, are fluorinated elastomers, thermoplastic elastomers, silicones, and sometimes reinforced PTFE. The seals are typically integrated into the flow field or bipolar plates.

5.7 Summary

In this chapter, we have introduced and discussed the following topics:

- Compared with low-temperature fuel cell using liquid electrolytes, a solid polymer electrolyte membrane provides mechanical strength and flexibility and therefore is simple and possible to construct all-solid cells with water as the only liquid in the system. The membrane can be used in small thickness with excellent resistance to gas reactant crossover. PEMFC can operate at high powder density with rapid load response and particularly suited for vehicular propulsion.
- Lower temperature fuel cells (AFC, PAFC, and PEMFC) are based on electrolytes conducting proton-containing species. The unique chemistry of protons enables the conductivity in two mechanisms: the proton transfer with a carrying species (the vehicle mechanism) or through a hydrogen bond chain by jumping from one site to another (the Grotthuss mechanism). These mechanisms determine to a large extent the features of cell construction and operation.
- Poly(perfluorosulfonic acid) membranes are commercially materials with high proton conductivity in fully hydrated states. It is water molecules that solvate protons and forming hydronium (H_3O^+) and other conducting species. The presence and movement of water are mandatory, and water management is a basic issue for PEMFC operation.
- Using short side chains or reinforced composite is an effective means to improve the membranes and fuel cell performance.
- For both anode and cathode, platinum is the most effective catalyst in terms of both activity and stability. The state-of-the-art catalysts are in form of nanoparticles

supported on high specific area carbons with a typical metal loading of 40–60 wt%.

- Alloying of platinum with transition metals such as nickel and cobalt enhances the ORR specific activity (exchange current density) by 3–5 times, attributable to modification of the surface electronic structure of Pt and hence weakening the binding of the reaction intermediates.
- Development of non-Pt catalysts has made significant progress based on the nitrogen (N)-coordinated iron single atoms embedded in carbon materials, i.e., FeN_x/C type composites though fabrication of high-performance electrodes as well as long-term stability is yet to prove.
- Variants of carbon-support materials are attempted to tailor the catalyst dispersion and improve the stability which are however limited to research activities.
- Catalysts in ionomer bonded thin layers can be applied on either membranes (catalyst-coated membrane, CCM) or gas diffusion layers (gas diffusion electrode, GDE) to create a three-phase boundary with the extended electrochemically active area. And membrane electrode assemblies (MEAs) are eventually constructed by attachment of membranes or gas diffusion layer materials.
- Bipolar plates (BPPs) are the fuel cell stacking component that contributes to the mass and volume, i.e., the power density as well as the cost of the stack. The main candidates for BPPs are graphite and stainless steel based materials.

5.8 Questions

5.1 Explain the following terms and materials

Hydrogen atom, hydrogen ion, proton and hydrated proton
Bare proton-conducting and hydrated proton-conducting processes
Proton conduction *via* the vehicle mechanism and/or Grotthuss mechanism
Ionomer and ionene
Electroosmotic drag of water
Composite membrane and reinforced composite membrane
Turnover frequency
ORR via 4-electron and 2-electron pathway
Pt/C and Pt alloy electrocatalysts
Non-precious metal catalyst
Carbon black, carbon nanotube, and graphene
Surface functionalization and self-assembly
Catalyst layer, gas diffusion layer and microporous layer
Hydrophobic and hydrophilic properties
Hot-press and catalyst-coated membrane, bipolar plates and gasket materials

5.2 Fuel cells can be constructed using an electrolyte conducting oxide anions ($O^{=}$) or molten carbonate ($CO_3^{=}$). Write anode and cathode half-cell reactions as well as the overall cell reactions of the two fuel cells and comment on the CO_2 management issue for the latter case.

5.3 Proton exchange membranes are often assumed as a bare proton (H^+) conductor. It is in fact a conductor of hydrated protons (e.g., H_3O^+). Write anode and cathode half-cell reactions as well as the overall cell reactions of fuel cells based on the two types of electrolytes and comment on the water management issue for the latter case.

5.4 An electrolyte is a substance conducting ions of different kinds. In a fuel cell the cations move from anode to cathode and anions move on the opposite direction, as shown in Figure below.

(a) Can the electrolytes that are conducting Li^+ or Na^+ cations be used to construct an H_2–O_2 fuel cell? If yes, write the anode and cathode half-cell reactions

(b) Can the electrolytes that are conducting Br^- or Cl^- anions be used to construct an H_2–O_2 fuel cell? If yes, write the anode and cathode half-cell reactions

(c) Can the electrolytes that are conducting NH_4^+ or $H_4PO_4^+$ cations be used to construct an H_2–O_2 fuel cell? If yes, write the anode and cathode half-cell reactions

(d) (d) Can the electrolytes that are conducting HCO_3^- or HSO_4^- anions be used to construct an H_2–O_2 fuel cell? If yes, write the anode and cathode half-cell reactions

5.5 Explain the vehicle mechanism and the Grotthuss mechanism of the proton conductivity in your own words and give examples of electrolyte materials of each type.

5.6 Explain differences of two types of ionic polymers, ionomers and ionenes in terms of mirco phase separation

5.7 In the lab you find two commercial membranes, Nafion® 211 and Aquivion®E79-09S. describe the differences of the two products.

5.8 A Gore-Select® composite membrane has a thickness of 20 micron and a proton conductivity of 50 mS/cm and a Nafion® 115 membranes has a thickness of 140 micron and a proton conductivity of 120 mS/cm. In a fuel cell operating at a current density of 1.5 A/cm^2, which membrane will have a larger ohmic loss?

5.9 The membrane degradation is often evaluated by measuring the H_2 crossover during a wet/dry humidity cycling. In a cell equipped with a Nafion 112 membrane with ambient pressure H_2 on one side and liquid water filled on the other. A hydrogen crossover current density of 1.6 mA/cm^2 is measured at 60 °C. Calculate the hydrogen permeability. (Answer: 4×10^{-11} mol H_2/cm s bar)

5.10 The CO-tolerant anode catalysts are of importance for both H_2-rich reformate and direct methanol as fuels. Give examples of the catalyst alloys and briefly explain the bifunctional mechanism

5.11 What are the dissociative and associative adsorptions of oxygen on the catalyst surface and explain the related ORR mechanisms

5.12 Commercial Pt/C catalysts contain platinum nanoparticles in a size range of 2–4 nm. Assuming the Pt particles are in form of spheres, calculate the specific surface area of platinum in m^2/g. The density of platinum is 21.5 g cm^3. (Answer: 70–140 m^2/g)

5.13 Platinum alloys with later transition metals, e.g., Ni. Co or Fe show enhanced ORR activity. With the help of terms the Pt-skin and Pt-skeleton structures explain the alloy effect.

5.14 Great progress has in last years been achieved in development of non-precious metal-based ORR catalysts. What are the major issues for applications of this type of materials in fuel cells?

5.15 Carbon blacks have been the base supports for catalysts while carbon nanotubes and graphene are the possible alternatives. What are the main barriers for the latter to replace carbon blacks?

5.16 A main requirement for gas diffusion layer materials is the electronic conductivity in both in-plane and through-plane. Why is the in-plane conductivity more important than the through-plane conductivity?

5.17 Explain the three-phase boundaries of a catalyst layer and the way of introducing ionic conductivity in the layer

5.18 For membrane electrode assembly fabrication the catalysts can be applied on either the gas diffusion layer or the membrane. The former is called gas diffusion electrode and the latter catalyst-coated membranes. Briefly discussion pros and cons of the two techniques.

5.19 Bipolar plates are the key stack materials made of either carbon or metals. What are the pros and cons of the two types of bipolar plates?

5.9 General Readings

1. Mauritz KA, Moore RB (2004) State of Understanding of Nafion. Chem. Rev. 104: 4535–4585.
2. Nie Y, Li L, Wei Z (2015) Recent advancements in Pt and Pt-free catalysts for oxygen reduction reaction. Chem. Soc. Rev. 44: 2168-2201.
3. Mathias MF, Roth J, Fleming J, Lehnert W (2003) Diffusion media materials and characterization, in Handbook of Fuel Cells, ed. W. Vielstich, A. Gasteiger, A. Lamm, Vol.3, Chapter 42, pp. 517-537, John Wiley & Sons.

References

1. Kreuer KD (2000) On the complexity of proton conduction phenomena. Solid State Ionics 136–137(1–2):149–160
2. Grubb WT, Niedrach LW (1960) Batteries with solid ion-exchange membrane electrolytes: II. Low-temperature hydrogen-oxygen fuel cells. J Electrochem Soc 107:131
3. Costamagna P, Srinivasan S (2001) Quantum jumps in the PEMFC science and technology from the 1960s to the year 2000 Part II. Engineering, technology development and application aspects. J Power Sources 102(1–2):253–269
4. Zawodzinski TA, Neeman M, Sillerud LO, Gottesfeld S (1991) Determination of water diffusion-coefficients in perfluorosulfonate ionomeric membranes. J Phys Chem 95(15):6040–6044
5. Li QF, He RH, Berg RW, Hjuler HA, Bjerrum NJ (2004) Water uptake and acid doping of polybenzimidazoles as electrolyte membranes for fuel cells. Solid State Ionics 168(1–2):177–185
6. Springer TE, Zawodzinski TA, Gottesfeld S (1991) polymer electrolyte fuel-cell model. J Electrochem Soc 138(8):2334–2342
7. Ghielmi A, Vaccarono P, Troglia C, Arcella V (2005) Proton exchange membranes based on the short-side-chain perfluorinated ionomer. J Power Sources 145(2):108–115
8. Kreuer KD, Schuster M, Obliers B, Diat O, Traub U, Fuchs A, Klock U, Paddison SJ, Maier J (2008) Short-side-chain proton conducting perfluorosulfonic acid ionomers: Why they perform better in PEM fuel cells. J Power Sources 178(2):499–509
9. Nakao M, Yoshitake H (2003). In: Vielstich W, Gasteiger H, Lamm A (eds) Handbook of fuel cells—Fundamental, technology and applications, vol 3. Willey, pp 412–419
10. Tang HL, Pan M, Wang F, Shen PK, Jiang SP (2007) Highly durable proton exchange membranes for low temperature fuel cells. J Phys Chem B 111(30):8684–8690
11. Tang H, Wan Z, Pan M, Jiang SP (2007) Self-assembled Nafion-silica nanoparticles for elevated-high temperature polymer electrolyte membrane fuel cells. Electrochem Commun 9(8):2003–2008
12. Gottesfeld S, Pafford J (1988) A new approach to the problem of carbon-monoxide poisoning in fuel-cells operating at low-temperatures. J Electrochem Soc 135(10):2651–2652
13. Norskov JK, Rossmeisl J, Logadottir A, Lindqvist L, Kitchin JR, Bligaard T, Jonsson H (2004) Origin of the overpotential for oxygen reduction at a fuel-cell cathode. J Phys Chem B 108(46):17886–17892
14. Wilson MS, Gottesfeld S (1992) High-performance catalyzed membranes of ultra-low Pt loadings for polymer electrolyte fuel-cells. J Electrochem Soc 139(2):L28–L30
15. Peuckert M, Yoneda T, Betta RAD, Boudart M (1986) Oxygen reduction on small supported platinum particles. J Electrochem Soc 133(5):944–947
16. Thompsett D (2003) Pt alloys as oxygen reduction catalysts. In: Vielstich W, Gasteiger H, Lamm A (eds) Handbook of fuel cells—Fundamental, technology and applications, vol 3. Wiley, pp 467–480
17. Toda T, Igarashi H, Uchida H, Watanabe M (1999) Enhancement of the electroreduction of oxygen on Pt alloys with Fe, Ni, and Co. J Electrochem Soc 146(10):3750–3756
18. Escudero-Escribano M, Malacrida P, Hansen MH, Vej-Hansen UG, Velazquez-Palenzuela A, Tripkovic V, Schiotz J, Rossmeisl J, Stephens IEL, Chorkendorff I (2016) Tuning the activity of Pt alloy electrocatalysts by means of the lanthanide contraction. Science 352(6281):73–76
19. Wang C, Chi MF, Li DG, Strmcnik D, van der Vliett D, Wang GF, Komanicky V, Chang KC, Paulikas AP, Tripkovic D, Pearson J, More KL, Markovic NM, Stamenkovic VR (2011) Design and synthesis of bimetallic electrocatalyst with multilayered Pt-skin surfaces. J Am Chem Soc 133(36):14396–14403
20. Gupta S, Tryk D, Bae I, Aldred W, Yeager E (1989) Heat-treated polyacrylonitrile-based catalysts for oxygen electroreduction. J Appl Electrochem 19(1):19–27
21. Lefevre M, Proietti E, Jaouen F, Dodelet JP (2009) Iron-based catalysts with improved oxygen reduction activity in polymer electrolyte fuel cells. Science 324(5923):71–74

22. Wu G, More KL, Johnston CM, Zelenay P (2011) High-performance electrocatalysts for oxygen reduction derived from polyaniline, iron, and cobalt. Science 332(6028):443–447
23. Chung HT, Cullen DA, Higgins D, Sneed BT, Holby EF, More KL, Zelenay P (2017) Direct atomic-level insight into the active sites of a high-performance PGM-free ORR catalyst. Science 357(6350):479–483
24. Kangasniemi KH, Condit DA, Jarvi TD (2004) Characterization of vulcan electrochemically oxidized under simulated PEM fuel cell conditions. J Electrochem Soc 151(4):E125–E132
25. Zhou JG, Zhou XT, Sun XH, Li RY, Murphy M, Ding ZF, Sun XL, Sham TK (2007) Interaction between Pt nanoparticles and carbon nanotubes—An X-ray absorption near edge structures (XANES) study. Chem Phys Lett 437(4–6):229–232
26. Tian ZQ, Jiang SP, Liang YM, Shen PK (2006) Synthesis and characterization of platinum catalysts on muldwalled carbon nanotubes by intermittent microwave irradiation for fuel cell applications. J Phys Chem B 110(11):5343–5350
27. Yuan WY, Lu SF, Xiang Y, Jiang SP (2014) Pt-based nanoparticles on non-covalent functionalized carbon nanotubes as effective electrocatalysts for proton exchange membrane fuel cells. RSC Adv 4(86):46265–46284
28. Fujigaya T, Nakashima N (2008) Methodology for homogeneous dispersion of single-walled carbon nanotubes by physical modification. Polym J 40(7):577–589
29. Wang S, Jiang SP, Wang X (2008) Polyelectrolyte functionalized carbon nanotubes as a support for noble metal electrocatalysts and their activity for methanol oxidation. Nanotechnology 19(26):265601
30. Yang DQ, Rochette JF, Sacher E (2005) Spectroscopic evidence for pi-pi interaction between poly(diallyl dimethylammonium) chloride and multiwalled carbon nanotubes. J Phys Chem B 109(10):4481–4484
31. Han M, Chan SH, Jiang SP (2006) Development of carbon-filled gas diffusion layer for polymer electrolyte fuel cells. J Power Sources 159(2):1005–1014
32. Raistrick ID (1989) 4876115. US Pat. No. 4876115
33. Tang HL, Wang SL, Jiang SP, Pan M (2007) A comparative study of CCM and hot-pressed MEAs for PEM fuel cells. J Power Sources 170(1):140–144

Chapter 6
Polymer Electrolyte Membrane Fuel Cells: Fabrication and Characterization

This chapter is devoted to an introduction of fabrication and characterization of key materials and components of PEMFC including proton exchange membranes, electrocatalysts, and membrane electrode assemblies. The characterization is primarily focused on electrochemical properties, including both ex situ and in situ techniques.

The chapter is written with a perfluorosulfonic acid-type membrane as an example but applies to other membrane types as well. Efforts are made to illustrate the processes and protocols with examples. Analysis of the characterization results and fuel cell performance will be further given in Chap. 7.

6.1 Proton Exchange Membranes

6.1.1 PFSA Membrane Fabrication

The perfluorosulfonic acids (PFSA) are copolymers synthesized from tetrafluoroethylene (TFE) and perfluorinated vinyl ether sulfonyl fluoride. While TFE is an often used building block for fluoropolymers, the structure of the sulfonyl fluoride monomers varies which determines the length of the perfluoroether pendant side chains with terminal sulfonic acid groups. As described in Chap. 5, the ratio between the two co-monomers decides the equivalent weight (EW) of the resulting polymer.

Figure 6.1 illustrates synthesis of the long side-chain (LSC) monomer, perfluoro sulfonyl fluroride ethyl propyl vinyl ether (PSEPVE), and the short side-chain (SSC) monomer, sulfonyl fluoride vinyl ether (SFVE). The synthesis of PFSA type materials requires highly dedicated process and equipment, and due to safety issues the use of TFE under conditions of high pressure and temperature, which easily forms explosive peroxides in contact with air, is strongly restricted by the authorities in many countries. Most of the research and development of PFSA materials are thus carried out in industrial laboratories, and the literature describing the fundamental polymer chemistry is rather scarce.

S. P. Jiang and Q. Li, *Introduction to Fuel Cells*,
https://doi.org/10.1007/978-981-10-7626-8_6

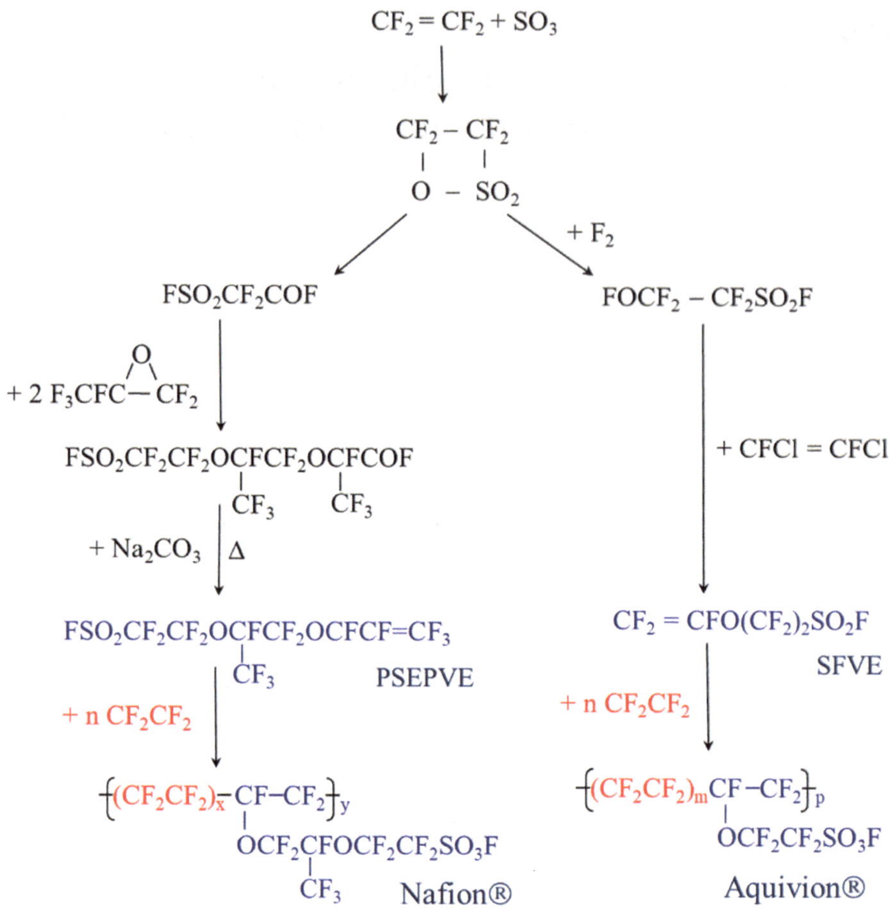

Fig. 6.1 Variations of perfluorinated vinyl ether sulfonyl fluoride monomers used for synthesis of perfluorosulfonic acid polymers

The perfluorinated vinyl ether monomers are relatively unreactive, and the polymerization is typically performed in perfluorocarbon solvents with a perfluorinated free radical initiator. The product is non-ionic precursor polymer in its sulfonyl fluoride form, which is stable in water and acids. It can be completely hydrolyzed in strong base, e.g., sodium hydroxide to give a Na^+ salt form that is further converted to the acidic form using nitric acid. Prior to the conversion to its salt form, the sulfonyl fluoride precursor polymer melts at above 200 °C and can be melt-extruded into membranes, a process suitable for mass production of homogeneous thin polymer films such as Nafion 1100 series. The extrusion-stretching mechanically reinforces the membrane through chain orientations, which makes the membrane tolerant toward the stresses induced in the membrane upon wet–dry cycling during fuel cell operation and enhances the long-term durability. The extruded membranes exhibit detectible, although small, anisotropy, i.e., the thickness swelling is higher than the swelling in the in-plane directions. Along the two in-plane directions, the membrane exhibits lower swelling and higher tensile strength in the machine (extrusion) direction (MD) than in the transverse direction (TD).

Another approach to membrane fabrication is by solution casting of the acid form of the ionomer. In fact, Nafion® or other PFSA ionomers do not form true solutions in low-boiling mixtures of alcohols and water. It is a dispersion of the ionomer aggregates in size of a few nanometers. PFSA dispersions of various EWs and ionomer concentrations are commercially available for both membrane casting and catalyst ink preparation.

Improvements in membrane chemistry and reinforcements have paved the way to reduce the thickness of membranes with chemical and mechanical stability. In 1990s Nafion 117 membranes of dry thickness 175 μm were widely used as a reference material in PEMFC studies, while recent developments are more focusing on use of membranes with 25 μm or smaller thickness (e.g., Nafion 211) or reinforced membranes of thickness as small as 5 μm (e.g., Nafion XL and Gore-Select). These thin membranes especially the PTFE-reinforced composite membranes are prepared by solution casting. In general, solution cast membranes are less crystalline than the extruded membranes, and high-temperature annealing is necessary in order to induce polymer chain reorganization for formation of semi-crystalline domains and therefore improves membrane mechanical properties.

6.1.2 Ion-Exchange Capacity and Equivalent Weight

Ion-exchange capacity (IEC) is defined as the number of active sites or functional groups responsible for ion exchange per unit mass of the dry polymer, usually expressed as milli-equivalent (meq) of ion-exchange groups contained in one gram dry polymer, in meq/g. Since the immobilized charge carrier of an ion-exchange membrane is always of mono-valence, this unit is equivalent to milli-moles per gram (mmol/g). In polymer chemistry another relevant term often used is the *equivalent weight* (EW). It is the inverse of the IEC, which should be understood as the mass of the polymer containing one mole of functional or reactive groups. For PFSA membranes with an EW of 800–1100 g/mol the corresponding IEC is 0.9–1.2 meq/g.

The IEC of membranes is a parameter that affects almost all other membrane properties. Determination of the IEC is therefore important in the membrane characterization. Experimentally, the ion-exchange capacity of a membrane can be determined by titration of the fixed ions. The membranes are first equilibrated in, e.g., a NaOH solution for cation exchange membranes or an HCl solution for anion exchange membranes. The membranes are then rinsed with deionized water to remove the excess ionic species. The ion-exchange capacity of the samples is then measured by back titration. The dry weight of the membrane samples can be determined before the ion exchange or after the titration. A critical step is to ensure a complete exchange of ions in the membrane before the titration. The IEC measurement is illustrated in Example 6.1.

Example 6.1

Measurement of Membrane Ion-exchange Capacity.

Preparation of membrane samples

Cut a 4 × 5 cm^2 piece of the Nafion 115 membrane and submerge it into 0.1 M sulfuric acid for 3 h at 30 °C. Rinse the sample thoroughly with deionized water.

Measurement

Submerge the sample into 100 mL 0.01M (C_{NaOH}) sodium hydroxide aqueous solution, and soak for 10 h at room temperature

Titrate the soaking solution using 0.01M (C_{HCl}) hydrochloric acid until it reaches the phenolphthalein (neutral) end point. The volume (V_{HCl}, ml) of the acid consumed is 5.6 ml.

Rinse the cation-exchanged membrane sample thoroughly with deionized water, and dry it in a vacuum (<100 mbar) oven for 4 h at 50 °C. The sample is then weighed and found to be $W_{dry} = 0.485$ g.

Calculation

The concentration of NaOH in the soaking solution is

$$C_{NaOH} = \frac{C_{HCl} \times V_{HCl}}{V_{NaOH}} = \frac{0.01\mathrm{M} \times 5.6\,\mathrm{ml}}{100\,\mathrm{ml}} = 0.00056\,\mathrm{M} \qquad (6.1)$$

That is to say, in 100 ml soaking solution the initial mole number of NaOH is 0.01 M × 0.1 L = 0.001 mol. After ion exchange, the remaining NaOH is 0.00056 mol, meaning that 0.001–0.00056 = 0.00044 mol NaOH is neutralized by H^+. This is the total mole number of H^+ or SO_3H^- contained in the 0.485 g polymer membrane. The membrane IEC is therefore

$$\mathrm{IEC} = \frac{0.00044\,\mathrm{mol}}{0.485\,\mathrm{g}} = 0.907\,\mathrm{mmol/g} \qquad (6.2)$$

This is corresponding to an equivalent weight of 1103 g/mol.

6.1.3 *Water Uptake and Swelling*

As described in Chap. 5, the water uptake and swelling of perfluorinated ionomeric membranes are essential to achieve a proton conductivity. The water uptake of a membrane is often reported in percentage, defined on the dry polymer basis.

The *water uptake* from the liquid phase can be obtained by submerging a pristine membrane in distilled water. After a certain period of time at room temperature or other specified temperatures, the membrane is taken out of the water, quickly blotted with a tissue, and weighed immediately. To determine the water uptake from the vapor phase at varied humidities and room temperature, a membrane sample is suspended above an aqueous solution of a salt, e.g., LiCl in a closed vessel. The relative humidity of the atmosphere above the salt solution is a function of the salt concentration, which is well-defined in the literature. An equilibrium time of a few days is suggested.

Determining the water sorption of a membrane seems simple. However, some precautions must be taken to achieve satisfactory measurement accuracy. For example, the surface water on a wet membrane must be removed without loss of water from the membrane bulk phase. This can be difficult with very thin membranes. To determine the drying weight all water must be removed. This often requires drying for an extended period of time by exposure to P_2O_5 at room temperature or under a vacuum <100 mbar at a mild temperature of 50 °C overnight. These measures may lead to some irreversible structural changes of the membrane. It is therefore a common practice to measure the dry weight and dimensions (size and thickness) after the water uptake and swelling measurement. Based on the measurement the following factors can be calculated:

$$\text{Water uptake}\,(\omega) = \frac{W_{\text{wet}} - W_{\text{dry}}}{W_{\text{dry}}} \times 100\% \tag{6.3}$$

$$\text{Area swelling} = \frac{A_{\text{wet}} - A_{\text{dry}}}{A_{\text{dry}}} \times 100\% \tag{6.4}$$

$$\text{Thickness swelling} = \frac{D_{\text{wet}} - D_{\text{dry}}}{D_{\text{dry}}} \times 100\% \tag{6.5}$$

$$\text{Volume swelling} = \frac{V_{\text{wet}} - V_{\text{dry}}}{V_{\text{dry}}} \times 100\% \tag{6.6}$$

The sulfonic acid groups in the membranes form the hydrophilic domains. Increase in the concentration of the ionic functionalities, as the EW of the polymer decreases, leads to a higher water uptake. This is shown in Fig. 6.2 where a clear correlation of the water uptake and ionic conductivity with the EW is seen. At very low EW values the membranes display excessive swelling (and ultimately complete dissolution), leading to dilution of the ionic concentration and therefore a decrease in the ionic conductivity. A set of water uptake and wet thickness data for Nafion® and Aquivion® membranes are listed in Table 6.1.

The water uptake of a proton exchange membrane is often expressed as the hydration number, the number of water molecules for each acid group, which can be calculated by the following equation:

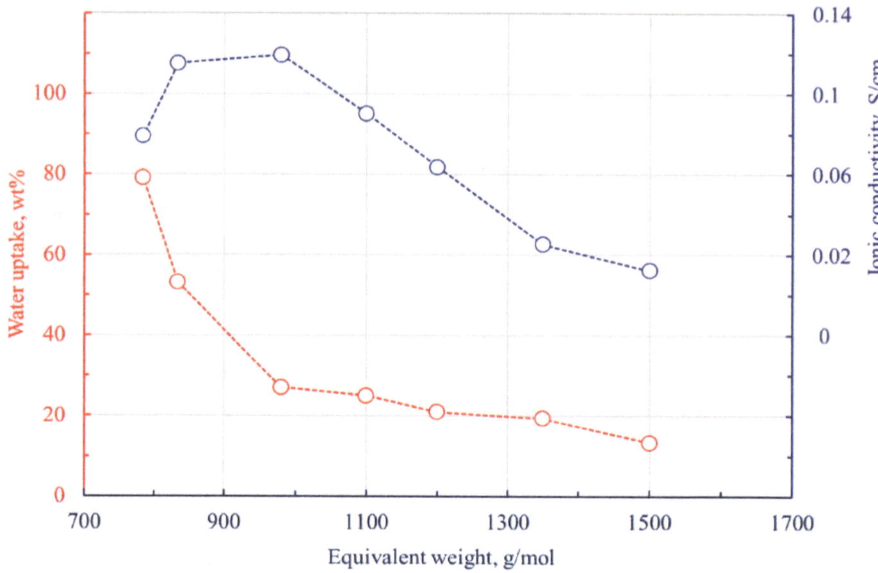

Fig. 6.2 Water uptake and ionic conductivity as a function of equivalent weight of Nafion membranes. Data were measured for membranes immersed in liquid water under ambient conditions, taken from Ref. [1]

Table 6.1 Water uptake and wet thickness of Nafion membranes [2]

Type of Nafion® membrane	Dry thickness (μm)	Wet thickness (μm)	Water uptake (wt%) on dry polymer basis	Hydration number (mol H_2O/mol SO_3H)
Nafion112	50	58–62	32–34	21–22
Nafion115	125	145–150	32–34	21–22
Nafion117	175	200–205	32–34	21–22
Nafion105	125	160–165	42–43	27–28

$$\lambda_{H_2O} = \frac{\text{mol}\,H_2O}{\text{mol}\,SO_3H} = \frac{(W_{wet} - W_{dry}) \times \frac{1\,\text{mol}\,H_2O}{18\,\text{g}} \times \frac{100}{100}}{W_{dry} \times \text{IEC}\frac{\text{mmol}}{\text{g}} \times 10^{-3}} = \frac{\omega}{\text{IEC}} \times \frac{10}{18} \tag{6.7}$$

A set of data for wet thickness, water uptake, and hydration number for Nafion and Aquivion membranes are listed in Table 6.1.

6.1.4 Proton Conductivity

For an object with a uniform cross section, for example, a strip of a film, its resistance (R) is proportional to its resistivity (ρ) and length (L) and inversely proportional to its cross-sectional area (A) according to Pouillet's law:

$$R = \rho\frac{L}{A} = \frac{1}{\sigma} \times \frac{L}{A} \tag{6.8}$$

Here σ is the conductivity or the reciprocal of the resistivity, having a unit of S/cm.

Two types of conductivity cells have been used, a two electrode cell for the through-plane conductivity ($\sigma_{\perp}$) and a four-probe cell for the in-plane conductivity ($\sigma_{\parallel}$) measurements. Both cells can be immersed in water or under atmosphere of known water vapor pressure. Figure 6.3 shows a four-probe conductivity cell. The membrane sample is fixed between two PTFE holders. The upper holder has two holes for locating the sensor and reference electrodes between which the potential is measured. The bottom holder has two platinum foils in contact with the lower side of the membrane sample serving as the working and counter electrodes. The holders and the fixed membrane sample are placed in a glass tube, which is equipped with endplates, through which four electrode lead wires and the humidified air inlet and outlet are connected with gas tight seals. The whole cell is kept in an oven for temperature control.

The conductivity cell assembly involves two interfaces between the platinum foils and the membrane electrolyte that the applied current should pass through. The platinum foils are an electronic conductor, while the membrane electrolyte is an ionic (H^+) conductor. The charger cannot simply pass through the interface where, in fact, a charged double layer is established in a similar manner as a capacitor with a capacitance C_{dl}. A capacitive current can only flow through a capacitor when the voltage across the capacitor is changing with respect to time:

$$i_C = C_{dl}\frac{\mathrm{d}E}{\mathrm{d}t} \tag{6.9}$$

This is illustrated in Fig. 6.3. Two types of alternating current are used for the conductivity measurement, i.e., the square-wave current (also called direct current interruption) or electrochemical impedance spectroscopy (EIS). For EIS, the AC frequency is often in a range from 0.1 to 1 MHz down to 1–10 Hz. The high-frequency intercept with the real impedance axis in a Nyquist plot gives the ohmic resistance (Fig. 6.4).

The most critical issue for the proton conductivity measurement is the control of water activity in the atmosphere under varying temperatures, which determines the water uptake and therefore the proton conductivity of the membrane. The equilibrium water activity can be controlled in a number of ways. The simplest approach is to immerse the membrane and confine the measurement in water or a salt solution. More

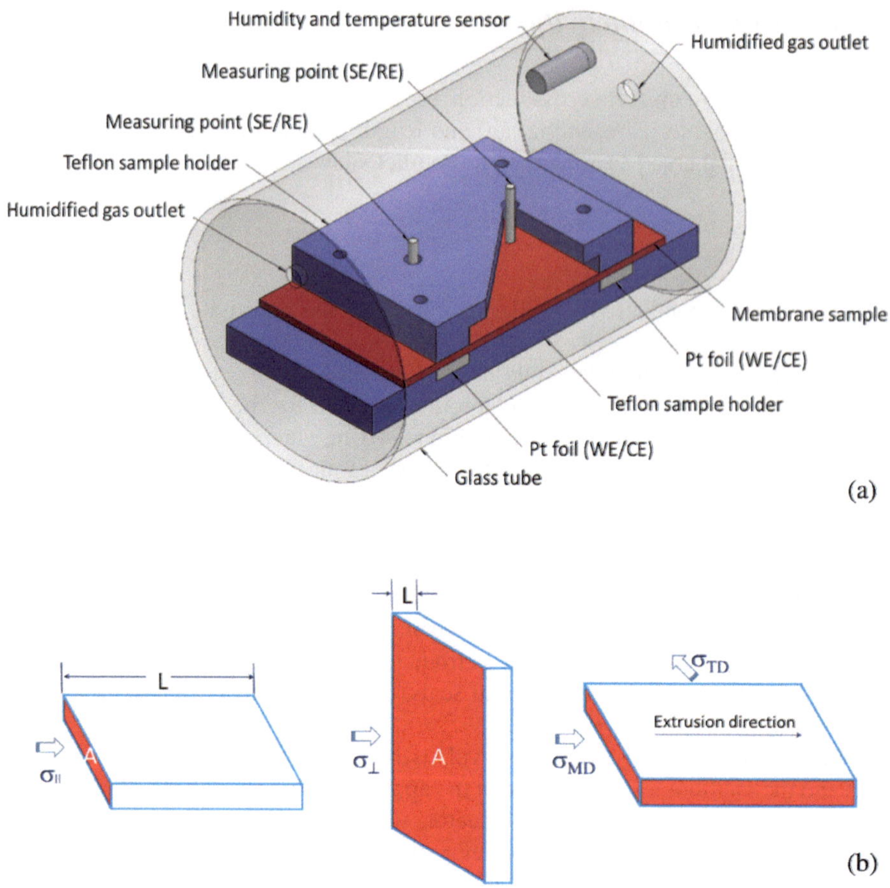

Fig. 6.3 **a** Construction of a four-probe conductivity cell and **b** illustration of in-plane ($\sigma_{\|}$), through-plane ($\sigma_{\perp}$) as well as machine direction (σ_{MD}) and transverse direction (σ_{TD}) conductivity

appropriate measurements are to tailor the water vapor pressure in the atmosphere. This can be done by equilibration of the gas phase with a salt solution of known activity or bubbling of a purging gas through a water bath at a certain temperature. For measurements at temperatures above 100 °C, humidification can be achieved through an evaporator into which water is pumped and mixed with a carrier gas. The temperature of the evaporator should be slightly higher than that of the conductivity cell in order to avoid any cold spot where condensation of water may occur.

In the in-plane conductivity measurements, it is always assumed that the current is homogeneously distributed throughout the entire membrane cross section. In case that a membrane shows an anisotropic behavior in the conductivity, for example, for extruded and particularly composite membranes, this is a problem. The directional extrusion and stretching is found to increase the conductivity in the machine direction (σ_{MD}) compared to the transverse direction (σ_{TD}). For Gore-Select membranes, for

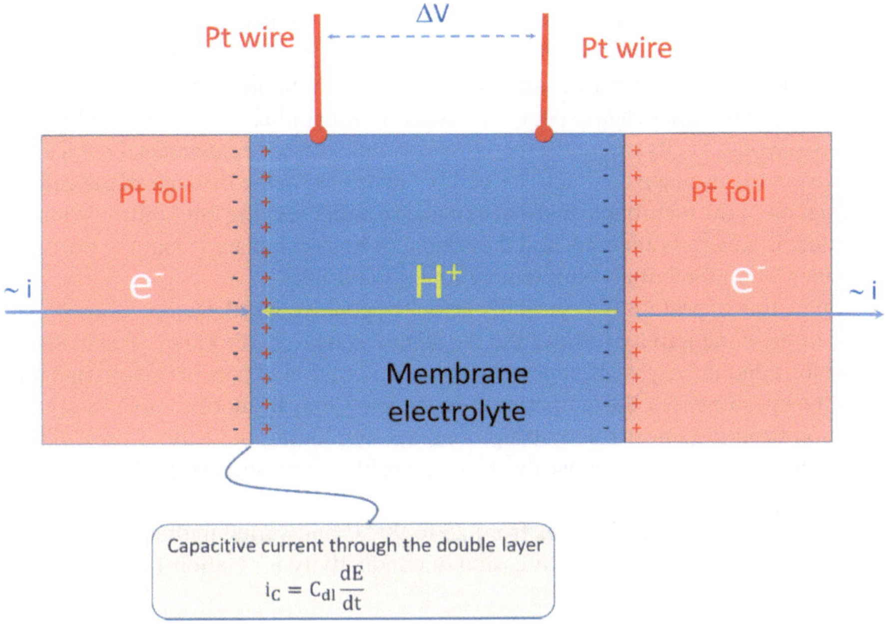

Fig. 6.4 Illustration of the double charged layer at the interface of the platinum foils and membrane electrolyte of the conductivity cell

example, it is reported that the in-plane conductivity is about 1.10–1.5 times higher than the through-plane conductivity, although this difference has not been observed for the dispersion cast membranes.

For through-plane conductivity measurements, the cell can be in a fuel cell configuration, in which the membrane is sandwiched between two electrodes. The electrodes typically consist of platinum disks or gas diffusion electrodes covering the entire membrane area. For this cell configuration, it is critical to minimize the polarization effect, which can be sufficiently achieved by using a high-frequency alternating current. A hydrogen flow, as a humidification carrier gas, is recommended on both electrode chambers, because its high electrochemical reversibility further minimizes the polarization. For a conductivity cell of 10 cm^2 active area with a membrane of 100 μm thickness, a conductivity of 0.1 S/cm corresponds to an overall resistance of 10 mΩ. This is very small resistance to measure. Any contact point through the cell components can be a major source of errors for the through-plane measurement. It is a good practice that the measured overall resistance is corrected by the short-circuit (with no membrane) cell resistance.

A series of measurements with membranes of different thicknesses allows for estimation of the contact resistance. Assuming that the contact resistance between two membrane layers is negligible, stacks of membrane samples can also be used instead of samples with different thickness. To some extent, conductivity measurements in a through-plane cell depend on the compressive forces on the membranes, and it is recommended to adjust the pressure to a fixed point by using a torque meter. The contact pressure should be enough to guarantee a good contact, but as low as possible.

With increasing temperature, dimensions of the membrane samples change due to thermal expansion, which is expected to have opposite influences on the conductivity of the two types of cells where the membrane sample is free to expand in the thickness direction in the in-plane conductivity cell but restricted in the through-plane cell. As information of the membrane thermal expansion under varying humidities is limited, this effect is generally ignored, and the conductivity is calculated from the measured resistance using the initial room temperature dimensions.

Conduction of protons is a thermally activated process, and the temperature dependence of the conductivity follows the Arrhenius equation (see Chap. 3). Physically the proton conductivity is influenced by the concentration of proton carriers and their mobility. In some cases, the temperature dependency may be more complex and cause deviation from the linear relationship obtained by plotting the natural logarithm of the conductivity (S/cm) against the inverse absolute temperature [$1/T$, K^{-1}] in an Arrhenius plot.

Based on data at temperatures from 25 to 90 °C and varied hydrations, Springer et al. suggested the following expression of conductivity for Nafion 117 membranes [3]:

$$\sigma\,(\mathrm{S/cm}) = (0.5139 \times \lambda - 0.326)\exp\left[1268 \times \left(\frac{1}{303} - \frac{1}{T}\right)\right] \tag{6.10}$$

It is seen that the conductivity is a linear function of λ, hydration number. For the membrane to exhibit a minimum conductivity, λ should be higher than 0.6, i.e., 0.6 water molecules per sulfonic acid site.

Figure 6.5 is another example of the in-plane conductivity for Aquivion E98-90S membranes which has an EW of 980 g/mol and thickness of 90 μm. The conductivity at 80 and 120 °C is measured under varied relative humidities. It should be noted that at 120 °C the saturated water vapor pressure is 1.96 bar. Under an atmosphere of pure water vapor the maximum relative humidity is 51%. This short side-chain membrane shows a conductivity of as high as 0.04 S/cm at 120 °C. The conductivity as a function of temperature is measured under 100% RH below 100 °C and under a constant water vapor pressure of 1 bar above 100 °C. A steady increase in the conductivity is observed until 100 °C. Further elevating temperature under the constant water vapor pressure of 1 bar leads to decrease in the relative humidity and evaporation of the water. As a result the conductivity decreases.

6.1.5 Electroosmotic Drag of Water

Water molecules that solvate the protons from the sulfonic acid groups form hydrated complexes such as hydronium cations (H_3O^+), Zundel cations ($H_5O_2^+$), or Eigen cations ($H_9O_4^+$). These proton-containing charge carriers are mobile and involved

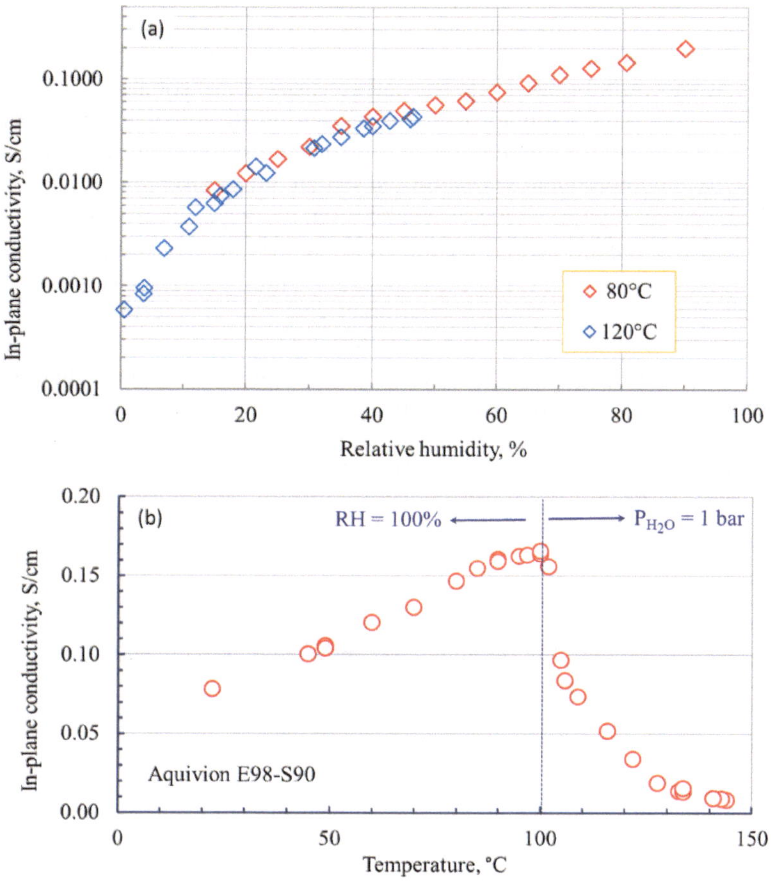

Fig. 6.5 In-plane conductivity of Aquivion E98-90S membranes under ambient conditions. **a** As a function of relative humidity at 80 and 120 °C and **b** as a function of temperature under 100% RH below 100 °C and under 1 bar water vapor pressure above 100 °C

in the charge transport, i.e., via the vehicle mechanism, as explained in Chap. 5. The water movement due to the proton transport is called the electroosmotic drag and therefore dependent on the current passing through the membrane. An electroosmotic drag coefficient of water is defined as the molar ratio of water moved with each proton in the absence of a water concentration gradient—any concentration gradient drives also the movement of water by diffusion as to be discussed in the next section.

In PEMFCs, the electroosmotic drag is of special interest as it, on one hand, directly impacts the hydration of membranes and therefore conductivity and ohmic losses of fuel cells. If the water content is too high, on the other hand, the electrodes are flooded, retarding kinetics and mass transport and hence lowering the cell performance.

A number of different techniques have been developed to determine the electroosmotic drag coefficient of water, and the most popular one is the electroosmotic drag cell, as shown in Fig. 6.6 [4]. The cell consists of two solution compartments, each with a horizontal capillary tube for measuring the small volume change of the electrolyte during the measurement. The attached syringes allow the electrolyte volume in each compartment to be adjusted so that the menisci remains in the capillary tubes. The membrane is sandwiched between two support screens to prevent the membrane deflection and sealed in a mount to separate the two compartments. Each compartment is equipped with a palladium–palladium hydride electrode allowing for proton transport upon passage of a direct current. The electrode reactions are simple consumption and production of protons, which are kinetically reversible and therefore minimize the electrode polarization and simplify the data analysis. The water drag can be determined by measuring, e.g., the water volume change in the different compartments.

The proton electroosmotic drag coefficient ξ of water is defined as the flux ratio of water to that of protons

$$\xi = \frac{j_{H_2O}}{j_{H^+}} \quad (6.11)$$

Here the water molar flux through the membrane due to the current is

$$j_{H_2O} = \frac{\Delta V}{\bar{V} \times A \times t} \quad (6.12)$$

$\bar{V}$ is the molar volume of water, ΔV is the volume change of either electrode chamber at time t, and A is the cross-sectional area of the membrane. Assuming that

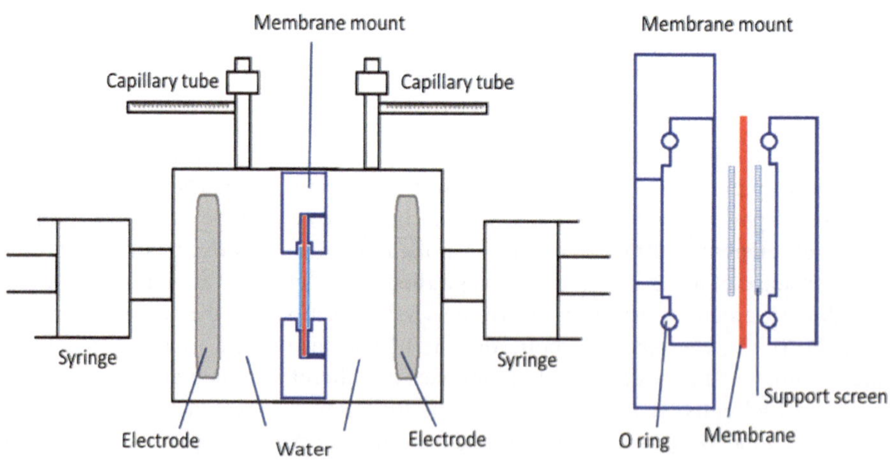

Fig. 6.6 Schematic of the experimental electroosmotic drag cell (left) and the membrane mount (right). Reproduced from Ref. [4] with permission from Elsevier, Copyright 2006

the cation transference number is unity, the proton molar flux through the membrane:

$$j_{H^+} = \frac{I}{F \times A} = \frac{i}{F} \tag{6.13}$$

where F is Faraday constant, I is the total current and i is the current density. Similar measurements can also be conducted in a fuel cell configuration operating in the hydrogen pumping mode. At the anode the humidified hydrogen is fed, generating protons that migrate toward the cathode where they are reduced to hydrogen. This H_2–H_2 mode eliminates the contributions of water production and water back diffusion. In case that the electrolyte concentration in the two compartments is different, water transport through the membrane by diffusion occurs, which is independent on the current and can be determined under OCV for correction.

For liquid water equilibrated Nafion membranes at room temperature, a coefficient of $\xi = 2.5$–3.0 H_2O/H^+ has been reported. For water vapor equilibrated membranes, however, the coefficient is much lower, between 1.0 and 1.4 H_2O/H^+ varying with the water vapor activity. Under fuel cell operating conditions, especially at higher temperatures and equilibrated with a water–methanol mixture, the presence of relatively large water-filled domains in membranes as well as the vehicle mechanism of proton conductivity results in even higher values of the drag coefficient.

It should be noted that, as a miscible liquid with water, methanol exhibits also electroosmotic drag phenomenon, which is a major mechanism of the methanol crossover, as to be discussed below.

6.1.6 Water and Methanol Diffusion

Any water concentration gradient, developed from the humidification of reactant gases, electroosmotic drag, or electrochemical formation at the cathode, is a driving force of water diffusion. This is the primary process to maintain the water balance and to ensure the dynamic response of the membrane and fuel cells to changes of, e.g., current density, humidification and temperature as well. In addition, the water diffusion can also be a major mechanism of the limiting current density.

The water diffusion coefficient in membranes can be determined by recording the transient water uptake curves of a dry membrane at varied temperatures. From the short-time slope of the water uptake curve, the diffusion coefficient as well as the activation energy could be estimated by assuming a constant diffusion coefficient over the course of the sorption measurement, which is in general not the case for water-swollen ionomers. Radiotracer and other labeling techniques, e.g., alkali metal cations have also been attempted. More often used is the pulsed gradient nuclear magnetic resonance spectroscopy (PG NMR) measurements to determine the ^1H self-diffusion coefficient, which can be converted into Fickian diffusion coefficients for purposes of comparison with other macroscopic measurements.

Methanol, as a liquid fuel in direct methanol fuel cells, is also of interest in the diffusion coefficient determination. Methanol diffusion is relatively easy to measure using a two-chamber cell with a methanol aqueous solution on one side and pure water on the other. The concentration gradient results in diffusion of methanol to the pure water side without interference from the electroosmotic drag. The change in the methanol concentration on the pure water side can be monitored by gas chromatography or changes in the refractive index or other analytic methods.

The water diffusion coefficient of Nafion membranes reported in the literature falls in a range of 10^{-6}–10^{-5} cm^2/s depending on the water content and temperature. Based on the NMR self-diffusion coefficient data and the water uptake isotherm measured at 30 °C for Nafion 115 membranes, the following expressions have been proposed for the Fickian diffusion coefficient in the low and high hydration ranges [5]:

$$D_{H_2O} = 3.10 \times 10^{-3} \times \lambda \times \left(e^{0.28\lambda} - 1\right) \times \exp\left[\frac{-2436}{T}\right] \quad \text{for } 0 < \lambda \leq 3 \qquad (6.14)$$

$$D_{H_2O} = 4.17 \times 10^{-4} \times \lambda \times \left(161 \times e^{-\lambda} + 1\right) \times \exp\left[\frac{-2436}{T}\right] \quad \text{for } 3 \leq \lambda < 17 \qquad (6.15)$$

where a discontinuity point with the maximum diffusion coefficient occurs at $\lambda = 3$.

6.1.7 Solubility and Diffusion of Oxygen

At the interface between the electrode and the membrane in fuel cells, the catalyst sites are covered by a water-swollen ionomer film. This means that the reactants and products need to dissolve and diffuse through the ionomer film to maintain the reaction. The solubility and diffusion of oxygen and hydrogen gases in electrolyte membranes are therefore important for electrode reactions and therefore fuel cell performance.

The solubility and diffusion coefficient of oxygen and hydrogen can be measured in solid-state electrochemical cells with a smooth platinum electrode, preferably a microelectrode, under conditions of controlled temperature, pressure and relative humidity, as schematically represented in Fig. 6.7. Typically, the working electrode is made of a 50 μm radius Pt wire sealed in glass and placed opposite to a large area Pt gauze counter electrode. In a close cell filled up with high-purity oxygen under well-controlled humidity, pressure and temperature, the reference is often a dynamic hydrogen electrode (DHE), consisting of a pair of Pt wires sealed in a glass rod and operating by passing a constant current of *ca.* 50 μA between the two wires by means of a battery.

Electrochemical techniques such as chronoamperometry and voltammetry are used for the measurements. As to be discussed in the catalyst characterization

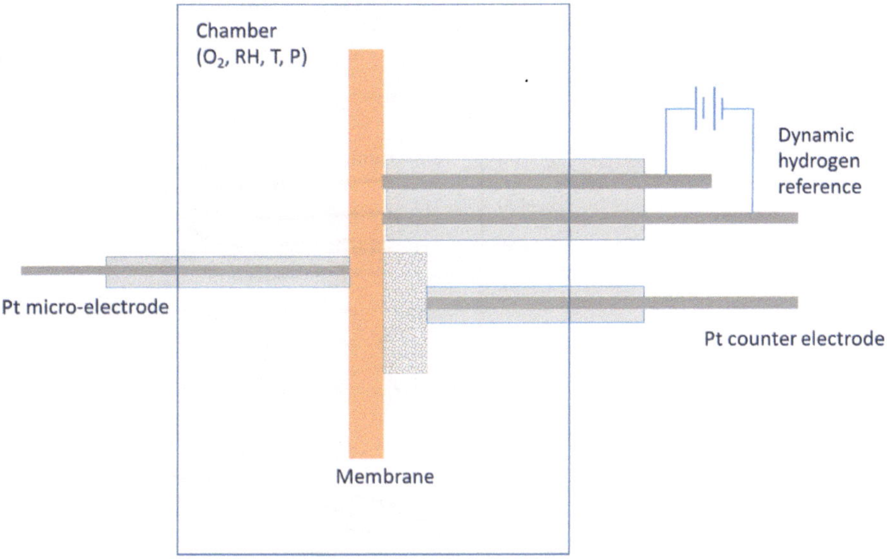

Fig. 6.7 Schematic representation of a solid-state electrochemical cell for measurements of the oxygen solubility and diffusion coefficient

sections, the voltammetry can be used to determine the electrochemical active surface (real) area of the electrode, which is used in kinetic studies, e.g., calculation of the exchange current density and Tafel slope from the polarization curves. An example of slow-sweep (5 mV/s) voltammograms is given in Fig. 6.8a for five temperatures with Nafion® 117 membranes under pure oxygen at pressure of 3 atm. The diffusion-limited current (i_L) on the microelectrode can be expressed [6]:

$$i_L = 4nFD_{O_2}c_{O_2}r \tag{6.16}$$

where $n = 4$ is the number of electrons transferred in the overall oxygen reduction reaction, F is Faraday constant, D_{O_2} and C_{O_2} are diffusion coefficient and solubility of oxygen in the membrane, and r is the radius of the microelectrode.

A set of chronoamperograms is shown in Fig. 6.8b. For large electrodes, the information obtained in a chronoamprometry is limited due to the Cottrellian decay with time. For microelectrodes, however, the current decays after a certain time reaching a diffusion-limited current. At short to intermediate times the current flow at the microelectrode resembles a quarter-sphere current flow and can be expressed by the modified Cottrell equation:

$$I(t) = \frac{nFAD_{O_2}^{1/2}c_{O_2}}{\pi^{1/2}t^{1/2}} + \pi nFD_{O_2}^{1/2}c_{O_2}r = \frac{B}{t^{1/2}} + C \tag{6.17}$$

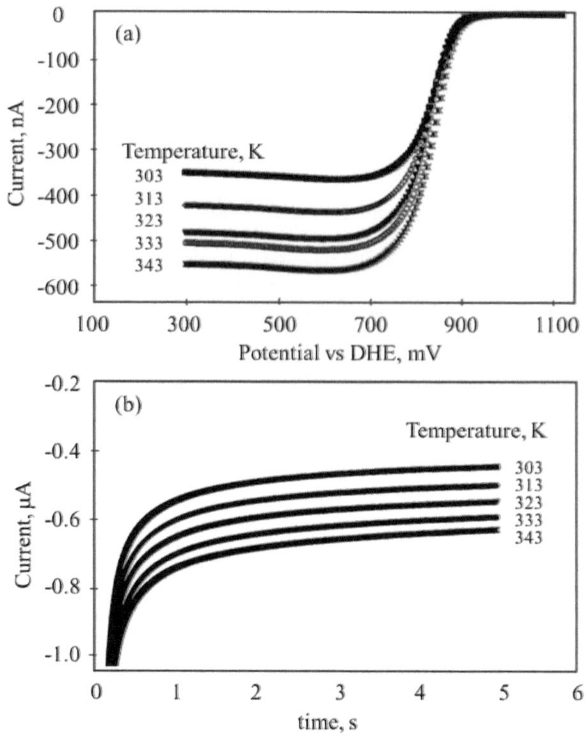

Fig. 6.8 **a)** Voltammograms (5 mV/s) and **b)** chronoamperograms for O_2 reduction at the Pt/ionomer interface at temperatures from 30 to 70 °C (303 to 343 K) under an O_2 pressure of 3 atm using microelectrode. Modified from Ref. [6] with permission from Elsevier, Copyright 1999

Here *A* is the geometric area of the microelectrode, and *B* and *C* are constants. By plotting the measured current against $t^{-1/2}$, *B* and *C* are the slope and intercept of the plot, from which the diffusion coefficient and solubility of oxygen in the polymer membranes can be determined. For Nafion 117 membranes, the oxygen diffusion coefficient is found to increase from 5.96×10^{-6} cm^2/s at 30 °C to 1.31×10^{-5} cm^2/s at 70 °C, while the oxygen solubility decreases from 9.19×10^{-6} mol/cm^3 at 30 °C to 7.81×10^{-6} mol/cm^3 at 70 °C.

6.1.8 Permeability of Hydrogen

In fuel cells, oxygen permeates from the cathode to the anode while hydrogen from the anode to the cathode through the electrolyte membrane. Due to the fast kinetics of hydrogen oxidation at the anode, the oxygen permeability has little effect on the anode potential, although the permeated oxygen chemically reacts with hydrogen. Of greater concern is the hydrogen permeability because it causes a significant decrease in the cathode potential.

The flux of a permeate is defined as the volume flow through the membrane per unit area per unit time, with an SI unit of m^3/m^2 · s. As the volume of a gas or vapor is very dependent on pressure and temperature, the volumetric flux should be always specified for pressure and temperature, e.g., $m^3_{STP}/m^2 \cdot s$, which refers to the volume under standard temperature and pressure (25 °C and 1 atm). Instead of this volumetric flux, the molar flux is often used for gases, in unit of mol/m^2 · s. The flux is a function of the driving force of permeation, e.g., pressure difference across the thickness of the membrane:

$$J = \kappa \frac{P}{\delta} \tag{6.18}$$

Here δ is the membrane thickness, κ is the permeability coefficient or simply permeability, defined as the flux of a permeate through a membrane per unit driving force per unit membrane thickness. With the pressure difference in cmHg, a common unit of gas permeability is Barrer, which is 1 Barrer = $10^{-10}\ \frac{cm^3_{STP} \cdot cm}{cm^2 \cdot s \cdot cmHg}$.

Henry's law links the partial pressure of a permeate in the gas phase and its concentration in the polymer phase:

$$C = S \times P \tag{6.19}$$

Here C is the gas solubility in the polymer phase (mol/cm^3), and S is the solubility coefficient (mol/cm^3 · atm). Through the membrane the gas permeation is described by Fick's diffusion law

$$J = D \frac{C}{\delta} \quad \text{mol/cm}^2 \cdot \text{s} \tag{6.20}$$

Here D is the diffusion coefficient of the permeate (cm^2/s). And the gas permeability can be written as

$$\kappa = D \frac{C}{P} \quad \text{mol/cm} \cdot \text{s} \cdot \text{atm} \tag{6.21}$$

Here the solubility is a thermodynamic term, and the diffusion coefficient is a kinetic term reflecting the effect of the surrounding medium on the molecular motion of the permeate.

The measurement of gas permeability can be carried out using a two-chamber cell separated by a membrane, with a high pressure on one side and a low pressure on the other. The gas permeation can be measured from changes of the pressure, volume or concentration with help of, e.g., gas chromatography. The atmospheric humidity can be tailored by humidification of the penetrating gas.

Electrochemical quantification of the permeated hydrogen is a convenient method using the hydrogen stripping in a fuel cell-like configuration. Here, one side of the membrane is a humidified hydrogen stream supplied which serves as the counter and reference electrodes (CE and RE). The other side is filled with liquid water or a humidified inert gas, e.g., N_2 or Ar and serves as the working electrode (WE).

Any permeated hydrogen through the membrane is oxidized by controlling the WE potential at a value above 0.4 V versus RHE. A critical issue is that traces of oxygen in the WE chamber should be completely avoided, otherwise oxygen will give a reduction current interfering the measurement. An indication of the residual oxygen is the resting potential between the working and reference electrodes, which otherwise follows the Nernst equation for a hydrogen concentration cell:

$$E = \frac{2.303RT}{nF}\log\frac{p_{H_2,\,RE}}{p_{H_2,\,WE}} \tag{6.22}$$

Here $p_{H_2,RE}$ is the partial pressure of the purging hydrogen on the reference electrode side, and $p_{H_2,WE}$ is the partial pressure of permeated hydrogen on the working electrode side. With a ratio of these two partial pressures being ~10^6, i.e., $p_{H_2,RE}$ is 1 atm and $p_{H_2,WE}$ in ppm level (10^{-6} atm), the calculated open circuit potential of this cell is about 180 mV.

The measurement of the stripping current can be done using linear sweep voltammetry, i.e., scanning the working electrode potential from the rest potential (~0.2 V) to ~0.5 V versus the counter/reference electrode. At potentials around 0.4 V the measured current is limited by the hydrogen diffusion through the membrane. It is optional that a reference scan is made with a nitrogen/nitrogen or nitrogen/water mode and the limiting current should be corrected by subtracting the background current, in order to eliminate the effect of the double layer charging.

Alternatively, the measurement can be done by chronoamperometry, recording a current–time curve under a constant potential of, e.g., 0.4 V. An example protocol for the electrochemical measurement of hydrogen permeability is given below. A set of typical current–time curves under a constant potential of 0.4 V versus RHE from hydrogen permeability measurements is shown in Fig. 6.9. The membrane is Aquivion E79-50S (dry thickness of 50 μm and wet thickness of 62 μm). The cell configuration is illustrated as the figure inset where liquid water is pumped through the working electrode chamber, ensuring full hydration of the membrane. The oxidation current of permeated hydrogen increases with the temperature, indicating the increase of permeability of hydrogen through the membrane.

The permeation current density can be translated into the hydrogen permeability. For example, according to Fig. 6.9, at 23°C, the limiting oxidation current density of permeated hydrogen is 0.625 mA/cm^2. Based on Eqs. (6.13) and (6.18), the hydrogen permeability can be calculated from the hydrogen oxidation current by:

$$\begin{aligned}\kappa &= \frac{\delta}{P}\times\frac{i}{F} = \frac{62\,\mu\mathrm{m}\times \mathrm{cm}/10000\,\mu\mathrm{m}}{1\,\mathrm{atm}}\times\frac{0.6\frac{\mathrm{mA}}{\mathrm{cm}^2}\times \mathrm{A}/1000\,\mathrm{mA}}{2\,\mathrm{mol\,e^-/mol\,H_2}\times 96485\mathrm{C/mol\,e^-}}\\ &= 1.9\times10^{-11}\mathrm{mol}H_2/cm\cdot s\cdot atm\end{aligned} \tag{6.23}$$

This leads to a hydrogen permeability of 1.9×10^{-11} mol H_2/cm·s·atm at room temperature.

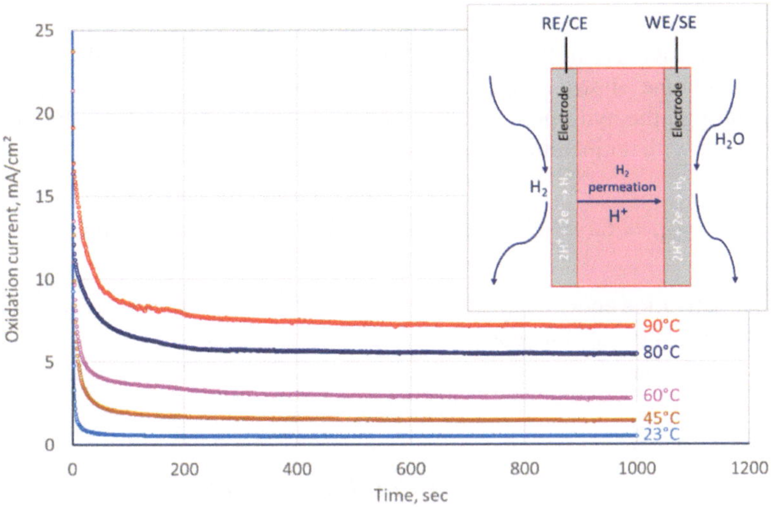

Fig. 6.9 Chronoamperograms for the hydrogen permeation at 100% RH and ambient pressure at 0.4 V versus RHE at different temperatures. The membrane is Aquivion E79-50S. The inset is an illustration of the cell configuration for hydrogen permeation measurements

Figure 6.10 represents hydrogen permeability data for Nafion® membranes collected by Kocha et al. [7]. Different symbols are results achieved by different groups using different methods. It can be seen that higher permeability coefficients are observed for wet membranes, which are approaching that in water whereas the dry membranes exhibit permeability coefficients similar to that of Teflon®. The increase

Fig. 6.10 Summary of hydrogen permeability data for Nafion® membranes. Origins of the data referred to [5]. Two dashed lines indicate the value of water (upper) and Teflon (lower). Redrawn after Ref. [7]

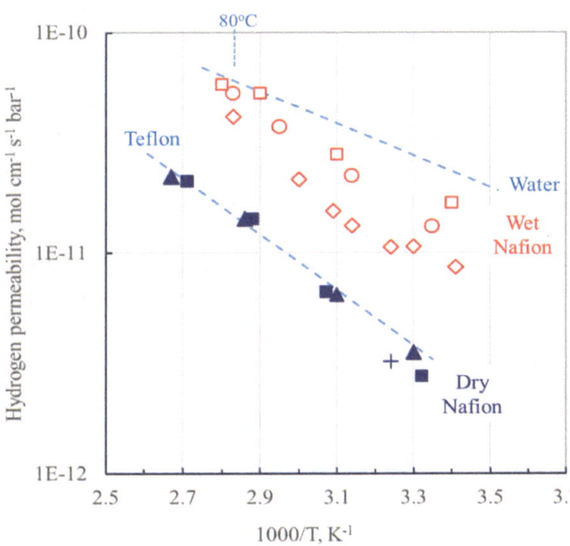

of the gas permeability coefficient with higher water contents of the membrane is due to the increase in the diffusion coefficient, while the solubility coefficient of the membrane shows insignificant change, indicating that the gas permeates mainly through the hydrophilic water-filled domains. The increase in the hydrogen permeability with temperature can be expressed:

$$\kappa_{H_2} = 6.6 \times 10^{-9} \exp\left(-\frac{21.30}{RT}\right) \quad \text{mol } H_2/\text{cm} \cdot \text{s} \cdot \text{atm} \tag{6.24}$$

where the activation energy is 21.03 kJ/mol. At 80 °C and 100% RH this gives a value of 5×0^{-11} mol/cm · s · atm, corresponding to a leaking current about 1 mA/cm^2 for an Nafion 115 membrane.

6.1.9 Methanol Crossover—Liquid and Vapor

The term *permeability* is specifically used to refer the permeation of a gas or liquid through a membrane via diffusion mechanism, i.e., driven by the concentration difference. The term *methanol crossover* is more often used to refer the overall permeation of methanol through a membrane driven by different transport mechanisms, e.g., the electroosmotic drag with the hydrated protons, diffusion due to the methanol concentration difference, and convection by the hydraulic pressure gradient between the anode and the cathode. Direct methanol fuel cells are normally operating under ambient conditions, and subsequently the hydraulic permeation can be ignored. When the fuel cell is idle, i.e., kept at OCV with zero current, the diffusion mechanism of methanol crossover dominates. The electroosmotic drag prevails when the cell operates particularly under high current densities.

6.1.9.1 Measurement of Methanol Permeability by Diffusion Cell

Liquid methanol diffusion can be measured in a two-chamber diffusion cell. The liquid permeation cell is composed of two compartments, separated by a permeable membrane. One compartment (cell A) is filled with methanol solution with designed concentration, and other compartment (cell B) is filled with deionized water. Before experiment, the membrane should be pre-treated and soaked in water for 24 h. The concentration-driven diffusion from cell A to cell B can be measured as function of time by, e.g., gas chromatography. Under the conditions of different methanol concentrations in cells A and B, there is a pseudo-steady-state conditions prevailing in the two compartments in the initial stage of the experiments. Consequently, the flux of methanol is constant, and the methanol concentration in cell B with time can be given as follows

$$V_B \frac{dC_B(t)}{dt} = A\frac{DK}{\delta}[C_A - C_B(t)] \quad (6.25)$$

where A is the membrane area, δ the membrane thickness, D the methanol diffusivity, K the methanol partition coefficient, V_B the volume of cell B compartment, C_A the methanol concentration in compartment A, and $C_B(t)$ the methanol concentration in compartment B at time t. The product, $P = DK$, is the permeability of methanol with unit of cm^2/s. C_B is measured as a function of time, and the methanol permeability can be calculated from the expression:

$$\ln\left(1 - \frac{C_B(t)}{C_A}\right) = -\frac{PA}{\delta V_B}(t - t_o) \quad (6.26)$$

where t_o is the start time of the measurement. Figure 6.11 shows a typical plot obtained for Nafion 112 membrane in a 5M methanol solution at 25 °C [8]. The methanol permeability can be obtained from the slope of the straight line. In this case, the methanol permeability of Nafion 112 membrane is found to $1.77 \pm 0.05 \times 10^{-6}$ cm^2/s at room temperature. The thickness used in the calculation is the membrane thickness after soaked in water for 24 h (62 μm). The nominal dry thickness of Nafion 112 is 50 mμ.

Methanol permeability can also be obtained from plots of methanol concentration in compartment B as a function of time by using following equation:

$$C_B(t) = \frac{PA}{\delta V_B} C_A(t - t_o) \quad (6.27)$$

For DMFCs, the ratio of proton conductivity to methanol permeability, σ/P, is often used to indicate the usefulness of a membrane. A high ratio indicates desirable membrane behavior, because the conductivity remains high while methanol crossover is low. For Nafion 112, the measured proton conductivity is 0.093 S/cm at room

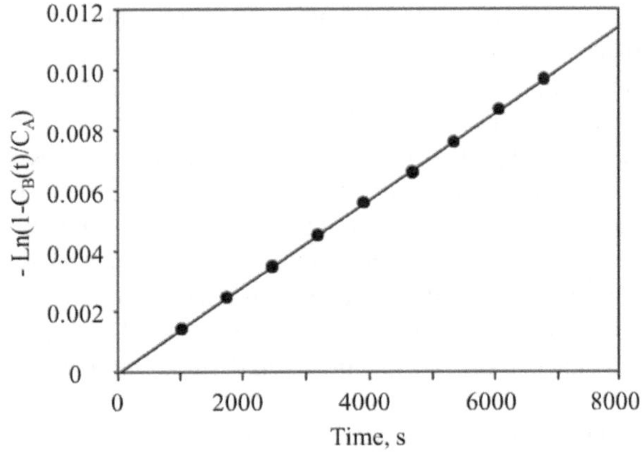

Fig. 6.11 Plot of-Ln(1 − $C_B(t)/C_A$) against measurement time for methanol permeability through Nafion 112 membrane at room temperature. Reproduced from Ref. [8] with permission from Elsevier, Copyright 2012

temperature using electrochemical impedance spectroscopy technique. This gives the ratio of 5.25×10^4 for Nafion 112 membrane.

Another method is to determine the CO_2 content of the cathode gas stream. When the permeated methanol reaches the cathode side, it reacts with oxygen forming CO_2. Chemically it is also possible to form intermediate products such as CH_xO_y and CO, but their amount is in general negligible. This method can be used to measure the methanol diffusion under OCV or sum of diffusion and electroosmotic drag under current. There are two issues in connection to the CO_2 method. One is that the methanol oxidation may not be complete. The other is the permeability of CO_2 through the membrane particularly under fuel cell operation at high current densities when a significant amount of CO_2 is generated on the anode side. The diffusion of CO_2 from the anode to the cathode will lead to an overestimation of the methanol crossover. Nevertheless methanol crossover in fuel cells operating at high current densities is of less importance [9].

6.1.9.2 Measurement of Methanol Permeability by Electrochemical Oxidation Method

Methanol permeability through the membrane can also be assessed by measuring the limiting oxidation current of crossover methanol through the MEA under fuel cell setup conditions at temperatures preferably below 80 °C under ambient pressures. The membrane is assembled in a standard test cell with attached gas diffusion electrodes on each side to form an MEA. In this case, a flow of methanol and water mixture, at specified concentration and flow rate, is applied on the anode, and humidified nitrogen is fed to the cathode side at controlled flow rate with or without back pressure. A potentiostat is used to sweep the potential in a similar way as the hydrogen permeability measurement. One difference is the reference electrode. For the hydrogen permeability measurement, the hydrogen purging CE is conveniently used as a reversible hydrogen reference. In the methanol cell, however, the methanol purging CE functions as a pseudo-dynamic hydrogen reference electrode.

The stripping of methanol, either by voltammetry or chronoamperometry, takes place typically at 0.9 V or higher. In this potential range, the platinum surface oxidation occurs, and the background current should be taken for the current correction. In the case of linear scan voltammetry, the potential is scanned from 0.1 to 0.9 V at a low scan rate, e.g., 0.5 mV/s, and the methanol that crosses over through the membrane will be oxidized at the Pt/C cathode. The methanol crossover rate is obtained by measuring the transport controlled limiting current.

A set of voltammograms for electro-oxidation of permeated methanol through Nafion 117 membrane in low methanol concentration solution (1M) at various temperatures is shown in Fig. 6.12. Steady limiting current densities were obtained at voltage above 0.8 V, which was 42 mA/cm^2 at 30 °C and increased to 117 mA/cm^2 at 70 °C.

Defining the methanol crossover rate as the methanol molar flux through the membrane (without taking into account of membrane thickness and methanol partial

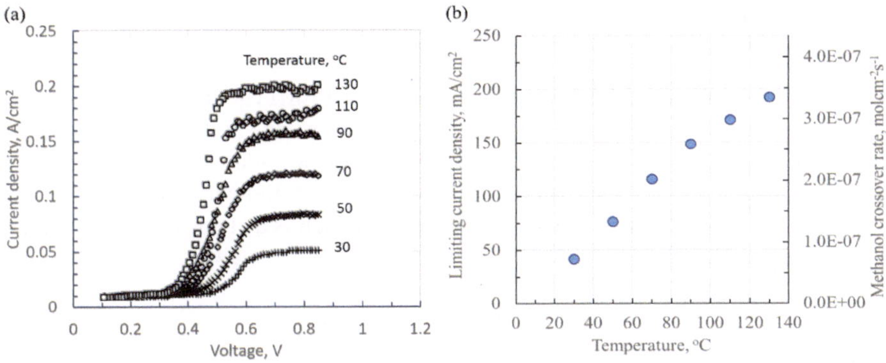

Fig. 6.12 **a** Voltammetric curves for oxidation of methanol permeating through Nafion 117 membrane exposed to 1M methanol solution at various temperatures. Reproduced from Ref. [10] with permission from IOP Publishing, Copyright 2000; **b** Reading of the limiting current density and calculated methanol crossover rate as a function of temperature

pressure), i.e.

$$j_{MeOH} = \frac{i}{6F} \tag{6.28}$$

Thus, the limiting current densities at 30°C and 70°C correspond to methanol crossover rates of 7.3×10^{-8} and 2.0×10^{-7} mol MeOH /cm^2·s, respectively.

These numbers are further increasing with the concentration of methanol solutions. For DMFC, a practical strategy to minimize the methanol crossover rate is to feed methanol at lower concentrations (<2M) and operate at relatively lower temperatures. For example, the methanol crossover rate could be reduced by *ca.* 50% by operating at a cell temperature of 60 °C compared to 90 °C. Membranes of larger thicknesses or/and higher EW are often used for the purpose of minimization of the methanol crossover, however, as expense of ohmic losses of the cell voltage. It can fit into some specific applications, for example, in relation to consumer electronics, which require very low current and power densities where the voltage loss is small. A large number of reports on methanol crossover measurements are available from literature with a varying magnitude of the results, which should be treated with care. Great efforts have been made to develop hydrocarbon polymer membranes as well as their composites with inorganic fillers, aiming at reduced methanol crossover. Also in this connection, a strong poisoning effect has been recognized on the ORR kinetics. Development of methanol-tolerant ORR catalysts has therefore been another focus of research.

6.1.10 Fenton Test

Membrane durability has become a major issue for PEMFCs. Reduction of oxygen involves generation of a small amount of hydrogen peroxide at the possible intermediates at the cathode. Also, the crossover of oxygen to the anode can lead to formation of active oxygen species such as peroxides, hydroperoxy radicals, and hydroxy radicals. These species aggressively attack polymers which is a chemical mechanism of polymer degradation.

$$O_2 + 2H^+ + 2e^- \rightarrow H_2O_2 \quad (6.29)$$

The H_2O_2 dissociates into hydroxyl or hydroperoxyl radicals with help of traces of transition metal cations as catalysts:

$$H_2O_2 \rightarrow 2HO^* \quad (6.30)$$

$$H_2O_2 + HO^* \rightarrow HOO^* + H_2O \quad (6.31)$$

Further reaction is pH dependent with $pK_a = 4.7$

$$HOO^* + H_2O \rightarrow O_2^- + H_3O^+ \quad (6.32)$$

At a low pH of fuel cells, the concentration of the superoxide radical O_2^- is very low, e.g., the concentration ratio of $[HOO^*]/[O_2^-] \approx 5000$, corresponding to a relative concentration of 99.98% HOO^* and 0.02% O_2^-.

In the presence of metal ions, the decomposition of hydrogen peroxide is accelerated via the so-called Fenton reaction:

$$Fe^{2+} + H_2O_2 \rightarrow HO^* + OH^- + Fe^{3+} \quad (6.33)$$

It is commonly accepted that degradation of PFSA membranes such as Nafion® is mediated by the HO^* as well as HOO^* radicals. Membrane fragmentation by attack of HO^* radicals can lead to chain unzipping, releasing fluoride ions [11]:

$$R_F - CF_2CF_2SO_3H + HO^* \rightarrow R_F - CF_2COOH + 2HF + H_2SO_4 \quad (6.34)$$

$$R_F - CF_2COOH + HO^* \rightarrow R_F - CF_2^* + CO_2 + H_2O \quad (6.35)$$

$$R_F - CF_2^* + HO^* \rightarrow R_F - CF_2OH \quad (6.36)$$

$$R_F - CF_2OH \rightarrow R_F - COF + HF \quad (6.37)$$

$$R_F - COF + H_2O \rightarrow R_F - COOH + HF \quad (6.38)$$

The degradation involves release of CO_2 and HF and can be detected by measurement of weight loss or the released F^- ions by selective electrodes or ionic chromatography. Correlation between the two species has been well established.

Attack of hydroxyl radicals on the Nafion® chain end groups leads to the unzipping polymer backbone and is considered a major degradation mechanism. The chemical stability of these membranes is ex situ characterized by Fenton test. In situ fuel cell test under open-circuit voltage is also conducted by measuring the fluoride ion content in the exhaust water, as to be discussed in Sect. 6.5.

In the Fenton tests, membrane samples are immersed in hydrogen peroxide solution containing a small amount of Fe^{2+}, e.g., in the form of $FeSO_4$, at typically 80 °C. A number of techniques have been successfully used in the qualitative studies, while measurements of the fluorine release rate, mass loss, or ion-exchange capacity (IEC) are typical methods for quantitative studies. Figure 6.13 shows an example of the microstructure change and chemical stability of Nafion membranes after treatment in a 30 wt.% H_2O_2 solution containing 12.3 ppm Fe, 6.1 ppm Cr, and 5.4 ppm Ni at 80–90 °C [12]. The metal ion solution was obtained by treating SS316L stainless steel, commonly used as bipolar plates in PEM fuel cells, in a 0.5 mol/L nitric acid. The solution was analyzed and replaced by fresh solution every 30 min. The attack by the radicals catalyzed by the metal ions in the solution results in loss of the polymer mass accompanied by the formation of small bubbles on the surface (Fig. 6.13a) and pores in the cross-section (Fig. 6.13b). The small bubbles gradually became pinholes during the chemical degradation process.

The structural stability of the membrane can also be evaluated by the measurement of fluoride and sulfuric contents of the treated solution using an atomic absorption spectroscopy (AAS). The change in the fluorine ion contents in the solution is also called the fluorine ion emission rate (FER), representing the C–F bonds loss ratios or

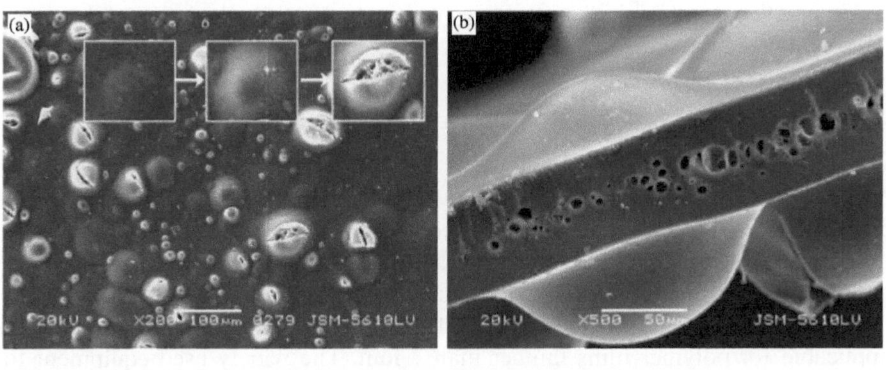

Fig. 6.13 SEM images of surface (**a**) and cross section (**b**) of a Nafion 111 membrane after treatment in H_2O_2/metal ion solution for 48 h. Reproduced from Ref. [12] with permission from Elsevier, Copyright 2007

the decomposition of C–F% of the membrane. In the Fenton test shown in Fig. 6.13, the FER as measured by AAS is 0.15 mg/h, which corresponds to 0.036 wt% of F released from this Nafion membrane per hour (the Nafion membrane before testing contains ca. 51wt% F). This is most likely due to the conversion of $–CF_2–$bond to –COOH under the chemical attack by the HO^* and/or HOO^* free radicals. On the other hand, the sulfuric content is closely related to the $-SO_3^-$ functional group loss ratio of the membrane. The loss of the repeating units containing proton-conducting functional $-SO_3^-$ groups will significantly decrease the proton conductivity of the proton exchange membranes.

6.1.11 Mechanical Properties

Shorter lifetime of PEMFCs is often caused by mechanical failures, for example, cracks, tears, punctures, pinholes, or delamination between the membrane and electrode. Any of these failures may result in reactant gas crossover and exothermic chemical reactions, which, in turn, generates hot spots. During the MEA preparation and operation, the membrane may experience local stresses at the reactant gas inlet or the edge of electrodes as well as the border of the channels, triggering tears and cracks. The resistance against this type of cracks is most commonly evaluated using tensile stress–strain curves. In this test a membrane specimen with initial dimensions is mounted between a fixed grip and a moveable grip. When an axial force F is applied, the sample is stretched from the initial length L_0 to length L. The strain or elongation (ε) is defined as

$$\varepsilon(\%) = \frac{L_o - L}{L_o} \tag{6.39}$$

The cross-sectional area of the sample during stretching is difficult to measure, and in practice it is the initial section area that is used in the calculation of the stress, the so-called engineering stress

$$\sigma_{\mathrm{E}} = \frac{F}{A_o} \tag{6.40}$$

The tensile strength of a membrane is the maximum tensile stress that a sample can be subjected to before failure. This corresponds, often but not necessarily, the tensile stress at break where the membrane ruptures. Also at the point when the membrane breaks under stress one measures the strain or elongation at break.

A standard procedure for testing thin films is given by ASTM D 882, which is applicable for polymer films thinner than 1 mm. The widely used equipment for measuring mechanical properties is the universal material strength testing machine. The standard sample geometry is of 2.53 cm broad and 15.24 cm long rectangular stripes; however, smaller sample sizes are often used. Mechanical properties depend on the crosshead rate, which is often chosen in the range of 5–10 mm/min. Sometimes an initial speed of as slow as 1 mm/min is used in order to obtain more reliable data of the Young's modulus.

A typical stress–strain curve for an Aquivion E98-09S membrane is shown in Fig. 6.14, from which the tensile strength at break, elastic modulus, and elongation at break can be extracted. The elastic modulus (or Young's modulus) is given by the initial slope of the linear part of the curve, showing the rigidness of the membrane. This is the region of the curve, in which the dislocation of the sample is practically reversible, i.e., the elastic deformation region. The standard deviation for tensile tests, particularly the stress and elongation at break, is relatively high, a number of specimens should be tested and error bars should always be presented. Tests should be excluded in the case that a sample is broken at the grips or other cracks originating from indentations or microscopic defects at the edge of the sample, preventing the test from a premature failure. Also for this reason the Young's modulus, the resistance against compression or stretching (the ratio of the stress to elastic strain in tension), is considered a more reliable measure. Most universal test machines do not provide an atmosphere control in terms of humidity and temperature. An additional climate chamber can be installed, in which the testing is conducted, and under controlled temperature by means of internal electric heaters and humidity by a purging gas, both can be monitored through a humidity/temperature sensor.

A set of typical values of tensile modulus, maximum strength, and elongation at break are listed in Table 6.2. The anisotropic values along the machine and transverse directions are also indicated for extruded membranes, while solution cast membranes behave isotropically. As the temperature and water hydration of the membrane increase, the tensile strength, modulus as well as elongation decrease.

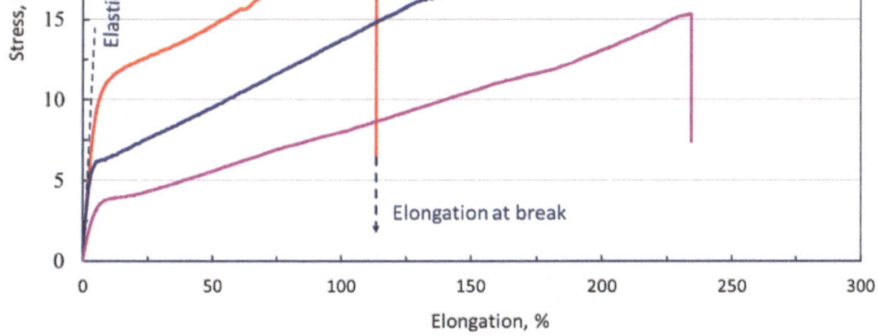

Fig. 6.14 Stress–strain curves of the Aquivion E98-09S membrane at varied temperatures and water vapor partial pressures with illustrations of terms

Table 6.2 Set of mechanical data for Nafion 1100 membranes, where MD indicates the machine direction and TD the transverse direction

Properties and test conditions	50% RH 23 °C	Water soaked at 23 °C	Water soaked at 100 °C
Tensile modulus (MPa)	249	114	64
Maximum tensile strength (MPa)	43 (MD) 32 (TD)	34 (MD) 26 (TD)	25 (MD) 24(TD)
Elongation at break (%)	225 (MD) 310 (TD)	200 (MD) 275(TD)	180 (MD) 240 (TD)

6.1.12 Thermal Stability Test

The thermal or thermo-oxidative stability of polymers can be tested by thermogravimetric analysis (TGA). In this method, a sample of about 10 mg is placed in a scale pan mounted in a furnace. TGA measurements are normally done in a temperature cycling mode at constant heating rate. A TGA graph is obtained by plotting the mass trace or most often normalized mass trace against temperature. Atmosphere should be specified as air leads to complete oxidation while nitrogen or argon to decomposition and residual soot. Heating rate is also an essential parameter as the TGA is not a thermodynamic but a kinetically controlled process.

The C–F bonds in the polymer backbone is so strong that only release of the adsorbed water takes place at temperatures of up to 200 °C. At higher temperatures until 280 °C, the side sulfonic acid groups begin to decompose with release of SO_2 and CO_2 in addition to water. Many other decomposition products have been detected before the polymer is totally decomposed at 550 °C, including carbonyl fluoride, thionyl fluoride, and hydrogen fluoride. Based on these thermal analysis thermal degradation mechanisms have been suggested as shown in Fig. 6.15 [13]. Changing atmosphere from 5% hydrogen in nitrogen to air and the presence of platinum nano-catalysts do not seem affect the polymer thermal stability.

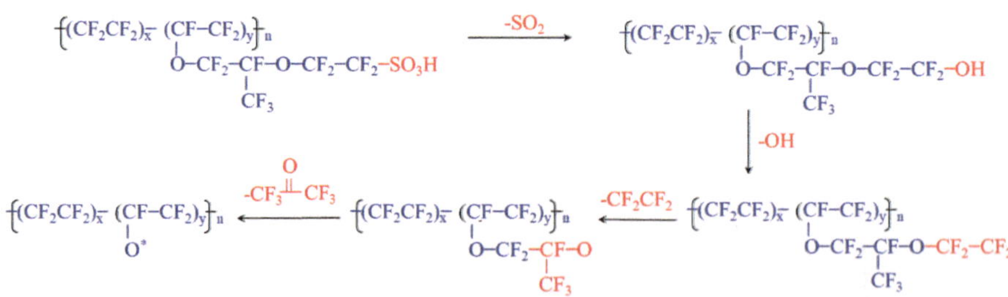

Fig. 6.15 Possible mechanisms of thermal decomposition for Nafion membranes. Reproduced from Ref. [13] with permission from IOP Publishing, Copyright 1996

6.2 Catalyst Synthesis

The standard reduction potentials of common Pt ions such as $[PtCl_6]^=$ and $[PtCl_4]^=$ are in the range 0.73–0.76 V vs SHE. Thus, Pt nanoparticles can easily be prepared at low temperatures using mild reducing agents such as hydrogen, sodium thiosulfate ($Na_2S_2O_3$), sodium borohydride ($NaBH_4$), $Na_4S_2O_5$, hydrazine (N_2H_4), formaldehyde (HCHO), ethylene glycol [$(CH_2OH)_2$], and formic acid (HCOOH). The most commonly used platinum precursors are hexachloroplatinic acid (H_2PtCl_6) and $Pt(NH_3)_4Cl_2$ complexes. One concern of using metal chloride precursors is the negative effect of possibly residual chlorides in the product catalyst. To avoid that issue, alternative precursor salts such as platinum sulphites, nitrates, or carbonyl complexes can be used instead. Pt carbonyl complexes are interesting as they can easily be prepared by bubbling carbon monoxide through the chloride solution, and the complexes undergo a thermal decomposition leading to the formation of nanoparticles with no need for reducing agents.

Bimetallic Pt catalysts can be made by co-reduction of two metal precursor salts. Carbon-supported catalysts are commonly prepared by the addition of carbon blacks before or after the Pt particles are formed. The main synthesis methods include the impregnation–reduction method and the colloidal adsorption method, as well as their variations.

6.2.1 Impregnation–Reduction Method

This method is a simple and straightforward process for preparation of supported Pt or Pt alloy nano-catalysts. The impregnation is typically done in aqueous solution media at room temperature, with no use of organic solvents. The process starts by ultrasonic mixing of the aqueous solution of metal precursors with a carbon support. After the impregnation, the powder can be collected by filtration and drying and reduced by a gaseous reducing agent typically hydrogen. The reduction can also be conducted in the slurry by addition of a solution-containing reducing agents. Commonly used liquid-phase reducing agents are solutions of sodium thiosulfate ($Na_2S_2O_3$), sodium borohydride ($NaBH_4$), $Na_4S_2O_5$, hydrazine (N_2H_4), formaldehyde (HCHO) and formic acid (HCOOH). Because the reduction occurs after the impregnation step, properties of carbon supports play a critical role in controlling the particle size of the Pt nanoparticles, as the pore of the carbon support limits the particle growth. A variety of methods have been developed to facilitate the impregnation, such as using surfactants or functionalizing the carbon support. The impregnation–reduction method is illustrated in Example 6.2.

Example 6.2

Preparation of 40%Pt/C catalysts by impregnation–reduction method.

Chemicals

0.1M H_2PtCl_6 solution
glacial acetic acid (CH_3COOH)
diethylamine ($(CH_3CH_2)_2NH$)
5% formic acid (HCOOH)
Pre-dried carbon black (Vulcan-XC 72R)
Deionized water

Impregnation

1. Add 400 mL deionized water in a 1L beaker, and add 600 g of dry carbon black (Vulcan-XC 72R).
2. Add 0.3 g of glacial acetic acid into the beaker.
3. Place the beaker on a magnetic stirrer, fix a thermocouple, start stirring, and slowly heat the slurry to 50 °C
4. Add 20.5 mL of 0.1 M H_2PtCl_6 solution to the slurry. Heat the slurry up to 95 °C at the rate of 10 °C/h.

Reduction

5. Add 5 mL of diethylamine
6. Add slowly 8 mL of 5% formic acid, leave the reaction to complete in 1 h at the temperature, and then cool it down to room temperature
7. Collect, filter, and wash the catalyst cake with 2 L of hot deionized water (>50 °C).
8. Dry the catalyst cake in a vacuum oven at 110 °C overnight. Cool the furnace to room temperature, and collect the catalyst.
9. Mill or grind the catalyst powder.

Notes

1. Carbon black may absorb moisture when stored in the air. Pre-drying is often recommended.
2. Glacial acetic acid is a low surface tension liquid and facilitates the wetting of carbon black.
3. Diethylamine is a week base and makes the acidic slurry a buffer within a stable pH range of 9–11.
4. Formic acid is the reducing agent with the following reduction reaction.

$$H_2PtCl_6 + 2HCOOH \rightleftharpoons Pt + 2CO_2 + 6HCl \quad (6.41)$$

5. It is critical to completely remove chloride ions during the washing. Residual chloride is tested by $AgNO_3$.

6.2.2 Colloidal Adsorption Method

A homogeneous mixture of a solute dissolved in a solvent is called a solution. A true solution is only one phase containing the dissolved species in the form of molecules or ions, typically in the size of subnanometers. When the solute is not fully dissolved, it forms a two-phase system, a continuous solvent phase, and a dispersed or suspended phase. Such a heterogeneous mixture is called a colloid, which can be an emulsion (a liquid–liquid mixture), a suspension (a liquid–solid mixture), or an aerosol (a gas–liquid mixture). A colloidal suspension normally contains dispersed particles in the size of 1 nm to 1 μm, and they do not settle in the suspension.

The first step of the colloidal method is to prepare metal nanoparticles in dispersion or suspension, followed by transferring the particles onto the porous carbon support by adsorption. The metal particles dispersed in the suspension tend to aggregate because of their very high surface energy in the nano-size and the attraction due to Van der Waals forces. In order to neutralize the Van der Waals interaction, stabilization of the particles is required, typically, by addition of a capping agent. A capping agent can be any molecules that adsorb on the particle surface to prevent nanoparticles from agglomeration. The capping effects can be either electrostatic or steric, the former being achieved by developing surface charges to repulse the aggregation and the latter through organic ligands, e.g., polymers, surfactants, and other small organic molecules rich in oxygen atoms such as citrate and oxalate. Compared to the impregnation–reduction method, the colloidal synthesis has been successfully tailored to allow for better particle size and composition control by a large variety of processes such as the polyol and microemulsion methods, among many others.

6.2.2.1 Polyol Method

A polyol or polyalcohol is an alcohol-containing multiple hydroxyl groups such as ethylene glycol, which is used as both the solvent and reducing agent. The method was developed by Fievet et al [14]. During the process, ethylene glycol is oxidized and forms glycolic acid or glycolate anions, depending on the solution pH, which are believed to act as a stabilizer for the metal nanoparticles. The size of the particles is typically in a range of 2–5 nm, with little dependence on the properties of the carbon supports and, therefore, suitable for preparation of catalysts with high metal loadings. The polyol method has been used to prepare both the single metallic and bimetallic nanoparticles.

The pH value of the solution is one of the crucial parameters of the polyol process and is often carefully controlled by the addition of NaOH. A steady decrease in the Pt particle size is observed with increasing the pH, likely due to the enhanced stabilization effect of the formed glycolate anions. The favorable pH range is above 7, where the concentration of glycolate anions is almost constant.

6.2.2.2 Microemulsion Method

An emulsion is a mixture of two immiscible liquids, such as water and an oil phase, which can be an oil-in-water emulsion where oil is the dispersed phase or a water-in-oil emulsion where water is the dispersed phase. When the dispersed phase is in the sizes of 1–100 nm or more often 10–50 nm, the system is called a microemulsion. It is a thermodynamically stable system typically made of water, oil, and surfactants. Since most metal precursors are inorganic salts that are soluble in water, not in oil, the most interesting systems are those with water droplets surrounded by the surfactant and dispersed in an organic solvent, which is also called inverted micelles. The water droplets with the dissolved metal salts exist as a large number of isolated nuclei, which upon reduction lead to the formation of nanoparticles. The size of water-in-oil droplets and, therefore, the size of the resulting metal particles can be tailored by the ratio of water to surfactant and the surfactant concentration. An example is a microemulsion system of water/n-heptane with sodium dioctyl sulfosuccinate as surfactant for the preparation of Pt–Ru nanoparticles[15].

Reducing agents also play a critical role in influencing the particle nucleation and hence the particle size. The reducing agent can be added to the microemulsion directly or as another microemulsion. Transfer of the formed nanoparticles can be achieved by stirring a carbon support in the microemulsion, with or without destabilizer. The residual surfactant molecules are typically removed by heat treatment. It is a relatively complicated and costly process. An illustration of the method is given in Example 6.3.

Example 6.3

Preparation of 50 wt% Pt/C catalysts by the colloidal adsorption method.

Chemicals

0.4M NaOH solution in ethylene glycol
0.04M H_2PtCl_6 solution in ethylene glycol
1M HCl aqueous solution
Pre-dried carbon black (Vulcan-XC 72R)
Acetone
Deionized water

Prepare the colloidal suspension and the reduction of Pt

1. Prepare a mixture of 60 mL of 0.4M NaOH solution and 55 mL of 0.04M H_2PtCl_6 solution in ethylene glycol in a 500 mL beaker.
2. Heat the mixture to 160 °C, and maintain the temperature for 10 min

   Precipitation and adsorption
3. Add 500 mL of 1M HCl solution to the colloidal Pt nanoparticle suspension for precipitation.
4. Centrifuge the mixture in portions, and discard the supernatant solvent. Repeat the washing and centrifugation step twice.
5. Collect the Pt nanoparticles, and disperse the powder in 100 mL of acetone.

6. Disperse 600 mg of pre-dried carbon black in 100 mL of acetone.
7. Mix the two acetone dispersions and evaporate the acetone in a rotary evaporator at 40 °C and 200 mbar.
8. Wash the obtained Pt/C catalyst powder with deionized water, and filter the power.
9. Dry the catalyst powder in a vacuum oven (110 °C, <100 mbar).

6.2.3 Self-assembly Method on CNT and Graphene Supports

Layer-by-layer self-assembly technology is one of the most effective synthesis approaches to fabricate highly ordered nanoscale structures or patterns with extended functionalities and activities and has been extensively adopted for the development of novel electrode and membrane modification in fuel cells [16]. The deposition at the surface is mainly controlled by charge compensation mechanism and can be used for the synthesis of stand-alone precious metal catalysts as well as precious metal catalysts on non-covalent-functionalized CNTs and graphene supports.

As shown in Chap. 5, non-covalent functionalization of CNTs and graphene supports via polyelectrolytes such as polydiallydimethylammonium chloride (PDDA) is effective to modify the surface of CNTs without detrimental effect on their intrinsic electronic and structure properties. One example of self-assembly methods is synthesis of Pt nanoparticles on PDDA-functionalized CNTs. In this process, Pt nanoparticles are in situ synthesized on the CNTs via the self-assembly between negatively charged Pt precursors, e.g., $PtCl_6^{=}$ ions, and positively charged functional groups of PDDA via electrostatic interaction. The anchored Pt precursors are then subsequently reduced by ethylene glycol, forming metallic Pt nanoparticles with uniformly distribution and high density. Because of the high density and positive surface charge of the non-covalent-functionalized CNTs, Pt catalysts on the functionalized CNTs are uniformly distributed with an average particle size of about 2 nm, and the Pt loading can be controlled in the range of 10–93 wt% [17]. Such a high Pt catalyst loading is difficult to achieve with other synthesis methods.

In addition to Pt catalysts, PtRu and PtSn alloys as well as Pd and Au nanoparticles can also be fabricated by the self-assembly method. Figure 6.16 shows an example of synthesis of phosphotungstic acid ($H_3PW_{12}O_{40}$, HPW)-modified Pd catalysts supported on PDDA-functionalized CNTs by this method [18]. First of all PDDA is irreversibly adsorbed on CNTs caused by the strong σ–π interactions between the polycationic molecules and the graphite-like basal planes of the carbon nanotubes. The negatively charged Keggin-type species are then self-assembled onto the surface of PDDA-functionalized CNTs. These self-assembled heteropolyanions prevent aggregation of the carbon nanotubes and promote formation of nanotube suspensions in the aqueous solution. The unsaturated cationic sites on these assemblies are subsequently utilized as anchoring sites to attract the $PdCl_4^{2-}$ complex

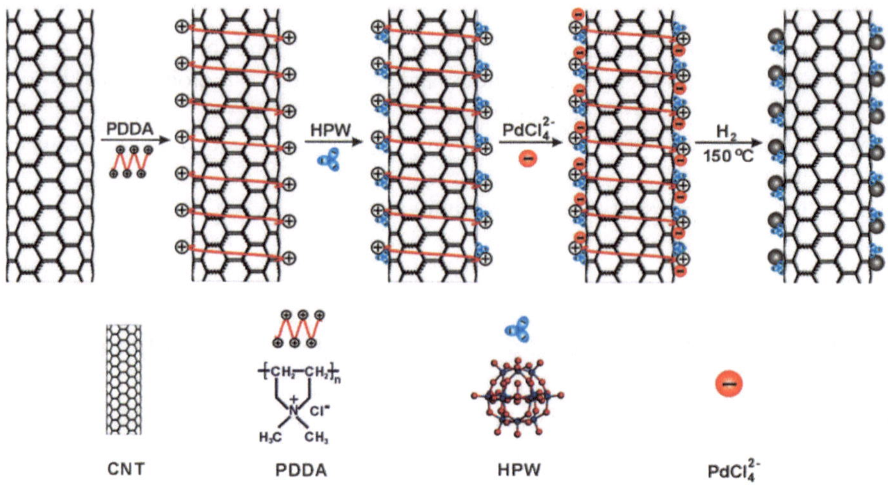

Fig. 6.16 Schematic representation of synthesis of 20 wt% Pd nanoparticles by the phosphotungstic acid (HPW) self-assembling method on polydiallydimethylammonium chloride (PDDA) modified surface of carbon nanotubes. Reproduced from Ref. [16] with permission from Royal Society of Chemistry, Copyright 2011

anions via electrostatic interactions. Finally upon reduction by hydrogen uniformly dispersed metal particles of ~2 nm size with a narrow particle size distribution are obtained. Most important, the electrocatalytic activity of Pd/HPW-PDDA-CNTs is highly active for the ORR in acid media, comparable to that of the state-of-the-art Pt/C electrocatalysts [18].

Highly dispersed and stand-alone electrocatalysts from nanoparticles to nanowire networks can also be synthesized by the self-assembly process with polyelectrolyte as the stabilizing agent. Polyelectrolytes are polymers bearing ionic groups, for example, quaternary ammonium cations (NR_4^+) in PDDA and sulfonic (SO_3^-) anions in Nafion or poly(sodium 4-styrenesulfonate). These ionic groups dissociate in aqueous solutions, making the polymer charged. Metal complex anions, e.g., $PtCl_6^=$ [19] are hence assembled into anisotropic networks with the charged polymers due to electrostatic attraction. When reduced by ethanol or $NaBH_4$ under tailored pH, uniform metal nanoparticles are obtained with a narrow size distribution.

6.3 Ex Situ Characterization of Catalysts

During the catalyst screening stage, the catalyst performance, including activity and stability, is commonly evaluated by the thin-film rotating disk electrode (TF-RDE) approach [20]. The technique is popular because it is simple and efficient, with no need for fabrication of membrane electrode assemblies (MEAs). This eliminates the influence of other fuel cell components on the catalyst performance. This technique allows for direct determination of the intrinsic activity information and kinetic parameters of the catalyst without resorting to complex mathematic modeling. An experienced operator may perform daily measurements for 1–2 inks or 5–10 electrodes. In addition, hardware and components such as glassy carbon (GC) electrodes, platinum or graphite counter electrodes, and a variety of reference electrodes are commercially available, also in reasonable prices. The testing results are, however, highly dependent on experimental practices, including dispersion of the catalyst ink, the homogeneity of the catalyst film, catalyst loading, contaminations from electrolyte and reference electrode, iR compensation, as well as data processing.

6.3.1 Catalyst Ink and Thin Catalyst Film

To prepare a thin catalyst film on the RDE tip, the catalyst is usually dispersed in a solvent, forming a well-dispersed catalyst ink. A certain amount of catalyst ink is then deposited on a glassy carbon electrode. A large number of ink recipes and coating tactics have been reported. A typical solvent is a mixture of deionized water and isopropanol, often with tailored pH value. The coating tactics include slow drying under a solvent vapor-saturated chamber or drying with rotation. Detailed information is given in Example 6.4.

Example 6.4

Preparation of catalyst ink and thin catalyst film.

1. Add 5.0 mg of 20 wt % Pt/C catalyst into a mixture of 85.7 μL of Milli-Q H_2O, 3299.0 μL of HCOOH, and 16.2 μL 5w% Nafion dispersion. The dry mass ratio of Nafion to carbon support is 0.33:1.
2. The obtained catalyst ink was ultrasonicated in a water bath at room temperature for about 30 min, yielding a uniform dispersion.
3. Polish a glassy carbon (GC) disk electrode (5 mm in diameter) with alumina paste and thoroughly cleaned in ethanol and Milli-Q H_2O. Before coating, 1–2 μL of isopropanol is first applied on the electrode surface.
4. Cast 10 μL of the catalyst ink on the tip, and dry it in air, optionally under the coverage of a beaker. After around 30 min, a uniform catalyst film is formed. The catalyst loading on the RDE is 15 $\mu g_{Pt}/cm^2$, which is within the typical range of 10–20 $\mu g_{Pt}/cm^2$ used for ORR measurements.
5. Examine the catalyst thin-film surface under a microscope.

6.3.2 Three-Electrode Cell Configuration

The configuration of a typical three-electrode cell is illustrated in Fig. 6.17. The cell is made of glass with a volume of about 150 mL, having two openings with standard

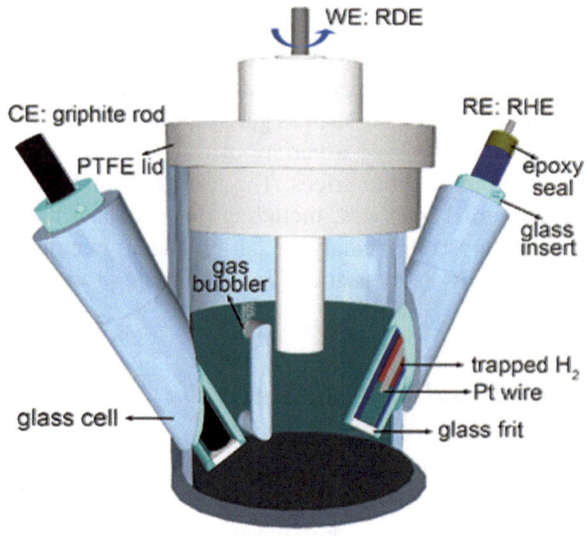

Fig. 6.17 Schematic illustration of a rotating disk electrode setup

conical joints for connection of a reference electrode and a counter electrode. The counter electrode (CE) is a graphite rod contained in a removable glass insert with a glass frit on the bottom separating the CE chamber from the working electrode (WE) chamber to avoid any gas crossover. The reference electrode is a reversible hydrogen electrode (RHE) with trapped hydrogen inside the electrode tube. The potential of the RHE was frequently monitored and calibrated by purging H_2 in the WE chamber. The shift of the reference electrode potential is typically less than 0.5 mV during the measurement and corrected during the data processing. The working electrode is described in connection to the thin-film preparation below. A bubbler for purging gases is embedded through the cell wall, with the outlet pointed to the cell wall to avoid severe disturbance to the electrolyte flow around the rotating disk electrode (RDE). The cell lid is made from polytetrafluoroethylene (PTFE). The cell hardware is, before use, cleaned by immersing in a Piranha solution for 1 h, followed by rinsing and boiling, together with other components, in Milli-Q water at least three times. Between electrochemical experiments, the cell and hardware are also kept in Milli-Q water.

Milli-Q water

Milli-Q water is high-purity water that has been purified using an ion-exchange cartridge. The purity of the water is monitored by measuring the specific resistance—the higher resistance, the fewer ions remaining in the water. A value greater than 18.2 MΩ/cm at room temperature is desirable.

6.3.3 Preconditioning and ORR Activity Measurement

The typical electrolyte solution is 0.1M $HClO_4$, prepared by diluting high-purity 70% $HClO_4$ with Milli-Q water. To start the test, the cell is filled with the electrolyte solution and equipped with the working, counter and reference electrodes in each compartment of the cell. Before the ORR measurement, the working electrode needs to be preconditioned in order to remove any adsorbed impurities on the Pt surface as well as reorganize the surface structure.

The measured I-V (or i-V) curves are always iR-corrected using the value of the electrolyte resistance across the space between the reference electrode tip and the surface of TF-RDE electrode. This resistance is determined from the high-frequency part of the electrochemical impedance spectra.

During the potential scan, the measured current contains contributions from the so-called background current, including the double layer capacitive current and current due to adsorption/desorption of electrode anions such as OH^- and ClO_4^-. These effects are assessed by conducting the cyclic voltammetry in the electrolyte saturated with Ar, using the same potentiodynamic parameters as the measurements for ORR curves. The procedure is illustrated in Example 6.5. The obtained polarization curve will be given and discussed in Sect. 6.3.5.

Example 6.5

Experimental procedure for the ORR activity measurement.

1. Purge the electrolyte with Ar gas. The electrode is scanned between 0.025 and 1.10 V at a scan rate of 0.1 V/s for around 130 cycles until a stable cyclic voltammogram is observed.
2. Replace the testing electrolyte with a fresh one, and purge the electrolyte with O_2.
3. Determine the electrolyte resistance between the working and reference electrodes by electrochemical impedance spectroscopy (EIS) under open circuit potential. An AC signal with 10 mV amplitude is applied over a frequency range from 100 kHz to 0.1 Hz, and the resistance is directly read out from the interception of the Nyquist plot with the *x*-axis.
4. Purge the electrolyte with Ar gas. Record a cyclic voltammogram (CV) as a background measurement. Sweep the potential between 0.05 and 1.10 V at a scan rate of 50 mV/s.
5. Record the ORR I-V curve. The electrode is scanned between 0.025 and 1.025 V in O_2 saturated electrolyte until a stable CV curve is achieved. The scan rate is 20 mV/s and the RDE rotation rate is 1600 rpm. The anodic scan of the CV curve is used later for calculating the ORR performance.

6.3.4 Determination of the ECSA

The electrochemically active surface area (ECSA) is a key parameter of fuel cell electrodes in order to assess the specific activity and investigate the utilization and degradation of the catalysts. Different to the specific surface areas measured by the N_2-adsorption isothermal, the ECSA represents the electrochemically active areas for electrochemical reactions in fuel cells. A number of ex situ methods are available to assess the ECSA, for example, by estimation of the average crystallite size of the metal particles and then calculate the surface area assuming the spherical shape. The most widely used method to quantify the ECSA is by underpotential hydrogen deposition and CO stripping by cyclic voltammetry (CV) technique.

6.3.4.1 Underpotential Hydrogen Deposition

Underpotential hydrogen deposition (H_{UPD}) on platinum refers to an electrocatalytic process that hydrogen is reduced and adsorbed on the electrode surface at potentials less negative than the reversible reduction potential of 0.0 V, indicating a strong interaction between hydrogen and platinum. It has been well established that the coverage of hydrogen atoms approaches unity at sufficiently negative potential and that one hydrogen atom binds per platinum surface atom. The ECSA of Pt-based electrocatalysts can be obtained from the charge of the atomic hydrogen desorption or adsorption peak in acid media after correcting for the capacitive charging current from the cyclic voltammograms. By using the charge passed for the H_{UPD}, Q_H, the ECSA can be calculated from the specific catalyst loading (L_{Pt}, $\mu g_{Pt}/cm^2_{Geo}$) and the geometric area of the electrode ($A_{Geo,}$ cm^2_{Geo}):

$$\mathrm{ECSA} = \frac{Q_H(\mu C)}{C_H(\mu C/cm^2) \times L_{Pt}(\mu g_{Pt}/cm^2_{Geo}) \times A_{Geo}(cm^2_{Geo})} \quad cm^2_{Pt}/\mu g_{Pt} \tag{6.42}$$

where C_H is the charge associated to the H_{UPD}, as an average over the three low index single crystal planes of Pt, being 210 $\mu C/cm^2{}_{Pt}$. This is equivalent to the charge required to oxidize a monolayer of H_2 on a bright Pt. Note that this value is determined for the polycrystalline Pt surface, and it may vary for Pt nanoparticles due to the different surface structures. The total charge Q_H is the stripping area in the potential range from +0.06 to +0.4 V, corrected for double layer capacitive contributions, see Fig. 6.18:

$$Q_H = \frac{1}{\nu}\int_{0.06}^{0.4} (i - i_C) \times dE \tag{6.43}$$

where ν is the scan rate, and i_C is the double layer charge current.

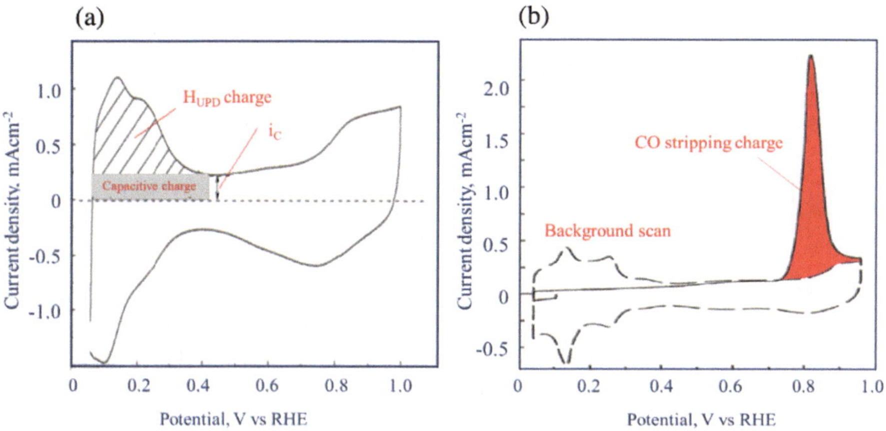

Fig. 6.18 Cyclic voltammogram with (**a**) the H_{UPD} and (**b**) CO stripping charge area marked

The CV method has been used to measure ECSA of PtRu electrocatalysts prepared on poly(ethyleneimine) (PEI) and acid-oxidized CNTs, PEI-CNTs and AO-CNTs, with the same metal loading of 20 wt % [21]. The average particle size of PtRu/PEI-CNTs and PtRu/AO-CNTs is ~2.4 nm and 3.2 nm, respectively. The CV curves of PtRu/CNTs electrocatalysts were obtained in a nitrogen-saturated 0.5M H_2SO_4 solution at a scan rate of 50 mV/s. Based on the area of the hydrogen adsorption peak, i.e., H_{UPD}, after correcting for the double layer charging current, mass-specific ECSA of PtRu/PET-CNTs is 113.2 m^2/g, which is higher than 82.5 m^2/g of PtRu/AO-CNTs. This indicates that PtRu nanoparticles supported on PEI-CNTs have a high electrochemical activity as compared to that supported on acid-treated CNTs.

6.3.4.2 CO Stripping

Another method to measure ECSA is the so-called CO stripping, i.e., by recording the anodic potential scan of the I-V curves after adsorbing CO at 0.05 V to the saturation coverage. The CO oxidation peak appears from around 0.8 V where the adsorption of electrolyte anions, e.g., OH^-, ClO_4^- is also possible. Contribution of these anions to the stripping charge should be corrected by recording and subtracting the background current in the potential range between 0.4 and 1.0 V from the following CV after the stripping. While one electron is involved during the oxidation of adsorbed H_{UPD}, the oxidation of CO to CO_2 involves two electrons, and consequently, the corresponding charge should be doubled. The total the value for the CO oxidation is calculated by

$$Q_{CO} = \frac{1}{2 \cdot \nu} \int_{0.4}^{1.0} \left(i - i_{bg}\right) \times dE \tag{6.44}$$

where i_{bg} is the background current measured under purging of an inert gas, e.g., argon. Based on the Q_{CO} value the ECSA is calculated from the specific catalyst loading (L_{Pt}, $\mu g_{Pt}/cm^2_{Geo}$) and the geometric area of the glassy carbon electrode (A_{Geo}):

$$\mathrm{ECSA} = \frac{Q_{CO}(\mu C)}{420\left(\mu C/\mathrm{cm}^2_{Pt}\right) \times L_{Pt}\left(\mu g_{Pt}/\mathrm{cm}^2_{Geo}\right) \times A_{Geo}\left(\mathrm{cm}^2_{Geo}\right)} \quad \mathrm{cm}^2_{Pt}/\mu g_{Pt} \tag{6.45}$$

CO stripping can also be used to measure the ECSA of Pd-based electrocatalysts because Pd nanoparticles have a poor hydrogen region due to hydrogen adsorption to form Pd/H alloy. In this case, the charge for oxidation of the CO monolayer on Pd is commonly taken as 405 $\mu C/cm^2$.

Example 6.6

Experimental procedure for underpotential hydrogen deposition and CO stripping.

1. Underpotential hydrogen deposition
   Preconditioning (see steps 1-4 in Example 6.5)
   Purge the electrolyte with Ar gas. The electrode is scanned between 0.025 and 1.10 V at a scan rate of 100 mV/s for around 130 cycles until a stable cyclic voltammogram is observed. Purge the electrolyte with Ar gas. Record a cyclic voltammogram (CV) as a background measurement. Sweep the potential between 0.05 and 1.10 V at a scan rate of 50 mV/s.
2. CO stripping
   Preconditioning (see steps 1–4 in Example 6.5)
   The electrode is held at 0.05 V for 10 min, while CO was purged into the electrolyte.
   The purging gas was then switched to Ar and purged for another 10 min, while the electrode potential is kept at 0.05 V.
   The CO stripping curve was obtained by scanning the electrode between 0.025 and 1.025 at the rate of 20 mV/s.
   Record the second cycle as background for correction.

6.3.5 Analysis of the Polarization Curves

Figure 6.19 shows a typical ORR polarization curve after the correction for the non-faradaic background current by subtracting the cyclic voltammogram recorded in the Ar-purged electrolyte. It should be noted that the current density for the *y*-axis is based on the geometric area of the glassy carbon electrode (A_{geo}). The ORR curve is always recorded in the anodic scan direction, i.e., from low potentials to high potentials. This

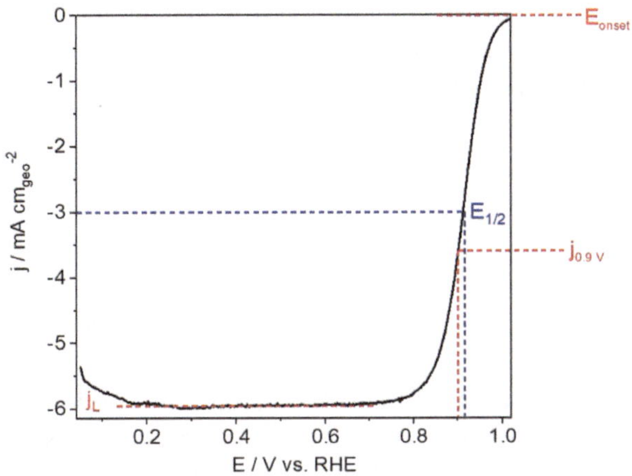

Fig. 6.19 Typical ORR polarization curve with illustrations of terms. These terms are explained in the text. The test is carried out with 20% Pt/C in 0.1M $HClO_4$ at room temperature with the scan rate of 20 mV/s

is because a Pt oxide layer is formed on the Pt surface at potentials above 0.8 V. If using the cathodic scan for the ORR measurements, i.e., scanning from high potential to low potential, experimental parameters such as the starting potential of the scan, the scan rate, and even the previous scan conditions can affect the amount of Pt oxide formed and thus the measured ORR activity, making it impossible to compare the results of different measurements.

To describe the catalyst performance based on the polarization curve, the following terms are commonly used:

Onset potential (E_{onset}) — The potential when the ORR current starts to rise from 0 and is normally defined as a potential at a fixed small current density, e.g., 0.1 $\text{mA/cm}^2_{\text{Geo}}$. Note that the E_{onset} is a technical term without a strictly defined scientific meaning.

Limiting current density (i_L) — The limiting current density is a function of the rotation speed expressed as the Levich equation:

$$i_{\text{L}} = 0.620\text{nF}A_{\text{Geo}}D_{O_2}^{2/3}\upsilon^{-1/6}\omega^{1/2}C_{O_2} \tag{6.46}$$

where n is the mole number of electrons involved in the reaction, A_{Geo} is the geometric electrode surface area, D the diffusion constant of O_2 in the electrolyte, C_{O_2} is the concentration of the oxygen, v is the kinematic viscosity of the electrolyte, and ω is the rotation speed of the rotating disk electrode.

In 0.1 M $HClO_4$ and at room temperature, the solubility of molecular O_2 is 1.26×10^{-3} mol/L, the diffusion coefficient is 1.93×10^{-5} cm^2/s, and the kinetic viscosity is 1.0×10^{-2}/cm$^2 \cdot$ s. For a glassy carbon electrode of the diameter 0.5 cm at the rotating speed of 1600 rpm at a scanning rate of 20 mV/s, the theoretical diffusion-limiting current density can be calculated to be $i_{\text{L}} = 5.8$ mA/cm^2.

Half-wave potential ($E_{1/2}$) — It is the potential at which the current density is equal to half of the diffusion-limiting current.

Kinetic current density — It is the current density at a given potential in the absence of mass transport limitation and can be calculated using the Koutecký–Levich equation:

$$\frac{1}{i} = \frac{1}{i_k} + \frac{1}{i_L} \tag{6.47}$$

or

$$i_k = \frac{i \times i_L}{i_L - i} \quad \mathrm{mA/cm^2} \tag{6.48}$$

where i_k is the kinetic current, i_L is the diffusion-limiting current, and i is the measured current. For catalyst evaluation the kinetic current is always based on the true or ECSA of the catalyst rather than the geometric surface area of the electrode substrate. The kinetic current at 0.9 V is used for comparing the ORR activity of different catalysts.

Roughness factor — The ORR activity of a catalyst should be normalized to the ECSA or the actual surface area (A_{ECSA}), on which the ORR reaction takes place. The roughness factor (RF) is defined as

$$\mathrm{RF} = \frac{A_{ECSA}\left(\mathrm{cm^2_{Pt}}\right)}{A_{Geo}\left(\mathrm{cm^2_{Geo}}\right)} \quad \mathrm{cm^2_{Pt}/cm^2_{Geo}} \tag{6.49}$$

For a smooth platinum metal surface, the RF is between 1.5 and 1.8 due to the steps, terraces, and edges arranged at different angles on the smooth surface. For the thin-film catalyst in RDE, the RF is about 10–20, while for gas diffusion electrodes in fuel cells, the RF is 500–800.

Area-specific activity or simply specific activity — Using the RF, the specific current density (j_{SA}) is the kinetic current density normalized to the ECSA of platinum (A_{ECSA}):

$$j_{SA} = \frac{I\ (mA)}{A_{ECSA}} = \frac{i_k}{\mathrm{RF}} = \frac{i \times i_L}{\mathrm{RF}(i_L - i)} \quad \mathrm{mA/cm^2_{Pt}} \tag{6.50}$$

This current density is the so-called specific activity, which is an intrinsic activity of the catalyst.

Mass-specific activity or mass activity — Mass activity is another intrinsic measure of the catalytic activity which can be defined as the kinetic current density per unit mass of platinum

$$j_{MA} = \frac{I\ (\mathrm{mA})}{\mathrm{Mass}\ (\mu g_{Pt})} = j_{SA}\ \left(\mathrm{mA/cm^2_{Pt}}\right) \cdot \frac{\mathrm{RF}\left(\mathrm{cm^2_{Pt}}/\mathrm{cm^2_{geo}}\right)}{L_{Pt}\left(\mu g_{Pt}/\mathrm{cm^2_{geo}}\right)} \quad \mathrm{mA}/\mu g_{Pt} \tag{6.51}$$

The mass activity of a catalyst is of engineering importance as it reflects the power/cost parameter in the economical assessment. In the catalyst evaluation, it also reflects the utilization of the catalyst. Typical values of the specific activity of the Pt/C catalyst are between 900 and 1200 $\mu A/cm_{Pt}^2$ and 400–600 mA/mg_{Pt}.

6.3.6 Stability Test

The stability is of special importance and is commonly evaluated through the accelerated stress test (AST) in an aqueous electrolyte. The most widely used AST protocol, one of those recommended by the U.S. Department of Energy, is by potential cycling between 0.6 and 1.0 V (*vs* RHE) in the Ar-saturated electrolyte solution. The conditions are to mimic the catalyst degradation process in fuel cells, which accelerates during the start-up/shutdown periods due to the dramatic change of electrode potentials. The cycling number is often 10,000, and the scan rate is 50 mV/s, as indicated in Fig. 6.20.

Example 6.7

Stability evaluation by the accelerated stress test with rotating disk electrode.

1. Preconditioning (see steps 1–4 in Example 6.5)
2. Record the initial ORR polarization curve. In electrolyte saturated with O_2, the potential is scanned between 0.025 and 1.025 V at 20 mV/s and rotation rate of 1600 rpm until a stable ORR curve is reached and recorded.
3. Record the initial background CV. The purging gas is switched from O_2 to Ar, and repeat the potential scan between 0.025 and 1.025 V until a stable CV curve is achieved and recorded.
4. Accelerated stress test. In Ar-saturated electrolyte solution, perform 10,000 potential cycles between 0.6 and 1.0 V at 100 mV/s.
5. Replace the testing electrolyte with a fresh solution and purge with O_2.
6. Record the end ORR curve by potential scan between 0.025 and 1.025 V until a stable CV curve is achieved and recorded. The scan rate is 20 mV/s. The RDE rotation rate is 1600 rpm.
7. Record the end background CV. The purging gas is switched from O_2 to Ar, and repeat the potential scan between 0.025 and 1.025 V until a stable CV curve is achieved and recorded.

An example protocol is given as Example 6.7. After conditioning by potential scans of ca. 100 cycles in the O_2-saturated electrolyte between 0.025 and 1.025 V in 0.1 M $HClO_4$, the I-V polarization and voltammetric curves are recorded as the initial performance. This is followed by an AST consisting of 10,000 cycle scans between 0.6 and 1.0 V at 100 mV/s in an Ar-saturated electrolyte, after which the I-V polarization and CV curves are recorded. The ORR polarization curves recorded at

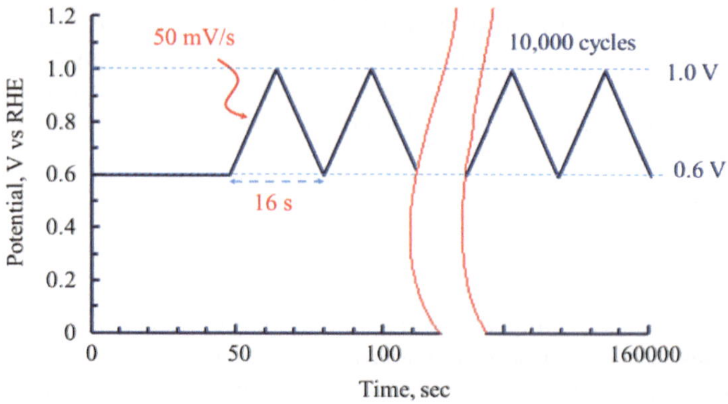

Fig. 6.20 Example of accelerated stress test protocol where an overall period of 160,000 s corresponds to 10,000 cycles

initial and after the AST are shown in Fig. 6.21a, with the region around the half-wave potentials ($E_{1/2}$) enlarged in the inset. The $E_{1/2}$ negatively shifted by 10.8 mV.

The corresponding CV curves are shown in Fig. 6.21b. The initial CV curve showed typical Pt features including the formation and reduction peaks of Pt oxide in a potential range from 0.5 to 1.0 V and the hydrogen adsorption and desorption region between 0.06 and 0.4 V. After the AST the Pt oxide formation and reduction peaks are significantly depressed, and the hydrogen adsorption peaks shifted to a slight lower potential. This should be due to the formation of very thick Pt oxide layers on the particle surfaces during the AST.

The catalyst activity is specified by the current at 0.9 V (*vs.* RHE) which can be converted to a specific and mass activity using the ECSA and Pt loading of the electrode. The ECSA-specific activities (SA at 0.9 V vs. RHE) and mass-specific activities (MA at 0.9 V vs. RHE) are shown in Fig. 6.21c–e. The initial values are 54 m^2/g-Pt of ECSA, 1.1 mA/cm^2_{ECSA} of SA, and 0.60 mA/mg_{Pt} of MA, which become 66%, 84%, and 50% remaining after the AST, respectively. The decrease in ECSA, SA and MA of the Pt/C catalyst after AST indicates the performance degradation of Pt/C catalysts for ORR.

6.3.7 *Half-Cell Characterization*

The widely used rotating disk electrode (RDE) testing is developed for catalyst evaluation in dilute acid solutions (e.g., 0.1M $HClO_4$ or 0.5M H_2SO_4) at room temperature. The reaction is for dissolved oxygen in the liquid electrolyte, which is different from the gas diffusion electrodes in fuel cells, where it is the gaseous oxygen that is reduced.

The RDE results can hardly give a good indication of the electrode performance. For PAFC and HT-PEMFC, concentrated acid is used as the electrolyte operating at 150–200 °C. In this type of electrolyte, the oxygen solubility is much lower. In addition, the viscosity of concentrated H_3PO_4 is two orders of magnitude higher than that of dilute acid solutions. Consequently, oxygen transport is the limiting factor, and the recorded ORR limiting current is very small. Figure 6.22a shows the ORR polarization curves of a Pt/C-coated RDE electrode at room temperature in 0.1M $HClO_4$ and H_3PO_4 solutions of different concentrations. The strong effect of the phosphate adsorption on the catalyst can be visualized by the negative shift of the kinetic region of ORR curves when switching from the dilute perchloric acid to 0.5 M H_3PO_4 and higher concentrations. The decrease in the limiting current is due to the reduced oxygen solubility and diffusion coefficient in concentrated H_3PO_4. When tested in 14.8M (85 wt%) H_3PO_4 the electrode exhibited a limiting current of only

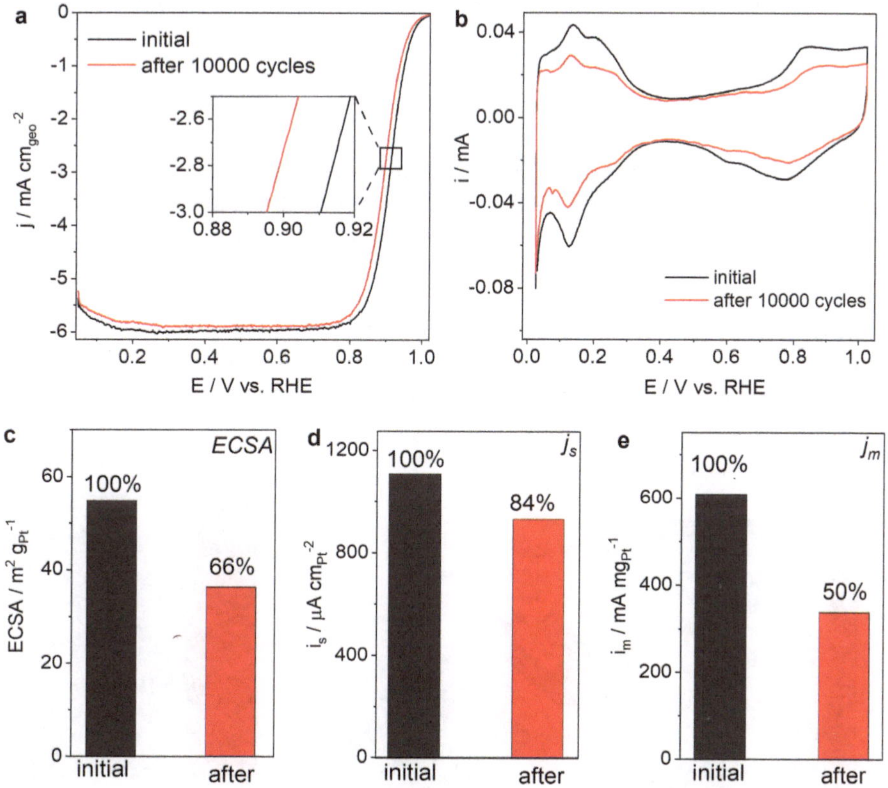

Fig. 6.21 Results of an accelerated stress test of Pt/C catalysts for ORR on a RDE. **a)** ORR polarization (anodic scans) and **b)** voltammetric curves before and after 10,000 potential cycles from which **c)** electrochemical surface area, **d)** surface activity, and **e)** mass activity of the catalysts before and after AST were obtained. The potential cycling was between 0.6 and 1.0 V in O_2-saturated 0.1 M $HClO_4$. Rotation speed: 1600 rpm; scan rate: 20 mV/s

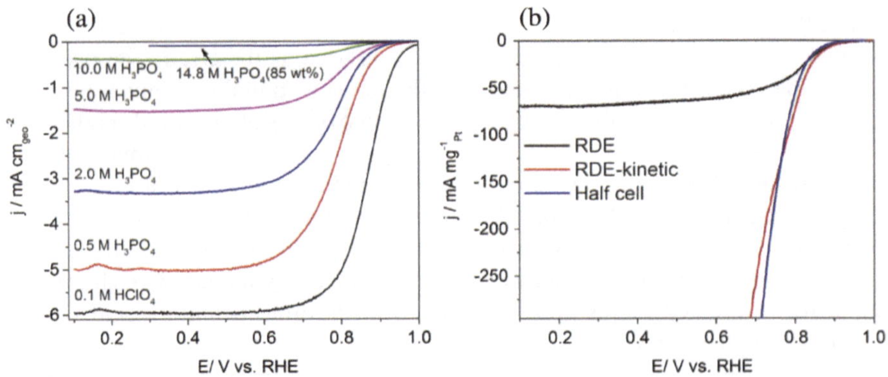

Fig. 6.22 **a)** ORR polarization curves of RDE tests at room temperature in 0.1M $HClO_4$ and H_3PO_4 of different concentrations. **b)** ORR polarization curves in 5.0M H_3PO_4 from half-cell and RDE (measured and kinetic currents). Reproduced from Ref. [22] with permission from Elsevier, Copyright 2017

0.08 mA/cm^2, which is even smaller than the double layer capacitive current of the carbon support.

Instead of being dissolved, reduction of gaseous oxygen via gas diffusion electrodes can be carried out in electrochemical half-cells for the catalyst characterization under practical conditions closer to fuel cell operation. Figure 6.23 shows the structure of a half-cell setup [22]. The cell body is made from a low thermal expansion borosilicate glass (Pyrex®) equipped with an oil heating jacket and a PTFE lid. The reference electrode is made of a RHE through the Luggin capillary tip close (*ca.* 5 mm) to the front surface of the working electrode, while a platinum mesh placed facing the WE is used as the counter electrode. The WE consists of a catalyst layer,

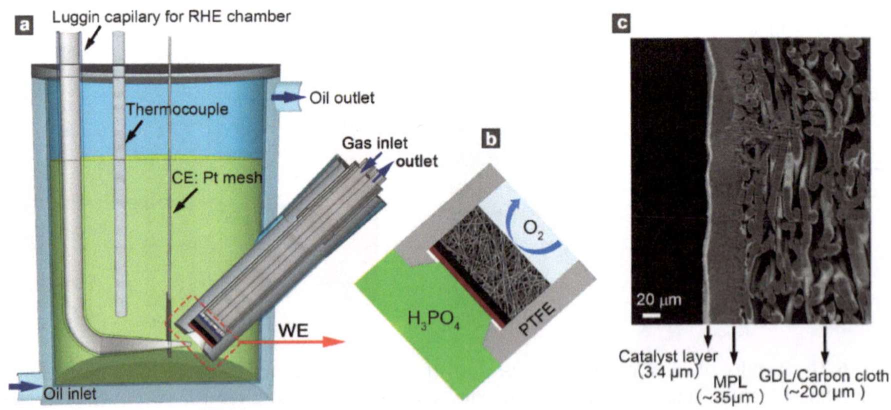

Fig. 6.23 Schematic illustration of the half-cell setup for gas diffusion electrode characterization in liquid electrolyte. **a)** Cross section of the cell, **b)** working electrode structure, and **c)** SEM image of an gas diffusion electrode

a microporous carbon layer (MPL), and a gas diffusion layer, as illustrated in the figure. It is embedded into a PTFE holder with a PTFE O-ring for sealing and also defining the active surface area of the electrode.

Figure 6.22b is a comparison of the polarization curves obtained by RDE and half-cell gas diffusion electrode test in 5.0M H_3PO_4 at room temperature. The same amount of Pt/C catalyst (3.9 μg Pt) is deposited on both the RDE and the half-cell electrodes. The ORR polarization curve obtained from the RDE shows an onset potential (at 1.0 mA/mg_{Pt}^2) of 0.95 V and a half-wave potential of 0.80 V. After correction of the mass transport resistance using the Koutecký–Levich equation, the obtained RDE kinetic current of ORR is very close to that directly measured in the half-cell test using gas diffusion electrode.

6.4 Fuel Cell Test for in Situ Characterization

6.4.1 Cell Assembling and Pretest Conditioning

Assembling a fuel cell is exemplified in Fig. 6.24. The flow fields are arranged perpendicular relative to one another, i.e., the gases are supplied in a cross flow orientation. Gaskets are placed between the MEA and the flow plates for gas sealing. A layer of expanded flexible graphite is used for gas sealing between the flow plates and the metallic current collectors. On the opposite side of the current collectors, a

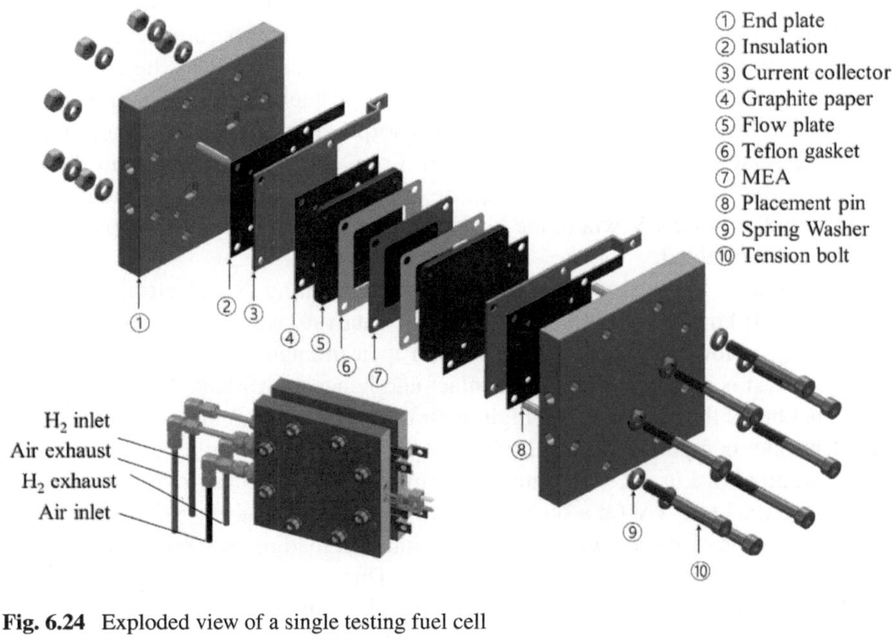

Fig. 6.24 Exploded view of a single testing fuel cell

thick layer of a fluoroelastomer is used for electrical insulation from the cell housing end plates. The end plates are used to clamp the MEA, flow plates, current collectors, and seals together. Clamping is done using bolts and nuts in each of the end plate corners. The end plate is equipped with heaters and thermocouples. Hydrogen and air are supplied through mass flow meters.

The constructed cell should before delivered to a test stand be tested for gas leak at elevated pressures and electric insulation. After the cell is mounted in a test stand, a series of diagnostics and conditioning should be conducted to ensure a baseline performance. OCV is an informative signal to record when H_2 and air are first supplied to the cell.

A simple method to condition a PEMFC is to operate the cell under H_2/air and cyclically scan the voltage over a limited current range. As soon as a current is drawn, the cell generates heat and water and hence conditions itself. Monitoring the ohmic resistance of the cell by current interrupt or high-frequency impedance is a common practice during the process of conditioning. A steady decrease of the cell resistance with time to eventually reach a constant value is an indication of the gradual and complete hydration of the membrane and the ionomer in the catalyst layer.

6.4.2 Determination of ECSA

For the ECSA determination, the fuel cell cathode is switched from operation with air to argon. Figure 6.25 shows an example of the measurement. Before the switch the fuel cell operates with H_2/air exhibiting an I-V curve as the red curve. After switching the cathode feed from air to argon the voltammograms are recorded within a potential range from 0.05 to 0.45 V, with the hydrogen anode as the counter and reference electrodes (CE and RE). As the oxygen is depleted, the reduction current is gradually decreased, while the hydrogen adsorption and desorption peaks appear, and steady-state plot is eventually obtained. From such a voltammogram the total charge for the hydrogen desorption can be obtained by integrating in the potential range from 0.05 to 0.4 V where the background current density of the double layer charging is subtracted.

It is similar to the ex situ measurement with the rotating disk electrode described in Sect. 6.3.4; however, the Pt catalysts in the catalyst layer of fuel cell electrodes have always a lower ECSA than the ex situ measurements, implying that not the entire catalyst is in electrochemical contact and hence unavailable for the electrode reaction. One of the reasons is the high platinum loading of 0.3–0.6 mg_{Pt}/cm^2 in fuel cell electrodes compared to typically 0.01–0.02 mg_{Pt}/cm^2 in rotating disk electrodes. In the example of Fig. 6.25 the integration of the current density over the potential range from 0.07 to 0.4 V (at a scan rate 20 mV/s) gives a total charge of 53 mC/cm^2 of the geometric electrode area. Using the platinum loading of 0.6 mg_{Pt}/cm^2 of the cathode the ECSA is estimated to be 42 m^2_{Pt}/g_{Pt}. This value is slightly lower than those reported in literature, apparently due to the high platinum-to-carbon ratio catalyst and the high platinum loading of the cathode as well as the presence of phosphoric acid

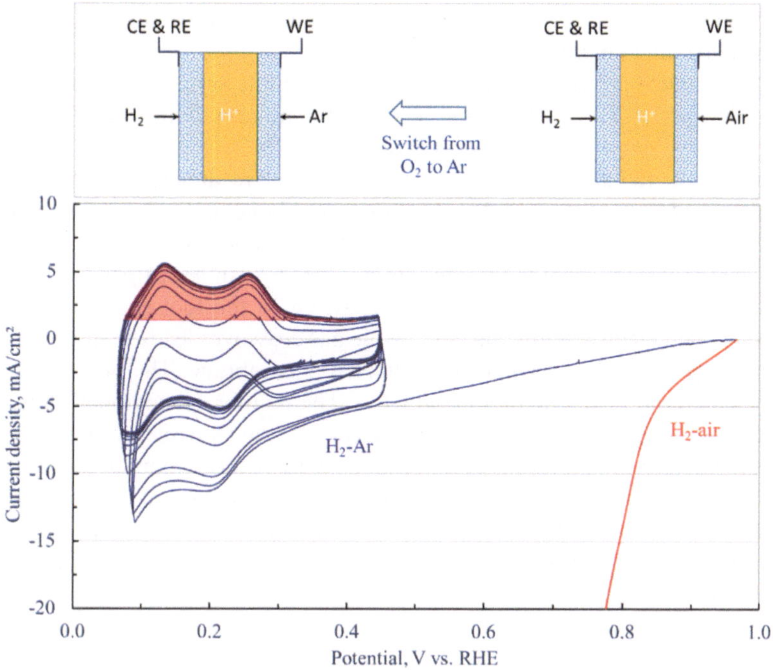

Fig. 6.25 Voltammograms of a 10 cm^2 fuel cell at room temperature during the cathode gas switching from oxygen to argon, as illustrated on the top. The cathodic catalyst was 39 wt% Pt-supported on carbon, and the platinum loading on the cathode is 0.6 mg_{Pt}/cm^2. The electrolyte is phosphoric acid doped polybenzimidazole membranes. The sweep rate was 20 mV/s. The ECSA is estimated from the desorption charge marked in red

in the electrolyte. The literature values are in a range of 60–70 m^2_{Pt}/g_{Pt} for catalysts of 30–50 wt% Pt on carbon. In general, the ECSA of Pt/C catalysts measured by in situ rotating disk electrodes has a relatively high value. The ratio of $ECSA_{in\text{-}situ}/ECSA_{ex\text{-}situ}$ is often used as a rough measure of the utilization of catalyst in the working fuel cell electrodes.

6.4.3 *Membrane Durability Test*

The membrane in fuel cells degrades primarily via mechanical and chemical mechanisms. The mechanical degradation includes membrane cracks, tears, punctures, and pinholes during the MEA fabrication process and particularly during operation in a wide range of humidity from ambient to 100% RH and temperature from −40 to 120 °C. The membrane swells at high relative humidities and shrinks as it loses water at low humidities. This swell–shrink cycling results in high mechanical stresses in the membrane and subsequent mechanical failure, leading to gas crossover through the membrane. There are accordingly two accelerated stress test protocols for mechanical and chemical stability of membranes.

The mechanical protocol is based on relative humidity cycling, e.g., from 0% RH to 90 °C dew point ($P_{H_2O} = 0.7$ atm). The membrane degradation can be monitored by the measurement of hydrogen crossover.

Chemical degradation is primarily due to oxidative attacks by peroxide radicals and results in breakage of the membrane backbone and side-chain groups and subsequent loss of mechanical strength and proton conductivity.

The chemical stability of membranes is assessed by using the open circuit voltage as a stressor under the hydrogen–air mode. At OCV, the fuel and oxidant are not being consumed electrochemically so that the gas crossover leads to more formation of hydrogen peroxide and radicals that attack the membrane. The protocol involves a steady-state hold of the membrane at OCV. The membrane degradation is, again, monitored by measuring the hydrogen crossover. Fluoride ion release or an equivalent is also an indicator of membrane deterioration for fluorinated membranes and is well correlated to the membrane lifetime. By the way combination of the two protocols, i.e., alternative cycles of the relative humidity and load, has also been proposed to provide additional insights; however, it does not allow isolation of the effects and degradation mechanisms.

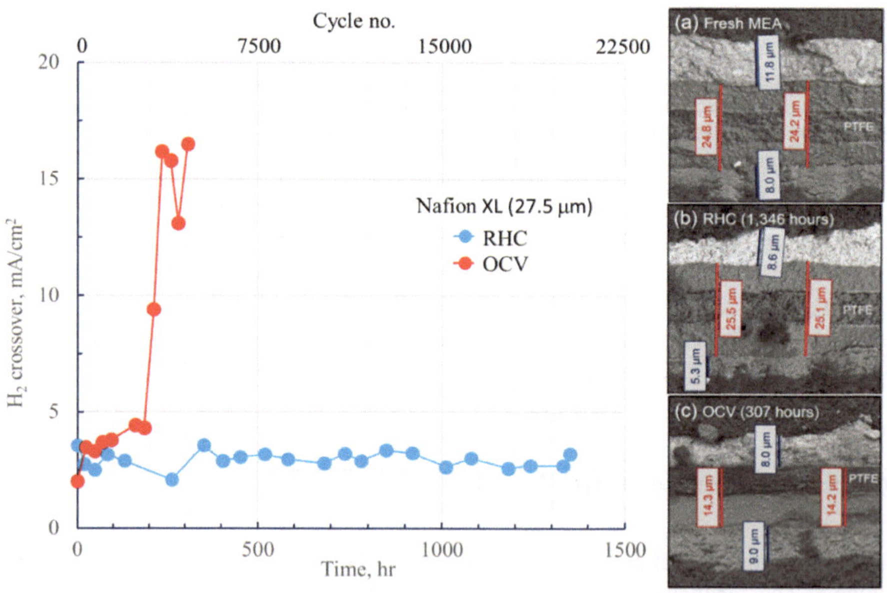

Fig. 6.26 Hydrogen crossover rates of Nafion XL membrane during the relative humidity cycling (RHC) and open circuit voltage AST (left). SEM micrographs of the MEA cross sections before (**a**) and after the humidity cycling (**b**) and OCV (**c**) testing (right). The cathode CL is near the top of each micrograph. The PEM and CL thicknesses are indicated by red and blue scale bars, respectively. The location of the PTFE reinforcement layer is also shown. SEM micrographs are reproduced from Ref. [23] with permission from IOP Publishing, Copyright 2018

Figure 6.26 shows an example of durability test for Nafion XL-reinforced composite membranes. The humidity cycling at 80 °C resulted in no observable increase in the hydrogen crossover after >20,000 cycles (>1300 h). This membrane contains PTFE reinforcement (as indicated in SEM images to the right of the figure) and is known to substantially improve PEM mechanical durability. Also in the SEM cross-sectional images are shown the membrane (red bar) and catalyst layer (blue bar) thicknesses before (a) and after the humidity cycling (b) and OCV (c) testing. The initial membrane thickness was about 25 μm (a), which remains unchanged after 1346 h of relative humidity cycling (b). The steady-state OCV test, on the other hand, resulted in failure of the membrane after ca. 250 h, as indicated by the sudden increase in the hydrogen crossover rate. This test, at high temperature of 90 °C and low RH of 30%, caused significant global membrane thinning of the MEA, from the original 25 μm to 14 μm.

6.4.4 Catalyst Stability Test

The catalyst layer (CL) consists of nanoscaled Pt or Pt alloy particles supported on high surface area carbon materials bonded together with an ionomer. The CL degradation relates to the stability of both the materials and the structure. Apart from the mechanical degradation such as cracks, delamination from the membrane and ionomer deterioration the major CL degradation involves Pt particle growth, Pt migration and carbon corrosion.

6.4.4.1 Catalyst Degradation Mechanisms

Several mechanisms have been proposed to account for the Pt particle growth. Collision of small Pt particles on the carbon surface may result in agglomeration, also called the coalescence effect. Small Pt particles may dissolve in the ionomer and electrolyte and redeposit on the surface of larger particles, a phenomenon called the Ostwald ripening [24]. Platinum may subsequently be migrating as ions through the ionomer and electrolyte in direction from cathode to anode and eventually reduced by the permeated hydrogen from the anode side. The Pt particle size growth leads to reduction of the catalytically active surface area and hence the catalyst activity. In addition, the electrochemical corrosion of carbon is thermodynamically favored in the potential range of fuel cell cathode though the carbon corrosion is in fact negligible due to the slow kinetics. The presence of Pt catalyzes the carbon oxidation reaction at potentials around the open circuit potential, i.e., when the fuel cell is idle or during start-up/shutdown cycling or when there is a fuel starvation. Carbon corrosion triggers the Pt growth and causes detachment of Pt particles. These mechanisms are illustrated in Fig. 6.27.

The catalyst degradation can be accelerated by potential control, undesirable temperatures and humidities, contaminants, and load cycling. In particular, at high

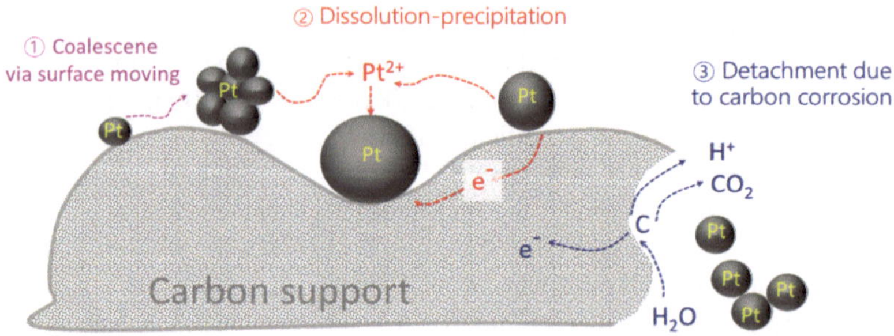

Fig. 6.27 Illustration of platinum degradation mechanisms in PEMFCs. (1) Coalescence via surface migration of Pt nanoparticles, (2) particle growth via the Ostwald ripening (dissolution and redeposition), (3) detachment of Pt nanoparticles from the carbon support

electrode potentials the durability of catalysts can be compromised by Pt sintering, particle growth, and dissolution. AST stressors can be any combination of these factors when testing is conducted in a complex environment. Because catalyst sintering/dissolution is accelerated under potential cycling, typical electrocatalyst AST is based on potential cycling.

6.4.4.2 Protocol for Catalyst Stability Test

Several institutions have explored different potential cycling protocols to evaluate the potential range and the scan profile for catalyst ASTs. Focusing on the platinum degradation a triangle-wave AST, similar to that used for RDE stability test, has been used consisting of potential cycling between 0.6 V -1.0 V at a scan rate of 50 mV/s (i.e., 16 s per cycle) for 30,000 cycles for a total test time of 133 h. An alternative protocol is a square-wave AST consisting of potential cycling between 0.6 V and 0.95 V with a dwell time of 3 s at each potential for 30,000 cycles for a total test time of 50 h. These protocols are exemplified in Exampe 6.8.

Example 6.8

AST protocol for catalyst stability test

Operating conditions	
Single cell	25–50 cm^2
Temperature	80 °C 80 °C
Relative humidity	Anode/cathode 100/100%

(continued)

(continued)

Fuel/oxidant	H_2/N_2 (200/50 sccm for 50 cm^2 cell)	
Pressure	Atmospheric	
Triangle-wave cycling		
Steps	Between 0.6 V and 1.0 V at scan rate 50 mV/s	
Number	30,000 cycles	
Cycle time	16 s	
Metric	*Frequency*	*Target*
Catalytic mass activity	At beginning and end of test	<40% loss on initial activity
I-V curve (0 − > 1.5 A/cm^2):	0, 1 k, 5 k, 10 k, and 30 k cycles	<30 mV loss at 0.8 A/cm^2
ECSA/cyclic voltammetry:	0, 1 k, 5 k, 10 k, and 30 k cycles	<40% loss of initial area
Square-wave cycling		
Steps	Between 0.6 V (3 s) and 0.95 V (3 s) with rise time ≤0.5 s	
Number	30,000 cycles	
Cycle time	6 s	
Metric	*Frequency*	*Target*
Catalytic mass activity	At beginning and end of test	<40% loss on initial activity
I-V curve (0- > 1.5 A/cm^2):	0, 1 k, 5 k, 10 k, and 30 k cycles	<30 mV loss at 0.8 A/cm^2
ECSA/cyclic voltammetry	0, 1 k, 5 k, 10 k, and 30 k cycles	<40% loss of initial area

A comparative study [25] of the triangle and square-wave AST protocols was performed, as shown in Fig. 6.28, where the reference is the drive cycle tests standardized for the long-term durability test for vehicular applications, called the wet drive cycle. With respect to the drive cycle durability reference, the square-wave AST (blue lines) exhibited a 20 times acceleration, on time basis, compared to the triangle-wave catalyst AST (red lines) and a 100 times acceleration compared to the wet drive cycle catalyst durability protocol (black line). Three lines of each color (red and blue) represent three types of Pt catalysts using different carbon supports. At the end of life the three Pt/C catalysts exhibited similar ECSA losses, indicating that both catalyst ASTs have minimized the effect of carbon support.

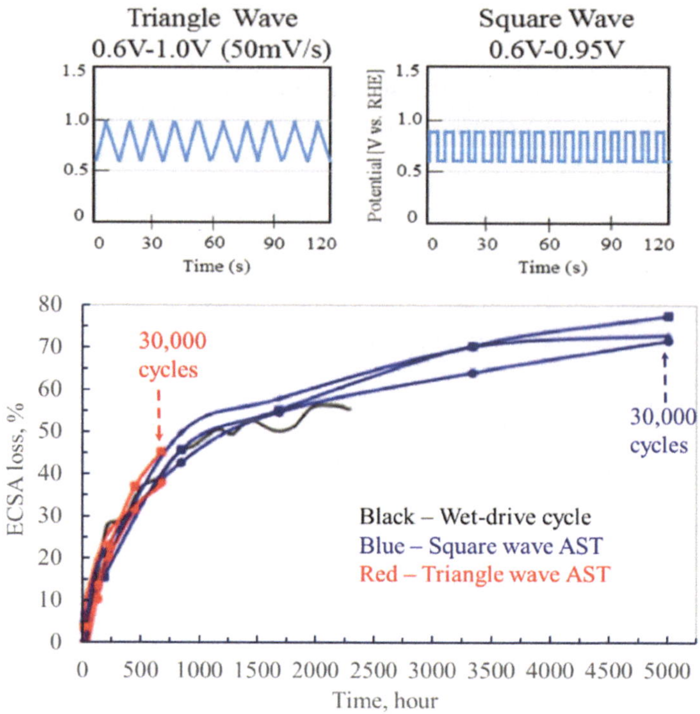

Fig. 6.28 Degradation rate (ECSA loss) during the square-wave and triangle-wave AST (illustrated on the top) as well as the wet drive cycle catalyst durability protocol with respect to total cycle time, where the actual testing time was multiplied with 5 for the triangle-wave and 100 for the square-wave AST. Modified from Ref. [25] with permission from IOP Publishing, Copyright 2018

6.4.4.3 Carbon Support Focused Stability Test

Carbon is the widely used catalyst support. Under prolonged conditions of high temperature, high water content, low pH, high oxygen concentration, and/or high potential, carbon corrosion is prone to occur. Corrosion of high surface area carbon supports at high electrode potentials is of significant concern, which is typically accelerated during fuel starvation and start-up/shutdown cycling. Evaluation of carbon support durability is thus performed at high voltage with hydrogen on the anode and nitrogen on the cathode in order to isolate electrochemical corrosion from chemical corrosion in the presence of molecular oxygen in air.

Two types of carbon corrosion ASTs have been recommended: cathode potential hold at 1.2 V and potential cycling from 1 to 1.5 V at 500 mV/sec. A study with three catalysts using high surface area carbon, Vulcan, and graphitized carbon support materials is shown in Fig. 6.29a. The performance of the three MEAs after 0 (initial), 20, and 400 h by the 1.2 V hold AST shows distinct differences. After only 20 h of the AST, the catalyst on high surface area carbon support already exhibits significant loss in performance. After 400 h the catalysts on graphitized support show only

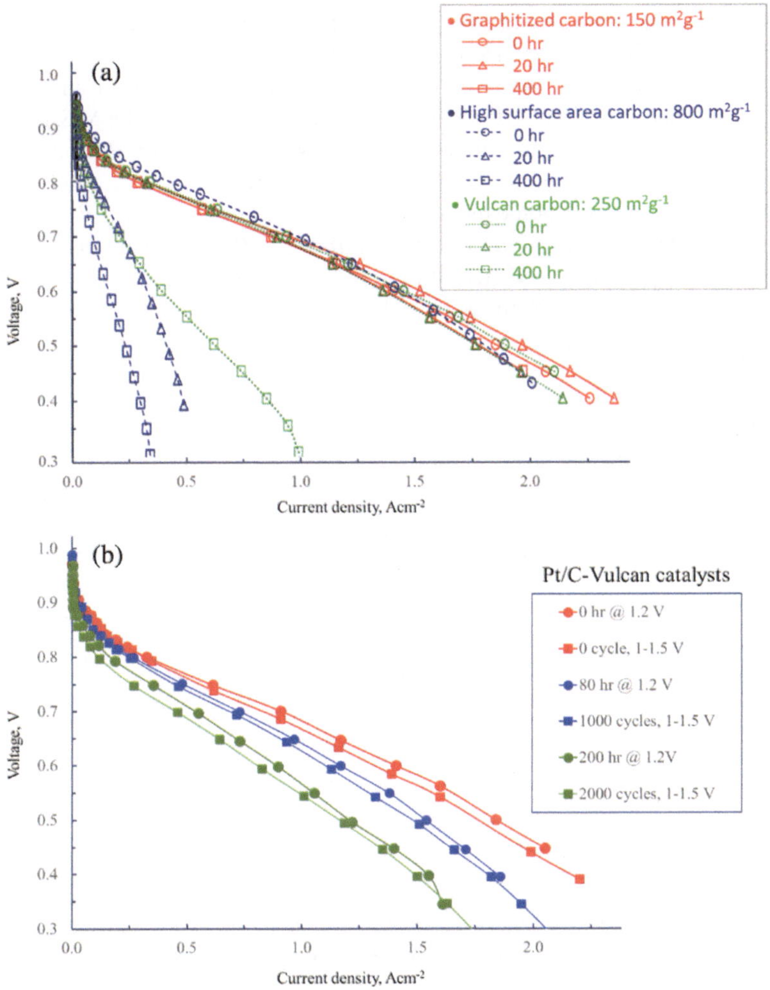

Fig. 6.29 **a** Carbon corrosion AST by potential hold at 1.2 V for Pt/C catalysts using three types of carbon supports as indicated in the figure; **b** Comparison of the potential hold and potential cycling ASTs for Pt/C-Vulcan catalysts. Modified from Ref. [26] with permission from IOP Publishing, Copyright 2018

40 mV performance loss at 1.5 A/cm^2, whereas the Vulcan-supported catalysts also has substantial performance degradation.

The potential hold AST needs a period of 400 h since the carbon corrosion is slow due to passivation. An alternative potentiodynamic method is by potential cycling from 1 to 1.5 V at a scan rate of 500 mV/s (**Example 6.9**). The protocol uses a rapid triangle-wave voltage cycle to accelerate catalyst support corrosion while minimizing degradation of the platinum catalyst itself. This potential range is similar to that which would occur in an unmitigated system start-up or shutdown. Each cycle takes 2 s, and a protocol of 5000 cycles needs less than 3 h.

A comparison study of the two ASTs is shown in Fig. 6.29b for an MEA with Vulcan-supported catalysts. After 80 h of potential hold at 1.2 V, the MEA exhibited a similar level of catalyst support degradation as occurred after 1000 cycles from 1.0 to 1.5 V which takes about 33 min. Similarly, 200 h at 1.2 V correspond to the same level of degradation achieved after 2000 cycles from 1.0 to 1.5 V which takes about 67 min. This potential cycling AST exhibits a 150 times faster decay than that for the constant potential hold AST. In terms of the Pt particle growth the potential cycling AST is faster by 100 times than the potential hold while retaining the same degradation mechanism [26].

Example 6.9

Protocol for catalyst support stability test

Operating conditions		
Single cell	25–50 cm^2	
Temperature	80 °C	
Relative humidity	Anode/cathode 100/100%	
Fuel/oxidant	H_2/N_2 (200/50 sccm for 50 cm^2 cell)	
Pressure	Atmospheric	
Square-wave cycle		
Steps	Between 1.0 and 1.5 V at 500 mV/s	
Number	5000 cycles	
Cycle time	2 s	
Metric	*Frequency*	*Target*
Catalytic mass activity	At beginning and end of test	$\leq$40% loss of initial activity
I-V curve (0- > 1.5 A/cm^2)	0, 10, 100, 200, 500, 1 k, 2 k, 5 k cycles	$\leq$30 mV loss at 1.5 A/cm^2 or rated power
ECSA/cyclic voltammetry	0, 10, 100, 200, 500, 1 k, 2 k, 5 k cycles	<40% loss of initial area

6.5 Summary

In this chapter, we have introduced and discussed the following topics:

- Perfluorosulfonic acid membranes, the state-of-the-art electrolyte material, are synthesized in industrial processes and cast by melt-extrusion and therefore may exhibit some anisotropy.
- Membrane characterization includes the following phenomena and properties:
  - Ion-exchange capacity or equivalent weight of the polymer is determined by titration which affects almost all other membrane properties.
  - Water uptake and swelling are essential to achieve the proton conductivity.
  - The optimal proton conductivity is always accompanied with compromised mechanical strength and permeability of hydrogen and oxygen.
  - The permeation of liquid alcohols such as methanol (crossover) is of significance for direct alcohol fuel cells.
  - In association with the proton conduction is the parasitic movement of water (electroosmotic drag) through the membrane.
  - Any concentration differences across the membrane cause diffusion of, e.g., water and methanol.
  - The electrolyte membrane provides a medium for the electrode reactions, and the solubility of reactants (oxygen and hydrogen) is of relevance.
  - The membrane resistance to oxidative dioxygen and radical oxidation is relative to the membrane durability and evaluated by Fenton test.
- Catalyst synthesis and evaluation are the most active area of the PEMFC research.
  - Pt and its alloy nanoparticles with carbon supports can be prepared by impregnation–reduction, colloidal adsorption, or self-assembly methods with many variations.
  - Ex situ characterization of the catalysts with respect to the activity and stability is commonly carried out by the thin-film rotating disk electrode, and the following parameters can be determined: the electrochemically active surface area (ECSA); the catalytic activities (in terms of onset potential, half-wave potential, and specific and mass activities)
  - The stability can be evaluated through the accelerated stress test (AST) with well-established protocols.
  - Gas diffusion electrodes as a fuel cell component can be evaluated by means of half-cell test.
  - Fuel cell tests focusing on in situ characterization of catalysts and membranes can be performed to determine
- Real electrochemical active surface area and utilization of the catalysts in catalyst layers
- Catalyst degradation and stability under specifically defined AST conditions
- Membrane durability under specifically defined AST conditions

6.6 Questions

6.1 Explain the following terms and materials:

Perfluorosulfonic acid
Short side-chain and long side-chain monomer
Ion-exchange capacity, equivalent weight, water uptake, and swelling
Through-plane and in-plane conductivity *via* machine direction and transverse direction
Ratio of proton conductivity to methanol permeability
Fenton reaction and fluorine ion emission rate
Roughness factor
Area-specific activity or specific activity and mass-specific activity or mass activity
iR-corrected polarization curve
Underpotential hydrogen deposition and electrochemically active surface area
Accelerated stress test (AST, RHC AST, OCV AST, potential hold AST and potential cycling AST)
Ostwald ripening process.

6.2 The equivalent weight (EW) of a proton exchange ionomer is the weight of the polymer for each mole of proton exchangeable sulfonic acid moieties, in g/mol. The lower the EW the high the concentration of sulfonic acid groups and hence the higher proton conductivity. In Fig. 6.2, however, a decrease in the proton conductivity is observed in the low EW range where a high water uptake occurs. Explain the phenomenon.

6.3 A Nafion 117 membrane has a water uptake of $\omega = 32\%$ on dry polymer basis which corresponds to a hydration number of $\lambda = 21$ mol H_2O/mol SO_3H. Use the ionomer density of 1.98 g/cm^3 and water density of 1.0 g/cm^3 and estimate the volume fraction of water in the hydrated ionomer (Answer: Water volume/total hydrated ionomer volume = 0.3).

6.4 Justify that for a four-probe conductivity cell (Fig. 6.3) the measurement of membrane resistance needs a high-frequency current or voltage signal. (Clue: a capacitive current through the double charge layer)

6.5 The water activity in the atmosphere is an essential factor for the proton conductivity of membranes which is often expressed as the relative humidity defined as $\mathrm{RH} = {}^{P_{H_2O}}/_{P_{H_2O}^{Sat}} \times 100\%$ where $P_{H_2O}^{Sat}$ is the saturated vapor pressure of water at a specific temperature. At 80 °C $P_{H_2O}^{Sat\,(80°C)} = 0.47$ atm. Calculate the oxygen partial pressure of the water saturated air at 80 °C. (Answer: $P_{air} = 0.53$ atm and hence $PO_2 = 0.11$ atm).

6.6 The proton conductivity for Nafion 117 membranes is proposed to be expressed by Eq. (6.10). Calculate the activation energy E_a (in kJ/mol) and explain the physical meaning of it.

6.7 A cell configuration similar to the one shown in Fig. 6.9 was used for determination the methanol permeability through an Aquivion® E89-09S membrane with a wet thickness of 112 μm. 1 M methanol solution was fed on the reference electrode (RE)/counter electrode (CE) side, while pure water was fed on the working electrode (WE) side. At 80 °C and a potential of 0.9 V an oxidation current of 83 mA/cm^2 was measured. Calculate the methanol permeability coefficient in mol CH_3OH/cm·s. (Answer: 1.6×10^{-9} mol CH_3OH/cm s).

6.8 By an impregnation–reduction method carbon-supported platinum nanocatalysts can be prepared using formic acid as the reducing agent. Look at reduction reaction (6.41) and explain the buffering effect of diethylamine during the process.

6.9 Of the two methods for determination of the electrochemical surface area of platinum-based catalysts, the underpotential hydrogen deposition method is more often used for pure Pt/C catalysts while the CO stripping method for Pt alloy catalysts. Why?

6.10 For determination of the ECSA by the CO stripping method (Fig. 6.18b) it is important that a background scan after the CO stripping is recorded and subtracted. Why?

6.11 What is the difference between the determined ECSAs by the RDE and fuel cell methods?

6.12 Look at Fig. 6.26 and explain the differences between the mechanical and chemical stability tests.

6.13 Explain mechanisms involved in the platinum particle growth and operational parameters that are affecting these mechanisms.

6.14 In Fig. 6.28 three catalysts with different carbon support materials are tested but shown similar degradations in both triangle and square-wave potential cycling tests. Why?

6.7 General Readings

1. Gasteiger HA, Kocha SS, Sompalli B, Wagner FT (2005) Activity benchmarks and requirements for Pt, Pt alloy, and non-Pt oxygen reduction catalysts for PEMFCs. Applied Catalysis B: Environmental 56:9–35.
2. Meyer Q, Zeng Y, Zhao C (2019). In Situ and Operando Characterization of Proton Exchange Membrane Fuel Cells. Advanced Materials, 31:1901900.
3. Yuan X–Z, Li H, Zhang S, Martin J, Wang H (2011) A review of polymer electrolyte membrane fuel cell durability test protocols. J. Power Sources 196:9107–9116.

References

1. Doyle M, Rajendran G (2003) In: Vielstich W, Gasteiger H, Lamm A (eds) Handbook of fuel cells—fundamental, technology and applications, vol 3. Wiley, pp 351–395
2. Buchi FN, Scherer GG (2001) Investigation of the transversal water profile in nafion membranes in polymer electrolyte fuel cells. J Electrochem Soc 148(3):A183–A188
3. Springer TE, Zawodzinski TA, Gottesfeld S (1991) Polymer electrolyte fuel-cell model. J Electrochem Soc 138(8):2334–2342
4. Pivovar BS (2006) An overview of electro-osmosis in fuel cell polymer electrolytes. Polymer 47(11):4194–4202
5. Motupally S, Becker AJ, Weidner JW (2000) Diffusion of water in Nafion 115 membranes. J Electrochem Soc 147(9):3171–3177
6. Beattie PD, Basura VI, Holdcroft S (1999) Temperature and pressure dependence of O_2 reduction at Pt vertical bar Nafion((R)) 117 and Pt vertical bar BAM((R)) 407 interfaces. J Electroanal Chem 468(2):180–192
7. Kocha SS, Yang JDL, Yi JS (2006) Characterization of gas crossover and its implications in PEM fuel cells. Aiche J 52(5):1916–1925
8. Jiang SP, Tang HL (2012) Methanol crossover reduction by Nafion modification via layer-by-layer self-assembly techniques. Colloid Surf A-Physicochem Eng Asp 407:49–57
9. Thomas SC, Ren XM, Gottesfeld S, Zelenay P (2002) Direct methanol fuel cells: progress in cell performance and cathode research. Electrochim Acta 47(22–23):3741–3748
10. Ren XM, Springer TE, Zawodzinski TA, Gottesfeld S (2000) Methanol transport through nafion membranes—electro-osmotic drag effects on potential step measurements. J Electrochem Soc 147(2):466–474
11. Curtin DE, Lousenberg RD, Henry TJ, Tangeman PC, Tisack ME (2004) Advanced materials for improved PEMFC performance and life. J Power Sourc 131(1–2):41–48
12. Tang HL, Shen PK, Jiang SP, Fang W, Mu P (2007) A degradation study of Nation proton exchange membrane of PEM fuel cells. J Power Sourc 170(1):85–92
13. Samms SR, Wasmus S, Savinell RF (1996) Thermal stability of Nafion(R) in simulated fuel cell environments. J Electrochem Soc 143(5):1498–1504
14. Fievet F, Lagier JP, Blin B, Beaudoin B, Figlarz M (1989) Homogeneous and heterogeneous nucleations in the polyol process for the preparation of micron and sub-micron size metal particles. Solid State Ionics 32–3:198–205
15. Godoi DRM, Perez J, Villullas HM (2007) Influence of particle size on the properties of Pt-Ru/C catalysts prepared by a microemulsion method. J Electrochem Soc 154 (5):B474–B479
16. Xiang Y, Lu S, Jiang SP (2012) Layer-by-layer self-assembly in the development of electrochemical energy conversion and storage devices from fuel cells to supercapacitors. Chem Soc Rev 41(21):7291–7321
17. Wang SY, Jiang SP, White TJ, Guo J, Wang X (2009) Electrocatalytic activity and interconnectivity of Pt nanoparticles on multiwalled carbon nanotubes for fuel cells. J Phys Chem C 113(43):18935–18945
18. Wang DL, Lu SF, Kulesza PJ, Li CM, De Marco R, Jiang SP (2011) Enhanced oxygen reduction at Pd catalytic nanoparticles dispersed onto heteropolytungstate-assembled poly(diallyldimethylammonium)-functionalized carbon nanotubes. Phys Chem Chem Phys 13(10):4400–4410
19. Tian ZQ, Jiang SP, Liu ZC, Li L (2007) Polyelectrolyte-stabilized Pt nanoparticles as new electrocatalysts for low temperature fuel cells. Electrochem Commun 9(7):1613–1618
20. Paulus UA, Schmidt TJ, Gasteiger HA, Behm RJ (2001) Oxygen reduction on a high-surface area Pt/Vulcan carbon catalyst: a thin-film rotating ring-disk electrode study. J Electroanal Chem 495(2):134–145
21. Cheng Y, Jiang SP (2013) Highly effective and CO-tolerant PtRu electrocatalysts supported on poly(ethyleneimine) functionalized carbon nanotubes for direct methanol fuel cells. Electrochim Acta 99:124–132

22. Hu Y, Jiang YL, Jensen JO, Cleemann LN, Li QF (2018) Catalyst evaluation for oxygen reduction reaction in concentrated phosphoric acid at elevated temperatures. J Power Sourc 375:77–81
23. Mukundan R, Baker AM, Kusoglu A, Beattie P, Knights S, Weber AZ, Borup RL (2018) Membrane accelerated stress test development for polymer electrolyte fuel cell durability validated using field and drive cycle testing. J Electrochem Soc 165(6):F3085–F3093
24. Watanabe M, Tsurumi K, Mizukami T, Nakamura T, Stonehart P (1994) Activity and stability of ordered and disordered Co-Pt alloys for phosphoric-acid fuel-cells. J Electrochem Soc 141(10):2659–2668
25. Stariha S, Macauley N, Sneed BT, Langlois D, More KL, Mukundan R, Borup RL (2018) Recent advances in catalyst accelerated stress tests for polymer electrolyte membrane fuel cells. J Electrochem Soc 165(7):F492–F501
26. Macauley N, Papadias DD, Fairweather J, Spernjak D, Langlois D, Ahluwalia R, More KL, Mukundan R, Borup RL (2018) Carbon corrosion in PEM fuel cells and the development of accelerated stress tests. J Electrochem Soc 165(6):F3148–F3160

Chapter 7
Polymer Electrolyte Membrane Fuel Cells: Performance and Operation

This chapter is devoted to analysis and discussion of performance and operation of the PEMFC. The state-of-the-art performance in the form of polarization curves is first presented. Based on the results of in situ and ex situ characterization from Chap. 6, analysis of *i-V* curves is made focusing at evaluation of kinetic, ohmic, and mass transport losses. Operational issues are then addressed including fuels and CO poisoning, water and thermal management. An example is given to illustrate the effect of operational parameters on the design and construction of fuel cell stacks. Examples of stack and system calculation based on the operating points are given and durability issues are briefly discussed at the end of this Chapter.

7.1 Performance and Analysis

Parameters characterizing a fuel cell performance under operation are current, voltage, and time. The characterization is typically made by recording a current–time (*i-t*) curve at a controlled voltage, a voltage–time (*V-t*) curve at a controlled current, or more often a current–voltage (*i-V*) curve at a specified time. The measurements of *i-t* and *V-t* curves, the so-called potentiostatic or galvanostatic measurements are performed by controlling the cell voltage while recording the responding current or controlling the load current and recording the responding voltage. The equilibrium time for each setting step is commonly a couple of minutes in order to reach a steady-state responding current or voltage. Figure 7.1 shows typical *i-V* curves obtained by operating with hydrogen and oxygen or air at 65 °C.

Due to the innovations in materials and processing technologies, the state-of-the-art performance of PEMFC reaches a cell voltage above 0.7 V at current density of 1.0 A/cm^2 under ambient pressure of humidified H_2 and air at an intermediate cathode Pt loading of 0.4 mg$_{Pt}$/cm^2. Owing to the high catalytic activity of Pt toward H_2 oxidation reaction (exchange current density on the order of 10^{-3} A/cm$_{Pt}{}^2$ based on the platinum surface area), the anode Pt loading for pure H_2 can be as low as

S. P. Jiang and Q. Li, *Introduction to Fuel Cells*,
https://doi.org/10.1007/978-981-10-7626-8_7

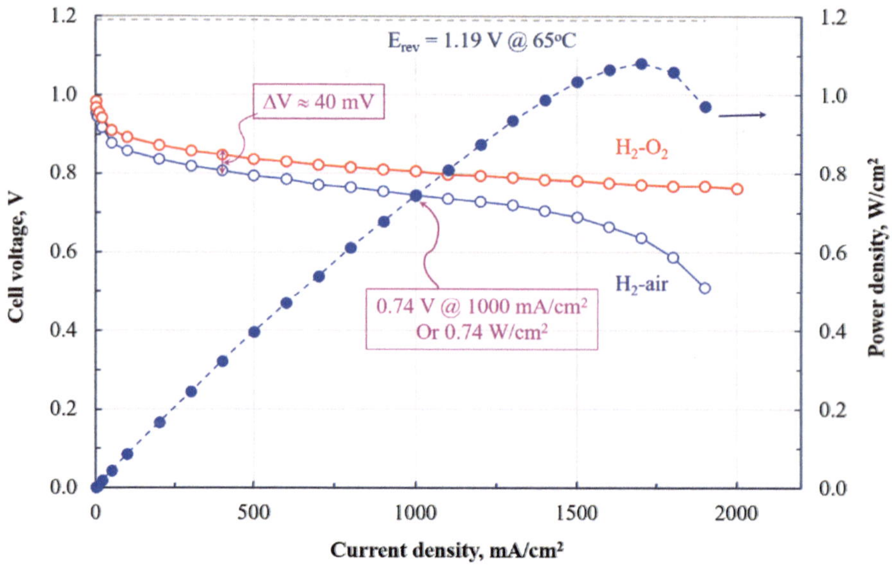

Fig. 7.1 Polarization curves of a PEM single cell operated using pure H_2 on the anode and 100% as well as 20.9% O_2 in N_2 on the cathode under ambient pressure. The cell temperature was 65 °C and reactants were fully humidified at the same temperature. High reactant flow rates were employed to minimize the effect of stoichiometry. The Pt loading on the cathode was 0.4 mg/cm^2 and Nafion 112 was used as electrolyte membrane. Power density curves is plotted for the H_2-air cell. Data were taken from Ref. [1] with permission from John Wiley & Sons, Copyright 2003

0.05 mg$_{Pt}$/cm^2 without any significant voltage loss. In practical studies, however, the used anode Pt catalyst loading is often higher. Under this assumption, the fuel cell performance of 0.7 V at 1.0 A/cm^2 can be achieved at a total Pt loading of 0.45 mg$_{Pt}$/cm^2 (0.05/0.40 mg$_{Pt}$/cm^2 anode/cathode loadings). At this rated point the power density based on the geometric electrode area is 0.7 W/cm^2, corresponding to a Pt mass-specific power density of 1.55 kW/g$_{Pt}$ or specific Pt consumption of 0.64 g$_{Pt}$/kW. Further reduction in the cathode Pt loading is desirable and challenging. At elevated pressures, for example, 1.5 bar a for hydrogen and air at 80 °C, a rated point of 810 mW/cm^2, i.e., 1.2 A/cm^2 at 0.67 V has been demonstrated.

In addition to the fuel cell specifications more information is summarized below based on which further analysis of the *i-V* data will be made in the following sections:

(1) At 65 °C the saturated water vapor pressure of $p_{H_2O} = 0.25$ atm. Under the total pressure of 1 atm the fully humidified (100% RH) hydrogen corresponds to a hydrogen partial pressure of $p_{H_2} = 0.75$ atm. Similarly for fully humidified air one has $p_{O_2} = 0.16$ atm.

(2) The hydrogen crossover is measured by the cyclic voltammetry and found to be $i_x = 1.8$ mA/cm ^{2} for Nafion ® 112 at 65 °C and 1 atm, corresponding to a hydrogen permeability of $4 \times 1\,0^{-11}$ mol/cm·s. In the low current density range, this hydrogen crossover current influences the cell voltage. In analysis of the cathode kinetics it should be taken into account by using an effective

current density in the low current density range, instead of the as-measured current density, in the analysis:

$$i = i_{\text{measured}} + i_x \tag{7.1}$$

(3) The electrochemically active surface area of the cathode platinum was determined by integration of the hydrogen desorption charge from cyclic voltammetry and found to be $\text{ECSA}_{(\text{Pt, MEA})} = 60\ \text{m}^2/\text{g}_{\text{Pt}}$. Considering the typical $\text{ECSA}_{(\text{Pt, RDE})} = 65–85\ \text{m}^2/\text{g}_{\text{Pt}}$ from the ex situ rotating disk electrode with low Pt loading thin-film catalyst layer, this value corresponds to a platinum utilization of

$$U_{\text{Pt}} = \frac{\text{ECSA}_{(\text{Pt,MEA})}}{\text{ECSA}_{(\text{Pt,RDE})}} = 70 \sim 86\% \tag{7.2}$$

(4) Taking the Pt electrode loading of $L_{\text{Pt}} = 0.4\ \text{mg}_{\text{Pt}}/{\text{cm}_{\text{Geo}}}^2$ one has an electrode roughness factor (RF):

$$\begin{aligned} \text{RF} &= 10 \times \text{ECSA}_{\text{MEA}}\left(\text{m}^2/\text{g}_{\text{Pt}}\right) \times \text{L}_{\text{Pt}}\left(\text{mg}_{\text{Pt}}/\text{cm}^2_{\text{Geo}}\right) \\ &= 10 \times 60\frac{\text{m}^2}{\text{g}_{\text{Pt}}} \times 0.4\frac{\text{mg}_{\text{Pt}}}{\text{cm}^2_{\text{Geo}}} = 240\,\text{cm}^2/\text{cm}^2_{\text{Geo}} \end{aligned} \tag{7.3}$$

7.1.1 *Analysis of i-V Curves*

The following equation has been shown to give very good fits to experimentally obtained polarization curves and is thus a simple tool to separate contributions from the ohmic, activation, and mass transport losses.

$$E_{\text{cell}} = E_{\text{rev}} - \eta_{\text{ORR}} - i\sum R_i - \eta_{\text{mt}} \tag{7.4}$$

Here E_{rev} is the reversible cell voltage, $\sum R_i$ is sum of ohmic resistances, η_{ORR} the cathode activation loss, and η_{mt} the mass transport loss. The ohmic resistance includes the area-specific resistance of the electrolyte membrane, electrode substrate and catalyst layer, bipolar plates, and leading wires. The anode activation loss, which is often assumed to be negligible, has also ohmic behavior and is in fact included in the total ohmic resistance.

The reversible cell voltage is a function of temperature and pressures of hydrogen and oxygen and can be expressed as [2]

$$E_{\text{rev}} = 1.23 - 0.9 \times 10^{-3}(T - 298) + \frac{2.303\ \text{RT}}{2F} \times \log\left[\left(\frac{p_{\text{H}_2}}{p^*_{\text{H}_2}}\right)\left(\frac{p_{\text{O}_2}}{p^*_{\text{O}_2}}\right)^{\frac{1}{2}}\right] \tag{7.5}$$

where $p^*_{H_2} = p^*_{O_2} = 1$ atm are the reference pressure of hydrogen and oxygen. At 65 °C the saturated water vapor pressure is 0.25 atm, and therefore the partial pressure of hydrogen and oxygen in fully humidified reactant streams is 0.75 atm. The calculated E_{rev} is hence 1.19 V at 65 °C.

The cathode overpotential is expressed by the Tafel equation:

$$\eta_{ORR} = b\log\left({}^{i}/_{i^0}\right) = a + b\log(i) \tag{7.6}$$

where b is the Tafel slope, and i^0 is the exchange current density. The concentration or mass transport overpotential can be expressed as

$$\eta_{mt} = c\log\left(1 - {}^{i}/_{i_L}\right) \tag{7.7}$$

where i_L is the limiting current density. For the H_2-O_2 cell performance in Fig. 7.1, the mass transport has little contribution to the voltage loss. In the following the analysis of the *i-V* data is made for the H_2–air cell where the limiting current is assumed to be, arbitrarily, however, 2.05 A/cm^2.

Taking Eqs. (7.6) and (7.7) into (7.4) one has

$$E_{cell} = A - iR - b\log(i) - c\log\left(1 - {}^{i}/_{i_L}\right) \tag{7.8}$$

Using i, $\log(i)$, and $\log(1\text{-}i/i_L)$ as independent variables, a multiple linear regression gives the following constants: $A = 0.96$ V, $R = 0.036$ Ωcm^2, $b = 0.055$ V/dec, and $c = 0.30$ V.

First of all the value of constant A approximately corresponds to the cell voltage at the hydrogen crossover current density (*ca.* 1–2 mA/cm^2), i.e., the OCV. Values of R, b, and c determine voltage losses of the ORR activation, ohmic, and mass transport, respectively, as shown in the lower part of Fig. 7.2. By subtraction of these losses from the measured *i-V* curves one obtains the η_{mt} and *iR*-corrected *i-V* curves. These losses are to be discussed below.

7.1.2 Ohmic Losses

The overall ohmic resistance estimated from regression consists of contributions of the ionic resistance of the electrolyte membrane and electronic resistance of gas diffusion layers, bipolar plates, and contact resistances. The electronic resistance of cell hardware can be determined in an ex situ measurement by assembling the cell hardware without placing the MEA, while the ionic resistance of the membrane can be calculated from the known membrane conductivity and thickness. In the above regression, the area-specific resistance (ASR) is found to be 35 mΩ cm^2. At an operational current density of 1.5 A/cm^2, for example, the ohmic voltage loss is

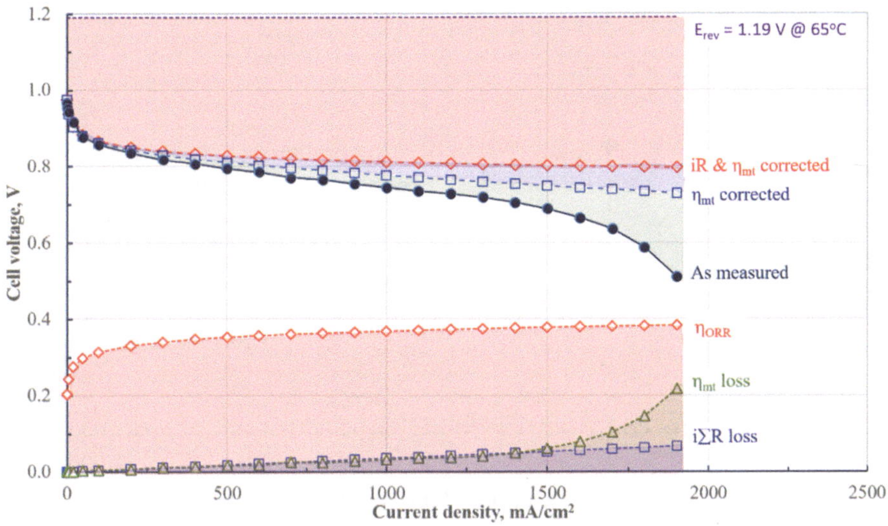

Fig. 7.2 Polarization curves of a H_2–air PEMFC (as measured, solid circle), voltage losses by ORR overpotential (η_{ORR}), ohmic ($i\sum R$) and mass transport (η_{mt}), and *i-V* curves corrected by subtraction of the mass transport and ohmic losses

about 50–60 mV. Using the conductivity of 0.15 S/cm and a wet thickness of 55 μm (Nafion 112) the ASR of the membrane is estimated to be

$$\mathrm{ASR} = \frac{\text{thickness, cm}}{\sigma,\ \text{S/cm}} = \frac{0.0055\,\text{cm}}{0.15\,\text{S/cm}} = 0.036\,\Omega\text{cm}^2 \qquad (7.9)$$

which seems to be the major contribution to the overall resistance. In addition to the membrane, other minor courses of the ohmic losses may include resistances of electrodes and bipolar plates.

7.1.3 ORR Kinetic Parameters

The exchange current density is a measure of the intrinsic activity of platinum catalysts and theoretically defined on basis of the active surface area of Pt nanoparticles. Based on the Pt loading and its utilization in the gas diffusion electrode, this can be translated into an apparent exchange current density on basis of the geometric active area of the electrodes. In the low current density range, say below 100 mA/cm^2, the ohmic and the mass transport losses are relatively small compared to the activation loss and can be ignored. The approximate activation loss can be obtained directly from the as-measured *i-V* data. If plotted on a semi-logarithmic scale, the as-measured *i-V* data show a nearly linear behavior, as seen from Fig. 7.3a.

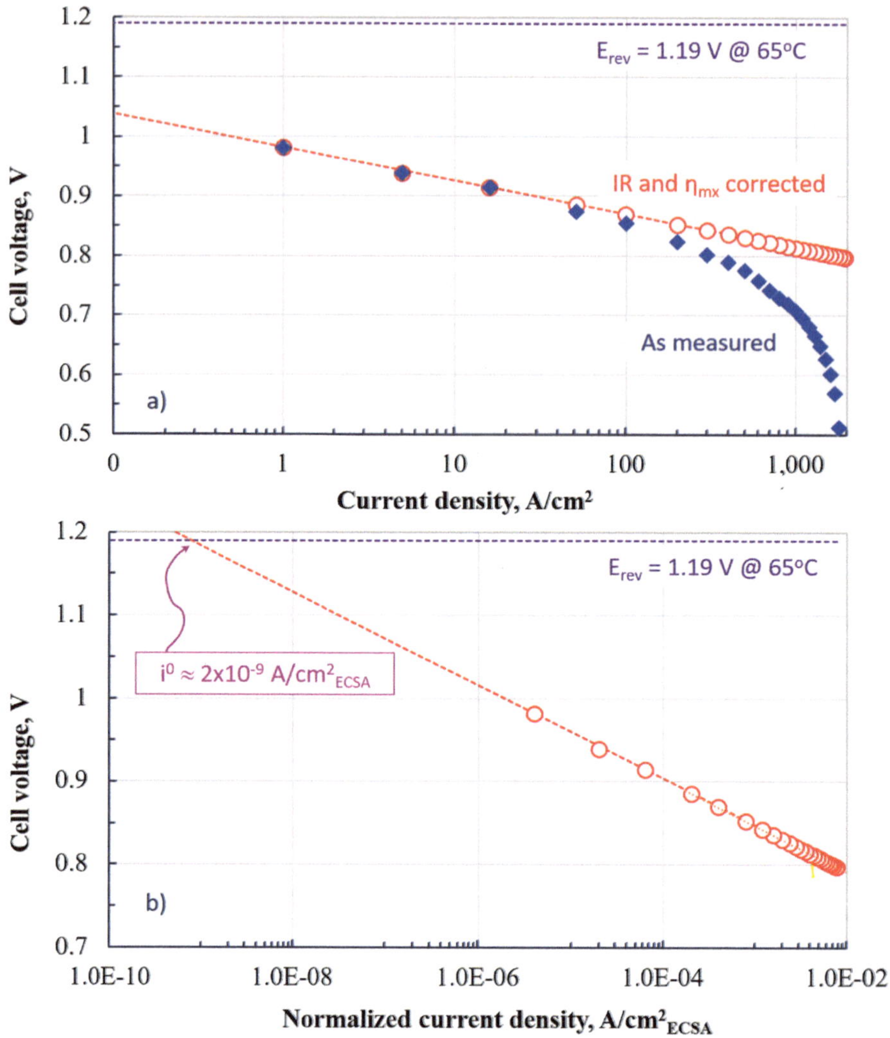

Fig. 7.3 **a** Tafel plot of the H_2–air cell without and with *iR* and η_{mt} correction with current density based on the electrode geometric area. **b** The *iR*- and η_{mt}-corrected Tafel plot based on the electrochemical surface area (ECSA) of Pt/C catalysts. The *i*-*V* data were taken from Fig. 7.1

ORR kinetic parameters such as the Tafel slope, exchange current density, specific activity, and mass activity of Pt can be derived from the *iR*- and η_{mt}-corrected *i*-*V* data. It should be noted that, for the kinetic studies, the catalytic activity of the cathode Pt should be defined using the electrochemical surface area (ECSA) of platinum, instead of the electrode geometric area (Geo). For this purpose, the geometric current density can be normalized to the ECSA-based current density i_{ECSA} using the platinum loading of $L_{Pt} = 0.4\ mg_{Pt}/cm^2$ and a specific surface area of 60 m^2/g or a roughness factor of 240 $cm^2{}_{ECSA}/cm^2{}_{Geo}$.

$$\begin{aligned} i_{ECSA} &= (i + i_x)\frac{A}{cm^2_{Geo}} \times \frac{1}{L_{Pt}}\frac{cm^2_{Geo}}{mg_{Pt}} \times \frac{1}{ECSA_{MEA}}\frac{g_{Pt}}{m^2_{ECSA}} \\ &= (i + i_x)\frac{A}{cm^2_{Geo}} \times \frac{1}{RF}\frac{cm^2_{Geo}}{cm^2_{ECSA}} \end{aligned} \quad (7.10)$$

A plot of the cell voltage against this ECSA-specific current density is shown in Fig. 7.3b. By extrapolating the plot to the reversible cell voltage of 1.19 V the extracted exchange current density is $i^0 = 10^{-9}$ A/cm^2_{ECSA}, a value in good agreement with literature. It should be remarked that this extrapolation of the Tafel plot over 7–8 orders of magnitude in current density is always associated with a large error in the obtained exchange current density. A practical parameter commonly used in characterizing catalytic activity is the current density at the iR-corrected voltage of 0.9 $V_{iR\text{-}free}$. This is understandable as this number is a direct reading from experimental curves with no extrapolation needed. In the above example, this current density is found to be 75 mA/cm^2_{Geo}, @ 0.9 $V_{iR\text{-}free}$ based on the electrode geometric area. This value has kinetic meaning only when it is converted into the Pt (area) specific activity (SA) or Pt mass specific activity (MA) as:

$$SA = 75\frac{mA}{cm^2_{Geo}} \times \frac{1}{240}\frac{cm^2_{Geo}}{cm^2_{ECSA}} = 0.31\ mA/cm^2_{ECSA} \quad (7.11)$$

$$MA = 75\frac{mA}{cm^2_{Geo}} \times \frac{1}{0.4}\frac{cm^2_{Geo}}{mg_{Pt}} = 187\ mA/mg_{Pt} \quad (7.12)$$

The DOE 2020 target mass activity of electrocatalysts is 440 mA/mgPt @ 0.9 $V_{iR\text{-}free}$ at a low platinum loading of 0.1 $mgPt/cm^2_{Geo}$, corresponding to a current density of 40 mA/cm^2_{Geo}. Both the exchange current density and specific current density at 0.9 $V_{iR\text{-}free}$ are dependent on temperature and partial pressure of oxygen. The above values are obtained at 65°C and $p_{O_2(air)} = 0.16$ atm for fully humidified air under ambient pressure.

Considering the two I-V curves in Fig. 7.1 obtained with pure oxygen and air, the oxygen pressure effect can be expressed as the *cell voltage gain* when the cathode feed is switched from humidified air to humidified oxygen. At current densities larger than the hydrogen crossover current, say, above 100 mA/cm^2, this cell voltage gain can be written as [3]

$$\Delta V_{O_2/air} = E_{O_2} - E_{air} = b\ \log\left(\frac{p^*_{O_2}}{p_{O_2(air)}}\right)^m \quad (7.13)$$

where b is the Tafel slope, $p^*_{O_2}$ is the reference partial pressure of oxygen in the humidified stream ($p_{O_2} = 0.75$ atm at 65 °C), and m is the kinetic order of ORR with respect to p_{O_2}. For the present example at 65 °C, the ratio $p^*_{O_2}/p_{O_2(air)} = 4.7$. For the reaction order $m = 1$ and Tafel slope $b = 60$ mV/dec, the oxygen gain is about ~40 mV. This value is close to the observed voltage difference in the low to intermediate current density range as indicated in Fig. 7.1. At higher current densities, more oxygen gain is obtained due to the mass transport impact. It should be mentioned, as a comparison, that the Tafel slope of ORR in the phosphoric acid doped PBI membrane or concentrated phosphoric acid is typically of 90–110 mV/dec, as to be discussed in Chaps. 8 and 14.

7.1.4 Mass Transport Losses

As expressed by Eq. (7.7) the mass transport loss is sensitive to the limiting current. Limiting current density is an important characteristic of the electrode structure when the *i-V* curve is restricted by the diffusion of reactants or products through a thin film of ionomer covering the catalyst surface. Such a thin film and therefore the limiting current density are often poorly defined.

As the kinetic (logarithmic) term is much better defined and the ohmic resistance can be independently determined, it is therefore practical to obtain the mass transport loss by subtraction of the ohmic and kinetic losses from the as-measured *i-V* curves. As seen from Fig. 7.2 this term has a value of more or less in the same range as the ohmic loss in the low to intermediate current range and becomes significantly larger at higher current densities. The mass transport loss can be minimized by optimization of design and water management in gas channels, diffusion layers, and catalyst structures.

Another parameter, often used as a measure of the quality of the electrode structure with respect to mass transport, is the effect of utilization (U) or stoichiometry (λ) of oxygen on the cell performance:

$$U = 1/\lambda \tag{7.14}$$

This is a parameter more engineering than fundamental. In practical stacks, a high utilization of air, as well as fuel, is desired in order to reduce the parasitic power consumption. It is also essential to manage the water balance. High utilizations of oxygen, on the other hand, sacrifice the overall cell performance due to the effect of further reduced oxygen partial pressure on both the electrode kinetics and the mass transport loss, as shown in Fig. 7.4. In a practical range of air utilization less than 50% (with $\lambda_{Air} > 2$) reasonable performance can be achieved in the low to intermediate current density range, say, below 1 A/cm^2. Further increase in the air utilization results in significant performance loss.

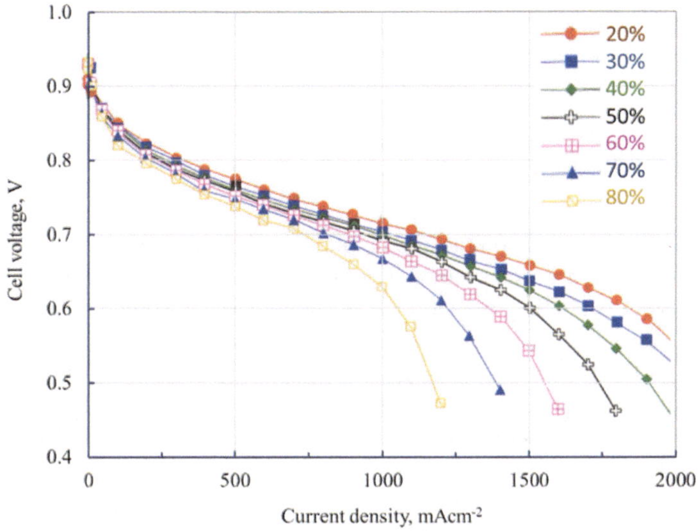

Fig. 7.4 *i-V* curves of a single cell operated under a fixed H_2 utilization of 80% and varied air utilizations indicated in the figure. The cell operated as 65 °C with fully humidified air under a total ambient pressure. The cathode Pt loading was 0.4 mg_{Pt}/cm^2. Modified from Ref. [1] with permission from John Wiley & Sons, Copyright 2003

7.2 Fueling Options and CO Poisoning

High-performance PEMFC requires high-purity hydrogen, which faces challenges including on board storage and establishment of hydrogen supply and filling infrastructure. Automotive fuel cell developers have made significant progress in high-pressure hydrogen tanks in reducing the volume and weight. The state-of-the-art technology is a carbon fiber-reinforced tank at a compression pressure of 700 atm storing 4–5 kg H_2. This H_2 storage capacity is sufficient for a driving range of 500–600 km which is also dependent on the efficiency of the fuel cell as well as the powertrain system. An alternative solution to avoid the complexities and compromises of hydrogen storage is a fuel processing unit based on reforming of alcohol or hydrocarbon fuels. Two most practical options of reforming use steam, air, or both as the oxidant, followed by a water gas shift reactor to convert carbon monoxide and steam into hydrogen and carbon dioxide. Carbon dioxide and unreacted hydrocarbon fuel in the fuel stream act as diluents at the anode, while the carbon monoxide content after the shift reactor at around 200 °C is still around 1000 ppm and needs to be further removed down to < 10 ppm by, for example, selective oxidation. Such a system currently covers 40–50% of the total system cost and considerably increases the system size, volume, weight, and time for start-up and transient response.

In addition, CO_2 as a major component of reformate gases has been shown to have negative effect beyond the simple dilution. It can react with hydrogen to form CO through a reverse water gas shift reaction:

$$CO_2 + H_2 \rightarrow CO + H_2O \tag{7.15}$$

7.2.1 CO Adsorption on Pt

The oxidation of hydrogen on the anodic platinum catalyst takes place via the dissociative chemisorption which requires two free adjacent sites of the platinum surface atoms. The electrochemical oxidation of the chemically absorbed hydrogen is kinetically fast, exhibiting a negligible anodic overpotential under fuel cell operational conditions. In the presence of CO in the hydrogen gas stream, CO competes with hydrogen for the adsorption sites of platinum. Two types of bonding modes of the adsorbed CO molecules on Pt have been suggested, i.e., the *linearly bonded and the bridge-bonded* (See Fig. 7.5). At low CO coverage, e.g., below 4%, almost all CO is bridge-bonded to platinum. This is the case either at high temperatures or at lower CO concentrations, resulting in a linear *i-V* polarization curve. In this manner, each absorbed CO species occupies one surface site of platinum. At lower temperatures or high CO concentrations, however, the adsorption is strongly favored. The increased number of absorbed CO species at the catalyst surface might lead to a mutual repulsion among the absorbed CO molecules, resulting in a higher fraction of the linearly bonding orientation, probably accompanied by the transfer of the bridge-bonded to the linearly bonded CO, especially at coverage over 70%. At a CO coverage around 90%, the fraction of the linearly bonded CO is close to 100% [4].

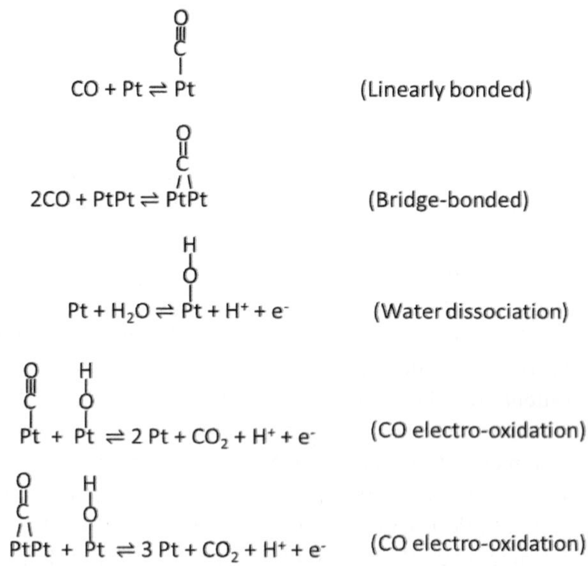

Fig. 7.5 Reaction schemes for CO bonding modes, water dissociation, and CO electro-oxidation

7.2.2 CO Poisoning Effect

Figure 7.6 shows polarization curves of an H_2-O_2 fuel cell operating with pure hydrogen and hydrogen containing CO at 80 °C. Figure 7.6a is the recorded *i-V* curves after switching from H_2 to a mixture of H_2 and 100 ppm CO with specified time intervals between which a constant current density of 250 mA/cm^2 was applied. The *i-V* performance showed a steady decrease, and a steady-state curve was only obtained after 210 min. Thus obtained steady-state *i-V* curves with a varied amount of CO are shown in Fig. 7.6b.

As seen from Fig. 7.6b, the steady-state *i-V* curves showed a significant poisoning effect when 25 ppm CO is present in H_2 at 80 °C. At this low content of CO, the current–voltage curve is linear. At higher CO contents the current–voltage curves exhibit two distinct slopes, larger in the higher voltage (400–700 mV) region and smaller in the low voltage (200–300 mV) region. The large slope at low current densities indicates the effect of the CO adsorption on the overvoltage of hydrogen oxidation at the anode. In the low cell voltage region, the potential of the anode reaches values at which the adsorbed CO is oxidized to CO_2 and the reaction rates for hydrogen adsorption and oxidation are increased. As shown in Fig. 7.5 the electro-oxidation of the adsorbed CO is achieved with help of the hydroxyl species from the water dissociation. This reaction needs an anode potential around 0.5 V versus RHE. Under operation of a PEMFC, the anode potential is within the region between 0 and 0.1 V versus RHE, which means that CO is hence an inert adsorbate on the platinum catalyst surface.

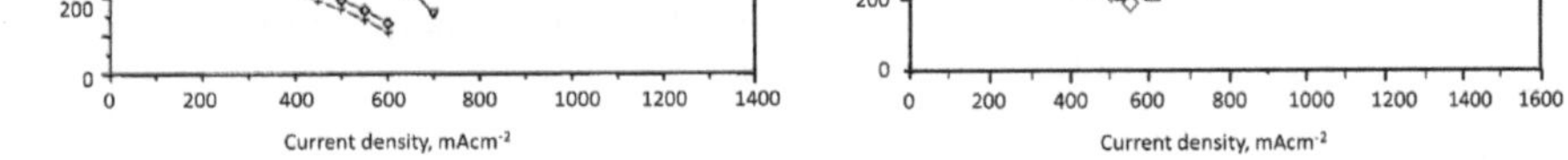

Fig. 7.6 Polarization curves of PEMFC **a** with H_2 containing 100 ppm CO during the poisoning period and **b** with H_2 containing 0 to 250 ppm CO. Anode and cathode were made on 30% Pt on Vulcan XC 72 at Pt loading of 1 mg/cm^2. Measurements were made at 80 °C with hydrogen partial pressure of 2.2 atm and oxygen partial pressure of 2.4 atm. Reproduced from Ref. [5] with permission from IOP Publishing, Copyright 1996

7.2.3 Relative Activity and Temperature Effect

To characterize the CO poisoning effect, the surface coverage of platinum catalysts by CO (θ_{CO}) has been determined. It is suggested that the CO coverage follows a Temkin isotherm at a fixed current density, i.e., the coverage can be expressed as a linear function of log [CO]/[H_2]. It is the fraction of the unblocked catalyst sites ($1-\theta_{CO}$) that is correlated to kinetics of hydrogen oxidation:

$$1 - \theta_{CO} = \left[\frac{i_{(H_2+CO)}}{i_{H_2}}\right]^n \tag{7.16}$$

where i_{H_2+CO} and i_{H_2} are the oxidation currents of hydrogen in the presence and absence of carbon monoxide, respectively, at a certain voltage. Electrochemical measurements of CO poisoning on both planar and gas diffusion electrodes in the phosphoric acid electrolyte suggest a constant of $n = ½$. This number implies that only adjacent free Pt sites are active for the hydrogen oxidation. Under conditions more appropriate to the operation of PEMFCs, it is suggested $n = 1$.

The most appropriate measure for the CO poisoning is the decrease in fraction of active catalyst surface sites available for the hydrogen oxidation under equilibrium conditions. This measure can be further defined as the ratio of the active surface site number for the H_2 oxidation in the presence of CO to the total surface site number available for the H_2 oxidation in the absence of CO, i.e., the simple ratio of poisoned H_2 oxidation current to the pure H_2 oxidation current, i_{CO+H_2}/i_{H_2}. The physical meaning of this current ratio is the relative activity of the catalysts for hydrogen oxidation in the presence of CO. A value of unity of the ratio indicates no change in the number of active catalyst surface sites for the hydrogen oxidation even in the presence of CO, i.e., no poisoning.

Using this simple current ratio, the relative activity of the Pt catalyst for hydrogen oxidation in the presence of CO was estimated as a function of temperature and CO concentration [6], as shown in Fig. 7.7. The relative activity of the platinum catalyst for 20 ppm CO at 55 °C is about 24%, close to that for 100 ppm CO at 80 °C or for 1% CO at 125 °C. As to be discussed in Chap. 8, the CO tolerance of up to 3% CO has been demonstrated in HT-PEMFCs at 160 °C.

7.2.4 CO Poisoning Mitigation and Bifunctional Mechanism

Several *operando* techniques have been developed to mitigate the CO poisoning in PEMFC. Niedrach et al. seem to be the first, who found that addition of Ru, Rh, and Ir elements into Pt could improve the CO tolerance of the catalysts [7]. Thereafter a number of alloy catalysts have been investigated. The presence of alloying elements is believed to have a beneficial effect on the oxidation of adsorbed CO via bifunctional mechanism of Pt alloy catalysts, i.e., by promoting the water dissociation and

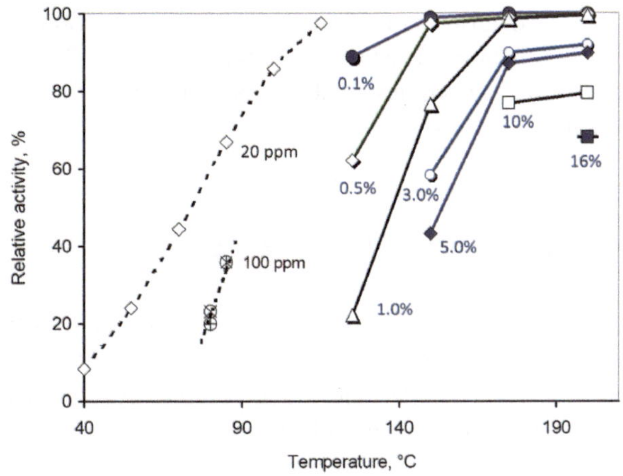

Fig. 7.7 Relative activity of Pt/C catalysts for hydrogen oxidation as a function of temperature at different CO concentrations indicted in the figure. Reproduced from Ref. [6] with permission from IOP Publishing, Copyright 2003

adsorption of hydroxyl groups as well as the electrooxidation of adsorbed intermediate species by the oxygen species of the hydroxyl groups. Among a long list of investigated binary and ternary alloys, Ehteshami et al. [8] reported that PtMo/C has the highest CO tolerance, followed by PtCoMo/C, PtRuMo/C, PtRuPb/C, PtRu/C, PtCo/C, PtFe/C, PtNi/C, and Pt/C, as shown in Fig. 7.8.

Alternative approaches including air-bleeding to the CO-containing anode feed stream or addition of hydrogen peroxide into an anode humidifer have been suggested. The bleed oxygen reacts chemically with CO and enhances its oxidation to CO_2. The method was first reported by Gottesfeld and Pafford [9], who demonstrated that introduction of 2–5% O_2 into the hydrogen steam can completely mitigate the poisoning effect of 100 ppm CO at 80 °C, though at expense of consumption

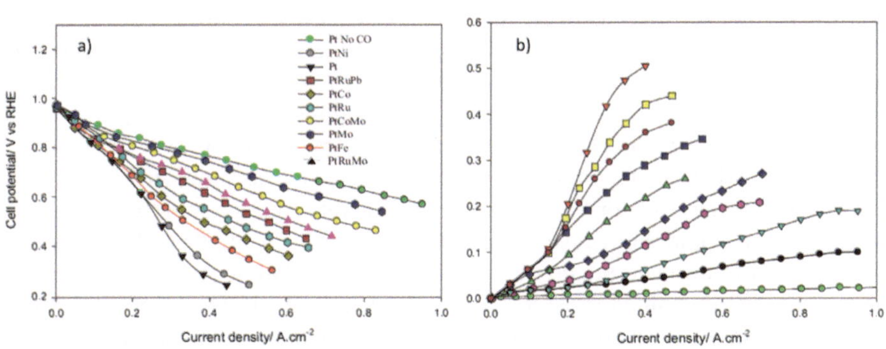

Fig. 7.8 **a** Polarization curves and **b** anode overpotentials for different anode catalysts Pt-M (M = Ru, Mo, Fe, Co, Ni) and PtRuMo/C, PtCoMo/C, and PtRuPb/C in PEMFC operating at 85 °C under 100% RH. Oxygen was used as the oxidant and reformate fuel including 100 ppm CO in a mixture of 50–50 H_2–CO_2. Reproduced from Ref. [8] with permission from Elsevier, Copyright 2013

of hydrogen as well as deterioration of the membrane due to formation of hydrogen peroxide and local overheating. The principle applies to internally permeated oxygen from the cathode when it becomes relatively significant at elevated pressures or minimized membrane thickness.

7.2.5 Catalysts for Methanol Electro-Oxidation

Thermodynamically the methanol electro-oxidation has an equilibrium potential close to that of the hydrogen oxidation. Direct oxidation of methanol needs, however, a significantly higher overpotential due to the complex reaction mechanism. This phenomenon is related to the strong adsorption of the intermediate CO, and it is the step of removing the strongly adsorbed CO from the catalyst surface that requires a relatively high potential. From kinetics point of view, a good methanol electro-oxidation catalyst should possess the ability to (1) dehydrogenate methanol to formation of intermediate CO^*; (2) activate the H_2O dissociation; and (3) weaken the binding of CO^* as well as OH^* to enable the CO_2 formation [10]. Pure Pt is a good catalyst for oxidization of methanol to CO^*, but it binds the CO^* too strongly and has also difficulty in activating the water dissociation and hence removing CO^* from its surface. Alloy catalysts have been proposed to overcome these issues, and the PtRu catalyst is among the most effective materials though both metals are expensive. Structured carbon supports, e.g., carbon nanotubes functionized with polyelectrolytes have also been shown enhanced activity for the methanol oxidation due to its high tolerance toward CO poisoning [11]. Other active and cost-effective electrocatalysts include binary and ternary alloys that have been discussed in above sections.

7.2.6 Other Effects of CO

It is also recognized that CO in the fuel stream may permeate through the membrane and affect the cathode performance. More interesting is the effect on the corrosion of carbon support of catalysts. During start-up or shutdown of a PEMFC the anode side is switched from a hydrogen flow to air. The transition front triggers a reverse current, which drives the cell between the oxygen reduction on the original anode side and the carbon oxidation on the original cathode side, as shown in Fig. 7.9. This reverse current may last for a short period but generate a peak voltage of up to 1.5 V, which leads to rapid and irreversible carbon corrosion. A number of patents have been granted to mitigate the problem including recirculation of the fuel gas or use of auxiliary loads or shunt resistors to limit the peak voltage. The presence of CO in the fuel stream on the anode side is reported to mitigate the cathode carbon support corrosion during start/stop operation.

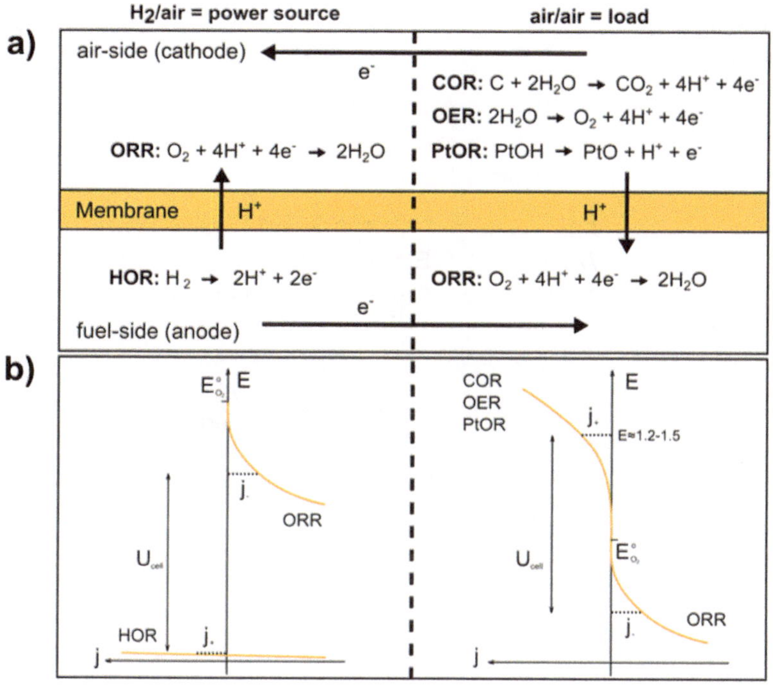

Fig. 7.9 **a** Start/stop degradation mechanism including the main reaction pathways and **b** schematic polarization curves within the different characteristic fuel cell regions caused by the gas front passing through the fuel compartment. Reproduced from Ref. [12] with permission from Springer Nature, Copyright 2016

7.3 Water Management

7.3.1 Water Balance and Transport

Water management is the most critical operational issue for PEMFC, arising from the fact that the membrane needs to be well hydrated to maintain the conductivity and durability. During operation of a fuel cell, water is formed at the cathode, which may diffuse toward the anode driven by the concentration gradient. At the same time, water moves from the anode to the cathode due to the electroosmotic drag effect as the membrane has a drag coefficient of 1–3.0 H_2O/H^+ depending on water content. Of the two mechanisms, the diffusion dominates at low current densities, while the drag effect prevails at high current densities. As a result, both hydrogen and air need to be humidified in order to manage the water balance and achieve the optimal fuel cell performance. Figure 7.10 schematically represents the water balance within a hydrogen–air fuel cell.

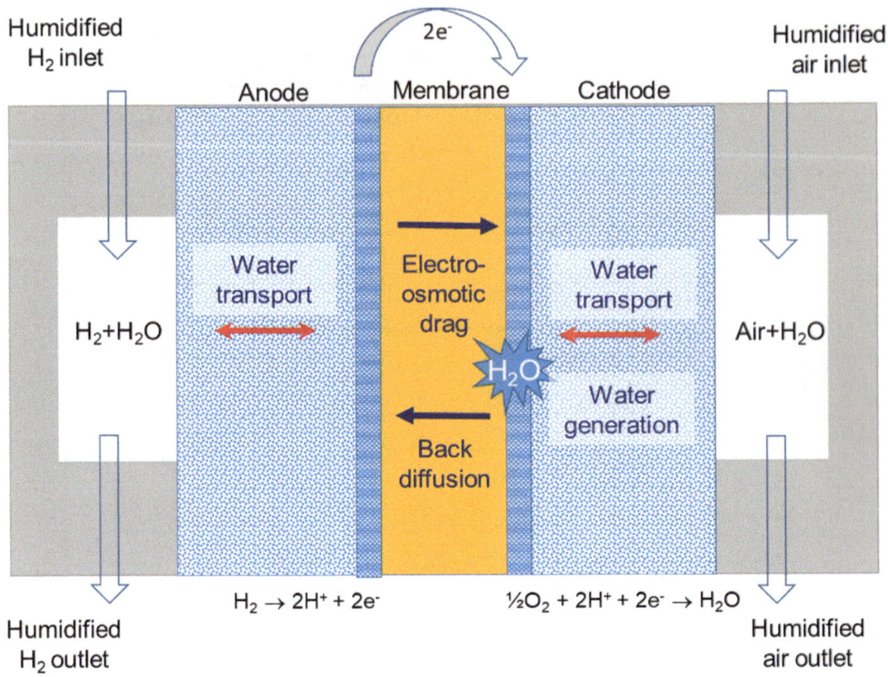

Fig. 7.10 Scheme of the water balance within a hydrogen–air fuel cell

7.3.2 Saturated Water Vapor Pressure and Relative Humidity

The *relative humidity* (RH) of an atmosphere containing water at a partial pressure of p_{H_2O} is defined as

$$\mathrm{RH}(\%) = \frac{p_{H_2O}}{p_{H_2O}^{Sat}} \times 100 \tag{7.17}$$

Here $p_{H_2O}^{Sat}$ is the saturated water vapor pressure which is an exponential function of temperature, as listed in Table 7.1. At room temperature the water content in a "fully humidified" atmosphere is about 3%, which is increased to about 25% at 65 °C and 47% at 80 °C. Another term often used to describe the water content is the *dew point*, the temperature when water contained in the atmosphere starts to condense. For example, air containing 25% water has a dew point of 65 °C. It should be noted that the term relative humidity can be misleading in cases at temperatures above 100 °C. For pure water vapor under the ambient pressure, i.e., $p_{H_2O} = 1$ atm, the relative humidity at 150 °C is only 21.3%.

Table 7.1 Saturated water vapor pressure and maximum relative humidity of pure water vapor as well as the fuel cell cathode exit at an air stoichiometry of 2, as a function of temperature

Temperature (°C)	Saturated water vapor pressure ($P^{Sat}_{H_2O}$) atm	Maximum RH under pure water vapor (%)	Maximum RH of cathode exit at $\lambda_{air} = 2$ (%)
10	0.008	>100.0	>100.0
20	0.023	>100.0	>100.0
30	0.042	>100.0	>100.0
40	0.073	>100.0	>100.0
50	0.122	>100.0	>100.0
60	0.196	>100.0	96.5
70	0.307	>100.0	61.7
80	0.467	>100.0	40.6
90	0.692	>100.0	27.4
100	1.00	100.0	19.0
120	1.959	51.1	9.7
130	2.667	37.5	7.1
140	3.567	28.1	5.3
150	4.698	21.3	4.0
160	6.100	16.4	3.1
170	7.817	12.8	2.4
180	9.895	10.1	1.9

7.3.3 *Self-Humidification*

For maintaining the water balance of a fuel cell a simple criterion is that the RH of the cathode exit is around 100%. Too low RH of the cathode exit means that the air flow absorbs water from the cell and the membrane will be drying out sooner or later. On the other hand, too high RH leads to water condensation in the cell—liquid water will flood the porous electrode structure and impede the access of gas reactant to the catalytic sites.

Assuming that all the water produced at the cathode is purged away in the cathode exhaust with the airflow, the water partial pressure in the cathode exit $p^{out}_{H_2O}$ is then a function of the air stoichiometry (λ_{air}) and the overall pressure at the cathode outlet p^{out}_{total} and can be calculated using the following equation:

$$p^{out}_{H_2O} = \frac{0.42}{\lambda_{Air} + 0.21} \times p^{out}_{total} \tag{7.18}$$

In the equation the number 0.21 is the molar fraction of oxygen in air, and each mole oxygen generates two moles of water (0.42). This equation can be used for estimation of airflow and temperature for self-humidification of fuel cells, i.e., fuel cells operating with dry inlet air without additional humidification ($\Psi = 0$). For $\lambda_{air} = 2$, i.e., the oxygen utilization of 50%, the water content in the airflow at the cathode exit corresponds to a partial pressure of 0.19 atm under the ambient pressure [13]. For a fuel cell operating at 60 °C ($p_{H_2O}^{Sat} = 0.196$ atm) this partial pressure can be translated into a RH of 97% at the cathode exit.

The RH *versus* the air stoichiometry at different fuel cell temperatures under ambient pressure is plotted in Fig. 7.11a. The grey area in the figure indicates the desired humidity range of the cathode exit for fuel cell operation — a cell will be flooded if the air exit humidity is higher and dried out if lower. When a cell operates at 40 °C, for example, the air stoichiometry should be around $\lambda_{air} = 6$. In a practical stoichiometry of $\lambda_{air} = 2$, a fuel cell can operate at a maximum temperature of 60 °C with unhumidified air inlet. At higher temperatures, e.g., 80 °C, this stoichiometry gives a RH of as low as 40% at the cathode exit, which is too low to maintain the water balance. At this temperature operation of a fuel cell with self-humidification can only be possible at elevated overall pressures. As seen from Fig. 7.11b, at 80 °C and $\lambda_{Air} = 2$, the fuel cell without external humidification needs to operate under a pressure of 2–3 atm for the purpose of water management.

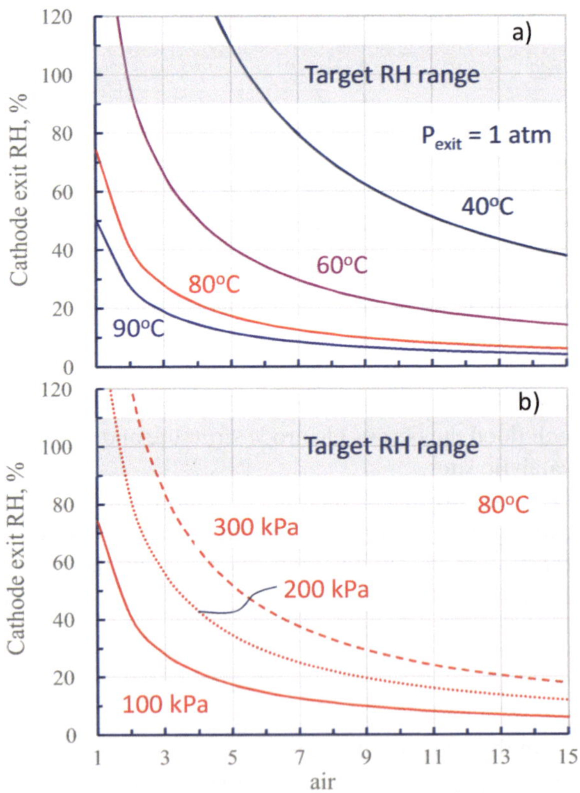

Fig. 7.11 Relative humidity of the airflow at the cathode exit of self-humidified fuel cells, **a** for $\lambda_{air} = 2$ at different fuel cell temperatures and **b** at 80°C at varied overall pressures

In addition researchers have attempted to modify the MEA for enhancing the performance under low humidity operation or self-humidification. A variety of hygroscopic additives to the catalyst layer have been explored including nano-oxides, e.g., SiO_2 or polyvinyl alcohol. More discussion will be made for membrane modification in Sect. 7.3.5.

7.3.4 External Humidification

With external humidification of the air stream, a partial pressure ratio of water to air at the inlet is defined as

$$\Psi = \frac{p^{\text{in}}_{\text{H}_2\text{O}}}{p^{\text{in}}_{\text{air}}} \tag{7.19}$$

Including this factor the water partial pressure at the cathode outlet can be expressed as (for detailed derivation see [14]):

$$p^{\text{out}}_{\text{H}_2\text{O}} = \frac{(0.42 + \psi \times \lambda_{\text{air}})}{(1 + \Psi)\lambda_{\text{air}} + 0.21} \times p^{\text{out}}_{\text{total}} \tag{7.20}$$

By bubbling the inlet airflow through a water bath at 80 °C, the saturated water vapor pressure $p^{\text{in}}_{\text{H}_2\text{O}} = 0.47$ atm. At a total inlet pressure of $p^{\text{in}}_{\text{total}} = 1$ atm the inlet air partial pressure is then $p^{\text{in}}_{\text{air}} = 0.53$ atm and therefore $\psi = 0.88$. For an air stoichiometry of $\lambda_{\text{air}} = 2$, one has the outlet water partial pressure from Eq. (7.20) $p^{\text{out}}_{\text{H}_2\text{O}} = 0.54$ atm. This is sufficient to maintain a good water balance if the fuel cell is also operating at 80 °C. In practice some losses of the water vapor are inevitable, and the humidifier temperature is therefore always 5–10 °C higher than the fuel cell temperature. If the fuel cell is operating at 90 °C where the saturated water partial pressure is 0.7 atm, however, this outlet water partial pressure corresponds to an outlet humidity of 0.54/0.70 =78%, a kind of low.

A variety of techniques has been developed for external humidification of both air and hydrogen. A simple method of bubbling the gases through a water bath at a controlled temperature is commonly used for laboratory investigations. For fuel cell stacks, direct injection of water into the feed gas as a spray is a straightforward method, which requires a pump to pressurize water, a fairly expensive method in terms of additional devices with increased volume and weight as well as parasitic energy consumption. The need for an external water supply makes the method more suitable for stationary applications. Alternatively water can be recycled from the cathode exhaust gas without condensation to liquid. The device is called an enthalpy wheel using a rotating wheel with capability of absorbing water from the exhaust gas and introducing it into the dry cathode inlet. Another method is by use of a membrane placed between the cathode exit and the cathode inlet gas. Water vapor in the exit condenses on the membrane and then passes through it to the dry inlet side.

7.3.5 *Modified and Thinner Membranes*

Technically the humidification of fuel cells can be enhanced in several ways. First of all, inclusion of Pt and hygroscopic oxide nanoparticles, e.g., SiO_2 and TiO_2 in membranes has been proposed. Pt particles catalyze the reaction of permeated hydrogen and oxygen inside membranes, and the generated water is then adsorbed and retained by the oxide particles. Design of fuel cells with a counter flow of hydrogen and air across the MEA is also beneficial for bettering the water distribution. Anyway for operation at higher temperatures and also for large stacks external humidification is employed.

Use of thinner membranes has demonstrated better water management characteristics due to the shorting of water diffusion distance. Figure 7.12 shows fuel cell voltages under different reactant humidities for MEAs using different membranes. The two Nafion membranes have EW 1100 and dry thicknesses of 50 (Nafion 112) and 25 (Nafion 111) μm, while the GorePrimea® uses a membrane of EW <1000 and thickness of 25 μm. At 0.8 A/cm^2 the cell voltage of the 50 μm thick membrane (Nafion 112) strongly depends on the reactant humidity. Switching from fully humidified to dry reactants leads to a drop of *ca.* 50 mV. This loss is about halved for thinner (Nafion 111) membranes, while the low EW GorePrimea® exhibits nearly no voltage losses upon reducing the reactant humidity.

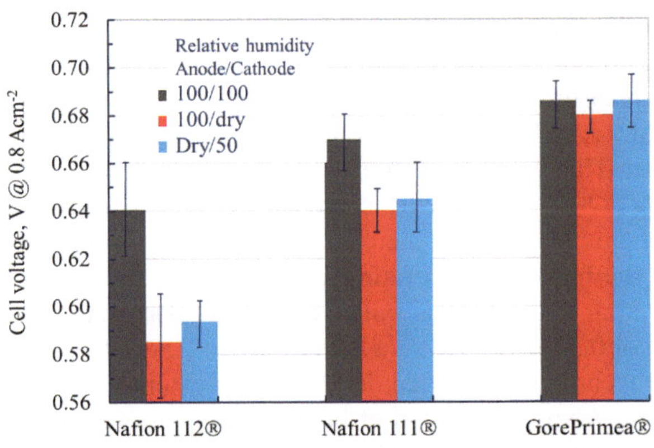

Fig. 7.12 Fuel cell voltages at 0.8 A/cm^2 for different membranes under varied anode (H_2) and cathode (air) humidities indicated in the figure. Temperature: 80 °C, stoichiometry: 2.0/2.0, pressure: 270 kPa, thickness: Nafion 112 (50μm, EW1100), Nafion 111 (25 μm, EW1100), and GorePrimea (25 μm, <EW1100). MEAs had an active area of 500 cm^2 with platinum loading of 0.40/0.40 mg_{pt}/cm^2 operated in a short stack (error bars represent cell voltage standard deviations for four nominally identical MEAs of each kind in the stack). Voltages were an average value between 10 and 15 min holding time at each current density. Reproduced from Ref. [3] with permission from John Wiley & Sons, Copyright 2003

7.4 Thermal Management

7.4.1 Heat Generation

The total energy released for water formation from hydrogen and oxygen, assuming water in the liquid form, is the high heat value $\Delta H_{HHV} = -286$ kJ/mol, corresponding to a thermal neutral voltage of 1.48 V. When water is released as vapor, the low heat value $\Delta H_{HHV} = -241$ kJ/mol, corresponding to a thermal neutral voltage of 1.25 V, is used instead. The respective electrical efficiency is therefore

$$\varepsilon_{HHV} = \frac{V_{cell}}{1.48} \tag{7.21}$$

$$\varepsilon_{LHV} = \frac{V_{cell}}{1.25} \tag{7.22}$$

For the cooling purpose it is always assumed that water is released as vapor, i.e., the condensation heat of the product water is not included. The cooling duty needs handle the amount of heat based on the low heat value.

A practical fuel cell system requires auxiliary components such as pumps, fans, instrumentation, control panel, and DC/AC voltage regulator that consume power. This parasitic power often counts for 10–12% of the rated power. As a result the fuel cell needs to deliver a gross power:

$$P_{gross} = P_{rated} + P_{parasitic} \approx 1.12 \times P_{rated} \tag{7.23}$$

In the calculation of total amount of heat generated the gross power (P_{gross}) should be used instead of the rated power:

$$Q_{LHV} = \left(\frac{1}{\varepsilon_{LHV}} - 1\right) P_{gross} = \left(\frac{1.25}{V_{cell}} - 1\right) P_{gross} \tag{7.24}$$

For a fuel cell stack of 80 kW rated power operating at a single cell voltage of 0.67 V, the heat generation rate is

$$Q_{LHV} = \left(\frac{1.25}{0.67} - 1\right) \times 80 \times 1.12 = 77.6\,\text{kW} \tag{7.25}$$

In consideration of a cooling system, a ratio of the total heat generation, Q, and the difference between the stack coolant exit temperature and the ambient temperature, ΔT, is often defined as a measure of the radiator size, called the *radiator constant*:

$$Q/\Delta T = m \times C_p \tag{7.26}$$

where m is the mass flow rate in kg/s and C_p is the specific heat capacity in kJ/kg K of the coolant. For an automotive fuel cell stack of rated power of 80 kW or a gross power 90 kW operating at a cell voltage of 0.67 V at the stack coolant exit temperature of 80 °C and ambient temperature of 40 °C, the radiator constant for the fuel cell stack is

$$\frac{Q}{\Delta T} = \frac{77.6\,\text{kW}}{80 - 40} = 1.94 \tag{7.27}$$

If the target value of the radiator constant is 1.45 instead of 1.94, this is only achievable at a fuel cell operating temperature higher than 95 °C at a fixed cell voltage of 0.67 V.

7.4.2 Cooling Duty and Radiator Area

The total heat generated in a fuel cell stack should be rejected in order to maintain the operating fuel cell temperature. Due to the low temperature of fuel cell operation, the fraction of heat removal by radiation and exhaust is small. A major part of the total heat, *ca.* 70% of the Q_{LHV}, must be rejected through the cooling loop, called the *cooling duty* Q_{duty}.

Here we estimate the cooling duty of the above-mentioned fuel cell stack as a comparison to a diesel engine. Figure 7.13 gives a comparison of the heat balance for a diesel engine and a fuel cell stack. For a 90 kW gross power diesel engine at an energy efficiency of, say, 30%, for the mechanical work output, the total amount of waste heat generated, Q_{waste}, is:

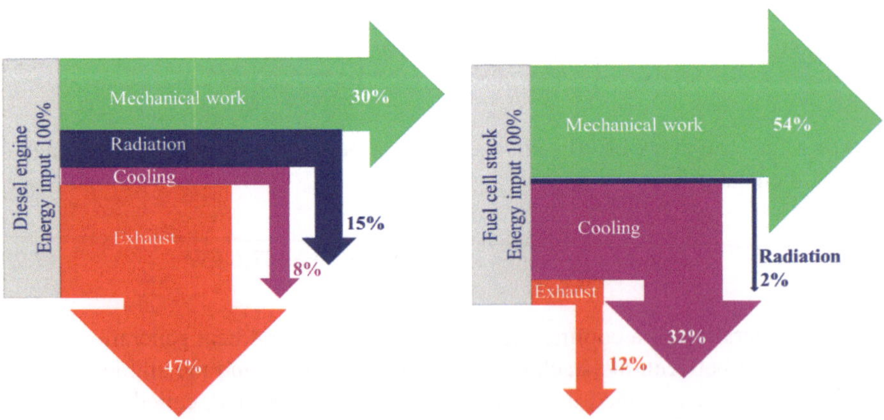

Fig. 7.13 Energy balance of (left) a diesel engine and (right) a fuel cell stack

$$Q_{\text{waste}} = P_{\text{total}} - P_{\text{gross}} = \frac{90}{0.3} - 90 = 300 - 90 = 210\text{kW} \quad (7.28)$$

It is assumed that a major part (say, 186 kW, i.e., 186/300 = 62%) of the heat is dissipated by exhaust and radiation and only 24 kW (i.e., 24/300 = 8%) is rejected through the cooling loop. This assumption is generally valid due to the high operating temperature of diesel engines. In addition the coolant temperature can be as high as 110 °C and therefore $\Delta T \approx 70$ °C when the atmosphere temperature is 40 °C during summer.

For a fuel cell stack operating at 80 °C, a high electrical efficiency of, e.g., 54% for mechanical work output, means that the total amount of waste heat is:

$$Q_{\text{waste}} = P_{\text{total}} - P_{\text{gross}} = \frac{90}{0.54} - 90 = 167 - 90 = 77\text{kW} \quad (7.29)$$

which is much less than 210 kW in the case of the diesel engine, but a major part of this heat, say, 53 kW, i.e., 53/167 = 32% of the total, needs to be rejected through the cooling loop, while only 14% of the total heat (24/167 = 14%) through the exhaust and radiation due to the much lower operating temperature of PEMFC. This is to say, the cooling duty for the fuel cell stack (i.e., 53 kW) is more than doubled in relation to that (24 kW) of a diesel engine. For a fuel cell stack operating at 80 °C the temperature difference ΔT of the coolant is only half of that for the motor engine. As a result the front area of the radiator for a fuel cell power unit should be at least four times larger than the one used in motor vehicles.

7.4.3 Air Cooling Versus Water Cooling

The cooling strategy depends greatly on the size of the fuel cell or stack. For small fuel cell stacks below about 100 W, sufficient cooling can be achieved via the evaporation of the product water and natural airflow. An open stack structure plus high cathode airflow, sometimes using a small fan, is also optional with a limited (a few percent) parasitic loss of the system power. Larger fuel cell systems require forced cooling using either air or water as coolant.

The specific heat capacity of water is 4200 J/kg K while that of dry air is 993 J/kg K, i.e., 4.2 times smaller. In addition, the density of water is 1000 kg/m^3 and that of air is about 800 times lighter. Roughly speaking, water has a mass-based cooling effect about 3000 times higher than air. For humid air, however, the difference will be smaller. Equation (7.26) tells that the airflow rate for the cooling purpose should be significantly higher than the cathode stoichiometry in order to maintain the water balance. As a result, air cooling is practically adopted for small stacks of a few kW size, with cooling channels separate from the cathode channels. Water cooling is more popular for large stacks. Advantages of water cooling include the compact design of small channels and also easy heat recovering for, e.g., stationary uses.

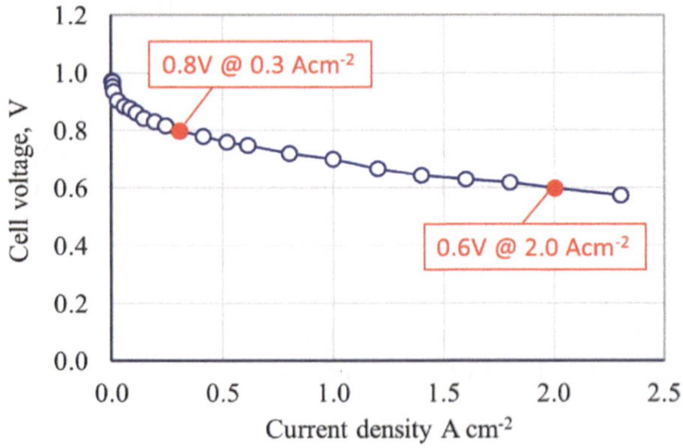

Fig. 7.14 Illustrative fuel cell *i-V* performance with an operation point of 0.8 V@ 0.3 A/cm^2 or 0.6V @ 2 A/cm^2 at 80°C and 2 atm

7.4.4 Operating Points and Stack Calculation

In this section an example of fuel cell stack calculation is given by choosing two operating points along a single cell polarization curve which is shown in Fig. 7.14. The electrodes have a platinum loading of 0.4 mgPt/cm^2 for cathode and 0.05 mg/cm^2 for anode. The cells operate with hydrogen ($\lambda_{H_2} = 1.2$) and air ($\lambda_{air} = 2$) at 80 °C and 2 atm. As comparison, two operation points are chosen at either 0.8 V@ 0.3 A/cm^2 or 0.6V @ 2 A/cm^2. Table 7.2 illustrates the calculation in the differences in construction and operation of a 50 kW stack having a voltage output of 336 V based on two operation points.

7.5 Durability Issues and Mitigation

7.5.1 Status and Target of Durability

The performance discussed so far is the power output of a fuel cell from the *beginning of life* (BoL). All fuel cells, more specifically, the components of fuel cells experience a gradual decline in the performance during operation. At the beginning the performance loss is so small that it is within an acceptable extent of the overall power range. As the degradation of fuel cell components continues, a point will be eventually reached when the required power cannot be supplied any longer. This point is called the *end of life* (EoL) of the fuel cell. During the *lifetime* between the EoL and BoL various modes of failures may occur which can be either temporary or permanent, recoverable or catastrophic.

In engineering terms, the ability of a fuel cell to recover its power loss during continuous operation is its *stability*, which concerns reversible changes of materials and components typically due to operational issues such as water management and CO adsorption. As long as the performance loss is within an acceptable level, a fuel

Table 7.2 System requirement calculation based on two scenarios

Option 1: Fuel cell operates at a rated point of 0.3 A/cm^2 @ 0.8 V	Option 2: Fuel cell operates at a rated point of 2.0 A/cm^2 @ 0.6 V
1. Cell number and size (a) All cells in the stack are connected in series. From the rated cell voltage (0.8 V), the number of cells (N_{cell}) needed to achieve the required stack voltage is $N_{cell} = 336$ V/ 0.8 V/cell = 420 cells (b) At the required voltage, the total current through each cell is $I_{cell} = P_e/V = 50{,}000$ W/ 336 V = 148.8 A (c) The active electrode area of each cell can be calculated from the rated current density (0.3 A/cm^2), and the electrode area is $A_{ele} = 148.8$ A / 0.3 A/cm^2 = 496.0 cm^2 (circa 22 cm × 22 cm in square)	**1. Cell number and size** (a) All cells in the stack are connected in series. From the rated voltage (0.6 V), the number of cells (N_{cell}) needed to achieve the required stack voltage is $N_{cell} = 336$ V/ 0.6 V/cell = 560 cells (b) At the required voltage, the total current through each cell is $I_{cell} = P_e/V = 50{,}000$ W/ 336 V = 148.8 A (c) The active electrode area of each cell can be calculated from the rated current density (2 A/cm^2), and the total electrode area $A_{ele} = 148.8$ A/ 2 A/cm^2 = 74.4 cm^2 (circa. 8.7 × 8.7 cm in square)
2. Platinum usage and cost assuming Pt price is 30 €/g (d) The total area of cathode or anode $A_{Cathode} = A_{Anode} = 496$ cm^2/cell×420 cells = 208,320 cm^2 Total Pt usage = (0.4+0.05) mg/cm^2×208,320 cm^2/1000 mg/g = 93.75 g The total Pt cost = 30 €/g ×93.75g = 2812.5 €	**2. Platinum usage and cost** assuming Pt price is 30 €/g (d) Tota area of cathode or anode $A_{Cathode} = A_{Anode} = 75$ cm^2/cell×560 cells = 42,000 cm^2 Total Pt usage = (0.4+0.05) mg/cm^2×42,000 cm^2/1000 mg/g =18.90 g The Pt cost =30 €/g×18.90 g = 567 €
3. Hydrogen supply (e) The hydrogen usage of the stack for $\lambda_{H2} = 1.2$ is $\frac{148.8\ C/s \times 1.2}{2\ \text{mol e}^-/\text{mol H}_2 \times 96485\ C/\text{mol e}^-} \times 420\ \text{cells} = 0.39\ \text{mol H}_2/\text{s}$ *Note* 148.8 A is for one of 420 cells, and all cells should be fed with H_2	**3. Hydrogen supply** (e) The hydrogen usage of the stack for $\lambda_{H2} = 1.2$ is $\frac{148.8\ C/s \times 1.2}{2\ \text{mol e}^-/\text{mol H}_2 \times 96485\ C/\text{mol e}^-} \times 560\ \text{cells} = 0.52\ \text{mol H}_2/\text{s}$ *Note* 148.8 A is for one of 560 cells, and all cells should be fed with H_2

(continued)

Table 7.2 (continued)

Option 1: Fuel cell operates at a rated point of 0.3 A/cm^2 @ 0.8 V	Option 2: Fuel cell operates at a rated point of 2.0 A/cm^2 @ 0.6 V
4. Air flow and humidification (f) The oxygen flow with a stoichiometry of $\lambda_{O2} = 2$ is $\frac{148.8\ C/s \times 2.0}{4\ mol/mol\ H_2 \times 96485\ C/mol} \times 420\ cells = 0.32\ mol\ O_2/s.$ The air flow rate is therefore $0.32 \frac{mol\ O_2}{s} \times \frac{1}{0.209} = 1.55\ mol\ Air/s$ (g) At 80°C the saturation water pressure is 0.47 atm. To reach a minimum inlet humidity of 90%, the water partial pressure in the inlet air should be higher than 0.43 atm. At a total inlet pressure of 2 atm, the water to air molar ratio is $\psi = 0.43/(2\text{-}0.43) = 0.43/1.57 = 0.27$. The amount of water needed to add into the inlet air flow is $1.55\ mol\ Air/s \times \frac{0.43\ mol\ H_2O}{1.57\ mol\ Air} = 0.42\ mol\ H_2O/s.$ This corresponds to 27.3 kg H_2O/h. *Note* This is the amount of water under a total pressure of 2 atm. If the system pressure is ambient, this number will be $1.55\ mol\ Air/s \times \frac{0.43\ mol\ H_2O}{0.57\ mol\ Air} = 1.53\ mol\ H_2O/s.$ This corresponds to 74.5 kg H_2O/h.	**4. Air flow and humidification** (f) The oxygen flow with a stoichiometry of $\lambda_{O2} = 2$ is $\frac{148.8\ C/s \times 2.0}{4\ mol/mol\ H_2 \times 96485\ C/mol} \times 560\ cells = 0.43\ mol\ O_2/s.$ The air flow rate is therefore $0.43 \frac{mol\ O_2}{s} \times \frac{1}{0.209} = 2.07\ mol\ Air/s$ (g) At 80°C the saturation water pressure is 0.47 atm. To reach a minimum inlet humidity of 90%, the water partial pressure in the inlet air should be higher than 0.43 atm. At a total inlet pressure of 2 atm, the water to air molar ratio is $\psi = 0.43/(2\text{-}0.43) = 0.43/1.57 = 0.27$. The amount of water needed to add into the inlet air flow is $2.07\ mol\ Air/s \times \frac{0.43\ mol\ H_2O}{1.57\ mol\ Air} = 0.57\ mol\ H_2O/s.$ This corresponds to 36.7 kg H_2O/h. *Note* This is the amount of water under a total pressure of 2 atm. If the system pressure is ambient, this number will be $1.55\ mol\ Air/s \times \frac{0.43\ mol\ H_2O}{0.57\ mol\ Air} = 1.53\ mol\ H_2O/s$. This corresponds to 99.3 kg H_2O/h.
5. Water production (h) The rate of water formation in the stack is $\frac{148.8\ C/s}{2\ mol\ e^-/mol\ H_2 \times 96485\ C/mol\ e^-} \times 420\ cells = 0.32\ mol\ H_2O/s.$ This corresponds to 21 kg/h, which is insufficient even all collected for the humidification purpose.	**5. Water production** (h) The rate of water formation in the stack $\frac{148.8\ C/s}{2\ mol\ e^-/mol\ H_2 \times 96485\ C/mol\ e^-} \times 560\ cells = 0.43\ mol\ H_2O/s.$ This corresponds to 28 kg/h, which is insufficient even all collected for the humidification purpose.

(continued)

Table 7.2 (continued)

Option 1: Fuel cell operates at a rated point of 0.3 A/cm^2 @ 0.8 V	Option 2: Fuel cell operates at a rated point of 2.0 A/cm^2 @ 0.6 V
6. Heat generation and cooling (i) At the rated power, the electric efficiency based on the low heat value (1.25 V) is $\varepsilon_{LHV} = \frac{V_e}{1.25} = \frac{0.8}{1.25} = 64\%$ The LHV is used in calculation of the cooling duty assuming that the product water is released, *i.e.*, condensation heat is not accounted in cooling. When 50 kW power is delivered, the amount of heat generated by the stack is then $50\ \text{kW} \times \left(\frac{1.25}{0.8} - 1\right) = 28.1\ \text{kW}$ (j) Assuming that 30% of the generated heat is removed by radiation and convention while 70% is rejected through the cooling loop, the cooling duty is 28.13 kJ/s ×70% = 19.69 kJ/s Water is used as the coolant with inlet temperature of 25°C and outlet temperature of, say, 60°C. The specific heat capacity of water is 4.2 kJ/kg K. The circulating rate of the cooling water is $m_{H_2O} = \frac{19.6\ \text{kJ/s}}{4.2\frac{\text{kJ}}{\text{kg}\cdot\text{K}} \times (333-298)K} = 0.13\ \text{kg/s}$ This corresponds to 482.14 kg/h *Note* The electric efficiency of the fuel cell stack should always be based on the HHV: $\varepsilon_{HHV} = \frac{V_{cell}}{1.48} = \frac{0.8}{1.48} = 54\%$ If water is supposed to be condensed before being released the cooling duty should be based on the HHV. The cooling duty becomes $50\ \text{kW} \times \left(\frac{1.48}{0.8} - 1\right) \times 70\% = 29.75\ \text{kW}$ And the cooling water flow is $m_{H_2O} = \frac{29.75\ \text{kJ/s}}{4.2\frac{\text{kJ}}{\text{kg}\cdot\text{K}} \times (333-298)\text{K}} = 0.20\ \text{kg/s}$ or 728.57 kg/h	**6. Heat generation and cooling** (i) At the rated power, the electric efficiency based on the low heat value (1.25 V) is $\varepsilon_{LHV} = \frac{V_e}{1.25} = \frac{0.6}{1.25} = 48\%$ The LHV is used in calculation of the cooling duty assuming that the product water is released, *i.e.*, condensation heat is not accounted in cooling. When 50 kW power is delivered, the amount of heat generated by the stack is then $50\ \text{kW} \times \left(\frac{1.25}{0.6} - 1\right) = 54.17\ \text{kW}$ (j) Assuming that 30% of the generated heat is removed by radiation and convention while 70% is rejected through the cooling loop, the cooling duty is 54.17 kJ/s ×70% = 37.92 kJ/s Water is used as the coolant with inlet temperature of 25°C and outlet temperature of, say, 60°C. The specific heat capacity of water is 4.2 kJ/kg K. The circulating rate of the cooling water is $m_{H_2O} = \frac{37.8\text{kJ/s}}{4.2\frac{\text{kJ}}{\text{kg}\cdot\text{K}} \times (333-298)K} = 0.26\ \text{kg/s}$ This corresponds to 928.57 kg/h *Note* The electric efficiency of the fuel cell stack should always be based on the HHV: $\varepsilon_{HHV} = \frac{V_{cell}}{1.48} = \frac{0.6}{1.48} = 41\%$ If water is supposed to be condensed before being released the cooling duty should be based on the HHV. The cooling duty becomes $50\ \text{kW} \times \left(\frac{1.48}{0.6} - 1\right) \times 70\% = 51.33\ \text{kW}$ And the cooling water flow is $m_{H_2O} = \frac{51.33\ \text{kJ/s}}{4.2\frac{\text{kJ}}{\text{kg}\cdot\text{K}} \times (333-298)\text{K}} = 0.35\ \text{kg/s}$ or 1257.14 kg/h
7. Summary and comments Operating at high voltage and low current (low power density 0.24 W/cm^2) the fuel cell stack: – has high electric efficiency (64% LHV or 54 HHV) – has large size (496 cm^2) but a small number (420) of cells – has a large total electrode area (208,320 cm^2), using more platinum (93.75 g) as well as all other construction materials consumes less hydrogen (0.39 mol/s) – has lower air flow (1.55 mol/s) and water demand (27 kg/h) for humidification – has low water production rate (21 kg/h) – generates less heat (28 kW) – handles less (482 kg/h) cooling water	**7. Summary and comments** Operating at high current and low voltage (high power density 1.2 W/cm^2) the fuel cell stack: – has low electric efficiency (48% LHV or 41% HHV) – has small size (75 cm^2) but a large number (560) of cells – has a small total electrode area (42,000 cm^2), using less platinum (18.90 g) as well as all other construction materials consumes more hydrogen (0.52 mol/s) – has higher air flow (2.07 mol/s) and water demand (36.7 kg/h) for humidification – has high water production rate (28 kg/h) – generates more heat (54 kW) – handles more (928 kg/h) cooling water

cell continues to operate until a point is reached when a catastrophic failure occurs and a replacement or a maintenance is needed. The ability of a fuel cell to perform without such a failure is its *reliability*, a measure of which is often defined as the *mean time between failures*. On the other hand, the *durability* of a fuel cell refers to the ability to resist permanent degradation in the performance typically associated with issues of material aging such as polymer degradation, catalyst sintering, carbon and bipolar plate corrosion.

Durability targets have been proposed for commercializing PEMFCs as electrical power sources for automobile as well as for stationary and portable power applications. Table 7.3 is a summary of the 2015 status, 2020 target, and ultimate numbers collected from the Hydrogen and Fuel Cell Technologies Office, DOE.

The durability of automotive fuel cell systems operating with direct hydrogen is measured in terms of service life until 10% performance degradation. Compared to present automotive engines, the durability should be at least 5,000 h including daily drive cycling. Fuel cells for stationary applications require at least 40,000 h of reliable operation. For portable power units in a power range of 5–50/100–200 watts and auxiliary power units of a kilowatter scale the durability target is usually defined as a performance degradation by 20%, though other levels of power degradation may be acceptable depending on applications. In addition, the mean time between failures has been specified including system component failures that render the system inoperable without maintenance. Also included in the table are targets for key PEMFC components, i.e., membranes, electrocatalysts, and membrane electrode assemblies (MEAs) for automobile applications.

A large number of durability investigations have been carried out. Accelerated stress test protocols to evaluate the stability and durability of membranes, and electrocatalysts have been outlined in Chap. 6. The durability decay rate under continuous operation has been reported to be in a range of 2–10 μV/h. Taking an average of 5 μV/h and a starting cell voltage of 0.8 V, the 10 and 20% degradation corresponds to a lifetime of 16,000 and 32,000 h, respectively. Operation parameters, e.g., fuel impurities, temperature, thermal and load cycling particularly start-ups and shutdowns further increase the performance decay. For automotive uses under drive cycles nearly 4000 h of fuel cell stacks/systems have been demonstrated, which is almost the halfway of the ultimate target.

7.5.1.1 Mitigation

Fundamental understanding and mitigation of key materials, e.g., dissolution and sintering of platinum nanoparticles, corrosion of carbon supports, polymer oxidation and membrane thinning have been well addressed, as discussed in Chap. 6.

A considerable part of the cell performance loss is due to the degradation of the catalyst layer. In addition to the metal dissolution and precipitation, the carbon corrosion is known to weaken the interaction of Pt nanoparticles with carbon support and eventually aggravate platinum agglomeration and the particle detachment, resulting in decreases in the electrochemically active surface area of Pt and therefore lowering

Table 7.3 DOE durability targets for PEMFC automotive, stationary, and portable power systems. For automotive applications information of fuel cell components is also included

Characteristics		Units	Status 2015	Targets 2020	Targets Ultimate
CHP systems on natural gas					
•1–25 kW_e:	Degradation with cycling	%/1000 h	<2	0.1	–
	Operating lifetime (20% degradation)	h	>12,000	60,000	–
•100 kW_e – 3 MW:	Operating lifetime (10% degradation)	h	>40,000	80,000	–
Portable power systems (5–50/100–200 watts):					
	Durability (20% degradation)	h	1500/2000	–	5000/5000
	Mean time between failures	h	500//500	–	5000/5000
APU on ultra-low sulfur diesel (1 – 10kW_e):					
	Degradation with cycling	%/1000 h	2.6	1.0	–
	Operating lifetime (20% degradation)	h	3,000	20,000	–
Backup power systems on direct H_2 (1–10kWe):					
	Lifetime	year	10	15	–
	Durability (10% degradation on backup power duty cycle)	h	8,000	10,000	–
	Mean time between failure	year	5	5	–
Automotive 80–k$W_{e(net)}$ systems on direct H_2:					
• System & stack	Durability in drive cycle (10% Degradation)	h	3,900	5,000	8,000
	Start-up/shutdown durability	cycles	–	5,000	5,000
• Membranes durability	Mechanical (cycling until 15 mA/cm^2 H_2 crossover)	cycles	>20,000	>20,000	–
	Chemical (time until 15 mA/cm^2 H_2 crossover)	h	>500	>500	–
• Electrocatalyst durability	Loss of initial mass activity	% mass activity loss	66	<40	–
	Loss in performance at 0.8 A/cm^2	mV	<30	<30	–
	Support stability	% mass activity loss	41	<40	–
	Loss in performance at 1.5 A/cm^2	mV	65	<30	–
• Membrane electrode assemblies (MEAs):	Durability with cycling	h	2,500	5,000	–
	Start–up/shutdown durability	cycles	–	5,000	–

the fuel cell performance. A long variety of approaches to mitigate the electrocatalyst degradation have been proposed to enhance the lifetime and increase the stability of the catalyst support. Thermal treatment of carbon black-based supports increases the graphitization extent to reduce the corrosion kinetics and improve the metal nanoparticle dispersion. New structures of carbon such as carbon nanotubes and graphene have shown greater stability than normal carbon blacks as support materials. Other support materials than carbon have also been explored, including metal oxides (e.g., indium tin oxides, doped titanium oxides, zirconium oxides, etc.), carbides, and nitrides. Few of these materials have been evaluated in terms of long-term stability and durability in fuel cells.

Of the two major degradation mechanisms of membranes, the mechanical degradation often causes early life failures via formation of perforations, cracks, tears, or pinholes during the MEA fabrication or humidification changes in operation. This can be mitigated by reinforcement of membranes by, e.g., the PTFE fibers. A more critical issue is the chemical degradation of membranes which is primarily due to reactive oxygen species, e.g., hydroxyl radicals (OH^*) generated in situ through both chemical and electrochemical pathways. The source of these species is hydrogen peroxide, a product of reactions between the crossover gases (H_2 and O_2) and/or an intermediate during electrochemical reduction of oxygen. Any measures to reduce the hydrogen crossover through the membrane are beneficial to lessen the chemical degradation but compromised by reduction of the membrane thickness. Use of free radical scavengers, typically cerium oxide nanoparticles embedded in either electrodes or membranes, has been suggested. Cerium oxide is able to scavenge the radical oxidative species by lattice oxygen vacancy sites:

$$OH^* + Ce^{3+} + H^+ \rightarrow Ce^{4+} + H_2O \tag{7.30}$$

The depletion of Ce^{3+} can be regenerated in acidic media, particularly with Pt as catalysts

$$2Ce^{4+} + H_2 \rightarrow 2Ce^{3+} + 2H^+ \tag{7.31}$$

$$4Ce^{4+} + 2H_2O \rightarrow 4Ce^{3+} + 4H^+ + O_2 \tag{7.32}$$

Figure 7.15 shows the fluoride emission rate (FER), a measure of the PFSA membrane degradation, determined for MEAs using composite Nafion® membranes containing CeO_2, Pt-supported CeO_2 as well as MnO_2 nanoparticles. The test was conducted with O_2 and H_2 under open circuit voltage at 90 °C and 30% RH for 24 h. It can be seen that the addition of CeO_2 and MnO_2 nanoparticles into Nafion® membranes results in a reduction in the fluoride emission rate by over an order of magnitude. The presence of Pt further promotes the free radical scavenging of CeO_2, and as a result a further reduction of *ca.* 50% has been observed.

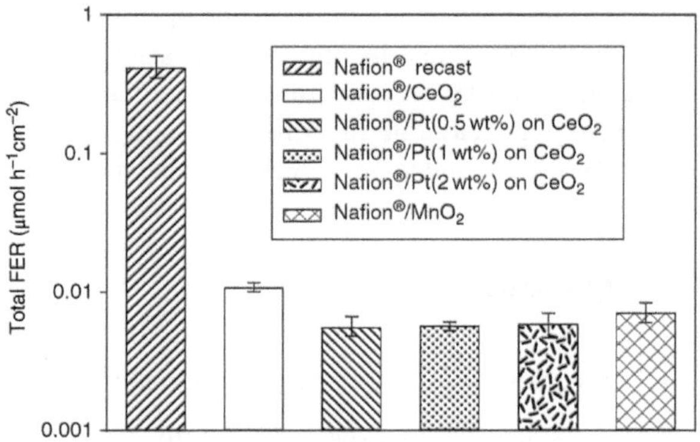

Fig. 7.15 Total fluoride emission rate (FER) measured for composite Nafion® membrane-based MEAs during accelerated durability tests at 90 °C and 30% RH. Reproduced from Ref. [15] with permission from Royal Society of Chemistry, Copyright 2011

7.6 Summary

In this Chapter, we have introduced and discussed the following topics:

- Typical PEMFC performance reaches about 0.7 V at 1.0 A/cm^2 at 65 °C under ambient pressure of humidified H_2 and air with a Pt loading of 0.4 mgPt/cm^2, i.e., a power density of 0.7 W/cm^2.
- The following set of data exemplify the PEMFC performance one can obtain from the *i-V* curves as shown in Fig. 7.1
  - An area-specific resistance (of the membrane) of 0.03 Ωcm^2 (Nafion 112)
  - A hydrogen crossover corresponds to an internal current of 1.8 mA/cm^2
  - A real ECSA around 60 m^2/g_{Pt} corresponding to a Pt utilization of 70–86%
  - An extracted exchange current density $i^0 = 10^{-9}$ A/cm$^2_{ECSA}$ and a Tafel slope of of 55–60 mV/dec.
- PEMFC performs well with pure hydrogen. When reformat hydrogen is used, CO poisons the Pt catalyst by strong adsorption. The CO coverage is temperature-dependent and can be expressed as a function of log[CO]/[H_2] indicating the competition between CO and H_2 to the active site. Pt alloys, e.g., PtRu/C and PtMo/C, improve the CO tolerance.
- Similar effect was observed for electro-oxidation of, e.g., methanol where CO is the intermediate species of the electrode reaction and can be removed by OH from the water dissociation. Bifunctional Pt alloy catalysts have been used as the anode catalysts.
- Water management is an issue for PEMFC operation, often addressed by either external or self-humidification. The issue is mitigated by use of modified and/or thinner membranes.
- Another important issue of PEMFC operation is the thermal management of heat generation and rejection. Intensive cooling is needed using either air or water as coolant.

- An example is given to illustrate the operating point (power density at low or high voltage/current) and its impact to the stack design and operation.
- Durability is of most concern in commercialization of the technology which shows typically a decay rate of 2–10 μV/h under steady-state operation. A lifetime of nearly 4000 h under drive cycles has been demonstrated.

7.7 Questions

7.1. Explain the following terms and materials:

Hydrogen crossover and effective current density
Activation, ohmic, and mass transport polarization losses
Cell voltage gain
CO coverage *via* the linearly bonded and bridge-bonded CO to Pt surface
Pt tolerance toward CO poisoning
Humidification, relative humidity, dew point, air stoichiometry and fuel stoichiometry
Thermal management, radiator constant, and cooling duty
Rated power, gross power, total waste heat
Stack and system operation requirement calculation
Stability, reliability, BoL, EoL and and mean time between failures.

7.2. Briefly comment on the state-of-the-art performance of the PEMFC technology.

7.3. A 5 kW PEMFC stack was constructed with two electrodes each containing 0.5 mgPt/cm^2. The cell has a single cell performance of 0.6 W/cm^2. Calculate the total amount of platinum that is used in this stack. (Answer: 4.2 g).

7.4. A 25 cm^2 fuel cell operates with an air flow rate of 500 scc/min. A constant current of 10 A is obtained. What is the air stoichiometry (λ) of the cell? (Answer: $\lambda = 3$).

7.5. A 50 cm^2 fuel cell operates with an air flow rate of 550 scc/min and a hydrogen flow rate of 300 mL/min. At 80 °C the fuel cell exhibits a limiting current of 35 A. Justify that it is the cathode, not the anode, that is experiencing a starvation at the limiting current.

7.6. It has been proposed that the CO poisoning on the anode can be mitigated by adding a small amount of oxygen into the fuel stream, the so-called air-bleeding. Comment on this method.

7.7. The calculation of the energy efficiency of a PEMFC is always based on the high heat value (ΔH_{HHV}), while in consideration of the cooling duty of a stack, the low heat value is used. Explain why.

References

1. Kocha SS (2003) In: Vielstich W, Gasteiger HA, Lamm A (eds) Handbook of fuel cells—fundamentals, technology and applications, vol 3. Wiley, pp 538–565
2. Bernardi DM, Verbrugge MW (1992) A mathematical model of the solid-polymer-electrolyte fuel cell. J Electrochem Soc 139(9):2477–2491
3. Gasteiger HA, Gu W, Makharia R, Sompalli B (2003) In: Vielstich W, Gasteiger HA, Lamm A (eds) Handbook of fuel cells—fundamentals, technology and applications, vol 3. Wiley, pp 593–610
4. Igarashi H, Fujino T, Watanabe M (1995) Hydrogen electrooxidation on platinum catalysts in the presence of trace carbon-monoxide. J Electroanal Chem 391(1–2):119–123
5. Oetjen HF, Schmidt VM, Stimming U, Trila F (1996) Performance data of a proton exchange membrane fuel cell using H_2/CO as fuel gas. J Electrochem Soc 143(12):3838–3842
6. Li QF, He RH, Gao JA, Jensen JO, Bjerrum NJ (2003) The CO poisoning effect in PEMFCs operational at temperatures up to 200 °C. J Electrochem Soc 150(12):A1599–A1605
7. Niedrach LW, McKee DW, Paynter J, Danzig IF (1967) Electrocatalysts for hydrogen/carbon monoxide fuel cell anodes. Electrochem Technol 5:318
8. Ehteshami SMM, Jia QY, Halder A, Chan SH, Mukerjee S (2013) The role of electronic properties of Pt and Pt alloys for enhanced reformate electro-oxidation in polymer electrolyte membrane fuel cells. Electrochim Acta 107:155–163
9. Gottesfeld S, Pafford J (1988) A new approach to the problem of carbon-monoxide poisoning in fuel-cells operating at low-temperatures. J Electrochem Soc 135(10):2651–2652
10. Rossmeisl J, Ferrin P, Tritsaris GA, Nilekar AU, Koh S, Bae SE, Brankovic SR, Strasser P, Mavrikakis M (2012) Bifunctional anode catalysts for direct methanol fuel cells. Energy Environ Sci 5(8):8335–8342
11. Cheng Y, Jiang SP (2013) Highly effective and CO-tolerant PtRu electrocatalysts supported on poly(ethyleneimine) functionalized carbon nanotubes for direct methanol fuel cells. Electrochim Acta 99:124–132
12. Engl T, Gubler L, Schmidt TJ (2016) Catalyst and catalyst-layers in HT-PEMFCs. In: Li Q, Aili D, Hjuler HA, Jensen JO (eds) High temperature polymer electrolyte membrane fuel cells—approaches, status and perspectives. Springer, New York, pp 297–313
13. Martin S, Garcia-Ybarra PL, Castillo JL (2017) Long-term operation of a proton exchange membrane fuel cell without external humidification. Appl Energy 205:1012–1020
14. Dicks AL, Rand DAJ (2018) Fuel Cell Systems Explained. 3rd Edition, Wiley, pp. 69–133
15. Trogadas P, Parrondo J, Ramani V (2011) Platinum supported on CeO_2 effectively scavenges free radicals within the electrolyte of an operating fuel cell. Chem Commun 47(41):11549–11551

Chapter 8
High-Temperature Polymer Electrolyte Membrane Fuel Cells

8.1 Why and How High Temperatures

Polymer electrolyte membrane fuel cells (PEMFCs) discussed in last three chapters are based on perfluorosulfonic acid (PFSA) membranes and operate at around 80 °C. A careful water management is mandatory by humidification of the cathode air to maintain the high proton conductivity. A PEMFC operating at above 100 °C is desirable but facing challenges from both materials science and technological points of view.

First of all, elevated temperatures tend to stress polymer materials in terms of thermal, chemical, and mechanical stabilities. Secondly, the proton conductivity usually involves a vehicular mechanism with help of water molecules, which are also an inevitable product of any fuel cell systems. Above 100 °C, water evaporates under ambient pressure. An increase of operating temperatures from 80 to 150 °C, for example, will lead to an increase of the saturated water vapor pressure from 47 to 500 kPa, resulting in great complexity of the system construction if water retention is an issue for the membrane conductivity. On the other hand, an operational temperature above the boiling point of water makes fuel cells a single phase system, i.e., only water vapor to handle.

Cooling of PEMFC is important for maintaining the fuel cell temperature within an operational range. The driving force of the heat transfer is the temperature difference between fuel cells and the surrounding. A higher working temperature of the fuel cell enables an efficient and compact cooling system. A temperature increase from 80 to 160 °C may mean 3–4 times reduction in the front area of radiators, which is also a key issue of the power system especially for automobile applications.

More importantly, platinum-based catalysts are very sensitive to traces of reactant impurities, e.g., carbon monoxide and sulfur-containing species. The poisoning effect is temperature dependent. The CO tolerance is dramatically enhanced from 10 to 20 ppm at 80 °C to over 100,000 ppm at, e.g., 160 °C. This high CO tolerance makes it possible for a fuel cell to use hydrogen directly from a methanol reformer with no

S. P. Jiang and Q. Li, *Introduction to Fuel Cells*,
https://doi.org/10.1007/978-981-10-7626-8_8

additional processors such as selective oxidation reactor or the membrane separator for CO cleanup.

Also in connection to the elevated temperature, the waste heat is of higher value. When a fuel cell operates at an energy efficiency of about 50%, there is about half of the energy of the fuel is released as heat. Making better use of the heat is an issue for a power system. A high-temperature PEMFC may deliver heat at 150 °C which can be used as steam, e.g., to evaporate alcoholic fuels for an onboard reformer. A fuel cell operating at temperatures slightly higher than 200 °C may enable internal methanol reforming or integration of fuel cell stack with reformer.

Complete integration of a fuel cell stack with a methanol reformer is desirable in several ways. The reforming is endothermic, while the fuel cell stack releases about three times more heat than the reformer needs. Integration is also expected to lead to simple and compact construction of the system and therefore fast load and thermal response. Use of the fuel cell waste heat for reforming is, however, only possible when the temperatures of the reformer and fuel cell stack match. The methanol steam reforming requires a temperature over 200 °C in order to reach sufficient kinetics and hydrogen generation rate. To develop such an integrated fuel cell and reformer power unit, the operation temperature of fuel cells needs to be above 220 °C [1].

In addition, fast kinetics for both electrode reactions is also expected at elevated temperatures, which is, however, not realized when phosphoric acid as electrolyte is involved as it strongly adsorbs on the platinum surface. However, non-precious metal (NPM) catalysts do not have the phosphoric acid adsorption problem and offer an opportunity to replace precious Pt-based electrocatalysts for PEMFCs at elevated temperatures.

In brief, high-temperature (HT-) PEMFC is a technology featured by simplified construction and operation with possible integration with, e.g., methanol reformers. The most successful system is phosphoric-acid-doped polybenzimidazole (PA/PBI) membranes which was first reported in 1995 [2] and has motivated extensive research activities with great progress in materials, components, stacks, and systems. This chapter is devoted to detailed discussion of the technology. Alternative membrane materials including non-PBI polymers and inorganic proton conductors are also discussed.

8.2 Polybenzimidazoles and Their Interactions with Phosphoric Acids

8.2.1 Synthesis and Molecule Weight Determination

PBI is a large family of aromatic heterocylic polymers containing benzimidazole units. The commercial product under the trademark Celazole® is poly 2,2´-*m*-(phenylene)-5,5´-bibenzimidazole (m-PBI). PBI was first synthesized in 1960s

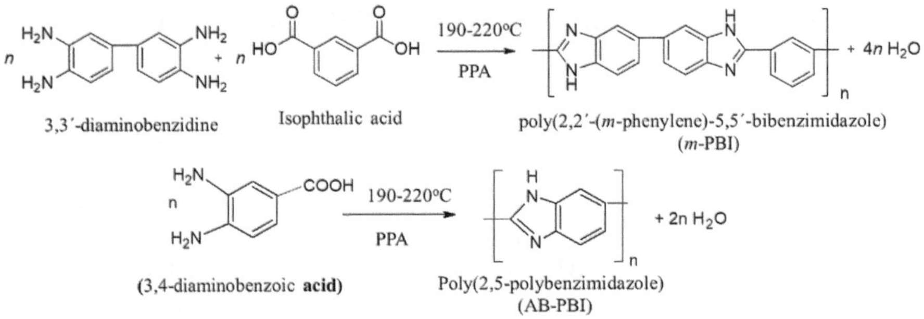

Fig. 8.1 Schematic illustration of synthesis of poly 2,2′-*m*-(phenylene)-5,5′-bibenzimidazole (m-PBI)

using melt condensation polymerization from aromatic tetraamines and aromatic dicarboxylic acid derivatives. On a laboratory scale, high molecular polymers can be prepared by homogeneous solution polymerization from 3,3′-diaminobenzidine and isophthalic acid. The most simple structure of polybenzimidazoles is poly(2,5-benzimidazole) (AB-PBI) which can be synthesized from a single and cheap monomer 3,4-diaminobenzoic acid containing both the diamine and the carbonyl group. Polyphosphoric acid (PPA) is typically used as the polycondensation solvent, which allows for dissolution of the monomers and growth of the polymer chain and removal of water formed from the polycondensation process (Fig. 8.1).

As an amorphous thermoplastic polymer, the aromatic nuclei of PBI provide the polymer high thermal stability (glass transition temperature, T_g = 425–436 °C), excellent chemical resistance, retention of stiffness and toughness. The synthesized polymer is usually characterized in terms of molecular weight by measurement of the inherent viscosity (η_{IV}, in dL/g) of a polymer solution, e.g., 0.5 g of the polymer in 100 ml 96 wt% H_2SO_4 at room temperature. The viscosity of polymer solutions can be measured with an Ubbelohde viscometer and is related to the weight-averaged molecular weight (MW, in g/mol) of the polymer by the Mark–Houwink–Sakurada equation:

$$\eta_{IV} = 1.94 \times 10^{-4} \times MW^{0.79} \tag{8.1}$$

The commercially available polymer (Celazole®) has a relatively low molecular weight in order to avoid the polymer insolubility, with M_W = 23,000–37,000 g/mol or an inherent viscosity of η_{IV} = 0.55–0.8 dL/g. Using the polyphosphoric acid method of synthesis at a solid concentration of 3–5% and temperature of 195–220 °C, the synthesized polymers may exhibit an inherent viscosity up to 1.5 dL/g as measured in sulfuric acid, which corresponds to a weight-averaged molecular weight of over 100,000 g/mol.

8.2.2 Membrane Casting from DMAc Solutions and Acid Doping

PBI dissolves in a few highly polar aprotic solvents, e.g., *N*,*N*-dimethylacetamide (DMAc), *N*,*N*-dimethylformamide (DMF), *N*-methyl-2-pyrrolidone (NMP), or dimethylsulfoxide (DMSO). It is also soluble in concentrated sulfuric acid or alkali metal hydroxide or alkoxide solutions in light alcohols. Traditionally, DMAc is a preferred organic solvent for PBI as the viscosity of the obtained solution is suitable for spinning of PBI fibers. A small amount of LiCl is often added to promote the polymer dissolution and stabilize the polymer solution from phasing out during the prolonged storage.

PBI membranes can be cast from DMAc solutions of different concentrations. The majority of the solvent can be slowly evaporated in a ventilated oven at temperatures from 60 to 120 °C. The membranes are then peeled off from the substrate, often a glass plate, and washed with hot water in order to remove the stabilizer (LiCl) if any. Strong interaction between PBI and DMAc molecules makes it necessary to remove the traces of the solvent at temperatures of up to 190 °C or under vacuum.

PBI membranes have a high affinity for moisture. By immersing a dry PBI membrane in distilled water at room temperature, up to 15–19 wt% of water can be absorbed, corresponding to about 3 water molecules per repeat unit of PBI. Chemically, PBI is a basic polymer ($pK_a = 5.23$ in its protonated form) and can readily react with phosphoric acid, which is thermally stable and highly conductive under anhydrous conditions at elevated temperatures. The procedure is schematically represented in Fig. 8.2.

During the acid doping, the weight gain of the membrane is due to both acid and water uptake. To distinguish between the contributions from water and acid,

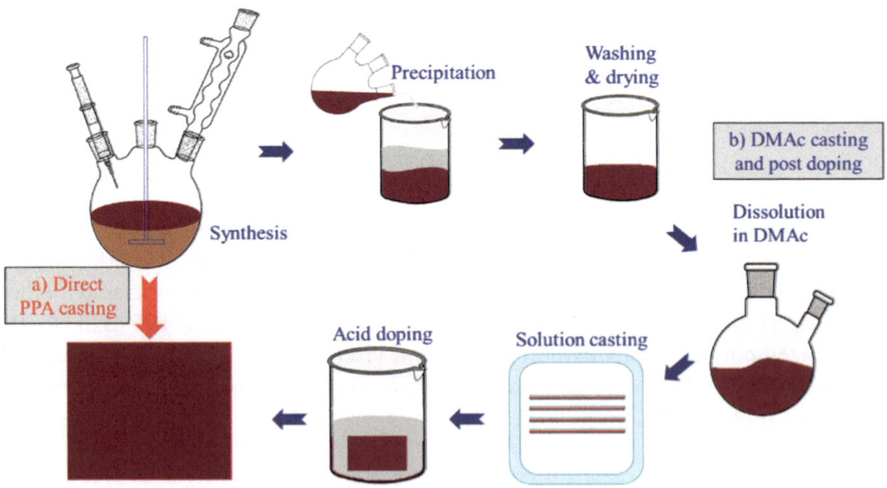

Fig. 8.2 Schematic representation of membrane preparation by **a** direct casting and **b** DMAc casting and post-doping

the membrane can be dried to remove the water or titrated with, e.g., NaOH, to determine the acid content. If the weight fraction of the polymer is unknown, the membranes can be dedoped by immersion in a large amount of water followed by drying *in vacuo*. The *acid uptake* (AU) of a membrane is often given in percent on a dry polymer basis, i.e., the ratio between the phosphoric acid content (W_{PA}) and the PBI content (W_{PBI}) of the membrane:

$$\mathrm{AU}(\%) = \frac{W_{PA}}{W_{PBI}} \times 100 \tag{8.2}$$

Sometimes the amount of acid is also reported as the acid content (AC) on the doped membrane:

$$\mathrm{AC}(\%) = \frac{W_{PA}}{W_{PBI} + W_{PA}} \times 100 \tag{8.3}$$

The term *acid doping level* (ADL) is widely used and is defined as the molecule number of phosphoric acid per polymer repeat unit

$$\mathrm{ADL} = \frac{W_{PA}/M_{PA}}{W_{PBI}/M_{PBI}} \times 100 \tag{8.4}$$

where M_{PA} is the molar mass of PA and M_{PBI} is the molar mass of the polymer repeat unit. It should be of note that the term ADL allows for a direct comparison of the phosphoric acid content of polymers of the same structure, i.e., with the same polymer repeat unit, otherwise the acid uptake or acid content is preferred to use for polymers of different structures.

8.2.3 Direct Cast Membranes

PBI is soluble in a mixture of trifluoroacetic acid (TFA) and phosphoric acid and has been suggested to cast membranes from the solution containing the desired amount of phosphoric acid. Another method is to utilize polyphosphoric acid as both the polymerization medium and the casting solvent for PBI polymers. After synthesis, no isolation or redissolution of the polymer is needed. Acid imbedded PBI membranes can be cast from the hot crude polymer solution in polyphosphoric acid, the so-called PPA method. Since both PBI and polyphosphoric acid are hygroscopic, moisture from atmosphere with or without humidification control will hydrolyze the polyacid to orthophosphoric acid. A sol to gel transition occurs as the polymer solubility is reduced in phosphoric acid as well as at decreased temperatures. Thus, obtained membranes may have acid doping levels as high as 20–40 mol PA per repeat unit of PBI and therefore high conductivity. For direct casting of membranes from acid solutions, high molecular weight polymers are preferred in order to obtain the improved mechanical strengths.

8.2.4 Membrane Characterization

The thermal stability of PBI has been extensively studied by thermogravimetric analysis (TGA). Pristine PBI shows an onset of thermal decomposition at around 500 °C in air. The observed weight loss is primarily due to the absorbed water, though oxidative degradation seems to start at temperatures above 350 °C. The chemical stability of membranes is of concern due to the oxidative attack of the in situ formed HO^* or HOO^* radicals. The initial oxidation occurs preferentially on the link between the nitrogen-containing heterocyclic and adjacent benzenoid rings. The presence of doping phosphoric acid and its acid–base interaction seems stabilizing the polymer from the radical oxidation.

The strong hydrogen bonding between –N = and –NH–groups is the dominant molecular force of the polymer and therefore good mechanical strength of membranes. The dry PBI membrane has a small elongation at break of around 1–3%, and an intermediate tensile strength of about 60–70 MPa at room temperature. When saturated with water, the elongation and tensile strength increase to about 7–10% and 100–160 MPa, respectively.

The doping with phosphoric acid results in volume swelling. For PBI membranes with an ADL of 10–12, the volume swelling is around 200%, which gives a considerable increase of the hydrogen and oxygen permeability [3]. It also means that the intermolecular van der Waals forces that are responsible for the superior mechanical strength of pristine PBI are strongly reduced. In other words, the phosphoric acid acts as a plasticizer. Further improvement can be achieved by increasing the linear molecular weight of the polymer as well as cross-linking.

The conductivity of phosphoric-acid-doped PBI membranes strongly depends on the acid doping level. At an acid doping level of lower than 2 where all acid molecules are likely involved in the acid–base interaction between the basic sites of benzimidazol (BIm) and phosphoric acid, forming the protonated benzimidazol ($BImH^+$) as the acid form:

$$BIm + H_3PO_4 \rightarrow BImH^+ + H_2PO_4^- \tag{8.5}$$

The membrane exhibits limited conductivity, being around 10^{-7} S/cm at room temperature. The presence of excess acid creates the acid anion chain ($H_2PO_4^- - H^+ \cdots H_2PO_4^-$) as well as the acid and H_2O ($H_2PO_4^- - H^+ \cdots H_2O$) chain depending on the water content. This significantly enhances the conductivity. At an acid doping level of 4–6 and temperatures close to 200 °C, the conductivity reaches about 0.04–0.07 S/cm [4]. The typical acid doping level of post-doped DMAc membranes for fuel cell uses is around 10–12, giving conductivities around 0.05–0.10 S/cm. The PPA membranes have acid doping levels as high as 20–40 and therefore higher

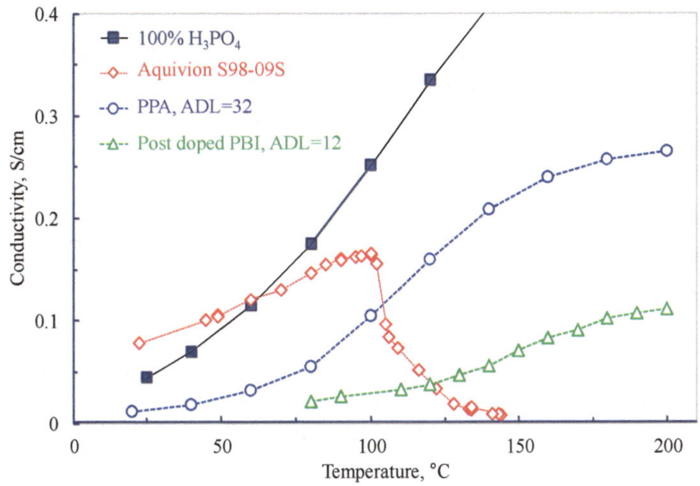

Fig. 8.3 Proton conductivity of 100% phosphoric acid, Aquivion S98-09S and acid-doped PBI membranes. The PPA-cast membrane was with an acid doping level of 32 and dry atmosphere. The DMAc-cast membrane was with an acid doping level of 12 and under dry atmosphere. The Aquivion membrane was under 100% RH below 100 °C and under $p_{H_2O} = 1$ atm above 100 °C

conductivity. For example, a PPA membrane having an acid doping level of 32, corresponding to 91 wt% phosphoric acid and 9 wt% PBI, exhibits a conductivity as high as 0.26 S/cm at 200 °C under dry conditions [5], as seen from Fig. 8.3. This conductivity is approximately half of that for 100% phosphoric acid at the same temperature. As a comparison, the conductivity of PFSA membranes was measured under 100% relative humidity (RH) below 100 °C and under $p_{H_2O} = 1$ atm at above 100 °C. The dehydration of the membrane at above 100 °C leads to decrease in the conductivity.

PFSA membranes are known to have water-mediated proton transport mechanisms and therefore with an electro-osmotic drag coefficient of water of about 2.5–3.0 H_2O per H^+. For phosphoric-acid-doped *m*PBI, on the other hand, the electro-osmotic drag coefficient of water is close to zero supporting the structure diffusion mechanism.

As a glassy polymer, the PBI membrane is dense with close chain packing (density of 1.34 g/cm^3), due to the rigidity of the structure and strong effect of hydrogen bonding. Very low gas permeability of hydrogen and oxygen has been reported for pristine PBI membranes, which is on a level 10^{-12} mol/cm·s·bar at 150 °C [3]. When doped with phosphoric acid, the membrane is swollen and as a result the gas permeability is about 2–3 orders of magnitude higher than for the pristine *m*PBI membrane.

The acid-doped PBI membrane was originally suggested as electrolyte for direct methanol fuel cells. Electro-osmotic drag for methanol was reported to be nearly zero, similar to that of water. Methanol crossover measurements in operating fuel cells showed a crossover rate corresponding to a leak current of 10 mA/cm^2 for vapor phase feeding at 150 °C. Test of direct methanol fuel cells has, however, showed a significant degradation within weeks [6]. The rapid degradation is, at least partly, connected to the chemical reaction between methanol and phosphoric acid, leading to the formation of methyl phosphoric acid ester derivatives with low conductivity and high volatility [7]. Reformate hydrogen instead of direct oxidation of methanol has therefore been used for fueling PBI cells.

8.2.5 PBI Variants

Much research effort has thus been devoted to the design and synthesis of PBI derivatives of varied backbone structures. Comparing with *m*-PBI membranes, *p*-PBI membranes containing the *para* phenylene structure show superior tensile strength and stiffness. The rigid structure of *p*-PBI makes it less soluble. Efforts have been made to improve solubility of the polymer in organic solvents by introducing flexible spacers though at expense of lowered thermo-oxidative stability. These variants of PBI can be synthesized using a variety of diacids with active groups such as pendant amino, carboxyl, sulfonic acid, hydroxyl or *tert*-butyl, as exemplified in Table 8.1. Typical examples are pyridine (PyPBI), ether (OPBI), sulfone (SO_2PBI), hexafluoropropylidene (F_6PBI), dihydroxy ($(OH)_2$PBI), etc., that have demonstrated the capability of improved acid retention and therefore high proton conductivity with sufficient mechanical strength [8, 9].

8.2.6 Cross-linking, Blends, and Thermal Curing

Cross-linking is an effective approach to reduce the softening of the membrane with increasing phosphoric acid content. The cross-linked membranes are mechanically stiff with increased elastic modulus and reduced elongation at break [10]. As a result cross-linked PBI membranes allow for enhanced ADLs and thus higher proton conductivity. In addition, the polymer resistance to the oxidative attack is improved as revealed by the reduced weight loss rate in the Fenton test, an indication of the long-term stability.

Chemically, the benzimidazole groups in PBI are nucleophilic, and the *N*-functionalization reactions are straightforward in the presence of strong base catalysts or if highly reactive electrophiles are used. They can be multiprotic inorganic or organic acids, e.g., perfluoroglutaric acid for ionic cross-linking. Akyl or other organo halides attached to a sp3 carbon atom are reactive toward primary or secondary amines forming the amides. Examples include α,α′-dibromo-*p*-xylene, dichloromethyl phosphinic acid, chloromethyl polysulfone, poly(vinylbenzyl chloride), etc. Cross-linked PBI can also be prepared via Michael-type addition using, e.g., divinyl sulfone and divinylbenzene, as shown in Fig. 8.4.

One potential issue with the cross-linking chemistry based on *N*-coupling is that the *N*-groups will be protonated by PA during doping. The resulting ammonium

Table 8.1 Structures of main-chain-modified PBI derivatives with different linkages of R and X

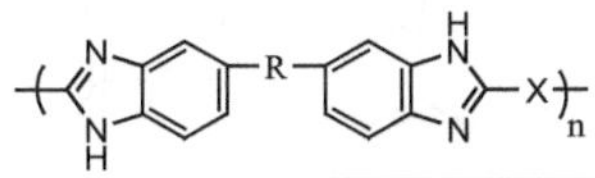

Polymers	R	X	Polymers	R	X
*m*PBI	–	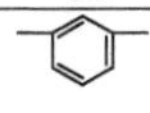	NH_2PBI	–	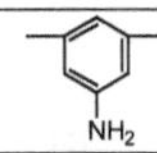
*p*PBI	–	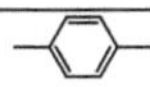	HPBI	–	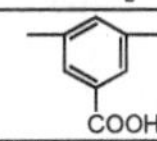
PyPBI	–	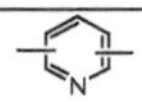	OHPyPBI	–	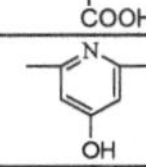
OPBI	–	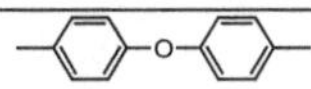	OHPBI	–	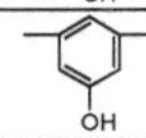
PBI-OO	–O–	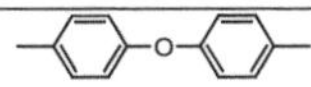	sPBI	–	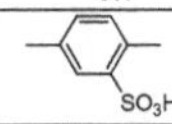
PBI-SO_2	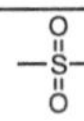	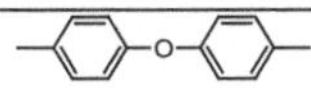	$(OH)_2$PBI	–	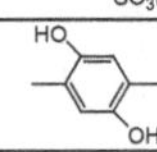
Py-PBI-BPDA	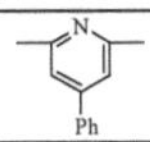	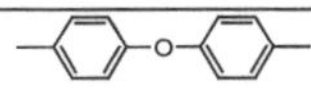	SO_2PBI	–	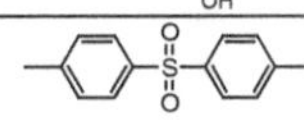
Py-PBI-HFIPA	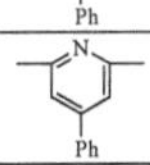	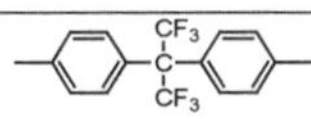	F_6PBI	–	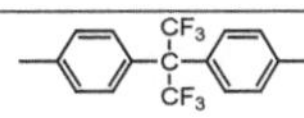
Py-PBI-BDA	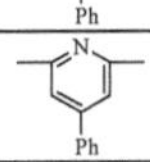		F_{14}PBI	–	$-(CF_2)_7-$
ImPBI	–	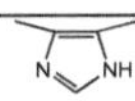	tert-butyl PBI	–	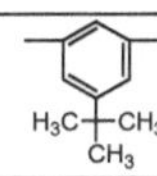

or imidazolium groups are good leaving groups in nucleophilic substitution reactions, which may trigger degradation. This can be mitigated by employing cross-linking methods that are not based on *N*-coupling, for example, via reactions between sulfonated polymers or cross-linking groups with electron-rich aromatic rings in the polymer backbone in a Friedel–Crafts sulfonylation [11]. The resulting sulfone bonds is stable in PA doped systems and at temperatures above 160 °C.

PBI as a basic polymer can be blended with another polymer-containing Brønsted acids to form the so-called ionically cross-linked membranes [12]. The acidic polymers are typically sulfonated hydrocarbons such as polysulfone and polyetherether-ketone (SPEEK). Heating a PBI membrane to a temperature of up to 400 °C, often in an inert atmosphere for a certain period of time, is a simple process without using any additional reagents [13], which is called thermal curing or thermal cross-linking. The mechanism is assumed to be via hydrolysis of imidazole rings by water molecules or probably via Friedel–Crafts reaction of aromatic rings. It seems that residual DMAc and/or moisture facilitates the curing at even lower temperatures. An obvious drawback is that it is a kinetically controlled process, which may lead to varying cross-linking degrees between batches.

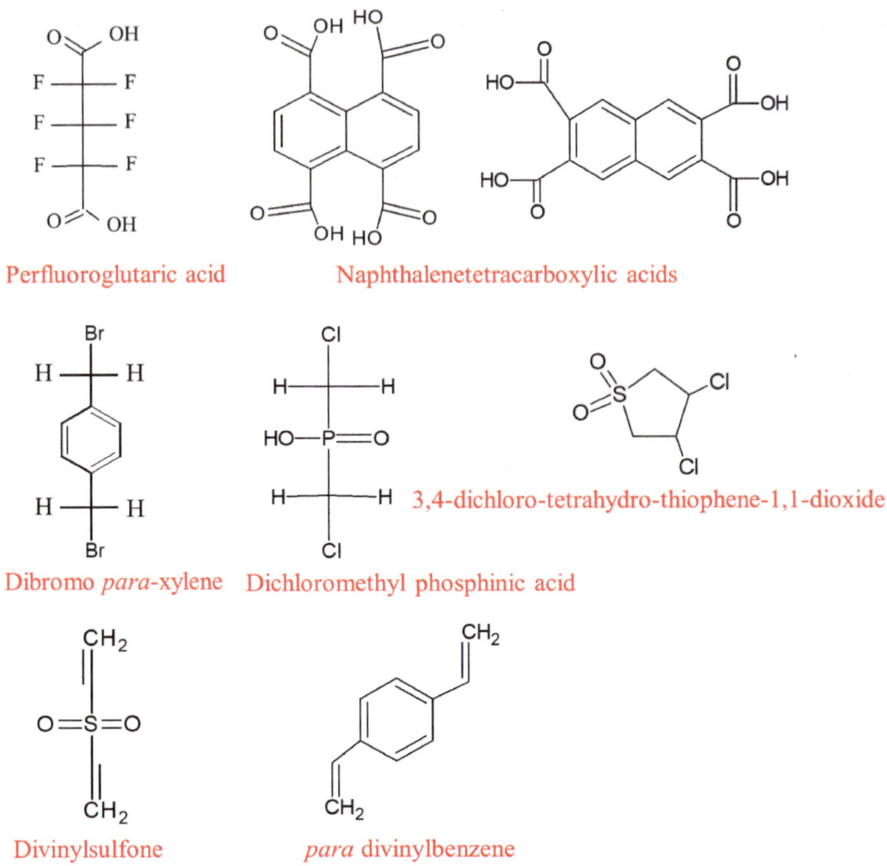

Fig. 8.4 Examples of cross-linkers for PBI

These cross-linked membranes need, in general, to be doped at elevated temperatures in order to achieve a sufficiently high acid doping level. On the other hand, the resultant volume swelling of membranes, which is a linear function of the acid doping level, shows a decreased slope, as seen from Fig. 8.5. At acid doping levels of 11–12, the volume swelling of linear PBI is about 200%, i.e., the doubled volume of the pristine membrane. For the variant polymers, e.g., SO_2PBI and F_6PBI membranes, particularly the cross-linked membranes, the volume swelling is smaller, i.e., the acid doping process results in less separation of the polymer backbones. As a consequence, the tensile strength of the cross-linked PBI membranes is higher than the linear membranes at high acid doping levels (See Fig. 8.6).

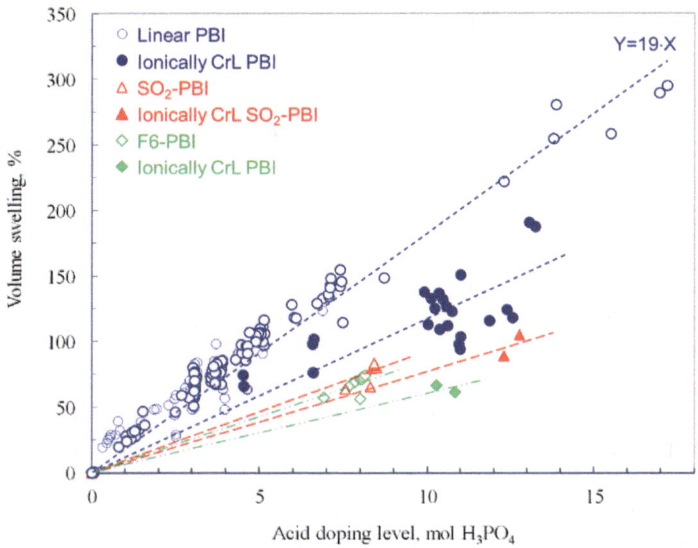

Fig. 8.5 Volume swelling due to acid doping. Modified from Ref. [14] with permission from Elsevier, Copyright 2010

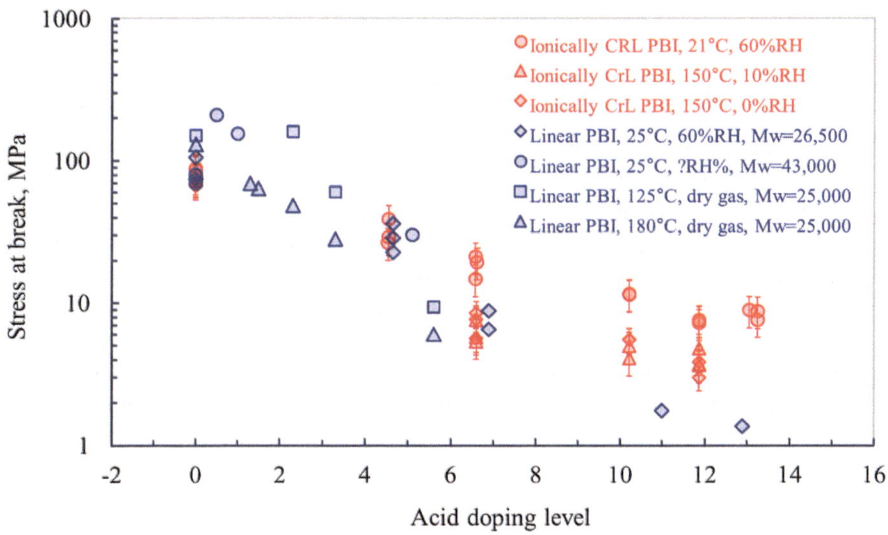

Fig. 8.6 Tensile strength at break of linear PBI of varied molecular weights and ionically cross-linked PBI membranes as a function of the acid doping level. Modified from Ref. [15] with permission from Springer Nature, Copyright 2008

8.2.7 PBI Composite Membranes

Introduction of inorganic fillers is a strategy to improve the acid retention capability, protonic conductivity, mechanical robustness, and thermal stability as well as reducing the gas crossover of PA/PBI membranes. Inorganic fillers used include oxide nanoparticles (e.g., SiO_2, TiO_2, ZrO_2, and mixed oxides), inorganic proton conductors (phosphate, pyrophosphate, and heteropolyacids), carbon-based materials, metal organic frameworks, and clays (aluminum silicates) [16]. Promising results have been demonstrated with the increased total acid content in the membrane with improved mechanical stability and therefore higher proton conductivity and fuel cell performance.

In general, there are two strategies to introduce the inorganic fillers into the polymer, one is by dispersing the filler in a polymer solution and co-casting a membrane, while the other is by *in situ* formation of the inorganic component inside the polymer from precursors introduced. The composite membrane can be fabricated from the pre-prepared nanoparticles by solution casting of a polymer solution with dispersed filler particles. A challenge is the agglomeration of the inorganic phase in the polymer phase. The nanoparticles are usually modified to improve the dispersity and compatibility in the polymer phase. To achieve compatibility with the PBI matrix, a number of modifiers with similar polarity, hydrogen bonding, or/and dispersive forces have been suggested [17].

Heteropolyacids, e.g., phosphotungstic acid (PWA) and silicotungstic acid, are known to have high intrinsic proton conductivity and can be supported on, e.g., nanosilica followed by surface modification. An example of preparation of PBI/SiO_2 composite membranes is shown in Fig. 8.7 [18]. In this process, SiO_2 nanoparticles are modified with surface coating of dequalinium chloride. Good dispersion is achieved in the polymer phase, as indicated by the transparent and uniform membranes containing up to 10 wt% SiO_2. At operating temperatures of fuel cells, the SiO_2 nanoparticles in the presence of doping acid are *in situ* converted into phosphosilicate nano-clusters, which are highly conductive and stable at high temperatures. It seems that the *in situ* formed phosphosilicate nano-clusters facilitate the immobilization of phosphoric acid. As a result, the composite membrane exhibits a high and stable proton conductivity in the range of 0.029 to 0.041 S/cm at temperatures as high as 250 °C.

8.3 Catalysts and Electrodes

8.3.1 Catalysts

8.3.1.1 Acid Adsorption Effect on Pt

The sluggish oxygen reduction reaction (ORR) in concentrated phosphoric acid, despite the elevation of operating temperatures, limits the fuel cell performance.

Fig. 8.7 Fabrication procedures for the synthesis of PA/PBI/phosphosilicate composite membranes with *in situ* formed phosphosilicate nano-clusters. Reproduced from Ref. [18] with permission from Elsevier, Copyright 2018

In terms of reaction rates, the ORR kinetics as well as the mass transport of O_2 behaves in a similar way to that in H_3PO_4. The reported ORR exchange current density is about 10^{-9} A/cm^2, which is two orders of magnitude lower than that in low-temperature PEMFC based on perfluorosulfonic acid membranes. This slow ORR kinetics in both PAFCs and HT-PEMFCs is associated to the strong adsorption of the acid molecules and anions. The adsorption is via the formation of Pt-O bonds involving undissociated acid molecules (H_3PO_4) in the low potential range and the acid anions ($H_2PO_4^-$) at high potentials. Two different configurations are identified for adsorption of the dihydrogen phosphate. At an electrode potential below 0.8 V, it is adsorbed through the two oxygen atoms that are not bonded to hydrogen, while above 0.9 V the adsorption changes to a single coordination. The adsorption shows a strong dependence on temperature. At lower temperatures, the phosphoric acid anions adsorb on the Pt surface over a wider potential range. It seems that the operating temperatures of PAFC and HT-PEMFC around 160–200 °C are not high enough to eliminate the phosphate anion adsorption.

A study using operando X-ray absorption spectroscopy on an HT-PEM MEA showed that on the Pt surface in contact with acid doped PBI membranes at temperatures of up to 170 °C, the adsorbate is hydrogen atoms (H_{ad}) at electrode potentials below 0.3 V and the oxide or hydroxyl species $O(H)_{ad}$ at potentials above 0.8 V. In the intermediate potential range, significant adsorption of the acid, most likely $H_2PO_4^-$ anions, occurs [19].

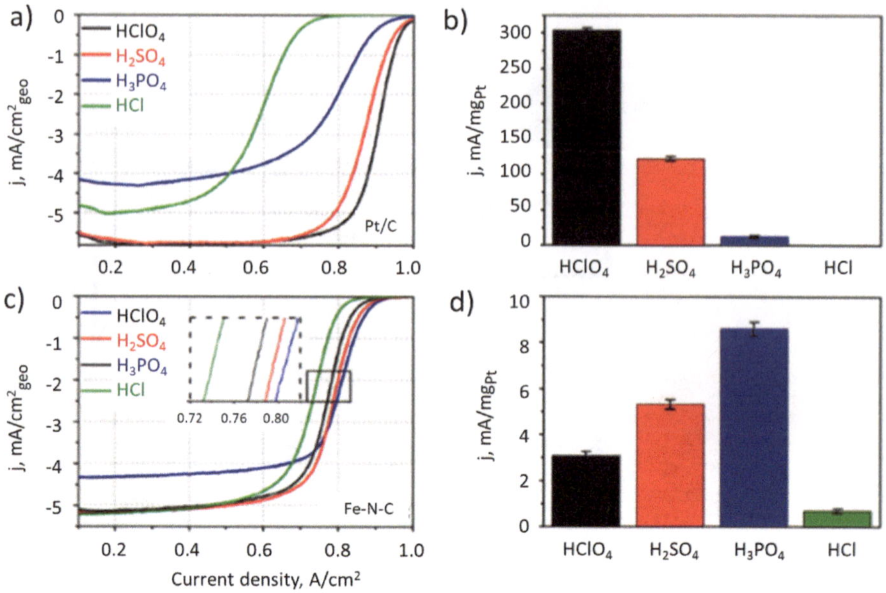

Fig. 8.8 ORR polarization curves based on geometric current density and mass specific current density at 0.9 V obtained in pH = 1.0 $HClO_4$, H_2SO_4, H_3PO_4, and HCl solutions for **a, b** 20 wt% Pt/C and **c, d** Fe–N-C catalyst. Rotation rate: 1600 rpm; scan rate: 10 mV/s. Reproduced from Ref. [20] with permission from Elsevier, Copyright 2018

Additives to the phosphoric-acid-based electrolyte such as the perfluoro alkane sulfonates or cyanide have been investigated, however, with no success in fuel cell applications. An interesting finding is that the strong poisoning effect by $H_2PO_4^-$ anions on the Pt surface is negligible on non-precious metal (NPM) catalysts like Fe-N-C catalysts. Of common anions of ClO_4^-, HSO_4^-, $H_2PO_4^-$, and Cl^- in acidic electrolytes at pH = 1.0, a minimum adsorption effect is observed for $H_2PO_4^-$, as shown in Fig. 8.8. In fact, the ORR activity was found to improve in the presence of the H_3PO_4 electrolyte compared with that in $HClO_4$, though the oxygen solubility is lower in H_3PO_4 [20].

Pt alloy catalysts have been extensively studied for the low-temperature PEM fuel cells, while for the application in HT-PEMFC, all Pt alloys with a non-noble metal will be de-alloyed to a certain degree under the harsh conditions in the cell. The de-alloying is believed to form a Pt overlayer and an underneath alloy core. The underneath alloy core is still able to affect the properties of the Pt overlayer including weakened adsorption of phosphates [21].

8.3.1.2 Non-precious Metal Catalysts

Non-precious metal catalysts (NPMC) have been developed to replace Pt electrocatalysts in particular for ORR and are of great significance to the commercial viability of

PEMFCs. Among various NPMC for ORR have the pyrolyzed iron–nitrogen–carbon (Fe–N–C) composites received great attention. The active sites of the catalysts are proposed to be consisting of nitrogen-coordinated transition metal ions embedded in the carbon support with general formula of MN_x/C in which M = Co, Ni, and Fe and $x \sim 4$, which are synthesized from a variety of precursors including polyaniline and metal organic framework. Iron single atoms coordinated in nitrogen doped carbon, FeNx/C have been found to be most active for ORR in acid conditions, but the role of iron in the active site structure has been a subject of debate. The metal content is generally less than 2 wt% and the low catalyst loading poses a significant challenge in the practical application of NPMC in fuel cells (see Chap. 5). An interesting route of synthesis has been reported, achieving Fe single atom catalyst content of up to 7.7% in the Fe–N–C composite [22].

NPMC has a promising potential as alternative and effective electrocatalysts in HT-PEMFCs. In 0.1 M $HClO_4$ electrolyte with the rotating disk electrode, the addition of 0.2 M H_3PO_4 into the electrolyte leads to a negative shift of the half-wave potential by 27 mV for the Pt/C catalyst due to the phosphate poisoning, as shown in (Fig. 8.9a). This potential shift is only 8 mV for the iron single atom catalysts embedded in nitrogen doped graphene (FeSA-G), indicating its resistance against the acid adsorption. The as-synthesized FeSA-G non-precious metal catalysts are tested in high-temperature PEMFC based on PA/PBI/SiO_2 composite membranes at temperatures of up to 230 °C under H_2/O_2 at atmospheric pressure without humidification. The reference cell is with a Pt loading of 1.0 mg/cm^2 at the cathode. At 160 °C, the fuel cell with the NPMC cathode showed an open circuit of 0.89 V and a peak power density of 276 mW/cm^2, lower than the observed 0.95 V and 373 mW/cm^2, respectively, for the Pt cathode (Fig. 8.9b). At 230 °C, the peak power density of the NPMC cathode increased to 325 mW/cm^2, higher than that for the Pt/C cathode. Most interestingly, the stability of the NPMC cell in term of current density at a constant cell voltage is better than that of the Pt/C cell at 230 °C. The better stability of the cell with FeSA-G cathode is most likely related to the strong coordination environment between iron single atoms and nitrogen in carbon matrix.

The activity of Fe-N-C composite catalysts can be further enhanced by incorporating another non-precious metal atoms, forming atomically dispersed bimetallic catalysts embedded on nitrogen doped carbon supports. An example is the bimetallic Fe and Cu atoms anchored on nitrogen doped carbon nanotubes (FeCu/N-CNTs) as NPMCs for ORR in the HT-PEMFCs [23]. The study indicates that in contrast to the adsorption and poisoning of phosphate to Pt/C electrocatalysts, phosphate enhances the electrocatalytic activity of FeCu/N-CNTs for ORR. The reason for the high and PA promoted activity and performance of FeCu/N-CNTs catalysts is most likely that Cu atoms adjacent to Fe atoms strongly bind the phosphate, which lowers the energy barrier of Fe active sites for oxygen double bond cleavage and provides local proton to assist the proton-coupled electron transfer ORR process. In general, the NPMCs based on atomically dispersed transition metal catalysts show a promising potential as a highly active, stable and new class of Pt-free electrocatalysts for HT-PEMFCs.

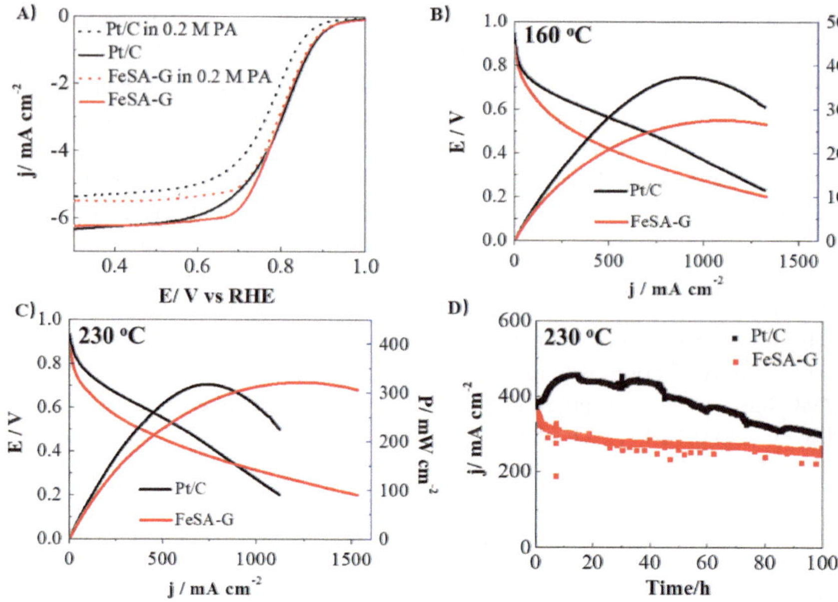

Fig. 8.9 Rotating disk electrode and HT-PEM fuel cell test of non-precious metal catalysts (FeSA-G) and Pt/C. **a** Linear scan voltammograms on rotating disk electrode in O_2-saturated $HClO_4$ solution with addition of 0.2 M H_3PO_4. **b** Polarization and power density curves of HT-PEMFCs at 160 °C and **c** at 230 °C; **d** Stability test of current density at 0.5 V for NPMC and 0.6 V for Pt/C. The fuel cells are fed with dry H_2 and O_2 with flow rate of 100 mL/min, the anode catalyst loading in both cells is 1 mg_{Pt}/cm^2, the cathode loading is 0.3 mg_{Fe}/cm^2 for NPMC and 1 mg_{Pt}/cm^2 for Pt/C as cathode, respectively. Reproduced from Ref. [22] under Creative Commons CC BY License

8.3.1.3 Carbon Supports

Carbon as the catalyst support material is susceptible to corrosion, which is a function of temperature, potential, water partial pressure, and the electrolyte acidity. In HT-PEMFCs, the high operating temperature and the harsh corrosive H_3PO_4 electrolyte lead to very pronounced corrosion of carbon supports, which triggers the agglomeration of platinum nanoparticles. Another consequence is the formation of surface oxides, causing a decrease in hydrophobicity of the catalyst layer and possible acid flooding. A well-known practice to mitigate the carbon corrosion is the heat treatment of carbon blacks at elevated temperatures in order to impart the character of graphite to reduce the number of structural defects where the corrosion starts. Alternatively carbon nanotubes, interestingly with a PBI wrapping layer, were found to improve the dispersion and stability of the Pt nanoparticles [24].

8.3.2 Gas Diffusion Electrodes

While low-temperature (LT-) PEMFC electrodes are optimized for easy removal of water, HT-PEMFC electrodes should facilitate the acid retention, i.e., preventing the acid loss via penetration through the macro porous layer (MPL) and gas diffusion layer (GDL). It seems that cracks within the MPL provide pathways, through which phosphoric acid flows into the GDL and further to the flow plates. Gas diffusion electrodes with a crack-free and thick MPL on a GDL substrate with small pores are considered beneficial to retaining the acid inside the catalyst layer during operation.

To optimize the catalyst layer structure, i.e., the triple-phase boundaries, the early HT-PEMFC effort was made to investigate the effect of the polymer binder, which attaches the catalyst particles as an integral layer on the GDL. Inert polymers as PTFE has been used as the binder to provide hydrophobicity to the catalytic layer and thus a balance between open pores and acid filled pores. PBI is another widely used binder, which assumes to absorb and stabilize the acid. It is interesting that recent attempt succeeds by complete elimination of the binder in the catalytic layer [25]. The proton conductivity of the catalytic layer is provided by the liquid phosphoric acid, which migrates from the membrane to the catalyst layer driven by the capillary force.

The state-of-the-art Pt loading for HT-PEMFCs is around 0.7–0.8 mg_{Pt}/cm^2. Great effort has been made during the last years to reduce the Pt loading of the electrodes while keeping a reasonable performance output. In laboratory cells, electrodes with Pt loading lower than 0.3 mg_{Pt}/cm^2, sometimes even lower than 0.1 mg_{Pt}/cm^2, have been reported with peak power over 0.40 W/cm^2. The feasibility of such electrodes in industrial cells operating with reformate fuels has not been verified. It should be mentioned that these low Pt loading electrodes are structured with either MPL-free [26] or binderless layers.

8.3.3 Membrane Electrode Assemblies

It is a common practice to assemble the HT-PEM MEA by hot-pressing two electrodes onto a membrane. The resulting MEAs are easy to handle and install in the fuel cell test hardware or stack. The hot-pressing parameters on the performance of the MEA have been optimized. The typical hot-press temperature is around 150 °C, which is around the operating temperature of the fuel cells. It seems that there is a practical tendency to omit the hot-pressing of MEAs in many recent studies and the assembling is made directly in the fuel cell test hardware. Figure 8.10 shows an SEM image of a PBI MEA cross section. The carbon fiber gas diffusion layers are coated with a macro porous layer of PTFE bonded carbon black. The cathode catalyst layer is thicker than the anode catalyst layer because different platinum loadings are used.

The acid transfer within MEAs is essential to obtain a stabilized fuel cell performance. The acid can be originating from the membrane or additionally introduced into the catalyst layer. No matter how the acid is introduced, the redistribution of the

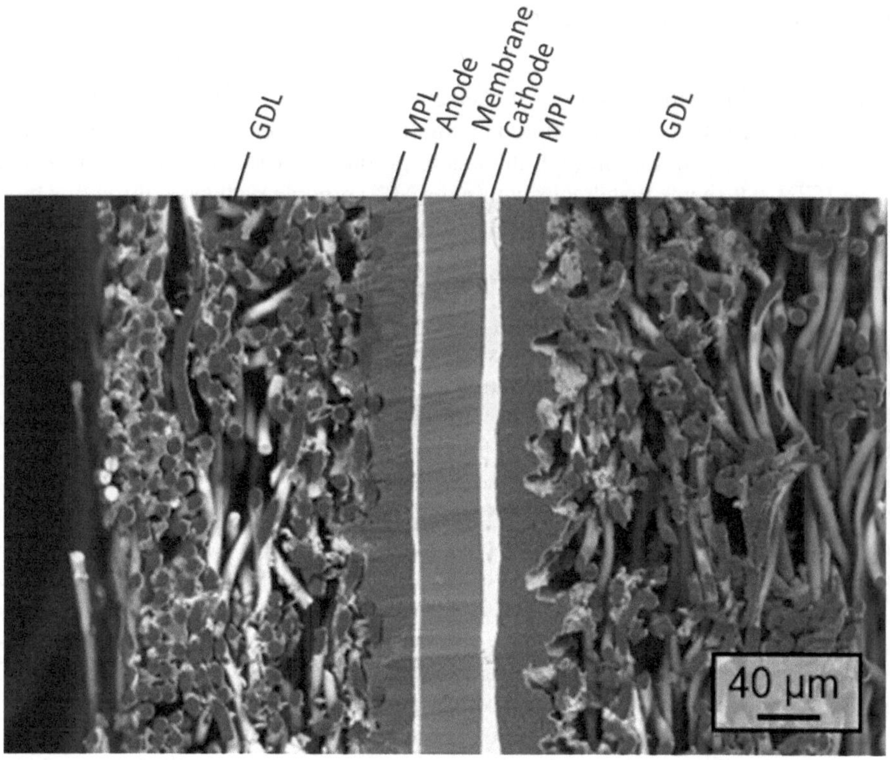

Fig. 8.10 A scanning electron microscopic image of an HT-PEMFC MEA cross section

acid between the membrane and catalyst layer occurs during the first hundred hours of operation, which is called the break-in period. The acid balance is determined by the microporous structures and hydrophobicity characteristics of the catalyst layer, the surface tension of the acid and acid retention capability of the membrane. A study reported that an equilibrium is eventually established an acid distribution in the membrane and catalyst layer with a mass ratio from 3 to 4 (13–16 mg H_3PO_4/cm^2 in membrane and 3–5 mg H_3PO_4/cm^2 in catalyst layer), which apparently depends on the amount of the polymer content in membrane and the thickness of the catalyst layer [27]. The discussed materials as well as operational features of PBI-based fuel cells are summarized in Table 8.2 in comparison with Nafion-based LT-PEMFC and PAFC.

Table 8.2 Technical features of PBI-based PEMFC compared with Nafion-based PEMFC and PAFC

	Nafion-based LT-PEMFC	PBI-based HT-PEMFC	PAFC
Catalyst and electrode	• Anode: Pt(Ru)/C, CO tolerance < 100 ppm • Cathode: Pt (alloy)/C, Fast kinetics; Corrosion • Ionomer-bonded catalyst layer	• Anode: Pt/C, CO tolerance > 100,000 ppm • Cathode: Pt/C, Slow kinetics (anion adsorption); Severe corrosion • Binder or binderless catalyst layer • Feasibility of replacing Pt/C by active and stable NPMC	• Anode: Pt/C, CO tolerance > 100,000 ppm • Cathode: Pt/C; Slow kinetics (anion adsorption) Severe corrosion • PTFE bonded catalyst layer
Electrolyte	• Essentially solid to handle • Mechanically supporting electrodes • Good separator - Low gas permeability • Small thickness (15–50 μm) • Hydrous conductivity (in hydrated state) • Water drag coefficient ~ 3 mol H_2O/H^+ • Tolerance to pressure difference (~ 300 kPa) • Chemically bonded acidic sites – No acid management – Critical water management • No creeping of electrolyte • High cost	• Essentially solid to handle • Mechanically supporting electrodes • Good separator - Low gas permeability • Intermediate thickness (50–100 μm) • Anhydrous conductivity • Nearly zero water drag • Tolerance to pressure difference (~ 300 kPa) • Absorbed doping acid – Acid loss issue – No water management • Less creeping of electrolyte • Low cost	• Immobilized liquid • Supported by electrodes • Gas separation by filling acid • Relatively large thickness (> 100 μm) • Anhydrous conductivity • Nearly zero water drag • Tolerance to pressure difference (~ 30 kPa) • Capillary force immobilized acid – Acid management (reservoir) – No water management • Creeping of electrolyte • Low cost
Other materials	• Bipolar plates: Metallic; composites • Stacking seals:	• Bipolar plates: Graphite; composites • Stacking seals:	• Bipolar plates; Graphite; composites(?) • Stacking seals: PTFE
Operating features	• Operating at 80 °C — narrow temperature range • High power density • Fast start-up and cycling • Critical water management (humidification) • High-purity fuel • Intensive cooling (large radiator) • Little value of heat	• Operating at 130–180 °C — wide temperature range • Medium power density — Low at low temperature • Fast start-up and cycling • No water management (no humidification) • Impure fuel (methanol reformate) • Less intensive cooling (small radiator) • Heat value for cogeneration	• Operating at 160–210 °C — wide temperature range • Low power density • Slow start-up and cycling • No water management • Impure fuel (natural gas reformate) • Less intensive cooling • Heat value for cogeneration

8.4 Fuel Cell Performance and Operation

8.4.1 Performance and Pt Loading

The fuel cell performance can be defined by a single point. It is a common practice that industrial manufacturers specify their MEA products by the cell voltage at a rated current density or a current density at a constant cell voltage. For HT-PEMFCs, an often used single point is 0.67 V at 0.2 A/cm^2 or a near peak power point of 0.5 V at 0.8 A/cm^2 (0.4 W/cm^2) when operating under reformate fuel and air under ambient pressure, though the total electrode platinum loading is usually about 2–3 times higher than that for LT-PEMFC.

Figure 8.11 shows the progress in the MEA performance of the HT-PEMFCs in terms of peak power and Pt loading in the latest years. The performance data correspond to MEAs operated under similar working conditions, i.e., H_2/air, ambient pressure, and cell temperatures between 150–180 °C. For each year, five publications exhibiting the highest peak power performances are selected as shown in the unfilled bars, while the gray bars and the square dots represent the average peak power density and cathode Pt loading calculated from all the works found in literature (~120 pieces in total) [28]. A slow yet steady increment of the performance is seen and a peak power of 450–500 mW/cm^2 at a Pt loading of 0.5–1.0 mg$_{Pt}$/cm^2.

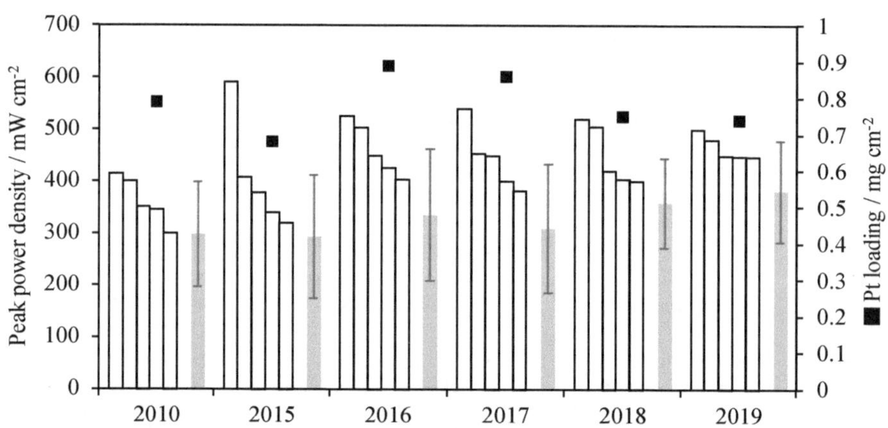

Fig. 8.11 Progress of performance and Pt loading of HT-PEMFCs. The unfilled bars are five selected works of the highest peak performance reported in the literature during a specific year, and the gray bars represent the average peak performance. The square dots is the average Pt loading for each year (evaluated from ~120 publications). Reproduced from Ref. [28] with permission from Springer Nature, Copyright 2020

8.4.2 *Impact of Fuel Impurities and Water*

Elevation of operating temperature is an effective way to alleviate the CO poisoning allowing for a simplified fuel processing. Operating at 150 °C - 180 °C, the HT-PEMFC can tolerate 1–3% CO in the fuel stream, which is about 1000 times higher than that for LT-PEMFCs (10–20 ppm) [29]. Studies have been devoted to evaluate the CO poisoning in reformate fuels, which are H_2-rich gases containing CO, CO_2, water vapor, and other impurities. The CO_2 is shown to have only diluting effect, while the water vapor is a multifaceted issue. HT-PEMFCs can operate with dry gases which significantly simplifies the system construction and operation. In practical operation, however, reformate fuels always involve excess water, which hydrates and ionizes phosphoric acid and hence increases the proton activity and enhances ORR kinetics in the catalyst layer.

The poisoning effect of CO is a strong function of operation temperature of fuel cells. A major advantage of HT-PEMFCs is their high CO tolerance enabled by the high operating temperatures. In most studies, however, the evaluation of the CO tolerance is carried out in H_2/CO gas mixtures. With a PA/PBI/SiO_2 composite membrane, the fuel cell operation at temperatures as high as 250 °C has been demonstrated [18]. At this temperature, the CO content of up to 16.7% has been tested, as shown in Fig. 8.12. The study shows that the PA/PBI/SiO_2 composite membrane cell displays a very high CO tolerance with a limited loss in performance at CO contents as high as 11.7% in H_2 at 240 °C. The cell delivers a peak power density of 283 mW/cm^2 and is stable at 240 °C for 100 h under a cell voltage of 0.6 V in 6.3% CO-containing H_2 fuel under anhydrous conditions. The reduced power performance of the cell at high CO contents in H_2 is partly due to the dilution effect of CO in the hydrogen fuel.

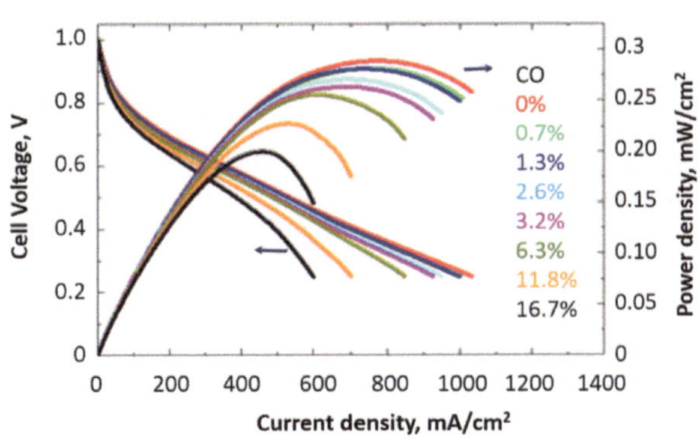

Fig. 8.12 Polarization and power density curves of HT-PEM fuel cell based on a PA/PBI/SiO_2 composite membrane operating at 250 °C with oxygen- and hydrogen-containing CO. The CO content is specified in the figure. Reproduced from Ref. [18] with permission from Elsevier, Copyright 2018

8.5 Alternative High-Temperature Membranes

8.5.1 Non-PBI Based High-Temperature Membranes

Great efforts have been made to develop basic polymers with various incorporated N-heterocycles that possess *p*Ka values high enough to be protonated by phosphoric acid for the purpose of acid doping. Triazole-functionalized polymers as well as their blends have been extensively explored with conductivity evaluation. Pyridine-containing aromatic polyethers with side-group modifications and cross-linking are a successful membrane system that is employed in industrial MEAs [30]. Another interesting example is polyvinylpyrrolidone (PVP) in its blend forms with polyether-sulfone (PES) or polyvinylidene fluoride (PVDF) where PVP contains functionality to absorb phosphoric acid, while the PES or PVDF provides the mechanical strength. The PA doping level of the 80%PVP-PES blend membranes reaches 9.1 mol of PA per mole of the pyrrolidone ring after acid doping in 85%H_3PO_4 at room temperature for a few hours. Such a membrane exhibits a proton conductivity of 0.21 S/cm at 180 °C [31]. Figure 8.13a shows the polarization performance of the PVP-PES blend membrane fuel cells operating with hydrogen and oxygen at 120–180 °C. The power performance of PA doped PVP-PES hybrid membrane cells increases with

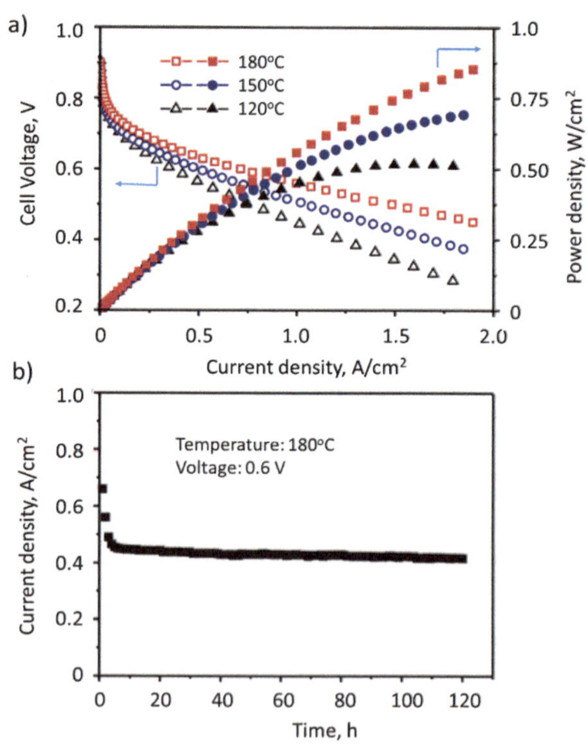

Fig. 8.13 **a)** Polarization and **b)** stability curves of 80wt%PVP-PES membrane fuel cells operating at different temperatures with ambient H_2 and O_2. The Pt loading for both anode and cathode was 0.5 mgPt/cm^2. Reproduced with permission from Ref. [32] with permission from Royal Society of Chemistry, Copyright 2015

the temperature, similar to that of PA/PBI based HT-PEMFCs. Under a constant cell voltage of 0.6 V, a short time test seems verifying the membrane stability (Fig. 8.13b).

8.5.2 *Inorganic High-Temperature Membranes*

Solid inorganic proton conductors have been developed for fuel cell applications at elevated temperatures of 100–300 °C. Two types of electrolytes are to be briefly discussed in this section: one is based on solid acids and pyrophosphates and the other is based on heteropolyacids impregnated into mesoporous silica matric.

8.5.2.1 Inorganic Proton Conductors

A large family of inorganic proton conductors have been extensively investigated for fuel cell and other electrochemical applications. The most interesting are solid acids and pyrophosphates. Solid acids particularly alkali metal or ammonia dihydrogen phosphates, e.g., cesium dihydrogen phosphate (CsH_2PO_4, CHP) are well-known solid proton conductors. This type of acidic salts is characterized by a superprotonic phase-transition temperature at about 230 °C. A dynamic disordered hydrogen bond network is developed above this temperature, leading to a jump in the proton conductivity by 4–5 orders of magnitude, i.e., 8.5×10^{-6} S/cm below 223 °C to 1.8×10^{-2} S/cm above 233 °C [33]. The superprotonic phase is stable at up to 250 °C, above which the dehydration takes place forming polymerized products such as $CsH_2P_2O_7$ or eventually $CsPO_3$ unless a humid atmosphere is maintained. The narrow temperature range of the superprotonic phase is a challenge for practical applications of the pure CsH_2PO_4, although heterogeneous doping of the acidic salt and other approaches have been explored to mitigate the issue.

Metal pyrophosphate is another interesting high-temperature proton conductor [34]. The cations are typically a tetra-valent metal such as tin, titanium, cerium, or zirconium. Among these, tin pyrophosphate (SnP_2O_7) is reported to exhibit the high proton conductivity at temperatures of up to 300 °C under unhumidified conditions. This conductivity can be further improved by doping with trivalent metal ions, e.g., Al^{3+}, Ga^{3+}, Gd^{3+}, or In^{3+} [35]. Alternatively, a combination of a tri- and a penta-valent metals, e.g., Fe(III) and Ta(V) have also been investigated as in $Fe_{0.4}Ta_{0.5}P_2O_7$ [36]. The synthesis of pyrophosphates is typically carried out by mixing metal oxides with an excess H_3PO_4 followed by heat treatment at up to 300 °C until a viscous paste is formed. The paste is then calcined at temperatures of 450–650 °C, air quenched to room temperature, and finally ground into fine powder. In form of uniaxially pressed pellets, the electrolyte has a proton conductivity above 10^{-2} S/cm, the minimum value for any technological applications. Fuel cell tests using the pyrophosphate electrolyte have been made at temperatures above 200 °C [36].

8.5.2.2 Functionalized Inorganic Mesoporous Proton Conductors

Functionalized inorganic mesoporous proton conductors belong to a special group of proton-conducting materials and are based on the principles that the highly ordered and confined mesoporous channels would facilitate the proton migration and diffusion processes, thus enhancing the proton conductivity of the materials. The mesoporous matrix can be made of TiO_2, silica, alumina, and other ceramic materials and the proton-conducting phase can be protonic acid such as HNO_3, HCl, H_3PO_4, or heteropolyacids (HPAs) [37].

Among the functionalized inorganic mesoporous proton conductors, HPAs functionalized mesoporous silica is extensively studied and is probably the most successful one with respect to the characterization and application in fuel cells, due to its high thermal stability and proton conductivity. HPAs are strong Brønsted acid with an unique Keggin unit which are linked by adsorbed water molecules via hydrogen bonds to form a secondary structure. Among HPAs, phosphotungstic acid ($H_3PW_{12}O_{40}{\cdot}nH_2O$, HPW) shows the highest proton conductivity of 0.18 S/cm in the fully hydrated state at room temperature.

HPW can be incorporated into the mesoporous silica matrix by an impregnation or a self-assembling method. The impregnation method without or with a vacuum facilitation is schematically shown in Fig. 8.14a [38]. The proton conductors, e.g., HPW, are incorporated into mesopores of silica materials, often facilitated by vacuum. Figure 8.14b shows typical TEM micrographs of mesoporous silica (type MCM-41) before and after the HPW impregnation. Mesoporous silica is characterized by highly ordered parallel channels in [110] direction and hexagonal channels in [100]

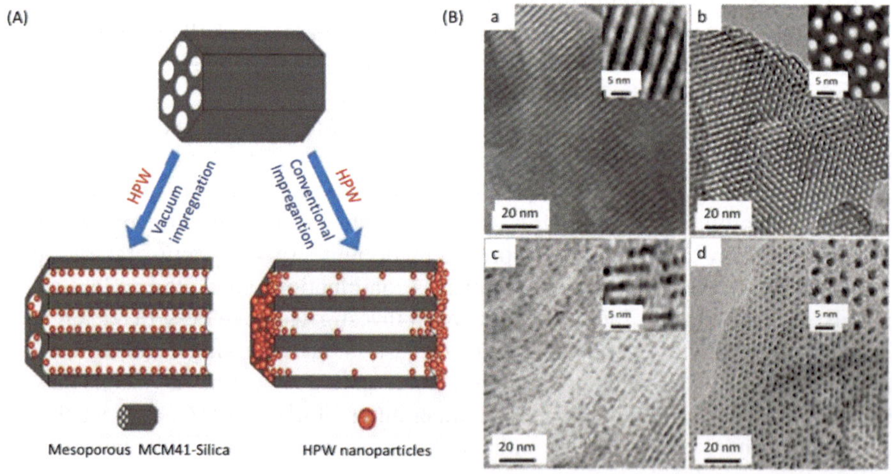

Fig. 8.14 **a)** Schematic representation of impregnation of HPW into *meso*-silica matric and **b)** TEM images of (**a, b**) pristine and (**c, d**) 30 wt% HPW impregnated MCM-41. Viewed (**a, c**) in [110] direction and (**b, d**) in [100] direction of the pore axis. Insets are high-resolution images. Modified from Ref. [38] with permission from Wiley–VCH, Copyright 2010

direction with a diameter of 3.0 nm and a wall thickness of 2.0 nm. The empty pore space is indicated by the white dot in the view of [100] direction, while the pore space changes to dark when HPW is impregnated into the channels of MCM-41. The distribution of HPW along the channels is indicated by the ordered but not continuous dark red dots. The higher loading of HPW particles inside the *meso*-silica, the lower activation energy barriers will be for the proton transportation in the *meso*-silica matrix.

The proton conductivity of HPW/*meso*-silica membranes depends strongly on the content and distribution of HPW in the mesopores. The threshold of the HPW content reaching a conductivity above 10^{-2} S/cm is about 10 wt%; however, the value remains nearly a constant of 7.0×10^{-2} S/cm at 25 °C and 100% RH when the HPW content is higher than 65 wt%. In addition to the HPW loading, the proton conductivity of HPW/*meso*-silica is also affected by the pore size and structure of the mesoporous silica matrix. In general, HPW/meso-silica nanocomposites with 3D mesostructures such as 3D face-centered cubic display a significantly higher proton conductivity and higher stability in comparison to 2D mesostructures such as 2D hexagonal. HPW/*meso*-silica composite membrane based fuel cells have been evaluated for both H_2 and alcohol fuels including methanol and ethanol. The applicability of HPW/*meso*-silica nanocomposite membranes was also demonstrated in short stacks [39].

8.6 Durability and Commercialization

Higher operational temperatures impose more challenges for the material durability as the carbon corrosion, platinum dissolution, and polymer oxidation are aggravated. In addition, the presence of phosphoric acid and its maintenance add also critical issues to the HT-PEMFC.

Oxidative degradation of the polybenzimidazoles seems to occur at the weak link of the benzenoid rings bearing nitrogen atoms. Possible attack by the peroxide radicals inside fuel cells is assumed and has been extensively investigated by means of accelerated aging test with the Fenton agent. Significant degradation of the polymers is identified and modification of the polymer macrostructures, covalently cross-linking and ionically blending of the polymers has been shown to improve the materials durability.

The phosphoric acid in HT-PEMFCs is mobile. When an MEA is assembled, redistribution of phosphoric acid takes place, i.e., the acid moves from the membrane to the catalyst layers. During operation of a fuel cell, acid migrates from the cathode to the anode and is balanced by the back diffusion. The loss of the acid out of the fuel cell is, however, dominated by the evaporation mechanism. The extrapolated acid vapor pressure to the fuel cell operational temperatures provides a guideline of the acid evaporation rate, which increases by 5 times from 160 to 190 °C. Following the acid vapor pressure trend, the acid loss rate measured by collection of the acid from anodic and cathodic off-gasses gives a consistent value of less than 1 $\mu g/m^2$

s at 150–160 °C. This can be translated into an acid loss of a few mg H_3PO_4 per square centimeter of electrode area over a duration of 10,000 h. For an MEA with an initial acid content of 36 mgH_3PO_4/cm^2 when operated at 160 °C at 0.6 A/cm^2 with a stoichiometry of $\lambda H_2/\lambda_{air} = 1.2/2$, it is estimated that the acid inventory is sufficient for a desired lifetime of 50,000 h [40]. At higher temperatures or higher current densities this lifetime target can hardly be achieved because of the expected higher acid loss.

Carbon corrosion and platinum dissolution in the acidic electrolyte at elevated temperatures are well recognized, both causing particle growth and therefore surface area reduction of platinum particles and being enhanced at higher temperatures and higher electrode potentials. It was reported that under steady-state operation for 17,800 h at 150 °C only 58% of the initial platinum surface area remained [41]. Using the equation below

$$\Delta V = b \log\left(\frac{\mathrm{SA}}{\mathrm{SA_o}}\right) \tag{8.6}$$

where $\mathrm{SA_o}$ and SA are the initial and remaining specific surface area of platinum catalysts, the $b = 100$ mV/dec is the Tafel slope of PBI cell polarization curves, and the fuel cell voltage loss (ΔV) caused by the platinum degradation can be estimated to be 24 mV.

Under steady-state operation, lifetimes of up to 18,000 h at 150–160 °C have been demonstrated showing a degradation rate of a few μV/h. A recent study [42] reported an average performance loss of 0.69 μV/h during a period of 17,500 h at 160 °C and 0.2 A/cm^2 with $\lambda_{H_2} = 1.2$ and $\lambda_{Air} = 2$. These studies indicate the great potential of PBI membranes for meeting the lifetime requirements of many devices, though increased cell temperature, current density, and reactant stoichiometries accelerated the degradation rate of PBI cells.

Fuel cells are much more stressed when operating under dynamic modes, i.e., with thermal, load, and shutdown-startup cycling. The shutdown-start-up or/and temperature cycling expose the cell to severe corrosion of carbon supports and sintering of noble metal catalysts. With a load cycling between OCV and 0.3 A/cm^2, the accelerated stress test with synthetic reformate fuel showed an averaged degradation of 31 μV/cycle [43].

Commercialization of key materials, components, MEAs, stack, and power units has started while automobile applications of HT-PEMFCs as a range extender are under active exploration. For micro-CHP applications, HT-PEMFCs have been assessed technically and economically with fueling option of steam reforming of natural gas. Power units with partially or thermally integrated methanol reformers are available in a power range from 5 to 15 kW. Complete integration of the HT-PEMFC stacks with internal methanol reforming needs membrane operational at above 220–230 °C.

8.7 Summary

In this chapter, we have introduced and discussed the following topics:

- Perspectives of PEMFC operating at temperatures above 100 °C are justified in consideration of fueling strategy, electrode kinetics, water management, cooling, and heat recovery.
- Phosphoric acid doped polybenzimidazole (PA/PBI)-based HT-PEMFC is so far the most successful system that has demonstrated the technical feasibility. It is represented by the simplicity of the power system where the humidifier, cooling, and fuel storage/processing are eliminated or minimized; however, the cell performance is low compared with the Nafion membrane-based LT-PEMFC technology.
- Polymer synthesis and membrane casting are presented in details. The membrane can be functionalized by direct casting from polyphosphoric acid solution or from an organic solution followed by acid doping. Acid-doped membranes are thermally stable and mechanical strong, showing an anhydrous conductivity above 0.05–0.15 S/cm at 160–180 °C under dry atmosphere.
- A large number of PBI derivatives of varied backbone structures have been synthesized aiming at improved processibility, stability, and particular capability of acid retention. Cross-linking, thermal curing, blends, and composites of the polymer membranes are effective approaches to further upgrade the membranes.
- Strong adsorption of acid anions on the platinum surface hinders the ORR kinetics and offsets the fuel cell performance expected from the temperature elevation. As a result, HT-PEM MEAs are often equipped with platinum loadings of about 1.0 mg_{Pt}/cm^2, about 2–3 times higher than that of LT-PEMFC, though low Pt loading electrodes have shown promise in laboratory MEA testing.
- Introduction of inorganic fillers can further increase the operating temperature of PA/PBI-based membranes. An example is the formation of PA/PBI/SiO composite membranes with in situ formation of phosphosilicate nanoclusters with high proton conductivity and stability at elevated temperatures of 200 to 250 °C. The elevated operating temperature offers an opportunity for the development and exploration of NPMCs, which are immune towards the acid adsorption.
- Operation with reformate fuels has been demonstrated with long-term durability, which remains a challenge for further evaluation and material development, in particular, under dynamic operation with thermal, load, or start/stop cycling.

8.8 Questions

8.1. Explain the following terms and materials:

High-temperature PEMFC and PA/PBI membrane characteristics
Acid uptake, acid content, and acid doping level
Acid–base interaction, acid anion chain, and acid water chain
Linear polymer and cross-linked polymer membrane

PA/PBI/SiO_2 composite membranes
Acid adsorption on Pt, acid tolerance of NPMC and enhanced activity and stability of NPMC
Solid acid and functionalized mesoporous silica proton conductor

8.2. A higher operating temperature is expected to enhance the kinetics of the ORR at the cathode. This is, however, not realized in the PBI-based HT-PEM fuel cells. Explain why.
8.3. One of the potential advantages of the HT-PEMFC is better use of the waste heat. Give examples of technological utilization of the waste heat in a temperature range from 120 to 200 °C.
8.4. The high CO tolerance of HT-PEMFC means that the CO can pass through the fuel cell stack with no significant deterioration the performance. The anode exit containing CO (as well as hydrogen) must be treated before it is released into atmosphere. How will the anode exit gas be cleaned in practice?
8.5. What are the similarities and differences between the HT-PEMFC and PAFC in terms of key materials and their characteristics?
8.6. What are the similarities and differences between the HT-PEMFC and Nafion-based PEMFC in terms of key materials and their characteristics?
8.7. Of the two types of acid-doped PBI membranes, the direct cast membranes have much higher conductivity than that of the DMAc cast and post-doped membranes (See Fig. 8.3). The performances of fuel cells based on these two kinds of membranes are however very close to each other. Why?
8.8. What are the most critical issues for further improvement of the HT-PEMFC durability?
8.9. The results have shown that the proton conductivity of HPW-functionalized mesoporous silica as shown in Fig. 8.14 is much less sensitive to RH as compared to Nafion membrane. Explain the reason.

8.9 General Readings

1. Aili D, Henkensmeier D, Martin S, Singh B, Hu Y, Jensen JO, Cleemann LN, Li QF (2020) Polybenzimidazole-Based High-Temperature Polymer Electrolyte Membrane Fuel Cells: New Insights and Recent Progress. Electrochemical Energy Reviews 3:793–845.

References

1. Zhang J, Xiang Y, Lu SF, Jiang SP (2018) High temperature polymer electrolyte membrane fuel cells for integrated fuel cell—Methanol reformer power systems: a critical review. Adv Sustain Syst 2(8–9):1700184
2. Wainright JS, Wang JT, Weng D, Savinell RF, Litt M (1995) Acid-doped polybenzimidazoles—a new polymer electrolyte. J Electrochem Soc 142(7):L121–L123
3. He RH, Li QF, Bach A, Jensen JO, Bjerrum NJ (2006) Physicochemical properties of phosphoric acid doped polybenzimidazole membranes for fuel cells. J Membr Sci 277(1–2):38–45

4. Ma YL, Wainright JS, Litt MH, Savinell RF (2004) Conductivity of PBI membranes for high-temperature polymer electrolyte fuel cells. J Electrochem Soc 151(1):A8–A16
5. Xiao LX, Zhang HF, Scanlon E, Ramanathan LS, Choe EW, Rogers D, Apple T, Benicewicz BC (2005) High-temperature polybenzimidazole fuel cell membranes via a sol-gel process. Chem Mat 17(21):5328–5333
6. Lobato J, Canizares P, Rodrigo MA, Linares JJ, Lopez-Vizcaino R (2008) Performance of a vapor-fed polybenzimidazole (PBI)-based direct methanol fuel cell. Energy Fuels 22(5):3335–3345
7. Aili D, Vassiliev A, Jensen JO, Schmidt TJ, Li QF (2015) Methyl phosphate formation as a major degradation mode of direct methanol fuel cells with phosphoric acid based electrolytes. J Power Sources 279:517–521
8. Yu S, Benicewicz BC (2009) Synthesis and properties of functionalized polybenzimidazoles for high-temperature PEMFCs. Macromolecules 42(22):8640–8648
9. Yang JS, Xu YX, Zhou L, Che QT, He RH, Li QF (2013) Hydroxyl pyridine containing polybenzimidazole membranes for proton exchange membrane fuel cells. J Membr Sci 446:318–325
10. Li QF, Pan C, Jensen JO, Noye P, Bjerrum NJ (2007) Cross-linked polybenzimidazole membranes for fuel cells. Chem Mat 19(3):350–352
11. Krishnan NN, Joseph D, Duong NMH, Konovalova A, Jang JH, Kim HJ, Nam SW, Henkensmeier D (2017) Phosphoric acid doped crosslinked polybenzimidazole (PBI-OO) blend membranes for high temperature polymer electrolyte fuel cells. J Membr Sci 544:416–424
12. Kerres JA (2005) Blended and cross-linked ionomer membranes for application in membrane fuel cells. Fuel Cells 5(2):230–247
13. Aili D, Cleemann LN, Li QF, Jensen JO, Christensen E, Bjerrum NJ (2012) Thermal curing of PBI membranes for high temperature PEM fuel cells. J Mater Chem 22(12):5444–5453
14. Li QF, Rudbeck HC, Chromik A, Jensen JO, Pan C, Steenberg T, Calverley M, Bjerrum NJ, Kerres J (2010) Properties, degradation and high temperature fuel cell test of different types of PBI and PBI blend membranes. J Membr Sci 347(1–2):260–270
15. Li Q, Jensen JO, Pan C, Bandur V, Nilsson MS, Schonberger F, Chromik A, Hein M, Haring T, Kerres J, Bjerrum NJ (2008) Partially fluorinated aarylene polyethers and their ternary blends with PBI and H_3PO_4. Part II. Characterisation and fuel cell tests of the ternary membranes. Fuel Cells 8(3–4):188–199
16. Sun X, Simonsen SC, Norby T, Chatzitakis A (2019) Composite membranes for high temperature PEM fuel cells and electrolysers: a critical review. Membranes 9(7):83
17. Plackett D, Siu A, Li QF, Pan C, Jensen JO, Nielsen SF, Permyakova AA, Bjerrum NJ (2011) High-temperature proton exchange membranes based on polybenzimidazole and clay composites for fuel cells. J Membr Sci 383(1–2):78–87
18. Cheng Y, Zhang J, Lu SF, Kuang HH, Bradley J, De Marco R, Aili D, Li QF, Cui CQ, Jiang SP (2018) High CO tolerance of new SiO_2 doped phosphoric acid/polybenzimidazole polymer electrolyte membrane fuel cells at high temperatures of 200–250 °C. Int J Hydrog Energy 43(49):22487–22499
19. Kaserer S, Caldwell KM, Ramaker DE, Roth C (2013) Analyzing the influence of H_3PO_4 as catalyst poison in high temperature PEM fuel cells using in-operando X-ray absorption spectroscopy. J Phys Chem C 117(12):6210–6217
20. Hu Y, Jensen JO, Pan C, Cleemann LN, Shypunov I, Li QF (2018) Immunity of the Fe-N-C catalysts to electrolyte adsorption: Phosphate but not perchloric anions. Appl Catal B-Environ 234:357–364
21. He QG, Shyam B, Nishijima M, Ramaker D, Mukerjee S (2013) Mitigating phosphate anion poisoning of cathodic Pt/C catalysts in phosphoric acid fuel cells. J Phys Chem C 117(10):4877–4887
22. Cheng Y, He S, Lu SF, Veder JP, Johannessen B, Thomsen L, Saunders M, Becker T, De Marco R, Li QF, Yang SZ, Jiang SP (2019) Iron single atoms on graphene as nonprecious metal catalysts for high-temperature polymer electrolyte membrane fuel cells. Adv Sci 6(10):e1802066
23. Cheng Y, Wang ME, Lu SF, Tang CJ, Wu X, Veder JP, Johannessen B, Thomsen L, Zhang J, Yang SZ, Wang SY, Jiang SP (2021) First demonstration of phosphate enhanced atomically dispersed bimetallic FeCu catalysts as Pt-free cathodes for high temperature phosphoric acid doped polybenzimidazole fuel cells. Appl Catal B-Environ 284:119717

24. Fujigaya T, Okamoto M, Nakashima N (2009) Design of an assembly of pyridine-containing polybenzimidazole, carbon nanotubes and Pt nanoparticles for a fuel cell electrocatalyst with a high electrochemically active surface area. Carbon 47(14):3227–3232
25. Martin S, Garcia-Ybarra PL, Castillo JL (2018) Ten-fold reduction from the state-of-the-art platinum loading of electrodes prepared by electrospraying for high temperature proton exchange membrane fuel cells. Electrochem Commun 93:57–61
26. Yao DM, Zhang WQ, Ma Q, Xu Q, Pasupathi S, Su HN (2019) Achieving high Pt utilization and superior performance of high temperature polymer electrolyte membrane fuel cell by employing low-Pt-content catalyst and microporous layer free electrode design. J Power Sources 426:124–133
27. Kwon Y, Kim TY, Yoo DY, Hong SG, Park JO (2009) Maximization of high-temperature proton exchange membrane fuel cell performance with the optimum distribution of phosphoric acid. J Power Sources 188(2):463–467
28. Aili D, Henkensmeier D, Martin S, Singh B, Hu Y, Jensen JO, Cleemann LN, Li QF (2020) Polybenzimidazole-based high-temperature polymer electrolyte membrane fuel cells: new insights and recent progress. Electrochem Energy Rev
29. Li QF, He RH, Gao JA, Jensen JO, Bjerrum NJ (2003) The CO poisoning effect in PEMFCs operational at temperatures up to 200 °C. J Electrochem Soc 150(12):A1599–A1605
30. Kallitsis JK, Geormezi M, Neophytides SG (2009) Polymer electrolyte membranes for high-temperature fuel cells based on aromatic polyethers bearing pyridine units. Polym Int 58(11):1226–1233
31. Xu X, Wang HN, Lu SF, Guo ZB, Rao SY, Xiu RJ, Xiang Y (2015) A novel phosphoric acid doped poly(ethersulphone)-poly(vinyl pyrrolidone) blend membrane for high-temperature proton exchange membrane fuel cells. J Power Sources 286:458–463
32. Guo ZB, Xu X, Xiang Y, Lu SF, Jiang SP (2015) New anhydrous proton exchange membranes for high-temperature fuel cells based on PVDF-PVP blended polymers. J Mater Chem A 3(1):148–155
33. Haile SM, Chisholm CRI, Sasaki K, Boysen DA, Uda T (2007) Solid acid proton conductors: from laboratory curiosities to fuel cell electrolytes. Faraday Discuss 134:17–39
34. Jin Y, Shen Y, Hibino T (2010) Proton conduction in metal pyrophosphates (MP_2O_7) at intermediate temperatures. J Mater Chem 20(30):6214–6217
35. Phadke SR, Bowers CR, Wachsman ED, Nino JC (2011) Proton conduction in acceptor doped SnP_2O_7. Solid State Ionics 183(1):26–31
36. Shen YB, Heo P, Pak C, Chang H, Hibino T (2012) Intermediate-temperature, non-humidified proton exchange membrane fuel cell with a highly proton-conducting $Fe_{0.4}Ta_{0.5}P_2O_7$ electrolyte. Electrochem Commun 24:82–84
37. Jiang SP (2014) Functionalized mesoporous structured inorganic materials as high temperature proton exchange membranes for fuel cells. J Mater Chem A 2(21):7637–7655
38. Lu S, Wang D, Jiang SP, Xiang Y, Lu J, Zeng J (2010) HPW/MCM-41 phosphotungstic acid/mesoporous silica composites as novel proton-exchange membranes for elevated-temperature fuel cells. Adv Mater 22(9):971–976
39. Zeng J, Jin B, Shen PK, He B, Lamb K, De Marco R, Jiang SP (2013) Stack performance of phosphotungstic acid functionalized mesoporous silica (HPW-meso-silica) nanocomposite high temperature proton exchange membrane fuel cells. Int J Hydrogen Energy 38(29):12830–12837
40. Eberhardt SH, Lochner T, Büchi FN, Schmidt TJ (2015) Correlating electrolyte inventory and lifetime of HT-PEFC by accelerated stress testing. J Electrochem Soc 162(12):F1367–F1372
41. Sondergaard T, Cleemann LN, Becker H, Steenberg T, Hjuler HA, Seerup L, Li QF, Jensen JO (2018) Long-term durability of PBI-based HT-PEM fuel cells: effect of operating parameters. J Electrochem Soc 165(6):F3053–F3062
42. Pingitore AT, Huang F, Qian GQ, Benicewicz BC (2019) Durable high polymer content m/p-Polybenzimidazole membranes for extended lifetime electrochemical devices. ACS Appl Energy Mater 2(3):1720–1726
43. Pinar FJ, Pilinski N, Rastedt M, Wagner P (2015) Performance of a high-temperature PEM fuel cell operated with oxygen enriched cathode air and hydrogen from synthetic reformate. Int J Hydrogen Energy 40(15):5432–5438

Part III
Solid Oxide Fuel Cells

Chapter 9
Solid Oxide Fuel Cells: Principles and Materials

9.1 Introduction

9.1.1 Operation Principles

Solid oxide fuel cell (SOFC) is also termed ceramic fuel cell due to the fact that cell components of a SOFC consist of only ceramic materials. The most common ceramic materials in a typical SOFC are dense oxide ion conducting yttria-stabilized zirconia (YSZ) electrolyte, porous lanthanum strontium manganite (LSM) cathode and nickel-YSZ (Ni-YSZ) cermet anode (see Fig. 9.1). The electrolyte needs to be dense to prevent the crossover of fuel and mixture of fuel and air. Porous electrodes are important for the transfer of reactants of fuel and oxidant to the reaction site and for the removal of products of the reaction.

The electrode reaction can bc illustrated using the most common H_2–O_2 fuel cell. At the cathode, the oxidant is oxygen from air and oxygen molecules are ionized to oxygen ion, $O^{=}$ by taking two electrons and incorporated into oxygen lattice site of an oxide ion conducting electrolyte by occupying an oxygen vacancy. Such oxygen reduction reaction (ORR) can be written as follows, using Kröger–Vink notation:

$$\frac{1}{2}O_{2(\text{cathode})} + 2e^{-} + V_O^{\bullet\bullet} \rightarrow O^x_{O\,(\text{electrolyte, cathode})} \tag{9.1}$$

where $V_O^{\bullet\bullet}$ is a vacant oxygen lattice site or oxygen vacancy and O_O^x is an oxygen ion on a regular oxygen site in the electrolyte lattice. At the same time, oxygen ions flow from the cathode side to the anode side through dense YSZ electrolyte via oxygen vacancies:

$$O^x_{O\,(\text{electrolyte, cathode})} \xrightarrow{V_O^{\bullet\bullet}} O^x_{O\,(\text{electrolyte, anode})} \tag{9.2}$$

At the anode, H_2 molecules react with the oxygen ions migrated from the cathode, forming H_2O, releasing two electrons and an oxygen vacancy back to the electrolyte:

S. P. Jiang and Q. Li, *Introduction to Fuel Cells*,
https://doi.org/10.1007/978-981-10-7626-8_9

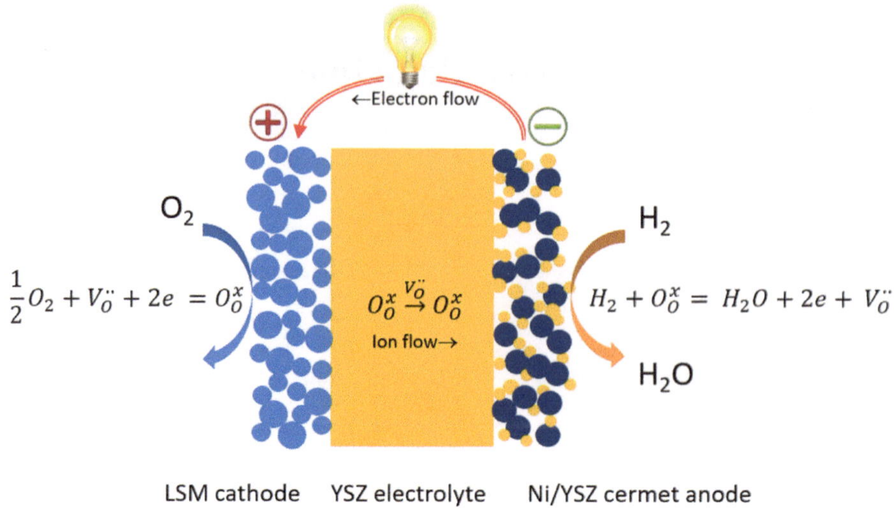

Fig. 9.1 Operating principle of a solid oxide fuel cell with common LSM cathode, YSZ electrolyte and Ni/YSZ cermet anode. The use of YSZ in the cermet anode extends the electrolyte phase into the anode

$$H_{2(\text{anode})} + O^{x}_{O(\text{electrolyte, anode})} \rightarrow H_2O_{(\text{anode})} + 2e + V^{\bullet\bullet}_{O(\text{electrolyte})} \tag{9.3}$$

The overall reaction is thus the combination of hydrogen from the anode and oxygen from the cathode via the oxygen vacancy migration through the electrolyte to form water on the anode side. In the meantime, the electrons flow from the anode (negative electrode) to the cathode (positive electrode) to do the electrical work, producing a power output. Thus, the overall reaction in SOFC can be written as:

$$H_{2(\text{anode})} + \frac{1}{2}O_{2(\text{cathode})} \rightarrow H_2O_{(\text{anode})} \tag{9.4}$$

The forward arrow in the reactions indicates the discharge mode of SOFC. In equilibrium, the forward and backward reaction rates in Eqs. (9.1–9.3) are equal and the above reaction becomes

$$H_{2(\text{anode})} + \frac{1}{2}O_{2(\text{cathode})} \rightleftharpoons H_2O_{(\text{anode})} \tag{9.5}$$

In SOFCs, the equilibrium potential or open-circuit voltage of the cell is a sole function of the partial pressure of oxygen at the cathode and anode sides because the force to drive the overall cell reaction is the difference between the chemical potentials of oxygen at the cathode and anode sides. If the oxygen ions migrated to the anode could be treated as oxygen species, the reaction (9.3) at equilibrium could also be expressed as follows:

$$H_{2(\text{anode})} + \frac{1}{2}O_{2(\text{anode})} \rightleftharpoons H_2O_{(\text{anode})} + 2e + V_{O(\text{electrolyte})}^{\bullet\bullet} \tag{9.6}$$

Thus, based on the partial pressure of oxygen at the cathode, i.e., Eq. (9.1), and partial pressure of oxygen at the anode, i.e., Eq. (9.6), the equilibrium potential can be described by the Nernst equation:

$$E = \frac{\text{RT}}{2nF}\ln\frac{p_{O_{2(c)}}}{p_{O_{2(a)}}} \tag{9.7}$$

where E is the *electromotive force* (emf) or the reversible (thermodynamic) potential or voltage and n is the number of electrons consumed in the reaction, $p_{O_{2(a)}}$ and $p_{O_{2(c)}}$ are the partial pressure of oxygen at the anode and cathode, respectively, R is the universal gas constant (=8.314 J/mol·K), T is the absolute temperature, and F is the Faraday constant (=96485 C/mol). For the H_2–O_2 fuel cell, the number of electrons involved in the reaction, n, is 2.

As the most common oxidant used in a SOFC is oxygen, $p_{O_{2(c)}}$ is either 1 or 0.21, depending on whether pure oxygen or air is used. However, the oxygen partial pressure at the anode, $p_{O_{2(a)}}$, depends on the type and composition of the fuel fed to the anode. In the case of hydrogen fuel, $p_{O_{2(a)}}$ is related to the equilibrium constant, K, of reaction (9.6):

$$K = \frac{p_{H_2O}}{p_{H_2} \times p_{O_{2(a)}}^{1/2}} \tag{9.8}$$

Substituting Eq. (9.8) for the anode oxygen partial pressure in Eq. (9.7) yields

$$E = E^0 + \frac{\text{RT}}{2nF}\ln p_{O_{2(c)}} + \frac{\text{RT}}{nF}\ln\frac{p_{H_2}}{p_{H_2O}} \tag{9.9}$$

where E^o is the equilibrium potential at standard temperature and pressure conditions (298.15 K and 1 atm) and is given as:

$$E^0 = \frac{\text{RT}}{nF}\ln K \tag{9.10}$$

Equation (9.9) is the Nernst equation (see Chap. 2) for the reaction (9.5). This in turn confirms the validity of Eq. (9.7). Equation (9.7) actually tells us that a SOFC can also be treated as a *concentration cell* electrochemically.

9.1.2 Charge Transport in Oxide Materials

As shown in Fig. 9.1, the operation of a SOFC is fundamentally based on the conduction or transport of charges in various ceramic cell components. There are two major types of charged particles in SOFCs: electrons (or electron holes) and oxide ions. The ionic species are transported through the electrolyte between two electrodes, and electrodes carry electrons to and from the reactive sites where the electrochemical reactions take place. The interconnect transfers electrons from the anode of one cell to the cathode of another cell to form a stack. Thus, a current that arises from the flow of electrons is termed *electronic conduction*, while a current due to the net motion of charged ions is termed *ionic conduction*. However, ceramic conductors have an intrinsic resistance to charge flow and resistance to charge transport results in a voltage loss for fuel cells. The voltage loss due to the resistance is generally termed as the ohmic loss or ohmic polarization. In fuel cells, the electrical conductivity, σ, is often used, which is the reciprocal of electrical resistivity, ρ:

$$\sigma = \frac{1}{\rho} \tag{9.11}$$

The conductivity for most materials can be expressed as the product of concentration and mobility of charge carrier:

$$\sigma = (nze)\frac{v}{\xi} = (nze)\mu \tag{9.12}$$

where n is the number of charge carriers, electrons (electron holes) or oxide ions in carriers/cm^3, ze is the charge per each carrier in which z is the valence and e is the electronic charge, v is the velocity (cm/s) of charge carriers and ξ is the electric field. The term v/ξ is defined as the mobility, μ. Thus, material conductivity is characterized by two major factors: the number of carriers available and the mobility of these carriers. For materials with more than one carrier, this can be expressed as:

$$\sigma_i = F\sum |z_i| c_i \mu_i \tag{9.13}$$

where c_i is the number of moles of charge carriers per unit volume, μ_i is the mobility of charge carriers within the material (cm^2/Vs), z_i is the charge number (or valence electrons) for the carrier ($|z_i|$ is always positive), and F is the Faraday's constant. The subscript i can be electrons, electron holes, ions, or vacancies.

In ceramic materials, electronic and ionic defects are the charge carriers that enable the conduction of electric current. These charge carriers can be formed within the material (intrinsic charge carriers), thus increasing entropy of the system, or can be introduced into the material by doping (extrinsic charge carriers). Examples for ionic point defects are vacancies and interstitials. Electronic defects are negatively charged *electrons* and positively charged *electron holes*, denoted as e' and $h^{\bullet}$, respectively, in the Kröger–Vink notation.

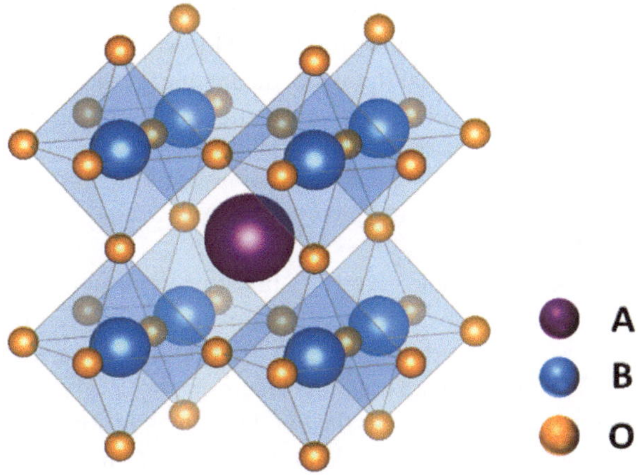

Fig. 9.2 A typical ABO_3 perovskite structure

9.1.2.1 Oxides with Perovskite Structure

Before the discussion of charge transport in specific oxide materials of SOFCs, it is useful to briefly introduce perovskite structure as majority of oxide materials used in SOFCs belong to perovskite and perovskite-derived structured oxides. A *perovskite* is any material with the same type of crystal structure as calcium titanate, $CaTiO_3$, with the general chemical formula of ABO_3, where "A" and "B" are two cations of very different sizes, and O is an oxygen anion. The ideal cubic structure has the B cation in sixfold coordination (coordination number or CN = 6), surrounded by an octahedron of oxygen anions, and the A cation in 12-fold cuboctahedral coordination (CN = 12) (Fig. 9.2).

The deviation from the ideal cubic perovskite structure is governed by the *Goldschmidt tolerance factor*, *t:*

$$t = \frac{(r_A + r_O)}{\sqrt{2}(r_B + r_O)} \tag{9.14}$$

where r_A, r_B and $r_O$are the ionic radii of cations *A* and *B* and oxygen anion, respectively. Cubic symmetry has been observed for $0.95 \leq t \leq 1.04$, while perovskite structure with lower-symmetry distorted versions, such as orthorhombic and rhombohedral symmetries are observed for tolerance factor in the range of 0.75–0.90. The ionic radii is related to CN.

9.1.2.2 Electronic and Ionic Conduction in Perovskite Oxides

In a metallic conductor, valence electrons associated with atoms of the metal become detached; i.e., they are excited and promoted into one of the empty and available energy state above the *Fermi level* of electrons and become free electrons that participate in the conduction process. This is due to the fact that for metals, the band structure is such that there are vacant energy states adjacent to the highest filled state at the Fermi level. Thus, very little energy is required to promote electrons to the low-lying empty states.

Electron hole

Hole is the name given to a missing electron in certain solids, especially semiconductors. According to the band theory of solids, electrons within a solid have energies only at certain discrete levels that combine into groups or bands. The valence band contains electrons that are bound into the atomic structure of the material, whereas the conduction band contains electrons at higher energies that are free to move. With the application of thermal energy, an electron can be promoted from the valence band across a band gap and into the conduction band, which leaves behind a hole. Since a missing electron is the same as an added positive electric charge, holes can carry a current—like that of electrons but in the opposite direction—under an electric field. Holes generally move more slowly than electrons, however, because they function within the tightly bound valence band rather than the conduction band.

Most ceramics or more precisely ionic ceramics are insulating materials at room temperature with typical electron energy band structures such that a filled valence band is separated from an empty conduction band by a relatively large band gap. Thus, at normal temperatures, only very few electrons may be excited across the band gap by the available thermal energy, resulting in a very low value of the conductivity. Increasing the temperature will increase the conductivity of oxide materials in general, however, in the case of oxide materials with metallic behavior, the conductivity will decrease with increasing temperature as shown late in this Chapter. For the ceramic oxide components used in SOFCs, there are ways to increase the electronic and ionic conductivity by playing with the *defect chemistry*.

The electronic conductivity of oxide materials of SOFCs can be either *n*- or *p*-type, with *p*-type materials being more common. Taking $LaCrO_3$ as an example. $LaCrO_3$ is a perovskite structured oxide with $A = La$ and $B = Cr$. $LaCrO_3$ is a *p*-type conductor from room temperature to high temperature (>1000 °C), and becomes nonstoichiometric through the reaction of oxygen with $LaCrO_3$, resulting in the formation of cation vacancies or defects:

$$\frac{3}{2}O_2 = V_{La}''' + V_{Cr}''' + 3O_O^x + 6h^\bullet \tag{9.15}$$

where V_{La}''' and V_{Cr}''' are La and Cr *cation vacancies*, respectively, and $h^\bullet$ is an electron hole. The negatively charged cation vacancies are electrically compensated by the concomitant appearance of positively charged electron holes. Electronic conduction in $LaCrO_3$ occurs by the small polaron hopping via transport of electron holes.

However, the electronic conductivity of undoped $LaCrO_3$ is very low. To improve the *p*-type electronic conductivity, we can substitute trivalent La cations with divalent cations such as strontium (Sr) and calcium (Ca), forming lanthum strontium chromite (LSCr) and lanthanum calcium chromite (LCCr):

$$\mathrm{SrO} \xrightarrow{\mathrm{LaCrO_3}} \mathrm{Sr}'_{La} + \mathrm{O}^x_O + h^\bullet \tag{9.16}$$

Or in Kröger–Vink notation,

$$\mathrm{SrO} \xrightarrow{\mathrm{LaCrO_3}} \mathrm{Sr}'_{La} + \mathrm{O}^x_O + \mathrm{Cr}^\bullet_{Cr} \tag{9.17}$$

Sr'_{La} is Sr^{2+} occupied on a La^{3+} lattice site and the valence change of Sr'_{La} is (+2)–(+3) = −1. Thus, divalent Sr on a trivalent La site creates a negatively charged defect. $\mathrm{Cr}^\bullet_{Cr}$ is Cr^{4+} on a Cr^{3+} site and the valence change of $\mathrm{Cr}^\bullet_{Cr}$ is (+4)–(+3) = +1, representing a positively charged electron hole localized on a Cr site. The excess negative charge of Sr'_{La} is compensated for by positively charged electron hole $\mathrm{Cr}^\bullet_{Cr}$ has the same function as $h^\bullet$.

The introduction of a positively charged electron hole on a cation site will lead to the attraction of anions and to the repulsion of cations at the same time. Such polarization of the surroundings of a localized electron hole due to the attraction and repulsion of ions is called a *polaron*. This is schematically shown in Fig. 9.3. For a small polaron, the polarization is of the size of one unit cell, while for large polaron, the polarized volume is much bigger than a unit cell.

The charge compensation can also occur by the formation of *oxygen vacancies*:

$$\mathrm{O}^x_O + 2\mathrm{Cr}^\bullet_{Cr} = V^{\bullet\bullet}_O + \frac{1}{2}\mathrm{O}_2 + 2\mathrm{Cr}^x_{Cr} \tag{9.18}$$

At equilibrium constant, we have

$$K^0 = \frac{\left[V^{\bullet\bullet}_O\right] \times \left[\mathrm{Cr}^x_{Cr}\right]^2 \times p^{1/2}_{O_2}}{\left[\mathrm{O}^x_O\right] \times \left[\mathrm{Cr}^\bullet_{Cr}\right]^2} \tag{9.19}$$

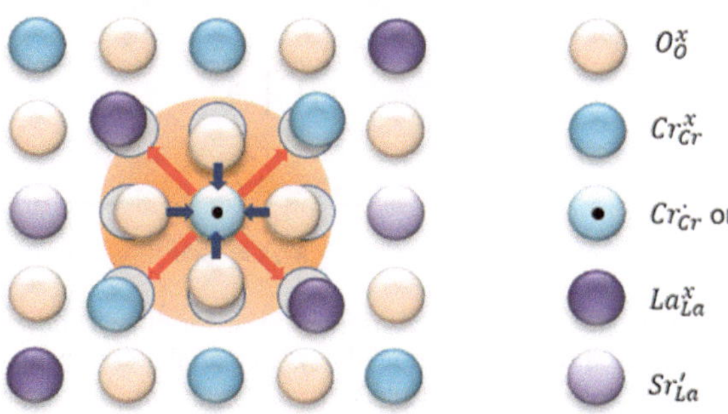

Fig. 9.3 Charge compensation of Sr^{2+} dopant in a La^{3+} site by the formation of positively charged electron holes *via* $Cr^{3+} \rightarrow Cr^{4+}$ transition in Sr-doped $LaCrO_3$. For simplicity, the arrangement of cations and anions is not to the scale

where K^o is the equilibrium constant of Eq. (9.18). This equation does not explicitly include the extrinsic point defect, but the concentration of point defect is involved in the charge neutrality condition:

$$\left[V_O^{\bullet\bullet}\right] + 2\left[\mathrm{Cr}_{Cr}^{\bullet}\right] = 2\left[\mathrm{Sr}_{La}'\right] \tag{9.20}$$

This means that under reducing conditions, the lattice oxygen transforms into a doubly charged oxygen vacancy consuming two electron holes simultaneously. Thus, for Sr-doped $LaCrO_3$, $La_{1-x}Sr_xCrO_3$ should be expressed as $La_{1-x}Sr_xCrO_{3-\delta}$ due to the presence of oxygen vacancies where subscript δ is used to represent the *oxygen nonstoichiometry*. The drop of electrical conductivity with decreasing oxygen partial pressure is a reflection of the reduced electron hole concentration $[h^\bullet]$. Nevertheless, under oxidizing environment ($p_{O_2} > 10^{-8}$ atm), oxygen vacancies are negligible. Therefore, all charge compensation occurs primarily via a $Cr^{3+} \rightarrow Cr^{4+}$ transition and the neutrality condition can be simply described as:

$$\left[\mathrm{Sr}_{La}'\right] = \left[\mathrm{Cr}_{Cr}^{\bullet}\right] \tag{9.21}$$

The *p*-type (hole) electrical conductivity increases with the concentration of Sr_{La}' and becomes independent of oxygen partial pressure beyond a pressure threshold. At low oxygen partial pressure, ionic compensation becomes a dominant mechanism through the formation of oxygen vacancies, resulting in a significant drop in electrical conductivity. Figure 9.4 shows the conductivity isotherms of Ca-doped $LaCrO_3$ at 1000 °C, showing the electronic compensation mechanism region at high $p_{O_2} > 10^{-10}$ atm ($[h^\bullet]$ is constant) and ionic compensation mechanism at low $p_{O_2} < 10^{-10}$–10^{-15} atm ($[h^\bullet]$ decreases). Nevertheless, the magnitude of pressure threshold scales with doping level.

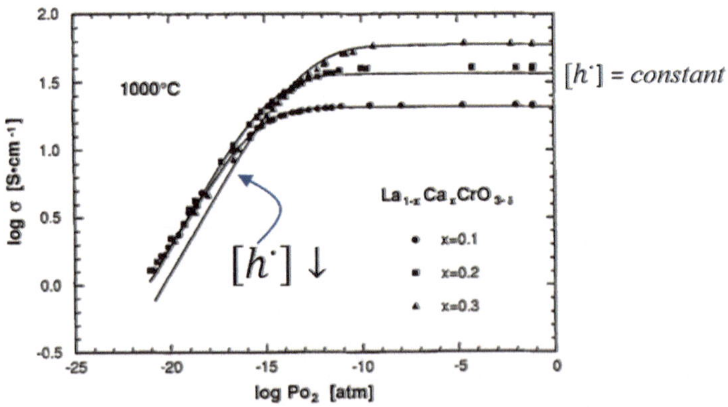

Fig. 9.4 Conductivity isotherms at 1000 °C for three different compositions: $x = 0.1$, 0.2 and 0.3 in $La_{1-x}Ca_xCrO_{3-\delta}$ as a function of oxygen partial pressure. Modified from Ref. [1] with permission from IOP Publishing, Copyright 1993

Electrical conductivity through hopping of small polarons is thermally activated and is generally expressed as

$$\sigma T = A_o \, exp\left(-\frac{E_a}{kT}\right) \tag{9.22}$$

where A_o is the pre-exponential constant and E_a is the activation energy, which can be obtained from the slope of the plots of log(σT) *versus* 1/*T*.

Strontium titanate perovskite, $SrTiO_3$, provides an example of an *n*-type conductor. In contrast to $LaCrO_3$ where a Sr^{2+} cation on a La^{3+} site creates a negatively charged defect, trivalent La on a divalent Sr site creates a positively charged defect, the excess charge being compensated for by negatively charged electrons:

$$La_2O_3 \xrightarrow{SrTiO_3} 2La^{\bullet}_{Sr} + 3O^{x}_{O} + 2e^{-} \tag{9.23}$$

Or in Kröger–Vink notation,

$$La_2O_3 \xrightarrow{SrTiO_3} 2La^{\bullet}_{Sr} + 3O^{x}_{O} + 2Ti'_{Ti} \tag{9.24}$$

$La^{\bullet}_{Sr}$ is La^{3+} on a Sr^{2+} lattice site and the valence change of $La^{\bullet}_{Sr}$ is (+3)–(+2) = +1. Thus, a trivalent La on a divalent Sr site creates a positively charged defect, while Ti'_{Ti} is Ti^{3+} on a Ti^{4+} site and the valence change of Ti'_{Ti} is (+3)–(+4) = −1, representing a negatively charged electron localized on a Ti site. The positive charge of $La^{\bullet}_{Sr}$ is compensated for by negatively charged electron Ti'_{Ti}. At low oxygen partial pressure and high temperature, the primary compensation mechanism is the reduction of Ti^{4+} to Ti^{3+}. Different to the electron holes located in the valence band of Cr in Sr-doped $LaCrO_3$, electrons formed in Ti'_{Ti} are located in the conduction band formed by the overlap of the Ti 3d orbitals. As expected, the electrical conductivity of doped $SrTiO_3$ increases with the decrease of oxygen partial pressure and decreases with the temperature, behaving like a metallic conductor.

Generally for the mixed ionic and electronic conducting (MIEC) ABO_3 perovskite oxides, the electrical conductivity behavior is closely related to the concentration of the high valence transition metal ions that are located in the B-site, which are charge carriers of electron holes. In the meantime, the generation of oxygen vacancies in the lattice will inhibit the electronic conduction. This can be explained by the conductivity behavior of the most representative MIEC material like $La_{0.6}Sr_{0.4}Co_{0.2}Fe_{0.8}O_{3-\delta}$ (LSCF).

LSCF is a typical *acceptor*-doped (acceptor dopants are cations with lower oxidation state than the host cation) cobaltite perovskite-type oxide and is characterized by enhanced lattice oxygen vacancy formation, which results in substantial departure from oxygen stoichiometry at increased temperatures. For example, $La_{0.8}Sr_{0.2}Co_{0.2}Fe_{0.8}O_{3-\delta}$ is stoichiometric (*i.e.*, $\delta = 0$) at room temperature and only becomes nonstoichiometric at high temperature and low p_{O_2}. At 800 °C, δ is 0.02 in air [2] and increases to 0.07 at $p_{O_2} = 1.4 \times 10^{-3}$ atm [3]. The high concentration of oxygen vacancies in conjunction with their relatively high mobility causes these materials to exhibit high oxygen ion conductivity. The electronc conductivity is even higher and becomes metallic at high temperatures.

For acceptor-doped perovskite oxides, the charge compensation occurs by both ionic and electronic defects. If the electronic charge compensation occurs preferentially by a valence change of iron ions over cobalt ions, the dissolution of SrO in the $La(Fe,Co)O_3$ pervskite lattice can be described by:

$$\mathrm{SrO} \xrightarrow{\mathrm{La(Fe,Co)O_3}} \mathrm{Sr}'_{La} + \mathrm{Fe}^{\bullet}_{Fe} + \mathrm{O}^{x}_{O} \tag{9.25}$$

where $Fe^{\bullet}_{Fe}$ is Fe^{4+} ion in a normal Fe^{3+} ion lattice site with a positive charge (similar to $Cr^{\bullet}_{Cr}$, functioning as an electron hole) and Sr'_{La} is Sr^{2+} in La^{3+} ion lattice site with a negative charge.

The charge compensation can also occur by the formation of positively charged doubly ionized oxygen vacancies:

$$2\mathrm{SrO} \xrightarrow{\mathrm{La(Fe,Co)O_3}} 2\mathrm{Sr}'_{La} + V^{\bullet\bullet}_{O} + 2\mathrm{O}^{x}_{O} \tag{9.26}$$

Since the ionic and electronic compensations occur simultaneously and compete with each other, the charge neutrality condition, assuming only the localized electronic charge carriers to be present and charge compensation also occurs by the valence change of cobalt ions, can be expressed as:

$$\left[V^{\bullet\bullet}_{O}\right] + \left[\mathrm{Fe}^{\bullet}_{Co}\right] + \left[\mathrm{Co}^{\bullet}_{Co}\right] = 4\left[\mathrm{Sr}'_{La}\right] \tag{9.27}$$

It has been known that the oxygen nonstoichiometry (δ) of LSCF varies almost linearly with the logarithm of p_{O_2}. This indicates that the predominant defects in LSCF are oxygen vacancies; $V^{\bullet\bullet}_{O}$, i.e., the charge compensation in LSCF is dominated by Eq. (9.26) rather than Eq. (9.25).

The electronic structure of LSCF is considered to be a combination of $La_{1-x}Sr_xCoO_{3-\delta}$ (LSC) and $La_{1-x}Sr_xFeO_{3-\delta}$ (LSF) perovskites, and the electronic conductivity has been explained by a mixed model; i.e., adiabatic hopping of small polarons (through Fe^{3+} and Fe^{4+} sites) occurs simultaneously with a more band-like conduction via a Co 3d/O 2p hybridization band [4]. LSC and LSF have similar crystal structure but differ in their electronic structure. In LSC, the conduction electrons are delocalized and occupy energy levels in a partially filled conduction band. This band is thought to represent the hybridization state between Co 3d and O 2p states and is responsible for the metal-like high conductivity of LSC [5]. In contrast, electrons and holes are localized in LSF with a hopping-type electrical conductivity [6]. Thus, the electronic conduction in LSCF is a mixed one with band-like conduction by Co 3d and O 2p hybridized orbital and hopping of small polarons with the charge disproportionation of Fe^{3+}. The conductivity of the hopping mechanism is almost constant and that of hybridized bands changes with oxygen vacancy introduction. Regardless of the temperature, the conductivity decreases with decreasing

p_{O_2} due to the decrease in the number of carrier holes with the increasing number of electrons in the Co 3d/O 2p hybridized orbital [7]. With the decrease in p_{O_2}, Co^{3+} is preferentially reduced to Co^{2+}, while Fe remains at higher valence above 3 [8]. This indicates that the charge compensation in LSCF is dominated by the formation of oxygen vacancies.

9.1.2.3 Oxide Ion Conductivity in Fluorite-Type Oxides

Fluorite-type oxides are the main ionic conductors and are used as the electrolyte materials in SOFCs. The fluorite-type structure is a face-centered cubic arrangement of cations, with anions occupying all the tetrahedral sites. The fluorite structure (CaF_2) has a general formula MO_2, where M is a large tetravalent cation, e.g., Zr^{4+}, Ce^{4+}, and Th^{4+} and anion is oxygen ion (see Fig. 9.5). To form the fluorite structure in MO_2, the minimum ionic radius ratio (i.e., the ratio of metal ion radius to oxygen ion radius) is 0.732. This structure has a rather open structure which is beneficial for rapid ion diffusion.

Oxide ion conductivity is generally related to the lattice defect structure in the metal oxides. There are three probable defect models in this case: (i) an oxygen ion vacancy model with all metal ions being fixed at their lattice sites, (ii) a cation

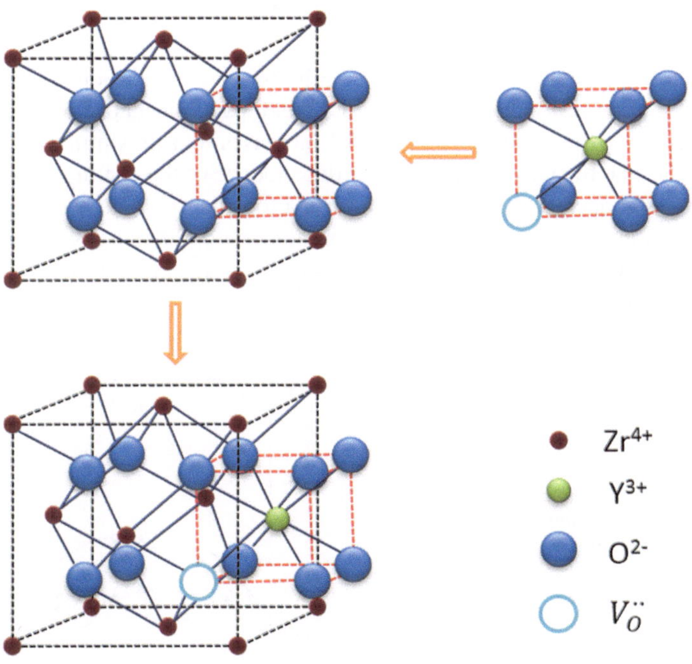

Fig. 9.5 Fluorite structure and scheme of substitution of Zr^{4+} by Y^{3+}, forming oxygen vacancies

interstitial model with all oxygen ions fixed at their lattice sites (Frenkel type), and (iii) a mixed model of (i) and (ii) (Schottky type). In the case of stabilized ZrO_2, the defect structure is the oxygen vacancy model. This also applies for stabilized ceria and other fluorite-type oxide ion conductors.

A typical example is the Y_2O_3-stabilized ZrO_2 or YSZ. Dissolution of Y_2O_3 in ZrO_2 results in incorporation of Y^{3+} on the Zr^{+4} cation sites with the concomitant formation of positively charged oxygen vacancies, $V_O^{\bullet\bullet}$, as charge compensation defects. The defect formation reaction can be written as:

$$Y_2O_3 \xrightarrow{ZrO_2} 2Y'_{Zr} + 3O_O^x + V_O^{\bullet\bullet} \tag{9.28}$$

where Y'_{Zr} is an Y^{3+} ion in an Zr^{+4} lattice site with a negative charge and O_O^x is a normal oxygen lattice site, see Fig. 9.5. A similar equation can also be written for Sc_2O_3 doping into zirconia:

$$Sc_2O_3 \xrightarrow{ZrO_2} 2Sc'_{Zr} + 3O_O^x + V_O^{\bullet\bullet} \tag{9.29}$$

The concentration of the oxygen vacancies is given simply by the electron neutrality condition. In this case, for every two Y^{3+} occupations of Zr^{4+} cation lattice sites, one oxygen vacancy is generated:

$$2[Y'_{Zr}] = [V_O^{\bullet\bullet}] \tag{9.30}$$

Thus, oxide ion conduction occurs through the oxygen vacancies. Oxygen ionic conductivity increases with temperature and the dopant concentration. As the Y_2O_3 content increases, the conductivity of YSZ increases to a maximum for YSZ with 8 mol% Y_2O_3 and then decreases. Similar dependence of the conductivity as a function of dopant concentration has been observed for other dopants (see Fig. 9.6) [9]. The Y_2O_3 content that corresponds to a maximum of conductivity is nearly independent of temperature. The decrease in conductivity is attributed to the formation of *dopant-vacancy pairs* or *defect clusters* according to

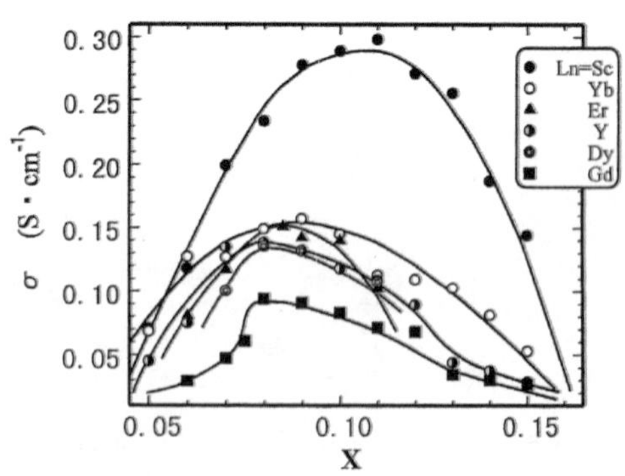

Fig. 9.6 Dependence of the conductivity of $(ZrO_2)_{1-x}(Ln_2O_3)_x$ as a function of dopant concentration, measured at 1000 °C. Reproduced from Ref. [9] with permission from Elsevier, Copyright 1999

$$Y'_{Zr} + V_O^{\bullet\bullet} = \left(Y'_{Zr} - V_O^{\bullet\bullet}\right)^{\bullet} \tag{9.31}$$

$$2Y'_{Zr} + V_O^{\bullet\bullet} = \left(Y'_{Zr} - V_O^{\bullet\bullet} - Y'_{Zr}\right)^{x} \tag{9.32}$$

in which the superscript • indicates the defect clusters with a positive charge and x the neutral defect cluster. These defect clusters trap oxygen vacancies, reducing the defect mobility and thus the conductivity.

The formation of such defect clusters or associates is due to Coulombic and elastic attractive forces between Y'_{Zr} and $V_O^{\bullet\bullet}$ and is affected by the binding enthalpy, E_a, which is more dominant at low temperatures. Based on Eq. (9.12), oxygen ion conductivity can be expressed by:

$$\sigma = (nze)\mu = ze\mu\left[V_O^{\bullet\bullet}\right] \tag{9.33}$$

The mobility μ is given by:

$$\mu = \frac{B}{T}exp\left(-\frac{E_m}{RT}\right) \tag{9.34}$$

where E_m is the enthalpy for the motion and B the constant. The activation energy for oxygen ion conduction E_i is thus:

$$E_i = E_m + E_a \tag{9.35}$$

At higher temperatures, defect clusters tend to be dissociated, forming free $V_O^{\bullet\bullet}$, while at lower temperatures, defect clusters or associations prevail, significantly affecting the population of free vacancies. This is reflected by the distinctive curvature of Arrhenius curves (activation energy plots) of the electrical conductivity of doped ZrO_2 systems [9]. A higher activation energy is observed at lower temperature, and the value of the low-temperature activation energy increases with dopant concentration, which are characteristics of dopant ions acting as nucleating agent for the formation of dopant–vacancy pairs or defect clusters. Such dependence of the conductivity on dopant concentration also exists for other dopants. Nevertheless, the concentration to reach the maximum conductivity changes with the dopants. In the case of YSZ, the maximum conductivity is obtained for YSZ is 8 mol% Y_2O_3, while for Sc-stabilized ZrO_2 (ScYZ), it is 10 mol% Sc_2O_3.

9.2 Electrolyte

The most important component of a SOFC is the electrolyte. The electrolyte conducts charged ionic species produced at one electrode to another electrode to balance the charge due to external electron flow, completing the electrical circuit in fuel cells.

Another function of the electrolyte is to separate the fuel from the oxidant in the fuel cell. Therefore, to be used as the electrolyte in SOFCs, materials should meet certain requirements in stability, conductivity, compatibility, and low cost:

- Must be chemically and structurally stable in both reducing and oxidizing atmosphere at operating temperatures of 500–1000 °C. In addition, the electrolyte should possess sufficient phase and dimensional stability, and mechanical strength during the fabrication and in the long-term operation and thermal cycling so as to prevent detrimental effect and damage to the cell structure and performance.
- Sufficiently high ionic conductivity, preferably 10^{-2} S/cm or higher in the working environment with negligible electronic conductivity.
- Must be chemically and thermally compatible with other cell components to minimize the chemical reactions and thermal stress with compatible thermal expansion coefficient (TEC).
- Low cost and abundant of raw materials for the synthesis of the electrolyte. The electrolyte materials can be easily fabricated in thin layer or film by scalable techniques and be easily densified at relatively low sintering temperatures, preferably $\leq$ 1400 °C.
- The *ionic transfer number*, t_i, of solid electrolyte is preferably 0.95 or higher in the temperature and oxygen partial pressure range relevant to SOFC operating conditions, which is given by:

$$t_i = \frac{\sigma_i}{\sigma_i + \sigma_e' + \sigma_h^{\bullet}} \tag{9.36}$$

where σ_i, σ_e', and $\sigma_h^{\bullet}$ are the ionic, *n*-type (electron), and *p*-type (hole) conductivities, respectively.

Any SOFC electrolyte should meet the requirements simultaneously as stated above. These requirements limit the choice of oxide ion conducting electrolyte materials to doped zirconia, doped ceria, and doped $LaGaO_3$. Some fluoride-type oxides such as Bi_2O_3 exhibit high ionic conductivity. However, Bi_2O_3 or doped Bi_2O_3 also show a high electronic conductivity, and these oxides are not very stable and are easy to be reduced under reducing atmosphere. Certain oxides, particularly perovskite oxides, such as doped $SrCeO_3$ and $BaCeO_3$, have been found to exhibit appreciably high proton conduction in steam-containing atmospheres at high temperatures, which will be discussed in protonic ceramic fuel cells in Chap. 16. Table 9.1 lists the conductivity and thermal properties of most common electrolyte materials in SOFCs.

Table 9.1 Ionic conductivity and thermal expansion coefficient of the most common electrolyte materials in SOFCs, measured in air

Composition	Conductivity, S/cm			E_a, kJ/mol		TEC, cm/cm·K	Sintering temp/°C
	1000 °C	800 °C	600 °C	400–500 °C	850–1000 °C		
$Y_{0.03}Zr_{0.97}O_2$ (TZ3Y or 3YSZ)	0.056	0.022	2.62×10^{-3}	77	92	10.8×10^{-6}	1350–1400
$Y_{0.08}Zr_{0.92}O_2$ (YSZ)	0.182	0.050	5.36×10^{-3}	°C	106	10.0×10^{-6}	1350–1400
$Sc_{0.093}Zr_{0.907}O_2$ (ScSZ)	0.35	0.128	1.3×10^{-2}	130	72	10.7×10^{-6}	1350–1400
$Gd_{0.2}Ce_{0.8}O_{1.9}$ (GDC)		7.49×10^{-2}	1.23×10^{-2}	86		11×10^{-6}	1400–1450
$Sm_{0.2}Ce_{0.8}O_{1.9}$ (SDC)		9.32×10^{-2}	1.64×10^{-2}	82–100		12.1×10^{-6}	1400–1450
$La_{0.8}Sr_{0.2}Ga_{0.8}Mg_{0.2}O_3$ (LSGM)		1.2×10^{-1}	1.7×10^{-2}	61–71		11.5×10^{-6}	1450–1500
$La_{10}Si_6O_{27}$ (LSO)		3.5×10^{-2}	6.62×10^{-3}	88		9.7×10^{-6}	1600–1700

9.2.1 Zirconium Oxides

Pure zirconium oxide, ZrO_2, is not an oxygen conductor. ZrO_2 exists as monoclinic crystal structure at room temperature and transforms to a tetragonal crystal structure at temperatures higher than 950 °C. The tetragonal structure is stable until 2370 °C; above this temperature, the ZrO_2 crystal structure becomes cubic or fluorite.

$$\text{monoclinic} \overset{950-1150\,^{\circ}\mathrm{C}}{\rightleftharpoons} \text{tetragonal} \overset{2370\,^{\circ}\mathrm{C}}{\rightleftharpoons} \text{cubic} \overset{2680\,^{\circ}\mathrm{C}}{\rightleftharpoons} \text{liquid} \tag{9.37}$$

The phase transformations will cause volume changes and decrease the mechanical stability and thermal shock resistance. For electrolyte application in SOFCs, ZrO_2 needs to be stabilized by adding dopants such as rare earth oxides like Y_2O_3 and Sc_2O_3 and alkaline earth oxides like CaO and MgO. In this case, lattice defects are created to preserve the electroneutrality condition in the solid solution, creating oxygen vacancies. In the ZrO_2-Y_2O_3 system, addition of Y_2O_3 reduces the tetragonal/monoclinic transformation temperature. In the composition range from 0–2.5 mol% Y_2O_3, tetragonal phase, *t*-ZrO_2, will transform on cooling to monoclinic phase, *m*-ZrO_2. At a higher Y_2O_3 content, a mixture of nontransformable *t*-ZrO_2 and cubic phase, *c*-ZrO_2, exists. Further increase in Y_2O_3 content leads to a homogenous cubic solid solution. The minimum amount of Y_2O_3 required to fully stabilize the cubic phase of ZrO_2 is 8–10 mol%. The composition of 8–9 mol% Y_2O_3 doped ZrO_2 yields the highest ionic conductivity, 0.14–0.16 S/cm at 1000 °C [9, 10]. In general, the ionic conductivity of *c*-ZrO_2 is independent of the oxygen partial pressure over several orders of magnitude ($1 < p_{O_2} < 10^{-20}$ atm), suggesting that the ionic transport number is close to unity in a wide p_{O_2} range.

The conductivity of stabilized ZrO_2 varies with dopant concentration, see Fig. 9.6. The minimum amount of dopant necessary to stabilize ZrO_2 in the cubic structure is close to the composition that gives the highest conductivity: 8 mol% Y_2O_3, 10 mol% Sc_2O_3, 15 mol% Nd_2O_3, 10 mol% Sm_2O_3, and 10 mol% Gd_2O_3. This maximum corresponds to the minimum dopant level required to fully stabilize the high-temperature cubic phase. Further increase in the dopant content will result in the formation of oxygen vacancy-dopant clusters as discussed. This behavior is characterized by a change in the slope of Arrhenius plots of $\ln(\sigma T)$ versus $1/T$ of zirconia-based electrolyte systems; a higher activation energy E_a is observed at lower temperatures, and the low-temperature value of E_a increases with dopant concentration. The conductivity is mainly determined by the amount of oxygen vacancies. On the other hand, conductivity as well as activation energy for oxygen ion conduction are strongly affected by the dopant ionic size, as reported by Arachi et al. [9] for the ZrO_2-Ln_2O_3 system (Fig. 9.7).

The Sc-doped ZrO_2 (ScSZ) shows the lowest ion migration enthalpy and highest conductivity. ScSZ with 9.0–9.3 mol% Sc_2O_3 has the conductivity of 0.34–0.36 S/cm at 1000 °C [11], higher than that of YSZ. The higher oxide ion conductivity of ScSZ is attributed to the smaller mismatch in size between Zr^{4+} (0.084 nm) and Sc^{3+} (0.087 nm) ions, as compared to that between Zr^{4+} (0.084 nm) and Y^{3+} (0.1019 nm)

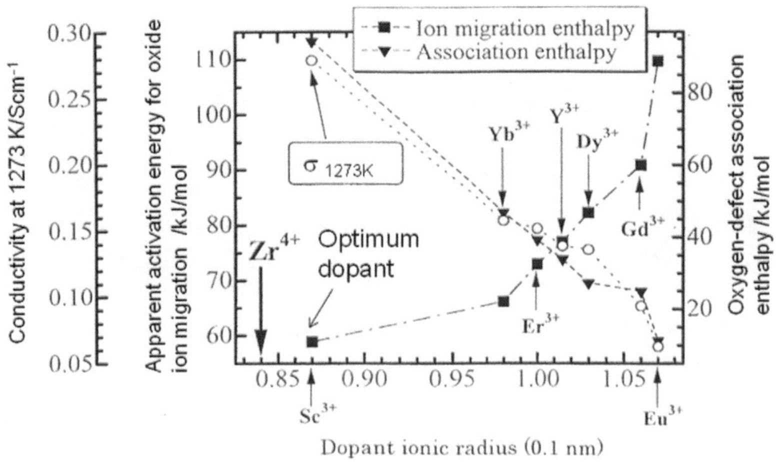

Fig. 9.7 Maximum conductivities, oxygen ion migration enthalpy, and association enthalpy of doped ZrO_2 at a specific concentration of dopant as a function of dopant ion radius. Reproduced from Ref. [9] with permission from Elsevier, Copyright 1999

ions. The ion migration enthalpy increases with increasing size of the dopant ion radius due to the elastic strain energy induced by the size difference on the cation lattice. This indicates that the diffusion of oxygen ion vacancy is affected by the elastic strain energy, which is related to the size mismatch between the host and dopant cations.

An important issue of zirconia-based electrolytes is the degradation in oxide ion conductivity due to the aging that occurs during operation at high temperatures. This phenomenon is commonly found in YSZ with Y_2O_3 content less than 9 mol%. Detailed TEM studies show that ZrO_2 is not perfectly stabilized in the *c*-ZrO_2 phase in YSZ, and the presence of metastable *t*-ZrO_2 contributes to the degradation of the ionic conductivity, 28% degradation in conductivity for 8 mol% Y_2O_3-stabilized zirconia (8YSZ) tested for 1000 h at 950 °C as compared to 5.7% degradation for 10 mol% Y_2O_3-stabilized zirconia (10YSZ) tested under similar conditions [12]. There is correlation between the formation and size distribution of the *t*-ZrO_2 phase and degradation of the ionic conductivity of YSZ. The oxide ion conductivity degradation decreases with increasing Y_2O_3 content and becomes small in the case of 10YSZ. The aging of ScSZ during operation has been attributed to the disappearance of a high conducting distorted fluorite phase, but such a degradation can be partly reduced by increasing the Sc_2O_3 content [11].

One particular attractiveness of YSZ electrolyte is its high mechanical strength. At room temperature, 8 mol% Y_2O_3-doped ZrO_2 or 8YSZ typically has a bending strength of about 150–250 MPa in the operation temperature ranges of SOFCs [13]. Adding Al_2O_3 can increase the mechanical strength of 8YSZ but at the expense of reduced conductivity. The mechanical properties of YSZ depend strongly on the yttria dopant level. For 3 mol% Y_2O_3-doped ZrO_2 or 3YSZ, the bending strength

Fig. 9.8 SEM images of **a** 8YSZ and **b** 3YSZ electrolyte surface after sintering at 1450 °C for 5 h in air

is ~1200 MPa, and the toughness can be as high as 6–9 MPa $m^{1/2}$. The main phase in 3YSZ is polycrystalline t-ZrO_2. Microstructurally, the t-ZrO_2 phase is also very different from the c-ZrO_2 phase. Figure 9.8 shows typical SEM images of surface of 3YSZ and 8YSZ electrolytes, sintered at 1450 °C for 5 h in air. The grain size of 3YSZ grains is in the range of 0.6 μm, significantly smaller than 8YSZ grain size of 1.7–5.7 μm. The much smaller grain size of 3YSZ is typically associated with t-ZrO_2 phase. The high mechanical strength of 3YSZ is due to toughening from the transformation of the tetragonal particles to the monoclinic phase. Consequently, 3YSZ is also called partially stabilized zirconia. It is important to consider grain boundary conduction in stabilized zirconia because the grain boundary contribution increases with decreasing temperature. This is particular the case for 3YSZ due to its high proportion of grain boundary areas. The thermal expansion coefficient of 3YSZ and 8YSZ is close, 10.0–10.8 × 10^{-8} K^{-1} [13].

High mechanical strength is important for electrolyte-supported cells as the electrolyte thickness can be reduced to ~50 μm. However, the conductivity of the 3YSZ phase is much lower than that of 8YSZ and is not thermodynamically stable [14]. At fuel cell operating temperature (600–1000 °C), stabilized zirconia generally exhibits little or no chemical interactions with other components such as LSM cathode, Ni/YSZ cermet anodes and $LaCrO_3$ or Fe–Cr metallic interconnect. However, zirconia-based electrolytes are not chemically compatible with cobaltite-type perovskite cathodes like LSCF. To use a LSCF cathode with a zirconia-based electrolyte, doped ceria is generally applied as barrier layer between the zirconia electrolyte and cobaltite-type electrode.

Stabilized zirconia electrolyte materials exhibit a minimum electronic contribution to the total conductivity in the oxygen partial pressure range most important for practical applications, from 1 atm down to 10^{-25}–10^{-20} atm.

9.2.2 Cerium Oxides

Pure cerium oxide ($CeO_{2-\delta}$) has the same fluorite structure as ZrO_2 with space group Fm3m over a wide temperature range from room temperature to the melting point of ~2477 °C. Due to the existence of the Ce^{3+}/Ce^{4+} redox couple in $CeO_{2-\delta}$, there is an appreciable variation of oxygen stoichiometry, δ, with the temperature and partial pressure of oxygen. However, pure $CeO_{2-\delta}$ does not possess an appreciable amount of oxygen ions and is not a good oxygen ion conductor, but in a manner similar to ZrO_2, substitution of alkaline earth or rare earth cations into the Ce^{4+} lattice can significantly increase the oxygen vacancies and stabilize the fluorite cubic phase, e.g., by dissolution of trivalent oxides such as Gd_2O_3:

$$Gd_2O_3 \xrightarrow{CeO_2} 2Gd'_{Ce} + 3O_O^x + V_O^{\bullet\bullet} \tag{9.38}$$

The oxide ion conductivity of doped ceria is generally higher than that of ZrO_2-based electrolytes, particularly at lower temperatures. Similar to the case of stabilized ZrO_2, the conductivity of doped CeO_2 electrolyte increases with dopant concentration to a maximum and then decreases with further increase in dopant concentration, see Fig. 9.9 [15]. The maximum in conductivity with the doping level is also due to the interaction between dopant and oxygen vacancies, leading to the formation of dopant–vacancy clusters, $(M''_{Ce} - V_O^{\bullet\bullet})$ or $(2M'_{Ce} - V_O^{\bullet\bullet})$, where M is dopant with divalent or trivalent charge.

Based on the critical dopant ionic radius (r_c) concept proposed by Kim [16], the ideal dopant is the one which does not cause any change in the volume of the host lattice. Thus, the smaller the difference between the dopant ionic radius and the r_c, the higher the conductivity. The ionic radius of Ce^{4+} in eightfold coordination (CN = 8) is 0.097 nm and r_c for a trivalent dopant in CeO_2 is calculated to be 0.1038 nm

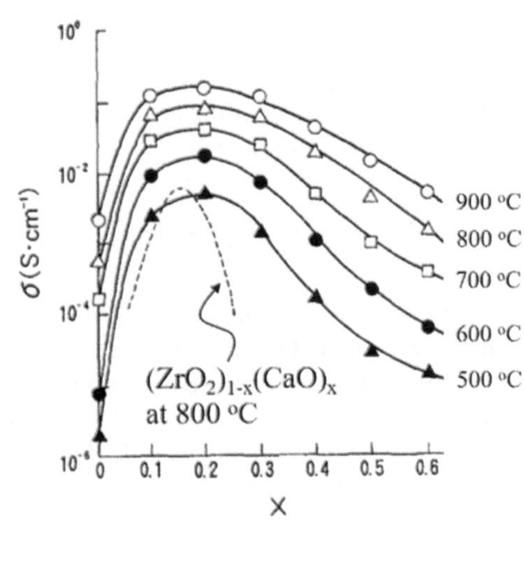

Fig. 9.9 Conductivity of $(CeO_2)_{1-x}(Sm_2O_3)_x$ as a function of doping level x at different temperatures. Reproduced from Ref. [15] with permission from Elsevier, Copyright 1996

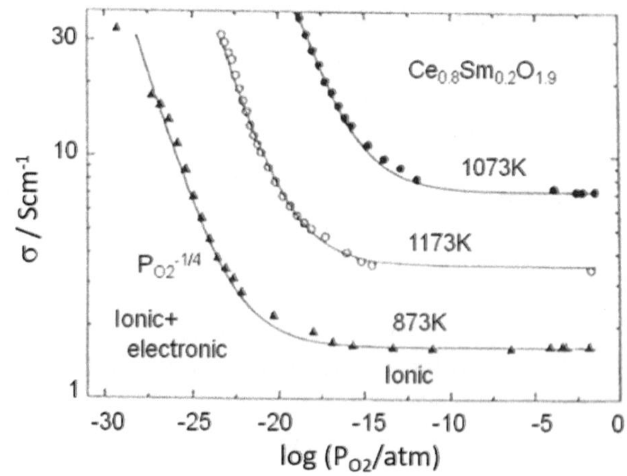

Fig. 9.10 Total electrical conductivity of $Ce_{0.8}Sm_{0.2}O_{1.9}$ as a function of oxygen partial pressure. Reproduced from Ref. [17] with permission from Elsevier, Copyright 1989

[16]. The highest conductivity occurs with the lowest size mismatch between the dopant ions and r_c and has been obtained with gadolinium (Gd^{3+}: 0.1053 nm) and samarium (Sm^{3+}: 0.1079 nm) in this case. The optimum dopant concentration is between 10 and 20 atomic % in general. Thus, $Gd_xCe_{1\text{-}x}O_2$ and $Sm_xCe_{1\text{-}x}O_2$ with $x = 0.1$ or 0.2 (GDC and SDC) are the most common ceria electrolytes in SOFCs [15]. In addition to gadolinium and samarium, other dopants include lanthanum (La^{3+}: 0.115 nm), yttrium (Y^{3+}: 0.1019 nm), ytterbium (Yb^{3+}: 0.0985 nm), and neodymium (Nd^{3+}: 0.1109 nm), but the conductivity of these doped cerias is much lower. The conductivity of GDC and SDC electrolytes is as high as 5.24–9.74 × 10^{-2} S/cm at 800 °C and 0.8–1.7 × 10^{-2} S/cm at 600 °C [15], making them attractive as electrolyte for low-temperature SOFCs.

Ceria-based electrolytes have pure ionic conductivity at high oxygen partial pressure and reach a plateau in the high oxygen partial pressure region, but it increases substantially with the decrease in oxygen partial pressure, see Fig. 9.10 [17]. This is due to the formation of free electrons resulted from the Ce^{4+} to Ce^{3+} transition. At low oxygen partial pressure environment that is prevalent on the anode side of fuel cells, the materials become partially reduced via the reduction of Ce^{4+} to Ce^{3+}. This can be expressed as follows:

$$O_O^x + 2Ce_{Ce}^x \rightarrow \frac{1}{2}O_2 + 2Ce_{Ce}' + V_O^{\bullet\bullet} \tag{9.39}$$

The interaction of the electrons localized at Ce_{Ce}' with neighboring oxygen ions constitute a small hopping polaron, and thus, the defects formed result in mixed ionic and electronic conductivity and reduce t_i, the efficiency of the system. The electronic current flows through the electrolyte even at open circuit, reducing the terminal voltage and efficiency of fuel cells. The increase of ionic radii from 0.097 nm in Ce^{4+} to 0.114 nm in Ce^{3+} also leads to the volume expansion of the lattice, which

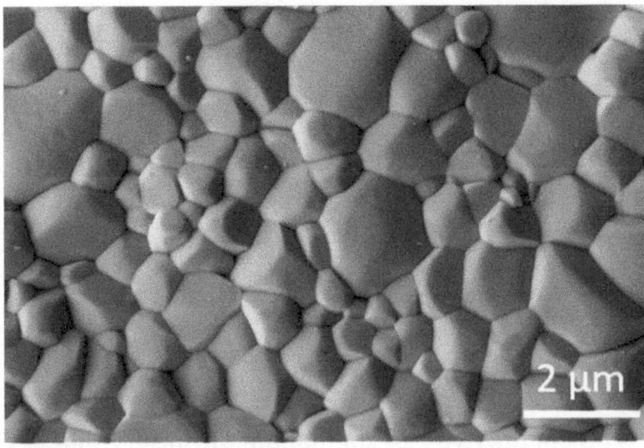

Fig. 9.11 SEM images of $Ce_{0.9}Gd_{0.1}O_{1.9}$ electrolyte surface after sintering at 1450 °C for 5 h in air

may cause mechanical failure. The backward reaction (9.39) occurs at high oxygen partial pressure conditions.

Ceria-based electrolytes can be densified at temperatures of ~1450 °C (see Fig. 9.11). Due to the small grain size of ceria-based electrolyte, the segregation of impurities such as silica to the grain boundaries is detrimental due to their blocking effect with regard to the ionic transport. Adding minor amount of oxide additives like FeO can scavenge impurities like SiO_2 and improve the conductivity. FeO can also function like sintering aid to decrease the densification temperature and promote sintering of doped ceria.

The TEC of ceria is ~12.1×10^{-6} K^{-1} from room temperature to 1000 °C. Mechanically, ceria is not as strong as zirconia and the bending strength of doped ceria is in the range of 115–220 MPa. However, ceria has a superior chemical compatibility as compared to zirconia. Doped ceria is stable with a wide variety of electrode materials including lanthanum manganite, lanthanum cobaltite, and lanthanum nickelate. Because of its excellent stability and compatibility with cobaltite-based cathode materials, doped ceria is commonly used as interlayer between the zirconia electrolyte and the cathode to prevent reaction between the two. However, the inter-diffusion at the YSZ-GDC interface can be an issue for the long-term stability of SOFCs.

CeO_2 and doped ceria are also extensively used as electrocatalysts in anodes and cathodes of SOFCs. The oxygen storage/release and redox properties of ceria-based materials play key roles in their catalytic activity, which can be mainly ascribed to the unique combination of high oxygen ion diffusivity coupled with the ability to shift easily between reduced Ce^{3+} and oxidized Ce^{4+} states.

9.2.3 Lanthanum Strontium Magnesium Gallate Perovskites

Lanthanum strontium magnesium gallate (LSGM) is the first perovskite-based oxide ion conductor, reported in 1994 by Ishihara et al. [18]. The high oxide ion conductivity in this oxide is achieved by double doping of lower valence cations into A- and B-sites of $LaGaO_2$ perovskite. Strontium with its ionic size almost the same as that of La^{3+} is the most suitable dopant for the La site in $LaGaO_3$, while it has been found that doping Mg at the Ga site is highly effective for increasing the oxygen vacancies and thus the conductivity. The dissolution of SrO and MgO in the $LaGaO_3$ lattice can be written as:

$$2SrO \xrightarrow{LaGaO_3} 2Sr'_{La} + 2O_O^x + V_O^{\bullet\bullet} \tag{9.40}$$

$$2MgO \xrightarrow{LaGaO_3} 2Mg'_{Ga} + 2O_O^x + V_O^{\bullet\bullet} \tag{9.41}$$

The conductivity of lanthanum strontium magnesium gallate (LSGM) depends on the dopant concentration and the highest conductivity has been found to be $La_{0.8}Sr_{0.2}Ga_{0.83}Mg_{0.17}O_{3-\delta}$ or $La_{0.9}Sr_{0.1}Ga_{0.8}Mg_{0.2}O_{3-\delta}$ [19]. The ionic conductivity of LSGM is 1.2×10^{-1} S/cm at 800 °C and exhibits fully ionic conduction over a wide range of oxygen partial pressure ($10^{-25} < p_{O_2} < 10^5$ atm) [20]. The conductivity of LSGM is higher than those of YSZ and ScSZ and similar, or comparable to that of doped ceria. However, LSGM does not have an easily reducible ion like Ce^{4+}, and thus is significantly better than doped ceria electrolytes at low partial pressure of oxygen conditions.

One potential issue is the propensity of loss of Ga_2O_3 in $LaGaO_3$ during high-temperature fabrication of LSGM materials and electrolyte via the vaporization reaction:

$$(Ga_2O_3)_{LSGM} \rightarrow Ga_2O_{(g)} + O_{2(g)} \tag{9.42}$$

Or during SOFC operation conditions:

$$2H_{2(g)} + (Ga_2O_3)_{LSGM} \rightarrow Ga_2O_{(g)} + 2H_2O \tag{9.43}$$

The loss of Ga would shift the single-phased composition to a two-phase region due to the very narrow single cubic-phase region of Sr- and Mg-doped $LaGaO_3$, resulting in a decrease in oxide ion conductivity [19]. The consequence of loss of Ga could also lead to the formation of impurity phases such as $LaSrGa_3O_7$ and $LaSrGaO_4$. Thus, very high-temperature processing of LSGM should be avoided. However, due to the complex cation composition, a number of intermediate phases may form if sintered or calcined at low temperatures. Full densification of LSGM electrolytes is required to be sintered at temperatures close to 1500 °C. Adding sintering aids such as different oxides of V, Zn, Si, Co, and Fe improves the densification and care should be taken to avoid the detrimental effect on the transport number. The high cost of gallium compounds and their low mechanical properties are also obstacles for using doped $LaGaO_3$ electrolyte in SOFCs. Therefore replacing Ga with inexpensive elements such as Al is desirable.

Because most common cathode materials have the perovskite structure, the interaction between LSGM electrolyte and SOFC cathode is different from that of zirconia or ceria. Rather than forming a separate phase, the interaction generally occurs trough inter-diffusion, for example, between LSM and LSGM and between LSCF and LSGM. The most common diffusion species is cobalt. Though small amounts of cobalt, iron, and nickel may be beneficial to the conductivity and performance of the electrolyte, excessive inter-diffusion is detrimental to the performance and activity of both cathode and electrolyte. Doping transition metal ions such as cobalt and iron could increase the electron hole conductivity, i.e., the leakage current, which is detrimental to cell performance.

Another issue with LSGM is its reactivity with a Ni-containing anode at high sintering temperatures, forming resistive and undesirable oxides, $LaNiO_3$, La_2NiO_4, etc., and degrading the cell performance. This problem can be solved by replacing the LSGM/Ni composite with ceria/Ni cermets, and by using La-doped ceria (e.g., $La_{0.4}Ce_{0.6}O_{2-\delta}$) buffer layer between LSGM electrolyte and anode. The increased chemical activity of La in the buffer layer with respect to that in LSGM effectively prevents the La diffusion from the LSGM electrolyte and thus stops the formation of impurities at the interface [21].

9.2.4 Apatite Oxides

Apatite-type oxides with a general formula $A_{10}(MO_4)_6O_{2\pm\delta}$ (M = Si or Ge and A is a rare earth or alkaline earth cation) are another class of oxygen ion conducting materials. Due to high cost and high volatilization of GeO_2, the silicate-based systems are preferred apatite electrolytes. As for the apatite structure, the oxygen ion transport in $Ln_{10}Si_6O_{27}$ (Ln = La, Pr, Nd, Sm, Gd, Dy) increases with increasing radius of Ln^{3+} with maximum conductivity for the La-containing phase [22]. The structure of lanthanum silicate apatite consists of isolated SiO_4 tetrahedral with La cations located in seven-coordinate and nine-coordinate cavity sites. Oxygen ions located in large conduction channels of the apatite structure suggest that this material should be appropriate for electrolyte applications for SOFCs. The conductivity of lanthanum silicate apatites increases with increasing oxygen stoichiometry, indicating an interstitial conduction mechanism. Lanthanum silicate apatites are usually prepared by a solid-state reaction route using La_2O_3 and SiO_2 as raw materials and require a very high temperature (1700 °C) and long dwelling time. Sintering and densification of lanthanum silicate oxyapatite, $La_{10}Si_6O_{27}$ (LSO) can be improved by a water-based gel-casting (WGC) technique [23]. Figure 9.12 shows the SEM micrographs of the surface of $La_{10}Si_6O_{27}$ prepared by solid-state reaction (SSR) and WGC methods, sintered at 1600 °C. There is a formation of holes in the SSR sample, but the one prepared by WGC is fully dense. The WGC process improves the relative density of apatite sample, achieving 95% relative density as compared to 91% prepared by SSR method.

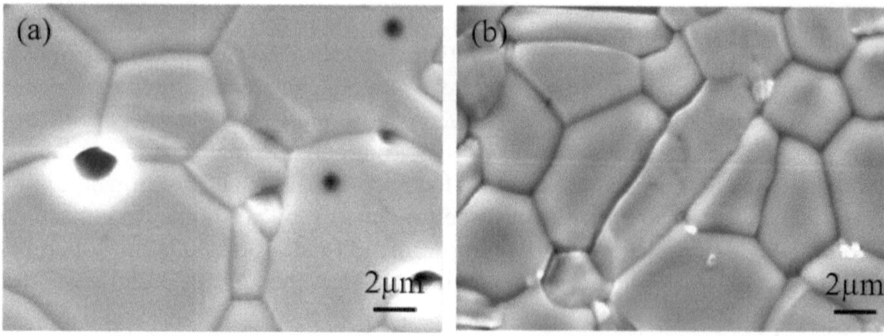

Fig. 9.12 SEM micrographs of surface of thermally etched $La_{10}Si_6O_{27}$ electrolyte samples prepared by **a** solid-state reaction (SSR) and **b** water-based gel-casting methods, sintered at 1600 °C

Incorporation of transition metal oxides such as FeO can improve the sinterability of apatite ceramics, but may also affect both the oxygen ionic and electronic transport properties. The presence of a secondary phase, $La_2Si_2O_7$, in the binary phase diagram of the La_2O_3-SiO_2 system indicates that it is difficult to obtain a homogeneous mixture of the oxide precursors prepared by the conventional solid-state reaction process. The TEC of the as-synthesized $La_{10}Si_6O_{27}$ apatite is in the range of 9.7×10^{-6} K^{-1} and conductivity is 3.46×10^{-2} S/cm at 800 °C. Doping at the lanthanum and/or silica sites with Ba, Sr, Ca, and Fe can further improve the conductivity.

In addition to the high sintering temperature, another potential issue with using lanthanum silicate apatite electrolyte is the inter-diffusion of silica at the electrode/electrolyte interface. This significantly inhibits oxygen ion migration into the lanthanum silicate apatite electrolyte at the interface and deteriorates the electrode electrochemical activity for O_2 reduction reaction.

9.2.5 Bismuth Oxides

Pure bismuth oxide, Bi_2O_3, exhibits a cubic δ phase from approximately 730 °C up to its melting point at 804 °C. When pure Bi_2O_3 is cooled below 730 °C, a transition from the δ-cubic phase to the metastable tetragonal β-phase, the body centered cubic (BCC) γ-phase, or the monoclinic α phase at room temperature takes place. The unit cell of δ-cubic Bi_2O_3 is face-centered cubic (FCC) with cations at the FCC sites and anions at the tetrahedral positions. The δ-cubic Bi_2O_3 has a defect fluorite structure with one-quarter of its oxygen sites randomly occupied by vacancies. The ionic conductivity of the high temperature δ-phase can be as high as 2.3 S/cm at 800 °C. The intrinsic vacancies of the material, the loose bonding of bismuth with oxygen, and the overall disorder of the structure allow for high oxygen ion mobility.

The stability range of δ-Bi_2O_3 can be extended to room temperature by doping with wide range of cations such as WO_3, Gd_2O_3, Er_2O_3, Dy_2O_3, Sm_2O_3, Y_2O_3,

Nb_2O_5, $Pr_2O_{3.66}$, $Tb_2O_{3.5}$, and La_2O_3. Doping with cations with ionic size smaller than that of Bi^{3+} will help to stabilize the high temperature δ-phase due to induced contract of the open structure of δ-Bi_2O_3 by the dopant and thus decreased diffusion of ions, while the Bi_2O_3 phase will exist as the rhombohedral phase when the lattice is doped with sufficiently large cations like La, Nd, Sm, and Gd. The doping of high valence cations (e.g., Nd^{+5}, W^{6+}) reduces the concentration of oxygen vacancy but increases the stability of the δ-phase.

Bismuth oxide-based materials, such as yttria and erbia-stabilized bismuth oxides (YSB and ESB), show high ionic conductivities. Ionic conductivity of $Bi_{0.75}Y_{0.25}O_{1.5}$ and $Bi_{0.8}Er_{0.2}O_{1.5}$ is quite high: 0.11 and 0.23 S/cm at 650 °C, respectively. However, the high oxygen mobility is a result of weak metal–oxygen bonds, and thus, Bi_2O_3-based materials have lower stability under reduced partial pressure of oxygen at the anode side. It decomposes to metallic bismuth at low oxygen partial pressure, e.g., $p_{O_2} < 10^{-13.1}$ atm at 600 °C [24]:

$$Bi_2O_3 \rightarrow 2Bi + \frac{3}{2}O_2 \tag{9.44}$$

This results in the instability and prevents it from being used as effective electrolyte in SOFCs. However, the stability of bismuth oxide-based electrolyte can be improved by construction of bilayer electrolyte with the ESB or YSB layer on the oxidizing side (cathode) and doped ceria layer on the reducing side (anode) [25]. The ceria layer at the anode side would help to stop the reduction of ESB or YSB by shielding it from a reducing atmosphere, while the ESB or YSB layer at the cathode side would reduce the leakage current from the ceria layer in a reducing atmosphere due to the nature of the pure oxide-on conductor of stabilized bismuth oxides. The combined benefit of bilayers would increase the transfer number, the open-circuit voltage, and thus the efficiency of fuel cells. Nevertheless, the performance and efficiency of the fuel cells depend strongly on the interface and structure between the two layers and the bilayer structure also increases the cost and complication of fabrication processes.

9.3 Anode

The anode in SOFCs is where the electrochemical oxidation of fuels such as hydrogen, methane and hydrocarbons takes place. However, as is the case for the anode materials for SOFCs, they must meet the basic requirements as listed below:

- Must be microstructure and phase stable in the reducing environment and be chemically and thermally compatible with other cell components during cell fabrication and operation.
- Must have sufficient electronic conductivity and preferably good oxygen ion conductivity. The electronic conductivity is preferably in the range of 200 S/cm or

higher under fuel reducing conditions. This will ensure there is the sufficient electrical conductivity for the porous electrode and supports for an anode-supported cell. The oxygen ionic conductivity is critical for high activity of the anode.

- Must have sufficient electrochemical activity for fuel oxidation and redox reactions and have a high resistance and tolerance toward carbon deposition, sulfur poisoning, and re-oxidation.
- Must be porous to transport gas-phase fuel to the reaction site and product from the reaction site. Finally, the fabrication process must be facile and raw materials for the fabrication of anodes must be abundant and cheap, which is particularly critical in the case of anode-supported cells due to the large proportion of the anode components used in the overall cell materials.

Because of the highly reducing environment, metals such as Ni, Co, Cu, Pd, Pt, and Ru are the natural choices for anode materials of SOFCs. Theoretically, the relative trend of electrochemical activity of metals can be evaluated by using ab initio *thermodynamics* calculations. This approach can estimate the relative activity or performance of the materials without the detailed knowledge of the charge-transfer and steps of the reaction. Figure 9.13 shows an example of the plots for the activity of different metals including W, Mn, Mo, Fe, Co, Ir, Ru, Rh, Ni, Cu, Pt, Pd, Ag, and Au toward three different fuels (H_2, CO and CH_4) using density functional theory calculation [26]. Based on the results, Ni, Ru, Rh, Ir, and Co show the highest electrode activity for the activation of methane and for the oxidation reaction of H_2 and CO. The volcano-type dependence of the metals' activity against the oxygen dissociative chemisorption energy (or oxide formation energy) underlines the critical role of *oxygen intermediates and species* in the oxidation reaction of fuels at high temperatures. Among metal families active for H_2, CO and CH_4, Ni is obviously the most abundant and easily available metal. An early study of the electrocatalytic activity of metal electrodes on YSZ electrolyte for H_2 oxidation *versus* the oxide formation energy of metal, $-\Delta H_f^0$, also shows the volcano-type dependence with Ni on the top of curve [27]. Ru also shows high electrocatalytic activity for H_2 oxidation

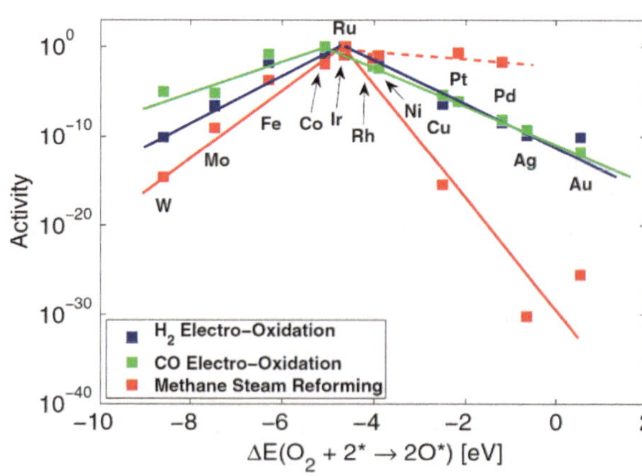

Fig. 9.13 Dependence of the activity of different metals for the oxidations of hydrogen and CO and methane steam reforming against the oxygen dissociative chemisorption energy based on ab initio thermodynamics calculations. Reproduced from Ref. [26] with permission from IOP Publishing, Copyright 2009

and methane steam reforming; however, its use as anode materials is limited by its high cost and the fact that it evaporates as RuO_x at high temperatures.

In addition to metals, a number of ceramic materials also show significant electrical conductivity and activity at high temperatures and can also be used as the anode of SOFCs.

9.3.1 Ni-Based Cermet Anode

In practice, Ni and Ni-based cermets are the most developed and widely utilized anode materials in SOFCs [28]. Metallic Ni is cost-effective and highly active for the oxidation of hydrogen- and hydrocarbon-based fuels. Most importantly Ni has a high electrical conductivity with electronic conductivity of 1.4×10^6 S/cm at 25 °C and ~2×10^4 S/cm at 1000 °C. The melting temperature of Ni is 1453 °C. However, direct use of a Ni anode in the SOFCs is not feasible due to its high TEC, 13.3×10^{-6} K^{-1}, higher than $10–12\times10^{-6}$ K^{-1} of zirconia- and ceria-based electrolytes. Such problems in TEC mismatch can be effectively solved by addition of the electrolyte component to form the cermet structure. With the cermet composition of 30vol.%Ni/70vol.%YSZ, the TEC is reduced to ~12.5×10^{-6} K^{-1}. The formation of a Ni/electrolyte cermet composite will also bring other benefits, including the improved interface contact between the anode and electrolyte, significantly reduced sintering and agglomeration of the Ni metal phase and the formation of ionic conducting paths in the cermet, and significantly increasing the three phase boundary (TPB) for hydrogen and fuel oxidation reactions. TPB is an important concept in SOFCs and will be described in details in Chap. 10.

A high sintering temperature (e.g., 1400 °C) is generally required to achieve a good bonding and form an intimate interface between Ni/YSZ cermet anode and YSZ electrolyte. Before the use as anode in fuel cells, NiO in the cermet has to be reduced to metallic Ni. The high-temperature sintering and reduction steps have a significant effect on the microstructure evolution of Ni-based cermets. Figure 9.14 shows such a microstructure evolution process for a Ni/YSZ cermet with 50/50 weight ratio [28]. The distribution of NiO and YSZ particles in the green state is very uniform, and the particle size of NiO and YSZ powder is very small, typically in the range of 0.03–0.06 μm. However, high-temperature sintering at 1400 °C in air leads to the significant agglomeration of both the NiO and the YSZ phases in the cermets. NiO particles grow to 1–5 μm, and YSZ particles are 0.5 μm. After being reduced at 1000 °C in hydrogen, Ni particles grow to as large as 7 μm in diameter, indicating the further grain growth of fine Ni particles during the chemical reduction of NiO to Ni. There is also an increase in the overall porosity of the anode resulting from the reduction of NiO to Ni with a reduction in volume of 25%. In general, the porosity increases with the increase in the Ni volume fraction, and in the case of 40vol.%Ni/60vol.%YSZ, the porosity is typically in the range of 30–40%. The porosity and pore morphology can also be controlled by adding pore-formers such as carbon fiber, graphite, and corn starch.

Fig. 9.14 Typical SEM micrographs of the cross section of Ni/YSZ cermet anodes at different states: **a** oxide state after sintered at 1400 °C in air, **b** reduced state after reduction in 97% H_2/3% H_2O at 1000 °C, and **c** green state. The scale bar applies to all SEM micrographs. Modified from Ref. [28] with permission from Springer Nature, Copyright 2004

For practical and industrial applications, Ni/YSZ and Ni/GDC cermets are generally made from commercially available Ni, YSZ, and GDC powders, and fabricated using conventional ceramic processes, i.e., mixing, coarsening, and firing as shown schematically in Fig. 9.15. Mechanical mixing and milling are used to homogenize the powder. The composition and phase distribution are important for the microstructure and performance of the Ni/YSZ cermet anodes. The parameters affecting the distribution, microstructure, and performance of Ni-based cermet anodes are summarized as follows, taking Ni/YSZ as an example:

- Raw materials—Characteristics of NiO and YSZ powders such as the average particle size, particle size distribution, and sintering behavior (e.g., shrinkage and shrinkage rate) have significant effect on the fabrication process and the electrochemical performance of the Ni/YSZ cermet anodes. This is relevant because the properties of NiO and YSZ powders are very different from different commercial suppliers.
- Matching shrinkage profile of the Ni/YSZ cermets and YSZ electrolyte is particularly important in the anode-supported cells. In practice, the sintering profile of the cermet powder can be adjusted in certain degrees by heat treatment of the starting NiO and YSZ powders. Coarsening of Ni, YSZ, and cermet powder is also effective to achieve a smooth- and crack-free electrode coating.
- The sintering temperature has a significant effect on the activity and performance of Ni/YSZ anodes. The formations of Ni-to-Ni electronic contact and YSZ-to-YSZ ionic contact networks are closely related to the sintering temperatures of the

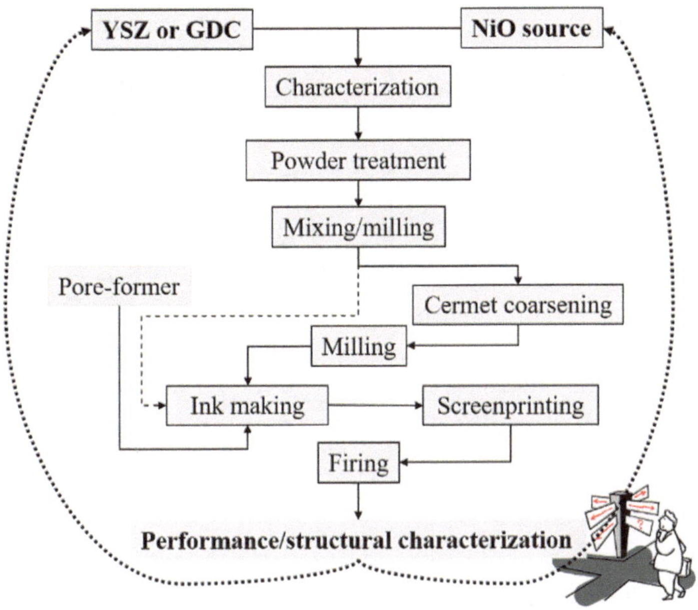

Fig. 9.15 Generalized processing route for the fabrication of Ni/YSZ cermet anodes

anode. With the increase of sintering temperatures, interconnectivity between Ni to Ni and YSZ to YSZ increases. This generally can be assessed by the reduction of the cell ohmic resistance with the sintering temperatures. The high sintering temperature is essential to create a good bonding between the YSZ phase in the cermet and the YSZ electrolyte, forming a rigid YSZ structure to support the Ni phase and the formation of Ni-to-Ni electronic contact network. In general, sintering at 1350–1400 °C is sufficient for the creation of a good interface between Ni/YSZ cermet anode and YSZ electrolyte.

In Ni/YSZ cermet materials, YSZ is an ionic conductor and Ni is an electronic conductor. In a simple diphase system the theory predicts a percolation threshold of ~30 vol.% of the higher conducting phase for the transition from dominant ionic conductivity to dominant electronic conductivity. This is generally true for the electronic conductivity of the composite Ni/YSZ cermet systems as shown in Fig. 9.16. Electronic conductivity of Ni/YSZ cermets is strongly dependent on the particle size and distribution of both Ni and YSZ phases, and thus, the threshold for the transition from the dominant ionic conductivity of the YSZ electrolyte phase to the dominant electronic conductivity of the Ni metal phase is related to the YSZ/NiO size ratio. In general, the threshold increases with the increase of Ni average particle size. For Ni/YSZ cermets with 30 vol.% (~40 wt%) or higher, the electrical conductivity of the cermets will be in the conductivity band with electrical conductivity equal or higher than 200 S/cm, meeting the requirement as the anode and anode supports for SOFCs.

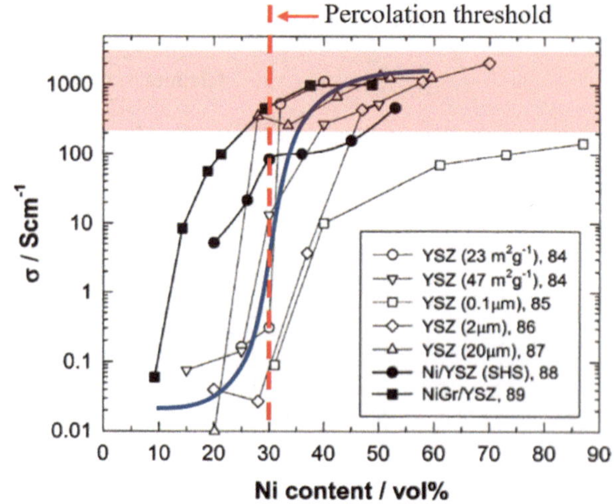

Fig. 9.16 Conductivity profile of Ni/YSZ as a function of Ni content. The conductivity values were obtained at 1000 °C in humidified H_2 and numbers are references cited in [28]. Modified from Ref. [28] with permission from Springer Nature, Copyright 2004

A Ni/YSZ cermet anode can be applied to an electrolyte by screen-printing in the case of electrolyte-supported cells. In the case of anode-supported cells, Ni/YSZ cermet is typically fabricated by tape casting a physical mixture of NiO and YSZ powders. A thinner YSZ tape is then joined with the green anode tape either by casting the two layers on top of each other or by laminating them together. Both layers are then co-sintered between 1300 and 1400 °C, yielding a thin, dense YSZ electrolyte (typically < 50 μm) supported on a Ni/YSZ anode substrate (typically ~500 μm). The cathode is then added to the free electrolyte side by screen-printing, forming an anode-supported cell. The porosity and thickness of the anode support have a significant influence on the cell performance. While tape casting is an industrially scalable process, it has limitations in the microstructure optimization of variations in the tape composition in terms of particle size, pore size, and distribution. In SOFCs, the reactive region is generally considered to be confined to several tens of micrometers from the electrode–electrolyte interface; that is, the oxygen ions migrating through the electrolyte are typically close to the anode–electrolyte interface. This has led to the development of functionally graded anodes, where the Ni/YSZ ratio, particle size, and pore size vary from the anode–electrolyte interface to the anode bulk. This is typically achieved by fabricating the anode with two distinct layers—an anode function layer (AFL) close to the anode–electrolyte interface and a thicker, more porous support layer. The AFL consists of smaller particles, with smaller pores to maximize TPB density, and hence electrochemical activity. AFL thickness is typically in the range of 10 μm.

Ni/YSZ or Ni/GDC cermet anodes show significant activity for the oxidation reaction of hydrogen and CO, but are not without disadvantages. The most significant of them is that the use of Ni places significant limitation in the fuel choice for SOFCs. Ni catalyzes the formation of graphitic carbon in dry hydrocarbon fuels. This problem can be partially surmounted by steam reforming of the hydrocarbon fuel to form H_2 and CO either feeding to the cell (external steam reforming) or on the Ni-based anode itself (internal steam reforming). The first solution increases cost by requiring an additional reactor in the system, and the second solution places large thermal stresses on the cells and stack. Steam reforming is highly endothermic while hydrogen oxidation is exothermic, and they do not necessarily occur in the same place. Another problem of Ni-based anodes is the low redox tolerance or instability toward reduction and oxidation cycling due to the large volume expansion accompanied by the oxidation of Ni to NiO. The potential damage to the microstructure and activity of Ni-based anodes by contaminants such as sulfur, generally existing as H_2S in common hydrocarbon fuels like natural gas is also a concern. These disadvantages and potential advantages of direct utilization of hydrocarbon fuels have led to significant research efforts to develop alternative anodes.

9.3.2 Cu/CeO_2/YSZ-Based Anode

Cu is used to replace Ni in the cermet anodes with the aim of overcoming the carbon cracking problems on Ni-based cermet anodes for direct hydrocarbon fuels [29]. Unlike Ni, copper (Cu) is inert to hydrogen or hydrocarbon oxidation and is catalytically inactive for cracking the hydrocarbons, thus suppressing the carbon deposition. The primary function of Cu in the cermet is to provide electronic conduction path. However, pure Cu/YSZ anode has very low electrochemical activity for both H_2 and CH_4 oxidation reactions. Ceria is added as an electrocatalyst to Cu/YSZ cermet to improve the electrode catalytic activity. The high electrocatalytic activity of ceria-based materials stems from the oxygen vacancy formation and diffusion associated with the reversible CeO_2/Ce_2O_3 or Ce^{4+}/Ce^{3+} transition. The cells have been operated with dry hydrocarbons including methane, butane, and synthetic diesel without any carbon deposition. Because Cu_2O and CuO melt at 1235 °C and 1326 °C, respectively, it is not feasible to prepare Cu/YSZ cermets by the conventional high-temperature sintering used for Ni/YSZ cermet anodes. An alternative method is to prepare Cu/CeO_2/YSZ anodes by infiltration of Cu and/or ceria into pre-formed YSZ or GDC skeletons on YSZ electrolyte. Cu-based anodes are generally used for operating temperatures < 800 °C. The addition of another solid phase also complicates the concept of an anode TPB. In this case, four phases, namely electrolyte, electrical conductor, electrocatalyst, and gas phase, must be in close proximity to create a functioning anode. Nevertheless, the microstructure degradation due to the low-melting point of Cu (1083 °C) is a limiting factor affecting the practical application of Cu/CeO_2/YSZ-based anodes in SOFCs.

9.3.3 Ceramic Oxide Anode

Electronic conducting and stable oxides under fuel environments have been explored for alternative anodes of SOFCs for the purpose of replacing Ni. This includes doped ceria, titanate-based perovskites ($La_{0.4}Sr_{0.6}TiO_3$), $La_{0.75}Sr_{0.25}Cr_{0.5}Mn_{0.5}O_3$ (LSCM), $La_{0.8}Sr_{0.2}Cr_{0.97}V_{0.03}O_3$ and double perovskites $Sr_2Mg_{1-x}Mn_xMoO_{6-\delta}$. Among the perovskite-based materials, LSCM is perhaps the most investigated ceramic oxide anode. Electrical conduction in LSCM occurs via thermally activated polaron hopping. The charge compensation for Sr^{2+} substitution and oxygen vacancy formation takes place via the Mn cation only with Cr maintaining a 3+ oxidation state and sixfold oxygen coordination [30]. This indicates that Mn is the catalytically active element in LSCM. Hence, the lattice oxygen is consumed to oxidize the reactant in a reduction step, before reoxidation in a subsequent step to complete the catalytic cycle. Thus, increase in the redox ability of the B-site transition metal cations could increase the catalytic activity of the perovskite anode. Pyrochlore structured oxides such as $Ga_2Ti_2O_7$ and $Pr_2Zr_2O_7$ are also considerd for use as anode materials of SOFCs.

However, the electrochemical activity of ceramic oxides such as LSCM is generally low for the oxidation reaction of H_2 and hydrocarbon fuels such as methane. The addition of YSZ or Pd catalytic nanoparticles substantially enhances the performance of LSCM-based anodes [31]. However, one major problem with ceramic oxide materials is their low electronic conductivity. For example, in reducing environment (10% H_2/N_2 to 100% H_2), the electrical conductivity of LSCM materials is 1.3–0.22 S/cm at 800 °C [31] and for $Sr_2MgMoO_{6-\delta}$, it is 0.8 S/cm at 800 °C [32]. This factor is a serious concern for practical SOFC electrodes, as low electronic conductivity will result in high polarization losses and high contact resistance between the electrode and current collector due to the constriction effect (see Chap. 11).

Example 9.1 iR drop or ohmic polarization of electrode coating

$La_{0.75}Sr_{0.25}Cr_{0.5}Mn_{0.5}O_3$ (LSCM) is stable in both oxidation and reducing environment and can be used as cathode and anode in symmetric solid oxide cells. Calculate (a) the ohmic loss of the cathode and anode if the thickness of the electrode coating is 50 μm and (b) the ohmic loss of the anode if LSCM is used as anode support with thickness of 1 mm. Current density is 1A/cm^2 and electronic conductivity of the electrode materials is 25 S/cm in air and 0.20 S/cm in 10% H_2/90% N_2 at 800 °C.

Solution:

a) If the thickness of both anode and cathode is 50 μm, we have

$$\eta_{\Omega,anode} = \frac{50 \times 10^{-4}\text{cm}}{0.2\,\text{S/cm}} \times 1\frac{\text{A}}{\text{cm}^2} = 0.025\,\text{V} = 25\,\text{mV} \tag{9.45}$$

$$\eta_{\Omega,cathode} = \frac{50 \times 10^{-4}\text{cm}}{25\,\text{S/cm}} \times 1\frac{\text{A}}{\text{cm}^2} = 0.0002\,\text{V} = 0.2\,\text{mV} \tag{9.46}$$

The iR loss or ohmic overpotential of anode and cathode is 25 and 0.2 mV, respectively.

b) If LSCM is used as anode support with substrate thickness of 1 mm, we have

$$\eta_{\Omega,anode} = \frac{1.0 \times 0.1\,\text{cm}}{0.2\,\text{S/cm}} \times 1\frac{\text{A}}{\text{cm}^2} = 0.5\,\text{V} = 500\,\text{mV} \qquad (9.47)$$

The iR loss of the LSCM anode substrate is 500 mV.

9.4 Cathode

In solid oxide fuel cells (SOFCs), the cathode is the material where pure oxygen or oxygen from air is reduced through the electrochemical reaction. As illustrated by Eq. (9.1), the oxygen reduction reaction (ORR) requires the presence of oxygen and electrons as well as the transfer of the generated oxide ions from the reaction site into the bulk of the electrolyte. This implies that as candidate materials for cathodes, they much meet following criteria:

- Must be a good electronic conductor and/or a mixed ionic and electronic conductor (MIEC). If the electrode material possesses only electronic conductivity, these requirements can only be fulfilled in the vicinity of the three phase boundary (TPB) between the electrode, electrolyte and oxidant gas. If the electrode material possesses MIEC, oxygen species may be transported through the bulk of the electrode and the reaction zone can be extended to areas away from the electrode–electrolyte interface.
- Must be electrochemically active for the ORR and tolerant toward volatile contaminants such as volatile chromium species from metallic interconnect and sulfur, water vapor, and CO_2 from air stream.
- Must be structurally stable under SOFC operating conditions with minimum surface segregation. Surface segregation of cations from the lattice of the structure has a significant implication for the electrochemical activity and stability of the cathodes.
- Must be chemically and thermally compatible with electrolyte to minimize chemical reactions during the fabrication and operation stages to prevent resistive phase formation. TEC is close to or compatible with the electrolyte in order to avoid cracking and delamination.

In SOFCs based on thin electrolyte films, the overall losses of the cell are generally dominated by the polarization losses for the ORR. This is partly due to the high activation energy and slow reaction kinetics for the ORR when compared with the

Table 9.2 Conductivity and thermal expansion coefficient of common electrode materials in SOFCs, measured in air

Composition	σ_e (S/cm)			E_a, kJ/mol	σ_i (S/cm)	TEC ($\times 10^{-6}$/K)
	1000 °C	800 °C	600 °C		(Temp °C)	
$La_{0.8}Sr_{0.2}MnO_3$	200	185	160	14.4	10^{-7}–10^{-8} (800)	11.5–12
$La_{0.6}Sr_{0.4}Co_{0.2}Fe_{0.8}O_{3-\delta}$	170	282	332	9–14	8×10^{-3} (800)	15–21
$Ba_{0.5}Sr_{0.5}Co_{0.8}Fe_{0.2}O_{3-\delta}$		29–40	30–40	6–31	1.8×10^{-2} (700)	20–23
$PrBaCoO_{5+\delta}$		418±270	515±328	8–12	na	20–24
$La_2NiO_{4+\delta}$		100	84	12–16	na	11–14
$(La_{0.73}Sr_{0.25})(Cr_{0.5}Mn_{0.5})O_3$		28	18	26	na	12

H_2 oxidation reaction (HOR). Consequently, the development of cathodes with high performance and high stability becomes increasingly critical for the development of intermediate temperature SOFC or IT-SOFC technologies. The cathode materials can be generally classified by the nature of conductivity, electronic conductivity-dominant materials represented by lanthanum strontium manganite (LSM) and MIEC materials represented by lanthanum strontium cobalt ferrite (LSCF). Table 9.2 lists electronic and ionic conductivity and TEC of most common cathode materials in SOFCs.

9.4.1 *Lanthanum Strontium Manganite*

Lanthanum manganite ($LaMnO_3$) and strontium substituted lanthanum manganite ($La_{1-x}Sr_xMnO_3$ or LSM) belong to the *Pbnm* space group with near-cubic ABO_3 perovskite phase and are the most common and investigated cathode materials of SOFCs [33]. Figure 9.17 shows the *Pbnm* near-cubic $La_{0.875}Sr_{0.125}MnO_3$ perovskite structure. In the perovskite structure, the larger rare earth ions (e.g., La, Sr) occupy the 12 coordinated A-sites and the transition metal ions (e.g., Mn) occupy the 6 coordinated octahedral B-sites.

Example 9.2 Calculate the Goldschmidt tolerance factor, *t*, of $La_{0.875}Sr_{0.125}MnO_3$ perovskite

Solution:

Using ionic radii of La^{3+} (0.136 nm, CN = 12), Sr^{2+} (0.144 nm, CN = 12), Mn^{3+} (0.0645 nm, CN = 6), and O^{2-} (0.140 nm), the Goldschmidt tolerance factor, *t*, can be calculated as,

$$t = \frac{(r_A + r_O)}{\sqrt{2}(r_B + r_O)} = \frac{(0.136 \times 0.875 + 0.144 \times 0.125 + 0.14)}{\sqrt{2}(0.645 + 0.14)} = 0.9578 \tag{9.48}$$

The tolerance factor $t = 0.9578$ and is within 0.95–1.04, confirming the near-cubic perovskite structure for the $La_{0.875}Sr_{0.125}MnO_3$ oxide.

Both the electrical conductivity and catalytic activity of $LaMnO_3$ are enhanced considerably when lanthanum is substituted partially with divalent ions such as Sr^{2+} and Ca^{2+}. Figure 9.18 shows an example of Arrhenius plots of LSM as a function of Sr-doping level [34]. The electrical conductivity varies significantly with the Sr-doping and activation energies determined from the slope decrease with the Sr-doping level, varying in the range of 24–3 kJ/mol. This suggests the predominant small-polaron hopping mechanism for electrical conduction, as discussed above. Maximum conductivity is for LSM with Sr-doping level at $x = 0.5$, but the minimum of TEC occurs for the Sr-doping level of $x = 0.1$–0.2 in the range of $11–12 \times 10^{-6}$ K^{-1}, close to that of most commonly used YSZ electrolytes. Compared to other cathode materials, LSM has an excellent microstructural stability and its long-term performance stability has been well established. $La_{0.8}Sr_{0.2}MnO_{3+d}$ is one of the most common compositions of LSM cathodes employed in SOFCs.

A detailed density functional theory (DFT) study by Pavone et al. showed that Sr substitution of A-site La leads to the formation of electron holes in the LSM electronic structure [35]. At SOFC operating temperatures, the electron holes are

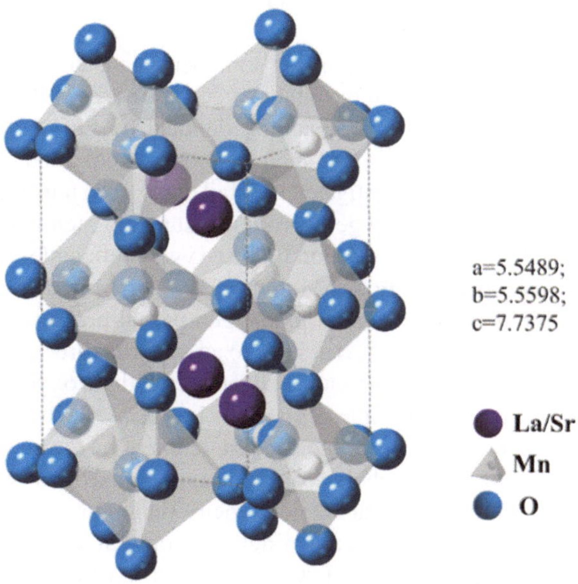

Fig. 9.17 *Pbnm* near-cubic $La_{0.875}Sr_{0.125}MnO_3$ perovskite structure

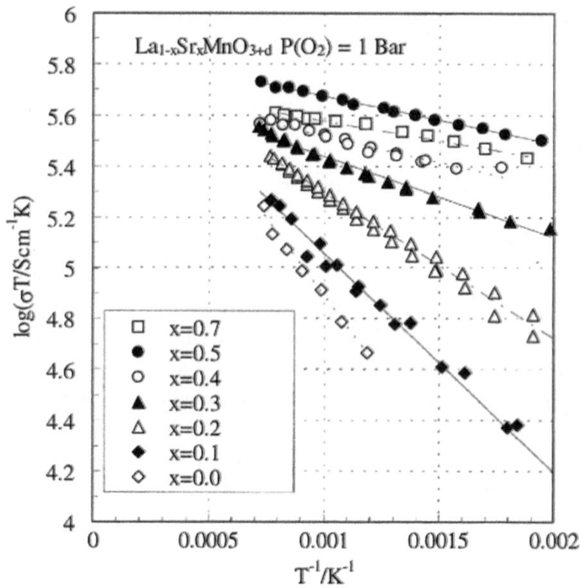

Fig. 9.18 Arrhenius plots of electronic conductivity (σ) of $La_{1-x}Sr_xMnO_{3+d}$ ($0 \le x \le 0.7$) at pure oxygen (p_{O_2} = 1 bar). Reproduced from Ref. [34] with permission from Elsevier, Copyright 2000

localized onto specific Mn ions, forming Mn^{4+} ions, $Mn_{Mn}^{\bullet}$ or electron holes, $h^{\bullet}$ and causing local distortion of the LSM lattice. Due to the high oxygen vacancy formation energy, oxygen vacancy concentration is too low to allow efficient oxygen ion diffusion in LSM. The neutrality condition can be simply described as:

$$SrO \xrightarrow{LaMnO_3} Sr_{La}' + Mn_{Mn}^{\bullet} + O_O^x \tag{9.49}$$

Lanthanum manganite is somewhat unusual in comparison to most perovskite oxides in that it can have the oxygen-excess as well as the oxygen-deficient nonstoichiometries. This is generally denoted by $La_{1-x}A_xMnO_{3\pm\delta}$ (A is divalent cation, such as Sr^{2+}, Ca^{2+}, "+" means oxygen-excess and "-" means oxygen-deficiency). Mizusaki et al. investigated in detail the oxygen nonstoichiometry of $La_{1-x}Sr_xMnO_{3\pm\delta}$ as a function of oxygen partial pressure, p_{O_2}, temperature and the composition, see Fig. 9.19 [36]. In addition to the oxygen-deficient region, the oxygen-excess region is also observed. For $LaMnO_3$ and $La_{1-x}Sr_xMnO_{3\pm\delta}$ with $x < 0.4$, the oxygen content exhibits two plateaus in its oxygen partial pressure dependence, one is around the oxygen excess (3 + δ) at high oxygen partial pressure and the other around the stoichiometric composition (δ = 0) at intermediate oxygen partial pressures (10^{-5}–10^{-10} Pa) (a typical example is given for $La_{0.9}Sr_{0.1}MnO_{3\pm\delta}$). At lower oxygen partial pressures, the oxides become oxygen deficient (3-δ) and the charge compensation of the positive effective charges of $V_O^{\bullet\bullet}$ is maintained by Mn reduction. $La_{1-x}Sr_xMnO_{3\pm\delta}$ is in the oxygen-excess nonstoichiometry region under normal fuel cell operation conditions and becomes oxygen-deficient only at very low partial pressures of oxygen, e.g., <10^{-10} Pa at 900 °C for x = 0.2. At high oxygen partial pressure, in order

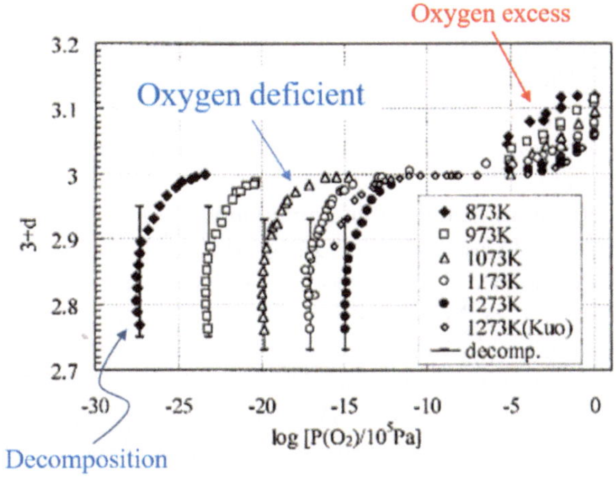

Fig. 9.19 Oxygen nonstoichiometry of $La_{0.9}Sr_{0.1}MnO_{3\pm\delta}$ as a function of oxygen partial pressure of oxygen, clearly showing the oxygen-excess and oxygen-deficient regions. Modified from Ref. [36] with permission from Elsevier, Copyright 2000

to compensate for the electrical neutrality raised by Mn oxidation, there can be two possibilities of defect formations: interstitial oxide ions, or cation vacancies. Neutron powder diffraction and high-resolution transmission electron microscopy (HRTEM) results indicate that cation vacancies, rather than oxygen interstitials, are responsible for the oxygen-excess nonstoichiometry. This is consistent with the close-packed nature of the perovskite lattice, which could not accommodate an excess of oxygen as an interstitial oxygen ion. Thus, apparent oxygen-excess can only be considered in terms of a Schottky-type disorder, which involves the formation of cation vacancies. The oxygen-excess nonstoichiometry of $La_{1-x}Sr_xMnO_{3+\delta}$ is most interesting as it is rarely observed in other perovskite-type oxides. This is probably the main reason for the dominant electronic conductivity of doped $LaMnO_3$ cathode materials under SOFC operation conditions.

In the oxygen-deficient region, the predominant ionic defect is the oxygen vacancy. The charge neutrality is maintained by the reduction of B-site Mn to the lower valence state. In the oxygen-excess region, electronic conductivity is constant, while it decreases sharply with the decrease in oxygen content in the oxygen-deficient region. Since the electronic conduction in $La_{1-x}Sr_xMnO_{3+\delta}$ is via a hopping-type *p*-type mechanism irrespective of the oxygen content ($\delta < 0$ and $\delta > 0$), disproportion of Mn ion into Mn^{2+}, Mn^{3+}, and Mn^{4+} can occur. This can be written as:

$$\frac{1}{2}O_2 + V_O^{\bullet\bullet} + 2Mn_{Mn}^x = 2Mn_{Mn}^{\bullet} + O_O^x \tag{9.50}$$

$$\frac{1}{2}O_2 + V_O^{\bullet\bullet} + 2Mn_{Mn}^{\prime} = 2Mn_{Mn}^x + O_O^x \tag{9.51}$$

where $Mn_{Mn}^{\bullet}$ is Mn^{4+} on a Mn^{3+} cation site and $Mn_{Mn}^{\prime}$ is Mn^{2+} on a Mn^{3+} site. The oxygen vacancy formation and the concomitant disproportionation of Mn ions into

Mn^{2+} close to the TPB region could also occur under cathodic polarization conditions due to the reduction in oxygen partial pressure.

The stability of the perovskite oxide, in general, increases with the size of A-site cations. Thus, the majority of work has been performed on lanthanum-based perovskites, in particular the $LaMnO_3$-based series oxides. Of the alkaline earth dopants, Sr substitution is preferred for SOFC applications because the resultant perovskite forms stable compounds with high conductivity in an oxidizing atmosphere. In the case of $La_{1-x}Sr_xMnO_3$, the electronic conductivity increases with increase in x; the maximum is observed for $x = 0.5$. However, extensive tests show that $La_{1-x}Sr_xMnO_3$ with $x = 0.1$–0.2 provides a high conductivity (~200 S/cm at 900 °C) while maintaining mechanical and chemical stability with YSZ. Compared to electronic conductivity, ionic conductivity of LSM is much smaller. The oxygen ionic conductivity of LSM materials is reported to be in the range of 5.9×10^{-8} S/cm in the oxygen partial pressure of 10^{-1}–10^{-3} atm at 800 °C.

As the incorporation and bulk diffusion of oxygen inside LSM cannot be expected to occur to a significant degree, the TPB region becomes the reaction sites for the O_2 reduction. The very low oxygen ionic conductivity of LSM is considered to be the main factor in the high polarization losses of LSM cathodes for the ORR in reduced SOFC operating temperatures. Due to the high activation energy of the ORR and negligible oxygen ionic conductivity of LSM-based electrodes, polarization resistance for the reaction on LSM-based cathodes increases dramatically with the decrease of SOFC operation temperatures. For example, for the O_2 reduction on a LSM electrode, R_E is 0.4 Ωcm^2 at 900 °C and increases dramatically to 56 Ωcm^2 at 700 °C. Thus, pristine LSM is not suitable for intermediate temperature SOFCs.

Various strategies have been developed to improve the electrocatalytic activity of LSM-based cathodes. The most used method is to form composite cathodes with the addition of YSZ and GDC electrolyte. For LSM/YSZ or LSM/GDC composite cathodes, the connectivity of the electronic conducting LSM and ionic conducting YSZ phases is crucial to produce a mixed conducting composite, because the ionic conductivity of LSM and the electronic conductivity of YSZ are negligible under normal SOFC operation conditions. The threshold for the ionic connectivity in the LSM/YSZ composite is between 50 and 70 wt% of YSZ phase depending on the porosity of the composite cathode. Optimization of an LSM-based cathode by doping, e.g., introducing scandium to the manganese site, $La_{0.8}Sr_{0.2}Mn_{1-x}Sc_xO_{3-\delta}$, to increase the oxide ion conduction as the high polarizability of Sc^{3+} would make it easier for the oxide ion to transport through the lattice and improve the electrode performance because of the suppressive effect of Sc^{3+} on surface SrO segregation. Introducing catalytically active components, such as doped CeO_2, Pd, into the LSM porous structure can also effectively enhance the electrocatalytic activity of LSM-based electrode materials for ORR.

9.4.2 Lanthanum Strontium Cobalt Ferrite

Another family of materials that has been extensively studied for SOFC cathodes is lanthanum strontium cobalt ferrite (LSCF) [37]. LSCF perovskite with a typical composition of $La_{0.6}Sr_{0.4}Co_{0.2}Fe_{0.8}O_3$ (LSCF6428) has been developed with desirable properties for IT-SOFC cathode due to the optimum tolerance factor and oxygen deficiency of both the A- and B-site substitutes involved. LSCF belongs to the $R\overline{3}c$ space group with near-cubic perovskite-type structure. Sr on the A-site lattice of LSCF acts as an acceptor, enhancing the oxygen vacancy formation, and Co ions on the B-site lattice have a smaller binding energy for oxygen than is the case with Fe ions, increasing the electronic conductivity. The oxygen self-diffusion coefficient of cobaltite-based materials is several orders of magnitude higher than that of the manganites.

The electrical conductivity of LSCF is critically related to the composition, as substitution of Fe for Co in the undoped or Sr-doped $LaCo_{1-y}Fe_yO_3$ alters not only the crystal structure but also the orbital configuration of valence electrons. The conductivity of LSCF increases with temperature through a maximum, then decreases (see Fig. 9.20) [38]. The increase in conductivity with temperature is typical of the semiconductor behavior; the electron holes located at the B-site are the main charge carriers. On the other hand, the decrease of the conductivity at high temperatures can be due to the metal-like conducting behavior as in $LaCoO_3$ and the increase of oxygen vacancies, which in turn reduces the p-type electronic carriers. In the case of LSCF6428, the conductivity at 900, 800, and 600 °C is 210, 282, and 332 S/cm, respectively, in air with TEC value of 15.3×10^{-6} K^{-1}. The conductivity behavior is

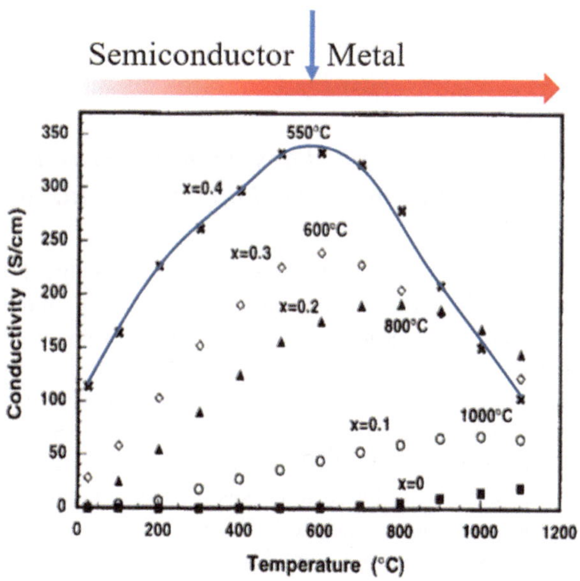

Fig. 9.20 Electrical conductivity of $La_{1-x}Sr_xCo_{0.2}Fe_{0.8}O_3$ as a function of temperature in air, showing the change from semiconducting to metallic conducting behavior with the maximum temperature change as a function of Sr dopant concentration. Modified from Ref. [38] with permission from Elsevier, Copyright 1995

due to the two competing compensation mechanisms, electronic, and ionic, respectively, that occur due to the partial replacement of the A-site La^{3+} cations with Sr^{2+} ions. The dependence of σ_e on p_{O_2} can be described by the equilibrium reaction:

$$O_o^x + 2h^\bullet = V_o^{\bullet\bullet} + \frac{1}{2}O_2 \quad (9.52)$$

As the p_{O_2} is reduced, the reaction tends to shift to the right. The concentration of oxygen vacancies increases, reducing the number of electronic charge carriers, and thus, the conductivity decreases.

The major advantages of lanthanum cobalt ferrite-based perovskites over lanthanum manganite-based perovskites are their significantly higher oxygen diffusion and exchange properties. The oxygen self-diffusion coefficient of cobaltite-based materials is several orders of magnitude higher than that of the manganite. For LSCF, increasing the substitution of Fe by Co will increase the ionic conductivity. For example, the oxygen ionic conductivity of $La_{0.8}Sr_{0.2}Co_{0.2}Fe_{0.8}O_{3-\delta}$ is 2.3×10^{-3} S/cm and increases to 8.1×10^{-2} S/cm of $La_{0.6}Sr_{0.4}Co_{0.8}Fe_{0.2}O_{3-\delta}$ at 800 °C in air.

The oxygen exchange process at the gas/electrode interface appears to become increasingly dominant with the decrease in operating temperature. The activation energy of ORR on LSCF electrodes is in the range of 135–150 kJ/mol, and the electrochemical activity of the LSCF electrode for the O_2 reduction decreases significantly with temperature. At temperatures below 600 °C, the reaction is increasingly driven to the electrode/electrolyte interface region due to the significant reduction in the ionic conductivity of LSCF materials below 600 °C. The addition of ionic conducting phase such as GDC can significantly promote the surface exchange kinetics and thus the electrocatalytic activity of LSCF electrodes. The promotion effect of the GDC is most effective at low operation temperatures due to the high ionic conductivity of the GDC phase at reduced temperatures.

9.4.3 Other Perovskites, Double Perovskites, and Ruddlesden-Popper Structured Oxides

The catalytic activity of perovskite oxides is directly related to the B-site cations, although the modification of the A-site cation could lead to a change in the valence state and redox properties of the B-site cations, thus altering the catalytic activity of the oxides. For example, substitution of La with Gd with the formation of $Gd_{1-x}Sr_xCoO_{3-\delta}$ oxides actually reduces the activity, while the replacement of La with Sm with the formation of $Sm_{1-x}Sr_xCoO_{3-\delta}$ (SSC) perovskites enhances the electrochemical activity. SSC materials show conductivity as high as 1000 S/cm at $x = 0.5$ with low polarization loses for ORR. Donor-doping at the B-site of cobaltite perovskites could generate both high oxygen vacancy concentration and electrical conductivity. Examples are Sb^{5+} doping at the cobalt site of $SrCoO_{3-\delta}$ perovskite and the electrical conductivity of $SrCo_{0.9}Sb_{0.1}O_{3-\delta}$ exhibits a maximum value of 300 S/cm at 400 °C and shows a low polarization resistances [39].

Replacing La with Ba at the A-site of LSCF substantially enhances its oxygen ion conductivity and thus the electrochemical activity for ORR at intermediate temperatures. Since the presence of Ba with large ionic radius (0.16 nm in 12 coordination number) is beneficial to form large lattice spacing and higher freedom of oxygen ionic movement, higher Ba content in perovskites has been associated with enhanced oxygen reduction reaction rate. The best example is $Ba_{0.5}Sr_{0.5}Co_{0.8}Fe_{0.2}O_3$ (BSCF) which has been reported to show a high activity on doped ceria electrolyte, achieving a power density of 402 mW/cm^2 at 500 °C under hydrogen and air [40]. BSCF shows a very high oxygen exchange kinetics with a high oxygen exchange coefficient and surface exchange coefficient. The oxygen ion conductivity is also high and estimated to be 0.018 S/cm at 700 °C [41]. The electronic conductivity of BSCF increases with temperature, reaching a value of 26 S/cm around 400 °C and leveling off with further increase in temperature, see Fig. 9.21a. Both the electronic conductivity and thermal expansion coefficient curves of BSCF show a distinct change in the slope around ~400 °C.

BSCF is characterized by high oxygen ion conductivity and high variable oxygen nonstoichiometry, i.e., (3-δ), which is closely related to the temperature and partial pressure of oxygen. Below 350–400 °C, the kinetics of oxygen exchange with the gas phase is very slow and can be considered to be virtually frozen. The nonstoichiometric oxygen content of BSCF would be constant at around 2.6–2.65, i.e., (3-δ) is constant. Thus, at temperatures below 400 °C, BSCF shows *p*-type electrical conduction via small polaron hopping mechanism due to the formation of Co^{4+} and Fe^{4+} derived from the charge disproportionation. The conductivity increases with the temperature. With the increase in temperature, the equilibration kinetics of oxygen exchange increases, and this will lead to an increase in nonstoichiometric oxygen content, i.e., (3−δ) decreases. The increase in the nonstoichiometric content indicates an increase in

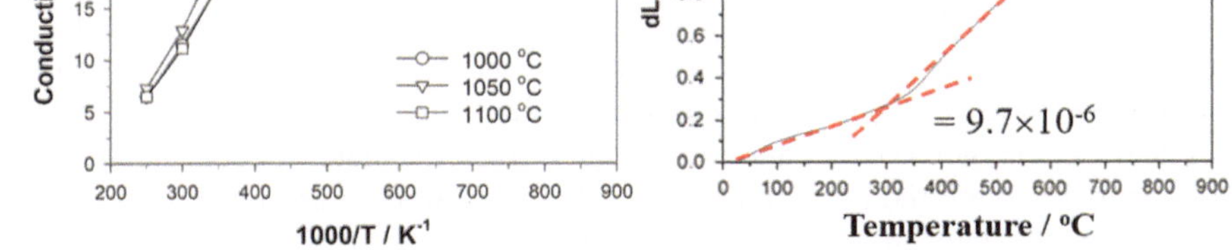

Fig. 9.21 **a** Arrhenius plots of electronic conductivity for $Ba_{0.5}Sr_{0.5}Co_{0.8}Fe_{0.2}O_{3-\delta}$ (BSCF) sintered at different temperatures in air and **b** thermal expansion curves of a BSCF sintered at 1050 °C for 6 h

oxygen vacancies and consequently a decrease in the electronic or hole conductivity, as shown above.

Two linear regions are observed for the electronic conductivity in the temperatures below 400 °C and between 400 and 700 °C with activation energy of ~31 and 6 kJ/mol in the low and high-temperature regions, respectively. The abrupt change in the activation energy indicates a transition in the conduction mechanism of the BSCF material. The fast oxygen exchange kinetics and oxygen ion transport properties contribute to the high activity of BSCF for ORR. The one sintered at 1050 °C shows a better conductivity due to the improved density of BSCF sintered at 1050 °C. At 800 °C the electrical conductivity of BSCF is ~29 S/cm. In the temperature range of 30–300 °C, TEC is 9.7×10^{-6} K^{-1} and increases to 23.2×10^{-6} K^{-1} between temperatures from 400 °C to 900 °C. As SOFCs are generally operated at temperatures higher than 400 °C, we should take 23.2×10^{-6} K^{-1} as the TEC value of BSCF, not the average TEC in the temperature range of 30–900 °C, 18.8×10^{-6} K^{-1}. Insufficient electrical conductivity may introduce a large contact resistance between the electrode and current collector. BSCF-based materials are also susceptible to CO_2 attack in the intermediate temperatures of 450–750 °C due to the formation of carbonates of Sr and Ba on the electrode surface. However, the thermal and electrical property and CO_2 tolerance of BSCF can be improved by surface modification, doping and composition optimization [41].

Layered cobaltites with general formula $LnBaCo_2O_{5+\delta}$ (Ln = Gd, Pr, Y, La, Sm, and Nd) have been considered as cathode materials of SOFCs [42]. These materials possess an ordered structure in which lanthanide and alkali-earth ions occupy the A-site sub-lattice, and the oxygen vacancies are localized into layers. Among them, Pr-based double compounds are one of the most popular double perovskites in SOFCs. Figure 9.22 shows the lattice structure of a typical A-site cation-ordered double perovskite oxide based on $PrBaCo_2O_{5+\delta}$. The crystal structure of these oxides can be regarded as a layered crystal $LnBaCo_2O_{5+\delta}$ (Ln = Pr in the case of $PrBaCo_2O_{5+\delta}$) by doubling the unit cell of the standard perovskite structure, consisting of consecutive layers $[CoO_2]$–[PrO]–$[CoO_2]$–$[BaO_\delta]$ stacked along the *c*-axis. The balance between electronic and ionic conductivity is related to the electron hopping along the Co–O–Co bonds. The transformation of a cubic perovskite to such a layered structure reduces the oxygen bonding strength in the BaO_δ layer and provides disorder-free channels for ion motion, which remarkably enhances oxygen diffusivity and opens the possibility for developing a new class of materials suitable for application as SOFC cathodes.

Stability of the double-layered perovskite structure of $LnBaCo_2O_{5+\delta}$ oxides is closely related to the ionic radius of the Ln^{3+} cation [43]. A stable layered structure is observed for Ln = Pr, Nd, Sm, and Gd, while it is in a metastable state for Ln = Y and La. Oxygen content and the nominal oxidation state of cobalt ions in the oxide increases with the ionic radius of Ln^{3+}. In addition, larger ionic radii for the Ln^{3+} cation translates into a higher oxygen mobility in the LnO_δ layer. Reasonable electrode performance has been reported for $LnBaCo_2O_{5+\delta}$, e.g., ASR of 0.213 Ω cm^2 at 600 °C for ORR on $PrBaCo_2O_{5+\delta}$, because of the fast exchange kinetics across the surface of the oxides. Layered cobaltites, $LnBaCo_2O_{5+\delta}$ oxides with Ln = Gd

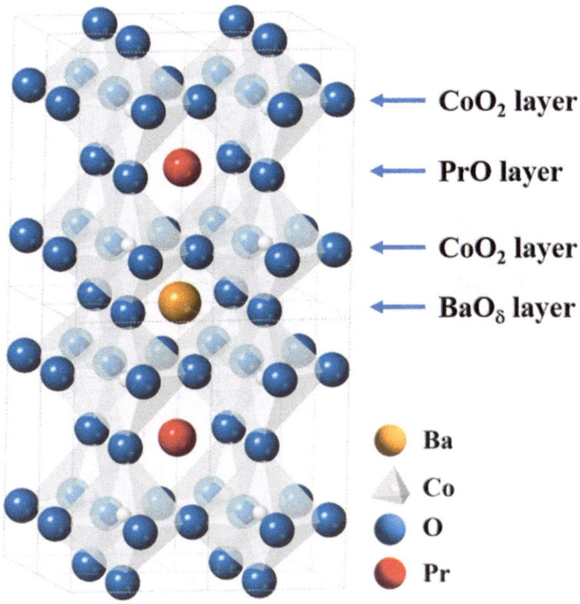

Fig. 9.22 Lattice structure of a typical A-site cation-ordered double perovskite oxide based on $PrBaCo_2O_{5+\delta}$

and Pr, could be one of the potential materials for IT-SOFC applications due to their excellent oxygen transport properties at low temperatures, i.e., high oxygen surface exchange coefficient and reasonable oxide ionic diffusivity, in combination with their high electronic conductivity. The overall conductivity of $PrBaCoO_{5+\delta}$ is typically in the range of 161–894 S/cm at 600 °C with TEC in the range of $20.4–24.6\times10^{-6}$ K^{-1} [42]. The significant variation in the electrical conductivity indicates the sensitivity of the materials to the synthesis and measurement conditions adopted by different research groups.

The Ruddlesden-Popper (RP) structured metal oxides are a series of oxide compounds with the general formula of $A_{n+1}B_nO_{3n+1}$ ($n = 1, 2, 3$) (see Fig. 9.23) and are also considered for cathodes of SOFCs [44]. The most simple one is K_2NiF_4-type structured oxides with general formula, $A_2BO_{4+\delta}$ (A = La, Nd, Pr; B = Co, Ni, Cu). The structure can be described as a stacking of ABO_3 perovskite layers alternating with AO rock-salt layers along the *c*-direction. There is sufficient space in the AO layer for oxygen ion migration. This structure allows for the accommodation of oxygen overstoichiometry as oxygen interstitial species with negative charges, which are balanced through the oxidation of the B-site cations. Studies have shown that these $A_2BO_{4+\delta}$ materials exhibit good electrical conductivity due to the mixed valence of B-site cation, good oxygen ionic transport properties due to the oxygen overstoichiometry and high oxygen mobility within the rock-salt layer, good activity for the ORR, and moderate thermal expansion properties.

$A_2BO_{4+\delta}$-type nickelates and cobaltites (e.g., $La_2NiO_{4+\delta}$) are *p*-type conductors with electrical conductivity in the range of ~100 S/cm at 800 °C and TEC of 11–13.7×10^{-6} K^{-1}. Introduction of cobalt into B-sites of nickelates can enhance the

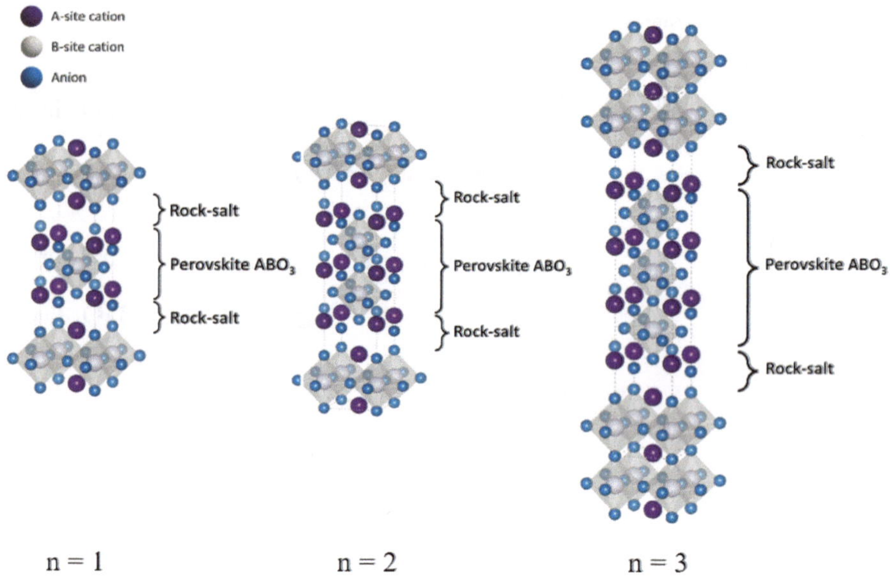

Fig. 9.23 Lattice structure of $A_{n+1}B_nO_{3n+1}$ (n = 1, 2, 3)

oxygen surface exchange process. Among $A_2BO_{4+\delta}$-type family, $Pr_2NiO_{4+\delta}$ shows a high performance and a much better stability as compared to LSCF and BSCF cathodes for ORR, tested at 750 °C and 0.8 V in wet H_2/air [45]. Partially replacing Pr with La and Ni with Cu and Ga can improve the oxygen diffusion properties, the electrocatalytic activity and stability for ORR.

High-order RP structured metal oxides of $La_3Ni_2O_7$ (n = 2) and $La_4Ni_3O_{10}$ (n = 3) are also potentially cathodes of SOFCs. However, there are difficulties in the synthesis of phase pure high-order RP-type materials due to the complex La-Ni-O phase relationships.

9.4.4 Composite Cathode

Similar to anode materials, it is also a common practice to mix single-phased perovskite oxides such as LSM, LSCF, BSCF with YSZ and doped ceria to form composite cathodes. The purpose is multiple; (1) increasing the ionic conductivity;(2) improving TEC and compatibility with the electrolyte; (3) increasing structure stability; and (4) possibility of synergetic effect of the composite to enhance the ORR.

In the case of a composite consisting of a pure electrical conducting phase like LSM or MIEC phase like LSCF and a pure ionic conducting phase like YSZ, LSGM, or doped ceria to a certain degree, the connectivity of the electronic conducting phase and ionic conducting phase is crucial to produce a mixed conducting composite. This is because the ionic conductivity of the electronic conducting phase (e.g., LSM) and the electronic conductivity of the ionic conducting phase (e.g., YSZ) are negligible under normal SOFC operation conditions. For porous LSM/YSZ composites, 50 wt% LSM and 50 wt% YSZ composition is commonly used.

Introducing an ionic conducting phase such as YSZ and GDC to form a composite cathode improves the electrochemical activities of not only electrically dominant electrodes like LSM, but also MIEC electrodes like LSCF and BSCF. This is generally indicated by the significantly reduced electrode polarization resistances for ORR on the composite cathodes as compared to the counterpart single-phased cathodes. The addition of the YSZ phase reduces the grain growth of LSM by the intervening YSZ phase and significantly extends the TPB areas [46]. Figure 9.24 shows AFM images of the YSZ electrolyte surface in contact with LSM/YSZ composite cathodes

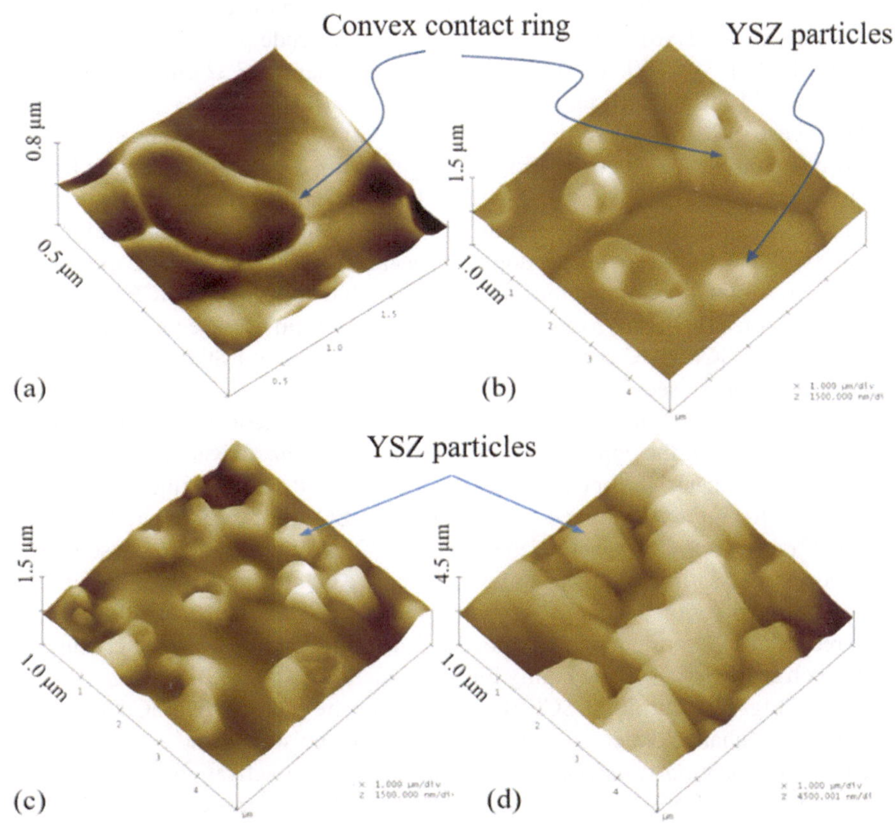

Fig. 9.24 AFM images of the YSZ electrolyte surface in contact with **a** LSM, **b** 90wt% LSM/10wt% YSZ, **c** 70wt% LSM/30wt% YSZ, and **d** 50wt% LSM/50wt% YSZ composite cathodes. LSM coating was removed by HCl treatment. Modified from Ref. [33] with permission from Springer Nature, Copyright 2008

after the removal of LSM by HCl treatment [33]. The formation of convex rings on the YSZ electrolyte surface are the contact areas between the LSM and YSZ, formed during the high-temperature sintering step at 1150 °C. The convex ring is actually the interface between LSM/YSZ for the oxygen reduction reaction, and this will be discussed in detail in Chap. 10. On the YSZ electrolyte surface in contact with the LSM/YSZ composite electrode, the isolated white particles are YSZ particles from the composites. The number of YSZ particles on the YSZ electrolyte surface increases with the increase in the YSZ content in the composite electrodes. As the YSZ content increases to 50 wt%, the number of YSZ particles on the YSZ electrolyte surface increases significantly and YSZ particles are interconnected (Fig. 9.24d). This indicates the formation of a three-dimensional network of ionic conducting YSZ phase in the 50wt% LSM/50wt% YSZ composite. Due to the formation of a LSM-to-LSM electrical network and a YSZ-to-YSZ ionic network, LSM/YSZ composite cathodes behave electrochemically like a MIEC cathode [46]. The composite cathodes can also be effectively formed by infiltration methods. In the case of a 5.8 mg/cm^2 GDC-impregnated LSM composite electrode, the electrode polarization resistance, R_E is 0.21 Ω cm^2 at 700 °C, substantially smaller than 11.7 Ω cm^2 measured on single-phased LSM cathode at the same temperature [47].

In the case of MIEC electrodes, the main function of adding an ionic conducting electrolyte phase is to reduce the TEC and improve the compatibility. This is largely due to the very high TEC values of MIEC electrodes, in particular, cobaltite-based electrode materials. In addition, the formation of composites with the addition of an ionic conducting phase such as doped ceria can significantly promote the surface exchange reaction kinetics and thus the electrocatalytic activity of LSCF electrodes. The enhancement in the performance is related to the ratio of the composite phases. In the case of LSCF/GDC composites, a ratio of 30–50 wt% LSCF and 70–50 wt% GDC is commonly used.

One potential benefit of the composite is the synergetic effect of the ionic conducting phase in the composite on the reduction of the activation energy of the parent electrode materials for ORR. This is particularly attractive for the development of cathodes for intermediate and low-temperature SOFCs. An early study showed that GDC can reduce the activation energy from 142 kJ/mol for a single-phased LSCF electrode to 122 kJ/mol for the LSCF/GDC composite electrode [48], although the reduction in activation energy is relatively small. Such a synergistic effect can be explored by using various ionic conducting phases, including a proton conducting phase such as $BaCeO_3$ and the highly ionic conducting Er doped Bi_2O_3, $Er_{0.4}Bi_{1.6}O_3$ (ESB), see Fig. 9.25 [49, 50]. LSCF/$BaCeO_3$ composite cathodes by $BaCeO_3$ infiltration show an enhanced tolerance toward sulfur (SO_2) poisoning and exhibit reduced electrode polarization resistance. Most significantly the activation energy for the ORR on LSCF/$BaCeO_3$ composite cathode is 96 kJ/mol, significantly smaller than 156 kJ/mol obtained for ORR on LSCF electrode. Addition of the ESB phase is also very effective in reducing the activation energy of the $La_{0.6}Sr_{0.4}Co_{0.2}Fe_{0.7}Nb_{0.1}O_{3-\delta}$ (LSCFNb) cathode for ORR. An LSCFNb/ESB composite with 60%LSCFNb and 40% ESB shows an activation energy of 98 kJ/mol, substantially smaller than

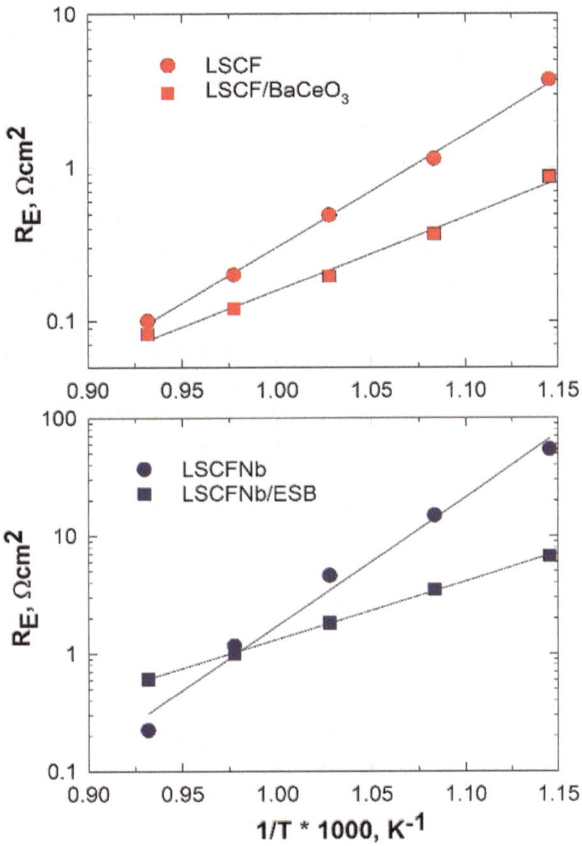

Fig. 9.25 Activation energy plots of LSCF and LSCF/$BaCeO_3$ composite cathodes, and LSCFNb and LSCFNb/ESB composite cathodes for ORR, measured in air in the temperature range of 600–800 °C

214 kJ/mol for the ORR on LSCFNb electrodes. This shows the promising potential in the development of cathode materials with reduced activation energies for intermediate and low-temperature SOFCs.

9.5 Interconnect and Sealant

9.5.1 Interconnect

In SOFCs, a single cell is rarely utilized; instead in order to increase the voltage output, multiple cells are stacked in series connected with interconnects. The interconnect is a critical component in SOFCs. Regardless of planar or tubular cell configuration, the interconnect provides electrical connection between cells, and at the same time, it acts as a physical barrier, protecting the air electrode from the reducing environment of the fuel on the anode side, and vice versa preventing the oxidation of anode materials from the oxidizing atmosphere of the cathode side. Interconnect can be classified as ceramic and metallic materials. In order to perform the intended function, interconnect materials need to fulfill the following conditions:

- Must exhibit excellent electrical conductivity with negligible ionic conductivity under the SOFC operating environments. A value of 10 S/cm is the minimum electrical conductivity for the usefulness of interconnect in SOFCs.
- Must have adequate stability in terms of dimension, microstructure, and composition at operating temperatures in both reducing and oxidizing atmospheres. In typical SOFC operation conditions, p_{O_2} varies from 0.21 to 10^{-21} atm. Thus, a significant p_{O_2} gradient builds up on both sides of the interconnect. Dimensional stability is important to minimize the mechanical stress and thus prevent the cracking or warping to the sealing.
- Must be chemically stable with no or minimum volatile species under reducing and oxidation conditions because the volatile species are potentially contaminants for the electrodes of SOFCs.
- Must have good thermal properties in terms of adequate TEC and good thermal conductivity. TEC of the interconnect should be in the range of 10–13 × 10^{-6} K^{-1} between ambient and operating temperatures to avoid failure caused by the thermal stresses developed during thermal cycles. Good thermal conductivity allows the fast transfer of heat generated from the cathode to the anode, facilitating heat uniform distribution within the stack. A value of 5 W/m·K is considered to be the low limit.
- Must possess good mechanical properties. High-temperature strength and creep resistance are particularly important for planar SOFCs where the interconnect would serve as a structural support.
- Be easy to prepare and fabricate. The formation of dense interconnect is critical to separate fuel and oxidant and to prevent the crossover during cell operation. Low cost of raw materials is important for the mass production of fuel cells.

9.5.1.1 Ceramic Interconnect

In ceramics, only a few oxide materials can be used as interconnect materials in SOFCs. Among them, lanthanum chromite ($LaCrO_3$) is the most common candidate for the interconnect materials because it is a *p*-type electronic conductor in oxidizing conditions, exhibits reasonable stability in both reducing and oxidizing environment and good compatibility with other cell components [51]. Electrical conduction in the undoped $LaCrO_3$ occurs by the small polaron mechanism via transport of electron holes. The electrical conductivity of pure $LaCrO_3$ is not sufficiently high for use as an interconnect in fuel cells, i.e., 0.34 S/cm at 700 °C and ~1 S/cm at 1000 °C. However, its conductivity can be increased through doping at the A- and/or B-site of $LaCrO_3$. The most common dopants for $LaCrO_3$ are Sr and Ca, which acts as electron acceptors on the trivalent La sites, and thus increase *p*-type conduction. The conductivity is also related to the size of the dopant. The smaller lattice distortion can result in a higher mobility, and thus a higher conductivity for a given defect concentration. For example, conductivity of $La_{0.8}Ca_{0.2}CrO_{3-\delta}$ is typically in the range of 30–40 S/cm at 1000 °C in air, higher than 20 S/cm measured on $La_{0.8}Sr_{0.2}CrO_{3-\delta}$. The higher conductivity of Ca-doped $LaCrO_3$ is due to the closer match of the radius of dopant Ca^{2+} (0.134 nm) with that of the host ion, La^{3+} (0.136 nm), as compared to the radius of Sr^{2+} (0.144 nm). Average TEC of doped $LaCrO_3$ is in the range of 10–11 × 10^{-6} K^{-1}, comparable to that of YSZ electrolyte. Calcium has a higher solubility limit, is more stable in oxidizing atmosphere, and is slightly more effective at increasing the conductivity at a high doping level. On the other hand, strontium is more effective at increasing TEC, reducing the expansion and maintaining strength in reducing atmosphere and leads to a lower chromium oxide activity. Doping at the B-site of $LaCrO_3$ such as Co, Fe, Ni, Mg, Cu, V not only alters the electrical conductivity, but also modifies the TEC and mechanical properties of $LaCrO_3$. In addition to aliovalent ions for increasing defect concentrations, La has been replaced with other lanthanide elements (Y, Nd, Pr, and others).

As shown early in this chapter, the electrical conductivity of doped $LaCrO_3$ behaves very differently in its compensation mechanism with oxygen partial pressure. Under oxidizing atmosphere at the cathode where the p_{O_2} ranges from 10^{-4} to 10^0 atm, the conductivity is enhanced via an electronic compensation mechanism (i.e., $Cr^{3+} \rightarrow Cr^{4+}$ transition). On the other hand, in the reducing environment such as with fuel gases at the anode where p_{O_2} is low (e.g., 10^{-18}–10^{-8} atm), the conductivity is significantly reduced due to the formation of oxygen vacancies via an ionic compensation mechanism. For example, for $La_{0.8}Ca_{0.2}CrO_{3-\delta}$, the electrical conductivity is ~37 S/cm in air at 1000 °C and reduces to ~6 S/cm at $p_{O_2} = 10^{-18}$ atm. Due to the strong dependence of electrical conductivity on the oxygen partial pressure, the electrical conductivity of doped $LaCrO_3$ in a reducing atmosphere like hydrogen is significantly lower than that in an oxidizing atmosphere like air. This leads to an establishment of a conductivity gradient across the doped $LaCrO_3$ interconnect when subjected to fuel on one side and oxidant on the other, resulting in the reduction of the overall conductivity of the doped $LaCrO_3$. This limitation renders that the practical operating temperature for the use of doped $LaCrO_3$ as interconnect should be higher than 800 °C. The significant decrease in their electrical and thermal conductivities with decreasing temperature is a major challenge in the developments of ceramic interconnects for IT-SOFCs.

Other issues concerning the use of doped $LaCrO_3$ interconnect in a reducing atmosphere include the expansion that originates from the oxygen-triggered lattice expansion. Under a fuel atmosphere, the concentration of oxygen vacancies will increase as the result of the ionic compensation mechanism. The lattice expansion occurs due to the increased cation size as Cr^{4+} (0.055 nm) transfers to Cr^{3+} (0.0615 nm) and/or the increased cation–cation repulsion resulting from removal of bridging oxygen ions. This will lead to the increase in thermal expansion at p_{O_2} below a critical value (e.g., $<10^{-10}$ atm). Such an oxygen vacancy-triggered expansion would create internal stress, reduce the mechanical strength, and thus degrade fuel cell performance.

The difficulty of sintering or densification of doped $LaCrO_3$ ceramics is also one of fabrication-related challenges. This is due to the refractory nature of $LaCrO_3$-based materials. The poor sinterability of $LaCrO_3$ has been attributed to the volatilization of CrO_3 species and formation of a thin solid layer of Cr_2O_3 at the interparticle neck, which inhibits the sintering of particles. Various approaches have been developed to address the densification issues, including using a Cr-deficient composition along with excess Sr or Ca doping, performing sintering in reducing atmosphere followed by oxidation treatment to restore its electrical conductivity and improving the sintering via a liquid-phase sintering process. However, the formation of a liquid phase could result in a second-phase formation, which can be detrimental during fuel cell operation. Formation of thin and mechanically strong ceramic interconnect plates by conventional ceramic processing techniques is also a challenge.

9.5.1.2 Metallic Interconnect

Metallic interconnects were originally developed to get around the difficulties encountered in the design of planar SOFCS, in particular the anode-supported planar SOFCs. The use of thin electrolyte in anode-supported SOFCs leads to a marked decrease of electrolyte ohmic resistance and thus the reduction in operating temperatures to 800 °C or below. The reduction in operating temperature, in principle, enables the doped $LaCrO_3$ interconnect to be replaced by a metallic one. Metallic materials show significant advantages as interconnect materials: higher electronic and thermal conductivity than the doped $LaCrO_3$, low cost, easy manufacture and good workability. The ohmic losses in the metallic interconnect are small. The excellent heat conduction allows the heat generated at the air electrode side to be easily transported to the fuel electrode, reducing the temperature gradient inside the cells. Metallic materials based on transition metal-based oxidation resistant alloys have been considered to be the primary candidates as the interconnect materials of SOFCs. These include Ni(-Fe)-Cr-based heat-resistant alloys, Cr alloys, and chromia-forming ferric stainless steels. The alloys with the formation of a protective and semi-conductive chromia scale (Cr_2O_3) to minimize further environmental attack during the high-temperature operation and with TECs of 11.0–12.5 $\times$ 10^{-6} K^{-1} are the preferred candidates. Most importantly, Cr_2O_3 has comparatively low electrical resistivity at elevated high temperatures (1 $\times$ 10^2 Ωcm at 800 °C in air) as compared to other metal oxides (Al_2O_3: 5 $\times$ 10^8 Ωcm at 700 °C in air; MgO: 1.8 $\times$ 10^7 Ωcm at 800 °C in air; SiO_2: 7 $\times$ 10^6 Ωcm at 600 °C in air) [51]. A good example in this category is Crofer 22 APU from ThyssenKrupp VDM, ferritic stainless steel AISI 441 or AISI 446 and ZMG232 from Hitachi Metal Inc. Table 9.3 lists the chemical compositions and TEC of the potential alloys for interconnect applications in SOFCs.

An oxide scale is inevitably formed on the surface of the alloys under IT-SOFC operating conditions. The oxidation of the alloys considered for metallic interconnect is usually a diffusion-controlled process that can be approximated by the parabolic law of Wagner's theory:

$$x^2 = k_p t \tag{9.53}$$

Table 9.3 Chemical composition (%) and TEC of commercially available ferritic stainless steels

Alloy	Fe	Cr	Mn	Si	C	La	Other	TEC ($\times 10^{-6}$/K^{-1})
Crofer22 APU	Bal.	22.8	0.45	–	0.005	0.06	Ti(0.08), P (0.016)	12.2
AISI 446	Bal.	25	1.5	1.0	0.02		P(0.04), S(0.03)	10.9
ZMG232	Bal.	21.96	0.05	0.4	0.02	0.04	Al(0.21), Ni(0.26), Zr (0.22)	12.5

or

$$\Delta w^2 = k_p t \tag{9.54}$$

where x is the thickness of the thermally grown oxide scale, Δw is the weight gain per area, and k_p is the parabolic rate constant or oxidation rate expressed in thickness (μm^2/s) or mass (g^2/cm^4·s). In practice, k_p is often used to describe the oxidation behavior of an alloy owing to the fact k_p is directly obtained from the experimental data with no need to know the exact chemical composition of the oxide scale formed. The oxidation constant k_p is primarily dependent on the alloying composition and is also affected by thermal history and surface conditions. The electrical resistance of the formed oxide scale is directly related to the scale thickness and conductivity.

One of the issues with ferritic stainless steels is the adhesion of scale to metal substrate. Thus, the formation of overgrown thick oxide scale of a metallic interconnect needs to be controlled to prevent subsequent spallation due to thermal stresses induced by thermal expansion mismatch between the scale and the substrate. The consequence of the spallation is serious as the regrowth of scale after spallation would lead to depletion of Cr, the scale-forming element, resulting in the rapid oxidation of base metal. Addition of reactive elements or oxide additives such as Y, Ce, La, Zr can reduce the oxidation rate and to promote the adherence of the oxide scale to the metallic substrate.

The growth rate, microstructure, and the composition of the thermally grown oxide scale are dependent on the alloy composition. The addition of trace amounts of Mn to the Fe–Cr or Ni-Cr alloys (e.g., Crofer 22 APU) leads to the formation of an electrically conductive $(Cr,Mn)_3O_4$ spinel top layer on the oxide scale. Due to the outward diffusion rate of Mn ions in Cr_2O_3 being faster than that of Cr in Cr_2O_3, the thermally grown oxide scale formed on Fe–Cr-Mn or Ni-Cr–Mn alloys belongs to $(Cr,Mn)_3O_4$ spinel with a duplex structure on top of the dense Cr_2O_3 sublayer (see Fig. 9.26). Generally, Cr_2O_3 and $(Cr,Mn)_3O_4$ spinel are the main phases in the oxide scale of the Fe–Cr alloys based metallic interconnects exposed to IT-SOFC operating environments. The formation of $(Cr,Mn)_3O_4$ spinel on top of the oxide scale significantly reduces the vaporization of volatile Cr-containing species. Because the electrical conductivity of $(Cr,Mn)_3O_4$ spinel is orders of magnitude higher than that

Fig. 9.26 **a** Surface and **b** cross-sectional view of oxide scale formation of Crofer 22 APU oxidized at 750 °C in air for 300 h

of Cr_2O_3, the formation of such a $(Cr,Mn)_3O_4$ spinel layer can also result in a higher electrical conductivity of the scale, compared with that only containing Cr_2O_3.

One main issue of chromium-based alloys is the formation of volatile Cr^{6+} species under operating conditions of SOFCs. The basic reason for the Cr volatility is the thermodynamic instability of the Cr_2O_3 scales formed on the chromia-forming alloys at high temperatures, forming gaseous species through the following reactions:

$$2Cr_2O_{3(s)} + 3O_{2(g)} \rightleftharpoons 4CrO_{3(g)} \tag{9.55}$$

$$2Cr_2O_{3(s)} + 3O_{2(g)} + 4H_2O_{(g)} \rightleftharpoons 4CrO_2(OH)_{2(g)} \tag{9.56}$$

$$Cr_2O_{3(s)} + O_{2(g)} + H_2O_{(g)} \rightleftharpoons 2CrO_2OH_{(g)} \tag{9.57}$$

The volatile Cr species, chromium oxides and chromium oxyhydroxides, from the chromia scale strongly depend on the partial pressure of oxygen, the water content and temperature. Due to the much lower p_{O_2} in the fuel side (10^{-7} and 10^{-18} Pa), the vaporization of Cr species occurs predominantly on the air side of SOFCs. Assuming a water vapor pressure of 3 kPa (i.e., the saturated water vapor pressure at room temperature) in air, the most representative vapor species over Cr_2O_3 are $CrO_2(OH)_2$ in wet air and CrO_3 in dry air. Figure 9.27 shows the temperature dependence of the partial pressure of these two dominant Cr species over Cr_2O_3 scale in wet and dry air, respectively. The partial pressure of CrO_3 decreases rapidly with the decrease of temperature while the change in the partial pressure of $CrO_2(OH)_2$ with the temperature is relatively small. For instance, the partial pressure of CrO_3 is 1.4×10^{-3} Pa at 900 °C and 3.4×10^{-7} Pa at 600 °C, a decrease by over 4 orders of magnitude. For $CrO_2(OH)_2$, the partial pressure is decreased from 9.2×10^{-2} Pa to 1×10^{-2} Pa over the same temperature range, a reduction in less than one order of magnitude. Water vapor pressure plays a more influential role than temperature in affecting the

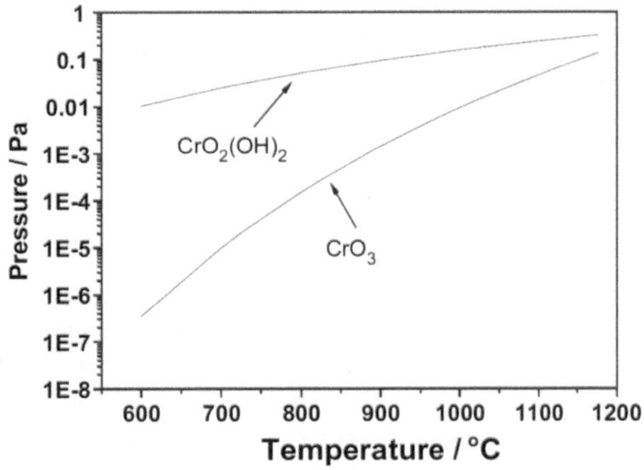

Fig. 9.27 Partial pressure of $CrO_2(OH)_2$ in wet air (3%H_2O) and CrO_3 in dry air over Cr_2O_3 scale at different temperatures

volatility of Cr^{6+} species. Thus, without effective protective coatings, the vaporization of chromium species from chromia scale poisons the cathodes and seriously degrades the cell performance.

To reduce the growth rate of the oxide scale and the vaporization of chromium species, a thin and dense oxide coating with high electrical conductivity such as CrN and CrAlN nitride, $(Mn,Co)_3O_4$ spinels, doped $LaCrO_3$, LSM and LSCF are often deposited on the metallic interconnect. The application of a dense and conductive coating can inhibit Cr volatilization, reduce the thickness of oxide scale, and improve the chemical compatibility with other SOFC components on the condition that the interfacial electrical resistance is small and does not increase significantly during the operation of fuel cells. The chromium volatility can also be suppressed by modification of the metallic interconnect materials. For example, Hua et al. [52] reported a new Ni–Mo–Cr alloy with a TEC value of 13.92×10^{-6} K^{-1} between 35 and 800 °C. After oxidation treatment at 750 °C for 1000 h, ASR of this alloy is 4.48 $m\Omega cm^2$. A poisoning study using LSM cathode indicates that the Cr deposition and poisoning of the Ni–Mo–Cr alloy is remarkably reduced as compared to the conventional Fe–Cr alloy [53].

The oxidation behavior of the Fe–Cr metallic interconnect under the reducing environment of the fuel is considered to be more complex, particularly when a hydrocarbon fuel is used. The presence of water, H_2 and carbon makes metallic interconnects susceptible to various forms of corrosion. Even though the oxygen partial pressure on the anode side is 10^{-18}–10^{-8} atm, significantly lower than that on the cathode side, oxides such as Cr_2O_3 and Mn–Cr spinel are still thermodynamically stable, but the surface morphology and properties of the scale can be significantly different. The presence of water vapor and hydrogen under the anode environment may affect the defect structure and hence change the diffusion behavior of metal and oxygen ions in oxides and thus have a significant effect on the oxidation mechanism and microstructure and properties of the scales formed on the surface of alloys. However, as the oxygen partial pressure in the fuel gas is in the range between 10^{-18}

and 10^{-8} atm, the effect of Cr vapor species on the performance of the anode can be ignored. Therefore, the Cr-poisoning caused by the metallic interconnets is mainly on the cathode.

9.5.2 Sealants

The sealing is an integral part of planar SOFCs. In planar SOFC designs, the places that need to be sealed include the edge between metallic interconnect and electrolyte, edge of the anode-supported cells as well as the interconnect and frame. Seals must be chemically stable and compatible with other SOFC components at high temperatures under both oxidizing and wet reducing atmospheres, with a lifetime greater than 40,000 h together with hundreds of thermal cycles for stationary applications, or a lifetime of at least 5000 h and not less than 3000 thermal cycles (start-up and shutdown) for transportation applications. The seal should be mechanically robust enough to accommodate the thermal stresses caused by thermal cycling as well as by temperature gradients induced by gas flowing and electrochemical reactions. Accordingly, the sealing material is regarded as one of the most significant technical challenges in the development of planar SOFCs.

The sealants can be broadly classified into rigid bonded seals, compressive seals, and compliant bonded seals. Each offers advantages and limitations. In rigid bonded sealing, the sealant forms a joint that is non-deformable at room temperature. Because the final joint is brittle, it is critical for the sealant to match the TEC of the adjacent components. High-temperature glasses and ceramic glasses such as alkali silica glasses and BaO-CaO-SiO_2 are among the most important rigid bonded sealants employed in joining planar SOFC stacks. Glass can wet the surface of adjacent components by viscous flow at high temperatures (sealing temperature) and then solidify at low temperature (working temperature). These materials have acceptable stability in the reducing and oxidizing atmospheres, are cheap, can be readily applied to the sealing surfaces as a powder paste or as a tape cast sheet, and can be tailored to cover a wide range by manipulating the glass composition and heat-treatment schemes. They are electrically insulating, and their TEC can be adjusted to those of the electrolyte and metallic interconnect. However, the brittle nature of glasses and ceramic glasses makes these seals vulnerable to cracking, and they tend to undergo phase transformations and react with the cell components and interconnect materials under SOFC operation conditions in the long run, due to their intrinsic thermodynamic instability.

In the case of metallic interconnects, sealing is best performed below 850 °C to prevent excess oxidation of the interconnect during the sealing process. Therefore, one of the primary goals of sealing glass design is to achieve a sufficient low viscosity to facilitate effective sealing at temperatures below 850 °C. The viscosity of a glass can be adjusted through composition, e.g., increasing the concentration of low melting point oxides like B_2O_3, P_2O_5, Bi_2O_3, etc., is helpful to decrease the viscosity of glass, while the viscosity increases with increasing the concentration of high melting point oxides like SiO_2, Al_2O_3, etc. For glass-ceramic sealants, the viscosity is dependent on both the resultant glass phase and the amount of crystalline phase. To match with other cell components, the TECs of glass sealants are best in the range of $11.0–13.5 \times 10^{-6}$ K^{-1} from room temperature to T_g. T_g is the glass-transition temperature corresponding to a glass viscosity of ~10^{12} Pa·s, at which the glass transforms from the solid state to the liquid state.

The chemical stability of glass seals containing B_2O_3 and P_2O_5 is of great concern, since B_2O_3 and P_2O_5 may evaporate significantly at high temperature, especially under a wet oxidation atmosphere. The evaporation of volatile species, in particular boron, not only degrade the long-term stability of the sealant glass but also poison the electrochemical activity of the electrodes of SOFCs. However, the boron volatility of boronsilicate-based glass can be reduced by manipulating the B–O network structures via the conversion of $[BO_3]$ to $[BO_4]$ units or crystallization of the boron-containing phases. For example, doping rare earth oxides such as Gd_2O_3 can significantly reduce the boron volatility and enhance the stability of the glass network structure, which in turn minimizes the poisoning effect of boron-containing glass sealant materials [54].

Compared with the YSZ electrolyte, reactions between glass-based seals and metallic interconnect are more problematic. The chemical interaction between glass-ceramic sealant and Fe–Cr alloy interconnect can cause severe corrosion of the interconnect along sealing rims, leading to short-circuiting and stack failure. Nevertheless, the interaction is dependent on the composition of both glass sealant and oxide scales. For example, in the case of barium silicate glass, the reaction between glass and chromia scale leads to the formation of $BaCrO_3$:

$$Cr_2O_3 + 2BaO + 1.5O_2 \rightarrow 2BaCrO_4 \tag{9.58}$$

Since $BaCrO_4$ have a very high TEC of $21–23 \times 10^{-6}$ K^{-1}, the excess formation of $BaCrO_4$ not only leads to depletion of barium in the glass-ceramic, but also causes cracking and separation of the glass-ceramic sealant from the interconnect matrix due to the thermal expansion mismatch. The addition of a barrier layer to the metallic interconnect can effectively prevent the formation of the $BaCrO_4$ phase and improve the thermal stability of glass sealing.

Compressive seals have been developed to avoid the disadvantages of the rigid bonded seals, with the merit of flexibility and compressibility, allowing the cells and interconnects to expand and contract freely during thermal cycles and operation. This type of sealing relies on the compressive load of the stack. So far, two kinds of compressive seals have been considered, i.e., the deformable metallic seals and the mica-based seals. The deformable metallic seals include ductile silver and corrugated or C-shaped superalloy gaskets. The most common compressive sealing material is based on mica. Mica belongs to a class of layered minerals known as phyllosilicates and is composed of cleavable silicate sheets. By incorporating a compliant interlayer such as a deformable metal or glass at the interface to form the hybrid mica-based seals, the sealing properties of mica can be significantly improved. Compliant bonded seals are based on metallic braze. Metallic materials have lower stiffness as compared to ceramics and can undergo plastic deformation, which allows for the accommodation of thermal and mechanical stresses. Silver and gold are stable in air and are commonly used as metal braze materials. A major challenge in obtaining a good metal–ceramic joint is adequate wetting of the ceramic by the braze metal.

9.6 Cell Structures and Stack Design

9.6.1 Cell Structures

There are number of cell support structures, and each is classified according to the layer that mechanically supports the cell. These include electrolyte-supported, anode-, or cathode-supported as well as porous substrate- or metal-supported structures, see Fig. 9.28.

Due to the thick electrolyte (typically in the range of ~100 μm) needed to mechanically support other cell components, the electrolyte-supported SOFCs are primarily developed for operation at high temperatures (~900 °C or above). However, the reduction in the operating temperature to an intermediate range (500–800°) greatly reduces the degradation of SOFC components, widens the materials selection, increases the thermal cycling ability, and enables the use of low-cost metallic interconnects and balance of plant (BoP) components. On the other hand, the overall performance of a SOFC may decrease because of the reduced ionic conductivity of the solid electrolytes and the increased polarization resistance of the electrodes. Another issue is that lower operating temperature requires high electrode electrocatalytic activity and high ionic conductivity of the electrolyte to compensate for the cell loss without penalty for the power density achieved at high temperatures. The most effective way to achieve this objective is to reduce the thickness of the YSZ electrolyte using anode or cathode-supported structures.

Up to now, the most popular and common-supported cell structure is the anode-supported thin film SOFCs. The state-of-the-art anode-supported SOFC is based on porous Ni/YSZ cermets as a support due to highly electrically and thermally conductive Ni cermet and the ability of co-firing with YSZ thin electrolyte layer at high temperature up to 1400 °C for the densification of the electrolyte with little reactivity between Ni and the YSZ electrolyte. Figure 9.29 presents a typical cross-sectional view of a Ni/YSZ anode-supported thin YSZ electrolyte cell. The cell

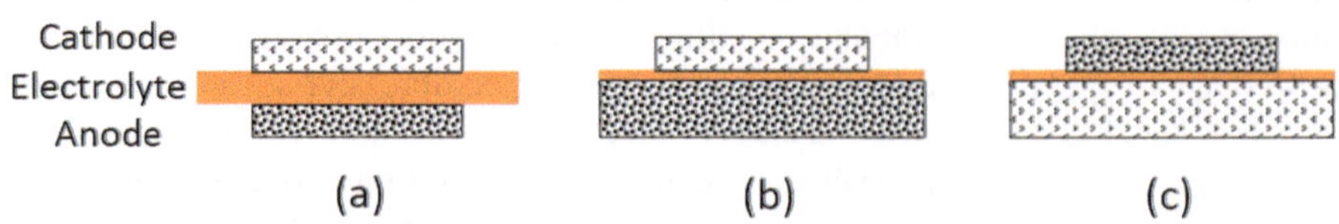

Fig. 9.28 Typical cell structures used in SOFCs: **a** electrolyte-supported, **b** anode-supported, and **c** cathode-supported

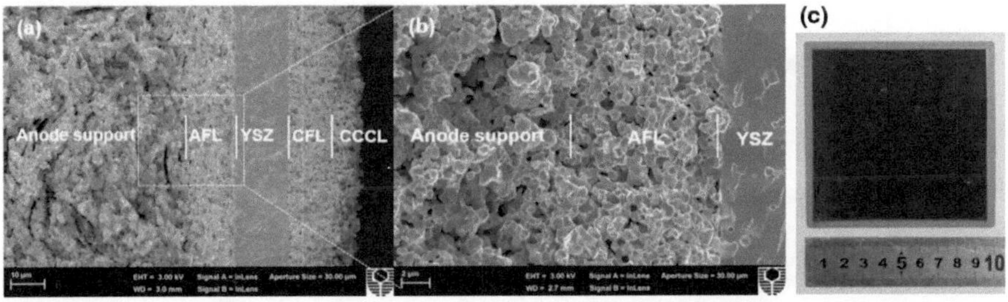

Fig. 9.29 A cross-sectional view of **a** a 5 layered a Ni/YSZ cermet-supported thin YSZ electrolyte cell consisting of a 800 μm Ni-YSZ anode support, a 13 μm Ni-YSZ anode functional layer (AFL), a 16 μm YSZ electrolyte, a 14 μm LSM-YSZ cathode functional layer (CFL) and a 8 μm LSM cathode current collection layer (CCCL), **b** magnified anode support/AFL/YSZ electrolyte triple layers, and **c** an optical photo of an anode-supported cell

consists of a relatively thick porous-supporting substrate (~500 μm) and a thin- and fine-structured electrode layer, the anodic functional layer (AFL). To reduce the electrolyte ohmic resistance and to enhance the cell efficiency, the electrolyte layer deposited should be as thin as possible. As a general role, the film thickness is inversely proportional to the pore size and/or propagated roughness of the surface, which means that the larger the pore size, the more difficult it is to get a thinner electrolyte. A LSM/YSZ cathode is used as a cathode functional layer (CFL) and a high electrical-conducting current collector layer (CCCL, in this case, LSM is used) is generally employed to reduce the contact resistance at the cathode side. If the CFL is a cobaltite-based electrode like LSCF, a doped ceria barrier layer is needed on the cathode side of the YSZ electrolyte to prevent the interaction between the cathode and YSZ electrolyte. Tape-casting and tape-calendering processes are the most common techniques in the fabrication of anode-supported structure for thin film SOFCs.

However, the porous composite anode support is relatively weak mechanically and can have difficulty withstanding the thermal and mechanical stresses generated by rapid temperature fluctuations. Moreover, Ni/NiO redox cycling induced by air diffusion into the anode compartment during the loss of fuel supply and other operational excursion can disrupt the anode microstructure, leading to irreversible degradation. For this reason, metal-supported structures have been developed to increase the mechanical flexibility, thermal shock resistance, and the robustness of SOFCs.

9.6.2 Stack Design

Several SOFCs designs have been developed, which include tubular, planar, banded, and corrugated structures. Each design has a different current path and cell and stack configuration. The cell and stack designs must satisfy the essential requirements such

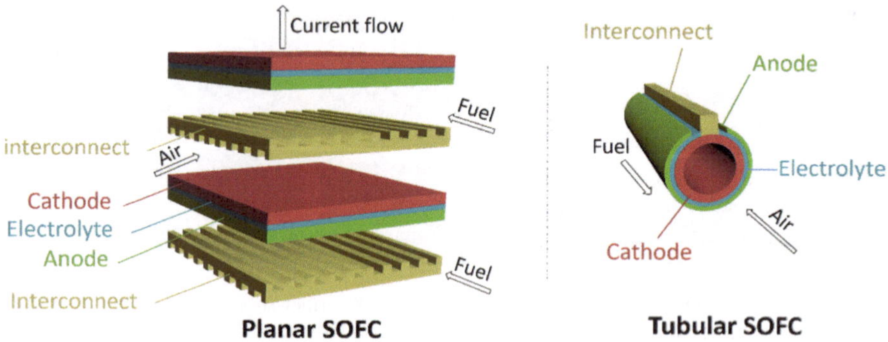

Fig. 9.30 Planar and tubular design of SOFCs

as thermal management, mechanical properties, and electrochemical and electrical performance to achieve high performance and high durability.

Planar SOFCs are the most preferred design (Fig. 9.30). The planar design is simpler as compared with tubular, monolithic, and segmented cells series designs. The performance of planar SOFCs is theoretically higher than that of tubular SOFCs because the thin planar SOFCs based on an anode-supported cell structure have a short current path. The reduced in-plane ohmic resistance of planar SOFCs enables operation at lower temperature. In addition, tape-casting, and other mass production techniques, for example screen-printing and plasma spray, can be easily be applied for planar SOFC component production, thus making possible a substantial reduction in production costs. On the other hand, the use of tubular geometry with the one-end-closed feature enables a seal-less design to be adopted. This is probably the strongest advantage over the planar geometry where gas seals are needed along the perimeters of the stack. In tubular SOFCs, cathode substrate-supported design is preferred because it allows cell-to-cell connections in a stack to take place in a reducing atmosphere where inexpensive metals such as Ni and Cu can be used. The tubular configuration, because of its geometry, is capable of solving the problems related to cracking, thermo-cycling, start-up time and sealing. High manufacturing cost and low power density appear to be challenging in tubular SOFCs. On the other hand, an anode substrate-supported design is a good choice for planar stacks. Anode-supported planar SOFCs permit a reduced operation temperature and high power density.

9.7 Variations of SOFCs

In addition to the most common planar and tubular designs, there are also variations in the design and arrangements in SOFCs. This includes single-chamber SOFCs (SC-SOFCs), metal-supported SOFCs (MS-SOFCs) and solid carbon fuel-based SOFCs or simply direct carbon fuel cells (DCFCs).

9.7.1 Single-Chamber Solid Oxide Fuel Cells

Single-chamber solid oxide fuel cells (SC-SOFCs) are developed probably for the purpose of solving the high-temperature sealing problems in conventional dual chamber SOFCs. In SC-SOFCs, both anode and cathode are exposed to the same mixture of fuel and oxidant gas. As a result, the gas-sealing problem can be avoided since no separation between fuel and air is required. In addition, carbon deposition becomes less problematic due to the presence of a large amount of oxygen in the mixture. Another advantage is the easy fabrication. Figure 9.31 shows the operating principle of SC-SOFCs.

The successful operation of SC-SOFCs requires the use of highly selective electrodes; (i) the anode has to be electrochemically active for the oxidation of the fuel but inert to ORR; and (ii) the cathode has to show the high activity for ORR but inert to the fuel oxidation reaction. The concept of SC-SOFC was first demonstrated by Hibino and Iwahara in 1993 [55]. Using YSZ as electrolyte, Ni/YSZ as anode and Au as cathode, the cell achieved an OCV of 350 mV and power density of 2.3 mW/cm^2 at 950 °C with a methane-air mixture (methane:oxygen = 2:1). Replacing the Au cathode with conventional LSM resulted in the significant increase in the OCV to 795 mV and maximum power density to 121 mW/cm^2 at 950 °C. Doped ceria can also be used electrolyte in SC-SOFCs. Using SDC as electrolyte, Ni/SDC as anode and $Sm_{0.5}Sr_{0.5}CoO_3$ as cathode, the cell achieved an OCV of 800 mV and a peak power density of 644 mW/cm^2 at 550 °C with a methane-air mixture [56]. SC-SOFCs also exhibit better performance with higher hydrocarbon fuels. Due to the fact that the gas at the electrodes in SC-SOFCs is a fuel-air mixture, the electrolyte does not need to be in a dense form. The use of a porous electrolyte allows for the preparation of thin electrolyte films by low-cost fabrication methods such as screenprinting. LSCF and BSCF can also be considered as cathodes in SC-SOFCs.

Though the heat generated by the partial oxidation of hydrocarbons at the anode is able to sustain the cell temperature without external heating, the efficiency of the fuel cell is generally low. This is partly due to the use of diluted fuel in order to avoid a fuel-air mixture explosion. The performance of SC-SOFCs depends strongly on the fuel/air molar ratio and flow rate. The catalytic oxidation of hydrocarbon fuels can also result in the substantial heating of the electrodes, temperature variation along the cells and reduced fuel utilization. Challenges still remain in the development of highly selective anode and cathode materials for fuel oxidation and ORR, respectively. However, due to the simplified cell structure (i.e., the compact design) and thus to the enhanced mechanical and thermal tolerance of the cells, SC-SOFCs can be used

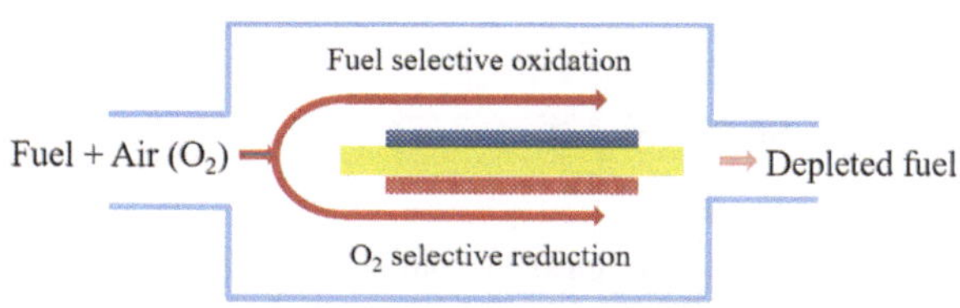

Fig. 9.31 Operating principle of SC-SOFCs

for the energy recovery from waste fuels such as engine exhaust, where the use of conventional SOFCs would be problematic.

9.7.2 Metal-Supported Solid Oxide Fuel Cells

Metal-supported solid oxide fuel cells (MS-SOFCs) have been developed to overcome the disadvantage of poor mechanical strength and poor thermal cycle ability of anode-supported SOFCs [57]. The use of metal supports can substantially reduce the thickness of electrode material supports (a further saving in material cost in SOFCs) so the electrode as well as electrolyte layers only need to be as thick as is necessary for the desired electrochemical function and performance. Due to the use of low-cost and inexpensive metals as supports, MS-SOFCs can provide advantages of reduced system cost, ease of manufacturing and more dynamic and thermal cycle operations. Figure 9.32 shows the general designs in MS-SOFCs.

In MS-SOFCs, the metal supports are generally used on the anode side, and thus, they must be porous for the diffusion and supply of fuel gas. The metal supports can be made from perforated ferritic stainless steel, porous metal powder, or metal meshes. The thickness of metal supports is generally in the range of 100–500 μm, depending on the mechanical strength of the material. Three layered cells with Ni cermet anode, GDC or YSZ electrolyte, LSM or MIEC type cathodes such as LSCF and BSCF are applied to the metal supports by screen-printing, tape casting, dip coating, vacuum plasma spray and other deposition techniques. Doped ceria can be added as buffer and/or diffusion barrier layers to avoid the interfacial reaction between the YSZ electrolyte and MIEC cathode and to improve the contact and bonding between the metal supports and Ni-based cermet anode. The use of ceramic glass sealant can be minimized by using metal-metal diffusion bonding and by welding together the metal support and metal interconnect plates.

High-temperature sintering is one of the critical challenges in the cell fabrication of MS-SOFCs. Sintering at high temperatures is necessary in SOFCs to increase the densification of electrolyte and barrier layer, form a good interface between electrode and electrolyte and enhance the microstructure of the electrode. However, it is unfortunate that the high-temperature sintering in air also induces the excess oxidation of metal supports, leading to the increased electrical resistance and reduced

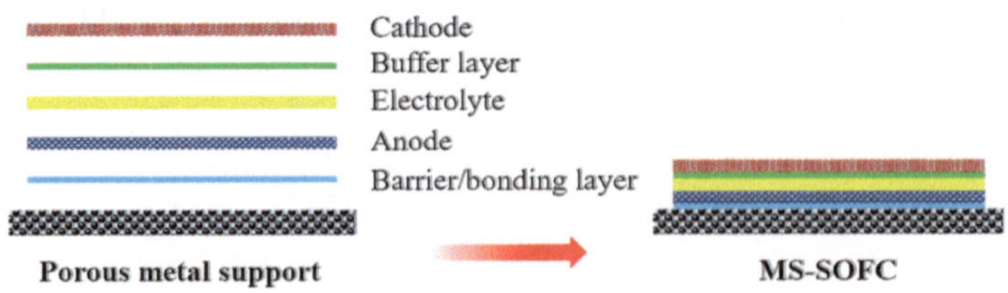

Fig. 9.32 General design and layout of MS-SOFCs

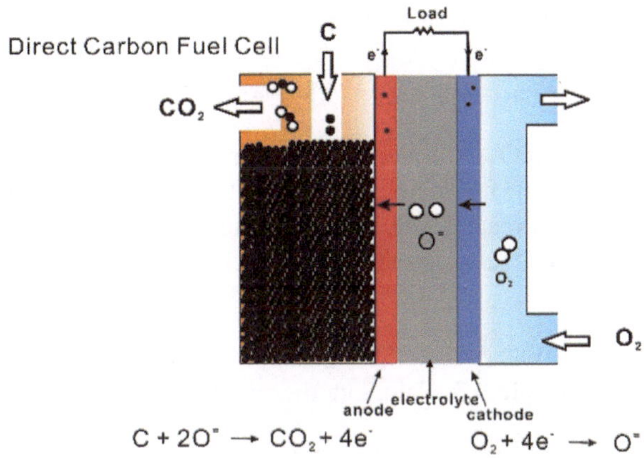

Fig. 9.33 Operating principle of a direct carbon fuel cell based on solid electrolyte

cell power output. Various approaches have been explored, including the use of high-temperature deposition techniques such as plasma spray, use of sintering aids such as cobalt oxide or copper oxide to improve the sintering of the doped ceria, in situ sintering, and sintering in a reducing atmosphere, etc. The metal-supported cells are particularly suitable for operating at 600 °C or lower as the corrosion constraints on the stainless steel materials would be far less severe.

9.7.3 Direct Carbon Fuel Cells

Direct carbon fuel cell (DCFC) is the only fuel cell type using solid fuel [58]. In DCFCs, solid carbon is directly fed into the anode compartment and electrochemically oxidized to CO_2 without the need for gasification, generating electricity. Figure 9.33 shows the operation principle of a DCFC based on solid electrolyte. The overall fuel cell reaction can be given by:

$$C + O_2 \rightarrow CO_2 \tag{9.59}$$

The reaction only produces pure CO_2 if high purity carbon is used. Due to the phases involved being solid, it is not difficult to see that only the carbon particles in direct contact with the anode and electrolyte will be electrochemically oxidized. The formed gaseous products will create a gap between the solid carbon fuel and reaction sites unless there is a way to push the solid fuel to the active sites. To improve the flow properties of solid carbon, molten metals such as Sb, Pb, In, Sn, and Bi are investigated as fuel electrodes in DCFCs. The driving force is the electrochemical oxidation of metals by oxygen ions from the electrolyte and then shuttling oxygen ions to a solid carbon fuel, reducing the metal oxides to metal. The most common molten metal is tin (Sn) due to its low metaling temperature (232 °C) and very high boiling temperature (2270 °C). In DCFCs, Sn is electrochemically oxidized in contact with the electrolyte to form tin oxide:

$$Sn_{liquid} + 2O^{=} \rightarrow SnO_2 + 4e^- \tag{9.60}$$

The SnO_2 is converted back to liquid Sn by the chemical reaction with solid carbon. The open-circuit voltage of the cells is related to the thermodynamics of the oxidation reaction of metals. Molten tin can also be used directly as liquid anodes in DCFCs. In this case, the electricity is produced by the direct oxidation of carbon, Eq. (9.59).

Example 9.3 Calculate thermodynamic efficiency of DCFCs

Solution:

We know the entropy change for the cell reaction is positive ($\Delta S = 1.6$ J/K·mol at 600 °C), which results in a slightly larger standard Gibbs free energy change ($\Delta G = -395.4$ kJ/mol at 600 °C) than the standard enthalpy change ($\Delta H = -394.0$ kJ/mol at 600 °C).

$$\varepsilon_{th} = \frac{\Delta G}{\Delta H} = 1 - \frac{T\Delta S}{\Delta H} = \frac{-395.4}{-394.0} = 100.4\% \tag{9.61}$$

In addition to solid electrolyte, molten carbonate (Li_2CO_3-K_2CO_3) and molten hydroxide (NaOH, KOH) are also used as the electrolyte due to their high ionic conductivity. However, for molten hydroxide, high corrosion and carbonate formation can lead to a high level of degradation. Molten carbonate electrolytes have the advantages of high stability in CO_2 but suffer corrosion problems similar to MCFCs. Despite the high thermodynamic efficiency, the performance reported for DCFCs is low, 5–120 mW/cm^2 at 700–800 °C [58], far below that of MCFCs and SOFCs. The low performance is partially related to the limited contact between the solid carbon, anode and the electrolyte and the sluggish oxidation kinetics of carbon. On the other hand, DCFCs can be used as secondary or primary battery types with or without manual feeding of carbon fuels.

Apart from the difficulties in continuous feeding of the solid carbon fuels to the reaction sites, the stacking and scaling-up of DCFCs are also significant challenges. Raw coal or biomass might not be able to be directly used as the fuel in DCFCs without cleaning as the impurities such as sulfur and chloride will be detrimental to the cell performance.

9.8 Summary

In this chapter, we have introduced and discussed the following topics:

- SOFC is a high-temperature fuel cell consisting of only ceramic materials. The most common ceramic materials in a typical SOFC are dense oxygen ion conducting yttria-stabilized zirconia (YSZ) electrolyte, porous lanthanum strontium manganite (LSM) cathode and nickel-YSZ (Ni-YSZ) cermet anode. Electrochemically, the cell potential of a SOFC can be obtained by the difference of the oxygen partial pressure at the cathode and anode, similar to the concentration cell. Nevertheless, there is a fundamental difference between the conventional electrochemical concentration cell and concentration cell in SOFC as discussed in this Chapter.
- The performance of a SOFC depends largely on the electrical conductivity properties of various cell components. There are two major charge species in SOFC materials, electronic charge carriers of negatively charged electrons or positively charged electron holes and oxide ions. The electronic conduction in perovskite-type oxides occurs through the cation vacancies or defects. A- or B-site doping is the most effective way to induce defects and charge compensation is the main mechanism for the formation of ionic and electronic defects. Depending on the type of electronic defects, we can have *n*- or *p*-type conductor.
- Fluorite-structured oxides are the most common ionic conducting materials in SOFC, and the typical examples are the Y_2O_3 and Sc_2O_3 stabilized zirconia, YSZ and ScSZ and Gd_2O_3 and Sm_2O_3 doped ceria, GDC and SDC. Oxide ion conductivity is mainly due to the formation of positively charge oxygen vacancies via the cation doping, e.g., dissolution of Y_2O_3 in ZrO_2 and Gd_2O_3 in CeO_2. The oxide conductivity depends on both the content and size of the dopant. In general, the smaller the mismatch between the dopant and host ions, the higher the oxide ion conductivity.
- Stabilized zirconia is the most stable oxide ionic conducting electrolyte materials with minimum electronic conductivity in the SOFC operating oxygen partial pressure range from 1 atm to 10^{-25}–10^{-20} atm. Doped ceria shows a high oxide ion conductivity as compared to stabilized zirconia, but develops appreciable electronic conductivity under reducing conditions due to the easily reducible Ce^{4+} ions. In addition to the fluorite-structured oxides, lanthanum strontium magnesium gallate (LSGM) perovskite and lanthanum silicate apatite oxides (LSO) have been investigated and developed. LSGM has shown the potential as promising alternative electrolyte materials for SOFC, while apatite-based oxides are facing the challenges in the high sintering temperature and silicate accumulated interface.
- Despite the significant advancement of SOFC technologies, Ni-based cermets are still the most common and adopted anodes of SOFC. Ni-based cermets have significant advantages of low-cost raw materials, high electronic conductivity, easy process ability, good interface, and high electrochemical activity for hydrogen oxidation reaction. The main limitation of N-based cermets is their low tolerance toward carbon deposition and sulfur poisoning. Thus, there is driving force for the development of alternative or Ni-free ceramic oxide-based anodes of SOFC.
- The development of cathode materials of SOFC is a totally different story. Due to the oxidizing environment, wide range of oxide materials can be used as cathodes of SOFC, ranging from electronic conducting LSM to mixed ionic and electronic

conducting LSCF and BSCF. The structures can be perovskite, double perovskite and Ruddlesden-Popper among others. However, to meet the requirement of high MIEC, high activity and structural stability, composite cathode becomes a viable approach with potential of synergetic effect.

- In the case of intermediate temperature SOFC, metallic interconnect is materials of choice due to its low cost, easy process ability, mechanical strength, and high electronic conductivity. Nevertheless, one critical challenge in the application of metallic interconnect is the chromium vaporization from the oxide scale, which poisons the cathode of SOFC. High-temperature ceramic glass is the most common sealant in SOFC, and in the case of boron-containing glasses, it is important to minimize the vaporization of boron species as it potentially deposits and poisons the cathode of SOFC.
- In addition to the planar and tubular design SOFC, there are other alternative designs. In this chapter, single-chamber SOFC, metal-supported SOFC and direct carbon fuel cells (DCFC) are briefly introduced. Although the detailed characteristics of the alternative SOFCs is not given due to the limited space, these alternative SOFCs have been designed to overcome some shortcomings of the conventional SOFC design and have specific advantages for targeted applications. It is important to keep in pace with the development of SOFC technologies in general.

9.9 Questions

9.1. Explain the following terms and materials:

Perovskite structured oxides and Goldschmidt tolerance factor.
Electron, electron holes, and polaron.
Cation vacancy, oxygen vacancy and charge compensation.
Oxide ion conductivity and dopant–vacancy pair or defect clusters.
Zirconia, ceria, lanthanum strontium magnesium gallate, apatite oxide, and bismuth oxide.
Dopant size and elastic strain energy.
Ni-based cermet, shrinkage profile, electronic conductivity, and percolation threshold.
Lanthanum strontium manganite and oxygen nonstoichiometry.
Lanthanum strontium cobalt ferrite and mixed ionic can electronic conductivity.
Double perovskite and Ruddlesden-Popper (RP) structured metal oxides.
Metallic interconnect, chromium oxide scale, and chromium vaporization.
Ceramic glass sealant, viscosity and boron volatility.
Electrolyte-supported and electrode-supported cell configuration.
Anode functional layer, cathode functional layer, and cathode current collection layer.

9.2. Explain the similarity and differences in the transportation properties/characteristics of electron holes and electrons in ceramic oxides.

9.3. Discuss the advantage and challenges in the use of apatite oxides based electrolyte materials in SOFC.

9.4. Referred to Figs. 9.18 and Fig. 9.20, LSM and LSCF show a very different dependence of electrical conductivity on temperature. Explain the reason.

9.5. In an H_2-air SOFC with Pt electrodes on both sides, the open-circuit voltage (OCV) has been measured to be 1.270 V in dry H_2, 1.128 V in 99.14% H_2 + 0.86% H_2O and 1.080 V in 98% H_2 + 2% H_2O at 1000 °C. Calculate the partial pressure of oxygen in the H_2 fuel.
(Answer: The oxygen partial pressure in the H_2 fuel is 1.64×10^{-16} Pa in dry H_2, 2.91×10^{-14} Pa in 99.14%H_2 + 0.86%H_2O and 1.68×10^{-13} Pa in 98%H_2 + 2%H_2O.)

9.6. Calculate the Goldschmidt tolerance factor, *t*, for $Ba_{0.5}Sr_{0.5}Co_{0.8}Fe_{0.2}O_3$ (BSCF), assuming Fe ions and 60% Co ions at B-site are in tetravalent state (Answer: $t = 1.05$).

Cation	Ionic radius (nm)
Ba^{2+}	0.161 (CN = 12)
Sr^{2+}	0.144 (CN = 12)
Co^{4+}	0.053 (CN = 6) HS
Co^{3+}	0.061 (CN = 6) HS
Fe^{4+}	0.0585 (CN = 6)
Fe^{3+}	0.0645 (CN = 6) HS
O^{2-}	0.14 (CN = 6)

CN: coordination number; HS: high spin

9.7. Is $Ba_{0.5}Sr_{0.5}Co_{0.8}Fe_{0.2}O_3$ (BSCF) a *n*- or *p*-type conductor? Give the charge compensation reactions for the Sr-doping.

9.8. Discuss the advantages and disadvantages of planar and tubular SOFCs, considering the power per unit area, power per volume, fabrication ability and cost, sealing ability, long-term stability, and thermal stability.

9.9. Discuss the similarities and differences between the SC-SOFC and conventional SOFC

9.10. Discuss the similarity and differences between the SC-SOFC and conventional SOFC with respect to characteristics, fuel utilization, energy efficiency and stack, and system construction.

References

1. Yasuda I, Hikita T (1993) Electrical-conductivity and defect structure of calcium-doped lanthanum chromites. J Electrochem Soc 140(6):1699–1704
2. Stevenson JW, Armstrong TR, Carneim RD, Pederson LR, Weber WJ (1996) Electrochemical properties of mixed conducting perovskites $La_{1-x}M_xCo_{1-y}Fe_yO_{3-\delta}$ (M = Sr, Ba, Ca). J Electrochem Soc 143(9):2722–2729
3. Mantzavinos D, Hartley A, Metcalfe IS, Sahibzada M (2000) Oxygen stoichiometries in $La_{1-x}Sr_xCo_{1-y}Fe_yO_{3-\delta}$ perovskites at reduced oxygen partial pressures. Solid State IonS 134(1–2):103–109
4. Lankhorst MHR, tenElshof JE (1997) Thermodynamic quantities and defect structure of $La_{0.6}Sr_{0.4}Co_{1-y}Fe_yO_{3-\delta}$ (y = 0–0.6) from high-temperature coulometric titration experiments. J Solid State Chem 130 (2):302–310
5. Mineshige A, Kobune M, Fujii S, Ogumi Z, Inaba M, Yao T, Kikuchi K (1999) Metal-insulator transition and crystal structure of $La_{1-x}Sr_xCoO_3$ as functions of Sr-content, temperature, and oxygen partial pressure. J Solid State Chem 142(2):374–381
6. Mizusaki J, Yoshihiro M, Yamauchi S, Fueki K (1985) Nonstoichiometry and defect structure of the perovskite-type oxides $La_{1-x}Sr_xFeO_{3-\delta}$. J Solid State Chem 58(2):257–266
7. Bouwmeester HJM, Den Otter MW, Boukamp BA (2004) Oxygen transport in $La_{0.6}Sr_{0.4}Co_{1-y}Fe_yO_{3-\delta}$. J Solid State Electrochem 8 (9):599–605
8. Mineshige A, Abe J, Kobune M, Uchimoto Y, Yazawa T (2006) Oxygen nonstoichiometry, mixed valency and mixed conduction in (La, Sr)(Co, Fe)$O_{3-\delta}$. Solid State Ionics 177(19–25):1803–1806
9. Arachi Y, Sakai H, Yamamoto O, Takeda Y, Imanishai N (1999) Electrical conductivity of the ZrO_2-Ln_2O_3 (Ln = lanthanides) system. Solid State Ionics 121(1–4):133–139
10. Badwal SPS (1992) Zirconia-based solid electrolytes—microstructure, stability and ionic-conductivity. Solid State Ionics 52(1–3):23–32
11. Badwal SPS, Ciacchi FT, Milosevic D (2000) Scandia-zirconia electrolytes for intermediate temperature solid oxide fuel cell operation. Solid State Ionics 136:91–99
12. Butz B, Kruse P, Stormer H, Gerthsen D, Muller A, Weber A, Ivers-Tiffee E (2006) Correlation between microstructure and degradation in conductivity for cubic Y_2O_3-doped ZrO_2. Solid State Ionics 177(37–38):3275–3284
13. Mori M, Abe T, Itoh H, Yamamoto O, Takeda Y, Kawahara T (1994) Cubic-stabilized zirconia and alumina composites as electrolytes in planar type solid oxide fuel-cells. Solid State Ionics 74(3–4):157–164
14. Jiang SP, Zhang JP, Foger K (2003) Chemical interactions between 3 mol% yttria-zirconia and Sr-doped lanthanum manganite. J Eur Ceram Soc 23(11):1865–1873
15. Inaba H, Tagawa H (1996) Ceria-based solid electrolytes—Review. Solid State Ionics 83(1–2):1–16
16. Kim DJ (1989) Lattice-parameters, ionic conductivity, and solubility limits in fluorite-structure Hf_4+O_2, Zr_4+O_2, Ce_4+O_2, Th_4+O_2, V_4+O_2 oxide solutions. J Am Ceram Soc 72(8):1415–1421
17. Yahiro H, Eguchi K, Arai H (1989) Electrical-properties and reducibilities of ceria rare Earth oxide systems and their application to solid oxide fuel-cell. Solid State Ionics 36(1–2):71–75
18. Ishihara T, Matsuda H, Takita Y (1994) Doped $LaGaO_3$ Perovskite-type oxide as a new oxide ionic conductor. J Am Chem Soc 116(9):3801–3803
19. Huang KQ, Tichy RS, Goodenough JB (1998) Superior perovskite oxide-ion conductor; strontium—and magnesium-doped $LaGaO_3$: I, phase relationships and electrical properties. J Am Ceram Soc 81(10):2565–2575
20. Ishihara T (2006) Development of new fast oxide ion conductor and application for intermediate temperature solid oxide fuel cells. Bull Chem Soc Jpn 79(8):1155–1166
21. Huang KQ, Wan JH, Goodenough JB (2001) Increasing power density of LSGM-based solid oxide fuel cells using new anode materials. J Electrochem Soc 148(7):A788–A794

22. Nakayama S, Kageyama T, Aono H, Sadaoka Y (1995) Ionic-conductivity of lanthanoid silicates, $Ln_{10}(SiO_4)_6O_3$ (Ln = La, Nd, Sm, Gd, Dy, Y, Ho, Er and Yb). J Mater Chem 5(11):1801–1805
23. Jiang SP, Zhang L, He HQ, Yap RK, Xiang Y (2009) Synthesis and characterization of lanthanum silicate apatite by gel-casting route as electrolytes for solid oxide fuel cells. J Power Sources 189(2):972–981
24. Takahashi T, Esaka T, Iwahara H (1977) Conduction in Bi_2O_3-based oxide ion conductors under under low oxygen pressure. 1. Current blackening of Bi_2O_3-Y_2O_3 electrolyte. J Appl Electrochem 7 (4):299–302
25. Wachsman ED, Jayaweera P, Jiang N, Lowe DM, Pound BG (1997) Stable high conductivity ceria/bismuth oxide bilayered electrolytes. J Electrochem Soc 144(1):233–236
26. Ingram DB, Linic S (2009) First-principles analysis of the activity of transition and noble metals in the direct utilization of hydrocarbon fuels at solid oxide fuel cell operating conditions. J Electrochem Soc 156(12):B1457–B1465
27. Setoguchi T, Okamoto K, Eguchi K, Arai H (1992) Effects of anode material and fuel on anodic reaction of solid oxide fuel-cells. J Electrochem Soc 139(10):2875–2880
28. Jiang SP, Chan SH (2004) A review of anode materials development in solid oxide fuel cells. J Mater Sci 39(14):4405–4439
29. McIntosh S, Gorte RJ (2004) Direct hydrocarbon solid oxide fuel cells. Chem Rev 104(10):4845–4865
30. Plint SM, Connor PA, Tao S, Irvine JTS (2006) Electronic transport in the novel SOFC anode material $La_{1-x}Sr_xCr_{0.5}Mn_{0.5}O_{3-\delta}$. Solid State IonS 177 (19–25):2005–2008
31. Babaei A, Zhang L, Tan SL, Jiang SP (2010) Pd-promoted (La, Ca)(Cr, Mn)O_3/GDC anode for hydrogen and methane oxidation reactions of solid oxide fuel cells. Solid State Ionics 181(25–26):1221–1228
32. Huang YH, Dass RI, Xing ZL, Goodenough JB (2006) Double perovskites as anode materials for solid-oxide fuel cells. Science 312(5771):254–257
33. Jiang SP (2008) Development of lanthanum strontium manganite perovskite cathode materials of solid oxide fuel cells: a review. J Mater Sci 43(21):6799–6833
34. Mizusaki J, Yonemura Y, Kamata H, Ohyama K, Mori N, Takai H, Tagawa H, Dokiya M, Naraya K, Sasamoto T, Inaba H, Hashimoto T (2000) Electronic conductivity, Seebeck coefficient, defect and electronic structure of nonstoichiometric $La_{1-x}Sr_xMnO_3$. Solid State Ionics 132(3–4):167–180
35. Pavone M, Munoz-Garcia AB, Ritzmann AM, Carter EA (2014) First-principles study of lanthanum strontium manganite: Insights into electronic structure and oxygen vacancy formation. J Phys Chem C 118(25):13346–13356
36. Mizusaki J, Mori N, Takai H, Yonemura Y, Minamiue H, Tagawa H, Dokiya M, Inaba H, Naraya K, Sasamoto T, Hashimoto T (2000) Oxygen nonstoichiometry and defect equilibrium in the perovskite-type oxides $La_{1-x}Sr_xMnO_{3+\delta}$. Solid State Ionics 129(1–4):163–177
37. Jiang SP (2019) Development of lanthanum strontium cobalt ferrite perovskite electrodes of solid oxide fuel cells—A review. Int J Hydrog Energy 44(14):7448–7493
38. Tai LW, Nasrallah MM, Anderson HU, Sparlin DM, Sehlin SR (1995) Structure and electrical-properties of $La_{1-x}Sr_xCo_{1-y}Fe_yO_3$. 2. the system $La_{1-x}Sr_xCo_{0.2}Fe_{0.8}O_3$. Solid State Ionics 76 (3–4):273–283
39. Aguadero A, Perez-Coll D, de la Calle C, Alonso JA, Escudero MJ, Daza L (2009) $SrCo_{1-x}Sb_xO_{3-\delta}$ perovskite oxides as cathode materials in solid oxide fuel cells. J Power Sources 192(1):132–137
40. Shao ZP, Haile SM (2004) A high-performance cathode for the next generation of solid-oxide fuel cells. Nature 431(7005):170–173
41. Zhou W, Ran R, Shao ZP (2009) Progress in understanding and development of $Ba_{0.5}Sr_{0.5}Co_{0.8}Fe_{0.2}O_{3-\delta}$-based cathodes for intermediate-temperature solid-oxide fuel cells: A review. J Power Sources 192 (2):231–246
42. Pelosato R, Cordaro G, Stucchi D, Cristiani C, Dotelli G (2015) Cobalt based layered perovskites as cathode material for intermediate temperature Solid Oxide Fuel Cells: A brief review. J Power Sources 298:46–67

43. Zhang K, Ge L, Ran R, Shao ZP, Liu SM (2008) Synthesis, characterization and evaluation of cation-ordered $LnBaCo_2O_{5+\delta}$ as materials of oxygen permeation membranes and cathodes of SOFCs. Acta Mater 56(17):4876–4889
44. Tarancon A, Burriel M, Santiso J, Skinner SJ, Kilner JA (2010) Advances in layered oxide cathodes for intermediate temperature solid oxide fuel cells. J Mater Chem 20(19):3799–3813
45. Zhou XD, Templeton JW, Nie Z, Chen H, Stevenson JW, Pederson LR (2012) Electrochemical performance and stability of the cathode for solid oxide fuel cells: V. high performance and stable Pr_2NiO_4 as the cathode for solid oxide fuel cells. Electrochim Acta 71:44–49
46. Zhen YD, Jiang SP (2006) Transition behavior for O_2 reduction reaction on (La, Sr)MnO_3/YSZ composite cathodes of solid oxide fuel cells. J Electrochem Soc 153(12):A2245–A2254
47. Jiang SP, Wang W (2005) Fabrication and performance of GDC-impregnated (La, Sr)MnO_3 cathodes for intermediate temperature solid oxide fuel cells. J Electrochem Soc 152(7):A1398–A1408
48. Hwang HJ, Ji-Woong MB, Seunghun LA, Lee EA (2005) Electrochemical performance of LSCF-based composite cathodes for intermediate temperature SOFCs. J Power Sources 145(2):243–248
49. He S, Zhang Q, Maurizio G, Catellani L, Chen KF, Chang QB, Santarelli M, Jiang SP (2018) In situ formation of $Er_{0.4}Bi_{1.6}O_3$ protective layer at cobaltite cathode/Y_2O_3-ZrO_2 electrolyte interface under solid oxide fuel cell operation conditions. ACS Applied Materials & Interfaces 10 (47):40549–40559
50. Wang CC, Luo DW, Jiang SP, Lin B (2018) Highly sulfur poisoning-tolerant $BaCeO_3$-impregnated $La_{0.6}Sr_{0.4}Co_{0.2}Fe_{0.8}O_{3-\delta}$ cathodes for solid oxide fuel cells. J Phys D-Appl Phys 51 (43):435502
51. Zhu WZ, Deevi SC (2003) Development of interconnect materials for solid oxide fuel cells. Mater Sci Eng A-Struct Mater Prop Microstruct Process 348(1–2):227–243
52. Hua B, Pu J, Zhang JF, Lu FS, Chi B, Jian L (2009) Ni-Mo-Cr alloy for interconnect applications in intermediate temperature solid oxide fuel cells. J Electrochem Soc 156(1):B93–B98
53. Chen XB, Hua B, Pu J, Li J, Zhang L, Jiang SP (2009) Interaction between (La, Sr)MnO_3 cathode and Ni-Mo-Cr metallic interconnect with suppressed chromium vaporization for solid oxide fuel cells. Int J Hydrog Energy 34(14):5737–5748
54. Zhang Q, Tan SW, Ren MY, Yang HW, Tang D, Chen KF, Zhang T, Jiang SP (2018) Effect of Gd_2O_3 doping on structure and boron volatility of borosilicate glass sealants in solid oxide fuel cells-A study on the $La_{0.6}Sr_{0.4}Co_{0.2}Fe_{0.8}O_{3-\delta}$ (LSCF) cathode. J Power Sources 383:34–41
55. Hibino T, Iwahara H (1993) Simplification of solid oxide fuel-cell system using partial oxidation of methane. Chem Lett 7:1131–1134
56. Hibino T, Hashimoto A, Yano M, Suzuki M, Yoshida S, Sano M (2002) High performance anodes for SOFCs operating in methane-air mixture at reduced temperatures. J Electrochem Soc 149(2):A133–A136
57. Tucker MC (2010) Progress in metal-supported solid oxide fuel cells: A review. J Power Sources 195(15):4570–4582
58. Giddey S, Badwal SPS, Kulkarni A, Munnings C (2012) A comprehensive review of direct carbon fuel cell technology. Prog Energy Combust Sci 38(3):360–399

Chapter 10
Solid Oxide Fuel Cells: Reactions

In SOFCs, due to the high temperature and solid electrode and solid electrolyte interface, the characteristics of the interface and electrode reactions in SOFCs are unique. In this chapter, we will start with the surface segregation of oxide electrodes as it is probably one of the most important factors influencing the O_2 reduction reaction (ORR) on the cathodes. This will be followed by the introduction and discussion on the electrode reactions at the cathode and anode, interfacial reactions between electrode and electrolyte, and the poisoning and deposition of various contaminants on electrodes in SOFCs. The issues related to the cell configuration, polarization performance analysis and specific electrochemical techniques in SOFCs will be addressed in Chap. 11.

10.1 Surface Segregation of Oxide Electrodes

Surface segregation is a common phenomenon of solid oxide materials and can be defined as the redistribution and accumulation of a specific cation or cations at the oxide surface due to a loss of symmetry at the surface relative to the bulk. The accumulation of specific cations on the surface will lead to the deviation of the surface composition from the nominal bulk composition, resulting in a change in the physicochemical and catalytic properties of the oxide materials. Surface segregation is particularly important and relevant in SOFCs due to the fact that all the electrochemical reactions are associated with the surface in one way or another and plays a critical role in the electrochemical activity of SOFC electrodes [1, 2]. Surface segregation also plays a dominant role in the performance degradation mechanism of SOFC cathodes due to the contaminant deposition and poisoning in addition to the thermally activated microstructural degradation processes.

The majority of the surface segregation issues of oxide electrodes in SOFCs are related to A-site cation segregation. Sr segregation in particular is frequently observed in Sr-containing perovskite-based electrodes such as lanthanum strontium

S. P. Jiang and Q. Li, *Introduction to Fuel Cells*,
https://doi.org/10.1007/978-981-10-7626-8_10

manganite (LSM) and lanthanum strontium cobalt ferrite (LSCF) under high temperatures in an oxidizing atmosphere. The segregated species occupy the active surface sites and thus degrade the electrochemical activity for the oxygen reduction reaction. The segregated SrO species are mobile and can diffuse across the doped ceria barrier layer, forming a $SrZrO_3$ insulating layer at the YSZ electrolyte interface. Surface segregation is also closely related to the performance degradation caused by the deposition and poisoning of contaminants such as chromium, sulfur and boron. Therefore, a fundamental understanding of the cation surface segregation is vital in the development of reliable and durable SOFCs technologies.

10.1.1 Surface Segregation Under Open Circuit

Surface segregation has been extensively studied under open circuit conditions on cathodes such as LSM and LSCF electrode materials. Experimentally, this is commonly performed by annealing dense bar samples under controlled temperature and atmosphere (usually air or oxygen) for a certain period and the extent of the segregation is measured by the observation of the surface using high-resolution microstructure and element analysis techniques. Figure 10.1 shows an example of surface segregation observed on the surface of $La_{0.6}Sr_{0.4}Co_{0.2}Fe_{0.8}O_{3-\delta}$ (LSCF) bar samples after annealing at 800 °C in air for 96 h. LSCF bar samples were prepared from LSCF powders, pressed into a rectangular bar at 300 MPa, and sintered at 1350 °C for 5 h in air [3]. The surface of freshly prepared LSCF sample before testing is dense, smooth, and clean with no impurity phases. However, the surface morphology of LSCF samples after heat treatment for 96 h in air changes significantly. There is a formation of submicron-sized particles (less than 1 μm) and dark colored particles as large as ~ 5 μm on the surface of the LSCF sample. EDS element mapping clearly identifies that fine particles are SrO, while large particles are CoO_x. The particles have very different crystalline domain or facets as compared to the surrounding LSCF grains. This indicates that the particles are segregated out of the LSCF perovskite structure.

The redistribution and accumulation of segregated cations are related to the nature of the segregated species. The extent of surface segregation of cations, in particular Sr, is subjected to various factors, including the stoichiometric composition, strain, temperature, oxygen partial pressure, nature of the oxide materials, type of cations in the B-sites and electrochemical polarization conditions. We need to consider all these factors when investigating the surface segregation phenomenon. In addition to the LSM and LSCF-based electrodes, surface segregation of cations also occurs on other perovskite oxides, layered perovskite oxides, and RP oxides.

There are various theories on the mechanisms of surface segregation, but the most accepted ones are elastic energy minimization and electrostatic interaction and [4, 5]. Here we only discuss the main points briefly.

Elastic energy minimization—In the case of Sr-doped lanthanum manganite perovskite oxides, the size of the Sr^{2+} dopant is 0.144 nm (CN = 12) and is 5.9%

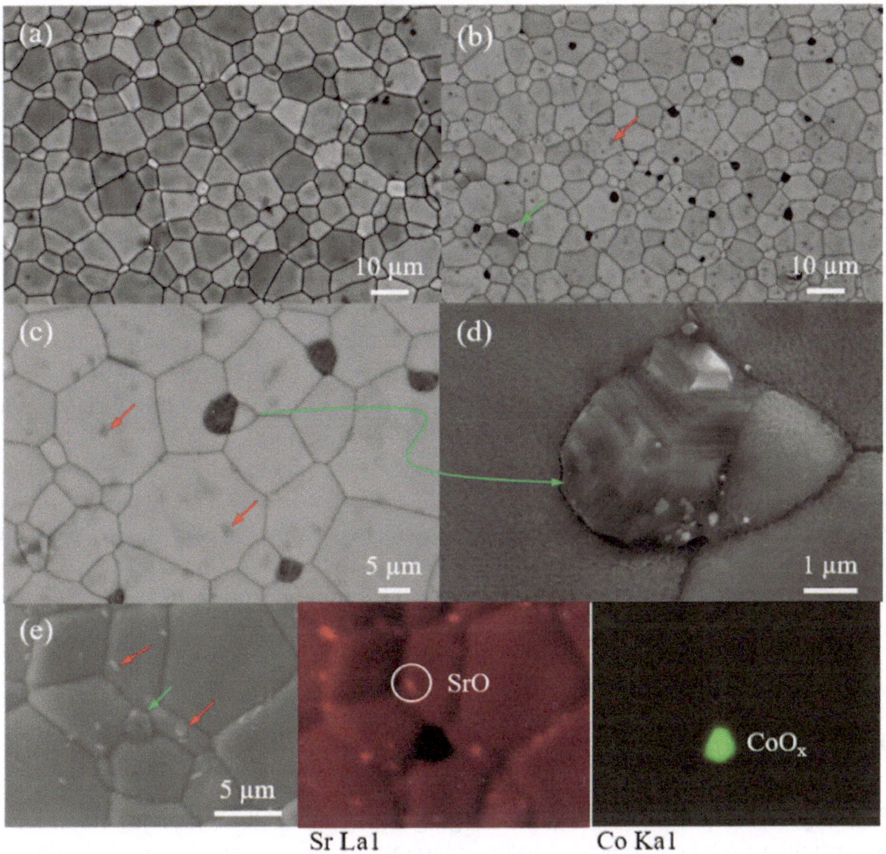

Fig. 10.1 SEM micrographs of **a** freshly prepared LSCF and **b** LSCF samples after heat treatment at 800 °C in air for 96 h; (**c**, **d**) enlarged image of the segregated micron-sized particle, and **e** EDS element mapping of Sr and Co. Arrows indicate segregated isolated submicron-sized particles and large particles. Modified from Ref. [3] with permission from Royal Society of Chemistry, Copyright 2014

larger than the ionic radius of the host cation La^{3+} (0.136 nm, CN = 12). Thus, a Sr^{2+} cation in an La lattice site, Sr'_{La}, is under a compressive strain state in the LSM lattice, which can be relaxed when it migrates to the surface. The high tendency of Sr segregation on the tensile-strained LSM surface is due to the large space available for accommodating the Sr cations with a reduced elastic energy on the surface compared to that in the bulk. The extent of dopant segregation has been shown to be proportionally dependent on the size of the dopant [4]. The elastic energy minimization has been proposed as a driving force for the segregation of SrO on the surface of Sr-doped perovskite oxides such as LSM and LSCF. We can also explain the observed increase of Sr surface segregation at a higher oxygen partial pressure by the elastic interaction mechanism. The high oxygen pressure incorporates oxygen into the lattice and contracts the lattice, and consequently, the shrinkage of LSM lattice under high oxygen partial pressure would lead to an increase of dopant strain energy, accelerating the Sr surface segregation.

Electrostatic interaction—We can understand from Chap. 9 that the concentration of oxygen vacancies, $V_O^{\bullet\bullet}$, on the electrode surface would increase as the oxygen partial pressure decreases. We can recall that $V_O^{\bullet\bullet}$ is positively charged, while the substitution of Sr^{2+} for La^{3+} in the LSM lattice will lead to a negative charged Sr'_{La} substitution. Thus, there would exist an interaction between the positively charged $V_O^{\bullet\bullet}$ and the negatively charged Sr'_{La} substitution in the LSM lattice. The large concentration of positively charged $V_O^{\bullet\bullet}$ on the surface may drive the substituted Sr cations from the bulk to the surface via electrostatic interaction in order to maintain the local electroneutrality [6]. The electrostatic interaction driving force may play a dominant role in Sr-doped cobaltite-based perovskite oxides like LSC and LSCF due to the easy formation of oxygen vacancies at the surface region.

However, the driving force for the B-site cations such as Co surface segregation is more complicated. For instance, Co substitution of Fe in LSCF would induce tensile strain (the Co cations are smaller than Fe cations) as well as increase the surface charge (Co substitution creates oxygen vacancies). Consequently, the Co surface segregation may be a compromise of the strain relaxation and surface charge minimization.

10.1.2 Surface Segregation Under Polarization

From a practical point of view, surface segregation under SOFC operating conditions, in particular under cathodic or anodic polarizations, is most significant and relevant to the activity and stability of the electrodes in SOFCs. However, surface segregation under polarization is a complex process with multiple factors in play. In addition to the polarization effect, surface segregation is also related to the nature of the electrode materials. We will examine this issue with two most common and studied electrode materials of SOFCs: LSM and LSCF.

10.1.2.1 Sr Segregation on LSM Electrode Under Polarization

The cation segregation on LSM electrode materials has been investigated by in situ probe techniques. One good example is given by Huber et al. [7] on the surface chemistry of LSM films under polarization by XPS and time of flight-secondary ion mass spectrometry (ToF–SIMS) techniques. They observed that cathodic polarization led to the depletion of Sr on the surfaces, i.e., inhibiting Sr segregation, while anodic polarization accelerated the Sr surface segregation.

LSM-based materials are predominantly electronic conductors with negligible oxygen ion conductivities. In fact, in the case of Sr-doped $LaMnO_3$, Sr substitution of A-site La leads to the formation of electron holes and at SOFC operating temperatures, the electron holes are localized onto specific Mn ions, forming Mn^{4+} ions, $Mn^{\bullet}_{Mn}$, i.e., electron holes, $h^{\bullet}$ [8]. The electroneutrality condition can be simply described as:

$$SrO \xrightarrow{LaMnO_3} Sr'_{La} + Mn^{\bullet}_{Mn} + O^{x}_{O} \quad (10.1)$$

Under cathodic polarization conditions, the generation of positively charged oxygen vacancies, $V^{\bullet\bullet}_{O,\mathrm{LSM,surface}}$, is a dominant process in LSM electrodes for ORR. The subscript *surface* indicates that the positively charged oxygen vacancies generated under cathodic polarization are primarily on the LSM surface in the vicinities of three-phase boundaries (TPB). Considering the high concentration of positively charged oxygen vacancies on the LSM surface, electroneutrality can also be established by incorporating the segregated SrO into the La lattice, forming negatively charged Sr'_{La}. Such incorporation of SrO through the electrostatic attractive interaction can be expressed as:

$$2SrO_{\mathrm{surface}} \xrightarrow{\text{Cathodic polarization, }LaMnO_3} 2Sr'_{La,\,\mathrm{surface}} + V^{\bullet\bullet}_{O,\,\mathrm{LSM\ surface}} + 2O^{x}_{O} \quad (10.2)$$

In this case, the driving force for the oxygen vacancy formation is the cathodic polarization and not the Sr substitution. As discussed in Chap. 9, the charge compensation through the oxidation of B-site Mn ions would mainly drive the B-site Mn cations to the surface via the elastic interaction, while incorporation of segregated SrO into La lattice sites would be the alternative charge compensation mechanism for the oxygen vacancy formation under cathodic polarization conditions. The incorporation of segregated SrO species can be a fast process due to the strong electrostatic attractive force between the positively charged $V^{\bullet\bullet}_{O,\,\mathrm{LSM,surface}}$ and negatively charged Sr'_{La}. The fast incorporation reaction of segregated Sr into LSM lattices is consistent with the observed activation effect and reduced Sr species at LSM electrode surfaces [9].

Under anodic polarization conditions, oxygen excess occurs, similar to the oxygen excess of LSM materials at high oxygen partial pressures [10]. The presence of oxygen excess in the form of negatively charged O''_{excess} ions (note: The oxygen excess could exist as negatively charged cation vacancies, e.g., V'''_{La} due to the compact nature of the LSM perovskite structure. Nevertheless, electronically O''_{excess} and V'''_{La} would function similarly due to the fact that both are negatively charged point defects) would push the negatively charged Sr'_{La} out of the lattice to the surface by repulsive electrostatic interaction. Sr segregation due to the excess oxygen ions in LSM under anodic polarization can be expressed as follows:

$$3Sr'_{La} + \frac{3}{2}O_2 \xrightarrow{\text{Anodic polarization, }O''_{\mathrm{excess}}\text{ or }V'''_{La}} 3SrO_{\mathrm{surface}} + V'''_{La} \quad (10.3)$$

This anodic polarization-induced electrostatic interaction can subsequently drive Sr segregation to the LSM surface. Accelerated surface segregation of SrO species under anodic polarization conditions has been supported by a careful study of Cr deposition on an LSM electrode under anodic polarization conditions of solid oxide electrolysis cells (SOECs) [11]. In this case, deposition and formation of $SrCrO_4$ and Cr_2O_3 occurs on the electrode surface due to the excess Sr segregation under anodic polarization. This is very different from the preferential Cr deposition at the electrode/electrolyte interface region and the dominant formation of $(Cr,Mn)_3O_4$ spinels for the LSM electrodes under SOFC operation conditions [12]. Therefore, the driving force for Sr segregation at LSM electrode surfaces under polarization is most likely the result of oxygen non-stoichiometry modulated electrostatic interaction on the LSM surface induced by cathodic/anodic polarization. The observed Sr incorporation rather than Sr segregation under cathodic polarization implies that elastic interactions due to the size mismatch between the dopant Sr and host La cations are compensated by the electrostatic interactions of the generated oxygen vacancies and are not key factors determining Sr surface segregation/incorporation in LSM under polarization conditions.

10.1.2.2 Sr Segregation on LSCF Electrode under Polarization

In the case of LSCF electrodes, the effect of polarization on the Sr segregation is very different from that observed on the LSM electrodes. We show this by giving an example in Fig. 10.2 for a directly assembled LSCF electrode on a YSZ electrolyte after anodic and cathodic polarization at a current density of 0.5 A/cm^2 and 750 °C for 40 h [13]. In the case of LSCF electrodes under the anodic polarization mode, there is a formation of a thin Sr-rich layer between the directly assembled LSCF electrode and YSZ electrolyte; however, the Sr-rich layer is isolated, not continuous and the thickness of the layer is 5–7 nm. The presence of a Sr-rich layer at the electrode/electrolyte interface may originate from Sr segregation/diffusion during

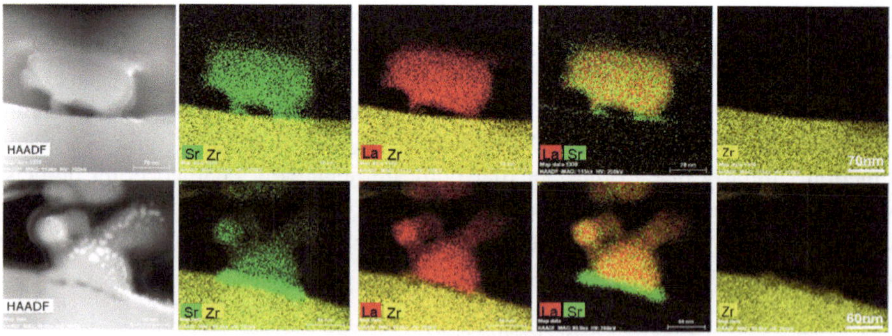

Fig. 10.2 STEM-EDS maps of electrode/electrolyte interface of directly assembled LSCF electrodes after polarization at 0.5 A/cm^2 and 750 °C under anodic current for 40 h (upper figure) and cathodic current for 40 h (lower figure). Reproduced from Ref. [13] with permission from Elsevier, Copyright 2018

the dwell of the electrode at 750 °C under open circuit conditions. However, after cathodic polarization for 40 h, formation of a pronounced and continuous Sr-rich layer with a thickness of ~ 19.5 nm is observed. There is also significant change in the microstructure of YSZ electrolyte surface in contact with the LSCF electrode, indicated by the roughness and holes formed at the interface (Fig. 10.2). The holes are the region for the reaction between Sr and Zr as there is no La in this reaction zone, indicating the formation of $SrZrO_3$. The accumulation of electrically insulating Sr-rich layer not only blocks the migration of oxygen ions at the electrode/electrolyte interface, but also deteriorates the electrocatalytic activity and electrical conductivity of LSCF electrodes due to the significant Sr deficiency in the A-sites of the perovskite structure. The significant differences in the deposition of Sr-rich layer at the electrode/electrolyte interface region clearly indicate that anodic polarization retards Sr segregation and cathodic polarization promotes Sr segregation.

Such polarization effects can be explained by considering the nature of LSCF electrodes. In contrast to LSM electrode materials, LSCF electrode materials are well-known MIECs with high oxygen ion conductivity (see Chap. 9). This will result in significant differences in the electrode behavior of LSCF and LSM. Cathodic polarization will cause an excess in oxygen vacancies on the electrode surface of LSCF. As a result, the increased accumulation of positively charged oxygen vacancies at the electrode surface would enhance electrostatic interactions. Combination with intrinsic elastic interactions subsequently drives Sr segregation. The observed Sr segregation rather than Sr incorporation under cathodic polarization is an indication of much stronger elastic interaction in the case of LSCF as compared to that in LSM. Under anodic polarization, oxygen vacancy concentrations at the electrode surface are significantly reduced and the substituted Sr at La sites would be held by the attractive force between negatively charged Sr'_{La} and positively charged $V_O^{\bullet\bullet}$ in the bulk, which will overcome the elastic interaction and result in a decreased driving force for Sr segregation, thus inhibiting Sr segregation. This in turn suggests the diminished role of elastic interactions on Sr segregation in LSCF electrodes under polarization conditions. As a result, the driving force for Sr segregation in LSCF electrodes under polarization is primarily related to the significant change in electrode surface properties, particularly the positively charged oxygen vacancy concentration at surfaces under cathodic and anodic polarization conditions. It should be emphasized that the surface segregation experiments of LSCF electrode on YSZ electrolyte as shown in Fig. 10.2 were performed using direct assembly technique (see Chap. 12) and not on pre-sintered LSCF electrodes. For pre-sintered LSCF electrodes on YSZ electrolyte, LSCF reacts with YSZ electrolyte, forming insulating $SrZrO_3$ phase as low as 800 °C. The facile interaction between LSCF and YSZ would change significantly the course of surface segregation behavior of LSCF electrodes.

10.1.2.3 Kinetics of Surface Segregation

It is important to remark that surface segregation is not a pure electrode material issue. Surface segregation is actually a system related issue in SOFCs. The reason for this is that we all know that Sr segregation is a predominant factor in the performance degradation of LSCF electrodes on GDC barrier layer protected YSZ electrolyte for ORR due to the diffusion and subsequent reaction of segregated Sr at the YSZ interface. In the case of doped ceria electrolytes, the LSCF electrode is very stable. Nevertheless, under catholic polarization conditions, Sr segregation would occur on both YSZ and GDC electrolytes, electrochemically. This indicates that we need to take into account the roles of electrolytes in the Sr surface segregation on LSCF electrodes [14]. To explain this, we examine the Sr segregation of directly assembled LSCF electrodes on YSZ and GDC electrolytes under identical polarization conditions, see Fig. 10.3 of the STEM-EDS mapping of directly assembled LSCF/GDC and LSCF/YSZ interface after polarization at 750 °C and 1000 mA/cm^2 for 100 h in air.

In the case of GDC electrolyte, the directly assembled LSCF electrode is in good contact with the GDC electrolyte surface with no voids or cracks at the interface. The EDS mapping analysis shows the clear interface between LSCF particles and the GDC electrolyte with no formation of a Sr-rich layer at the interface between LSCF and GDC. On the other hand, B-site cation Co segregation is observed, but, differently from the preferential deposition of Sr species at the interface (see Fig. 10.2), segregated Co species agglomerate and accumulate in the isolated areas or pockets of the electrode. Very different to the LSCF electrode on GDC electrolyte, a reaction

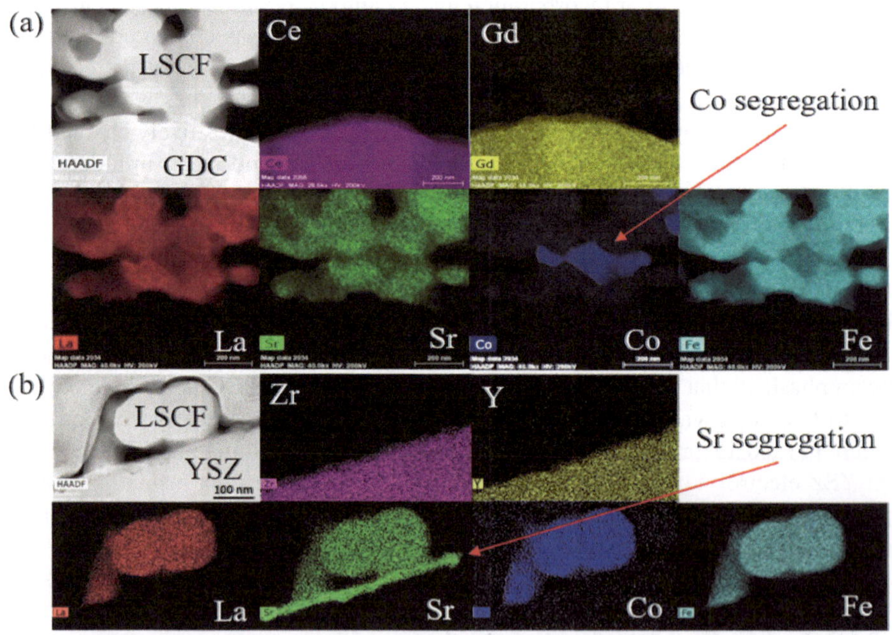

Fig. 10.3 STEM-EDS mapping of **a** directly assembled LSCF/GDC interface and **b** directly assembled LSCF/YSZ interface after polarization at 750 °C and 1000 mA/cm^2 for 100 h. Modified from Ref. [14] with permission from Elsevier, Copyright 2020

layer as thick as 20–30 nm is formed at the YSZ interface. The element mapping indicates the predominant presence of Sr and Zr and to a lesser extent, Fe, La, and Y in the reaction layer. Further detailed analysis indicates the formation of a $SrZrO_3$ phase at 750 °C under the polarization condition. This shows that the nature of the electrolyte plays a critical role in the interface microstructure stability under fuel cell polarization conditions.

Based on the comparison study, we can discuss the kinetics of the surface segregation and its relationship with the system parameters and the environment. In the case of LSCF/GDC cells, the segregation of Sr is much slower as compared with that on LSCF/YSZ cells under identical cathodic polarization conditions. The much slower kinetics of Sr segregation in the LSCF/GDC system is evidently due to the chemical stability between LSCF and GDC electrolyte, in contrast to the accelerated Sr segregation in the case of LSCF/YSZ cells induced by the facile reaction between segregated SrO and YSZ, forming a $SrZrO_3$ phase. This indicates that although the surface segregation of Sr is an inherent phenomenon of LSCF-based electrode materials, the kinetics of segregation is largely determined by the subsequent reaction between the segregated SrO species and the catcher in the system (e.g., Zr in the case of YSZ). In the absence of a catcher of the segregated species, the segregation process of Sr will be kinetically retarded and slow. On the other hand, significant segregation and accumulation of Co at isolated areas/pockets of the LSCF electrode are observed in the case of LSCF/GDC cells, but not in LSCF/YSZ cells. This indicates a shift of the segregation process from the A-site Sr cation to the B-site Co cation in the case of kinetically inhibited segregation of A-site Sr cations under the influence of the cathodic polarization conditions. This may indicate a compromise in the driving force or electrostatic interaction on the surface segregation under cathodic polarization in general.

Thus, based on the observations, we can illustrate the kinetics of the surface segregation reaction at the LSCF/YSZ and LSCF/GDC cells. In both cases, surface segregation of Sr species occurs under oxidizing conditions and under the influence of cathodic polarization at high temperatures:

$$La_{0.6}Sr_{0.4}CO_{0.2}Fe_{0.8}O_{3-\delta} \xrightarrow{i_{cathodic}} La_{0.6}Sr_{0.4-x}CO_{0.2}Fe_{0.8}O_{3-\delta} + xSrO \quad (10.4)$$

The segregated Sr species diffuses from the surface of LSCF to YSZ or GDC electrolyte surface:

$$SrO_{electrdoe} \xrightarrow{diffusion} SrO_{electrolyte} \quad (10.5)$$

In the case of LSCF/YSZ cells, the segregated SrO reacts with Zr, forming $SrZrO_3$,

$$SrO_{YSZ} + Y_{0.08}Zr_{0.92}O_2 \rightarrow SrZrO_3 + Y_2O_3 \quad (10.6)$$

In this case, Zr in the YSZ electrolyte is the catcher of the segregated Sr species. The reaction accelerates the Sr segregation and leads to the disintegration of the LSCF structure at the interface.

$$La_{0.6}Sr_{0.4}CO_{0.2}Fe_{0.8}O_{3-\delta} \xrightarrow{i_{\text{cathodic}}} SrO + La_2O_3 + CO_3O_4 + FeO \tag{10.7}$$

However, in the case of LSCF/GDC cells, no such catchers for the segregated SrO exist. Therefore, kinetically there is no driving force for the further and continuous segregation of the A-site Sr cations. The surface segregation process of the A-site Sr cation under the cathodic polarization conditions becomes a very slow process and an equilibrium could be reached:

$$La_{0.6}Sr_{0.4}CO_{0.2}Fe_{0.8}O_{3-\delta} \overset{i_{\text{cathodic}}}{\rightleftharpoons} La_{0.6}Sr_{0.4-x}CO_{0.2}Fe_{0.8}O_{3-\delta} + xSrO \tag{10.8}$$

The equilibrium of the surface segregation reaction implies that excess Sr segregation and disintegration of LSCF perovskite structure would not occur for the LSCF/GDC cells, due to the fact that segregation of the A-site Sr cation is kinetically inhibited. This is in general consistent with the observed excellent and high polarization performance and stability of LSCF electrodes on GDC electrolyte.

On the other hand, significant segregation of B-site Co cation occurs in the case of LSCF/GDC cells, but not in the case of LSCF/YSZ cells. This appears to indicate that cathodic polarization promotes the surface segregation of both the A-site cations and B-site cations from the LSCF perovskite structure. Nevertheless, the degree of the preferential segregation of A-site or B-site cations to the surface seems to be related to the chemical environment or the presence of chemical catchers of segregated species. In the presence of Sr catchers, e.g., Zr in the case of LSCF/YSZ cells, the segregation of the A-site cation, Sr is predominant. On the other hand, the segregation of B-site cations such as Co will take place in the absence of chemical catchers for the segregated Sr species:

$$La_{0.6}Sr_{0.4}CO_{0.2}Fe_{0.8}O_{3-\delta} \xrightarrow{i_{\text{cathodic}}} La_{0.6}Sr_{0.4-x}CO_{0.2-y}Fe_{0.8}O_{3-\delta} + xSrO + yCoO \tag{10.9}$$

The segregated Co species could be in the form of Co_3O_4. Figure 10.4 shows the schemes of the surface segregation of Sr and/or Co and interface reaction between the segregated species and electrolyte. Limited and isolated accumulation of Co species may not be as detrimental as the A-site Sr cation to the electrocatalytic activity of LSCF cathode, but the excess segregation of B-site cations such as Co should be avoided.

The significance of the above discussion of the chemical catchers induced driving force is that the presence of chemical catchers for the segregated species such as Sr in particular will accelerate the surface segregation of the LSCF electrodes and consequently the performance degradation under SOFC operation conditions. In addition to the electrolyte, impurities or contaminants can also function as catchers

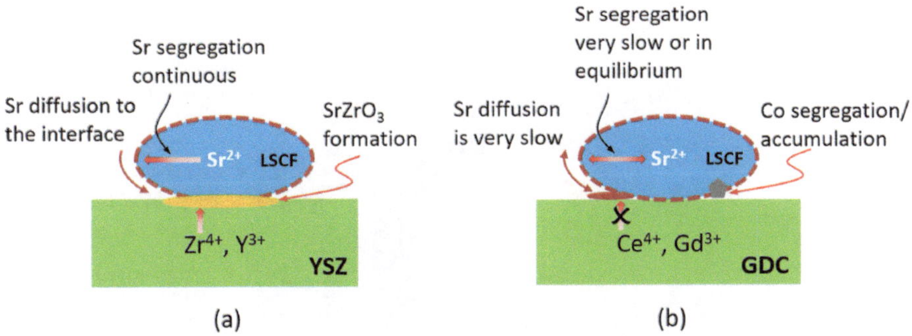

Fig. 10.4 Schemes of the surface segregation and interface reaction in **a** LSCF/YSZ and **b** LSCF/GDC cells. Red arrow indicates the diffusion and reaction direction. Reproduced from Ref. [14] with permission from Elsevier, Copyright 2020

for the segregated species; in particular Sr. We will have more discussion in relevant sections of the deposition and poisoning of contaminants in relation to the surface segregation on the cathodes of SOFCs.

10.1.3 A Note

Surface segregation is a complex process. Therefore, it is quite normal that significant discrepancies exist in the influential factors, mechanism, and kinetics of the surface segregation and segregated A-site and/or B-site cations on SOFCs cathodes as reported in the literature. One of the reasons may be related to the fact that as the Sr segregation is critically influenced by a variety of factors, the differences in sample preparation processes, surface chemistry, surface strain, surface morphology, atmospheres, impurities, pre-heat treatment, and polarization conditions most likely to have a critical effect on the observed phenomenon of surface segregation. Thus, it is important to conduct the surface segregation studies under carefully controlled conditions. Segregated cations, particularly Sr and Mn species, can further interact chemically with volatile impurities such as Cr, S and B in SOFC stacks, leading to contaminant deposition and poisoning as well as significant performance degradation. Because of this, the inhibition and significant reduction of surface cation segregation is critical in the enhancement of SOFC electrochemical performance and durability. Here, the manipulation of operating regimes as well as the modification of stoichiometric compositions and oxygen electrode surfaces are effective approaches to mitigate the detrimental effects of surface segregation on operational stability. In addition, phase migration and in situ deposition and formation of highly active and protective layers on the electrode/electrolyte interfaces have shown particular potential in the mitigation and inhibition of the detrimental effects of surface segregation in cobaltite-based electrodes of SOFCs. Figure 10.5 shows the interrelationship between the driving forces and surface segregation on SOFCs cathode

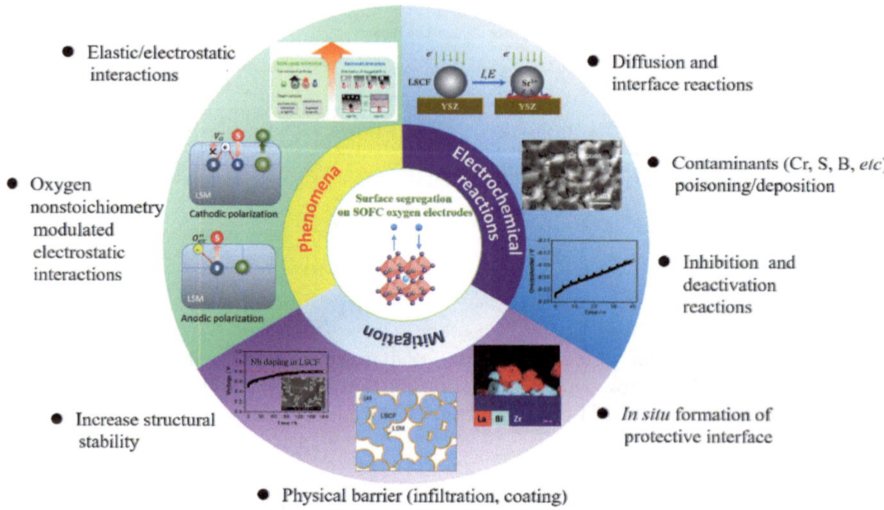

Fig. 10.5 Phenomena, electrochemical properties and mitigation strategies of cation surface segregation of SOFC oxygen electrodes. Reproduced from Ref. [2] under Creative Commons CC BY license

materials and electrochemical properties and mitigation strategies of cation surface segregation of SOFC cathodes [2].

It should be emphasized again here that the presence of catchers for the segregated species as demonstrated for the Sr segregation processes on LSCF/YSZ and LSCF/GDC cells has significant effect on the kinetics and behavior of surface segregation processes. This can be illustrated from the interrelationship between the poisoning and deposition of contaminants and the surface segregation, as shown in Sect. 10.5.1.

10.2 Reactions at the Cathode

10.2.1 Activation Process

There are various forms of activations in fuel cells, e.g., the activation processes associated with the initial start-up of low temperature PEMFCs. Such activation phenomena are generally associated with the stabilization of the physicochemical states of the electrocatalysts and catalyst layer and are conventionally referred to as conditioning. In high-temperature SOFCs, we generally define the *activation process or behavior* as the electrochemical phenomena associated with the significant improvement in the electrochemical performance of cathodes under the influence of cathodic polarization [15]. The best and probably the first known activation behavior is the observed substantial enhancement in the electrode polarization performance of LSM-based electrodes under cathodic polarization or current passage [9]. The activation behavior is characterized by a rapid decrease in the cathodic polarization potential and/or the *electrode polarization (interface) resistance* (R_E) or *area specific resistance* (ASR) after application of a cathodic polarization or cathodic current passage treatment. We should make a note here that both R_E and ASR are used in SOFCs communities, so they will be used interchangeably in this book. For ORR on a freshly prepared LSM electrode, the initial polarization losses are very high and decrease rapidly with the cathodic polarization/current passage. Consistent with the polarization potential, the impedance responses measured at open circuit decrease rapidly with the application of the cathodic current passage. Analysis of the impedance responses indicates that the effect of cathodic polarization is primarily on the reduction in the low-frequency impedance. However, the activation process can be reversed by the anodic polarization. With the application of anodic polarization current, both the polarization potential and R_E increase with the anodic polarization time, although the rate of increase is much slower [16]. The increase in the electrode polarization resistance of LSM for ORR with the anodic polarization is also called the *deactivation process*. Figure 10.6 gives an example of the activation and deactivation processes on a LSM electrode for ORR as a function of cathodic and anodic polarization processes, measured at 800 °C and 200 mA/cm^2 in air. From the R_E curve, we can see that the deactivation process is much slower than the activation process.

The first direct evidence of the intrinsic relationship between the activation process and surface segregation is demonstrated in the polarization behavior of an acid-treated as-prepared LSM electrode in 2001 [9]. In this experiment, an as-prepared $La_{0.72}Sr_{0.18}MnO_3$ electrode was washed with a 1 M HCl acid solution prior to the cathodic polarization. The nominal composition of the powder from the inductive coupled plasma-atomic emission spectrometry (ICP-AES) analysis is $La_{0.74}Sr_{0.15}MnO_3$, very close to the targeted composition. The ICP-AES analysis of the solution collected after the HCl treatment of the LSM electrode coating shows that the atomic ratio of the La/Sr/Mn based on the La concentration is 0.74/8.4/15.8 for the treated solution, which is significantly differently from the measured nominal ratio of 0.74/0.15/1 for the LSM coating. The error in the ICP-AES analysis is ±2% for La and Mn and ±5% for Sr. The high content of Mn is most likely due to A-site sub-stoichiometry with a small Mn excess. However, the high content of Sr on the LSM surface is an indication of the Sr segregation and enrichment on the LSM surface. Electrochemically, the activation process for ORR on the acid-treated LSM electrode disappears; first, the initial R_E is significantly smaller as compared to the as-prepared LSM without acid treatment; secondly, the reduction of R_E under the cathodic polarization treatment is also very small and can be neglected.

The initial state of LSM lattices can be examined to explain the activation phenomenon. The segregated SrO species on the LSM surface occupies the active sites and significantly inhibits the dissociation and diffusion of oxygen, leading to a very high electrode polarization resistance. Here, the removal of SrO species can be achieved through chemical processes such as weak acid etching as well as with

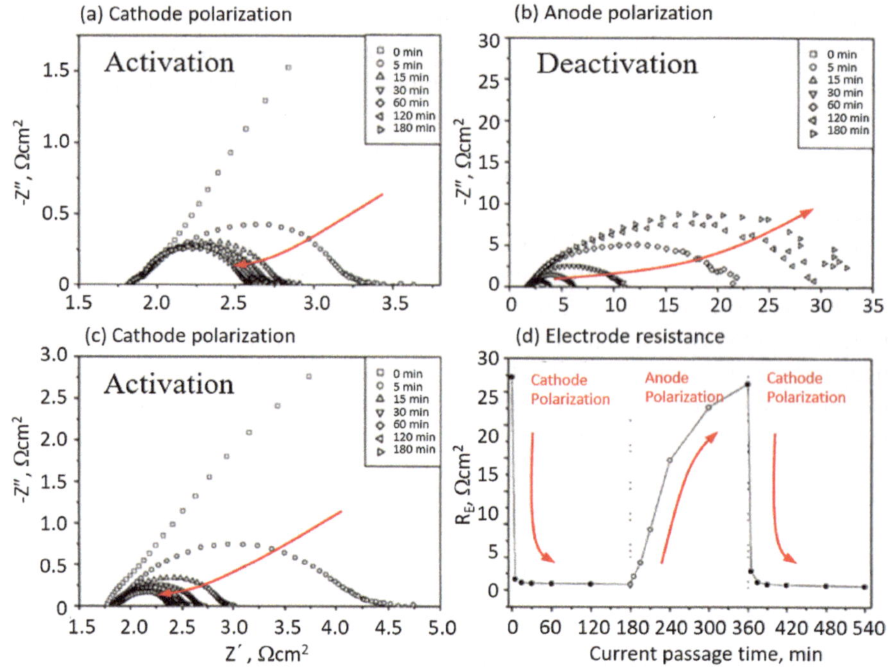

Fig. 10.6 Impedance behavior of a freshly prepared $La_{0.72}Sr_{0.18}MnO_3$ electrode for O_2 reduction as a function of **a** cathodic, **b** anodic, and **c** cathodic polarization time at 200 mA/cm^2 and 800 °C in air. The change of electrode polarization resistance, R_E, as a function of polarization processes is given in (d). Arrows show the change of the impedance arcs or R_E with the polarization time. Modified from Ref. [16] with permission from Elsevier, Copyright 2006

the application of cathodic polarization/current passage through the incorporation of segregated SrO species into LSM perovskite lattices due to the charge compensation between the positively charged oxygen vacancies formed and the incorporated negatively charged Sr'_{La} defects as discussed above. The incorporation of SrO under cathodic polarization conditions is kinetically favorable and a fast process due to the strong oxygen vacancy modulated electrostatic interactions. Under anodic polarization, the oxygen vacancies are consumed by the recombination reactions, followed by the oxygen excess and/or cation vacancy formation. As the cation vacancy formation would result in the Sr segregation on the LSM surface, the process would be highly energetic and is much slower as compared to the corresponding Sr incorporation process. This explains the much slower deactivation process for the reaction on the LSM electrode under anodic polarization in comparison to the activation process for the reaction on the LSM electrode under cathodic polarization.

The activation effect of the cathodic polarization/current passage on LSM/YSZ composite cathodes and MIEC electrodes such as LSCF is much smaller than that on pure LSM electrodes. The fundamental reason is due to the fact that the activation process is essentially a surface dominant process and is most effective on electrical dominant electrodes like LSM and Pt. For composite and MIEC electrodes, the reaction kinetics are mainly limited by the ionic diffusion in the electrode bulk. Therefore, the surface dominant activation process becomes less significant. Nevertheless, the activation process is a very important phenomenon and is of great practical relevance as the electrode performance not only depends on the fabrication process (i.e., the microstructure), but also on the polarization history.

10.2.2 Oxygen Reduction Reaction

As in low-temperature fuel cells, the *oxygen reduction reaction* or ORR is also the most important and fundamental reaction in high-temperature SOFCs. Different to the aqueous electrolyte systems, ORR at the cathode of SOFCs occurs at the solid/solid interface, and thus the surface chemistry of the solid phase, either electrode or electrolyte, is very critical for the reaction mechanism and kinetics. This has been shown very clearly from the activation process as discussed above. For the purpose of simplicity, in the discussion of the ORR in this section, we assume that the surface chemistry or state of the electrodes does not change.

As mentioned in Chap. 9, the overall ORR can be written by using Kröger–Vink notation:

$$\frac{1}{2}\mathrm{O_2} + 2\mathrm{e}^- + V_O^{\bullet\bullet} \rightarrow \mathrm{O_O^x} \tag{10.10}$$

where $V_O^{\bullet\bullet}$ is a vacant oxygen lattice site or oxygen vacancy and O_O^x is an oxygen ion on a regular oxygen site in the electrolyte lattice. As is well known in the fuel cell communities, ORR does not proceed in one step, instead ORR occurs in many individual steps. Consequently, there are many reaction models and mechanisms proposed for ORR at cathodes of SOFCs [17]. There are also significant disagreements on the kinetics of the reaction whether ORR is controlled by a chemical or electrochemical process [18]. Despite the differences in the proposed models, there are common steps for the ORR at the reactive sites of the cathode/electrolyte interface. As the resistance for the diffusion of reactant gas (air or O_2) from the bulk gas phase to the active or adsorption sites through the pores of the cathode would be very small, the gas diffusion step through the pores of the cathode is generally ignored in the overall contribution to the kinetics of ORR. Thus, the reaction steps can be expressed as:

Step 1. Adsorption of an oxygen molecule at the active sites or adsorption sites and dissociation into adsorbed oxygen atoms:

$$\mathrm{O_2} + 2\mathrm{ad} \xrightarrow{\text{adsorption and dissociation}} 2\mathrm{O_{ad}} \tag{10.11}$$

Step 2. The adsorbed oxygen atoms take the charge to become the charged O species (the charge transfer step):

$$O_{ad} + e^- \rightarrow O^-_{ad} \text{ or } O_{ad} + 2e^- \rightarrow O^=_{ad} \tag{10.12}$$

Step 3. Migration of the charged O species either through the surface or bulk of the electrode to the reaction sites (or TPB):

$$O^-_{ad} \text{ or } O^=_{ad} \xrightarrow{surface/bulk\,migration} O^-_{ad,TPB} \text{ or } O^=_{ad,TPB}(\text{electrode}) \tag{10.13}$$

Step 4. The $O^-_{ad,TPB}$ will take another electron to become oxygen ions and transfer from electrode (e.g., LSM or LSCF) to ionic conducting electrolyte (e.g., YSZ, GDC) by consuming an oxygen vacancy:

$$O^=_{ad,\,TPB}(\text{electrode}) + V_O^{\bullet\bullet}(\text{electrolyte}) \rightarrow O^x_{O,\,TPB}(\text{electrolye}) \tag{10.14}$$

Different models have been proposed with different sequences of the reaction steps. For example, the charge transfer step can be at the TPB prior to the ion transfer (step 4) or be combined with step 4. However, for the oxygen migration/diffusion (step 3), two main pathways can be identified: oxygen adsorption and dissociation at the electrode surface, followed by the surface diffusion pathway (pathway I) to the TPB area and/or by the bulk diffusion pathway II to the two phase area at the cathode/electrolyte interface, as shown in Fig. 10.7. The surface diffusion or bulk diffusion pathways are dependent on the nature of the electrode materials, i.e., the electronic conducting materials or MIEC materials. ORR can be dominated by one of the pathways or occur in parallel through the pathways. Among the species involved in the reaction, oxygen vacancies are the key factor that influences the ORR mechanism. In the bulk diffusion pathway, transport within the cathode occurs via

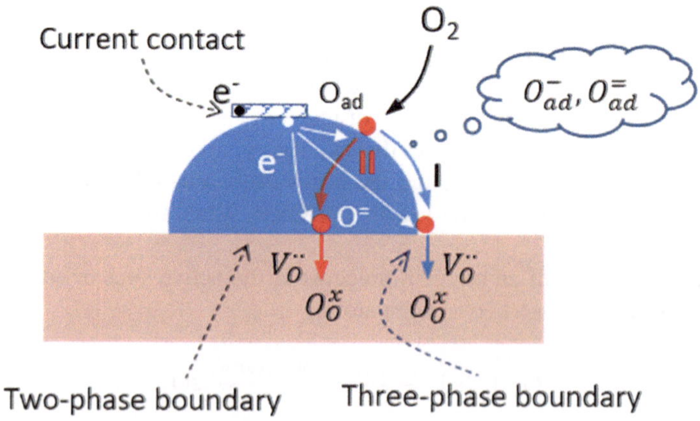

Fig. 10.7 Reaction pathways for ORR via surface diffusion process and bulk diffusion process

oxygen ionic conductivity until the ions are transferred into the electrolyte material. Oxygen vacancies act as sinks and active sites for the adsorbed oxygen, O_{ad} and at the same time, they are actively involved in the electrocatalysis. Kinetics of the reaction are related to the oxygen dissociation, diffusion or migration and exchange at the electrode surface and the ionic transportation properties of the cathodes.

Three-phase boundary (TPB)

The three-phase boundary (TPB) is particularly relevant to the electrochemical reactions in SOFCs. The three phases consist of the solid electrode phase to take or deliver electrons, solid electrolyte to deliver oxygen ions and the gas phase to deliver reactants. Electrochemical reactions can only occur in locations or regions where all three phases meet. TPB can be measured by the length of TPB, i.e., the TPB length and by the area, i.e., the TPB area.

Thus, for electronic dominant LSM electrodes on the YSZ electrolyte, the TPB is the region at which LSM, YSZ, and O_2 reactant phases are spatially close to each other and usually in the vicinities of the electrode/electrolyte interface. In the case of the MIEC LSCF electrode on GDC electrolyte, TPB can be the region at the LSCF electrode surface in contact with O_2 reactant phase in addition to the LSCF electrode, GDC electrolyte and O_2 reactants in the vicinities of the interface because LSCF functions as a mixed electronic and ionic conducting phase. In Ni-based cermet anodes, TPB is the region where Ni, electrolyte are (YSZ or GDC), and reactant gas meet.

In the case of MIEC which is a single-phase material having both ionic and electronic conductivity, the electrochemical reactions can take place at the two-phase area between the electrode and electrolyte.

The study of reaction mechanisms and kinetics of ORR is generally performed by electrochemical impedance spectroscopy (EIS). If there are significant differences in the reaction rate, it would be possible to separate the electrode reaction steps on the frequency domain. However, such separation in the electrode reaction steps is dependent on the nature of electrode materials, the test conditions like temperature and partial pressure of oxygen. Figure 10.8 shows an example of EIS curves for ORR on LSM and LSCF electrodes measured at 700 °C [19]. Obviously, the overall ORR on a LSM electrode is controlled at least by three electrode steps: low frequency, medium frequency and high frequency, while for the ORR on LSCF electrodes, only one arc can be identified. With respect to the equivalent circuit, the reaction on LSM can be presented by $LR_o(R_1Q_1)(R_2Q_2)(R_3Q_3)$ with three RQ units and on LSCF by $LR_o(R_1Q_1)$ with one RQ unit. In the circuit, L is the inductance, R_o is the electrolyte resistance, R_n and Q_n is the electrode resistance (R_E) and constant phase element (CPE) of the reaction step associated with individual impedance arc *n* (more details on equivalent circuit treatment are given in Chap. 11). Although the EIS of ORR on LSM electrode can be perfectly fitted with the equivalent circuit with three

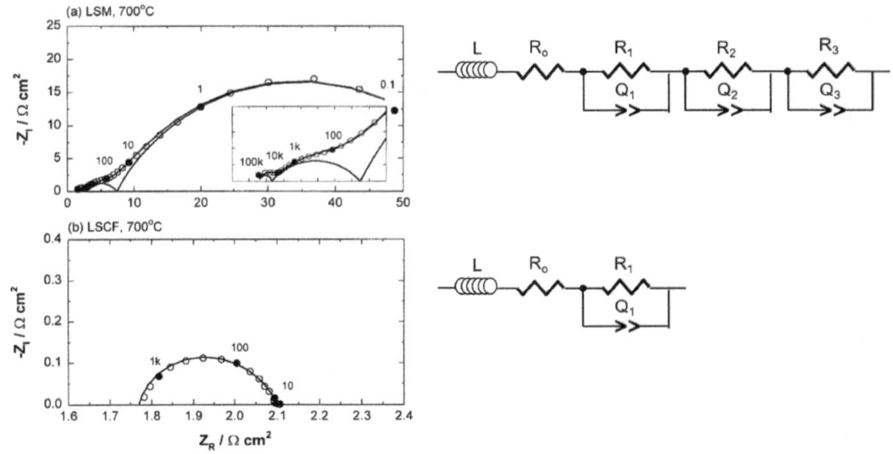

Fig. 10.8 Impedance responses and corresponding equivalent circuits for ORR on LSM and LSCF electrodes, measured at 700 °C in air. The numbers are frequencies in Hz. The symbols are experimental data, and lines are fitted data based on the equivalent circuit. Modified from Ref. [19] with permission from Elsevier, Copyright 2002

sub-circuit RQ, the questions remain to the physical significance of the individual electrode steps associated with frequency domains.

For the ORR on LSM electrode, the electrode polarization resistance associated with the high frequency arc (R_1Q_1) is generally very small and disappears at temperatures higher than 700 °C and does not significantly contribute to the reaction. In most cases, ORR is characterized by the electrode processes at low and medium frequencies, R_3 and R_2. The main question is what are the predominant electrode processes associated with the medium and low frequencies?

The essential physical significance of the electrode steps associated with specific impedance arcs can be identified by performing simple DC bias and temperature tests (see Chap. 11 for more details), although it would take much more efforts to identify elementary steps. The basic principle is that for the electrode process involving charge transfer, e.g., Step 2 of the reaction, the electrode resistance of the process will be affected by the applied dc bias, while the magnitude of the activation energy can be distinguished with certain processes. Figure 10.9 shows an example of the electrode polarization resistance as a function of dc bias and the corresponding activation plots for ORR on a LSM electrode [19]. The total electrode resistance, R_E, shows a parabolic dependence on the dc bias at 700 and 800 °C but disappear at 900 °C. This implies that the effect of charge transfer process on the ORR diminishes with the increase in temperature. Most significantly, we can see that the R_{E3} changes with the dc bias, while R_{E2} is in general independent of DC bias. This indicates that the electrode process associated with the low frequencies involves charge transfer, while the electrode process associated with the medium frequencies does not involve charge transfer. The appearance of a maximum of the change of the R_{E3} with DC bias simply indicates that the electrode process at low-frequency arc is not a pure charge

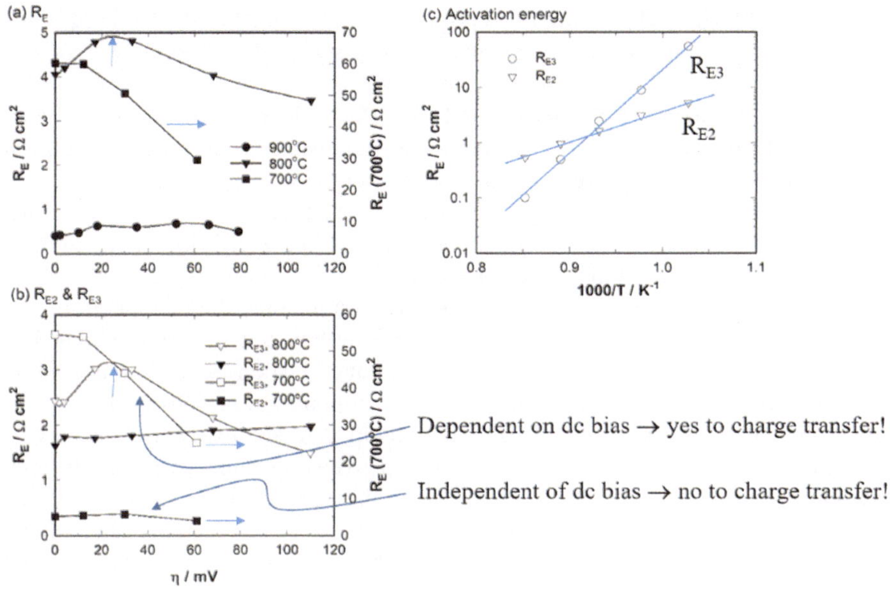

Fig. 10.9 Dependence of **a** the total electrode resistance (R_E) and **b** individual electrode resistance of the low- and high-frequency arcs (R_{E3} and R_{E2}) for ORR on LSM electrode and the DC bias at different temperatures, and the activation energy plots of R_{E3} and R_{E2}. Modified from Ref. [19] with permission from Elsevier, Copyright 2002

transfer process but involve other processes. Therefore, the processes associated with low-frequency arc can be identified with the adsorption, dissociation, and/or migration of charged oxygen species (i.e., Steps 1, 2, and/or 3). Polarization can promote the migration of the charged oxygen species to the TPB regions for ORR on LSM electrodes.

The measured activation energy of the electrode process associated with the medium-frequency arc, R_{E2}, is ~ 110 kJ/mol, while the activation energy for the electrode process associated with the low-frequency arc, R_{E3}, is 290 kJ/mol. The high activation energy of R_{E3} indicates that the oxygen dissociation and diffusion process is a kinetically demanding process with high energy barrier. This is probably due to the breaking of the double bonds of oxygen molecules. The activation energy of R_{E2} is actually close to that of the YSZ electrolyte (106 kJ/mol, see Chap. 9), indicating that the electrode process associated with this frequency arc is the oxygen ion migration from the LSM electrode to the ionic conducting YSZ electrolyte. It is not difficult to understand, as the oxygen is incorporated into the YSZ electrolyte by migration through the electrode/electrolyte interface and the energy barrier for such migration would be at least the activation energy of the electrolyte material. This implies that the electrochemical activity of cathode materials is also dependent on the ionic properties of the electrolyte. To separate the R_{E3} into elementary steps of adsorption, dissociation, and diffusion, the impedance behavior could be examined under controlled partial pressure of oxygen and temperatures [20]. In this case, it is possible to further identify the elementary steps associated with the low-frequency impedance arcs.

The step occurring at the gas/cathode interface (Steps 1–3) can be an electrochemical process involving the charge transfer of electrons across the current collector/cathode interface (e.g., on LSM electrodes) or a pure chemical process with no charge transfer (e.g., on MIEC electrodes like LSCF). The oxygen exchange process at the gas/electrode interface becomes increasingly dominant with the decrease in operating temperature, indicating that the rate-limiting step of ORR on MIEC electrodes like LSCF strongly depends on the operating temperature.

The main limitation in the application of EIS in the study of electrode reactions like ORR is the ambiguous nature of the impedance responses, particularly in the case of overlap or dispersion in the frequency domain among physical processes governing electrode reactions. This makes them difficult to resolve entirely by frequency domain. Another difficulty is that for a given impedance response, different mechanistic models can be fitted depending on the number of parameter used. This is particularly the case where the separation of impedance responses is generally not possible. More detailed discussions related to the equivalent circuit approaches are given in Chap. 11.

The electrocatalytic activity of cathodes for ORR also depends strongly on the properties of the electrolyte. This is due to the fact that oxygen ion transfer at the interface is related to the oxygen ion conductivity of the electrolyte.

10.2.3 Effect of Oxygen Vacancies

From the discussions above, it is clear that oxygen vacancies are the most important factors influencing the ORR mechanism and kinetics. On one hand, $[V_O^{\bullet\bullet}]$ is the key species to facilitate the oxygen ion transfer between the electrode and electrolyte (Step 4), and on the other hand, $[V_O^{\bullet\bullet}]$ provides sink and sites available for the adsorbed oxygen species (O_{ad}^{-} *and* $O_{\mathrm{ad}}^{=}$), promoting the adsorption and fast diffusion path for ORR. Thus the high concentration of oxygen vacancies, $[V_O^{\bullet\bullet}]$, in the vicinities of the reactive sites and regions enhances the kinetics of ORR by providing a high density of active sites available for oxygen incorporation. Such an intrinsic property depends strongly not only on the ionic conducting property of the materials but also on the microstructure of the electrode. It is difficult to actually measure the value of $[V_O^{\bullet\bullet}]$, but electrochemically, this can be illustrated by examining the impedance responses for ORR on purposely designed experiments [21].

Figure 10.10 shows such an example of the relationship between the impedance responses and microstructure of a nano-structured LSM/GDC composite cathode as a function of the infiltrated GDC nanoparticles (NPs) [21]. The impedance responses for ORR on a GDC-infiltrated LSM electrode heat-treated at 700 °C are characterized by a high-frequency capacitive arc and a distinctive low-frequency inductive loop, forming an impedance circle with a very small R_E value of 0.04 Ω cm^2. A

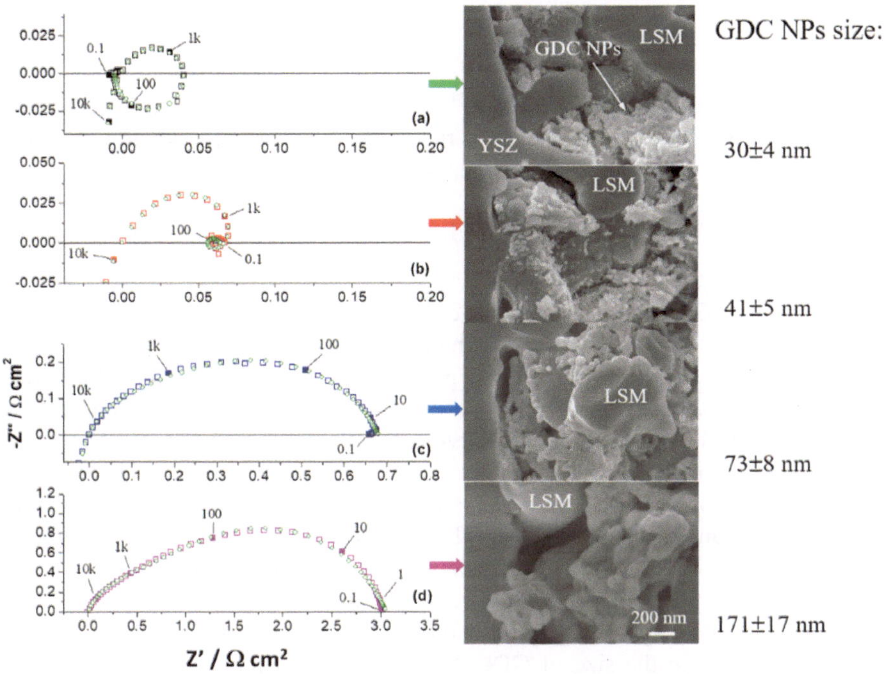

Fig. 10.10 Impedance responses for ORR on GDC ($Gd_{0.2}Ce_{0.8}O_{1.9}$)-infiltrated LSM composite electrodes heat-treated at (a) 700, (b) 850, (c) 950, and (d) 1100 °C for 2 h, measured at 800 °C. The infiltrated GDC loading was 1.2 ± 0.2 mg/cm^2. SEM images of the electrodes are shown on the right. Modified from Ref. [21] with permission from Elsevier, Copyright 2016

semicircle is called inductive if it has positive imaginary values. As the electrode heat treatment temperature increases to 850 °C, the size of low-frequency inductive loop is significantly reduced. The low-frequency inductive loop disappears and converts to a capacitive arc as the heat treatment temperature rises further to 950 and 1100 °C. Both the characteristics and size of the impedance responses change significantly with the heat treatment temperatures; the disappearance of the low-frequency inductive loop occurs with the concomitant increase in the size of the impedance arcs. This is accompanied by the increase in the GDC nanoparticles (NPs) size from 30 nm for the composite cathode calcined at 700 °C to 171 nm of the composite cathode calcined at 1100 °C.

The distinct inductance loop and very high electrochemical activity of LSM/GDC composite cathodes calcined at 700 °C can be explained based on the fact that ceria has the capacity to rapidly take up and release oxygen. Under open circuit and/or slightly reduced conditions, the intrinsic oxygen vacancies for pure ceria are formed by the partial reduction of Ce^{4+} to Ce^{3+} as described by the following non-stoichiometric reaction:

$$2\mathrm{Ce}_{Ce}^{x} + \mathrm{O}_{\mathrm{O}}^{x} \rightleftharpoons V_{O}^{\bullet\bullet} + 2\mathrm{Ce}_{Ce}' + \frac{1}{2}\mathrm{O}_2 \quad (10.15)$$

The transition reaction of Ce^{4+} to Ce^{3+} for oxygen vacancy formation and migration is reversible, largely depending on the partial pressure of oxygen. Gd doping can introduce additional extrinsic oxygen vacancies and can be described by

$$\mathrm{Gd_2O_3} \rightarrow 2\mathrm{Gd}_{Ce}' + V_{O}^{\bullet\bullet} + 3\mathrm{O}_{\mathrm{O}}^{x} \quad (10.16)$$

where Ce_{Ce}^{x} is a Ce^{4+} ion in the Ce lattice site, Ce_{Ce}' is a Ce^{3+} ion in the Ce lattice site, and Gd_{Ce}' is a Gd^{3+} ion occupying the Ce lattice site. In case of GDC-infiltrated LSM, GDC NPs would act as the oxygen source for the ORR due to the high ionic conductivity and high surface exchange rate and their ability to store and release oxygen, significantly enhancing the activities for the ORR on LSM. When the size of GDC NPs is small (e.g., 41 nm or smaller), surface area specific $[V_{O}^{\bullet\bullet}]$ is high due to the high specific area of GDC NPs, substantially promoting the dissociative adsorption and diffusion process. The O_2 reduction reaction is thus essentially controlled by the ion migration process at the electrode/electrolyte interface. This is supported by the presence of the low-frequency inductive loop instead of the low-frequency capacitive arc for the reaction on LSM/GDC cathode with fine GDC NPs. With the increase in the size of GDC NPs to 73 nm or above, $[V_{O}^{\bullet\bullet}]$ is reduced, thus the rate of oxygen supply decreases. The dissociative adsorption and diffusion processes of oxygen species become dominant, indicated by the disappearance of the low-frequency inductive loop with the concomitant increase of R_E. The gradual transfer of the low-frequency inductive loop to a capacitive arc for the reaction on LSM/GDC with the calcination temperature is clearly due to the increase of the GDC NPs size and thus the reduced capacity of GDC as an oxygen supply source for the reaction.

10.3 Reactions at the Anode

10.3.1 Hydrogen Oxidation Reaction

10.3.1.1 On Ni and Ni-Based Cermet Anodes and Spillover Mechanism

The *hydrogen oxidation reaction* (HOR) is the other important electrochemical reaction in SOFCs. The overall reaction can be generally formulated as:

$$\mathrm{H_2} + \mathrm{O}_{\mathrm{O}}^{x} = \mathrm{H_2O} + 2\mathrm{e} + V_{O}^{\bullet\bullet} \quad (10.17)$$

In the case of the Ni/YSZ cermet anode, HOR occurs at the TPB between the Ni, YSZ, and gas phases. Oxygen anions are transported to the reaction site through the

YSZ electrolyte, where they react with hydrogen absorbed on the Ni or YSZ surface to form H_2O. The released electrons flow through the Ni phase to the external circuit and the product H_2O diffuses out of the anode. The individual steps of this reaction include gas diffusion of reactants and products, adsorption/desorption of species onto the catalytically active sites, surface diffusion of adsorbed species and intermediates, and charge transfer at anode/electrolyte interface. Its overall rate can be described empirically as a function of experimental parameters such as the partial pressures of H_2 gas, partial pressure of H_2O, temperature, etc. The kinetics and mechanism, or description of the different processes involved in the HOR have been studied extensively on Ni and Ni/YSZ cermet anodes.

In the study of the HOR, experimentally obtained electrochemical impedance data are usually interpreted using an appropriate model of the electrochemical system, and parameters are obtained by fitting simulated impedance spectra to the experimental data. To avoid the complex microstructure of porous Ni/YSZ cermet anodes, modeling studies on HOR are generally carried out using geometrically well-defined pattern and point anodes. Pattern and point electrodes offer the advantages of easily accessible active sites, no porous transport effects, well-defined and measurable two-phase and three-phase lengths or areas. There are various modeling approaches including atomistic modeling, elementary kinetic modeling and global kinetic modeling [22].

Understanding the mechanism and kinetics of HOR on technically relevant Ni/YSZ cermet anodes is very important for the development of high performance and durable anodes for SOFCs. It has been well established that the electrochemical activity of Ni anodes increases almost proportionally with the TPB length [23], indicating that TPB or active areas are extremely important for Ni-based cermet anodes. In addition to the TPB, the mechanism and kinetics of HOR are also significantly affected by the content of H_2, partial pressure of H_2O, and the distribution of Ni and YSZ phases in the cermet.

Spillover Mechanism

Spillover significantly enhances the diffusion process of active species and is defined generally as the transport phenomena of an active species adsorbed or formed on a surface onto another surface that does not adsorb or form the active species under the same conditions. Two different surfaces, e.g., metal and metal oxide, are generally involved in the spillover processes. For example, the adsorption of gaseous hydrogen onto alumina is made possible by the presence of platinum, although the adsorption process of hydrogen from gas phase onto alumina is kinetically hindered. In particular the diffusion and migration of hydrogen atoms from the metal catalyst surface onto the non-metal oxide support surface is called the *hydrogen spillover*.

Water vapor plays an important role in HOR. To explain the role of water, hydrogen and/or oxygen *spillover mechanisms* have been proposed. As a spillover pathway requires direct contact of the metal oxide surfaces, the oxygen spillover and/or hydrogen spillover pathways are considered to occur at the YSZ electrolyte and Ni anode interface as it is generally assumed that there is no other phase in the system. In this case, H atoms adsorbed on the Ni phase move over the TPB toward the YSZ phase while simultaneously releasing their electrons to the Ni anode. In the presence of water, spillover of H from the Ni surface to hydroxyl ions (OH^-) on the YSZ surface occurs.

There is strong evidence that hydrogen spillover also occurs on the surface of the Ni anode if there is sufficient amount of water in the hydrogen fuel [24]. Figure 10.11 shows impedance responses of the effect of water vapor on the HOR on pure Ni and Ni/YSZ with (70wt% Ni and 30wt% YSZ) cermet anodes, measured at 1000 °C [25]. In dry H_2, the measured open circuit voltage (OCV) is 1.27 V and the H_2O content in H_2 can be calculated to be ~ 0.07%. The impedance behavior of HOR on both Ni and Ni/YSZ cermet anodes is characterized by a very large and depressed impedance arc in dry H_2, although the magnitude of the impedance responses for the reaction on Ni/YSZ cermet is relatively much smaller. Because the electrode process associated with low frequencies is generally related to the adsorption, dissociation and diffusion on the surface of electrodes, this indicates the difficulties of hydrogen dissociation and diffusion on Ni and Ni/YSZ cermet surface under dry H_2. In the presence of 2% H_2O, the impedance arc, i.e., the area specific resistance (ASR), is substantially reduced. For example, for HOR on the Ni anode, the ASR is estimated to be 23.6 Ωcm^2 in dry H_2 (by extrapolation) and reduces dramatically to 3.3 Ωcm^2 in 98%H_2/2%H_2O. For the HOR on a Ni/YSZ cermet electrode, in addition to the significant reduction in the ASR, two relaxation processes are distinguished in the low and high frequencies in wet H_2. In this case the reaction can be approximated by the equivalent circuit with two RQ units, i.e., $LR_o(R_1Q_1)(R_2Q_2)$.

The substantial reduction in the ASR of the electrode process associated with low frequencies can be explained by the hydrogen spillover mechanism. First let us examine the interaction of water with metallic Ni. The interaction of water with a metal surface can proceed via a completely or partially dissociative adsorption or a molecular adsorption step. The primary driving force for dissociation is the formation of a metal–oxygen or a metal-hydroxyl bond. In the case of Ni, the enthalpy change for metal-O bond formation, ΔH_f^0, is − 240 kJ/mol. Thus, the interaction of water with Ni would lead to the complete dissociation at high temperatures:

$$H_2O \rightarrow O_{ad} + H_2 \tag{10.18}$$

and/or

$$H_2O \rightarrow OH^-_{ad} + H \tag{10.19}$$

The subscript "*ad*" denotes the free adsorption site on a Ni surface. The formed O_{ad} and/or OH^-_{ad} can be thermodynamically stable on the Ni surface due to the negative enthalpy change of the metal-O formation. This would lead to the formation of suboxide in the form of ($Ni - O_{ad}$) and/or ($Ni - OH^-_{ad}$). The presence of a suboxide

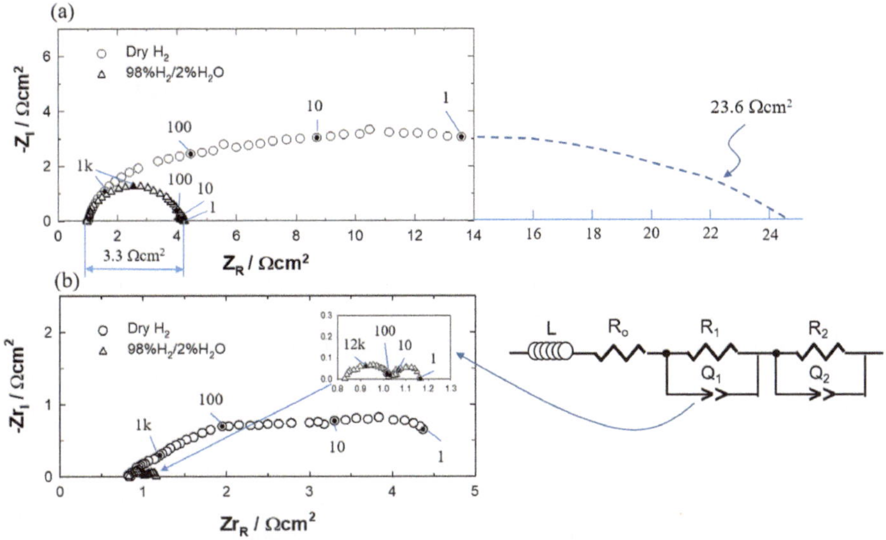

Fig. 10.11 Impedance spectra for H_2 oxidation on **a** Ni and **b** 70vol%Ni/30vol%YSZ cermet electrodes at 1000 °C in dry H_2 and 98%H_2/2%H_2O under open circuit conditions. Numbers are frequencies in Hz. Modified from Ref. [25] with permission from Elsevier, Copyright 1999

phase makes the direct contact of two surfaces of metal and oxide possible, a key condition for the occurrence of a spillover mechanism. The formation of suboxide on Ni is also consistent with the theoretical calculation of the activity of various metals against the oxygen dissociative chemisorption energy (or oxide formation energy) [26]. The reaction steps involving dissociative adsorption/surface diffusion and charge transfer facilitated by a hydrogen spillover from a Ni surface to a YSZ surface can be expressed as follows.

For HOR, the first step is the adsorption and dissociation of molecular H_2 on the surface of metallic Ni:

$$H_2 + 2Ni_{ad} \leftrightharpoons 2H_{ad,Ni} \tag{10.20}$$

With no presence of oxide or a second phase, diffusion of the formed adsorbed H_{ad} species on a bare metallic Ni surface would be very difficult. This is clearly indicated by the very high ASR of HOR on both Ni and Ni/YSZ cermet anodes in dry H_2. In the presence of the $(Ni - O_{ad})$ and/or $(Ni - OH^-_{ad})$ phase, spillover of the H_{ad} from the metallic Ni phase to the neighboring suboxide phase occurs, significantly promoting the diffusion of adsorbed H_{ad} species. As the concentration of suboxide phase on the Ni surface increases with the partial pressure of water, the rate of such dissociative adsorption/diffusion of hydrogen increases significantly with the water content in H_2. Therefore, hydrogen dissociatively adsorbed on a Ni metal site can effectively diffuse to an active site at TPB via the spillover mechanism:

$$H_{ad,\,Ni} + (Ni - O_{ad})/(Ni - OH_{ad}^{-}) \xrightarrow{\text{diffusion via spillover mechanism}} H_{ad,\,TPB} \quad (10.21)$$

This is followed by the combination with oxygen anions emerging from the YSZ electrolyte to form water, releasing electrons to the Ni anode. This is a charge transfer process:

$$2H_{ad,\,TPB} + O_{O,\,YSZ}^{x} \rightarrow H_2O + V_O^{\bullet\bullet} + 2e_{Ni}^{-} \quad (10.22)$$

Dissociative adsorption/diffusion reactions of hydrogen through the spillover mechanism are fast processes and occur favorably on metals such as Ni with sufficiently low Gibbs free energy for oxide formation. For inert metals such as Pt and Au, the hydrogen spillover would be more difficult due to the difficulty in the formation of metal-O bonds at high temperatures. Therefore, the electrochemical activity of Pt for HOR is much lower than that of Ni in high temperature SOFCs.

In the case of Ni-based cermet anodes, the addition of the ionic conducting YSZ phase significantly enhances the electrocatalytic activity of Ni for HOR. This is generally indicated by the substantially reduced electrode ASR. In some cases, the impedance behavior can be visibly separated into two impedance arcs in the frequency domain, indicating that the overall electrode reaction is controlled by at least two rate-limiting processes (see Fig. 10.11b). Similarly to ORR, the electrode processes associated with low- and high-frequency arcs can be identified by DC and temperature tests, see Fig. 10.12a for the HOR on a Ni/YSZ cermet anode, measured at 1000 °C in 98%H_2/2%H_2O. The electrode process associated with the low-frequency arc is less affected by the anodic DC bias, while the electrode process of the high-frequency arc decreases continuously with the DC bias. Two pieces of information can be obtained from this experiment: (1) the electrode process associated with the high-frequency arc (R_1) is most likely dominated by the charge transfer

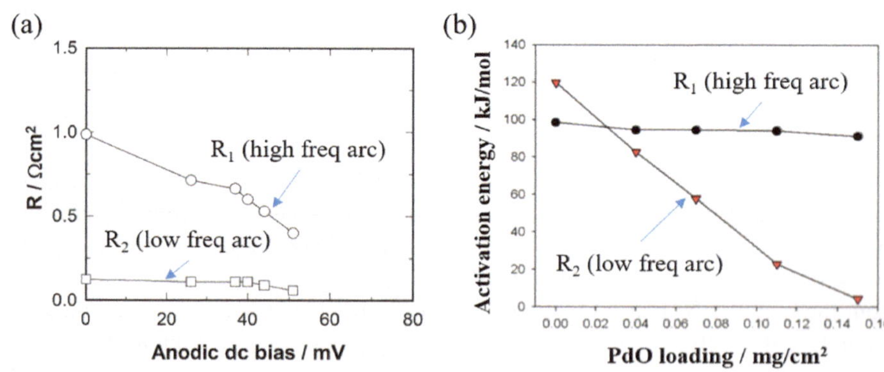

Fig. 10.12 Plots of electrode polarization resistance associated with high and low frequencies, R_1 and R_2 of **a** Ni/YSZ cermet anodes as a function of anodic DC bias (i.e., η) at 1000 °C in 98%H_2/2%H_2O, and **b** PdO-infiltrated Ni/GDC cermet anodes as a function of PdO loading measured in 97%H_2/3%H_2O

process of HOR and the electrode process associated with the low-frequency arc (R_2) is most likely related to the adsorption and diffusion of hydrogen on Ni/YSZ cermet anode, and (2) the hydrogen species involved in the hydrogen spillover promoted diffusion process are not charged hydrogen species. This in turn implies that the charge transfer most likely occurs at TPB of the electrode/electrolyte interface and not in the adsorption and dissociation stages of HOR.

The addition of redox couples can also substantially enhance the hydrogen spillover mechanism for HOR on Ni-based cermet anodes. An example is given in Fig. 10.12b for the HOR on Pd-infiltrated Ni/GDC cermet anodes [27]. In this case, Pd is infiltrated to Ni/GDC cermet anodes as Pd NPs and the HOR on Pd-infiltrated Ni/GDC anodes are characterized by two separated impedance arcs. Both the ASR and activation energy of the electrode process associated with the low-frequency arc (R_2) decrease significantly with the increase of the infiltrated Pd NPs, while the effect of infiltrated PdO has little effect on the activation process of the electrode process associated with the high-frequency arc (R_1). The significantly enhanced adsorption and diffusion processes of the HOR on Pd-infiltrated Ni/GDC anode are considered to be related to the significantly promoted hydrogen spillovers over the Pd/PdO redox couple or Ni/PdO couple. The significantly reduced activation energy of R_2 indicates the hydrogen spillover mechanism reduces the energy barrier for the hydrogen adsorption, and dissociation and diffusion process on the cermet anodes. In contrast to R_2, the activation energy for R_1 is almost independent on the infiltrated Pd loading. The activation energy of the electrode process associated with high-frequency impedances is generally in the range of 97–140 kJ/mol, also compatible with the activation energy of oxygen ion conductivity of YSZ electrolyte. All this information indicates that the electrode process at low frequencies can be attributed to the hydrogen dissociative desorption/diffusion on the surface of the anode, and the electrode process associated with the high frequencies is predominantly the charge transfer process. The physical significance of the individual electrode processes can also be obtained by tests under controlled concentrations of H_2O and H_2 [25].

Numerous evidence indicates that the spillover mechanism is an important process for the HOR in Ni and Ni-based cermet anodes. However, the important question regards the place or region most likely for the occurrence of the spillover mechanism. If the spillover process mainly occurs at the Ni/YSZ interface, the presence of water should not make such a dramatic impact on the ASR and the infiltration of Pd would be expected to inhibit the HOR as Pd is actually a poor catalyst for HOR. The most reasonable explanation is that spillover of hydrogen or oxygen to lesser extent occurs preferentially on the surface of Ni to facilitate the dissociative adsorption and diffusion of hydrogen toward the TPB region, followed by the charge transfer process at TPB, as shown in Fig. 10.13.

10.3.1.2 On Ceramic Oxide Anodes

Ceramic oxide materials such as ceria, titanate, and lanthanum chromite have been explored as alternative and potential anodes for SOFCs. The purpose is to overcome

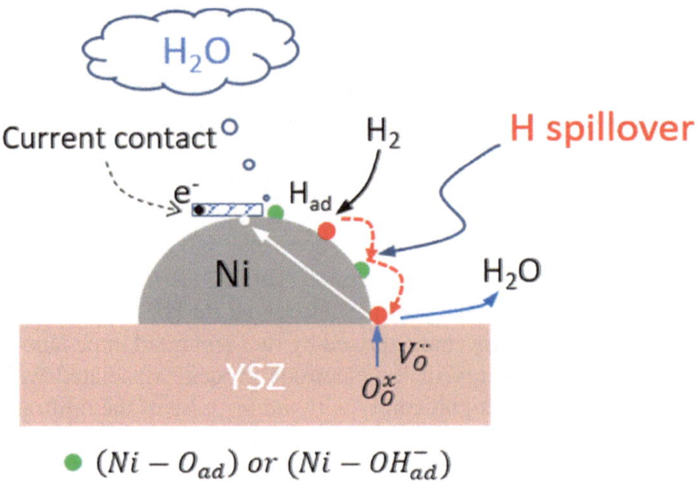

Fig. 10.13 Hydrogen spillover mechanism promoted HOR on Ni electrode. Similar reaction mechanism also applies to the HOR on Ni-based cermet anodes in general

the disadvantages of the traditional Ni-based cermet anode such as thermal instability and low resistances toward carbon deposition and sulfur poisoning. However, very different from the generally high activity of perovskite-based ceramic oxide cathodes for ORR, there are very few ceramic oxides processing good electrochemical activities for HOR at high temperatures. HOR on ceramic oxide anodes shows a significant dependence on the microstructure of the electrodes and may proceed with a different mechanism as compared to the reaction on Ni-based cermet anodes [28]. Figure 10.14 shows an example of impedance responses of HOR on the perovskite oxide $(La_{0.75}Sr_{0.25})(Cr_{0.5}Mn_{0.5})O_3$ (LSCM)/YSZ composite electrodes fabricated by an aqueous gel-casting technique and solid-state reaction (SSR) methods, measured at 750 °C. Impedance responses of the LSCM/YSZ composite anodes for HOR are characterized by two clearly separated arcs at low and high frequencies. As the temperature increases, the impedance arc associated with high frequencies disappears. HOR is dominated by the electrode process associated with low frequencies. This indicates that the reaction is limited by the hydrogen adsorption, dissociation, and diffusion processes on the LSCM electrode surface. The high electrode resistance for HOR on LSCM/YSZ composite anodes can be explained by the fact that very differently from the Ni-based cermet anodes, a spillover mechanism may not be possible in this case to facilitate the HOR due to the lack of a secondary metallic phase.

ASR is 2.2 Ωcm^2 at 750 °C for HOR on a LSCM/YSZ composite electrode with LSCM powder fabricated by gel-casting technique, while it is 4.8 Ωcm^2 for the reaction on the LSCM powder prepared by SSR method. This shows that the electrochemical activity of oxide materials depends strongly on the microstructure and properties of the oxide powders, which in turn are critically affected by the powder

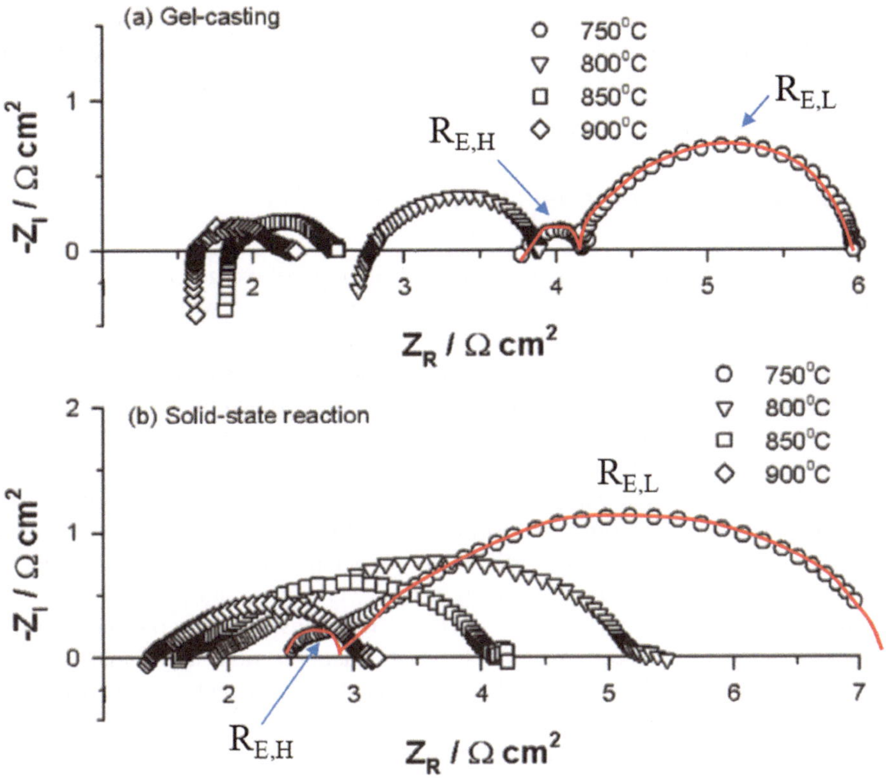

Fig. 10.14 Impedance spectra of the LSCM/YSZ composite anodes for the H_2 oxidation reaction, measured at different temperatures under open circuit. LSCM powders were prepared from gel-casting and solid-state reaction techniques, respectively. Reproduced from Ref. [28] with permission from Royal Society of Chemistry, Copyright 2007

synthesis process and conditions used. In addition to the activity, the impedance responses are also very different. For the HOR on LSCM/YSZ composite anodes prepared by gel-casting LSCM powders, the impedance could be analyzed based on an equivalent circuit with two RQ units, while for the same reaction on the anode from LSCM powder prepared by SSR, the fitting of the impedance data may require an equivalent circuit with three to four RQ units. The significant variation in the electrochemical performance of the ceramic oxide anodes is potentially a problem in the establishment of a performance bench-mark for the comparison of the ceramic oxide materials reported by different research groups. The generally low electronic conductivity of the ceramic oxide materials, particularly under reducing conditions, is also a concern for the development of anode-supported cells. In anode-supported cells, the electrical conductivity of the anode substrates must be high enough to minimize the overall cell resistance.

Nevertheless, the electrochemical activity of ceramic oxide anodes can be enhanced by the addition of metallic nanoparticles such as Ni, Pd, etc. The addition of metal NPs can be performed by infiltration or by in situ exsolution.

10.3.2 Hydrocarbon Fuel Oxidation Reaction

Due to the high operating temperature and availability of oxygen transported from the cathode to the anode, SOFCs can in principle utilize a wide range of fuels, including alcohols and hydrocarbons. Fuel cells that directly use hydrocarbon fuels such as natural gas, without first reforming those hydrocarbons to hydrogen, will have enormous advantages over conventional hydrogen-based fuel cells. Removing the need either to supply hydrogen to the fuel cell or to include a hydrocarbon-reforming system greatly decreases the complexity, size, and cost of the fuel cell system. Methane is the most important hydrocarbon fuel for SOFCs. However, unlike H_2 which can easily be oxidized to form H_2O, hydrocarbons can undergo a number of electrochemical and catalytic reactions in the anode including partial oxidation, total oxidation, steam reforming, dry reforming, cracking, and polymerization to form tars. Although only the electrochemical reactions directly contribute to the current generated by the cell, the non-electrochemical reactions may dictate the gas composition at the TPB. Some possible reactions of methane under SOFC operating conditions are exemplified as follows.

Total electrochemical oxidation reaction:

$$CH_4 + 4O_O^x \rightarrow CO_2 + 2H_2O + 8e^- + 4V_O^{\bullet\bullet} \tag{10.23}$$

Partial electrochemical oxidation reaction:

$$CH_4 + O_O^x \rightarrow CO + 2H_2 + 2e^- + V_O^{\bullet\bullet} \tag{10.24}$$

$$CH_4 + 3O_O^x \rightarrow CO + 2H_2O + 6e^- + 3V_O^{\bullet\bullet} \tag{10.25}$$

Steam reforming reaction:

$$CH_4 + H_2O \rightarrow CO + 3H_2 \tag{10.26}$$

Dry reforming reaction:

$$CH_4 + CO_2 \rightarrow 2CO + 2H_2 \tag{10.27}$$

Cracking or dissociation:

$$CH_4 \rightarrow C_{(s)} + 2H_2 \tag{10.28}$$

CO formed can react further with water to produce additional H_2 through water gas shift (WGS) reaction at lower temperatures (see Chap. 4). Among these reactions, only the electrochemical oxidation of methane contributes to the fuel utilization in SOFCs. The primary issue with Ni-based anodes in hydrocarbon fuels is the propensity of Ni to catalyze the formation of graphite from hydrocarbons, blocking the active sites for the fuel oxidation reaction. This limits the operating conditions of the cell. Under the high carbon activity environment, iron, nickel, cobalt, and their alloy-based electrodes could corrode by a process known as metal dusting. Metal dusting involves the disintegration of bulk metals and alloys into metal particles at high temperatures (300–850 °C) in environments that are supersaturated with carbon. Such degradation can be avoided by providing a high enough steam-to-carbon (S/C) ratio in the feed (external reforming), addition of steam to the feed (internal reforming), addition of oxygen to the feed, or maintaining a relatively high oxygen ionic flux to produce sufficient oxidation products within the anode itself. The goal is to supply enough oxygen to the anode compartment to avoid carbon cracking. Among them, external steam reforming is the most commonly adopted approach for industrial applications as it removes these complications by separating the reactions, enabling optimization of the steam reforming reactor system independent of the fuel cell system. This can then yield a full or partially reformed feed mixture of H_2 and CO for the SOFC stack. The external reforming can also enable the removal of higher hydrocarbons which may be present in the natural gas feed. These higher hydrocarbons contribute significantly toward carbon fouling in the anode. However, steam needs to be removed before entering the fuel cells as it lowers the electrical efficiency of the fuel cells by diluting the fuel.

Another option is the internal reforming of methane inside the anode of SOFCs. One of the potential advantages of internal reforming is the efficient use of the thermal energy and steam generated from the exothermic fuel oxidation reactions as the energy source for the endothermic steam reforming reactions and it being a simple system with no separate external reformer. The SOFC system operating on hydrocarbon fuels instead of hydrogen is more suitable for the residential power generation application than other types of fuel cells because it can reduce the complexity of fuel processing. However, due to the difference between the reaction rates of the endothermic methane reforming reaction and the exothermic electrochemical oxidation of hydrogen produced, cooling effects arise resulting in high temperature gradients inside the cell. The steep thermal gradients are potentially capable of mechanically damaging the cell stack, particularly in the hydrocarbon-rich inlet section of the fuel cell stack where the rates of these two reactions differ significantly. Nevertheless, the internal reforming activity of the anodes is important as it converts methane into more electrochemically active species of H_2 and CO. Unlike polymer electrolyte fuel cells, CO is also a fuel for SOFCs.

The operating conditions such as current load have a significant effect on carbon cracking and deposition on the Ni/YSZ cermet anodes in the hydrocarbon fuel. The water produced during the fuel oxidation reaction would reduce the carbon deposition via an in situ or a localized internal reforming reaction. This is particularly the case for anode-supported cells as the water produced at the interface would eliminate

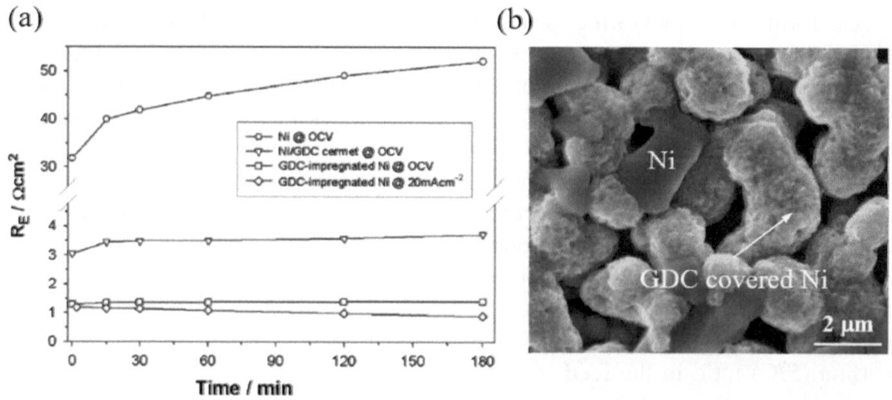

Fig. 10.15 **a** Comparison of electrode polarization resistance of Ni anode, Ni/GDC cermet anode and GDC-impregnated Ni anode in 97% CH_4/3% H_2O at 800 °C and **b** SEM image of a GDC-infiltrated Ni anode. Modified from Ref. [29] with permission from Elsevier, Copyright 2006

the carbon deposition at the interface region but may not be effective for the carbon cracking on the surface. Thus, carbon cracking is generally more pronounced at open circuit or low current density, as compared to that under a high current density. In general, the activity of Ni/YSZ cermets toward CO oxidation is much lower than for HOR.

Doped ceria such as GDC is a common additive to promote the electrocatalytic activity and stability of Ni-based cermet anodes and ceramic anodes for the oxidation of hydrocarbon fuels [29, 30]. Figure 10.15 shows an example of the polarization behavior of a Ni anode, Ni/GDC cermet anode and GDC-impregnated Ni anode in 97% CH_4/3% H_2O at 800 °C. The infiltrated GDC NPs loading is ~ 1.5 mg/cm^2. The significant increase in ASR or R_E for the reaction on a Ni anode is clearly due to the carbon cracking. For both the Ni/GDC cermet anode and the GDC-infiltrated Ni anode, R_E is stable at OCV. This indicates the high tolerance of the anodes for carbon cracking in 97% CH_4/3% H_2O. In this case, the steam/carbon ratio is 3/97 = 0.03, far lower than the threshold for the carbon deposition on a Ni phase. The specific promotion effect of doped ceria is most likely related to its capability to store and release oxygen, which reacts with methane to produce H_2 and CO via the redox reaction:

$$CH_4 + O^{x}_{O,\,\mathrm{GDC}} \rightarrow 2H_2 + CO + V^{\bullet\bullet}_{O,\,\mathrm{GDC}} + 2e^- \tag{10.29}$$

The above reaction is most effective on nano-sized GDC particles with high specific surface area. This reaction is also called the dry reforming process. In the dry reforming process, we need to add CO_2 to the methane stream to react with reduced ceria to replenish the lattice oxygen:

$$CO_2 + V^{\bullet\bullet}_{O,\,\mathrm{GDC}} \rightarrow CO + O^{x}_{O,\,\mathrm{GDC}} \tag{10.30}$$

However, under SOFC operation conditions, the lattice oxygen in ceria can be quickly replenished by the oxygen ions transported from the cathode side through the YSZ electrolyte:

$$O^{x}_{O,\,\mathrm{YSZ}} + V^{\bullet\bullet}_{O,\,\mathrm{GDC}} \rightarrow O^{x}_{O,\,\mathrm{GDC}} + V^{\bullet\bullet}_{O,\,\mathrm{YSZ}} \tag{10.31}$$

The reforming products, H_2 and CO, are removed from the system by the electrochemical oxidation reaction on Ni-based anodes and ceramic oxide anodes.

The promotion effect of infiltrated GDC NPs on the carbon tolerance of Ni-based anodes is similar to that on the ORR on the GDC-infiltrated LSM cathodes, see Fig. 10.10.

10.3.3 Redox Reaction

Metallic Ni in the Ni cermet anode may experience oxidation reaction under certain conditions such as imperfect sealing, failure of fuel supply or very high fuel utilization, and redox reactions in the cooling and heating thermal cycles during the lifetime of the cell. Ni redox reaction is accompanied by a large volume change (i.e., the volume increase from Ni to NiO is 25%), and this will cause a significant mechanical stress, resulting in crack formation or delamination of anode. In general, anode-supported thin electrolyte cells are more susceptible to redox cycles, as the thick Ni-based cermet supports would induce large mechanical stress and result in electrolyte cracking in fuel cell. Malzbender et al. studied the microstructural change of an anode-supported YSZ electrolyte cell during redox treatments [31]. In repeated re-oxidized stages, NiO particles were fragmented and the electrolyte fracture and microcracks in the anode substrate were clearly visible. Thus, the redox stability is an important property of Ni-based cermet anodes.

The main operation parameters affecting redox stability are temperature, humidity and initial anode microstructure. The temperature for the redox reactions has a strong impact on stability and performance of the cell. In general, reduction in the redox reaction temperature, i.e., thermal cycling, would increase the redox stability due to the reduced strain at lower temperatures for both oxidation and reduction reactions. A decrease in the operating and redox reaction temperature can result in a decreased degradation. The redox stability can be enhanced by operation control, the design of cell and stack geometry, the optimization of the anode microstructure and composition and rational materials selection such as functionally graded anode (FGA). The FGA can improve the redox stability by using less nickel content in the region close to the electrolyte to alleviate stress concentration and by applying a more dense diffusion barrier layer at the backside of the anode support to slow down re-oxidation kinetics.

Ni-free perovskite oxides such like $La_{0.75}Sr_{0.25}Cr_{0.5}Mn_{0.5}O_3$ (LSCM), $SrTiO_{3-\delta}$ based oxides are generally more robust, as compared to Ni-based cermet anodes.

10.4 Interface and Interface Reactions

10.4.1 LSM and YSZ (Doped Ceria) Systems

10.4.1.1 Interface Between an LSM Electrode with YSZ and GDC Electrolytes

The phase relationship between LSM and YSZ is one of the most important phase relations in SOFCs and has been extensively studied by theoretical thermodynamic calculations and experimental evaluation [32]. The phase relationship is important in order for us to understand the compatibility of different components or phases under elevated temperatures. Figure 10.16 gives an example of a ternary phase diagram of the (Zr, Y)O_2–La_2O_3–Mn_3O_4 system, obtained at 1400 °C based on the powder mixture experiments [33]. In the diagram, symbols " + " are the experimental data, (Zr,Y)O_2 represents the 3YSZ phase and La_yMnO_3 the perovskite phase due to its non-stoichiometry on the A-sites. Based on this (Zr, Y)O_2–La_2O_3–Mn_3O_4 phase diagram, we can identify that the *c*-ZrO_2 phase, rather than *t*-ZrO_2 phase, is in equilibrium with LSM perovskite at high temperatures. The phase diagram basically shows that the *t*-ZrO_2 (i.e., 3YSZ or TZ3Y materials) cannot be in equilibrium with LSM perovskite at high temperatures, while fluorite-type cubic zirconia solid solution phase, *c*-(Zr, Mn, La, Y)O_2, can be in equilibrium with the LSM perovskite

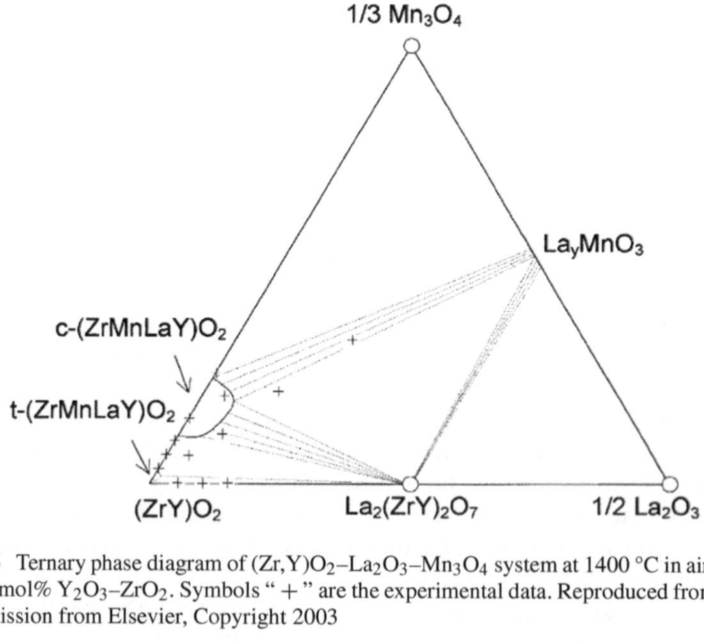

Fig. 10.16 Ternary phase diagram of (Zr,Y)O_2–La_2O_3–Mn_3O_4 system at 1400 °C in air. (Zr,Y)O_2 denotes 3 mol% Y_2O_3–ZrO_2. Symbols " + " are the experimental data. Reproduced from Ref. [33] with permission from Elsevier, Copyright 2003

phase. Thus, from the phase stability point of view, partially stabilized zirconia such as 3YSZ may not be an optimum choice as the electrolyte materials in SOFCs.

We will look at the interface between an LSM electrode and YSZ and GDC electrolytes [34]. Because of the solid LSM electrode/solid YSZ (GDC) electrolyte interface, the formation of intimate contact and interface can only be achieved by sintering at high temperatures, e.g., ~ 1150 °C. Sintering at 1150 °C leads to the formation of convex contact rings on the YSZ and GDC electrolyte surface. Figure 10.17 shows an example of a contact convex ring and interface in the case of LSM/YSZ. Similar convex rings and interface formation also occur in the case of LSM/GDC. This typical contact ring has a dimension of ~ 18 nm high and 600 nm in diameter, but the size of the contact rings on YSZ or GDC electrolyte is actually closely related to the particle size of LSM. There is a significant cation interdiffusion, particularly Mn^{2+} into the YSZ electrolyte in the convex contact ring region, and Mn, La, and Y cation diffusion along the grain boundary of the YSZ electrolyte also occurs. Very

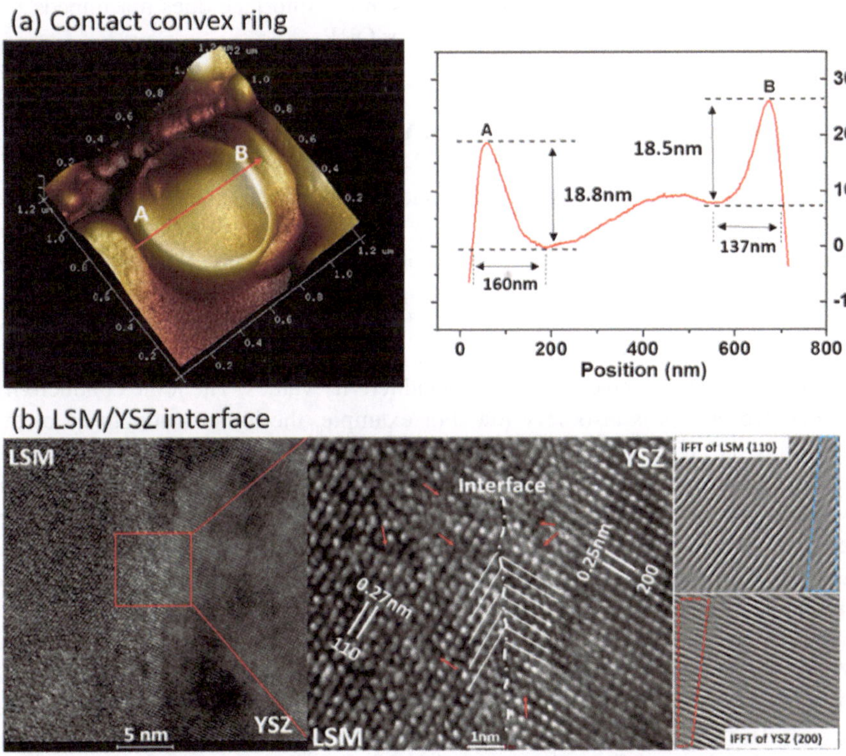

Fig. 10.17 **a** Contact convex ring on a YSZ electrolyte surface after removal of LSM cathode coating by acid treatment using Atomic Force Microscopy and **b** LSM/YSZ interface by HRTEM. The lattice planes of $\{110\}_M$ and $\{200\}_{YSZ}$ in an area adjacent to the interface are reconstructed by inverse fast Fourier transform (IFFT). The LSM electrode was removed by adhesive tape. The dislocations and defects around the interface regions are indicated by the arrows

different to the LSM/YSZ interface, no cation interdiffusion is observed between the LSM electrode and GDC electrolyte. The contact convex ring formation between the LSM particles and YSZ or GDC electrolyte is very similar to the neck formation phenomena in conventional ceramic sintering processes, which would diminish the curvature of the interface as well as the surface free energy. Therefore, cation interdiffusion is not an essential requirement for the formation of heterointerfaces between LSM electrodes and oxygen ion conducting electrolytes such as YSZ and GDC.

The formation of a semi-coherent interface, i.e., an atom-by-atom matching only exists locally in regions along the interface, but the orientation relationship and lattice mismatch factor between electrode and electrolyte crystallographic planes at the LSM/YSZ and LSM/GDC interfaces vary significantly. Lattice disorientation and misfit of the crystallographic planes occur within a narrow region of 0.10–0.5 nm wide with no amorphous or solid solution formation. The electrochemical performance of LSM electrodes on the YSZ or GDC electrolyte indicates that the disorientation and dislocation of the crystallographic planes at the interface does not impede the oxygen diffusion and incorporation process for ORR.

10.4.1.2 Interaction Between LSM and YSZ

The reactivity between the LSM and YSZ electrolyte varies with the stoichiometric composition of LSM, the La/Sr ratio at the A-site and the temperature and atmosphere of the heat treatments. The main products of the interaction between LSM and YSZ are the resistive pyrochlore phases, $La_2Zr_2O_7$ and $SrZrO_3$. The electrical conductivity of the $La_2Zr_2O_7$ layer is ~ $10^{-4} - 10^{-5}$ S/cm at 1000 °C, much lower than that of the LSM and YSZ. Strontium or praseodymium zirconates, $SrZrO_3$, Sr_2ZrO_4 and $Pr_2Zr_2O_7$ have similar low conductivity values. The ionic conductivity of pyrochlore phases is also very low. For example, the ionic conductivity of the $SrZrO_3$ phase is 1.87×10^{-6} S/cm at 800 °C [35], more than four orders of magnitude lower than GDC and YSZ at the same temperature. In addition, the electrochemical activity of the resistive pyrochlore phases for O_2 reduction is negligible. Therefore, in general the formation of the resistive pyrochlore phases at the Sr-doped lanthanum or praseodymium manganite electrode/YSZ electrolyte interface is detrimental to the electrochemical activity and performance of the electrode.

The interaction and interface phase formation between an LSM electrode and the YSZ electrolyte mainly depend on the stoichiometric composition of LSM and the sintering temperature. The driving force for the high activity of LSM with YSZ is the excess lanthanum oxide within the perovskite, which reacts immediately with YSZ to form a resistive $La_2Zr_2O_7$ phase at the interface. A-site overstoichiometric composition will lead to the excess La. However, excess La can also occur for the stoichiometric LSM due to the inhomogeneous phase formation during the synthesis process [36]. Figure 10.18 shows an example of a YSZ electrolyte surface in contact with a stoichiometric $La_{0.80}Sr_{0.2}MnO_3$ electrode after sintering at 1150 °C. The

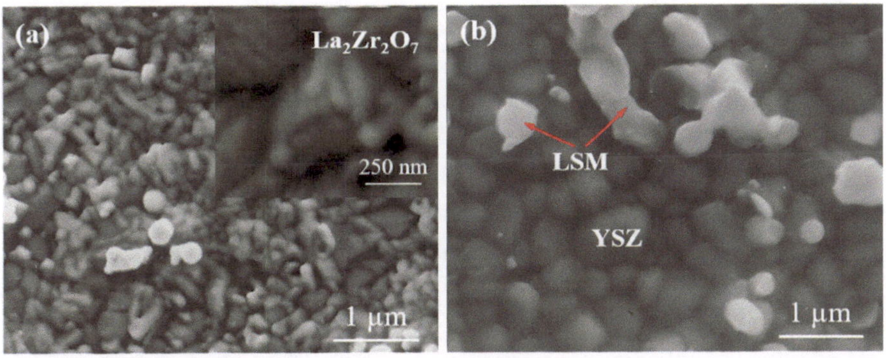

Fig. 10.18 SEM micrographs of a YSZ electrolyte surface in contact with **a** $La_{0.80}Sr_{0.2}MnO_3$ and **b** $(La_{0.80}Sr_{0.2})_{0.9}MnO_3$ after removal of the LSM coating by stick tape. The LSM electrodes were sintered at 1150 °C in air for 2 h. Modified from Ref. [36] with permission from Springer Nature, Copyright 2000

LSM electrode coating is removed by sticky tape. The surface of the YSZ electrolyte is almost fully covered by column-type particles, which are identified as the $La_2Zr_2O_7$, but no such phase formation is observed between the A-site non-stoichiometric $(La_{0.80}Sr_{0.2})_{0.9}MnO_3$ and YSZ sintered at the same temperature. The formation of $La_2Zr_2O_7$ indicates that the excess La in stoichiometric $La_{0.80}Sr_{0.2}MnO_3$ electrode is most likely due to the poor compositional homogeneity of the LSM electrode powder. In this case, the LSM powder is prepared by the solid-state reaction (SSR) method. In general, A-site non-stoichiometry or slightly lanthanum deficient manganite, $(La_{0.80}Sr_{0.2})_{1-x}MnO_3$ (with $x = 0.05$ to 0.1) is recommended in order to suppress La excess and to avoid the potential interfacial reaction.

The excess lanthanum oxide can also occur by the unidirectional diffusion of Mn into YSZ since manganese has a high solubility in YSZ, ranging from ~ 11.4% at 1300 °C to 5.1% at 1000 °C. The manganese ions in YSZ mainly exist as Mn^{2+} ions in the YSZ lattice. The diffusion of Mn from the B-site of LSM into the YSZ surface leads to the excess La and Sr in the A-sites, which will react with YSZ to form resistive pyrochlore phases $La_2Zr_2O_7$ and $SrZrO_3$ at the interface. In the case of Sr-doped praseodymium manganite (PSM), the formed pyrochlore phase is $Pr_2Zr_2O_7$. The reactivity between zirconia and manganites is also related to the yttria-stabilized zirconia phase. The tetragonal stabilized zirconia (3 mol% Y_2O_3–ZrO_2, 3YSZ or TZ3Y) is more active with manganite as compared to fully stabilized zirconia (8 mol% Y_2O_3–ZrO_2, yttria fully stabilized zirconia or YSZ) in the formation of the pyrochlore phase [33, 37]. Figure 10.19 shows the formation of pyrochlore phases at the interface of a LSM/TZ3Y and PSM/TZ3Y with stoichiometric, $(Pr_{0.8}Sr_{0.2})_{1.0}MnO_3$, and non-stoichiometric, $(Pr_{0.8}Sr_{0.2})_{0.9}MnO_3$, compositions after heat treatment at high temperatures. In the case of a LSM/TZ3Y sample, two distinct reaction layers are formed between the LSM coating and TZ3Y electrolyte, marked as "L1" and "L2" in the figure. The "L1" reaction layer is the formation of a $La_2Zr_2O_7$ phase and the "L2" layer is the cubic zirconia solid solution.

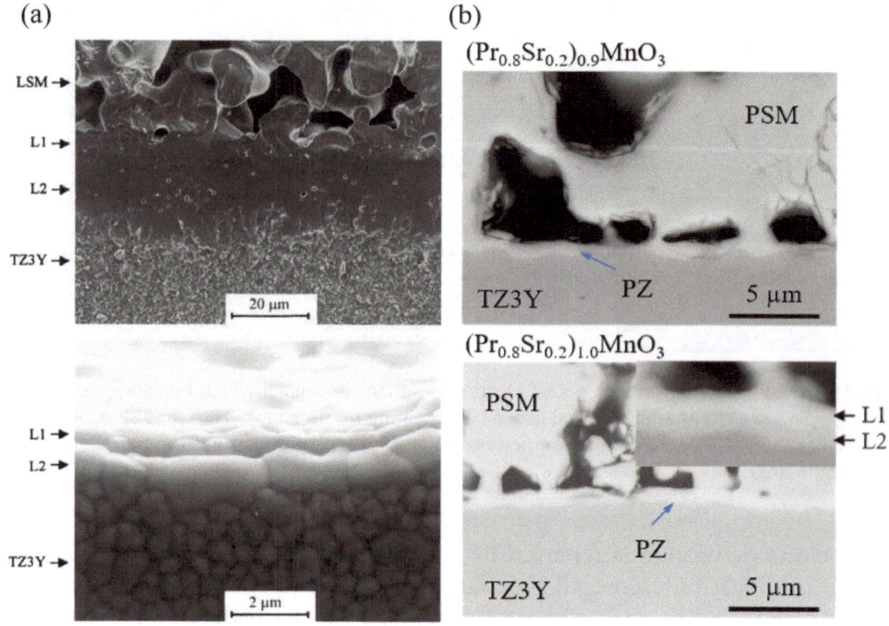

Fig. 10.19 **a** SEM micrographs of fractured and polished and thermally etched cross sections of a $La_{0.8}Sr_{0.2}MnO_3$ (LSM)/TZ3Y specimens after heat treatment at 1400 °C for 4 h. **b** The backscattered electron micrographs of the polished cross sections of $(Pr_{0.8}Sr_{0.2})_{0.9}MnO_3$/TZ3Y and $(Pr_{0.8}Sr_{0.2})_{1.0}MnO_3$/TZ3Y samples after heat treatment at 1200 °C for 168 h. The inset shows the two-layer formation at the interface. Modified from Ref. [37] with permission of Royal Society of Chemistry, Copyright 1998

The formation of the fluorite-type zirconia phase is due to the dissolution of Mn ions into TZ3Y, causing the phase transformation from tetragonal zirconia to cubic zirconia. In the case of PSM/TZ3Y samples, the chemical interaction between PSM and TZ3Y leads to the formation of $Pr_2Zr_2O_7$ (PZ as shown in the figure). The use of A-site non-stoichiometry will not deter the pyrochlore phase formation but will delay the reaction. This is indicated by the thinner pyrochlore layer formed on the interface with A-site non-stoichiometric PSM composition. Similarly to LSM/TZ3Y, two distinct layers are also formed.

The activity of manganite is affected by the A-site cations. For example, substitution of Ca for La in $LaMnO_3$ suppresses the manganese migration, thus reducing the reactivity with YSZ.

10.4.2 *Interface and Interaction Between LSCF and YSZ*

$La_{1-x}Sr_xMO_3$ (M=Co, Ni and Fe) perovskites are relatively unstable compared to their manganite counterparts. They readily react with zirconia electrolytes, leading

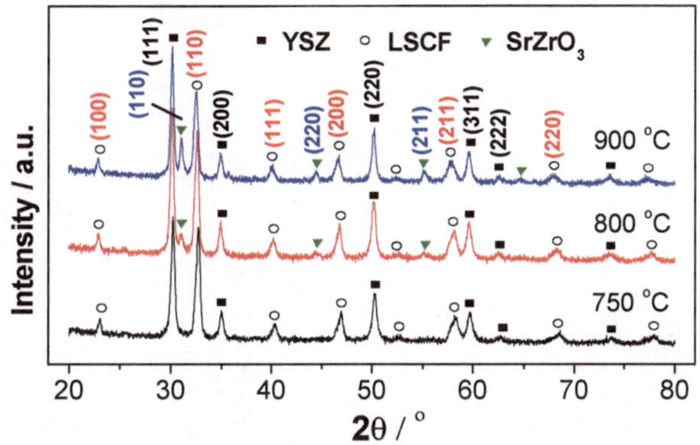

Fig. 10.20 XRD patterns of $La_{0.6}Sr_{0.4}Co_{0.2}Fe_{0.8}O_{3-\delta}$ (LSCF) and YSZ oxide couples calcined at different temperatures. Reproduced from Ref. [38] with permission from IOP Publishing, Copyright 2018

to the formation of secondary pyrochlore phases at temperatures as low as 800 °C in air. Figure 10.20 shows the XRD result of the LSCF/YSZ oxide couples calcined at different temperatures in air [38]. The oxide couple sintered at 750 °C only shows the presence of LSCF and YSZ phases. However, as the calcination temperature increases to 800 °C or above, diffraction peaks at 30.8°, 44.7°, and 54.9° are observed, which can be assigned to the $SrZrO_3$ (110), (200), and (211) planes, respectively. This indicates that low limit of the chemical reaction temperature between LSCF and YSZ is 800 °C.

However, the interaction between cobaltite-based perovskite electrodes such as LSCF and YSZ electrolyte is not as straightforward as we would normally expect from the high activity of cobaltite perovskite materials. In fact, like in any other perovskite oxide materials, there is a complex relationship between the interface formation, the surface segregation and interaction between the LSCF electrode and the YSZ electrolyte. The best technique to study such complex relationship between the highly active components is the use of directly assembled electrodes without the prior high-temperature sintering steps to bypass the potential chemical reaction (the direct assembly technique is given in Chap. 11). Figure 10.21 shows an example of the interface formation and Sr segregation of a directly assembled LSCF electrode on a YSZ electrolyte as a function of polarization at 750 °C and 1000 mA/cm^2 for 1 and 12 h [38]. The high resolution TEM (HRTEM) images are taken along the $[001]_{LSCF}$ zone axis. After polarization at 750 °C and 1000 mA/cm^2 for 1 h, LSCF electrode particles are in an intimate contact with the YSZ electrolyte surface with no voids or cracks at the interface. The atomic geometry of the interface is characterized by a high level of periodicity and symmetry. The diffractogram of the interface is presented by the fast Fourier transform (FFT) micrograph, showing the overlapping of the cubic perovskite structure of LSCF and the cubic fluoride structure of YSZ. $\{110\}_{LSCF}$ and $\{111\}_{YSZ}$ lattice planes are identified with an interplanar spacing of 0.27 nm and 0.29 nm, respectively, and meet at the interface with no Sr segregation and no other secondary phases (Fig. 10.21a). The orientation relationship between $\{110\}_{LSCF}$ and $\{111\}_{YSZ}$ planes is 48.8°, and the lattice mismatch factor (f) between $\{110\}_{SCF}$ and $\{111\}_{YSZ}$ planes is 6.9% in this case. However, we should point out

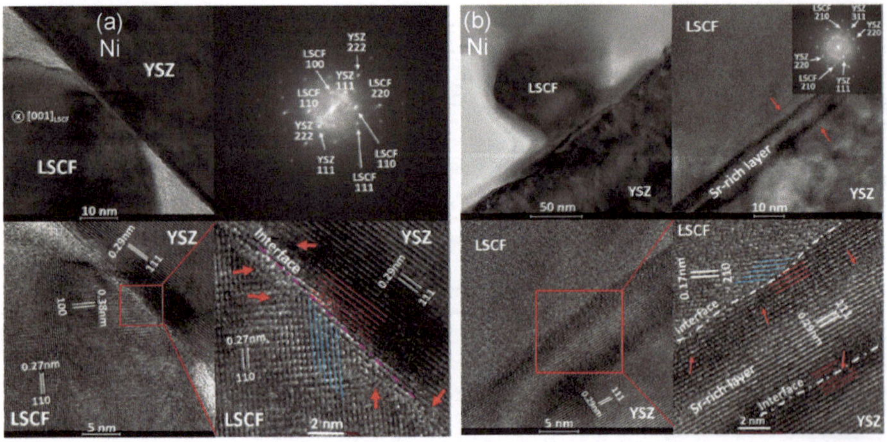

Fig. 10.21 HRTEM and the corresponding FFT micrograph of directly assembled $La_{0.6}Sr_{0.4}Co_{0.2}Fe_{0.8}O_{3-\delta}$ (LSCF)/YSZ interface after polarization at 750 °C and 1000 mA/cm^2 **a** for 1 h observed along $[001]_{LSCF}$ zone axis and **b** for 12 h observed along $[1\bar{1}\bar{2}]_{YSZ}$ zone axis. The inset presents the FFT diffractogram over the three-phase interface. The lattice distortion is indicated by red arrows. Reproduced from Ref [38] with permission from IOP Publishing, Copyright 2018

that the orientation relationship between LSCF and YSZ lattice planes would be different at different contact regions due to the polycrystalline nature of both phases.

Due to the crystallographic structure mismatch of the two heterogeneous phases, misfit dislocations occur (indicated by red arrows in Fig. 10.21a), in order to relax the mismatch strain and to accommodate the lattice misfit at the interface. Nevertheless, such misfit dislocations are limited within several nano-meters around the interface, without propagating into the electrode or electrolyte bulk. This indicates excellent interface formation between the LSCF cathode and the YSZ electrolyte, very similar to that between the LSM electrode and the YSZ electrolyte. Clearly such distortion and misfit at the LSCF/YSZ heterointerface would not impede the oxygen transfer process of ORR.

After the LSCF/YSZ cell was polarized at 750 °C and 1000 mA/cm^2 for a longer period of 12 h, the interface changed significantly; see Fig. 10.21b. A thin layer at the electrode/electrolyte interface is observed. This thin layer primarily contains Sr but no La. The existence of Sr-rich layer (SRL) at the interface and on the YSZ electrolyte surface is clearly due to the significant Sr segregation and diffusion under the influence of cathodic polarization, as discussed in Sect. 10.1.2.1. The deposition of Sr on the YSZ electrolyte surface beyond the interface contact region indicates the high mobility and volatility of segregated Sr species. The SRL is sandwiched between LSCF and YSZ, forming distinct LSCF/SRL and SRL/YSZ interfaces. The abrupt change at the interface regions is most likely due to the orientation change at the interfaces during the deposition of the SRL and the strain effect resulting from the lattice mismatch of the heterogeneous phases, i.e., LSCF/SRL and SRL/YSZ. SRL exhibits symmetry and crystallography similar to the YSZ electrolyte, characterized by the $\{111\}_{SRL}$ lattice planes with an interplanar spacing of 0.29 nm. The diffractogram at the SRL/YSZ interface shows a single YSZ cubic phase pattern, along $[1\bar{1}\bar{2}]_{YSZ}$ zone axis. This indicates that the SRL grows along the crystallographic orientation of the YSZ electrolyte. This is supported by the observed orientation relationship, $\theta_{\{111\}SRL/\{111\}YSZ} = 0°$, while in the case of SRL/LSCF interface, the orientation relationship is 44.3°. With further polarization, the thickness of the reaction layer increases and an orthorhombic $SrZrO_3$ secondary phase is formed with no particular orientation relationship with YSZ electrolyte. The excess Sr segregation leads to the disintegration of the LSCF perovskite structure, forming a Fe, La, Y-containing $SrZrO_3$ solid solution layer at the interface, as discussed in the Sect. 10.1.2.1.

The overall interaction between an LSCF cathode and the YSZ electrolyte under SOFC operation conditions can be presented in Fig. 10.22, summarizing the processes of interface formation, Sr segregation, and interfacial reaction of segregated Sr with YSZ to form $SrZrO_3$ at the interface. The surface segregation and subsequent reaction between the segregated species and catcher play a critical role in the interfacial reaction and subsequent disintegration of cobaltite-based electrodes. The surface segregated Sr species is more active with YSZ, resulting in the formation of $SrZrO_3$ insulation phase at reduced temperatures.

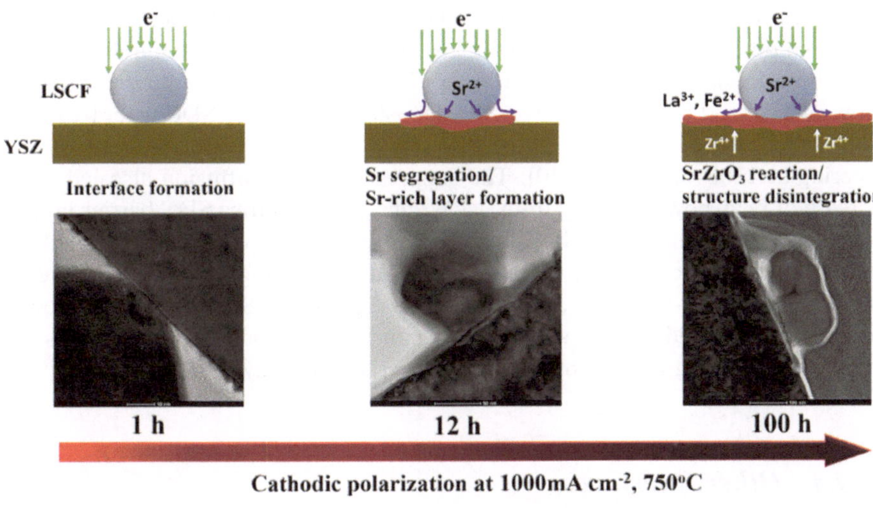

Fig. 10.22 Scheme of the effect of polarization time at 1000 mA/cm^2 and 750 °C on the interface formation, Sr segregation and diffusion, formation of Sr-rich layer and (Fe, La, Y)-containing $SrZrO_3$ reaction layer at LSCF/YSZ interface region. Reproduced from Ref. [38] with permission from IOP Publishing, Copyright 2018

10.4.3 LSCF/GDC/YSZ Interface

LSCF is ready to react with YSZ electrolyte. Therefore, to use LSCF electrodes, a doped ceria such as GDC or SDC barrier layer between the LSCF electrode and YSZ electrolyte is commonly used, so as to avoid the chemical reaction at the LSCF/YSZ interfaces. However, numerous reports have shown that even in the presence of the doped ceria barrier layer, the $SrZrO_3$ phase can be formed at the YSZ interface after the electrode sintering processes [39].

The diffusion of Sr across the GDC layer is mainly due to the high reactivity of segregated Sr species. The segregated Sr species are volatile and mobile under the SOFC operating conditions. Therefore, the volatile Sr species can diffuse across the pores or grain boundaries of a GDC barrier layer via gas- or solid-phase diffusion to reach the YSZ surface, forming $SrZrO_3$. Lu et al.[40] investigated the effects of sintering temperature of LSCF electrodes and observed the formation of a visible $SrZrO_3$ layer at the SDC/YSZ interfaces at a sintering temperature of 1150 °C and above. At a lower sintering temperature of 1050°s°C, EDS analysis revealed the presence of a significant amount of Sr at the SDC/YSZ interfaces. However, the actual reaction temperature for the significant Sr diffusion through the ceria barrier layer is very much dependent on the surface chemistry of LSCF and the physical states of the ceria barrier layer.

As the pores of the doped ceria barrier layer facilitates the diffusion of volatile Sr, dense and pore-free doped ceria barrier layers are adopted to block the Sr diffusion. It has been reported that in contrast to the significant Sr diffusion to the interfaces between YSZ and porous SDC layer, the deposition of a dense SDC layer by pulsed laser deposition (PLD) led to no detection of Sr at YSZ electrolyte surfaces after sintering at 1200 °C for 2 h [40]. The suppression of Sr diffusion also led to the increase of peak power density from 0.7 W/cm^2 for the porous SDC barrier layer to 1.2 W/cm^2 for the densified SDC layer. However, there is evidence that the dense GDC barrier layer cannot eliminate the Sr diffusion during an extended sintering step and under long-term operating conditions. Segregated Sr diffusion can proceed through grain boundary and surface diffusion paths in the GDC layer. This is a concern for the long-term stability of doped ceria barrier layer-based SOFCs.

10.4.4 Other Interfaces

In addition to the YSZ electrolyte, the interaction and diffusion between a SOFC cathode and other electrolytes are also reported. Huang et al. [41] studied the chemical reactions between the $La_{0.84}Sr_{0.16}MnO_3$ (LSM) and $La_{0.5}Sr_{0.5}CoO_3$ (LSC) cathodes, and $La_{0.9}Sr_{0.1}Ga_{0.8}Mg_{0.2}O_3$ (LSGM) electrolyte. Significant interdiffusion of Co into LSGM electrolyte and Ga into LSC was found at the LSC/LSGM interface at relatively low fabrication temperatures (e.g., 1050 °C for 2 h). In contrast, only small interdiffusion of Mn into LSGM and Ga into LSM was detected at the LSM/LSGM interface even though it was fired at a much higher temperature of 1470 °C. The Mn diffusion into the LSGM electrolyte is less pronounced. The substitution of Co for Ga in the LSGM electrolyte increases the ionic conductivity but also introduces the mixed conducting behavior. The interdiffusion of cations between LSCF electrode and LSGM electrolyte can be prevented by using a doped ceria barrier layer.

The performance of cells based on apatite electrolytes appears to be unsatisfactory; most likely due to the poor electrocatalytic activity of the electrode. For example, in the case of a $(La_{0.8}Sr_{0.2})_{0.9}MnO_3$ (LSM) electrode on $La_{9.5}Si_6O_{26.25}$ (LSO) apatite electrolyte, the activation effect of cathodic polarization on the electrochemical activities of the LSM cathode for ORR is significantly lower, as compared to that on the LSM/YSZ interface [42]. The migration and diffusion process of oxygen ions from TPB into an LSO apatite electrolyte is also much slower as compared to that in the case of YSZ electrolyte. The relatively low performance is most likely due to spreading and surface diffusion of silica from the apatite-type electrolyte.

10.5 Reaction with Contaminants

In SOFCs, the electrochemical activity of electrodes not only depends on the nature of the materials themselves and interactions with the electrolyte, but also is critically affected by the deposition and poisoning of contaminants in the cell and stack. There are various contaminants in high temperature SOFCs systems, either from the volatile species of the components or from the reactant stream or environment. The most common contaminants on the cathode and anode sides are listed below.

On the cathode sides:

- Volatile chromium species from chromia-forming interconnect and manifold;
- Volatile boron species from boron-containing sealing glasses;
- Sulfur from the air stream;
- CO_2 from the air stream;
- Water vapor from the air stream.

On the anode sides:

- Carbon deposition and cracking from the use of hydrocarbon fuels;
- Sulfur from the hydrocarbon fuels;
- Impurities in coal gasification syngas.

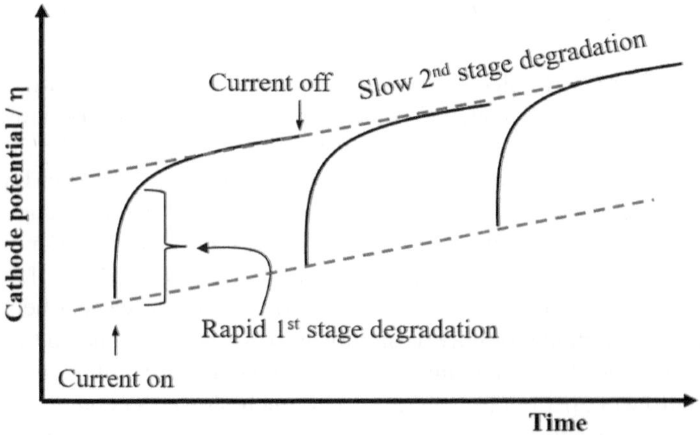

Fig. 10.23 Typical Cr poisoning effect on the polarization potential or overpotentials (η) of the ORR on cathode of SOFCs under cathodic polarization conditions. Gaseous Cr species exist in the system due to the use of Fe–Cr metallic interconnect

10.5.1 Reactions Between Cathode and Contaminants

10.5.1.1 Chromium

10.5.1.1.1 Phenomena of Chromium Deposition and Poisoning

Metallic interconnect such as chromia-forming alloys (e.g., stainless steel) is commonly used in intermediate temperature SOFCs or IT-SOFCs. However, Cr–Fe-based metallic interconnect produces volatile Cr-containing species at high temperatures in oxidizing atmospheres. The most common vapor species over Cr_2O_3 scale are $CrO_2(OH)_2$ in wet air and CrO_3 in dry air (see Chap. 9 for details). Without effective protective coatings, the gaseous chromium species can deposit and poison the electrochemical activity of SOFC cathodes and result in the degradation of SOFC performance. Deposition and poisoning of volatile chromium species is one of the most significant degradation mechanisms in SOFCs [43].

The Cr poisoning of ORR on SOFC cathodes is generally characterized by a two-stage degradation behavior, see Fig. 10.23. The electrode polarization potential or overpotential, η, shows a sharp increase with the application of a current initially, followed by a much slower increase in the polarization potential. The initial increase in η is largely reversible after switching off and switching on the polarization current, but the degradation occurring at the slow second stage cannot be reversed. The significant increase in the polarization potential and irreversible degradation behavior is a clear indication of the chromium deposition and poisoning on the activity of the cathode. The magnitude of the degradation is related to the electrocatalytic activity (i.e., electronic and ionic conductivity, surface chemistry, segregation) of the cathode material and the environment (i.e., the humidity, temperature, polarization).

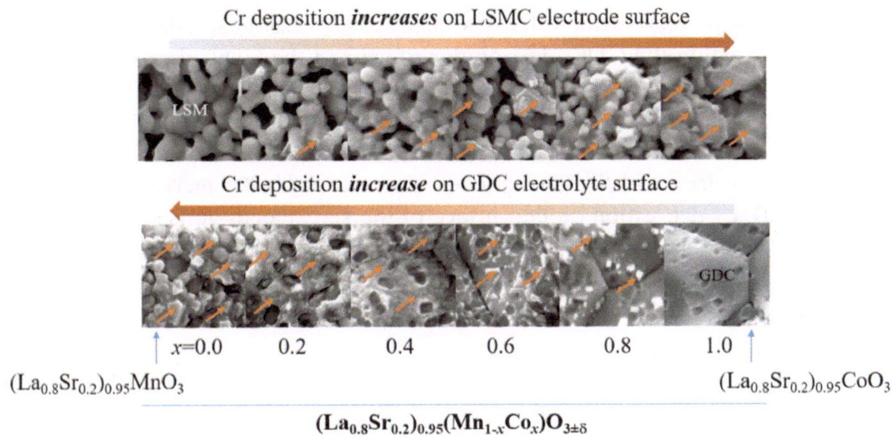

Fig. 10.24 SEM micrographs of chromium deposition on the surface of an LSMC electrode and the surface of GDC electrolyte after current passage in the presence of a Fe–Cr alloy metallic interconnect at 200 mA/cm^2 and 900 °C for 20 h. To examine the GDC electrolyte surface, the LSMC electrode coating was removed by acid treatment. Red arrows indicate the Cr deposits. Modified from Ref. [45] with permission from Elsevier, Copyright 2011

The largely reversible polarization degradation at the rapid 1st stage is related to the poisoning effect of adsorbed Cr species on the electrode process associated with surface processes such as adsorption, dissociation, and diffusion, while the irreversible loss of the activity at the slow 2nd stage degradation is more related to the deposition of solid Cr species either on the electrode surface and/or at the TPB areas, blocking the active sites and gas diffusion pores for ORR. Thus, the initial increase in η is generally much smaller for the ORR on MIEC-type electrodes such as LSCF as compared to that on electronic conducting electrodes like LSM and Pt [44].

Cr poisoning is characterized by Cr_2O_3 deposition and formation of Cr-containing solid phases in cathodes. Cr deposition and poisoning depend on the type and composition of the cathode as well as a number of operation parameters, such as the type and composition of cathode materials, polarization, air flow rate, and humidity in the air stream. Humidity in the air has a significant detrimental effect on the Cr deposition, as predicted from the thermodynamic equilibrium of chromium species. Air flow will affect the amount of Cr deposits but not the mechanism of Cr deposition. Among them, the nature or the type and composition of the cathode material are most important. Figure 10.24 shows one of the classic Cr deposition studies on $(La_{0.8}Sr_{0.2})_{0.95}(Mn_{1-x}Co_x)O_{3\pm\delta}$ (LSMC) with $x = 0$, 0.2, 0.4, 0.6, 0.8, and 1.0 [45]. With $x = 0$, the electrode material is a dominantly electronic conducting electrode, $(La_{0.8}Sr_{0.2})_{0.95}MnO_3$ (LSM) and with $x = 1.0$, the electrode material is an excellent mixed ionic and electronic conducting electrode, $(La_{0.8}Sr_{0.2})_{0.95}CoO_{3-\delta}$ (LSC). LSMC changes from a pure electrically conducting material to an excellent MIEC material with an increase in the value of x. In the case of LSM cathode, Cr is predominantly deposited on the GDC electrolyte surface and the GDC electrolyte surface is almost completely covered by the Cr deposits, Cr_2O_3 and $(Cr,Mn)_3O_4$ spinel. On the other hand, very little or no Cr deposition occurs on the LSM electrode surface. Cr deposition starts to occur on the surface of LSMC with $x = 0.2$. The Cr deposition on the surface of LSMC electrodes increases with the increase in x value. With $x = 1.0$, Cr deposition on the surface of the LSC electrode is substantial and completely covers the surface, forming a dense layer of Cr_2O_3 and $SrCrO_4$ phases with a scheelite structure. The opposite is observed for the Cr deposition on the GDC electrolyte surface with a decrease in the x value in LSMC.

10.5.1.1.2 Nucleation Theory and Electrochemical Deposition Theory

There are two main theories proposed in the literature for the mechanisms of the deposition and poisoning of chromium on SOFC cathodes: *Electrochemical Deposition Theory* and *Nucleation Theory*. Vaporization of chromium oxide scale, Cr_2O_3, from the interconnect surface forms CrO_3 or $CrO_2(OH)_2$ as the major gaseous species with chromium in the 6+ oxidation state. In the electrochemical deposition mechanism, the deposition of chromium is considered to be controlled by the electrochemical reduction of gaseous Cr species CrO_3 and $Cr(OH)_2O_2$ to solid-phase Cr_2O_3. The electrochemical reduction reaction of high valence chromium species is thermodynamically compatible with the ORR and is thus in direct competition with the ORR; the processes are written as follows [46]:

$$2CrO_{3(g)} + 6e^- \rightarrow Cr_2O_{3(s)} + 3O^= \tag{10.32}$$

$$2Cr(OH)_2O_{2(g)} + 6e^- \rightarrow Cr_2O_{3(s)} + 2H_2O_{(g)} + 3O^= \tag{10.33}$$

The formed Cr_2O_3 would react with cathodes such as LSM or LSCF to form $(Cr,Mn)_3O_4$ or $SrCrO_4$ phases. For example, the reaction between Cr_2O_3 and LSM may be written as follows:

$$\begin{aligned} &1.5Cr_2O_{3(s)} + 3(La, Sr)MnO_3 \rightarrow 3(La, Sr)\left(Mn_yCr_{1-y}\right)O_3 \\ &\quad + \left(Cr_yMn_{1-y}\right)O_4 + 0.25O_2 \end{aligned} \tag{10.34}$$

However, there are significant fundamental problems with the electrochemical deposition theory. Even in the case of LSM electrodes, the electrochemical deposition theory could not give satisfactory answers to the early stage behavior of Cr deposition observed on LSM electrodes [47]. Systematic studies of the chromium deposition and poisoning processes on a wide range of electrode materials including LSM, LSM/YSZ, LSCF, BSCF and (La,Ba,Sr)(Co,Fe)O_3, $LaNiO_3$, (La,Sr)(Mn,Co)O_3 and Pt show that there is no intrinsic relationship between the Cr deposition and the oxygen activity or the O_2 reduction at the electrode/electrolyte interface [43]. To explain the Cr deposition on SOFC cathodes, a chemical deposition mechanism based on nucleation theory is proposed [12]. In this nucleation theory, Cr deposition is essentially initiated by the nucleation reaction between the gaseous Cr species and nucleation agents in the electrode/electrolyte cells, resulting in the formation of solid nuclei at the electrode surface, electrode/electrolyte interface or electrolyte surface. As in any gas–solid reaction, the presence of nuclei will significantly accelerate the further deposition and reaction of gaseous Cr species, leading to the formation and grain growth of solid Cr_2O_3 and other Cr deposits. The deposition process induced by nucleation agents is essentially chemical in nature and can be expresses in general forms as follows:

$$N + \mathrm{CrO}_{3(g)} \rightarrow \mathrm{Cr} - N - \mathrm{O}_{x(\mathrm{nuclei})} \tag{10.35}$$

$$\mathrm{Cr} - N - \mathrm{O}_{x(\mathrm{nuclei})} + \mathrm{CrO}_{3(g)} \rightarrow \mathrm{Cr}_2\mathrm{O}_{3(s)} \tag{10.36}$$

$$\mathrm{Cr} - N - \mathrm{O}_{x(\mathrm{nuclei})} + \mathrm{CrO}_{3(g)} + N \rightarrow \mathrm{Cr},\, N - \text{containing deposits}_{(s)} \tag{10.37}$$

In the process, N is a nucleation agent and Cr–N–O_x is nucleus. The deposited Cr_2O_3 particles can also be nuclei for the deposition and grain growth of Cr,N-containing deposits. The presence or formation of nucleation agents mainly depends on the nature of the electrode materials and is closely related to the surface segregation phenomena as discussed in this chapter. On the other hand, the structure, composition, and the place for the deposition and growth of the Cr,N-containing deposits are very much related to the nature and mobility of the nuclei. Here are examples of the Cr deposition on the most common cathodes of SOFCs.

LSM electrode—Under cathodic polarization (i.e., SOFC operation conditions), Mn in the LSM lattice can be electrochemically reduced to Mn^{2+} ions due to the charge compensation mechanism with the concomitant formation of oxygen vacancies, see Chap. 9. On the other hand, Sr segregation is retarded under cathodic polarization. This leads to segregated Mn species rather than Sr species under cathodic polarization, which function as nucleation agents for the Cr deposition. The interaction between the segregated Mn species, e.g., Mn^{2+} ions, and gaseous CrO_3 species leads to the formation of Cr–Mn–O nuclei. The formation of stable Cr–Mn–O nuclei phases in turn accelerates the crystallization and grain growth of Cr_2O_3 and $(Cr,Mn)_3O_4$ solid phases. The segregation of Mn species occurs predominantly at the TPB region. This is due to the fact that ORR mainly occurs at TPB. Thus the deposition of Cr_2O_3 and $(Cr,Mn)_3O_4$ solid phases preferentially occurs at the electrode/YSZ (or GDC) electrolyte interface areas. The deposited solid Cr phases block the reaction sites at the TPB and deteriorate the electrochemical activity of LSM electrodes for ORR.

$$\mathrm{Cr} - \mathrm{Mn} - \mathrm{O}_{x(\mathrm{nuclei})} + \mathrm{CrO}_{3(g)} + \mathrm{Mn}^{2+} \rightarrow (\mathrm{Cr},\, \mathrm{Mn})_3\mathrm{O}_{4(s)} \tag{10.38}$$

As the Mn segregation generally occurs under the influence of cathodic polarization, the deposition of $(Cr,Mn)_3O_4$ and Cr_2O_3 usually takes place under polarization, rather than under open circuit conditions.

LSCF electrode—Sr segregation to the surface of LSCF electrodes is very facile under both open circuit and cathodic polarization conditions. The segregated Sr species, existing as SrO, will act as nucleation agents and reacts with gaseous Cr species, forming Cr–Sr–O nuclei. Thus, due to the existence of nuclei on the electrode surface, the deposition of Cr species will occur preferentially on the surface of the LSCF electrode region rather than at the TPB areas, forming $SrCrO_4$ as the main products.

$$\mathrm{Cr-Sr-O}_{x(\mathrm{nuclei})} + \mathrm{CrO}_{3(\mathrm{g})} + \mathrm{SrO} \rightarrow \mathrm{SrCrO}_{4(\mathrm{s})} \tag{10.39}$$

The Cr deposition process is accelerated under cathodic polarization due to the increased Sr segregation. The degradation of LSCF cathodes by Cr poisoning is generally slower as compared to that on LSM due to the fact that the ORR on MIEC electrodes like LSCF is not dominated by the surface adsorption and diffusion of oxygen. However, the formation of Cr_2O_3 and Sr_2CrO_4 deposits blocks on the pores on the electrode surface and deteriorates the electrode electrochemical activity. Also, the formation of $SrCrO_4$ phase accelerates the Sr segregation and LSCF gradually becomes Sr-deficient. Sr deficiency significantly deteriorates the activity of LSCF for ORR due to the reduced oxygen ionic conductivity property of the material.

Cr deposition on the LSCF electrode surface is also selective. As shown in Fig. 10.1, both Co and Sr are segregated out of the LSCF lattice, forming SrO and CoO_x particles. However, only the deposition of $SrCrO_4$ and not $CoCr_2O_4$ is observed [48]. This suggests that the Cr deposition on LSCF preferentially takes place on the segregated SrO rather than on the segregated CoO_x.

BSCF electrode—$Ba_{0.5}Sr_{0.5}Co_{0.8}Fe_{0.2}O_{3-\delta}$ (BSCF) is an interesting material for Cr deposition [49]. In the case of BSCF, segregation of both Sr and Ba takes place. The volatile Cr gaseous species would react with segregated Sr and Ba species to form Cr–Sr–O and/or Cr–Ba–O nuclei with subsequent crystallization and grain growth of Cr_2O_3 and $BaCr_2O_4$ or $BaCrO_4/SrCrO_4$ solid phases.

$$\mathrm{Cr-Ba-O}_{x(\mathrm{nuclei})} + \mathrm{CrO}_{3(\mathrm{g})} + \mathrm{BaO} \rightarrow \mathrm{BaCrO}_4 + \mathrm{BaCr}_2\mathrm{O}_{4(\mathrm{s})} \tag{10.40}$$

The rate of the deposition and formation of barium chromites, $BaCrO_4$ or $BaCr_2O_4$, as a result of reaction between the gaseous chromium species and segregated BaO is much lower as compared to the deposition of $SrCrO_4$. This may be related to the relatively low stability of $BaCrO_4$ in comparison with that of $SrCrO_4$. The predominant Cr deposit on the BSCF cathode is $SrCrO_4$. This shows that the Cr deposition process is also kinetically affected by the thermal stability of the Cr deposit products. Cr deposition on the BSCF in turn blocks the active sites and result in the high ASR of the electrodes for the O_2 reduction.

10.5.1.2 Boron

In high-temperature SOFCs, we use glass and glass ceramics as sealant materials to seal the components of the cells, e.g., between the cell and metallic interconnect, due to their tailorable structure and properties. However, the volatility and reactivity of some components of sealing glasses can be significant at the high operating temperatures of SOFCs. For instance, the use of boron oxide (B_2O_3) in glasses is an established means of tailoring physical properties such as viscosity and softening temperature. One possible issue regarding the use of boron-containing sealing glasses for SOFC is the high volatility of boron species such as BO_2 under dry conditions and $B_3H_3O_6$ under wet, reducing conditions [50]. The gaseous boron species can poison the electrochemical activity of cathode materials for ORR and is a contaminant for SOFCs.

Boron species shows a high activity with the commonly used electrolyte and cathode materials of SOFCs such as YSZ, GDC, LSM, LSCF, BSCF, $SrCo_{0.2}Fe_{0.8}O_3$ (SCF), etc. under open circuit and under SOFC cathodic operation conditions in air. YSZ and GDC react with boron, forming YBO_3 and $GdBO_3$, respectively. In the case of LSM, LSCF, and BSCF electrode materials, the chemical reaction with boron causes the disintegration and decomposition of the perovskite structures. The driving force for the high chemical activity between selected materials and boron or boron oxide is most likely due to the high stability of borates. Different to gaseous Cr species, boron shows a higher reactivity with lanthanum oxide (La_2O_3) as compared with strontium oxide (SrO). On the other hand, the nano-sized electrode particles are more susceptible by boron attack as compared to large particles.

In the case of an LSM cathode, boron deposition occurs randomly on the LSM electrode surface under open circuit but is driven to the electrode/electrolyte interface region under cathodic polarization conditions, resulting in the formation of $LaBO_3$ and Mn_2O_3 and the disintegration of the LSM perovskite structure [51]. Under open circuit conditions, $LaBO_3$ rather than strontium borate is formed on the LSM grain surface, indicating that gaseous boron species are highly active with La instead of Sr. As lanthanum does not segregate on the surface of LSM, the reaction must take place between the boron species and La in the LSM lattice sites. This can be expressed by:

$$La_{0.8}Sr_{0.2}MnO_3 + BO_2 \xrightarrow{\text{open circuit}} La_{0.8-x}Sr_{0.2}MnO_3 + xLaBO_3 \qquad (10.41)$$

Under cathodic polarization condition, the LSM phase near the TPB region would become more active and energetic in comparison with the LSM in the electrode bulk or on the electrode surface. This is also related to that fact that cathodic polarization would enhance the formation of oxygen vacancies at the electrode/electrolyte interface. The presence of positively charged oxygen vacancies would make the surrounding atoms polarized toward the vacancies [52]. Thus, under cathodic polarization, the boron deposition is driven to the interface region due to the increased activity of La at the LSM lattice sites. The preferential reaction between boron and LSM at the electrode/electrolyte interface under cathodic polarization is written as follows:

$$La_{0.8}Sr_{0.2}MnO_{3,\,TPB} + BO_{2(g)} \xrightarrow{\text{polarization}} LaBO_3 + SrB_xO_y + Mn_2O_3 + O_2 \quad (10.42)$$

The significant extraction of lattice La will lead to the complete decomposition of LSM perovskite at the electrode/electrolyte interface region, forming $LaBO_3$ and Mn_2O_3 nanoparticles. The boron deposition and reaction under open circuit and under cathodic polarization condition are schematically shown in Fig. 10.25 [53]. The presence of volatile boron species degrades the electrical conductivity and electrode polarization activity of LSM electrode for ORR.

Similar to LSM, there is also a preferential reaction between boron and LSCF, forming $LaBO_3$ [54]. The interaction between LSCF and volatile boron can lead to a significant reduction in the surface exchange properties of LSCF. LSCF is more active toward the boron as compared to LSM and the formation of $LaBO_3$ is kinetically favorable, leading to the quick disintegration of the LSCF perovskite structure. The high reactivity between LSCF and boron is most likely related to the fact that lanthanum cobaltite-based perovskites are less stable structurally as compared to their lanthanum manganite counterpart.

For BSCF electrodes, the volatile boron species reacts with the segregated Sr and Ba, forming B/Ba/Sr-containing borates and Ba(Co,Fe)O_x or $BaFeO_x$ phases. The segregation of Sr and Ba to the surface accelerates the excess depletion of A-site cations and could cause substantial non-stoichiometry and thereby decomposition of the BSCF perovskite structure. The reaction between BSCF and volatile boron species such as BO_2 in dry air can be expressed by following equation:

$$Ba_{0.5}Sr_{0.5}Co_{0.8}Fe_{0.2}O_3 + BO_2 \rightarrow \text{Ba, Sr} - \text{Borates} + BaFeO_x + Ba(Fe, Co)O_x \quad (10.43)$$

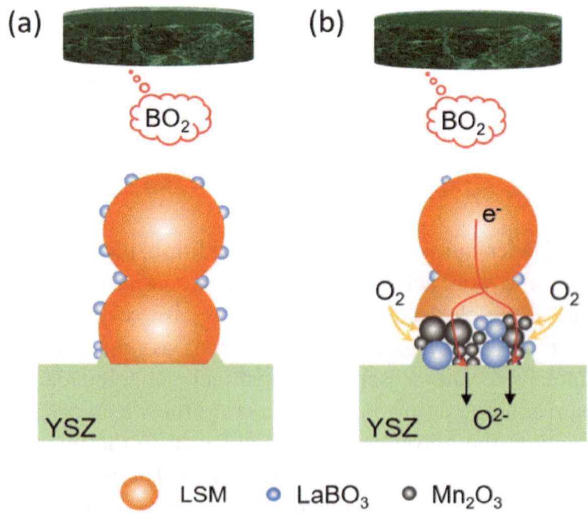

Fig. 10.25 Scheme of boron deposition on LSM cathode under **a** open circuit and **b** cathodic polarization conditions. Reproduced from Ref. [53] with permission from IOP Publishing, Copyright 2015

However, BSCF cathodes have a much better tolerance toward boron deposition and poisoning. The fundamental reason for the much higher stability of BSCF electrodes as compared to LSM and LSCF is most likely related to the much slower kinetics of the formation of strontium and barium borates as compared to the formation of lanthanum borates.

In practical SOFC stacks, the contact between the volatile boron species and cathodes of SOFCs is generally much smaller as compared to that between the cathode and volatile Cr species. Also, there is no direct contact between the glass sealant and cathode. The exposure of the glass seal can also be minimized by the stack design. Nevertheless, any direct contact with volatile boron species should be avoided to minimize the detrimental damage of boron poisoning on the stability and durability of SOFCs. The chemical interaction between glass ceramic sealant and metallic interconnect can cause a severe corrosion of the interconnect along sealing rims, leading to short-circuiting and stack failure. Silicon from the sealants can also migrate and deposit at the electrolyte/electrode interface of Ni-based fuel electrodes.

In the study of the poisoning effects of gaseous boron or chromium species, it is important to consider the ways the contaminant species are introduced. More detailed discussion on the experimental conditions of contaminant studies is given in Chap. 11.

Boron species are highly active and are not chemically compatible with almost all the electrode and electrolyte materials currently used in SOFCs. Thus, in SOFCs, the best way to avoid the boron deposition and poisoning is to use boron-free glass sealant materials.

10.5.1.3 Sulfur

The air that is used in SOFCs is not pure and may contain impurities such as carbon dioxide (CO_2) and sulfur dioxide (SO_2). The main source of SO_2 in the air is from industrial activities that process materials or fuels that contain sulfur, e.g., the generation of electricity from coal, oil, or natural gas. Although the concentration of SO_x may be low (0.02 to 0.20 ppm, depending on the location and time of day), the accumulation of sulfur is shown to be detrimental to the cathode performance of SOFCs. Thus, sulfur in the air stream is a contaminant for SOFC cathodes.

Similar to the deposition and poisoning of Cr, the segregated species, and in particular Sr, play an important role in the deposition and poisoning of sulfur species on the electrocatalytic activity of SOFC cathodes such as LSM and LSCF [55, 56]. Sulfur poisoning of the electrochemical activity of LSM electrodes can also be described by two-stage degradation processes: initial rapid degradation by the adsorbed SO_2 on the surface of LSM electrodes, inhibiting the dissociation, adsorption and diffusion processes of oxygen, followed by a slow degradation due to the deposition and formation of $SrSO_4$ on the LSM electrode surface and at the TPB region. The degradation due to the adsorption of SO_2 is largely reversible and is characterized by the initial rapid increase of polarization potential. Such rapid increase in the polarization potential depends on the temperature. The degradation in the electrochemical activity of

LSM at the second slow stage is related to the formation of a solid $SrSO_4$ phase. The formation of $SrSO_4$ accelerates the Sr segregation and consequently depletes the Sr content at the A-site of LSM, leading to significant degradation of the electrochemical activity of LSM electrodes.

Deposition of and poisoning by SO_2 can also be explained using nucleation theory. The first step is the interaction between SO_2 and segregated SrO, forming nuclei on the active sites of the LSM surface:

$$SO_{2(g)} + SrO \overset{\text{active sites}}{\leftrightharpoons} S - Sr - O_{x(\text{nuclei})} \tag{10.44}$$

In the presence of nuclei, the interaction between SO_2 and segregated SrO will lead to the deposition of S-containing compounds such as $SrSO_4$ and SrS,

$$S - Sr - O_{x(\text{nuclei})} + SO_{2(g)} + SrO \rightarrow SrSO_4 + SrS \tag{10.45}$$

but the formation of $SrSO_4$ and/or SrS compounds may depend on the temperature. Formation of $SrSO_4$ occurs most favorably at high temperatures (i.e., $\geq$ 700 °C) and at low temperatures (i.e., < 700 °C), the product would be mainly in the form of SrS. The effect of temperature on the deposition and poisoning by SO_2 is more complicated. This is due to the fact that the concentration of sulfur contaminant usually does not change with the temperature. Therefore, increasing temperature would increase the desorption of the adsorbed sulfur species on the LSM surface, i.e., the reverse process of the nucleation reaction, Eq. (10.44). This will reduce the rate of the sulfur deposition. On the other hand, the Sr segregation in general increases at high temperature, and this will increase the nucleation reaction of sulfur deposition. Therefore, depending on the circumstances, the increase in temperature does not necessarily increase the deposition and poisoning by SO_2 on LSM electrodes.

The reversibility of sulfur poisoning effects depends strongly on the electrode materials. In the case of LSCF electrodes, the presence of SO_2 has a significant poisoning effect on the electrochemical activity for ORR and the degradation in polarization performance cannot be 100% recovered by polarization treatment in SO_2-free and clean air. Figure 10.26 shows a typical example of the polarization behavior for the ORR on an LSCF electrode as a function of polarization time at 200 mA/cm^2 in 1 ppm SO_2-containing air for 20 h, followed by polarization in pure air for 20 h at 800 °C. The E_{Cathode} is measured between the LSCF electrode and Pt reference electrode. In SO_2-free air, LSCF electrodes are generally stable in air (e.g., the one showing in Fig. 10.26 for a typical polarization performance of a LSCF electrode measured at 200 mA/cm^2 at 700 °C for 40 h). The presence of SO_2 has a significant effect on the performance stability of LSCF electrodes. At 800 °C and in the presence of 1 ppm SO_2, the E_{Cathode} increases continuously with the polarization for 20 h. After switching to SO_2-free air, the E_{Cathode} decreases initially and finally stabilizes after a further polarization for 20 h. However, the stabilized value is still significantly higher than the initial value before the introduction of SO_2. The gap

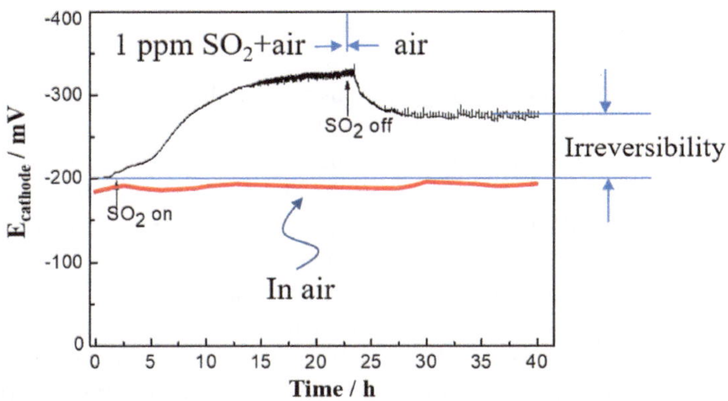

Fig. 10.26 Typical example of the polarization behavior for the ORR on an LSCF electrode, measured at 200 mA/cm^2 in the 1 ppm SO_2-containing air for 20 h, followed by the polarization in pure air for 20 h at 800 °C

between the initial and stabilized value is due to the irreversibility of sulfur poisoning on the electrochemical activity of LSCF electrodes.

Similarly to that described for the sulfur deposition on LSM electrodes, the deposition of SO_2 can also be described by the nucleation theory, i.e., the deposition is initiated kinetically by the nucleation reaction between gaseous SO_2 and segregated SrO on the surface of LSCF particles, followed by the crystallization and growth of sulfur deposits, $SrSO_4$. The formation of $SrSO_4$ would deplete the Sr content at the A-site of the perovskite structure, leading to the decomposition of the LSCF perovskite structure. There is no evidence to indicate the preferential deposition of sulfur at the LSCF electrode/GDC electrolyte interface. However, sulfur deposition and poisoning is more pronounced at low temperatures. The reason may be because of the thermodynamically favorable interaction between LSCF and SO_2 [56]. Thermodynamically, Sr from the LSCF perovskite can react with SO_2, leading to the decomposition of the LSCF perovskite structure and in turn forming $SrSO_4$, $CoFe_2O_4$, Fe_2O_3 and to a lesser degree La2$(SO_4)_3$. The interaction between LSCF and SO_2 can be written as follows:

$$La_{0.6}Sr_{0.4}Co_{0.2}Fe_{0.8}O_{3-\delta} + SO_{2(g)} + O_{2(g)} \rightarrow SrSO_4 + CoFe_2O_4 + Fe_2O_3 + La_2(SO_4)_3 \quad (10.46)$$

The mechanism and kinetics of sulfur poisoning and deposition have a significant implication in the development of high sulfur-tolerant cathodes for intermediate and low-temperature SOFCs.

10.5.1.4 CO_2

Carbon dioxide (CO_2) is known as greenhouse gas in the atmosphere and the main source of CO_2 in air due to the increased use of fossil fuels in the electricity generation and transportation sectors. In addition to its effect on the environment, CO_2 in air has also a detrimental and poisoning effect on the electrochemical activities of certain cathode materials of SOFCs. Among them, barium-containing cathode materials such as BSCF are more susceptible to CO_2 attack [57].

Interaction between CO_2 and BSCF does not occur at room temperature, but the reactivity of CO_2 with the BSCF electrode increases with increasing temperature. The interaction between CO_2 and the BSCF electrode surface is generally considered to be started by CO_2 adsorption, followed by the reaction with segregated Sr and/or Ba species on the BSCF surface to form carbonates on the electrode surface. Thermodynamically, both barium and strontium oxides would be segregated on the BSCF surface, and the formed strontium and/or barium carbonates occupy the active sites and deteriorate the electrochemical activity of the BSCF electrode for ORR. The reaction can also cause the decomposition of the BSCF perovskite structure. The interaction between CO_2 and BSCF can be written as follows:

$$CO_{2(g)} + S_{\mathrm{ad,\,BSCF}} \rightarrow CO_{2,\,\mathrm{ad}} \tag{10.47}$$

This is followed by the reaction with segregated SrO and BaO to form carbonates on the BSCF surface:

$$CO_{2,\,\mathrm{ad}} + SrO + BaO \rightarrow SrCO_3 + BaCO_3 \text{ or } Ba_xSr_{1-x}CO_3 \tag{10.48}$$

Kinetically, adsorption of CO_2 molecules on the surface of a perovskite oxide is related to the oxygen vacancies. BSCF is a material with a very high oxygen non-stoichiometry and its non-stoichiometry, i.e., oxygen vacancies, increases significantly with temperature (see Chap. 9). Thus, we can expect that CO_2 poisoning on the BSCF electrode will become significant and pronounced at high temperatures. This can also explain the relatively high stability of LSCF electrodes in the presence of CO_2. Although Sr segregation is also significant on the LSCF surface, the much lower concentration of oxygen vacancies on the LSCF surface limit the CO_2 adsorption, making the formation of strontium carbonate kinetically unfavorable.

10.5.2 Reaction Between Anode and Contaminants

10.5.2.1 Carbon Deposition and Coking Reaction

In any reaction involving hydrocarbon fuels, carbon deposition or the coking reaction on the metal catalyst surface is one of the main deactivation causes of fuel electrodes and is thermodynamically favored in the operating conditions of SOFCs

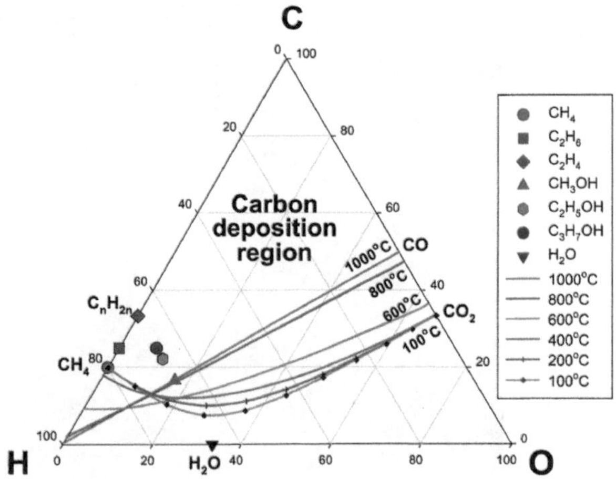

Fig. 10.27 Carbon deposition limit lines in the C–H–O phase diagram. Reproduced from Ref. [59] with permission from IOP Publishing. Copyright 2003

[58]. Figure 10.27 shows a phase diagram of CH–O, showing that all common hydrocarbon fuels used in SOFCs are in the carbon deposition region below 1000 °C [59]. The factors which can make the carbon deposition thermodynamically unfavorable include higher temperature (this is not always feasible as in SOFCs, we are actually trying to decrease the operating temperature of SOFCs), lower carbon/oxygen ratio and high oxygen fluxes (or adding O-containing species to the reaction).

The principle reactions leading to carbon deposition include the disproportion reaction of carbon monoxide (CO) and the decomposition reactions of hydrocarbons.

$$2CO_{(g)} \rightarrow C_{(s)} + CO_{2(g)} \tag{10.49}$$

$$C_nH_{m(g)} \rightarrow nC_{(s)} + \frac{m}{2}H_{2(g)} \tag{10.50}$$

Reaction (10.49) is commonly known as the Boudouard reaction and is exothermic at all temperatures. Due to the reduction in entropy, the reaction becomes more favorable at lower temperatures, e.g., below 700 °C. Reaction (10.50) is endothermic with an increase in entropy, thus the decomposition of hydrocarbon fuels is generally favored at high temperatures. For example, decomposition of methane occurs thermodynamically at temperatures above 500 °C.

Carbon deposition is strongly affected by the presence of sulfur and aromatic compounds in the fuel. The effect of sulfur is more complicated, but the presence of aromatics in the fuel tends to increase carbon deposition as they are more susceptible to non-catalytic thermal cracking. Once formed, these aromatic deposits serve as nucleation sites for the formation of polynuclear carbon compounds. Hydrocarbon decomposition and cracking would lead to the formation of solid carbon particles or fibers on the surface of Ni-based anodes, hindering the fuel transport, blocking active reaction sites and as a consequence degrading the cell performance. Figure 10.28 gives an example of the formation of carbon fibers on a Ni/YSZ cermet anode surface

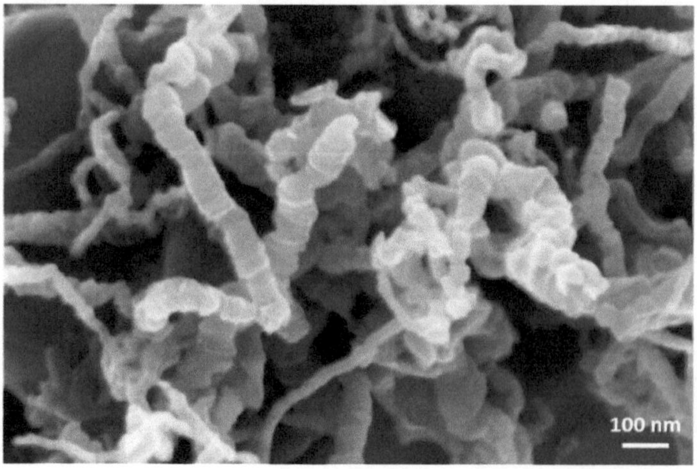

Fig. 10.28 Formation of carbon deposits in the form of carbon fibers on the surface of a Ni/YSZ cermet anode after tested in bio-oil fuels at 800 °C. Bio-oil fuels contain large amount of aromatics

after the electrochemical oxidation of bio-oil fuels at 800 °C, which contain a large amount of aromatic compounds.

The mechanism and kinetics of carbon deposition or coking reactions is a complex function of electrode materials (metal or metal oxides, type of metals, particle size, etc.) and extrinsic conditions (the temperature, current density, gas composition, oxygen supply and source, etc.). Carbon deposition has been studied using both ex situ and in situ techniques. The ex situ techniques such as microstructure and elementary examination are performed on the initial and final stage of the electrode materials after the dismantling of the experiments. Some critical information can be obtained by in situ techniques. One example is the use of in situ Raman spectroscopy in combination with scanning electron microscopy [60]. This in situ technique allows the characterization of coke morphologies and distribution over time and the mapping and monitoring of coke formation on Ni surface at high temperatures.

Carbon deposition on the Ni anode can be avoided by increasing the steam to carbon (S/C) ratio and the deposited carbon could be removed by the oxygen fluxes that are electrochemically driven through the electrolyte. Thus, the applied current density or overpotential has a significant effect on the coking formation, as discussed early. Carbon deposition is also affected by the composition of the cermet anode. Ni/GDC or Ni/SDC cermet anodes are more stable in hydrocarbon fuels such as propane than the Ni/YSZ cermet anode. The better carbon tolerance property of the Ni/doped CeO_2 anodes over the Ni/YSZ anodes is due to the strong interaction between Ni and doped ceria and the specific capacity of doped ceria for storage and releasing oxygen, as compared to YSZ. Reducing the size of Ni particles also enhances the resistance to coke formation due to the higher saturation concentration of carbon in the small Ni particles. Modification of Ni-based cermets with alkaline earth metal oxides such as BaO and MgO can enhance the resistance to carbon deposition although these oxides essentially have no electrochemical activity. Addition of precious metals like Pd and Ru into the Ni-based cermets can promote the electrocatalytic activity of the anode and suppress the carbon formation.

Perovskite-based oxides such as $La_{0.75}Sr_{0.25}Cr_{0.5}Mn_{0.5}O_{3-\delta}$ (LSCM), $Sr_2MgMoO_{6-\delta}$, $BaZr_{0.4}Ce_{0.4}Y_{0.2}O_3$, and $PrBaMn_2O_{5+\delta}$ exhibit a good carbon deposition tolerance, as compared to Ni-based cermet. However, the power densities of the cells with the oxide anodes are not as high as those with the Ni-based anodes, due to the relatively low electronic conductivity of the perovskite-based oxides under fuel-reducing conditions and/or the low electrochemical activity of the ceramic oxides for the oxidation reaction of H_2 and CH_4. Nevertheless, the oxide-based anode performance can be enhanced by infiltration or in situ exsolution of electrocatalytic active nano-scale particles such as Ni to enhance the electronic conductivity and electrocatalytic activity of the oxide anodes.

10.5.2.2 Sulfur

For hydrocarbon fuels ranging from natural gas to diesel, the most abundant impurity is sulfur. Sulfur is a major impurity in coal, so sulfur tolerance is a major issue for SOFC power plants designed to utilize gasified coal. Sulfur can exist as SO_2 and H_2S, but in the conditions occurring in an anode compartment of SOFCs, all sulfur compounds will be transformed into gaseous hydrogen sulfide, H_2S. The poisoning effect of H_2S in the fuel stream for SOFCs is well known. Under typical SOFC operating conditions with low sulfur concentration, H_2S quickly dissociates into hydrogen and atomic sulfur, S^*. The strongly adsorbed S^* species block the active sites for electrochemical oxidation of the fuel, leading to a rapid increase in the anode polarization. Further interaction between sulfur and Ni will lead to the formation of nickel sulfides, Ni_3S_2, NiS, Ni_3S_4, etc., accompanied with significant change in the morphology of the anode. The nickel sulfide formation degrades the activity and performance of the anode, although the degradation rate may vary significantly depending on the sulfur concentration, temperature, anode composition, and other fuel cell operating parameter. Once nickel sulfides are formed, the degradation in performance can no longer be reversed.

The effect and reversibility of sulfur poisoning on the Ni-based cermet anodes are affected by the nature of the electrode material and by the operational conditions, such as temperature, H_2S concentration, current density, H_2/CO ratio, and water content. On Ni/YSZ cermet anodes, the reversibility of sulfur poisoning is very low. Figure 10.29 shows a typical polarization behavior of a Ni/YSZ cermet anode measured in alternative H_2 and H_2S-containing H_2 fuels at 200 mA/cm^2 and 800 °C. The anode is first poisoned in H_2S-containing H_2 for 20 h, followed by regeneration in pure H_2 for 20 h. The performance or activity reversibility is measured by the anode potential loss between the initial anode potential and the one after regeneration. Even for the H_2 with 5 ppm H_2S, the electrochemical activity of Ni/YSZ cermet anode is not completely recovered after the regeneration in H_2 and the loss in the polarization performance is 20 mV. The irreversibility increases with the concentration of H_2S in

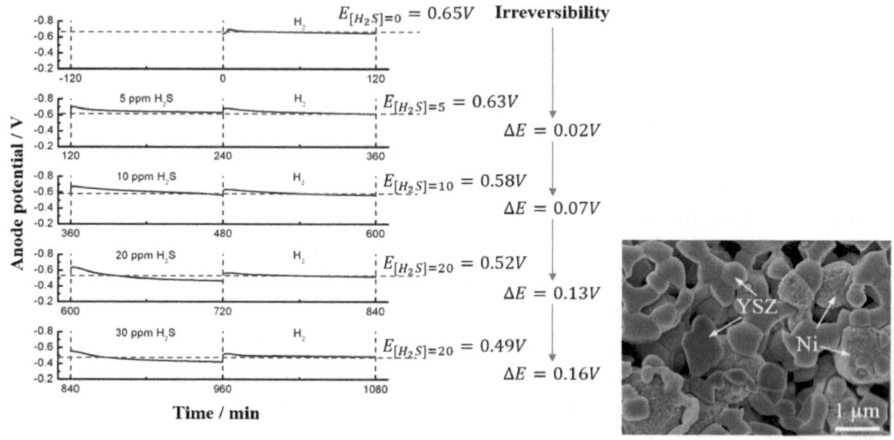

Fig. 10.29 Polarization behavior of a Ni/YSZ cermet anode measured in H_2 and H_2S-containing H_2 fuels at 200 mA/cm^2 and 800 °C. The H_2S concentration was increased from 5 to 30 ppm and the duration of the polarization in alternate H_2 and H_2S-containing H_2 fuels was 2 h. The SEM micrograph is taken on the surface of a Ni/YSZ anode after polarization treatments in alternative H_2 and H_2S-containing H_2 fuels at 200 mA/cm^2 and 800 °C. Modified from Ref. [61] with permission from Elsevier, Copyright 2010

H_2 fuel. The surface of the Ni/YSZ cermet, and in particular the Ni phase, changes significantly after the sulfur poisoning. This indicates the formation of nickel sulfides, e.g., Ni_3S_2, NiS_2. However, in the case of Ni/GDC cermet anodes, the performance can be fully recovered after exposed to H_2S-containing H_2 up to H_2S concentration of 30–50 ppm [61], indicating the significantly enhanced sulfur resistance of Ni/GDC cermet anodes.

Electrochemical oxidation of H_2S

In SOFCs, H_2S itself is a fuel and can be electrochemically oxidized under SOFC operation conditions:

$$H_2S + 3O_O^x \rightarrow H_2O + SO_2 + 6e^- + 3V_O^{\bullet\bullet} \qquad (10.51)$$

The reaction can occur similarly to HOR, but requires three times more oxygen ions (and an equal number of electrons) being used per mole of H_2S. As the result, the electrochemical reaction requires the continuous supply of a large amount of oxygen ions and the release of electrons at the TPB. Insufficient or slow supply of oxygen or release of electrons would lead to the competitive H_2S dissociation reaction.

Ni-free ceramic oxides have also been investigated as alternative anode materials for H_2S-containing fuels. However, similar to that of carbon tolerant oxide anodes, the relatively low electrical conductivity of the oxide materials under fuel reducing conditions is a major consideration in SOFCs.

10.5.2.3 Impurities in Coal Gasification Syngas

Coal is one of the most abundant fossil fuels. Coal syngas (H_2, CO, CO_2, N_2 and H_2O) produced by coal gasification is a promising fuel for SOFCs. On the other hand, multiple minor and trace impurities are present in the coal syngas even after gas cleanup processing and could affect the SOFC performance and durability. Among them, phosphorus is a trace element but potentially is the most detrimental. The concentration of phosphorus in the syngas produced from the coal gasification is typically in the range of a few ppm. Phosphorus readily reacts with Ni to form Ni phosphides (Ni_xP_y) under SOFC operation conditions, significantly degrading the performance of Ni-based anodes. The presence of phosphine causes Ni agglomeration and densification of the outside anode layer, the growth of Ni particles reduces the active TPB sites, and the densification of the outside anode layer hinders fuel diffusion in the porous anodes, while the formation of Ni_xP_y phases leads to decreased electrical conductivity of the anode.

Chlorine is one of the impurities in coal syngas, existing in the form of HCl. Cell degradation has been observed in the presence of HCl, but the effect of chloride on the performance degradation is much smaller as compared to that of phosphorus. The cell performance degradation due to the presence of chloride has been reported to be recoverable [62]. Other impurities in the coal gas such as arsenic (As), antimony (Sb), cadmium (Cd), and selenium (Se) could also cause degradation of the anode, due to the formation of the secondary phases in the Ni cermet anodes.

The mitigation of impurities in coal gasification fuels is less investigated, but the strategies discussed for carbon deposition and sulfur tolerance anodes can also be employed in this case.

10.5.3 Contaminant-Tolerant Electrodes

In addition to the electrochemical activities of the electrodes, the resistance or tolerance toward contaminants is also an important consideration in the development of cathodes and anodes in SOFCs. Depending on the mechanism of poisoning and deposition of contaminants, there are various strategies in the development of contaminant-tolerant electrodes. In this section, the discussion is divided into two parts—contaminant-tolerant cathodes and contaminant-tolerant anodes.

10.5.3.1 Contaminant Tolerant Cathodes

Based on the nucleation theory, reduction or elimination of nucleation agents or nuclei can effectively inhibit or decrease the deposition and poisoning of Cr contaminants on SOFC cathodes. Therefore, Cr deposition on LSM/YSZ and LSM/GDC composite cathodes is significantly smaller as compared to that on pure LSM [63]. The reason for the significantly reduced Cr deposition in the LSM-based composite cathodes is most likely due to the decrease of Mn^{2+} ions generated under cathodic polarization as the results of the addition of the ionic conducting electrolyte phase. This leads to a significant reduction in the nucleation and grain growth reaction for the Cr deposition. Reducing the Sr content and/or enhancing Sr stability at the A-site of LSCF by proper doping can significantly improve the chromium tolerance of the cathodes.

One of the best examples is the development of $(La_{0.24}Sr_{0.16}Ba_{0.6})(Co_{0.5}Fe_{0.44}Nb_{0.06})O_{3-\delta}$ (LSBCFN) perovskite oxides prepared by a novel direct mixing synthesis of $(La_{0.6}Sr_{0.4})(Co_{0.2}Fe_{0.8})O_{3-\delta}$ (LSCF) and $Ba(Co_{0.7}Fe_{0.2}Nb_{0.1})O_{3-\delta}$ (BCFN) as Cr-tolerant cathodes for ORR [64]. BCFN has been used as oxygen permeation membrane due to its high oxygen flux. However, similarly to BSCF, the electronic conductivity of BCFN is low, less than 20 S/cm at 800 ºC. Different to BCFN, LSCF possesses high electrical conductivity, but is low in the tolerance toward Cr deposition and poisoning. The new LSBCFN electrodes combine the structural stability and activity of BCFN and high conductivity of LSCF and show a better electrochemical activity for ORR than either LSCF or BCFN in the temperature range of 600–900 °C; see Fig. 10.30. For example, ASR is 0.38 Ωcm^2 at 600 °C for BCFN, significantly smaller than 0.97 Ωcm^2 for LSCF; however, the electrode ohmic resistance of BCFN is 6.7 Ωcm^2, significantly larger than 2.4 Ωcm^2 measured on LSCF. On the other hand, LSBCFN showed the lowest ASR and its electrode ohmic resistance is compatible with that of LSCF (Fig. 10.30a). Most importantly, LSBCFN shows an excellent stability and tolerance toward chromium deposition and poisoning under SOFC operation conditions (Fig. 10.30b). Significant Cr deposition occurs on the surface of both LSCF and BCFN due to the formation of $SrCrO_4$ on LSCF and $BaCr_2O_4$ and/or $BaCrO_4$ on BCFN. Cr deposition also occurs on the LSBCFN electrode surface, but the surface of the LSBCFN electrode is much cleaner, indicating the much slower kinetics of Cr deposition processes on the LSBCFN electrode surface as compared to its constituent LSCF and BSCF electrodes (Fig. 10.30c). The high tolerance of the LSBCFN electrode is closely related to its inhibited surface segregation processes due to the doping of niobium at the B-site of the perovskite structure.

Infiltration of BaO NPs is also effective in substantially enhancing the chromium tolerance of LSCF electrodes [65]. The infiltrated BaO NPs inhibit the formation of $SrCrO_4$ by forming the thermodynamically more stable and conducive $BaCrO_4$, thus preventing the excess segregation of Sr at the A-sites of the LSCF perovskite structure. This in turn mitigates the poisoning effect of Cr on the electrocatalytic activity of LSCF. In this strategy, the infiltrated BaO phase primarily functions as a scavenger to the gaseous Cr contaminants. Once the BaO NPs are exhausted in

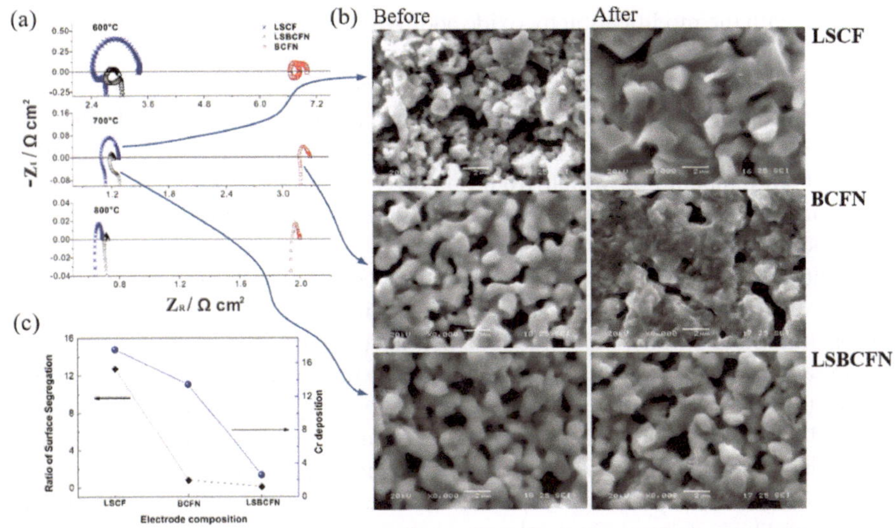

Fig. 10.30 **a** Nyquist plots of LSCF, BCFN, and LSBCFN electrodes at 600, 700, and 800°C, **b** SEM micrographs of the LSCF, BCFN, and LSBCFN electrode surface before and after the test at 900 °C under a current passage of 200 mA/cm^2 in the presence of the metallic interconnect, and **c** plots of Cr deposition and ratio of surface segregation of Sr and Ba of LSCF, BCFN, and LSBCFN electrodes. Modified from Ref. [64] with permission from Royal Society of Chemistry, Copyright 2013

forming $BaCrO_4$, the infiltrated BaO phase may lose its protective properties for LSCF cathodes. However, in practical applications, Fe–Cr interconnects are protected by the dense and electronic conductive protective layers, which substantially reduces the Cr volatility. Therefore, the increased Cr tolerance and resistance of the cathode materials would significantly contribute to the overall stability of the SOFCs based on metallic interconnect. The scavenger effect of infiltrated BaO NPs is also effective in enhancing the sulfur tolerance of LSCF cathodes [66]. In this case, the infiltrated BaO reacts with SO_2 to form $BaSO_4$, inhibiting the formation of $SrSO_4$. This in turn effectively prevents segregation of Sr from the LSCF perovskite structure and thus enhances the tolerance of LSCF electrodes toward the contamination of SO_2.

10.5.3.2 Contaminant-Tolerant Anodes

There are various strategies to increase the carbon deposition tolerance and resistance of Ni-based cermet anodes [67], and some of them have already been discussed in Chap. 9. Studies show that adding noble metals such as Au, Ru, Pd, and Pt into the Ni/YSZ porous anode addition promotes reforming activity and suppresses carbon deposition. In addition, alloying nickel with other metals such as Cu, Sn, Fe, etc. is another way to enhance the chemical stability of Ni-based cermet anodes in hydrocarbon fuel. The carbon deposition tolerance of Ni-based cermet anodes can also be enhanced with the modification by oxide additives, such as SrO, BaO, La_2O_3, SnO_2, CeO_2 and Al_2O_3. Among them, BaO appears to be most effective. For instance, Ni/YSZ cermets with BaO additive demonstrated an enhanced coking tolerance under the fuels of propane, wet CO and gasified carbon at 800 °C [68]. In the presence of BaO, a thin $Ba(OH)_2$ layer would form on the Ni surface in wet hydrocarbon fuels at high temperature, which would oxidize the formed coke into CO or CO_2, thus significantly enhancing the coking resistance of BaO modified Ni anodes via a water-mediated process.

Manipulation of the operational parameters such as polarization currents can also suppress the carbon deposition and cracking in Ni-based cermet anodes. Under fuel cell operating conditions, the oxide ions that have migrated through the YSZ electrolyte can spill over from the Ni/YSZ interface to the Ni surface and can effectively oxidize the deposited carbon on the Ni surface. Figure 10.31 shows an example of the SEM images of cross section of a Ni/YSZ cermet anode supported YSZ cell after testing in biofuel and bio-oils[69]. The Ni/YSZ cermet anode consisted of a Ni/YSZ support of ~ 500 μm thickness and a Ni/YSZ anode functional layer of 10 μm. Biofuels and bio-oils were obtained as products of the fast pyrolysis of mallee wood and contained ~ 30 different compounds consisting of water and various $C_mH_nO_x$

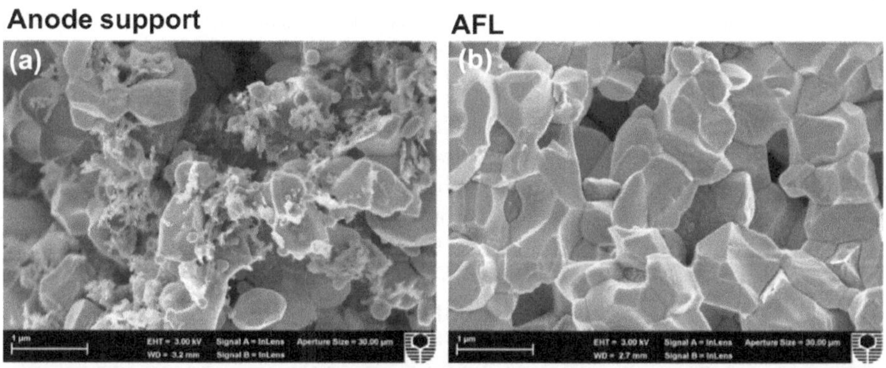

Fig. 10.31 SEM images of cross section of Ni/YSZ cermet anodes supports and anode functional layer (AFL) of Ni/YSZ cermet anodes after test in (**a**, **b**) biofuel and (**c**, **d**) bio-oil at 800 °C. Reproduced from Ref. [69] with permission from Elsevier, Copyright 2016

compounds ($2 \leq m \leq 14$, $4 \leq n \leq 14$ and$1 \leq x \leq 5$). Carbon particles and fibers are formed on the anode support, but not on the AFL. The Ni/YSZ anode functional layer close to the YSZ electrolyte is clean with no carbon deposition, indicating that carbon deposition does not occur at the region close to the electrode/electrolyte interface. This clearly demonstrates that high oxygen fluxes to the AFL and the water produced can react or catalytically reform the deposited carbon. Adding an anode catalyst layer is proposed to be effective in increasing the activity and stability of Ni-based cermet anodes [70].

Similar strategies can be applied to the development of sulfur-tolerant anodes by introducing a proton conducting phase. Yang et al. [71] replaced the oxygen ion conductor YSZ in a Ni/YSZ cermet anode by a mixed proton and oxygen ion conductor like $BaZr_{0.1}Ce_{0.7}Y_{0.2-x}Yb_xO_3$ (BZCYYb). This Ni/BZCYYb cermet anode shows superior coking and sulfur tolerance at 750 °C for up to ~ 20 ppm H_2S using a cell based on a BZCYYb electrolyte and ~ 50 ppm H_2S using a cell based on SDC electrolyte. Its ability to resist deactivation by sulfur and coking is considered to result from the enhanced catalytic activity of BZCYYb for sulfur oxidation as well as enhanced water adsorption capability to facilitate the oxidation of H_2S or elemental sulfur to SO_2 at the Ni active sites. Infiltration of proton conducting oxides such as $BaCe_{0.9}Yb_{0.1}O_{3-\delta}$ (BCYb) is also effective at enhancing the sulfur poisoning resistance of Ni/GDC cermet anodes [72]. Figure 10.32 shows the polarization performance of Ni-GDC anode-supported GDC electrolyte cells with $NdBa_{0.75}Ca_{0.25}Co_2O_{5+\delta}$ double perovskite cathodes. BCYb is infiltrated in the Ni-GDC anode substrate to form BCYb + Ni-GDC anode. The infiltration of BCYb nanoparticles substantially improves the electrochemical performance and stability of the Ni-GDC anode in both H_2 and 500 ppm H_2S containing H_2 fuels. For Ni-GDC cermet anodes, the sulfur deposition and poisoning causes significant change in the microstructure of the Ni phase. The holes formed on the Ni surface are most likely due to the formation and decomposition of the nickel sulfides, e.g., Ni_3S_2, NiS_2. On the other hand, no such morphological change is observed on BCYb-infiltrated Ni-GDC cermet anodes.

10.6 Summary

In this chapter, we have introduced and discussed the following topics:

- Because the fuel cell reaction occurs predominantly on the surface and at the electrode/electrolyte interface, the surface and surface chemistry of electrodes play critical roles in the electrocatalytic activity and stability of electrode materials of SOFC. This is particularly relevant to the cathode of SOFC due to the fact that majority of cathode materials are oxide-based and surface composition is inherently different from that in the bulk due to the loss of symmetry at the surface relative to the bulk. This causes the redistribution and/or accumulation

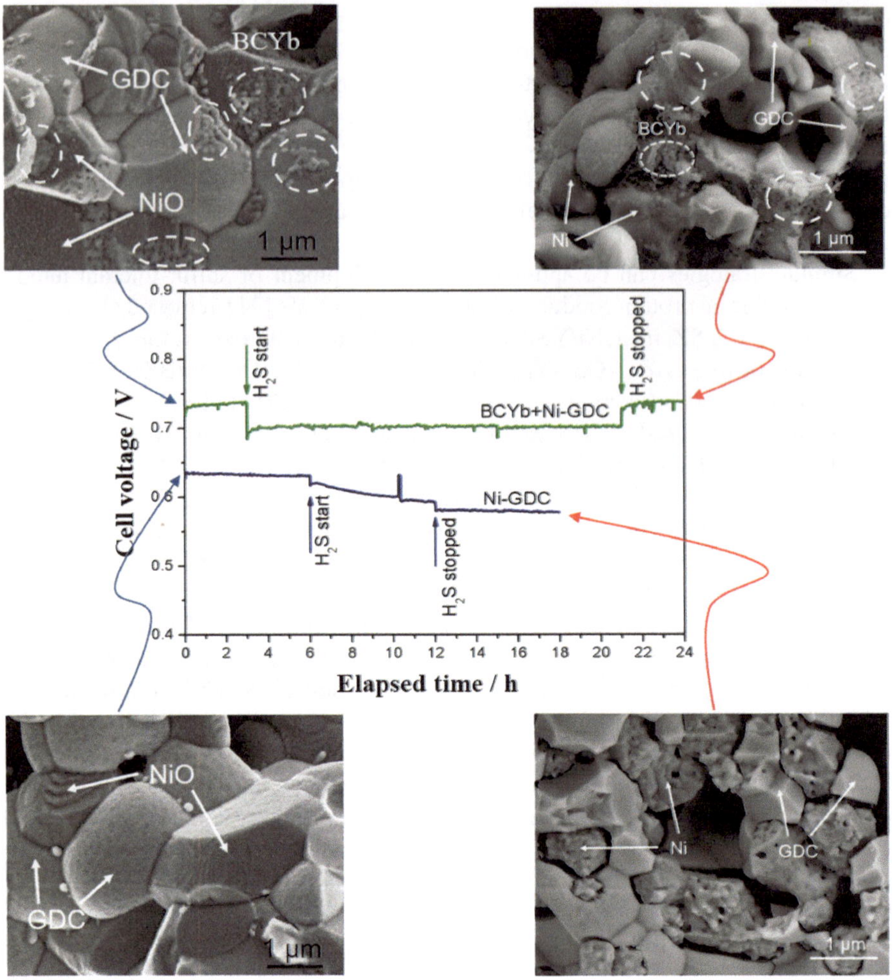

Fig. 10.32 Cell voltages as a function of time for the cell with Ni-GDC cermet anode and BCYb-infiltrated Ni/GDC (BCYb + Ni-GDC) cermet anode, measured at a constant current density of 640 mA/cm^2 and 650 °C in H_2 and H_2 containing 500 ppm H_2S. The SEM images show the anode before and after the stability test in H_2 containing 500 ppm H_2S fuel. The circles indicate the infiltrated BCYb particles. Modified from Ref. [72] with permission from American Chemical Society, Copyright 2016

of a specific cation or cations at the oxide surface, a phenomena called surface segregation.

- Surface segregation is generally related to the A- and B-site cation segregation and is frequently observed in the Sr-containing perovskite-based electrodes such as LSM and LSCF. The segregated species occupy the active sites and degrade the electrocatalytic activity of the electrodes. The segregated species could also react with gaseous species or contaminants in the air electrode sides, leading to the contaminant deposition and poisoning. The interaction between the contaminants and segregated species in turn accelerates the surface segregation process and ultimately results in the decomposition and disintegration of the perovskite structure of the electrodes. In general, surface segregation needs to be carefully controlled and monitored during the operation of SOFC.

- The generally accepted theory of elastic energy minimization and electrostatic interaction is quite successful in the explanation of surface segregation phenomena under open circuit conditions. However, there are significant discrepancies or controversy in the mechanism and kinetics of the segregation processes under polarization conditions, which is in fact most critical for the fundamental understanding of the electrode process under SOFC operation conditions. One of the reasons is the lack of in situ techniques in the identification of surface segregation process and the difficulties in the separation of the interfacial reaction and surface segregation under high temperature particularly for cobaltite-based oxide materials. By the use of direct assembly technique, it is possible to separate the interface reaction and surface segregation. This has been discussed in detail for the surface segregation processes under polarization conditions for LSM and LSCF electrodes.
- Activation process observed for the O_2 reduction reaction under cathodic polarization conditions in high-temperature SOFC is essentially related to the incorporation and/or cleaning of the surface segregated species. Deactivation process is a typical example of the surface segregation of cations under anodic polarization conditions for ORR. However, due to the fact that surface reaction is most critical for electronic predominant electrode materials, the activation process is most pronounced for ORR on LSM electrode and in much less extent on MIEC LSCF electrode.
- ORR is the most important reaction in SOFC due to its high activation energy and kinetic barrier. Thus, development of electrode materials with high electronic conductivity and oxide ion conductivity is essential for the high-performance SOFC. One of the effective routes is to increase the oxygen vacancies of the electrodes through the nano-structured approaches using materials with high oxygen storage and release capabilities such as doped ceria and doped bismuths.
- H_2 oxidation reaction (HOR) on Ni-based cermet anodes is facile and proceeds fast in the presence of small amount of water vapor due to the promotion effect of spillover mechanism. For the hydrocarbon oxidation reaction, e.g., methane oxidation, the reaction is much more complicated due to multiple reaction steps of both electrochemical and chemical reactions and possibilities of carbon cracking and deposition. Nano-structured approach via composite and functional nanoparticles is effective for the development of highly active and stable anodes for hydrocarbon fuel oxidation reaction in SOFC.
- It is important to note that the interface formation between electrode and electrolyte induced under high-temperature sintering does not necessarily involve the cation interdiffusion. The interface formation, characterized by the formation of

convex rings in the case of LSM electrode and YSZ or GDC electrolyte, is essentially related to the minimization of curvature and surface energy, similar to the neck formation phenomena in conventional ceramic sintering process. The lattice disorientation and misfit of crystallographic planes do exist at the interface but such atom-to-atom mismatch of the crystallographic planes at the interface does not impede the oxygen migration processes of ORR.

- One of the limitation of stabilized zirconia-based electrolyte is the interface reaction with cathodes of SOFC during the high-temperature sintering steps and operating conditions, and this is particular the case for the cobaltite-based electrodes such as LSCF and BSCF. The interface reaction leads to the formation of insulating phases such as lanthanum and strontium zirconates at the interface and deteriorates the performance. In order to use cobaltite-based cathodes, a barrier layer such as doped ceria is usually applied to the surface of zirconia electrolyte.
- In addition to the interfacial reaction, the interaction between the electrode and contaminants either in air or fuel streams is generally detrimental to the performance and stability of SOFC and by all means should be avoided and minimized. In the case of cathode, the dominant degradation mechanism related to the contaminants is the deposition and poisoning of chromium species due to the wide use of chromia-forming metallic interconnect in intermediate to low temperature SOFC. Among various deposition and poisoning mechanisms proposed, the nucleation theory is considered the most viable as it satisfactorily explains the observed Cr deposition phenomena on wide range of cathode materials and offers the effective strategy for the development of Cr tolerance cathodes of SOFC. The nucleation theory is also applicable for the deposition and poisoning of other contaminants like sulfur. For the anodes, the main degradation mechanism is related to the sulfur deposition and carbon cracking.
- Different strategies have been discussed for the development of contaminant tolerance electrodes, including the composition modification to increase the structural stability and thus to reduce the surface segregation, nano-scale electrodes by introducing Cr resistance nanoparticles, introduction of proton conducting phase for the in situ formation of water phase to mitigating the carbon cracking, sulfur adsorption and deposition on Ni-based cermet, etc.

10.7 Questions

10.1 Explain the following terms and materials:

Surface segregation, elastic energy minimization, and electrostatic interaction

Activation, deactivation process, and ORR under influence of polarization

ORR and three-phase boundary

HOR, effect of water vapor, and spillover mechanism

Convex ring contact ring, cation interdiffusion, LSM/YSZ interface, LSM/GDC interface and LSCF/YSZ interface
Interface formation, surface segregation, and interfacial reaction
Cr deposition, poisoning, electrochemical depostion theory and nucleation theory
Boron deposition and lanthanum borates
Reversibility and irreversibility of sulfur poisoning
Carbon cracking and metal dusting
Contaminant-tolerant cathode and anode

10.2 What is the role of oxygen vacancy formation under the cathodic polarization conditions on the Sr surface segregation on LSM and LSCF electrodes?

10.3 In the literatures, there are significant discrepancies in the surface segregation phenomena of SOFC cathodes like LSM and LSCF under polarization conditions. In your opinion, what are the reasons for such discrepancies and in some cases contradictory results?

10.4 Activation phenomena are most pronounced for ORR on LSM electrode as compared to that on LSCF electrode under identical experimental conditions. Explain the reason.

10.5 Figure 10.10 gives an example of the occurrence of low-frequency inductance loop as a function of impregnated GDC NP size, an indication of effect of oxygen capacity of GDC phase on ORR. Are there other examples of the occurrence of low-frequency inductance loop for ORR of SOFC?

10.6 For hydrogen oxidation reaction on Ni/YSZ cermet anodes, both hydrogen spillover and oxygen spillover mechanisms are proposed. In your opinion, which one is more relevant and important? Provide the evidence for the argument.

10.7 What is the main difference between the hydrogen oxidation reaction on Pt in low-temperature PEMFCs and high-temperature SOFCs? Explain the reaction mechanism and kinetics of the reaction in PEMFCs and SOFCs.

10.8 In the oxygen reduction reaction on perovskite oxide cathodes such as LSM and LSCF, is the oxygen spillover mechanism important? Explain the reason behind your answer.

10.9 For the methane oxidation reaction, if the oxidation occurs by the full electrochemical oxidation reaction, what is methane fuel mass flow rate in L/min to produce 1 A current under atmospheric pressure, assuming 100% fuel utilization efficiency? (Answer: 1.74×10^{-3} L/min).

10.10 For an industrial natural gas, if the oxidation occurs by the full electrochemical oxidation reaction, what is natural gas mass flow rate in L/min to produce 1 A current, assuming 100% fuel utilization efficiency? What is the minimum air mass flow rate at the cathode? (Answer: 1.51×10^{-3} L/min for natural gas and the minimum air flow rate at the cathode is 1.66×10^{-2} L/min).
The natural gas composition: 88%CH_4, 6%C_2H_6, 4%C_3H_8, 2%C_4H_{10}.

10.11 List the advantages and disadvantages of using an internal reforming process in methane fuel-based SOFCs.

10.12 In addition to the nucleation theory, electrochemical deposition theory has been proposed for the Cr deposition on cathode of SOFC, see Eq. (10.32) and (10.33). Can you use the electrochemical deposition theory to explain the Cr deposition as shown in Fig. 10.24?

10.13 Read relevant papers and discuss the role of temperature in the sulfur poisoning and deposition on cathodes of SOFCs.

10.14 Referred to Fig. 10.32, explain the role of impregnated $BaCe_{0.9}Yb_{0.1}O_{3-\delta}$ (BCYb) phase on the promotion of sulfur tolerance of Ni-GDC cermet anode in H_2S-containing H_2 fuel.

10.15 If H_2S can, in principle, be electrochemically oxidized on Ni-based cermet anodes like Ni/YSZ, why is Ni/YSZ still poisoned by the presence of H_2S in hydrogen fuel?

References

1. Li Y, Zhang W, Zheng Y, Chen J, Yu B, Chen Y, Liu M (2017) Controlling cation segregation in perovskite-based electrodes for high electro-catalytic activity and durability. Chem Soc Rev 46(20):6345–6378
2. Chen K, Jiang SP (2020) Surface segregation in solid oxide cell oxygen electrodes: phenomena, mitigation strategies and electrochemical properties. Electrochem Energy Rev 3:730–765
3. Zhao L, Drennan J, Kong C, Amarasinghe S, Jiang SP (2014) Insight into surface segregation and chromium deposition on $La_{0.6}Sr_{0.4}Co_{0.2}Fe_{0.8}O_{3-\delta}$ cathodes of solid oxide fuel cells. J Mater Chem A 2(29):11114–11123
4. Lee W, Han JW, Chen Y, Cai Z, Yildiz B (2013) Cation size mismatch and charge interactions drive dopant segregation at the surfaces of manganite perovskites. J Am Chem Soc 135(21):7909–7925
5. Koo B, Kim K, Kim JK, Kwon H, Han JW, Jung W (2018) Sr segregation in perovskite oxides: why it happens and how it exists. Joule 2(8):1476–1499
6. Fister M (2008) In situ characterization of strontium surface segregation in epitaxial $La_{0.7}Sr_{0.3}MnO_3$ thin films as a function of oxygen partial pressure. Appl Phys Lett 93 (15):151904.
7. Huber AK, Falk M, Rohnke M, Luerssen B, Amati M, Gregoratti L, Hesse D, Janek J (2012) In situ study of activation and de-activation of LSM fuel cell cathodes—electrochemistry and surface analysis of thin-film electrodes. J Catal 294:79–88
8. Pavone M, Munoz-Garcia AB, Ritzmann AM, Carter EA (2014) First-principles study of lanthanum strontium manganite: insights into electronic structure and oxygen vacancy formation. J Phys Chem C 118(25):13346–13356
9. Jiang SP, Love JG (2001) Origin of the initial polarization behavior of Sr-doped $LaMnO_3$ for O_2 reduction in solid oxide fuel cells. Solid State Ionics 138(3–4):183–190
10. Mizusaki J, Yonemura Y, Kamata H, Ohyama K, Mori N, Takai H, Tagawa H, Dokiya M, Naraya K, Sasamoto T, Inaba H, Hashimoto T (2000) Electronic conductivity, Seebeck coefficient, defect and electronic structure of nonstoichiometric $La_{1-x}Sr_xMnO_3$. Solid State Ionics 132(3–4):167–180
11. Chen KF, Hyodo J, Dodd A, Ai N, Ishihara T, Jian L, Jiang SP (2015) Chromium deposition and poisoning of $La_{0.8}Sr_{0.2}MnO_3$ oxygen electrodes of solid oxide electrolysis cells. Faraday Discuss 182:457–476

12. Jiang SP, Zhang JP, Apateanu L, Foger K (2000) Deposition of chromium species at Sr-doped $LaMnO_3$ electrodes in solid oxide fuel cells I mechanism and kinetics. J Electrochem Soc 147(11):4013–4022
13. Ai N, He SA, Li N, Zhang Q, Rickard WDA, Chen KF, Zhang T, Jiang SP (2018) Suppressed Sr segregation and performance of directly assembled $La_{0.6}Sr_{0.4}Co_{0.2}Fe_{0.8}O_{3-\delta}$ oxygen electrode on Y_2O_3–ZrO_2 electrolyte of solid oxide electrolysis cells. J Power Sources 384:125–135
14. Sun Y, He S, Saunders M, Chen K, Shao Z, Jiang SP (2021) A comparative study of surface segregation and interface of $La_{0.6}Sr_{0.4}Co_{0.2}Fe_{0.8}O_{3-\delta}$ electrode on GDC and YSZ electrolytes of solid oxide fuel cells. Int J Hydrog Energy 46:2606–2616
15. Jiang SP (2007) Activation, microstructure, and polarization of solid oxide fuel cell cathodes. J Solid State Electrochem 11(1):93–102
16. Wang W, Jiang SP (2006) A mechanistic study on the activation process of (La, Sr)MnO_3 electrodes of solid oxide fuel cells. Solid State Ionics 177(15–16):1361–1369
17. Adler SB (2004) Factors governing oxygen reduction in solid oxide fuel cell cathodes. Chem Rev 104(10):4791–4843
18. Li YH, Gemmen R, Liu XB (2010) Oxygen reduction and transportation mechanisms in solid oxide fuel cell cathodes. J Power Sources 195(11):3345–3358
19. Jiang SP (2002) A comparison of O_2 reduction reactions on porous (La, Sr)MnO_3 and (La, Sr)(Co, Fe)O_3 electrodes. Solid State Ionics 146(1–2):1–22
20. Jiang SP, Love JG, Ramprakash Y (2002) Electrode behaviour at (La, Sr)MnO_3/Y_2O_3-ZrO_2 interface by electrochemical impedance spectroscopy. J Power Sources 110(1):201–208
21. Chen KF, Ai N, Jiang SP (2016) Origin of low frequency inductive impedance loops of O_2 reduction reaction of solid oxide fuel cells. Solid State Ionics 291:33–41
22. Bessler WG, Vogler M, Stormer H, Gerthsen D, Utz A, Weber A, Ivers-Tiffee E (2010) Model anodes and anode models for understanding the mechanism of hydrogen oxidation in solid oxide fuel cells. Phys Chem Chem Phys 12(42):13888–13903
23. Mizusaki J, Tagawa H, Saito T, Kamitani K, Yamamura T, Hirano K, Ehara S, Takagi T, Hikita T, Ippommatsu M, Nakagawa S, Hashimoto K (1994) Preparation of nickel pattern electrodes on YSZ and their electrochemical properties in H_2–H_2O atmospheres. J Electrochem Soc 141(8):2129–2134
24. Jiang SP, Badwal SPS (1997) Hydrogen oxidation at the nickel and platinum electrodes on yttria-tetragonal zirconia electrolyte. J Electrochem Soc 144(11):3777–3784
25. Jiang SP, Badwal SPS (1999) An electrode kinetics study of H_2 oxidation on Ni/Y_2O_3-ZrO_2 cermet electrode of the solid oxide fuel cell. Solid State Ionics 123(1–4):209–224
26. Ingram DB, Linic S (2009) First-principles analysis of the activity of transition and noble metals in the direct utilization of hydrocarbon fuels at solid oxide fuel cell operating conditions. J Electrochem Soc 156(12):B1457–B1465
27. Babaei A, Jiang SP, Li J (2009) Electrocatalytic promotion of palladium nanoparticles on hydrogen oxidation on Ni/GDC anodes of SOFCs via spillover. J Electrochem Soc 156(9):B1022–B1029
28. Jiang SP, Zhang L, Zhang Y (2007) Lanthanum strontium manganese chromite cathode and anode synthesized by gel-casting for solid oxide fuel cells. J Mater Chem 17(25):2627–2635
29. Wang W, Jiang SP, Tok AIY, Luo L (2006) GDC-impregnated Ni anodes for direct utilization of methane in solid oxide fuel cells. J Power Sources 159(1):68–72
30. Jiang SP, Chen XJ, Chan SH, Kwok JT (2006) GDC-impregnated, $(La_{0.75}Sr_{0.25})(Cr_{0.5}Mn_{0.5})O_3$ anodes for direct utilization of methane in solid oxide fuel cells. J Electrochem Soc 153 (5):A850-A856.
31. Malzbender J, Steinbrech RW (2007) Advanced measurement techniques to characterize thermo-mechanical aspects of solid oxide fuel cells. J Power Sources 173(1):60–67
32. Yokokawa H (2003) Understanding materials compatibility. Ann Rev Mater Res 33:581–610
33. Jiang SP, Zhang JP, Foger K (2003) Chemical interactions between 3 mol% yttria-zirconia and Sr-doped lanthanum manganite. J Eur Ceram Soc 23(11):1865–1873
34. He S, Chen KF, Saunders M, Li J, Cui CQ, Jiang SP (2017) A FIB-STEM study of $La_{0.8}Sr_{0.2}MnO_3$ cathode and Y_2O_3-ZrO_2/Gd_2O_3-CeO_2 electrolyte interfaces of solid oxide fuel cells. J Electrochem Soc 164 (13):F1437-F1447.

35. Poulsen FW, Vanderpuil N (1992) Phase-relations and conductivity of Sr-Zirconates and La-Zirconates. Solid State Ionics 53(6):777–783
36. Jiang SP, Zhang JP, Ramprakash Y, Milosevic D, Wilshier K (2000) An investigation of shelf-life of strontium doped $LaMnO_3$ materials. J Mater Sci 35(11):2735–2741
37. Zhang JP, Jiang SP, Love JG, Foger K, Badwal SPS (1998) Chemical interactions between strontium-doped praseodymium manganite and 3 mol% yttria-zirconia. J Mater Chem 8(12):2787–2794
38. He S, Saunders M, Chen KF, Gao HF, Suvorova A, Rickard WDA, Quadir Z, Cui CQ, Jiang SP (2018) A FIB-STEM study of strontium segregation and interface formation of directly assembled $La_{0.6}Sr_{0.4}Co_{0.2}Fe_{0.8}O_{3-\delta}$ cathode on Y_2O_3–ZrO_2 electrolyte of solid oxide fuel cells. J Electrochem Soc 165 (7):F417–F429.
39. Wang F, Nishi M, Brito ME, Kishimoto H, Yamaji K, Yokokawa H, Horita T (2014) Sr and Zr diffusion in LSCF/10GDC/8YSZ triplets for solid oxide fuel cells (SOFCs). J Power Sources 258:281–289
40. Lu Z, Darvish S, Hardy J, Templeton J, Stevenson J, Zhong Y (2017) $SrZrO_3$ Formation at the interlayer/electrolyte interface during $(La_{1-x}Sr_x)_{1-\delta}Co_{1-y}Fe_yO_3$ Cathode Sintering. J Electrochem Soc 164(10):F3097–F3103
41. Huang KQ, Feng M, Goodenough JB, Schmerling M (1996) Characterization of Sr-doped $LaMnO_3$ and $LaCoO_3$ as cathode materials for a doped $LaGaO_3$ ceramic fuel cell. J Electrochem Soc 143(11):3630–3636
42. Cao XG, Jiang SP (2013) Identification of oxygen reduction processes at (La,Sr)MnO_3 electrode/$La_{9.5}Si_6O_{26.25}$ apatite electrolyte interface of solid oxide fuel cells. Int J Hydrog Energy 38 (5):2421–2431.
43. Jiang SP, Chen XB (2014) Chromium deposition and poisoning of cathodes of solid oxide fuel cells—a review. Int J Hydrog Energy 39(1):505–531
44. Jiang SP (2001) Use of gaseous Cr species to diagnose surface and bulk process for O_2 reduction in solid oxide fuel cells. J Appl Electrochem 31(2):181–192
45. Chen XB, Zhang L, Liu EJ, Jiang SP (2011) A fundamental study of chromium deposition and poisoning at $(La_{0.8}Sr_{0.2})_{(0.95)}(Mn_{1-x}Co_x)O_{3+/-\delta}$ (0.11 <= x<1.0) cathodes of solid oxide fuel cells. Int J Hydrog Energy 36 (1):805–821.
46. Hilpert K, Das D, Miller M, Peck DH, Weiss R (1996) Chromium vapor species over solid oxide fuel cell interconnect materials and their potential for degradation processes. J Electrochem Soc 143(11):3642–3647
47. Jiang SP, Zhang S, Zhen YD (2005) Early interaction between Fe-Cr alloy metallic interconnect and Sr-doped $LaMnO_3$ cathodes of solid oxide fuel cells. J Mater Res 20(3):747–758
48. Chen XB, Zhang L, Jiang SP (2008) Chromium deposition and poisoning on $(La_{0.6}Sr_{0.4-x}Ba_x)(Co_{0.2}Fe_{0.8})O_3$ ($0 \leq x \leq 0.4$) cathodes of solid oxide fuel cells. J Electrochem Soc 155(11):B1093–B1101
49. Kim YM, Chen XB, Jiang SP, Bae J (2012) Effect of Strontium Content on Chromium Deposition and Poisoning in $Ba_{1-x}Sr_xCo_{0.8}Fe_{0.2}O_{3-\delta}$ ($0.3 \leq x \leq 0.7$) Cathodes of Solid Oxide Fuel Cells. J Electrochem Soc 159(2):B185–B194
50. Zhou XD, Templeton JW, Zhu Z, Chou YS, Maupin GD, Lu Z, Brow RK, Stevenson JW (2010) Electrochemical performance and stability of the cathode for solid oxide fuel cells. III. Role of Volatile Boron Species on LSM/YSZ and LSCF. J Electrochem Soc 157(7):B1019–B1023
51. Chen KF, Ai N, Zhao L, Jiang SP (2013) Effect of Volatile boron species on the electrocatalytic activity of cathodes of solid oxide fuel cells I. (La, Sr)MnO_3 based electrodes. J Electrochem Soc 160 (2):F183–F190.
52. Kotomin EA, Mastrikov YA, Heifets E, Maier J (2008) Adsorption of atomic and molecular oxygen on the $LaMnO_3$(001) surface: ab initio supercell calculations and thermodynamics. Phys Chem Chem Phys 10(31):4644–4649
53. Chen K, Liu S-S, Guagliardo P, Kilburn MR, Koyama M, Jiang SP (2015) A Fundamental study of boron deposition and poisoning of $La_{0.8}Sr_{0.2}MnO_3$ cathode of solid oxide fuel cells under accelerated conditions. J Electrochem Soc 162 (12):F1282–F1291.

54. Chen KF, Ai N, Zhao L, Jiang SP (2013) Effect of volatile boron species on the electrocatalytic activity of cathodes of solid oxide fuel cells II. (La,Sr)(Co,Fe)O_3 based electrodes. J Electrochem Soc 160 (3):F301–F308.
55. Wang CC, Chen KF, Jiang SP (2016) Mechanism and kinetics of SO_2 poisoning on the electrochemical activity of $La_{0.8}Sr_{0.2}MnO_3$ cathodes of solid oxide fuel cells. J Electrochem Soc 163 (8):F771–F780.
56. Wang CC, He S, Chen KF, Rowles MR, Darvish S, Zhong Y, Jiang SP (2017) Effect of SO_2 poisoning on the electrochemical activity of $La_{0.6}Sr_{0.4}Co_{0.2}Fe_{0.8}O_{3-\delta}$ cathodes of solid oxide fuel cells. J Electrochem Soc 164 (6):F514–F524.
57. Yan A, Cheng M, Dong YL, Yang WS, Maragou V, Song SQ, Tsiakaras P (2006) Investigation of a $Ba_{0.5}Sr_{0.5}Co_{0.8}Fe_{0.2}O_{3-\delta}$ based cathode IT-SOFC-I. The effect of CO_2 on the cell performance. Appl Catal B-Environ 66 (1–2):64–71.
58. Boldrin P, Ruiz-Trejo E, Mermelstein J, Menendez JMB, Reina TR, Brandon NP (2016) Strategies for carbon and sulfur tolerant solid oxide fuel cell materials, incorporating lessons from heterogeneous catalysis. Chem Rev 116(22):13633–13684
59. Sasaki K, Teraoka Y (2003) Equilibria in fuel cell gases—II. The C–H–O ternary diagrams. J Electrochem Soc 150 (7):A885–A888.
60. Blinn KS, Abernathy H, Li XX, Liu MF, Bottomley LA, Liu ML (2012) Raman spectroscopic monitoring of carbon deposition on hydrocarbon-fed solid oxide fuel cell anodes. Energy Environ Sci 5(7):7913–7917
61. Zhang L, Jiang SP, He HQ, Chen XB, Ma J, Song XC (2010) A comparative study of H2S poisoning on electrode behavior of Ni/YSZ and Ni/GDC anodes of solid oxide fuel cells. Int J Hydrog Energy 35(22):12359–12368
62. Marina OA, Pederson LR, Thomsen EC, Coyle CA, Yoon KJ (2010) Reversible poisoning of nickel/zirconia solid oxide fuel cell anodes by hydrogen chloride in coal gas. J Power Sources 195(20):7033–7037
63. Zhen YD, Jiang SP (2006) Transition behavior for O_2 reduction reaction on (La, Sr)MnO_3/YSZ composite cathodes of solid oxide fuel cells. J Electrochem Soc 153(12):A2245–A2254
64. Chen XB, Jiang SP (2013) Highly active and stable $(La_{0.24}Sr_{0.16}Ba_{0.6})(Co_{0.5}Fe_{0.44}Nb_{0.06})O_{3-\delta}$ (LSBCFN) cathodes for solid oxide fuel cells prepared by a novel mixing synthesis method. J Mater Chem A 1(15):4871–4878.
65. Chen KF, Ai N, O'Donnell KM, Jiang SP (2015) Highly chromium contaminant tolerant BaO infiltrated $La_{0.6}Sr_{0.4}Co_{0.2}Fe_{0.8}O_{3-\delta}$ cathodes for solid oxide fuel cells. Phys Chem Chem Phys 17 (7):4870–4874.
66. Wang CC, Luo DW, Hou BX, Jiang SP (2019) Effect of BaO impregnation on sulfur tolerance of $La_{0.6}Sr_{0.4}Co_{0.2}Fe_{0.8}O_{3-\delta}$ cathodes of solid oxide fuel cells. Materials Research Express 6 (7):075504
67. Yue WX, Li YF, Zheng Y, Wu T, Zhao CH, Zhao J, Geng G, Zhang WQ, Chen J, Zhu JX, Yu B (2019) Enhancing coking resistance of Ni/YSZ electrodes: In situ characterization, mechanism research, and surface engineering. Nano Energy 62:64–78
68. Yang L, Choi Y, Qin W, Chen H, Blinn K, Liu M, Liu P, Bai J, Tyson TA, Liu M (2011) Promotion of water-mediated carbon removal by nanostructured barium oxide/nickel interfaces in solid oxide fuel cells. Nat Commun 2(1):357
69. Chen KF, Zhang L, Gholizadeh M, Ai N, Hasan MM, Mourant D, Li CZ, Jiang SP (2016) Feasibility of tubular solid oxide fuel cells directly running on liquid biofuels. Chem Eng Sci 154:108–118
70. Wang W, Qu JF, Juliao PSB, Shao ZP (2019) Recent advances in the development of anode materials for solid oxide fuel cells utilizing liquid oxygenated hydrocarbon fuels: a mini review. Energ Technol 7(1):33–44
71. Yang L, Wang SZ, Blinn K, Liu MF, Liu Z, Cheng Z, Liu ML (2009) Enhanced sulfur and coking tolerance of a mixed ion conductor for SOFCs: $BaZr_{0.1}Ce_{0.7}Y_{0.2-x}Yb_xO_{3-\delta}$. Science 326 (5949):126–129
72. Li M, Hua B, Luo JL, Jiang SP, Pu J, Chi B, Li J (2016) Enhancing sulfur tolerance of Ni-based cermet anodes of solid oxide fuel cells by ytterbium-doped barium cerate infiltration. ACS Appl Mater Interfaces 8(16):10293–10301

Chapter 11
Solid Oxide Fuel Cells: Techniques and Characterization

Characterization of the activity and performance of the anode, cathode, and electrolyte is one of the most fundamental requirements in fuel cell diagnostics. However, the electrochemical techniques which can be used in SOFCs are limited due to the high operating temperature (600–1000 °C) and solid-state nature of the electrolytes. This limits the construction and placement of a reference electrode. For example, in contrast to the aqueous electrolyte systems, the reference electrode cannot be easily incorporated into the electrolyte and instead can only be placed on an external surface of the solid electrolyte. Therefore, the cell configuration, electrode placement, and electrochemical characterization techniques are of unique features for SOFCs. In addition, in this chapter, we will also discuss the constriction effect, the performance scalability, etc., typically associated with solid state electrolyte cells.

11.1 Electrode Arrangement and Test Stations

11.1.1 Test Station Design and Arrangement

The design and configuration of test cells are very important for the accurate measurement and deconvolution of the electrochemical performance of the anode, cathode and electrolyte of SOFCs. There are a number of practical and technical considerations in the electrode arrangement and cell configuration design, which are summarized as follow:

- Easy assembly and dismantle of the test cells. This includes adequate space and position for the sealing of the cell if required and placement of thermocouples for temperature monitoring and easy access of electrical connection for current and voltage measurements.
- Reliable contact between the current collector (e.g., Pt or Au mesh) and working, counter and reference electrodes. Maintaining a good contact is critical during the

S. P. Jiang and Q. Li, *Introduction to Fuel Cells*,
https://doi.org/10.1007/978-981-10-7626-8_11

electrochemical test for reliable and accurate measurements over a wide range of SOFC operating conditions (currents, temperatures, partial pressure of oxygen, etc.) in order to clarify the elementary reaction steps of the electrode/electrolyte system.
- Pt (Ag or Au) wires used as current and voltage probes should be as short as possible in the hot zone to reduce the electrical resistance. The conducting wires should be straight and avoid crossover to minimize the effect of the interference of current conducting wires on the electrochemical signal, in particular the electrode impedance responses.
- The cell configuration should be so designed that the individual electrode performances can be monitored with minimum distortion, which may arise due to non-uniform distribution of the electrical field between the working and the counter electrodes.

The technical consideration is mainly related to the position and placement of the reference electrode (RE). The placement of the RE is a complicated issue and is closely related to the equipotential line distribution in a solid electrolyte cell. We will discuss this important issue in a separate section. The first step in the electrochemical characterization of a SOFC cell is to design the electrode cell with a proper arrangement of current collectors, current probe, voltage probe, sealant, etc. in the test rig. Specific requirements of the cell arrangement are, however, related to the electrode cell configuration to be tested: electrolyte-supported cells or anode-supported cells.

Figure 11.1 shows an example of the electrode and current collector arrangements for small electrolyte-supported and anode-supported button cells. The button cells can be in a square or circular shape. For electrolyte-supported cells, the size of the cells should be 2 cm or larger in diameter for the convenience of the electrode arrangement and placement of the RE. For button cell testing, fine Au, Ag or Pt mesh is usually used for the current collector and electrical contact. In the case of low conductivity of the electrode coating, additional Au, Pt, or Ag paste may be used in order to eliminate the effect of the electrical conductivity of the electrode materials on their electrochemical performance. However, care should be taken because the metallic paste may contribute to the electrochemical activity or performance of the electrode materials. In this case, a background test with the metallic paste electrode coating should be carried out and used as the base line of the electrochemical test results.

Pt or Ag wire is generally used for the current and voltage probes. The addition of voltage probes to the working electrodes is necessary in order to reduce the contribution of the ohmic resistance of the Pt wires. At high temperatures, the resistance of Pt wire is quite high and cannot be ignored. For example, at 1000 °C, the total resistance of a Pt wire with diameter of 0.5 mm and total length of ~ 80 mm in the hot zone can be as high as about one ohm. The measurement of cathode performance can be made in a single gas environment with a symmetrical electrode arrangement. In the case of fuel electrodes, it is necessary to seal the anode side by a Pyrex glass ring, ceramic glass putty or ceramic bond. The air and fuel flow rates are generally in the range of 100 to 500 ml/min for small button cells. Pure hydrogen can be humidified

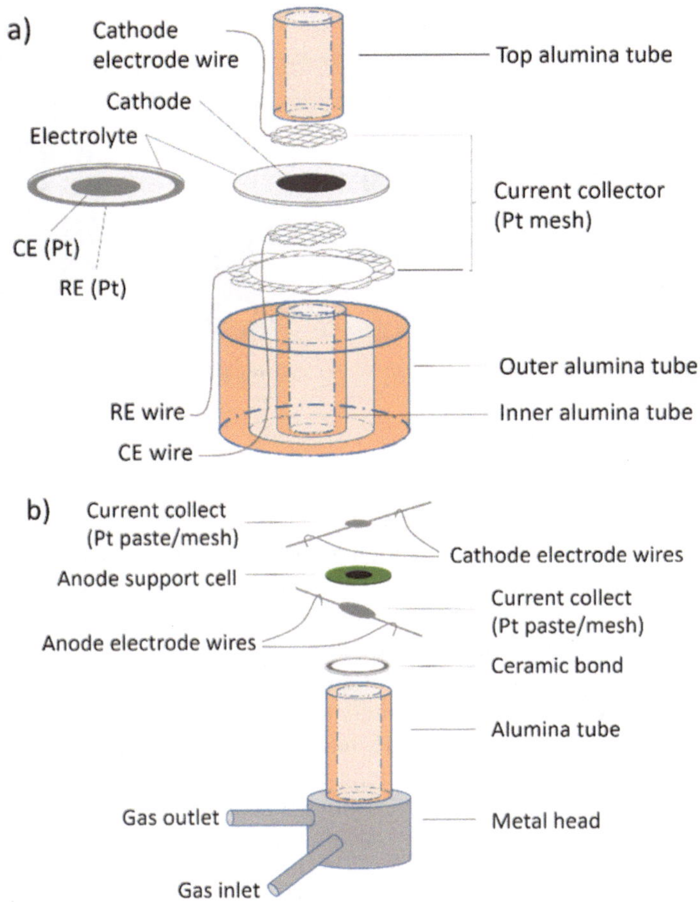

Fig. 11.1 Cell and electrode arrangements for **a** electrolyte-supported cell and **b** anode-supported cell

via bubbling through a water bath. At room temperature (25 °C), the water content in hydrogen fuel is ~ 3%. The quality of the seal can be assessed by the stability of open circuit voltage (OCV) responses at the cell testing temperature.

The cell and electrode arrangement needs to be hosted inside a high-temperature furnace with control panel for electrochemical data collection and fuel and air flow. Figure 11.2 shows a typical SOFC test station with upper and base connectors for the gas flow and support of the cell arrangements and electrical connection panel. An electrochemical instrument with electrochemical impedance spectroscopy and polarization control capability is used to control and record the electrochemical data of the cell. The contact between the support tubes and electrode cell components is maintained by a spring load to ensure the good contact between the working electrode

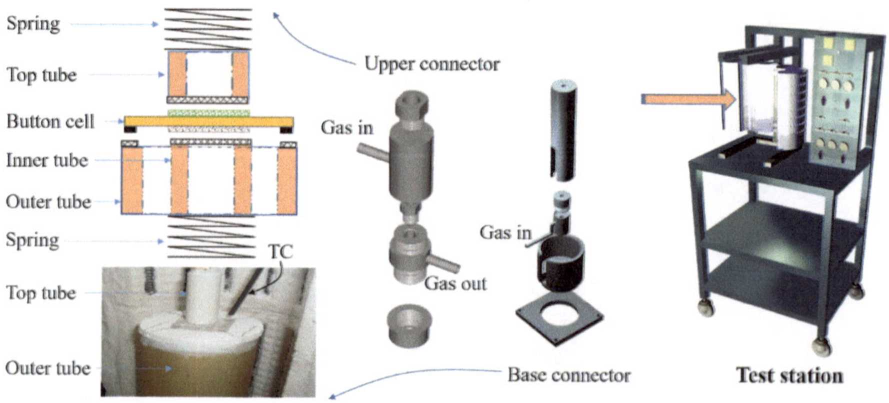

Fig. 11.2 SOFC test station with furnace and movable rack

(WE), counter electrode (CE), RE, and current collector. If sealing is required, an outer tube can be added to the top section of the cell arrange to seal the WE.

In SOFCs, different gases are required, depending on the reactions and test conditions. In the case of the studies involving CO and sulfur-containing fuels, we will also need to consider the treatment of the exit gas. Figure 11.3 shows a typical cell configuration and schematic diagram of an experimental apparatus for the sulfur-poisoning study on anodes in sulfur-containing H_2. The cell is fixed with an alumina tube and sealed via a ceramic sealant. H_2 and H_2S-containing H_2 with different H_2S concentrations are humidified at room temperature (~3% H_2O) as fuel while the CE and RE are exposed in static air. The humidification of the fuel gas can also be turned

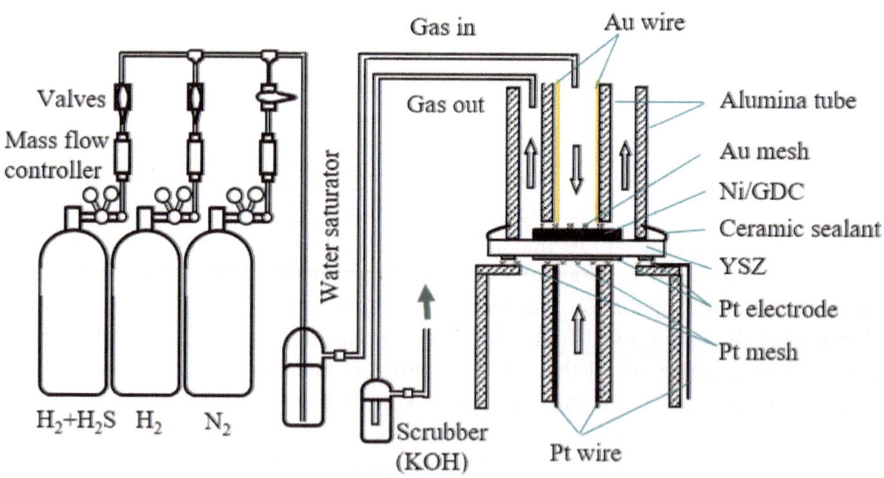

Fig. 11.3 Typical gas line arrangement and experimental apparatus for the sulfur-poisoning study on anodes

off. The fuel flow rate is controlled by a mass flow controller. Due to the toxicity of the unreacted sulfur in the H_2, the exit gas is treated by going through a KOH scrubber to cleanse the unreacted H_2S before disposing to atmosphere. By rearrangement and change of the gas bottles, various electrode reactions at the anode or cathode can be investigated.

11.1.2 Electrode Arrangements

It is generally straightforward to study the electrode reactions of pure reactants such as H_2, O_2 or other reactive species in terms of the gas supply and test cell arrangement. However, if the electrode reaction involves contaminants, it needs to consider the way to introduce the contaminants to the cell unless the contaminants are part of the gas components such as H_2S in H_2 fuel gas and CO_2 in air. The way to introduce the contaminants to the electrodes is important for the critical evaluation of the mechanism and kinetics of contamination reactions at the electrodes. Depending on the objectives of the study, different methods are used for introducing the contaminants. We will illustrate this with the examples of the investigation of deposition and poisoning of gaseous Cr species on a SOFC cathode.

Because volatile chromium species can only be produced at high temperatures, a high-temperature and oxidizing environment is necessary for introducing the gaseous Cr species. The deposition and poisoning of Cr species can be done under a chemical environment or under the SOFC operation conditions. Figure 11.4 shows two different arrangements for the introduction of Cr species for the purpose of study of Cr deposition and poisoning on SOFC cathode; *static arrangement* (a) and *dynamic or in situ arrangement.*

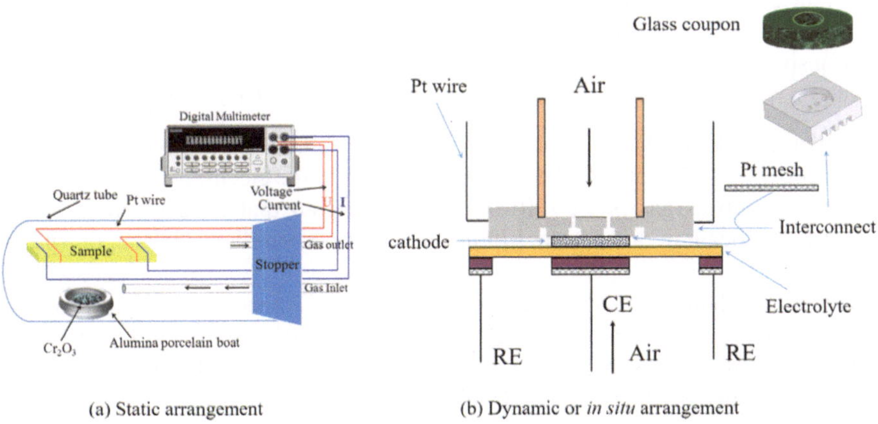

Fig. 11.4 Introducing volatile Cr species by **a** static arrangement using chromium oxide powder and **b** dynamic and in situ arrangement using chromia-forming alloy interconnect

In the static arrangement, the electrode sample, e.g., $La_{0.6}Sr_{0.4}Co_{0.2}Fe_{0.8}O_{3-\delta}$ (LSCF), is prepared from LSCF powder and then sintered at high temperature (e.g., 1350 °C) in air to obtain dense LSCF bar samples with a high relative density (e.g., > 97%). Cr_2O_3 oxide is used as the source for gaseous Cr species. The LSCF bar samples and Cr_2O_3 powders are housed in a quartz tube with controlled air flow or other desired atmosphere. The air can be dry or humidified by passing through a water bath. The bar samples are exposed to the Cr-containing atmosphere and heat treated at different temperatures for certain periods. Such an arrangement is often used to study the effect of gaseous Cr species on the surface properties of cathode materials such as the surface segregation and surface exchange properties by the electrical conductivity relaxation (ECR) measurement [1, 2]. The static arrangement can also be used to investigate the effect of sulfur, boron, or other contaminants on the electrochemical performance and chemical compatibility of SOFC electrodes and electrolytes [3, 4].

However, in order to study the effect of Cr contaminants on the electrochemical activity of the electrode, we need to introduce the species to the electrode in situ under the SOFC operating conditions. Thus, the cell arrangement used for the study of the deposition and poisoning of contaminants on SOFC cathodes should be as close as possible to the stack arrangement in practical planar cells. In the dynamic arrangement as shown in Fig. 11.4b, a chromia-forming alloy such as high chromium ferritic stainless steel is used as the Cr source for the study. The alloy is machined into coupons (e.g., 12 mm × 12 mm × 5 mm thick). One side of the coupon is cut with channels (1.2 mm × 1.2 mm deep) with holes in the center. Air is distributed to the electrode coating through channels by an alumina tube, simulating the air flow situation in an actual fuel cell stack of planar-type SOFCs. Pt wires are spot-welded to the coupon to serve as voltage and current probes. There is no additional Pt mesh placed between the chromia-forming alloy and the electrode. The chromia-forming alloy coupon is in direct contact with the cathode coating and also acts as a current collector in this case. Due to the channel and rib contacts between the electrode coating and alloy coupon, Cr deposition and poisoning will be related to the transportation processes of chromium species, depending on the contact. Thus, we will have two different situations for the deposition of Cr species:

- Under the ribs of the interconnect coupon, the cathode is in direct contact with the alloy. Consequently, Cr species can transport to the electrode coating via solid state as well as gas phase diffusion. It is expected that Cr deposition will be most significant in the vicinities of the contact between rib of the coupon and the electrode.
- Under the channels of the interconnect coupon, Cr species can only reach the electrode surface via gas phase diffusion. This would result in less Cr deposition in the electrode areas under the channels of the coupon. The differences of the Cr deposition under rib and channels of the coupon could indicate the roles of solid and/or gas diffusion of Cr species on the deposition and poisoning of Cr species on the electrode.

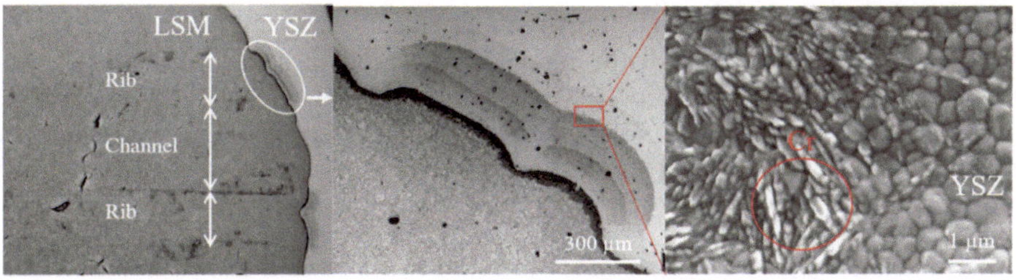

Fig. 11.5 SEM micrographs of a LSM electrode coating after cathodic current passage at 200 mA/cm^2 and 900 °C for 4 h, showing the contact marker of the rib and channel areas of the interconnect and the wave-like bands on the YSZ electrolyte surface

To completely avoid the solid diffusion process of Cr on the deposition and poisoning, Pt mesh can be placed between the chromia-forming coupon and cathode. In this case, volatile chromium species can only travel to the surface of cathode via the gas diffusion route. Due to the air flow from the center of the electrode to the edge of the cell, the mobility and deposition of the Cr species on different electrode areas away from the reactive zone can also be examined [5]. Figure 11.5 shows an observation of the Cr deposition on the YSZ electrolyte surface under the rib area of a Fe–Cr interconnect coupon on a LSM electrode after polarization at 200 mA/cm^2, 900 °C for 4 h. On the YSZ electrolyte surface, wave-like bands are observed, which is due to the deposition of solid Cr species under air flow. In areas close to the LSM electrode edge, the deposits are mainly $(Cr,Mn)_3O_4$ spinels, while it is Cr_2O_3 on the YSZ electrolyte surface far away from the LSM electrode edge. Such deposition of Cr species on the YSZ electrolyte surface away from the active zone has a significant implication for the evaluation of the mechanism and kinetics of the Cr deposition in SOFCs. Air flow can also be humidified to study the effect of water vapor on Cr deposition processes.

The dynamic or in situ arrangement can be used to study other contaminants such as volatile boron species from borosilicate glass-based sealants on the electrochemical activity of cathode for ORR. In this case, the chromia-forming alloy coupon is replaced with a boron-containing glass coupon [6]. The glass coupon made from borosilicate glass powder (e.g., 70% SiO_2, 10% B_2O_3, 8% Na_2O, 8% K_2O and 1% CaO) is used as the source of boron. The glass powder is pre-sintered into a pellet or coupon with a hole in the center and heat treated at 800 °C for 2 h. The size of the glass pellet is similar to the button cells and the thickness is around 4–5 mm. The glass pellet is placed on top of the cathode, similar to the metallic interconnect coupon. However, the glass pellet should not be in direct contact with the electrode coating due to its high activity. This is also due to the fact that in SOFC stacks, glass sealants are not in direct contact with electrode coatings. To avoid the direct contact between the glass coupon and cathode coating, a Pt mesh is placed between the glass coupon and electrode coating. In this way, volatile boron species can only deposit on the cathode via gas phase diffusion. In practical planar SOFCs, glass sealants are generally used on the edge of the cells to seal the interface between the electrode and metallic interconnect. The contact between the sealant materials and cathode can be very small or minimum and volatile boron species could only escape from the edge of the sealant. This is very different from the face-to-face contact between the electrode and interconnect. Thus, the arrangement as suggested in Fig. 11.4b represents the accelerated study for the deposition and poisoning of boron.

11.1.3 Comparative Experimental Approach

A comparative experimental approach can be very effective in the diagnosis of electrode and electrolyte process in fuel cells if the conditions are selected properly and correctly. To illustrate the point, one of the best examples is the study of the reaction mechanism and kinetics of the Cr deposition and poisoning on the ORR of cathodes under SOFC operating conditions [7]. There are two main mechanisms of Cr deposition and poisoning on SOFC cathodes reported in the literature; electrochemical deposition mechanism of Cr species due to the thermodynamic compatibility of the Gibbs free energy changes for the electrochemical reduction reactions of high valence Cr species (e.g., CrO_3) to Cr_2O_3, and nucleation theory of Cr deposition initiated by the presence of nuclei in the electrode/electrolyte systems. The nucleation theory implies that Cr deposition is essentially a chemical reaction in nature.

To identify the dominant reaction mechanism for Cr deposition, the experiment should be designed in such way that the results will contradict each other if the set mechanism prevails. For example, if the driving force for the Cr deposition is electrochemical, Cr deposition would not occur under both cathodic and anodic polarization or under open circuit conditions. For the same reason, Cr deposition phenomena would be very similar for the reaction on LSM and Pt electrodes as both LSM and Pt are electronically dominant electrode materials. Figure 11.6 is the list of conditions for the identification of predominant Cr deposition mechanisms; electrochemical or chemical processes.

It turns out that Cr deposition occurs under cathodic as well anodic polarization conditions, and the Cr deposition on LSM and Pt electrodes is very different despite the fact that both Pt and LSM are electronic conducting electrodes. Extensive experimental data indicate that Cr deposition is essentially a chemical process initiated by a nucleation reaction between high valence Cr species and nucleation agents such as Sr, Mn, Ba, etc. More detailed discussion can be found in Chap. 10.

Comparative experimental approaches have also been used successfully in the study of the hydrogen oxidation reaction on Ni and Pt electrodes under dry and humidified H_2 because Ni and Pt have very different oxidation behavior under dry and humidified H_2 [8] and oxygen reduction reactions on LSM and LSCF cathodes [9].

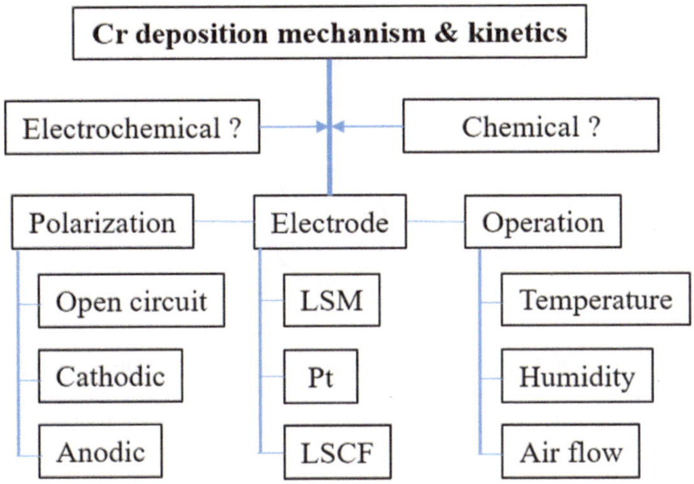

Fig. 11.6 Experimental parameters in the identification of mechanism and kinetics of chromium deposition and poisoning on cathodes of SOFCs

11.2 Cell Configurations and Performance Scalability

11.2.1 Cell Configurations

In SOFCs, *2-electrode configuration* and *3-electrode configuration* are the two main cell configurations used to measure the electrode performance. Symmetrical two-electrode cells with the same electrode and gas composition on both sides are often used to obtain impedance data unambiguously assignable to a cathode or anode. However, to accurately isolate electrode performance, and in particular the polarization behavior of the cathode or anode, the three-electrode cell configuration with a RE needs to be used. In solid electrolyte cells, in addition to the alignment of the WE and CE, the position and placement of the RE is ultimately critical. A proper RE should possess well defined and stable potential as well as low impedance. The most common design used in the evaluation of electrode and cell performance in SOFCs is the planar disk or plate-type cells with symmetric WE and CE and point or ring-shaped RE due to their simple geometrical arrangement and easy fabrication. Thick electrolyte pellet cells with Luggin-type RE [10] are also used, which is considered to be analogous to using a Luggin capillary in an aqueous electrolyte. Nevertheless, this is a complex design and thick pellets with an internal channel for Luggin-type RE are not easy to fabricate. The thick electrolyte (~20 mm) may also limit the practical and technologically meaningful current density that can be achieved. Figure 11.7 shows a typical thick electrolyte cell with Luggin-type RE and planar cell configurations commonly used in the performance measurement and isolation in SOFCs. Button cells with circular WE and CE with a diameter of 2–3 cm or relatively large planar

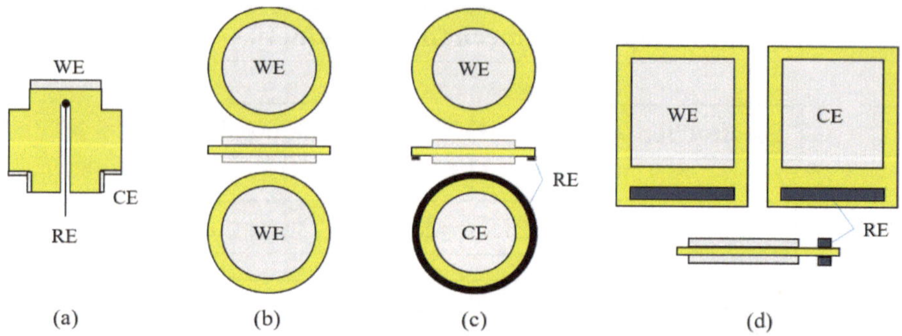

Fig. 11.7 Cell configurations used in performance measurements in SOFCs. **a** Thick electrolyte pellet cells with Luggin-type RE, **b** two-electrode symmetrical cell, **c** three-electrode cell with RE, and **d** three-electrode cell with two RE for large planar cells. WE—working electrode; CE—counter electrode; RE—reference electrode

cells with large size (e.g., 5 × 5 cm) are commonly used. In SOFCs, RE can be located on the electrolyte surface, either on the WE or CE side of the cell. However, in disk-type cells, the ring-shaped RE is better to be placed on the electrolyte surface of the CE side to minimize the effect of gas partial pressure changes at the WE on the RE potential. Due to the relatively high-electrolyte resistance and location of a RE on the electrolyte surface, the incorrect use of a RE can lead to inaccurate and misleading results. This will be discussed in detail later.

On the other hand, 2-electrode configuration with only anode and cathode is commonly used for the cell polarization performance evaluation, where the separation of the electrode polarization at anode and cathode is not possible, e.g., in the case of anode-supported thin electrolyte cells or tubular cells, or the separation of individual electrode polarization is not required.

11.2.2 Symmetric Cell

In symmetric cell configurations, both WE and CE are exposed to the same atmosphere, and electrochemical performance or activities are measured at equilibrium or close to the equilibrium conditions. Assuming the complete reversibility of the reaction at WE and CE under equilibrium conditions, the electrochemical impedance behavior of the electrode for the individual reaction processes would be equal to the half of that measured on the symmetric cell. Figure 11.8 shows the principle of the symmetric cell configuration and the ideal electrochemical polarization and impedance behavior [11]. To be reversible, the reaction should be in the linear region of polarization curves in both the cathodic and anodic polarization directions. The electrochemical impedance spectroscopy curves measured at the WE and CE should be identical in characteristics and size. In high-temperature solid oxide cells (i.e., including SOFCs and SOECs), the reaction occurs at the solid electrode and solid

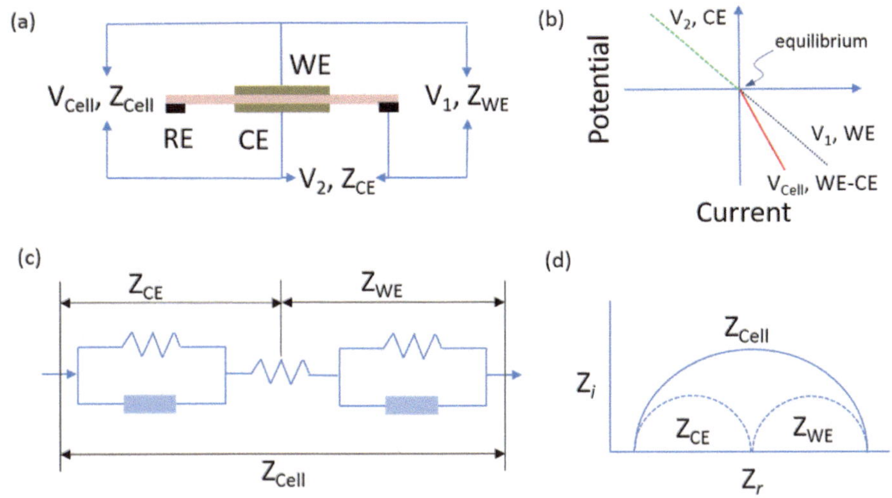

Fig. 11.8 **a** Symmetric cell with a reference electrode, **b** polarization curves of a symmetric cell where the overpotential of CE and WE is measured between CE and RE and, WE and RE, respectively, **c** equivalent circuit for symmetric cell, and **d** impedance responses of ideal symmetric cell. Reproduced from Ref. [11] with permission from Elsevier, Copyright 2020

electrolyte interface. Therefore, the reversibility of the electrode reaction depends essentially on the reversibility of the interface involved in the individual reaction steps, which may not be the same for the reaction at the symmetric electrodes, i.e., WE and CE. The electrochemical measurement of symmetric cells is typically performed at open circuit voltage conditions or at low overpotentials in order to maintain the electrode reactions at both sides in the reversible or equilibrium region. The experimental setup for a two-electrode cell configuration is simple and is particularly useful for the initial identification of the electrochemical activity of cathode materials. However, care should be taken to verify the reversibility of the electrode reactions in symmetric cells.

However, the reversibility of a symmetric cell configuration is related to the magnitude of the DC bias or overpotential. Figure 11.9 shows an example of electrochemical impedance responses of the WE and CE on a symmetric LSM electrode on a YSZ electrolyte-supported cell with a ring-type RE. At open circuit, the electrochemical impedance responses measured from WE and CE are identical and the sum of impedance measured from WE and CE is equal to that measured between WE and CE, i.e., the cell impedance. This indicates the reversibility of the reaction at WE and CE at equilibrium. With the application of a DC bias (even as low as 10 mV), the electrode polarization resistance starts to deviate. The electrode polarization resistance, R_p or R_E, for the oxygen reduction reaction (ORR), i.e., the reaction on WE, increases while R_E for the oxygen evolution reaction (OER), i.e., the reaction on CE, decreases. The reason is most likely related to the differences in the interface for the reaction, for ORR the reaction occurs on three phase boundaries (TPB), while

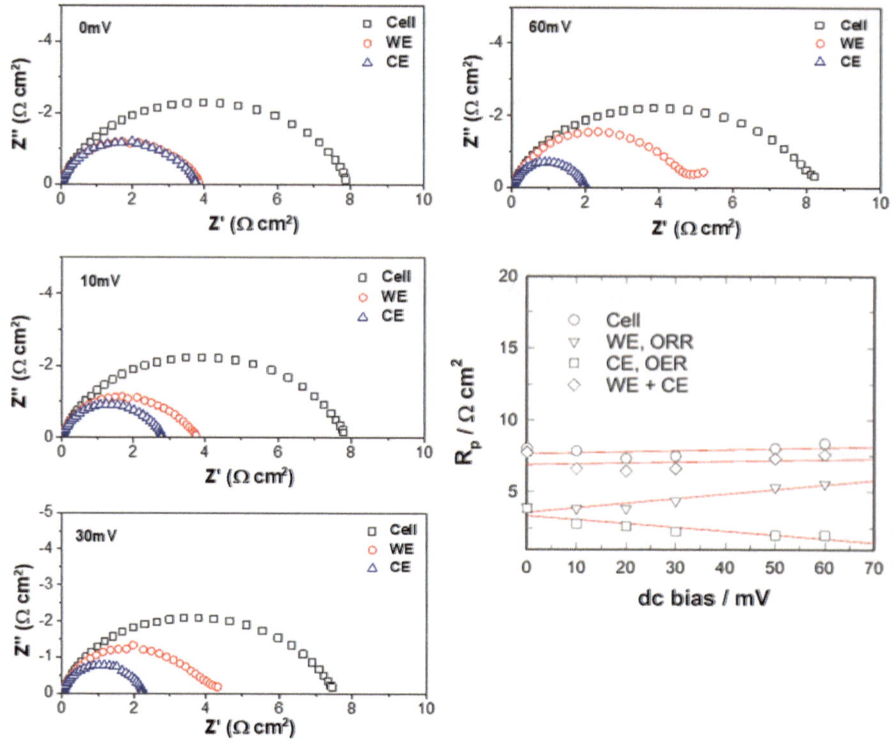

Fig. 11.9 Impedance spectra of the full cell, WE and CE for a LSM symmetric cell with DC bias (η) of 0, 10, 30 and 60 mV and the electrode polarization resistance, R_p (i.e., R_E) for ORR, OER and whole cell at 800 °C. Modified from Ref. [11] with permission from Elsevier, Copyright 2020

for OER, the reaction takes place at TPB as well as two phase regions between the electrode and electrolyte, although on a molecular level, the individual reaction steps are reversible for ORR and OER on WE and CE, respectively. This indicates that the assumption of reversibility of the electrode reaction, ORR and OER, in this case on LSM electrode and YSZ electrolyte, is valid only under equilibrium conditions and not under DC bias.

The reversibility of electrode reactions in solid electrolyte cells is related to the reversibility at the molecular level (the individual reaction steps) as well as at the interface level (the regions where the individual reaction steps occur). The reversibility of the symmetric cells also depends strongly on the nature in particular the ionic conductivities of electrode and electrolyte materials. MIEC electrodes such as LSCF show a high reversibility in symmetric cells, as compared to LSM. On the other hand, Pt electrodes do not show reversibility even under open circuit conditions [11]. Another point is that the electrochemical information obtained from symmetric cells under open circuit (i.e., with no DC bias or overpotential) is in reality presenting the electrode reaction of O_2 under reversible conditions for ORR and OER. This needs to be taken into account in the evaluation of the reaction mechanism and kinetics, unless the electrochemical tests are done under either cathodic or anodic DC bias.

11.2.3 Polarization Performance Analysis

11.2.3.1 Polarization losses

The performance and power output of a SOFC strongly depend on the operating temperature, the air and H2 fuel flow rates, electrochemical activity and properties of electrode and electrolyte materials, etc. Due to the limitation of solid electrolyte and high temperatures, electrochemical impedance spectroscopy and polarization curves are the two most important electrochemical techniques (see Sects. 11.4 and 11.5) in the evaluation of electrochemical activity of electrode materials and performance of cells. The polarization losses in a SOFC can be divided into activation polarization, concentration polarization, and ohmic polarization, as discussed in Chap. 3. Thus, the total polarization or overpotential, η_{total}, is expressed as:

$$\eta_{total} = \eta_{electrode} + \eta_{\Omega} = \eta_a + \eta_c + \eta_{\Omega} = (\eta_{a,act} + \eta_{a,conc}) + (\eta_{c,act} + \eta_{c,conc}) + \eta_{\Omega} \tag{11.1}$$

The ohmic polarization, η_{Ω}, is the overpotential due to the ohmic resistance of electrolyte, electrodes, electrode/electrolyte interface, and the contact between electrode and current collector (i.e., contact resistance). $\eta_{electrode}$ is the sum of polarization losses at the anode and cathode, η_a and η_c, which can be further separated into activation and mass transport or concentration polarization losses, $(\eta_{a,act} + \eta_{a,conc})$ and $(\eta_{c,act} + \eta_{c,conc})$, respectively. An efficient operation of SOFCs requires both ohmic polarization and electrode polarization losses as small as possible. Parameters such as ionic conductivity of electrolyte, electronic, and ionic conductivities of anode and cathode materials, thickness of the electrode coating and electrolyte, porosity of electrode and contact layers and electrode/electrolyte interface all affect the polarization performance of a SOFC. The main contributor to the total ohmic resistance, R_{Ω}, of a SOFC is considered to be ionic conductivity and thickness of the electrolyte. In addition, the interface between electrode and electrolyte and between electrode and current collector is also important for η_{Ω} as shown later in the sections related to the constriction effect and contact resistance in this chapter.

The activation and concentration polarization at anode and cathode are dependent on the current at the interface and distribution of reactants and products, i.e., partial pressure of hydrogen and water vapor at the anode, p_{H_2} and p_{H_2O}, and partial pressure of oxygen, p_{O_2} at the cathode. Figure 11.10 shows a schematic diagram of the cross section of anode-supported cell with the changes in the partial pressure of various gaseous species in a H_2-O_2 cell [12]. In the case of humidified hydrogen at room temperature, $p^o_{H_2} = 0.97$ and $p^o_{H_2O} = 0.03$. The partial pressure of H_2 and H_2O at the anode support/AFL is related to a current density, i, given by $p_{H_2(i)}(i)$ and

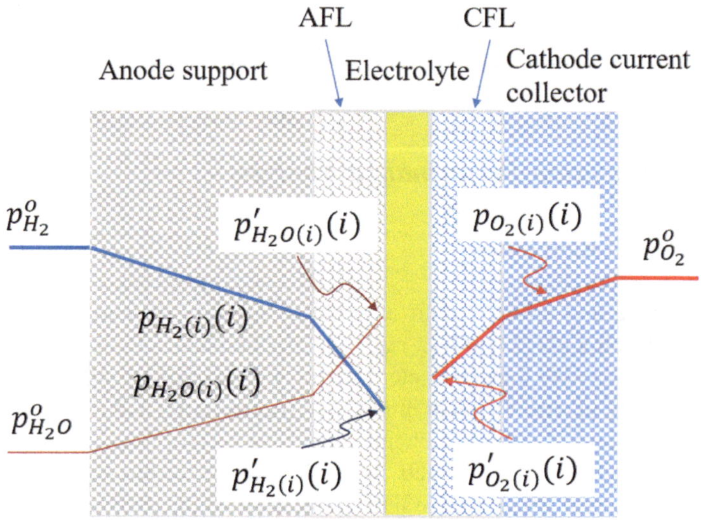

Fig. 11.10 Schematic diagram of an anode-supported cell with five layers and variations of partial pressures of reactants and products in a H_2-O_2. In the figure, AFL and CFL are the anode functional layer and cathode functional layer, respectively. The superscript o and ′ indicates position of partial pressure of species outside of the electrode and at the interface, respectively. Adopted from Ref. [12] with permission from Elsevier, Copyright 2005

$p_{H_2O(i)}(i)$, and the partial pressure of O_2 corresponding to a current density is given by $p_{O_2(i)}(i)$. It is assumed that gaseous transport through the porous electrode can be described by binary diffusion and microstructure of electrode and current collector layers is uniform. Under these conditions, the partial pressure of gaseous species within a given region would be expected to vary linearly with the distance, as shown in Fig. 11.10. With the input of various performance parameters, analytical models can be established to predict the polarization performance of a SOFC as a function of various operation parameters such as temperature, partial pressure of oxygen and hydrogen, and material parameters such as electrode and electrolyte thickness, the porosity of electrode, and contact layers.

Similar to PEMFCs, the polarization performance of a SOFC also shows an activation effect; i.e., the performance of a freshly prepared cell improves with the polarization current treatment, characterized by the reduction of both cell ohmic resistance and cell area-specific resistance. Figure 11.11a shows an example of the activation effect of Ni/YSZ anode-supported cell with LSM/YSZ composite cathode and thin YSZ electrolyte (45 μm) [13]. The initial impedance of the cell before current treatment is characterized by a large arc at low frequencies and a much smaller one at high frequencies. The ASR of the cell, measured by the differences between the low- and high-frequency intercepts, is ~1.80 Ωcm^2, and cell ohmic resistance, R_Ω, is ~0.2 Ωcm^2. After polarization treatment at 200 mA/cm^2 for 5 min, ASR decreases to 1.35 Ωcm^2 and is further reduced with increasing polarization treatment. After polarization treatment for 4 h, ASR is reduced to 0.9 Ωcm^2, half of the as-prepared cell. The reduction in the cell impedance is primarily on the low frequencies, indicating that such activation or conditioning effect is largely related to the activation effect of polarization on the LSM/YSZ composite cathode as shown in Chap. 10. The reduction in the cell impedance is also evident by the change of cell voltage measured at 200 mA/cm^2 with the polarization time. The cell voltage increases from initial 0.59 V to 0.82 V after polarization current passage for 2 h, an increase of 230 mV in cell voltage. This corresponds to an increase of 38% in power density of the cell! However, the cell activation or conditioning effect is very much dependent on the nature and activity of the electrode materials. For instance, in the case of MIEC cathodes, the effect of polarization treatment on the cell activation is much smaller.

Similar to that in the case of PEMFCs, the performance of SOFCs is kinetically limited by the sluggish oxygen reduction reaction (ORR). Thus, the oxygen supply rate has a significant effect on the power output of the cell particularly in the case of using air as oxidant. The effect of air flow rate on the cell performance is most evident via the polarization curves, see Fig. 11.11b. The overall cell impedance decreases slightly with the increase in air flow rates. In contrast, the cell polarization curves change significantly with the air flow rates. The effect of air flow rates on the cell voltage versus current density characteristics is most significant at high

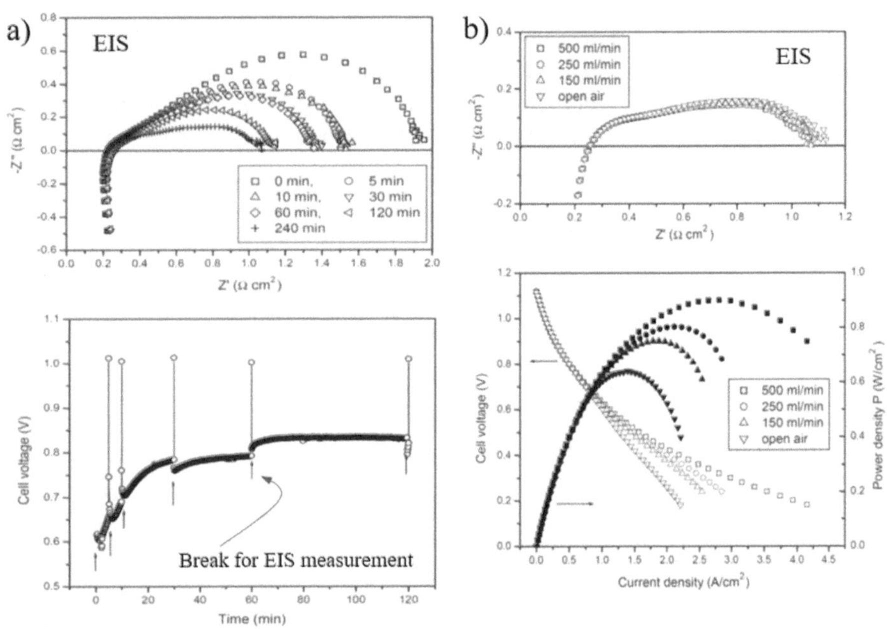

Fig. 11.11 **a** Cell activation or conditioning effect as shown by the change of impedance and cell voltage at 200 mA/cm^2 of an as-prepared anode-supported cell as a function of polarization treatment time at 200 mA/cm^2 and 800 °C and **b** effect of air flow rate on impedance and cell performance of the same cell at 800 °C. Modified from Ref. [13] with permission from Elsevier, Copyright 2004

current densities. This is a clear indication of the dominance of mass transport in the polarization performance of this specific cell. In this case, the peak power density (PPD) increases from 0.64 W/cm^2 in open air to 0.90 W/cm^2 with air flow rate of 500 mL/min. This example also shows that it is important to evaluate the performance of SOFCs through both EIS and polarization techniques.

11.2.3.2 Deviation from ideal *i-V* curves

With the combination of EIS and polarization techniques, the total ohmic resistance and electrode polarization components of cells can be separated. Figure 11.12 shows the typical *i-V* curves with the separation of ohmic drop and electrode polarization of an anode-supported cell [13]. The *i-V* curves were obtained using linear sweep voltammetry at low scan rate, e.g., 1 mV/s. The total cell ohmic resistance, R_Ω, can be obtained from the high-frequency intercept of the impedance measured under OCV. To increase the accuracy of the cell resistance, R_Ω can also be obtained by the impedance responses measured under dc bias. For the cell shown in the figure, $R_\Omega = 0.198\ \Omega cm^2$ obtained from the high-frequency intercept from the impedance measurement at OCV. The OCV, E_{OCV}, is 1.11 V, very close to the theoretically calculated value of 1.05 V at 800 °C under 97% H_2/3% H_2O as fuel and air as oxidant. The *i-V* curve shows the activation loss and ohmic loss regions but no concentration polarization losses in this case probably due to the high air flow rate. The transition from activation region to ohmic loss region is rather gradual, different from the instantaneous and fast transition as observed in the case of low temperature

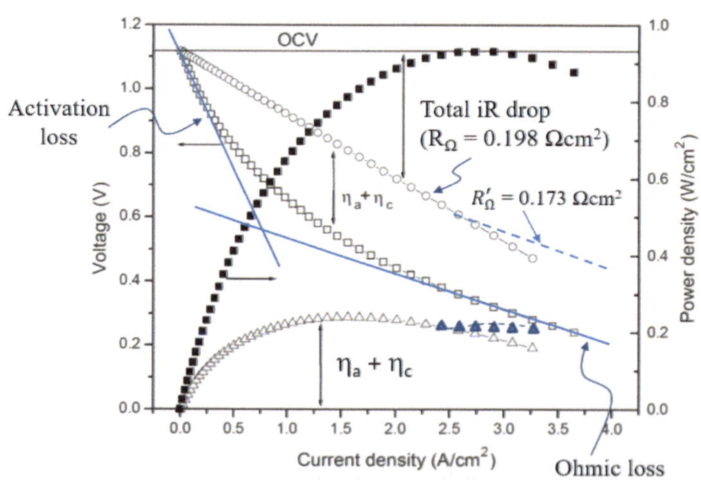

Fig. 11.12 *i-V* curve of an anode-supported cell with separation of cell iR drop and cell polarization losses, $\eta_a + \eta_c$, measured at 800 °C in 97% H_2/3% H_2O at hydrogen flow rate of 250 mL/min and air at a flow rate of 500 mL/min. The YSZ electrolyte thickness of the Ni/YSZ anode-supported cell is 45 μm. Modified from Ref. [13] with permission from Elsevier, Copyright 2004

PEMFCs. This is probably due to the high kinetics of the electrode reactions at high temperatures. Based on the R_Ω of the cell, the electrode polarization losses, $\eta_a+\eta_c$, can be calculated:

$$\eta_a + \eta_c = E_{OCV} - E_{cell} - iR_\Omega \tag{11.2}$$

where E_{cell} is the measured cell voltage. The electrode polarization for both oxygen reduction and hydrogen oxidation reactions increases initially with the polarization current and accounts for 48% of overall losses of the cell at a current density of 1.5 A/cm^2. With the increase of current density to 2.5 A/cm^2, the electrode polarization losses decrease to 33% of overall cell polarization losses. The decrease of the electrode polarization losses at high current densities indicates the deviation of polarization behavior of SOFCs from ideal *i-V* curves as shown in Chap. 3.

For SOFCs, heat can be produced locally at the electrode/electrolyte interface region due to the effect of Joule heating particularly at high current densities. The heat produced would increase the cell temperature and thus decreases the cell ohmic resistance. This is indicated by the reduced slope of the ohmic loss region of the *i-V* curve, as compared to that based on the R_Ω obtained under OCV. From the slope of the ohmic loss region, the overall cell resistance can be calculated, $R'_\Omega = 0.173$ Ωcm^2, lower than cell ohmic resistance of 0.198 Ωcm^2 obtained at OCV. As R'_Ω also includes the ASR, the actual cell ohmic resistances under high current load could be even smaller. Using R'_Ω obtained in the high current density region, the electrode polarization losses at high current density region can be calculated, see dotted line and solid triangle symbols in Fig. 11.12. The revised electrode polarization losses increase with the current density in general, consistent with that expected from the ideal *i-V* curves. Nevertheless, the studies show that the deviation from ideal *i-V* curves is most significant in cells with high ohmic resistance because Joule heating effect is most prominent at high ohmic resistance and high current densities. The deviation from ideal polarization curve will lead to the underestimation of cell electrode polarization losses at high current densities.

The performance of SOFCs is primarily evaluated by the measurement of impedance of the cell at OCV and the measurement of *i-V* curves. However, there are distinctions between the cell impedance responses measured at OCV and the polarization performance or power output of the cell (see Fig. 11.11); the former reflects the cell performance only at OCV or in the region of very low current densities, while the latter is directly governed by the cell ohmic resistance and electrode polarization resistance, which are correlated closely with the current load.

11.2.3.3 Performance Scalability

In SOFCs, one important issue which requires special attention is that the electrochemical performance is sometimes dependent on the size or area of the electrode [14]. This is particularly important because for the development of electrode materials for SOFCs it is a common practice to evaluate the performance and activity of the electrode materials and cells on small button cells due to the convenience and easiness in the test rig setup. However, the question is, can the electrode performance obtained from small button cells (e.g., for an electrode area of 8 mm in diameter, the electrode area is ~ 0.5 cm^2) be scalable to the large cells, e.g., cells with active electrode areas of 10 or 100 cm^2? The variation of performance as a function of the cell area can be measured by a parameter termed *performance scalability*.

We can define the scalability as the ratio of the performance or activity measured on small button cells and large plate cells based on the overpotential, η:

$$\varphi = \sum_{i=1}^{i=n} \frac{\eta_{i,\ \text{large plate cell}}}{\eta_{i,\ \text{button cell}}} \tag{11.3}$$

For practical electrodes, φ is equal to or greater than 1. The higher φ value, the poorer the scalability. The scalability of the electrode performance depends on number of factors, including the conductivity of the porous electrode coating, the electrical contact and contact area between the current collector and electrode, and the test conditions such as the mechanical load used in the test station. Figure 11.13 compares the electrode performance of LSM cathodes and Ni/YSZ cermet anodes measured on small button cells with an active electrode area of 0.44 cm^2, and on 50 × 50 mm planar cells with an active electrode area of ~ 10 cm^2 under similar test conditions [14]. The overpotentials are measured at 250 mA/cm^2 and 900 °C in air and in 97%H_2/3%H_2O. The small button cells and 50 × 50 mm planar cells are fabricated at the same time, using the same LSM and Ni/YSZ inks. Pt and Ni woven meshes are used as current collectors for the LSM cathode and the Ni/YSZ

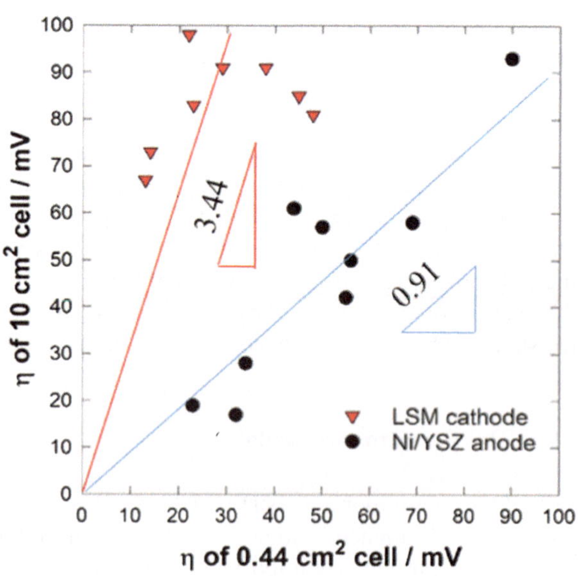

Fig. 11.13 Plots of overpotentials (η) measured at 250 mA/cm^2 and 900 °C for ORR on LSM cathode and HOR on Ni/YSZ cermet anodes on small button cells (0.44 cm^2) and planar cells (10 cm^2) under identical test conditions. Modified from Ref. [14] with permission from Elsevier, Copyright 2003

cermet anode, respectively. Despite the similar contact and identical test conditions, there are significant differences in the electrochemical performance of LSM cathodes measured on small button cells and large planar cells. However, for Ni/YSZ cermet anodes, the differences are much smaller.

Based on the data given in Fig. 11.13, we can calculate the scalability for LSM cathodes and Ni/YSZ cermet anodes:

$$\varphi_{\mathrm{LSM}} = 3.44 \pm 1.71 \tag{11.4}$$

$$\varphi_{\mathrm{Ni/YSZ}} = 0.91 \pm 0.16 \tag{11.5}$$

For LSM electrodes, φ_{LSM} is 3.44 ± 1.71. This indicates that the performance losses measured on large plate cells with an electrode area of 10 cm^2 are more than three times higher than that measured on small button cells with an electrode area of 0.44 cm^2. In the case of Ni/YSZ cermet anodes, $\varphi_{\mathrm{Ni/YSZ}}$ is 0.91 ± 0.16, very close to unity, indicating an excellent scalability and correlation between the performance measured on small button cells and on 50 × 50 mm large cells. The effect of cell size on the performance scalability is most pronounced on LSM cathodes but is negligible on Ni/YSZ cermet anodes.

The significantly reduced performance of LSM cathode on large cells is due to the low conductivity of the porous LSM electrode. In this experiment, the electronic conductivity of porous LSM electrode coating and Ni/YSZ cermet anode coating is 45 and 254 S/cm at 800 °C, respectively [15] (please note the electronic conductivity of a porous electrode coating is generally much lower than the corresponding material conductivity). The low electronic conductivity will lead to a high contact resistance between the electrode coating and current collector. With the increase in the cell size, the load to the cell will be reduced and the pressure between the electrode coating and current collector becomes increasingly uneven. This again will increase the contact resistance between the electrode coating and the current collector, which in turn will lead to the increase in the polarization loss at the electrode/electrolyte interface [15]. The fundamental reason for the effect of the electrical contact between electrode and current collector on the activity of the electrode is due to the constriction effect. This will be discussed in following sections. The much higher electronic conductivity of Ni/YSZ cermet anodes improves the contact between the cermet and current collector, and substantially reduces the constriction effect on the polarization performance of the anode.

The scalability can also be defined as the ratio of the peak power density (PPD) measured on small button cells and large plate cells:

$$\varphi = \sum_{i=1}^{i=n} \frac{\mathrm{PPD}_{i,\ \text{button cell}}}{\mathrm{PPD}_{i,\ \text{large plate cell}}} \tag{11.6}$$

Clearly, the absolute value of scalability is closely related to the size of the electrodes or cells used and is case sensitive.

11.2.3.4 Accelerated stress test

For commercially viable SOFC technologies, the useful lifetimes for SOFC stacks are expected to be 40,000 to 80,000 h for both residential and stationary combined heat and power (CHP) applications. In addition to the development of reliable system control and better SOFC stack, the intrinsic degradation of the cell/stack needs to be adequately addressed and known. Clearly, it is not practical to proven the durability of every cell materials and stack modifications under fuel cell operation condition in real-time tests due to the extreme time consuming and high cost involved, though real-time durability is required for verification of key materials and cells. A simple comparison of cell voltage reduction rates at different temperatures does not necessarily represent the degree of degradation. This is due to the fact that performance degradation at high temperatures is driven by various factors and proceeds through complicated pathways, including the deterioration of electrode, electrolyte and interconnect materials, interfacial reactions between various components and responses to various external and internal contaminants. Accelerated stress test (AST) is also used in SOFCs to enable rapid analysis of realistic degradation rates and enable the degradation models that can be used to extrapolate realistic performance of the system to end of useful life.

The establishment of AST protocol is critical and needs to be defined that characterize the degradation behavior of a SOFC toward end of life (EoL) in much shorter time. To establish AST protocol, one would need to have an extensive database with various accelerating factors including operation conditions such as temperature, current, fuel utilization, thermal cycle, load cycle, redox cycle, and materials characteristics such as microstructure sintering and agglomeration and surface segregation and reaction. This will offer the guidelines for the quantitative identification and prediction of critical degradation mechanisms. Based on the proposal by Polverino et al. [16], a general procedure for AST protocol design can be written as follows:

- Identify degradation processes and mechanism and design/select representative microscale models;
- Identify key parameters influenced by degradation and design/select related mesoscale models;
- sign/select macroscale performance models;
- Combine the different models through a multiscale modeling approach;
- Perform a parametric analysis to investigate the impact of the operating conditions on degradation rate;
- Design mathematical transfer functions to correlate accelerating factors and degradation rate;
- Develop guidelines/tools for lifetime estimation and AST protocol design.

Based on the procedure, Polverino et al. [16] developed a multiscale modeling framework to correlate the microscale level (i.e., Ni agglomeration) to the macroscale level (i.e., degradation in cell voltage) for the performance durability assessment of SOFCs.

11.3 Equipotential Line and Constriction Effect

11.3.1 Equipotential Line

In SOFCs, the accurate separation and determination of polarization losses of various processes at the electrolyte, cathode, and anode are essential for the optimization of electrode microstructure and the performance of SOFCs. By using a RE, one can in principle obtain the polarization performance and kinetic parameters of the anode or cathode independently. However, due to the high operating temperature of 600–1000 °C and the all-solid components, the polarization measurement of individual electrodes of a SOFC is constrained by the cell/electrode geometry and position of the RE.

In an idealized thin, flat-plate cell consisting of a dense oxygen ion conducting electrolyte with active electrodes on each side and a RE placed on the surface of the inactive portion of the electrolyte, the equipotential line or surface would be present somewhere between the two active electrodes, i.e., anode and cathode, see Fig. 11.14. To satisfy the requirement of a RE on an inactive surface of electrolyte, the RE needs to be placed at a distance greater than three electrolyte thicknesses from the edge of the active electrodes. This can be characterized by an aspect ratio, which is defined as the distance between the RE and the edge of the active electrode divided by the electrolyte thickness. For plate-type geometry, the electrolyte is generally approximately equipotential at aspect ratios higher than 3.

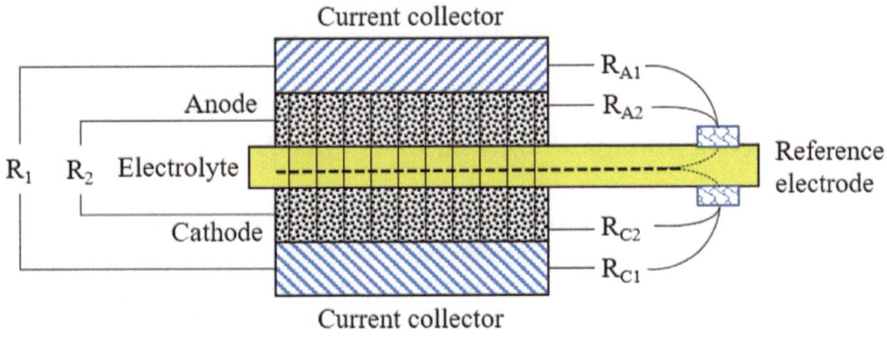

Fig. 11.14 Equipotential line/surface location (dotted line) and partitioning of cell resistance in a thin electrolyte cell with symmetric contacts at the anode/electrolyte and cathode/electrolyte interface

Let us to start to look at the resistance distribution and measurement in a SOFC. In a solid electrolyte cell based on Fig. 11.14, the total cell resistance, R_1 can be written as:

$$R_1 = R_{A1} + R_{C1} \tag{11.7}$$

where R_{A1} and R_{C1} are the anodic and cathodic partitioning resistance of R_1 measured by a conventional voltage probe against RE, respectively. R_{A1} includes the contact resistance between the anode and current collector, anode resistance and part of the electrolyte resistance. R_{C1} include the contact resistance between cathode and current collector, cathode resistance and the remaining electrolyte resistance. As reference potential samples an average of the equipotential surface in the cell, the division or partitioning of the cell resistance is an indication of the location of the equipotential line/surface. Under idealized conditions, the equipotential line lies within the electrolyte, and thus the cell voltage (V_{cell}) can be divided into V_A and V_C. V_A contains the overpotential of the anode plus some fraction of the electrolyte *iR* drop and V_C contains the overpotential of the cathode plus the remaining fraction of the electrolyte *iR* drop. The total cell impedance can thus be separated into two half-cell impedances.

However, the polarization and impedance separation of a solid electrolyte cell is based on an ideal symmetric cell with perfect alignment and similar characteristics (resistance and capacitance) of the active electrodes. In reality, such an idealized cell may not exist. Perfect alignment is difficult to achieve, particularly for small button cells. In addition to the alignment and characteristics of active electrodes, the equipotential line/surface in a cell also depends on the contact between the porous electrode coating and current collector [15]. In practical cells, particularly with a thin electrolyte, the equipotential line/surface may be shifted away from the electrolyte, leading to inaccuracy or misleading results. Thus, it is fundamentally important to understand the intrinsic relation between the location of the equipotential line, the microstructure and electrochemical characteristics of the electrode and current collector. This essentially requires the accurate measurement of cell resistance partition and thus an alternative design with specific voltage probes.

11.3.2 Special Voltage Probe and Resistance Distribution

Conventional voltage probes (VP) would not be able to separate the anodic and cathodic partitioning resistance accurately due to the large contribution of the contact resistance between the electrode and current collector. In order to accurately assess the resistance partition of the electrolyte, a specific voltage probe needs to be used [17]. Figure 11.15 shows the cell configuration with special voltage probes. In the configuration, VP1 is the conventional voltage probe attached to the current collector (CC) and VP2 is the special voltage probe. Different to VP1, *special voltage probes*

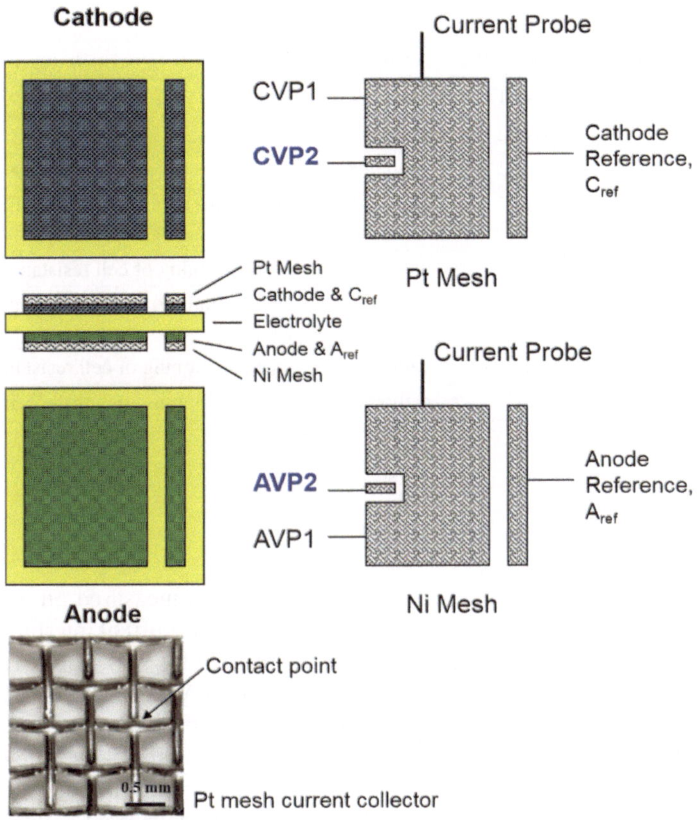

Fig. 11.15 Cell configuration and arrangement of conventional and special voltage probes (VP1 and VP2) with cathodic and anodic reference electrodes (C_{ref} and A_{ref}). A and C represent anode and cathode, respectively, and the cell has typically an area of 5 × 5 cm. An optical picture of the Pt woven mesh current collect is also shown in the figure. Reproduced from Ref. [18] with permission from IOP Publishing, Copyright 2017

are electrically isolated from the current collector. The function and resistances measured by the voltage probes and reference electrodes are given in Table 11.1 [18].

From the resistance distribution measured by various VPs and RE, we have following equations:

$$R_1 = R_{A/C} + R_2 + R_{C/C} = R_{A1} + R_{C1} \tag{11.8}$$

$$R_2 = R_{A2} + R_{C2} \tag{11.9}$$

$$R_{A2} = R_{A1} - R_{A/C} \tag{11.10}$$

Table 11.1 Resistances measured by VP1, VP2 and RE

Resistance	Between probes/RE	Assigned resistances
R_1	CVP1 and AVP1	total cell resistance
R_2	CVP2 and AVP2	electrolyte resistance
$R_{A/C}$	AVP1 and AVP2	anode/CC contact resistance (including anode resistance)
$R_{C/C}$	CVP1 and CVP2	cathode/CC contact resistance (including cathode resistance)
R_{A1}	AVP1 and C_{ref}	anodic resistance partitioning of cell resistance
R_{A2}	AVP2 and C_{ref}	anodic resistance partitioning, minus the anode/CC contact resistance
R_{C1}	CVP1 and A_{ref}	cathodic resistance partitioning of cell resistance
R_{C2}	CVP2 and A_{ref}	cathodic resistance partitioning, minus the cathode/CC contact resistance

$$R_{C2} = R_{C1} - R_{C/C} \tag{11.11}$$

Figure 11.16 shows examples of polarization curves measured on a YSZ electrolyte cell with Pt anode and Pt cathode by a galvanostatic current interruption (GCI) technique at 900 °C. The total cell resistance measured by VP1, R_1, is 1.23 Ωcm^2, much higher than the electrolyte resistance measured by VP2, $R_2 = 0.61\ \Omega cm^2$. This indicates the existence of significant contact resistance between electrode and electrolyte. GCI curves measured between voltage probes on the same electrode sides are dominated by *iR* losses with negligible overpotential contribution: characteristics of the contact resistance [16]. From GCI curves, the anode/current collector contact resistance and the cathode/current collector contact resistance, $R_{A/C}$ and $R_{C/C}$, are 0.34 and 0.29 Ωcm^2, respectively. The very high polarization losses, ~ 230 mV at 25 mA/cm^2 measured between voltage probes are actually due to the fact that Pt is a very poor electrode material for both H_2 oxidation and O_2 reduction reactions (HOR and ORR) at high temperatures. The cathodic resistance as measured between CVP1 and A_{ref} is very small, $R_{C1} = 0.08\ \Omega cm^2$. On the other hand, the anodic resistance measured between AVP1 and C_{ref}, R_{A1} is 1.17 Ωcm^2. The most interesting observations are the GCI curves measured between CVP2 and A_{ref} on the Pt cathode side, showing an unusual and depressed GCI curve (indicated by the circle, Fig. 11.16b, c). However, from the instantaneous potential drop immediately after the current interruption, a resistance between CVP2 and A_{ref} can be measured, R_{C2} is 0.24 Ωcm^2. Nevertheless, as R_{C2} is obtained from the depressed GCI curve, R_{C2} could be regarded as depressed or *reversed ohmic resistance* between the special cathodic voltage probe and RE. On the other hand, a normal GCI curve is observed between AVP2 and C_{ref}, giving $R_{A2} = 0.82\ \Omega cm^2$. Table 11.2 lists the resistances measured on the cell with Pt cathode and Pt anode. From the resistance values, other resistance parameters can also be obtained. For example, the difference between R_{A1}

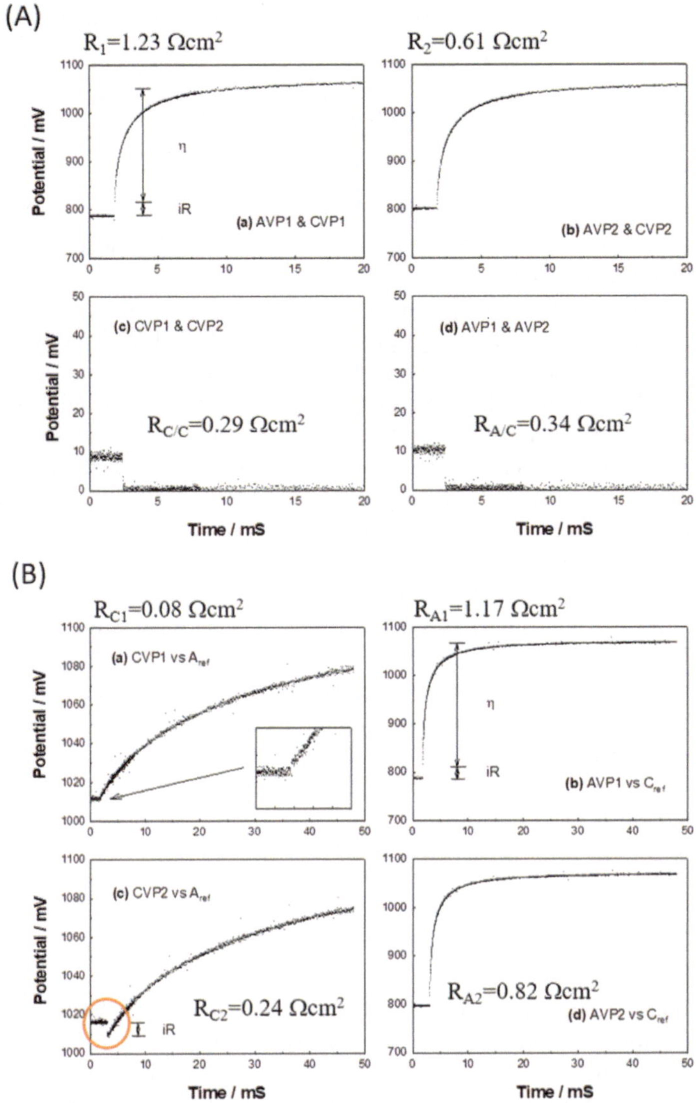

Fig. 11.16 GCI curves between (A) the voltage probes and (B) the voltage probes and reference electrodes on a cell with Pt anode and Pt cathode, measured at 25 $mAcm^{-2}$, 900 °C. The thickness of 3 mol%Y_2O_3 doped ZrO_2 (3YSZ) electrolyte plate is 108 ± 4 μm. The circle indicates the depressed polarization decay

Table 11.2 List of resistances measured on cells with Pt cathode and Pt anode in Ωcm^2, measured at 900 °C and with LSM cathode and Ni/YSZ anode, measured at 800 °C

Cell	R_1	R_2	$R_{A/C}$	$R_{C/C}$	R_{A1}	R_{A2}	R_{C1}	R_{C2}
Pt/YSZ/Pt	1.23	0.61	0.34	0.29	1.17	0.82	0.08	0.24
LSM/YSZ/Ni/YSZ	1.41	0.77	0.34	0.32	0.03	0.32	1.37	1.07

and R_{A2} is the anode coating resistance and the contact resistance between anode and current collector, which is 0.35 Ωcm^2.

Let us briefly discuss the physical significance of resistances measured by the conventional and special voltage probe couples (R_1 and R_2) and by the voltage probes on the same electrode side ($R_{A/C}$ and $R_{C/C}$). R_1 is the total cell resistance measured by VP1 and includes the resistive contribution from the current collector, the electrode and the electrolyte. On the other hand, R_2 is primarily determined by the electrolyte resistance due to the fact that VP2 is isolated from the current collect with no current passing through the probe. $R_{A/C}$ and $R_{C/C}$ are the resistive component of electrode and contact resistance between electrode and current collector.

As $R_{A/C}$, R_2 and $R_{C/C}$ are connected in series, AVP2 is located between $R_{A/C}$ and R_2 and CVP2 between $R_{C/C}$ and R_2. For the same reason, AVP1 and CVP1 are located outside $R_{A/C}$ and $R_{C/C}$, respectively. R_{A1} and R_{C1} represent the anodic and cathodic partitioning resistance of the cell resistance (R_1), respectively. As the reference potential samples in average the equipotential line in the cell, the division or partitioning of the cell resistance is an indication of the location of the equipotential line. Based on Table 11.2, the resistance distribution in the cell with a Pt anode and a Pt cathode with an estimated location of the equipotential line can be obtained, and this is shown in Fig. 11.17.

From the resistances measured (Table 11.2) and Eq. (11.8), the total resistance of the cell with a Pt anode and a Pt cathode is:

$$R_1 = 0.34 + 0.61 + 0.29 \approx 1.17 + 0.08 = 1.23\,\text{cm}^2 \tag{11.12}$$

which is equal to the measured total cell resistance, R_1. The measured anodic and cathodic fraction resistances are far from equal and R_{A1} (= 1.17 Ωcm^2) is larger than the electrolyte resistance, R_2 and is almost equal to the total cell resistance, R_1. This indicates that the partitioning of the cell resistance is not in the electrolyte region. R_{C1} (= 0.08 Ωcm^2) is much smaller than the cathode contact resistance, $R_{C/C}$ (= 0.29 Ωcm^2), indicating that the equipotential line of the cell is located inside the cathode. In another words, the equipotential line is located inside the Pt cathode side near the Pt electrode/YSZ electrolyte interface region rather than within the YSZ electrolyte. This is fundamentally the reason for the depressed GCI responses measured between the special voltage probe (CVP2) against RE on the Pt cathode side. This in turn indicates that the cathodic resistance fraction moves away from the electrolyte region and becomes negative, R_{C2} = -0.24 Ωcm^2 with respective to the electrolyte direction. The depressed ohmic resistanceis an indication of the measured

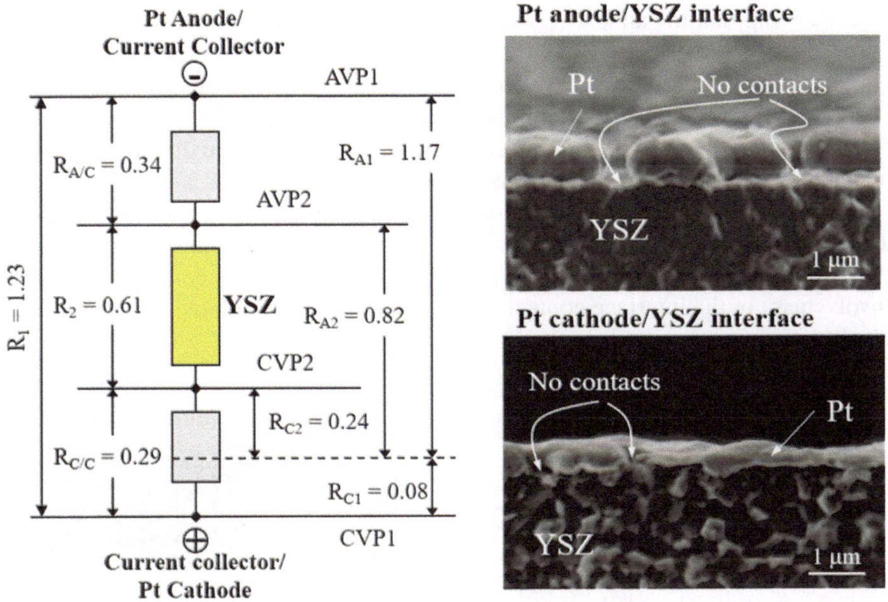

Fig. 11.17 Resistance distribution and partitioning of cell resistance in a YSZ electrolyte cell with Pt cathode and Pt anode at 900 °C, showing the equipotential line location (dotted line) within the Pt cathode. The SEM micrographs show the asymmetric contacts at the electrode/electrolyte interface, showing that the Pt anode/electrolyte interface contact is much smaller than that at the Pt cathode/electrolyte interface. Reproduced from Ref. [18] with permission from IOP Publishing, Copyright 2017

resistance partitioning outside the electrolyte region. By definition of Eq. (11.9), we can also have

$$R_2 = R_{A2} + R_{C2} = 0.82 + (-0.24) = 0.58\,\text{cm}^2 \tag{11.13}$$

This is very close the measured value of R_2 (0.61 Ωcm^2).

The shift of the equipotential line to the Pt cathode side can be explained by the microstructure of the interface of the cell. As shown in Fig. 11.17, for the cell with Pt electrodes, Pt anode is much more porous compared to that of the Pt cathode. Thus, the contact areas at the Pt anode/YSZ electrolyte interface are much smaller than the contact areas between the Pt cathode and YSZ electrolyte. This will push the equipotential line toward the Pt cathode side. Resistance distribution and location of the equipotential line have also been measured on the cell with Ni/YSZ anode and LSM cathode, see Fig. 11.18 [18]. The resistance distributiuon for the cell with Ni/YSZ anode and LSM cathode is given in Table 11.2. Similar to the cell with Pt electrodes, R_{C1} and R_{A1} are also far from equal, but R_{C1} is much higher than R_{A1} and is close to the total cell resistance, R_1. Different to the cell with Pt electrodes, the depressed GCI curves were obtained between AVP2 and C_{ref} on the Ni/YSZ anode side, showing that the anodic resistance fraction, becomes negative with respect to the electrolyte direction, R_{A2} = -0.32 Ωcm^2. This indicates that the location of the equipotential line is inside the Ni/YSZ cermet anode rather than inside the YSZ electrolyte. This is consistent with the much higher interfacial contact between the Ni/YSZ cermet anode and YSZ electrolyte, as compared to the discrete interfacial contact between the LSM cathode and YSZ electrolyte.

Although the estimation of the location of equipotential line/surface as shown above is purely based on the resistance measurement, the consistency between the resistance partitioning and electrode/electrolyte interface contact implies the involvement of the interface contact on the location of the equipotential line in solid electrolyte cells in general.

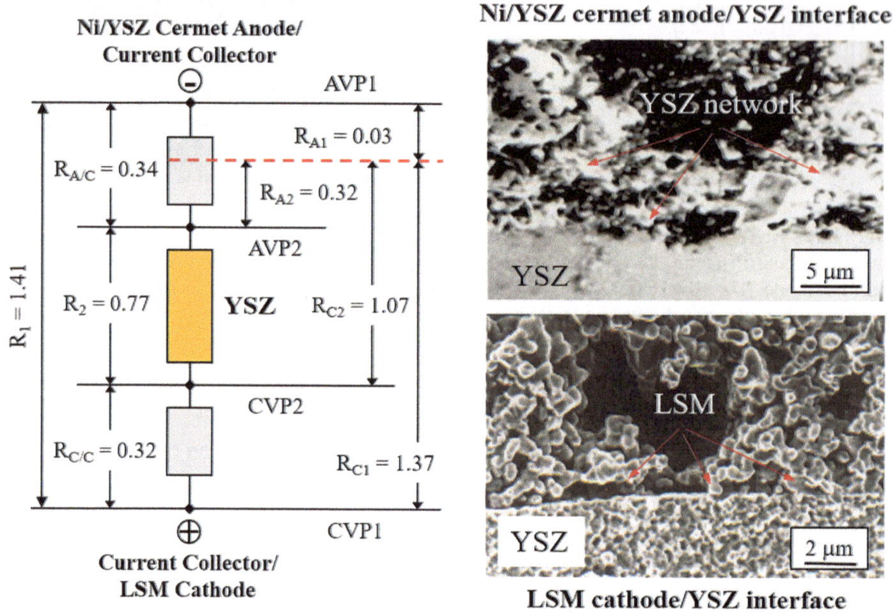

Fig. 11.18 Resistance distribution and partitioning of cell resistance in a YSZ electrolyte cell with a LSM cathode and Ni/YSZ cermet anode at 800 °C, showing the equipotential line location (dotted line) within the Ni/YSZ cermet anode. The SEM micrographs show the asymmetric contacts at the electrode/electrolyte interface, showing Ni/YSZ cermet anode/electrolyte interface contact is much higher than that at the LSM cathode/electrolyte interface due to the presence of YSZ-to-YSZ networks in the cermet. The Ni phase in the Ni/YSZ cermet was removed by acid treatment. Reproduced from Ref. [18] with permission from IOP Publishing, Copyright 2017

11.3.3 Relationship Between Validity of Reference Electrode and Electrolyte Thickness

It is a common knowledge that it is not reliable to separate the electrode polarization from the total cell polarization by reference electrodes in thin solid electrolyte cells. Numerous modeling studies show that distortion to the electrochemical impedance measurement of electrode reactions on thin electrolyte cells is unavoidable due to the misalignment of anode and cathode and the differences in their electrochemical characteristics. Experimentally, we also show that asymmetric contact at the electrode/electrolyte interfaces can cause the significant shift of the equipotential line away from the electrolyte region, leading to significant distortion in the polarization measurements. The distortion in the polarization behavior is demonstrated by the depressed GCI curves [18]. On the other hand, studies demonstrate that in thick electrolyte cells, the polarization, and impedance responses at the electrode/electrolyte interface can be accurately separated and measured [19, 20]. The question is how thick is the thick electrolyte cells or what is the limit in the thickness of the electrolytes in the applicability of RE in the separation of the polarization and impedance responses at the electrode/electrolyte interface in solid electrolyte cells. This question can be approached by experimental studies.

As the distortion is in nature related to the misalignment or asymmetric contacts at the anode and cathode interfaces, such asymmetric contacts can, in principle, be measured by the anodic and cathodic fraction or partitioning resistances, R_{A1} and R_{C1}, of the cell. Therefore, the relative change of the R_{A1} and R_{C1} values can be used as an indicator of the change of the position of the equipotential line. This can be done under controlled conditions. Figure 11.19 shows the results of the dependence of the partitioning of cell resistance as a function of YSZ electrolyte thickness in the range of 65 to 247 μm. In this experiment, large planar cells with an electrode area of 10.88 cm^2 are used with model LSM cathodes and Ni/YSZ cermet anodes [21]. The use of cells with large areas is necessary to increase the accuracy. The same Pt mesh current collectors are used on the cathode and anode side to ensure there is equal contact resistance at both sides. For the cell with 65–100 μm thick electrolytes, the relative cathodic fraction with respect to the total resistance is 80–95%, indicating the significant distortion and shifting of the equipotential line to the cathode/electrolyte interface side. This indicates that for thin electrolyte cells with electrolyte thickness less than ~ 125 μm, the use of RE is no longer reliable. However, the differences between R_{C1} and R_{A1} become smaller with the increase of electrolyte thickness. This implies the shift of the equal potential line to the middle of the electrolyte region with the increase of electrolyte thickness. The effect of the misalignment at the electrode/electrolyte interface contact on the potential distribution decreases with the increase of the electrolyte thickness, effectively diminishing the distortion of misalignment on the polarization and impedance measurements of solid electrolyte cells. For cells with electrolyte thickness larger than ~ 125 μm, the anodic and cathodic partitioning resistance, R_{C1} and R_{A1}, fall within the range of 20–80%. In

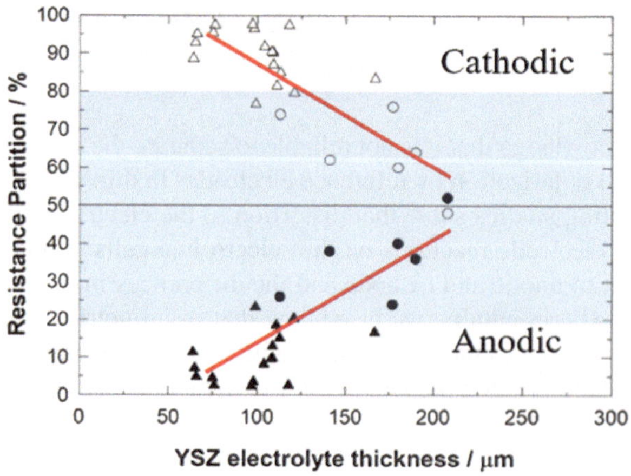

Fig. 11.19 Resistance partitioning measured by relative anodic and cathodic fraction resistances ($\%R_A + \%R_C = 100\%$) of overall cell resistance of 50 × 50 mm cells with LSM cathode and Ni/YSZ anode as a function of YSZ electrolyte thickness. Lines are for the guide only. Symbols: (•) R_{A1} at 1000 °C, (o) R_{C1} at 1000 °C, (▲) R_{A1} at 900 °C, and (Δ) R_{C1} at 900 °C. Modified from Ref. [21] with permission from Springer Nature, Copyright 2004

this case, the anodic and cathodic polarization losses can be accurately separated experimentally under the conditions of the study.

Despite the fact that the cell partitioning resistances are obtained on cells with significant asymmetric electrode/electrolyte contacts with model LSM cathode and Ni/YSZ anode, the clear trend of the partitioning of cell ohmic resistance with the electrolyte thickness has a significant implication for the change of the equipotential line in electrolyte-supported cells with geometric symmetrical electrodes. With an electrolyte thickness equal to or higher than 250 μm, the equipotential line moves to the middle region of the electrolyte, indicating the substantially reduced effect of the physical misalignment between anode and cathode on the distortion of the potential distribution in the electrolyte cells. This in turn indicates that the reliability of RE in the polarization and impedance measurement of the electrode/electrolyte interface is a strong function of the electrolyte thickness. As the misalignment between WE and CE increases with the decrease in the electrode area, therefore, as a general guidelines, it is recommended to use electrolytes with thickness of 250 μm at least for planar cells with electrode areas of ~ 10 cm^2 for the accurate separation of cathode and anode polarization. In the case of small button cells with electrode areas of ~ 1 cm^2, the minimum electrolyte thickness should be 0.5 mm or higher with the ring-shaped or point RE for accurate electrochemical activity measurement.

11.3.4 Constriction Effect

In solid electrolyte cells, it is often observed that the measured electrolyte resistance is larger than the one based on the calculated electrolyte resistance from the electrolyte thickness and electrolyte ionic conductivity. Taking the previous YSZ electrolyte cell with Pt electrodes as an example, the YSZ electrolyte used is 3YSZ and its conductivity at 900 °C is 0.039 S/cm. For a cell with 108 μm thick 3YSZ electrolyte, the calculated resistance is:

$$R_{\text{electrolyte}} = \frac{\delta}{\sigma_i} = \frac{108 \times 10^{-4}}{0.039} = 0.28\,\text{ohmcm}^2 \tag{11.14}$$

The measured electrolyte resistance, R_2, via the special voltage probe is 0.61 Ωcm^2. Thus, the measured electrolyte resistance is more than two times higher than the calculated electrolyte resistance. The higher resistance value of the measured resistance is evidently related to the non-uniform contact at the interface, as shown in the SEM micrographs of the Pt electrode/YSZ electrolyte interface. In solid electrolyte cells, the contact between a porous electrode and a dense electrolyte is usually not continuous and the electrode (e.g., Pt and LSM) is only active at some discrete contact points, leading to the loss of the effective cross-section for the current flow through the electrolyte. The asymmetric contact at the electrode/electrolyte interface on the anode and cathode sides will lead to the asymmetric distribution of current line across the cell, resulting in the higher measured resistance as compared to the calculated resistance based on electrolyte thickness. Such phenomena of the increased resistance of the electrolyte due to a reduction in the active contact/transport area at the interface are generally called the *current constriction* or *constriction effect*. The constriction effect can be represented by a cell/electrolyte resistivity ratio:

$$\gamma = \frac{\rho_{\text{cell}}}{\rho_{\text{electrolyte}}} \tag{11.15}$$

$$\rho_{\text{cell}} = \frac{R_{\text{cell}}}{\delta} \tag{11.16}$$

where $\rho_{\text{electrolyte}}$ is the resistivity of the electrolyte material, ρ_{cell} is the measured cell resistivity, R_{cell} is the measured cell resistance, and δ is the electrolyte thickness. As $\rho_{\text{electrolyte}}$ is the property of an electrolyte and does not change with the electrolyte thickness, the change of λ with the electrolyte thickness is an indication of the constriction effect. λ is equal to or greater than 1.

The constriction effect is most pronounced with the solid electrolyte cells such as SOFCs due to the solid/solid interface. The constriction effect in SOFC depends strongly on the electrolyte thickness, electrochemical activities of the electrodes and the operating temperature [22]. Figure 11.20 shows the results of a carefully designed experimental study of the constriction effect as a function of electrolyte thickness and electrochemical activities of the electrodes. In this experiment, planar cells with

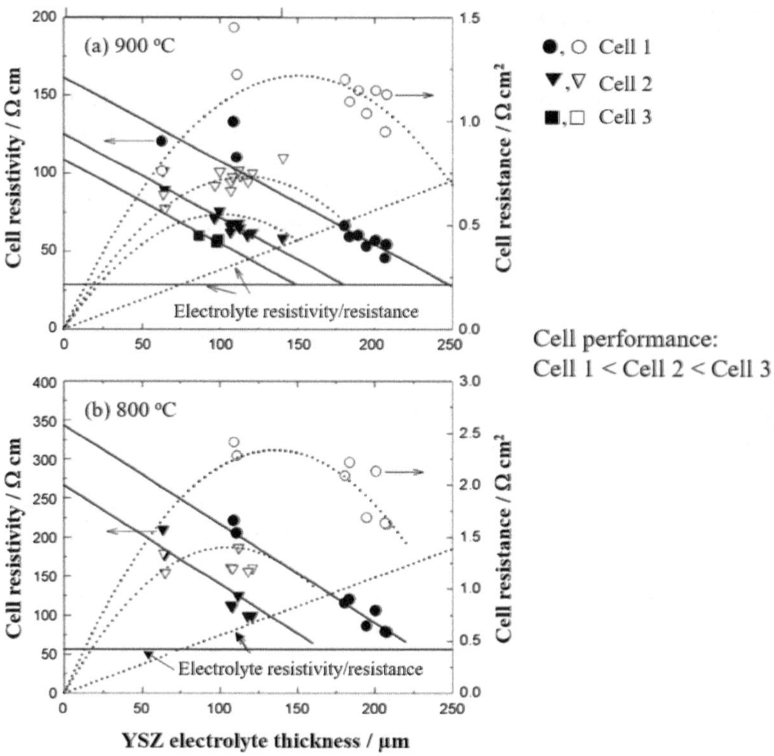

Fig. 11.20 Plots of cell resistivity and resistance of Cell 1, Cell 2, and Cell 3 groups as a function of YSZ electrolyte thickness measured at 900 °C and 800 °C. Modified from Ref. [22] with permission from Elsevier, Copyright 2008

cell area of ~ 10 cm^2 are used under consistent and carefully controlled testing conditions. YSZ electrolyte is prepared from 3 mol% Y_2O_3-doped ZrO_2 (3YSZ) powder with thickness in the range 70 to 200 μm. LSM and Ni/YSZ cermet are used as cathode and anode, respectively, but prepared by three different processes. Based on the performance, three groups of cells are obtained. In a typical test condition, the overpotential of Cell 1, 2, and 3 groups is 186 ± 13 mV, 130 ± 18, and 89 ± 6 mV, measured at 900 °C and 250 mA/cm^2. The electrochemical activity of the cells follows the sequence of Cell 1 < Cell 2 < Cell 3. However, within each group, the activities of the LSM cathodes and Ni/YSZ cermet anodes are similar.

The measured cell resistivity and resistance are a significant function of the electrolyte thickness. In contrast to the constant electrolyte resistivity, the measured cell resistivity increases almost linearly with the decrease in the electrolyte thickness regardless of the cell group. The slope in the increase in the cell resistivity increases with the decrease in operating temperatures. The measure cell resistance is significantly higher than that calculated based on the electrolyte thickness, showing a parabolic dependence on the electrolyte thickness as the electrolyte becomes thinner.

This indicates the constriction effect is more serious with the reduction in cell operating temperature and becomes increasingly important in thinner electrolyte cells. With the enhancement in the electrochemical activities of cells, the cell resistivity line moves to the left with the significant reduction in cell resistance. Improvement in the cell performance is due to the increased electrochemical activities of the LSM cathode and Ni/YSZ cermet anodes and the improvement in the interfacial contact at the electrode/electrolyte interface. This indicates that the most effective way to reduce the constriction effect on thin electrolyte cells is to increase the electrochemical activity of the electrodes and the uniform and symmetric contacts of both anode and cathode/electrolyte interfaces.

The constriction effect on the cell resistance is also related to the contact between the current collector and electrode coatings. This is due to the fact that non-uniform contacts at the current collector/electrode can affect the current and potential distribution in the solid electrolyte cells, and thus the internal resistance [15]. The constriction effect is a significant function of electrolyte thickness and the effect decreases in general with the increase of electrolyte thickness. This implies that the constriction effect is most detrimental for thin electrolyte cells: therefore, good contact and high electrical conductivity of the electrode coatings are particularly critical to minimize the constriction effect on the performance of thin electrolyte based IT-SOFCs.

In principle, the constriction effect can be significantly reduced or eliminated completely through the development of high electrochemically active electrodes with uniform electrode/electorlyte interface and high electrical contacts between the electrode and current collector layers.

11.4 Equivalent Circuit Analysis in Electrochemical Impedance Spectroscopy

Electrochemical impedance spectroscopy (EIS) is a powerful technique for the electrochemical characterization of a wide range of electrochemical systems and for determining the contribution of individual electrode or electrolyte processes. This is particularly true for the solid electrolyte cells. EIS is properly the most common and popular electrochemical technique in diagnosis of the electrode and electrolyte processes and breaking down the losses within a SOFC.

SOFCs are complicated systems containing multiple mass transfer, chemical, electrical, and electrochemical processes. When an excitation or perturbation signal (voltage or current generally in the sinusoidal form) is imposed on a SOFC, a corresponding output signal determined by the coupling of multiple processes gives a characteristic impedance of the system:

$$Z(j\omega) = Z' + jZ'' \tag{11.17}$$

where Z' is the real part, j is the imaginary unit, Z'' is the imaginary part, $Z(j\omega)$ is the transfer function of the system and reflects the amplitude and the phase shift relationship between the excitation and response signal as a function of angle frequency, ω ($\omega = 2\pi f$, where f is the frequency in Hz). The impedance responses can be presented in Nyquist form by plotting Z' against Z'' or in Bode diagrams by plotting impedance against the log of frequency. A Bode plot shows explicitly the frequency information.

There are different models developed for the interpretation of measured EIS data, including both numerical approaches such as electrochemical impedance model, transmission line model, state-space model, and equivalent circuit model approach [23]. Among them, the equivalent circuit model (ECM) approach is phenomenological, easy to use, and is thus most popular in the interpretation of the experimental impedance data in SOFCs. The fundamental laws governing the relationship between the charge and potential and the properties of linear systems are similar in passing from electronic to ionic materials. Therefore, as a first approximation, a close connection exists between SOFC physicochemical processes and the operation of an electrical circuit consisting of discrete components (resistors, capacitors, inductors). Thus, various processes in an electrochemical system could be approximated by an electrical equivalent circuit which would depict the way in which various conducting elements in the electrochemical cell are connected. The equivalent circuit approach in EIS data analysis is popular in the study of solid electrolyte cells like SOFCs, as detailed microscopic models of all possible electrode and electrolyte processes are generally not available or are extremely complex.

However, in reality, the impedance of solid electrolyte systems cannot be well approximated by the impedance of an equivalent circuit involving a limited or finite number of ideal resistors and capacitors, which give rise to perfect semicircular arcs with discrete time constants for each specific electrochemical reaction or process. The commonly observed electrochemical behavior of a solid electrolyte system in the impedance plane is rarely in the form of semicircular arcs with their origin on the real impedance axis, but instead consists of depressed arcs. These arcs can often be skewed on the low or high-frequency end of the impedance spectrum. This type of behavior often indicates heterogeneity in the system and a distribution of time constants and is commonly simulated by including distributed or *constant-phase elements* (CPE) in the equivalent circuit. The empirical fractional power law for CPE is given by:

$$Z_{\mathrm{CPE}} = A(j\omega)^{-n} \tag{11.18}$$

where A is a constant and is independent of ω, and n is frequency power with $1 \geq n \geq 0.5$. With $n = 1$, Z_{CPE} is a classic capacitance and for $n = 0.5$, it is a Warburg-type diffusion impedance. The physical origin of CPE has not been fully established. However, it has been shown that the surface roughness and/or strong adsorption/desorption are important contributing factors to the observed frequency dispersion.

11.4.1 Equivalent Circuit for Capacitance Impedance

In the simplest representation, a solid electrolyte cell can be considered to consist of a series network of several resistors and capacitors in parallel, with each sub-circuit representing a different electrode or an electrolyte process. It can be shown that for each capacitor in parallel with a resistor, the response in the complex plane is a semicircle with its center on the real or x-axis. The difference between the low and high-frequency intercepts of the semicircle on the real axis gives the value of the resistor. The value of the capacitor and the time constant of the circuit can be calculated from the frequency (f_o) at the top of the semicircle ($\omega_o = 2\pi f_o = 1/\tau = 1/\mathrm{RC}$). If each sub-circuit (or an electrochemical process represented by it, denoted by RQ where Q can be capacitance C or CPE) has a different time constant associated with it, then the response of the equivalent circuit in the complex impedance plane will consist of several semicircles each corresponding to a sub-circuit. In electrochemical cells, the ohmic resistance of the electrolyte and electrode is represented as the resistance, R_0 in series with the circuit and can be obtained at the high-frequency intercept. The inductance in solid electrolyte cells can occur at high or low frequencies. If the time constants associated with each process are not at least two orders of magnitude different, a considerable overlap of the individual frequency domains over which each process relaxes will occur, leading to overlapping arcs in the complex impedance plane.

Analysis of impedance arcs can give valuable information on individual electrode processes. This is particularly true if the analysis is accompanied by the microstructure/microanalysis results. Figure 11.21 shows an example of electrochemical impedance responses measured on a LSM electrode before and after Cr poisoning, measured at 700 °C [9]. The electrochemical reactions on the LSM cathode can be presented by an equivalent circuit with three sub-circuits and fitted by using a non-linear-least-square (NLLS) analysis program. The significant change of the impedance responses of the reaction on LSM electrodes is consistent with the deposition of Cr at the LSM electrode/YSZ electrolyte surface [24]. Before Cr deposition, the contact between LSM and YSZ is clean and characterized by the convex contact rings (Fig. 11.21b). After Cr-poisoning tests, Cr deposits form at the LSM/YSZ interface, leading to the significant increase in the electrode resistance associated with the oxygen ion transfer and migration from the LSM electrode to the YSZ electrolyte, indicated by the significantly increased R_2. This indicates a significant effect of Cr deposition and poisoning on the mechanism and kinetics of the ORR on LSM electrodes.

Table 11.3 gives the results of the impedance parameters obtained by the equivalent circuit fitting for the ORR on LSM electrodes before and after Cr poisoning, measured at 700 °C in air. The frequency power (n) for the low and high-frequency arcs is between 0.6 and 1, indicating that Q is not a pure capacitive component. For ORR on a fresh LSM electrode, R_3 is much higher than R_2, while for the reaction on the Cr-poisoned LSM electrode, R_3 is much smaller than R_2. The significantly increased polarization resistance associated with medium frequencies (R_2) is a clear

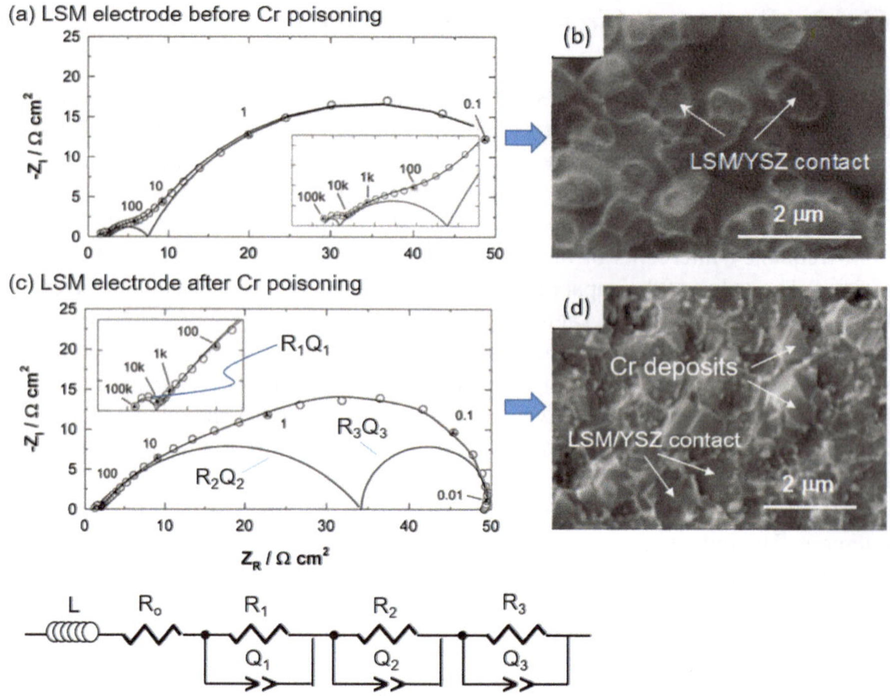

Fig. 11.21 Electrochemical impedance curves measured at 700 °C for O_2 reduction on a $La_{0.72}Sr_{0.18}MnO_3$ (LSM) electrode **a** before and **c** after Cr poisoning. Symbols are experimental data, and lines are the NLLS-fitted results using the equivalent circuit. Cr poisoning on LSM was carried out at 200 mA/cm^2 and 900 °C for 129 h. SEM micrographs of the YSZ electrolyte surface in contact with LSM electrode **b** before and **d** after Cr poisoning. LSM electrode coating was removed by HCl treatment. Modified from Ref. [9] with permission from Elsevier, Copyright 2002

Table 11.3 Impedance parameters evaluated by NLLS program for ORR on the LSM electrode before and after Cr poisoning, measured at 700 °C in air

	R_1Q_1			R_2Q_2			R_3Q_3		
	R_1 Ω cm^2	Q_1 Ω cm^{-2} s^{n1}	n_1	R_2 Ω cm^2	Q_2 Ω cm^{-2} s^{n2}	n_2	R_3 Ω cm^2	Q_3 Ω cm^{-2} s^{n3}	n_3
LSM	0.7	4.0 x 10^{-6}	1	5.2	3.4 x 10^{-2}	0.6	54.5	1.2 x 10^{-2}	0.7
LSM/Cr	0.7	5.3 x 10^{-6}	1	32.3	7.9 x 10^{-3}	0.6	16.3	3.3 x 10^{-2}	1

indication of the inhibiting effect of deposited Cr on the oxygen ion transfer at the LSM/YSZ interface region, consistent with the observed significant Cr deposition at the LSM/YSZ interface.

The *n* value for the R_1Q_1 sub-circuit associated with high frequencies is equal to 1, indicating the pure capacitive nature of the electrode process. With the observed capacitance of 4–5 μF/cm^2, the time constant of the process or reaction at this high-frequency arc can be calculated to be ~ 4 × 10^{-6} s. Such time constant is in the range associated with the charge transfer reaction, e.g., the one for HOR on Ni/YSZ cermet anodes [25]. This indicates that the high-frequency arc is most likely associated with the charge transfer process. With the additional information from the microstructure and other analysis, the middle frequency arc, R_2Q_2, is associated with the transfer and migration of oxygen ions from electrode to electrolyte, while the low frequency arc, R_3Q_3, is related to the adsorption, dissociation and diffusion of oxygen species on the electrode surface (see Chap. 10).

11.4.2 Equivalent Circuit for Inductance Impedance

In SOFCs, inductance is frequently observed. The high-frequency inductance, $L1$ is generally considered to be associated with the Pt probes and heating elements of the furnace of the test station. But the situation with the low-frequency inductance loop is much more complicated. The inadequate positioning of the RE can cause the appearance of low-frequency inductance loop [26]. However, such interference related to the placement of the RE can be identified by the microstructure test, as described below. The interference or error induced a low-frequency inductance loop should not be sensitive to the microstructure change of the electrodes.

Due to the fact that the low-frequency inductance loop occurs most frequently for the reactions on highly active and/or nanostructured electrodes, the low-frequency inductive loop is most likely a result of significant adsorption and dissociation of oxygen species on the electrode surface, enhanced by the nanostructure and the presence of catalytic promoters. Figure 11.22 gives an example of an equivalent circuit with low-frequency inductance and fitting results for the ORR on mixed ionic and electronic conducting $La_{0.6}Sr_{0.4}Co_{0.2}Fe_{0.8}O_{3-\delta}$ (LSCF) and a $Gd_{0.2}Ce_{0.8}O_{1.9}$ (GDC) infiltrated $La_{0.8}Sr_{0.2}MnO_3$ (GDC-LSM) electrodes measured at 800 °C [27]. The LSCF particles are in a wide size range of 0.1–1.5 μm and are in good intimate contact with the GDC interlayer. In the case of a GDC-LSM composite electrode, the infiltrated GDC particle size is 30 ± 4 nm. In both cases, the impedance response is characterized by a high-frequency capacitive arc and a low-frequency inductive loop, forming an impedance circle. An equivalent circuit with a low-frequency inductance, L_2, is used. The replacement of low-frequency capacitance with low-frequency inductance implies that the low-frequency capacitive arc and low-frequency inductive loop do not co-exist. Table 11.4 gives the fitting results of equivalent circuit for ORR on a LSCF and GDC-LSM electrodes, measured at 800 °C.

Based on the equivalent circuit fitting results, the low-frequency inductive loop has an inductance, L_2 in the order of 10^{-4} H, two orders of magnitude higher than 10^{-6} H of L_1, the high-frequency inductance. This indicates that the low-frequency inductive loop, L_2, has a very different origin as compared to that of L_1, which in turn implies that the high-frequency inductance originates from the Pt leads or the high-frequency phase shift of the heating element of the high-temperature furnace

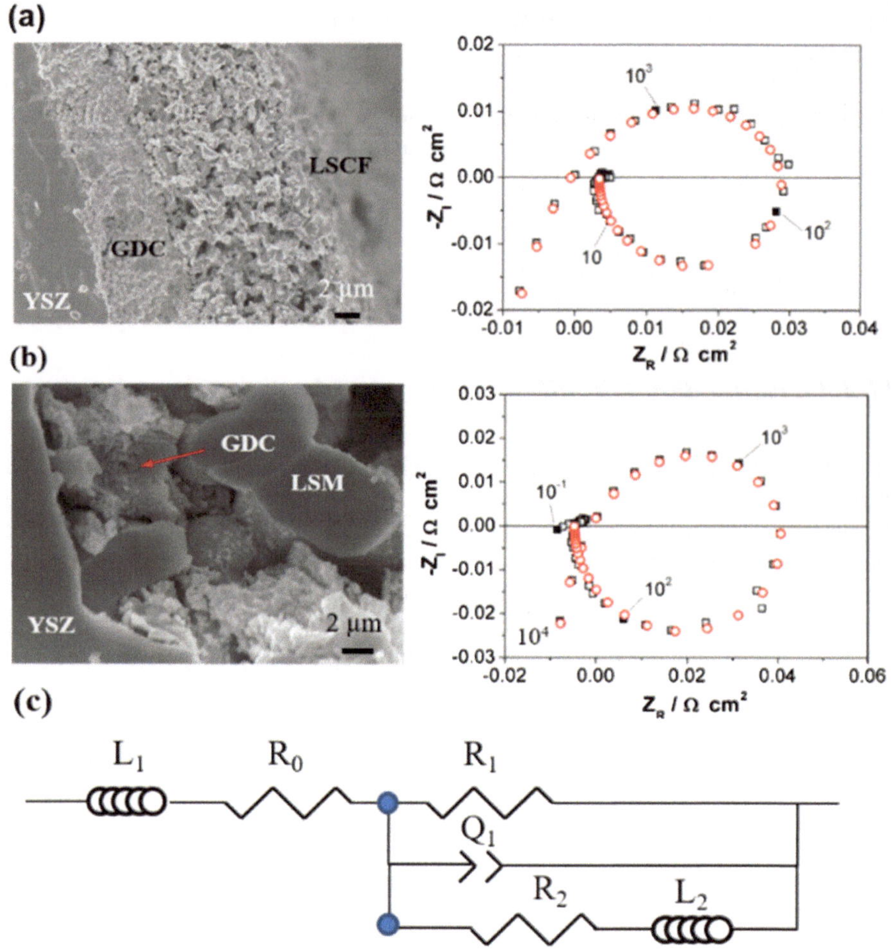

Fig. 11.22 SEM micrograph of cross section and initial impedance responses of (a) a LSCF cathode and (b) GDC-LSM cathodes, measured at 800 °C in air. The equivalent circuit used for fitting the experimental data is given in (c). Numbers are frequencies in Hz; the rectangular symbols (black) are experimental data and the circular symbols (red) are fitted data. Modified from Ref. [27] with permission from Elsevier, Copyright 2016

Table 11.4 Impedance parameters evaluated by NLLS program for ORR on a LSCF and GDC-LSM electrode, measured at 800 °C in air

	L_1 H	R_o Ω cm²	R_1 Ω cm²	Q_1 Ω cm^{-2} s^{n1}	n_1	L_2 H	R_2 Ω cm²
LSCF	1.89×10^{-6}	1.34	0.056	4.72×10^{-2}	0.7	6.99×10^{-4}	0.041
GDC-LSM	1.42×10^{-6}	2.14	0.062	6.52×10^{-3}	0.9	1.13×10^{-4}	0.012

used in the test station. The physical significance of the low-frequency loop is most likely related to the easy and fast oxygen transfer and supply of nano-sized GDC and LSCF particles for the ORR at the electrode/electrolyte interface (for detailed evaluation of the transformation of the low-frequency inductance to low-frequency capacitance, see the ORR section in Chap. 10). Thus, the electrode process associated with dissociative adsorption and diffusion becomes negligible for the ORR, indicated by the disappearance of the low-frequency capacitance arc. Nevertheless, the fundamental understanding of such low-frequency inductance loop requires the systematic investigation of the electrochemical impedance responses as a function of the electrode microstructure [27].

11.4.3 Physical Significance of Equivalent Circuit Elements

We should emphasize here that the equivalent circuit model or ECM approach with each suppressed semicircle ascribed to a specific reaction or process in an electrochemical system is seldom unique. It is not uncommon to fit the same impedance responses with two or more different equivalent circuits. An important consideration with the ECM approach is that an equivalent circuit involving three or more circuit elements can often be rearranged in various ways and still produce similar impedance responses. Figure 11.23 shows an example of the fitting of three different equivalent circuits for the impedance responses measured at 850 °C and oxygen partial pressure of 0.01 atm for the ORR on a LSM cathode[20]. The fittings between the experimental and the fitted data are generally good. Nevertheless, a good fitting is not a sole verification of equivalent circuit used. The equivalent circuit approach to the impedance response of the electrochemical system must be applied carefully as there are no well-defined rules for the selection of an equivalent circuit to best represent the electrochemical system under the interest of investigation.

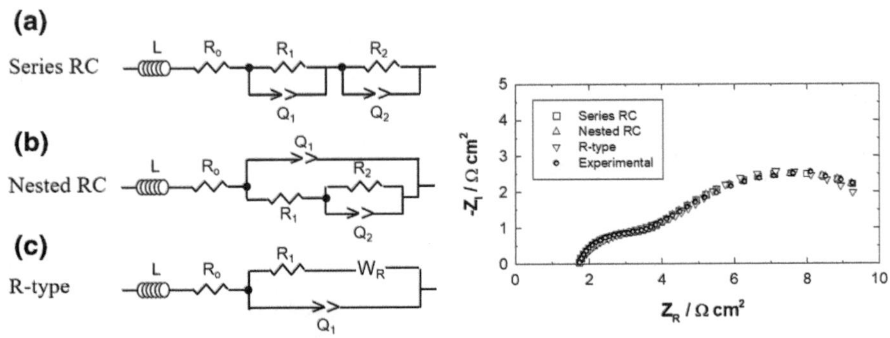

Fig. 11.23 Fitting of the equivalent circuits of **a** series RC, **b** nested RC and **c** R type to the experimental data of ORR on a LSM electrode, measured at 850 °C and O_2 partial pressure of 0.01 atm. Modified from Ref. [20] with permission from Elsevier, Copyright 2002

The equivalent circuit approach, in which the circuit model is constructed based on the prior understanding of electrical, chemical, and electrochemical processes in a solid electrolyte system, is straightforward. However, in general, a good understanding of the system and multiple processes involved in the reaction is required in order to avoid an ambiguity in the selection of an equivalent circuit. An equivalent circuit model should only be evaluated in combination with other experimentally supported evidence. In this respect, both intrinsic factors such as materials, composition, microstructure, and component ratio, and operational factors such as temperature, pressure, fuel/air flow rate, concentration, and polarization potential should be considered. To verify the validity of the selected circuit model, it is necessary to establish the relationship between the measured impedance, $Z(j\omega)$, and overlapped or coupled responses from both intrinsic and operational factors.

However, there are tests which can be performed to determine and differentiate the physical significances of impedance arcs, including:

- DC bias test—if the electrode process involves charged species or charge transfer steps, the associated impedance arc or electrode polarization resistance would change with the applied DC bias. It should be pointed out that DC bias should be expressed in overpotential, η, as η is a direct measure of the electrochemical driving force for the reaction. The magnitude of η should be in the linear polarization region and less than 100 mV as a high η value will change the local p_{O_2}, leading to interference and distortion of the system and thus error in the measurement.
- Partial pressure test—the change of the electrode polarization resistance with partial pressure or content of reactants such as p_{O_2} and p_{H_2} will give information on the molecular diffusion or atomic surface diffusion. An example of the dependence of the impedance arcs on the p_{O_2} is given in Fig. 11.24. It shows that the partial pressure of O_2 has the most significant effect on the electrode process associated with low-frequency arcs. From the dependence of the individual arcs on the p_{O_2}, the reaction order can be obtained and based on the reaction order, it is also possible to separate the gas diffusion and gas conversion processes in porous electrodes.

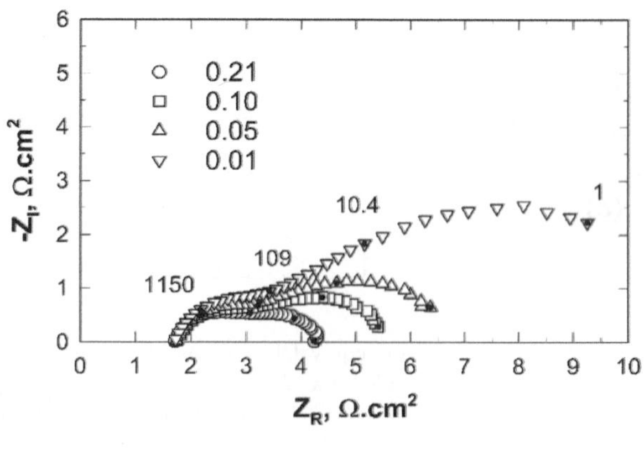

Fig. 11.24 Impedance responses of ORR on a LSM electrode measured at 850 °C at different partial pressure of 0.21, 0.10, 0.05, and 0.01 atm. Numbers in the figure are frequency in hertz. Modified from Ref. [20] with permission from Elsevier, Copyright 2002

From the impedance data, it is also possible to get the reaction of order of the individual electrode processes associated with impedance arcs as a function of the partial pressure of reactants.

- Helium–oxygen shift test—the method of helium–oxygen (He-O_2) shift can be used to determine the concentration polarization contribution of a cathode. The principle of the method is based on the fact that the O_2 diffusivity in a He-O_2 mixture is much faster than that in a N_2-O_2 mixture. Thus, a change in the impedance responses from N_2-O_2 mixture (e.g., air) to He-O_2 mixture (e.g., 20%He-80%O_2) is an indication of concentration polarization change.
- Activation energy (E_a) test—the magnitude of the activation energy will tell the energy barrier of the process. If the process is independent of the temperature ($E_a =$ 0), this implies that the process is not thermally activated like gas phase diffusion. If the magnitude of E_a is close to that of the electrolyte, this may indicate the electrode process associated with the impedance arc is affected by or involving the electrolyte phase, e.g., the oxygen ion transfer at the cathode/electrolyte interface region of ORR.
- Inhibitor/promotor test—one can purposely introduce an inhibitor or promoter to the system of interest to inhibit one process, making the process of interest more pronounced and visible by the electrochemical techniques. For example, at high temperatures, gaseous chromium species are a contaminant for ORR. However, at the initial stage of Cr poisoning, the effect of Cr species is predominantly on the electrode processes associated with surface adsorption, dissociation and diffusion. Thus, in the presence of gaseous Cr species, the surface process, but not the bulk diffusion process, is inhibited. This has been used to distinguish between the roles of surface and bulk processes on the kinetics for ORR under SOFC operation conditions [28].
- Microstructure test—changes in the microstructure of the electrode can have a significant effect on the electrode processes associated with the surface and/or bulk properties of the electrode materials. Examples include the transition of the impedance responses from the low-frequency inductance loop to low-frequency capacitance arc for GDC infiltrated LSM electrodes [27] and significant impedance responses for the H_2 oxidation reaction on Ni and Ni/YSZ cermet anodes [25].

11.4.4 Impedance Measurement in SOFCs

There is various and well-established commercially available equipment for the data acquisition and operation in the impedance measurement in SOFCs. The issues affecting the accuracy and reliability of EIS in solid electrolyte cells like SOFCs are related to the experimental parameter selection. This includes the frequency range, impedance signal amplitude, reference electrode, and interference from the test station.

The frequency range used in SOFCs is generally between 0.01 Hz to 1 MHz. This will allow the simultaneous analysis of multiple and overlapped responses from adsorption, dissociation and diffusion, to ionic and electron charge transfer. The amplitude of the perturbation signal should be small and meet the linearity requirement of the impedance transfer function. In SOFCs, an amplitude of the voltage signal of 10–40 mV is recommended. Too high amplitude will lead to the deviation of the voltage or current responses from the linear region of the reaction.

The reliability issue of RE in solid electrolyte cells has been discussed in Sect. 11.3.2. Another frequently accounted phenomenon associated with impedance responses in SOFCs is the interference and/or noises of the test cells and system. The interference and noise can be reduced to a certain degree by shielding, but that related to the test cell and test station is not avoidable. Fortunately, such interference is generally shown as inductance at high frequencies and tends to be more pronounced in larger cells and stacks. Figure 11.25 shows an example of impedance responses measured between two voltage probes, VP1 and VP2 (see Fig. 11.15 for explanation) at 800 °C on a large plate cell. Despite the significant differences in the high-frequency inductance behavior, the impedance responses measured by two VPs are similar. This in fact implies that the inductances at high frequencies have little effect on the electrode processes. Nevertheless, the high inductance will contribute

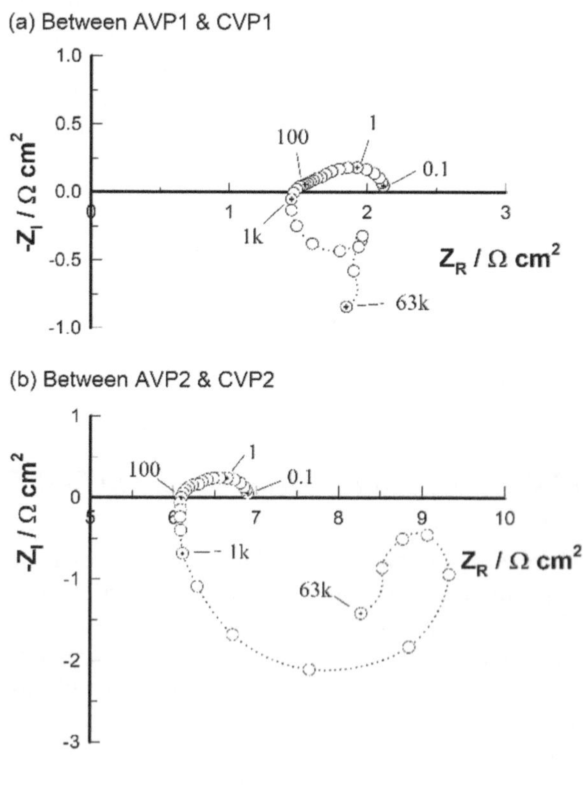

Fig. 11.25 Impedance responses measured between voltage probe couples of **a** AVP1 and CVP1 and **b** AVP2 and CVP2 on a YSZ cell with Ni/YSZ cermet anode and LSM cathode at 800 °C under open circuit. The numbers are frequencies in hertz

to the error in the ohmic resistance measurement of electrolytes or cells. In general, the high-frequency inductance would cause an underestimation of the ohmic resistance, which is taken as the intercept at the high frequencies. In this case for accurate measurement of the electrolyte/cell resistance, the GCI technique, as shown below, should be used.

11.5 Galvanostatic Current Interruption Technique

The study of the electrode kinetic and transport properties in solid-state electrochemical systems by the impedance technique can be complemented by the use of a *galvanostatic current interruption* (GCI) technique. This technique is exceptionally useful for studying the electrode kinetic behavior in solid electrolyte systems where the ohmic resistance of the system is usually significant. In solution electrochemistry, the ohmic contribution is minimized by the correct placement of a Luggin probe near the working electrode. GCI, however, can separate the ohmic potential contribution from the total cell potential where the placement of the reference electrode is geometrically limited as in the case of SOFCs.

The GCI technique allows separation of electrolyte and electrode contributions in the time domain and is useful for determining the contribution of electrode and electrolyte processes in the dynamic mode (that is at typical operating current densities of up to several hundreds mA/cm^2). Also, EIS measurements under cell operating conditions, where the system is perturbed from equilibrium in the potentiostatic or galvanostatic modes, are inherently slow and useful only for systems that are stable over the time frame of the experiment. The GCI method is fast but cannot provide the detailed information available from an electrochemical impedance spectrum.

The GCI technique involves analyzing potential decay over a short time scale (of the order of milliseconds or less) after interruption of a steady constant current flowing through the electrochemical system. The time-dependent (or transient) behavior consists of ohmic losses due to finite resistance of the electrolyte and overpotential losses due to various superimposed electrochemical processes at the electrode/electrolyte interface.

The GCI technique can be described in terms of an equivalent circuit representing basic polarization processes in a solid electrolyte system with working, counter and reference electrodes. The current is passed between the WE and CE, and the voltage is measured across WE and RE. The use of RE through which no current passes allows the cathodic and anodic processes to be studied separately. On application of a constant current to an electrochemical device, the potential rises to a steady-state value corresponding to the potential drop across all resistors in the measuring circuit (WE and RE).

In general, capacitances associated with solid electrolyte processes (capacitance associated with intragrain and intergrain processes) are several orders of magnitude smaller than the interfacial or double layer capacitance (C_{dl}) and the behavior is almost that of a pure resistor. On interruption of a constant current with a fast switch

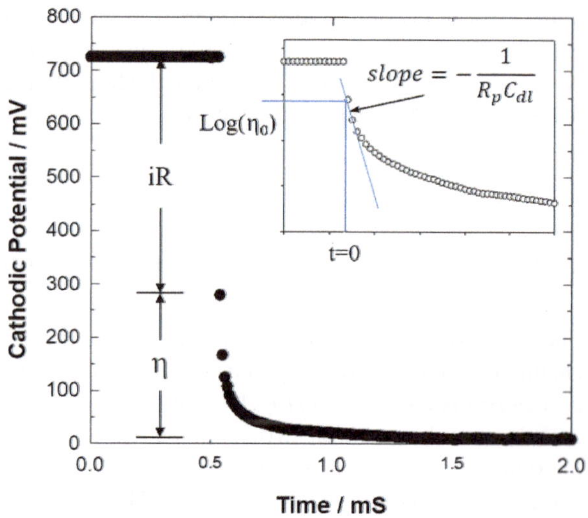

Fig. 11.26 GCI curve recorded for ORR on a LSM electrode at 500 mA/cm^2 and 800 °C

(switching time < < 0.1 μsec) the ohmic potential drops and disappears virtually instantaneously (usually in less than 0.1 μsec) because of the very small time constant associated with these processes (unless an associated capacitance exists as in oxide and passive layers). The electrode processes are much slower, have much longer time constants, and the slow part of the potential—time transients usually corresponds to overpotential losses across the electrode/electrolyte interface. Appropriate data acquisition equipment with resolution on a microsecond scale allows the ohmic contribution to be separated, see Fig. 11.26. In this case, the overpotential (η) is 279 mV at 500 mA/cm^2 for ORR on the LSM electrode and the overall ohmic resistance of the electrolyte pellet is 0.89 Ωcm^2. The slow part of the transient for which data is recorded is due to the potential difference across both the double layer capacitor C_{dl}, and the complex nonlinear impedance Z.

Therefore in the time domain of GCI both the electrode and the electrolyte contributions can be easily separated. The analysis of these potential-time transients for different current densities and for a variety of experimental conditions yields information about the kinetics of the interfacial processes.

The time dependent overpotential decay can be given by:

$$\eta(t) = \eta_o \exp\left(\frac{-t}{\tau}\right) \text{or}(t) = \eta_o \exp\left(\frac{-t}{R_{\text{p}}C_{\text{dl}}}\right) \quad (11.19)$$

where τ is the time constant and η_{o} is the overpotential at time zero (the instant of current interruption) and equates to the overpotential of the charge transfer process at the supplied constant current (see the inset in Fig. 11.26). These overpotential values can be used subsequently in generating polarization (η_{o} versus current density) curves and for calculation of the kinetic parameters and elucidation of the reaction mechanism. The current interruption curves could be analyzed on the basis of equivalent circuits using a standard R(RC) circuit. In practice for solid electrolytes the equivalent circuit is much more complex and requires the inclusion of CPE. The analysis of the time dependent polarization decay curves obtained by the GCI technique is complex as the potential decay curves often contain contributions from both Faradic and non-Faradic processes with overlapping time constants.Thus it is seldom to use equivalent circuit approach to fit GCI data in SOFCs.

The GCI technique requires data to be collected over a short period of time following current interruption. The basic experimental requirements are a constant current source for passing current, an electronic switch with fast switching time (preferably less than 0.1 μS) and capability to take desired current loads, and a fast data recording device with resolution in the microsecond range such as an oscilloscope or a transient recorder. The sharp change in the voltage responses corresponds to the ohmic loss and slow change is related to non-ohmic polarization losses. Care must be taken before the GCI to ensure the attainment of steady state. The accuracy of the ohmic resistances obtained can be verified by plotting the iR drop as a function of applied current. A linear relation between the iR drop and applied current passing the origin suggests the accurate caption of the ohmic portion of the polarization in the GCI experiment. Various commercial electrochemical instruments are available to construct a GCI facility.

11.5.1 Comparison of EIS and GCI Techniques

With EIS, the contribution of various electrode processes is determined in the frequency domain and usually performed under open circuit or near equilibrium conditions. The GCI method separates the electrode processes in the time domain and is useful for determining the contribution of the electrode and electrolyte processes under different current densities. In principle, EIS and GCI measurements should generate equivalent information on the electrode behavior for the reactions occurring at the electrode/electrolyte interface. For a simple electrode reaction with either exponential or non-exponential behavior, EIS and GCI data are mathematically transferable.

However, sometimes there are differences in the reaction parameters as evaluated from GCI and EIS measurements, which are fundamentally related to the mechanism and kinetics of the reaction. For example, in the case of the H_2 oxidation reaction (HOR), there is a significant difference in the reaction orders and activation energies obtained by GCI and EIS. The reaction order with water content, p_{H_2O}, and activation energy for HOR are generally smaller in the case of the GCI technique compared to those measured by the EIS technique. For example, from the polarization data obtained by GCI, the activation energy is ~ 76 kJ/mol for the HOR on Ni/YSZ cermet electrodes, while based on the EIS measurements, the activation energy is in the range of 170 kJ/mol, see Fig. 11.27. The much lower activation energy for HOR

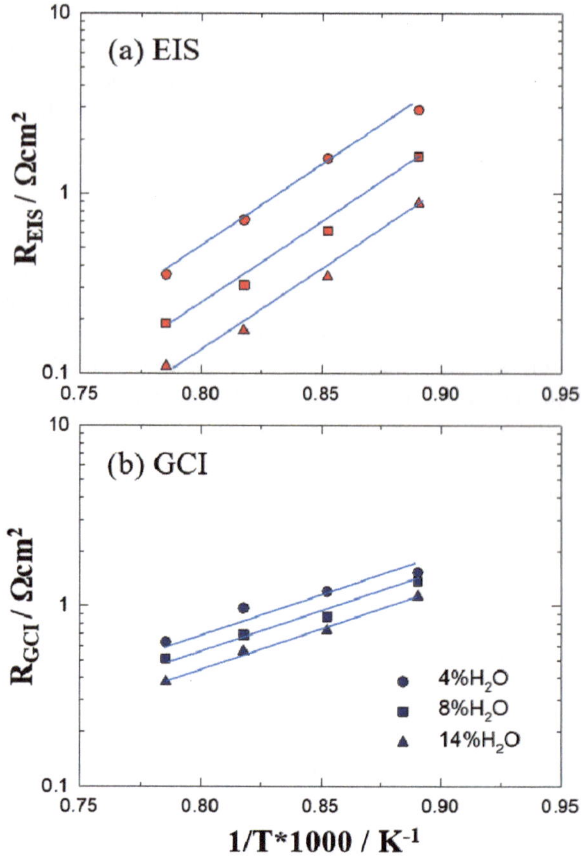

Fig. 11.27 Activation energy plots of electrode polarization resistance of a Ni/YSZ cermet anode at different H_2O contents, measured by EIS and GCI techniques. R_{GCI} is measured from the slope of the polarization curves in the linear region, while R_{EIS} is measured from the differences of the low and high-frequency intercepts on the impedance curves. Modified from Ref. [19] with permission from Elsevier, Copyright 1999

on Ni/YSZ cermet anodes measured by GCI technique is related to the significant promotion effect of produced water on the reaction kinetics [19]. On the other hand, as the EIS measures the reaction responses at open circuit, the effect of the produced water on the reaction is negligible. In the case of ORR, the differences in the reaction parameters measured by EIS and GCI are expected to be small due to the negligible effect of the intermediates or reaction products on the reaction kinetics.

11.6 Conductivity Measurement of SOFC Components and Materials

11.6.1 Two-Probe and Four-Probe Methods

Electronic and/or ionic conductivity is one of the most important properties of SOFC materials and high conductivity is vital to the high performance and stable operation of SOFCs. The two-probe and four-probe configurations are two of the most commonly used techniques for the conductivity measurement of SOFC materials with a well-defined geometry such as a rectangular or a cylindrical bar as shown in Fig. 11.28. By varying the current (i) across the specimen, the voltage drop (V) produced across a given length is proportional to the resistance of the sample. The conductivity in unit S/cm of the specimen is then calculated as follows:

$$\sigma_i = \frac{i \times d}{V \times S} \tag{11.20}$$

where d in cm and S in cm^2 are the separation distance of the probes and the cross-section area of the specimen, respectively. The electrical conductivity of a material is generally measured by two-probe methods at room temperature due to the fact that at room temperatures, the iR losses due to the current probe (e.g., Pt wires) are very small and can be ignored. In the two-probe method, both voltage and current probes are attached to the specimens. The two-probe method is simple to use and various specimen shapes and geometries such as disk and bar can be used. However, it is not recommended to use the two-probe configuration for the conductivity measurement at high temperatures (500 °C or higher) as the resistance of the Pt wires is no longer negligible as compared to the conductivity of the specimen.

To eliminate the resistive interference of the shared current and voltage probe, the four-probe method is often used (Fig. 11.28b). As there is no current flow in the voltage probe, the voltage measured is purely due to the resistivity loss of the specimen. All probes must be reversible to charge carrying species within the ionic conductor and physically isolated from each other. With proper care in the selection

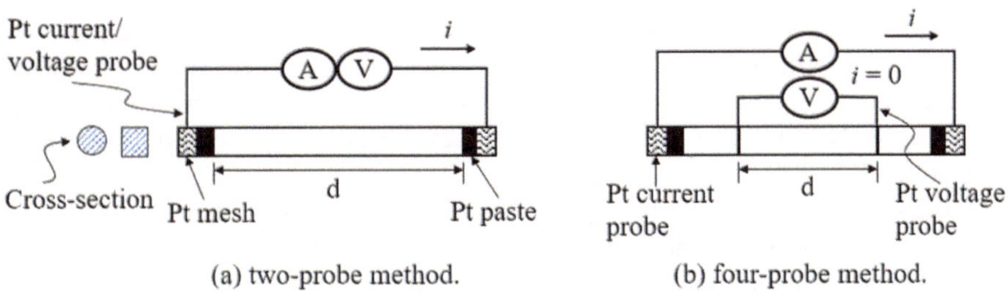

Fig. 11.28 Electrode arrangement of the two- and four-probe electrical and ionic DC techniques

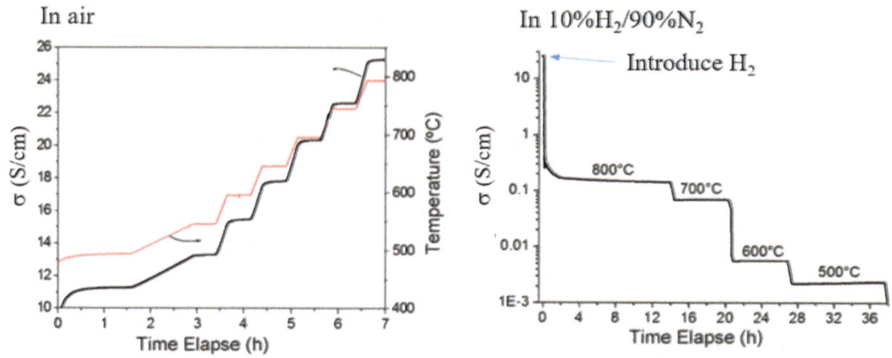

Fig. 11.29 Conductivity-time plots of a $(La_{0.73}Sr_{0.25})(Cr_{0.5}Mn_{0.5})O_3$ (LSCM) bar sample, measured in air and in 10% H_2/90%N_2 at different temperatures. The relative density of LSCM is 94%. Modified from Ref. [30] with permission from Elsevier, Copyright 2008

of the probe material and positioning of the probe, accurate measurements can be made. The four-probe method can be used for the ionic conductivity measurement using electron-blocking techniques. Pure oxygen ion conducting YSZ can be used as electron-blocking probes placed between the specimen, and Ag or Pt paste on both ends of the sample in order to collect pure ionic current. To further improve the accuracy of the measurements, two ionic probes, which have the same composition as the ionic current collector but are sharpened at the edge, can be mechanically contacted to voltage probes placed at the sides of the specimen [29]. The experimental setup for the conductivity measurements on bar samples is similar to the one shown in Fig. 11.4a with controlled atmosphere.

The electrical conductivity of perovskite oxides commonly used in SOFCs depends strongly on the atmosphere of the test environment. This is due to the fact that the charge compensation mechanism of perovskite oxides is related to the partial pressure of oxygen, p_{O_2}. In general, the electrical conductivity would decrease with decreasing p_{O_2} due to the reduced electron hole concentration. Figure 11.29 shows typical conductivity responses of a $(La_{0.73}Sr_{0.25})(Cr_{0.5}Mn_{0.5})O_3$ (LSCM) bar sample in air and in 10% H_2/90%N_2 at different temperatures [30]. The responses are relatively fast, indicating a fast establishment of the equilibrium in conductivity with the temperature. However, the conductivity drops significantly when the atmosphere is changed to 10% H_2/90%N_2. The relatively slower responses as compared with that in air are probably due to the slow kinetics of the surface reaction with H_2. The measured conductivity of LSCM is 0.20 S/cm in reducing environment (e.g., 10% H_2/90%N_2) at 800 °C, much lower than 25.3 S/cm in air measured at the same temperature.

If the as-prepared ceramic bar sample is not fully dense, the measured conductivity is also termed apparent conductivity, σ_{app}, and needs to be corrected by taking into account of the effect of pores on the true conductivity value of the material. The low density of the sample as compared to the theoretical density is often due to the poor sinterability of ceramic oxides. The effect of porosity on the conductivity values can be taken into account by correcting the measured σ_{app} as follows:

$$\sigma_{corr} = \frac{\sigma_{app}}{2\left(\frac{d_{rel}}{100} - 0.5\right)} \tag{11.21}$$

where σ_{corr} is the porosity-corrected conductivity and d_{rel} is the relative density. In the case of LSCM, the relative density is 94%. After porosity correction, σ_{corr} is 28.8 and 0.22 S/cm for LSCM material at 800 °C in air and 10% H_2/90%N_2, respectively [30]. The conductivity of LSCM increases with increase in temperature, indicating a thermally activated process. The activation energy can thus be calculated according to:

$$\sigma T = A\exp\left(-\frac{E_a}{RT}\right) \tag{11.22}$$

From a plot of log(σT) against 1/T, a straight line will be expected. From the slope of the plot, the activation energy of the conductivity, E_a, can be calculated. In the case of LSCM materials, activation energy of the electrical conductivity is 26 kJ/mol and 68 kJ/mol in air and 10% H_2/90%N_2, respectively [30].

11.6.2 Conductivity Measurement of Porous Electrode Coating

Porosity has a significant effect on the conductivity of a material and this is particularly true for the porous electrode coating in SOFCs, because the conductivity of porous electrodes has direct implications for the actual performance of the practical electrodes as well as the cells. Thus, it is of technical importance to measure the conductivity of porous electrode coatings under SOFC operating conditions. Figure 11.30 shows a modified four-probe configuration for the conductivity measurement of porous electrode coating using electrolyte-supported cells [15]. The

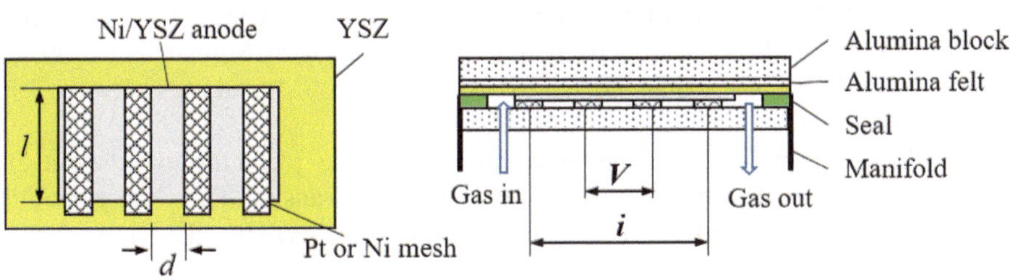

Fig. 11.30 Four-probe configuration for the conductivity measurement of an electrolyte-supported porous electrode coating. Modified from Ref. [15] with permission from Elsevier, Copyright 2003

area of the electrolyte substrates should be large enough to host the mesh-type current and voltage probes. It is recommended to employ electrolyte plates with 50 × 50 mm in lateral dimension and 100–200 μm thickness. For example, to measure the conductivity of a porous anode such as Ni/YSZ cermet electrode, Ni/YSZ cermet is prepared by a screen-printing method onto a YSZ electrolyte plate, followed by sintering at 1400 °C, similar to the process for conventional Ni/YSZ electrode coating. There is no need to have a cathode coating on the other side of the electrolyte. The anode side is sealed with the hydrogen gas inlet and outlet and the gas is distributed across the cell, similar to that in a fuel cell test. On the reverse side, the plate is open to air with an alumina felt to evenly distribute the load. The load to the cell is kept constant, e.g., ~1 kg. The electrical contact for the current and voltage probes is made by Pt or Ni mesh along the length of the Ni/YSZ cermet electrode coating with 2–3 mm in width. The average distance (e.g., 5 mm) between the edges of the voltage probes is kept constant. Current is passed through the two outside current probe meshes and the voltage is measured using the two inner voltage probe meshes. Measurement is carried out in reducing environment (e.g., 97%H_2/3%H_2O) at SOFC operating temperatures. The conductivity of the porous electrode coating on the electrolyte is obtained using the formula:

$$\sigma_{\text{coating}} = \left(\frac{i}{V}\right)\left(\frac{d}{l \times \delta}\right) \tag{11.23}$$

where δ is the thickness of the coating, l the contact length of the Ni mesh (i.e., the width of the coating), d the distance between the Ni meshes, i and V the applied current and measured voltage, respectively.

In the case of cathode coating, the same cell configuration can be used but with no requirement of sealing of the electrode sides.

The cell configuration as shown above has been used for the conductivity measurement of practical electrode coatings. For example, for a YSZ-supported Ni/YSZ (50:50 in vol%) with thickness of ~ 50 μm and ~ 50–60% porosity, the measured σ_{coating} is ~ 155 S/cm at 800 °C. In the case of tape-cast porous Ni/YSZ tapes with thickness of ~ 600 μm, the electrical conductivity is in the range of 400 to 600 S/cm at 800 °C [15]. The higher σ_{coating} for the thicker Ni/YSZ anode substrates is most likely due to the reduced porosity (40–45%) of the tape-casted tapes. In the case of a porous LSM electrode coating used in SOFCs, $La_{0.72}Sr_{0.18}MnO_3$ (LSM) coatings sintered at 1150 °C show a conductivity in the range of 34 to 56 S/cm in air at 800 °C. This is considerably lower than the 175 to 300 S/cm for LSM materials between 700 to 1000 °C.

The electron transfer on the electrode coating can take place through the out-of-plane transfer as well as through the in-plane electron transfer. Thus, the generally much lower conductivity of porous electrode coatings needs to be taken into account in the design of the current collector or current collector layer in fuel cells.

11.6.3 Contact Resistance Between Electrode and Current Collector

The reduced conductivity of porous electrode coatings has two important implications for the performance evaluation of practical cells; the high contact resistance between the electrode coating and the current collector, and the poor scalability of the electrode performance obtained on small button cells, as discussed in Sect. 11.2.2 in this chapter.

In SOFC stacks, maintaining a decent and reliable electrical contact is necessary in order to achieve good overall stack performance. On the anode side, the nickel mesh that serves as a contact layer provides excellent contact with the interconnector and the anode/anode substrate. Due to the fact that the electrical conductivity of cathode materials is generally much lower than that of a Ni-based cermet anode, the electrical contact resistance between the current collector and porous cathode coating can be a significant issue. In general, contact resistance is closely related to the contact areas between current collector and electrode coating [15]. Figure 11.31 shows an example of Ni/YSZ cermet anode-supported cells with $Pr_{0.80}Sr_{0.20}MnO_3$ (PSM) cathode as a function of the contact areas between the current collector and PSM cathode, measured at 800 °C. Both the cell resistance and polarization performance depend strongly on the contact area between the PSM cathode and current collector. For instance, in the case of a low contact area of ~ 5%, the overall cell resistance is 1.43 Ωcm^2 and the cell overpotential is 225 mV at 250 mA/cm^2. As the contact area increases to 27%, both cell resistance and overpotential decrease dramatically

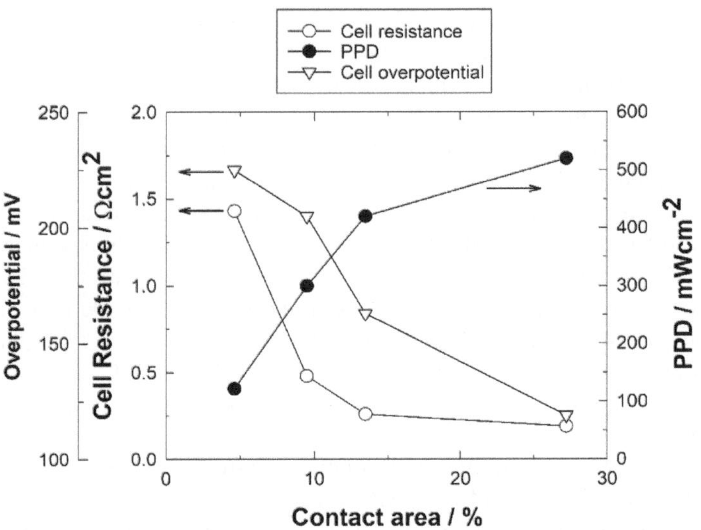

Fig. 11.31 Plots of cell resistance, cell overpotential and peak power density (PPD) of Ni/YSZ cermet-supported cells with PSM cathode as a function of contact areas between current collector and PSM cathode at 800 °C in 97%H_2/3%H_2O and air

to 0.19 Ωcm^2 and 119 mV, respectively. The peak power density (PPD) of the cell reaches 520 mW/cm^2. The high sensitivity of the cell performance to the contact area can be ascribed to the constriction effect as described above (Sect. 11.3.4). The reason for the high constriction effect is due to the low conductivity of the porous PSM electrode, 3–4 S/cm at 800 °C in this case.

Thus, to reduce the contact resistance particularly at the cathode side in SOFC stacks, an additional cathode contact layer or cathode current-collecting layer (CCCL) can be added or deposited to ensure satisfactory electrical contact. The CCCL is a critical component in SOFC stacks. This layer not only ensures electrical contact between the interconnect and cathode but also serves as compensation for the dimensional tolerances of the various parts of the stack. This layer should have a high electrical conductivity, an acceptable thermal expansion coefficient and no reaction with the chromium barrier layer on the interconnect. LSM and some cobaltite based materials can be used as CCCL.

Consistent and good contact is critical for the reliability and reproducibility of the electrochemical performance and good scalability of the electrode and stack performance in SOFCs.

11.6.4 Ionic Conductivity of Electrolyte by EIS

In addition to the four-probe technique, EIS is the most common technique in the measurement of the ionic conductivity of the solid electrolyte materials of SOFCs. Both the particle size and porosity of the electrolyte samples affect the conductivity of the electrolyte. A single crystal sample will show only a bulk and an electrode semicircle. Polycrystalline samples consist of a number of grains and grain boundaries and the grain boundaries are a critical part of the microstructure. For polycrystalline samples, the impedance responses are generally characterized by three semicircles or arcs. The high-frequency arc is attributed to bulk or grain interior resistance, R_B, the medium frequency arc to grain boundary resistance, R_G, and the low-frequency arc to electrode contributions, R_E. The R_G of acceptor-doped zirconia and ceria can be two to three orders of magnitude higher than the corresponding bulk values. The high grain boundary resistance can be related to the existence of grain boundary impurities such as a siliceous phase, which blocks the diffusion of oxide ions. In the case of high-purity materials, such a phenomenon could be due to the fact that oxygen vacancies are severely depleted in the space-charge layer [31]. From an electrical point of view, a grain boundary consists of a grain boundary core and two adjacent space-charge layers.

Figure 11.32 shows an example of the impedance studies of YSZ electrolyte pellets made from two different kinds of YSZ powders measured at 350 °C in air, nano-sized YSZ and micron-sized YSZ powders [32]. The nano-sized YSZ powder has an average particle size of 50–60 nm and specific surface area of 120–130 m^2/g, while for micron-sized YSZ powder the average particle size is 500 nm and specific surface

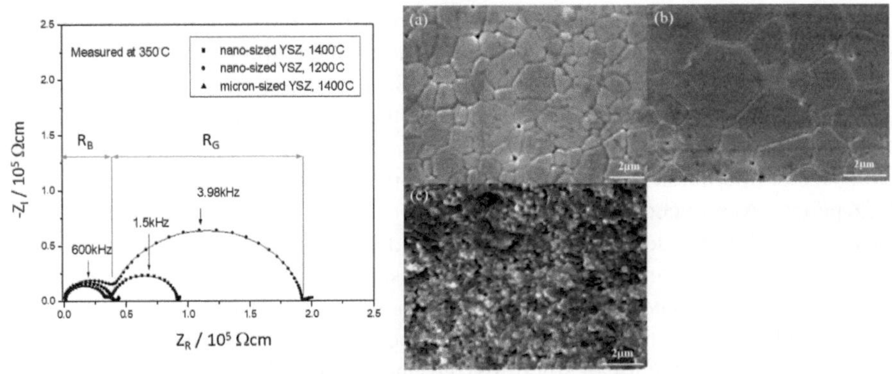

Fig. 11.32 Normalized impedance spectra of sintered YSZ pellets with Ag electrodes measured at 350 °C (sintering time: 2 h, solid line is fitting curve), and SEM images of the surface of a YSZ pellet. (a) micron-sized YSZ, sintering at 1400 °C for 2 h; (b) nano-sized YSZ, sintering at 1400 °C for 2 h; (c) nano-sized YSZ, sintering at 1200 °C for 2 h. Modified from Ref. [32] with permission from Elsevier, Copyright 2003

Table 11.5 Bulk resistance and grain boundary resistance, R_B and R_G, of YSZ measured at 350 °C

YSZ electrolyte powder	Relative density	Grain size, μm	R_B, Ωcm	R_G, Ωcm
Micron-sized YSZ @ 1400 °C, 2 h	96%	1.3	3.32×10^4	7.24×10^3
Nano-sized YSZ @ 1400 °C, 2 h	96%	1.7	3.72×10^4	5.46×10^4
Nano-sized YSZ @ 1200 °C, 2 h	92%	0.3	3.78×10^4	1.56×10^5

area is 6.1 m^2/g. The sintered YSZ pellets have a relative density of 92–96%. Two semicircles at high and low frequencies can be clearly identified, which are associated with the resistance of the bulk and grain boundary. The value of the resistance is determined by curve-fitting the impedance data with two RC circuits in series and the results are given in Table 11.5. The bulk resistance of the nano-sized YSZ sintered at 1400 °C and 1200 °C is close to that of micron-sized YSZ sintered at 1400 °C. This indicates that the bulk conductivity of the YSZ is not affected significantly by the grain size. However, there is a significant difference in grain boundary resistance among the YSZ specimens studied. The grain boundary resistance of micron-sized YSZ sintered at 1400 °C is 7.24×10^3 Ω cm, much smaller than that measured on nano-sized YSZ powders. From the sintered density measurement, the porosity of the pellets made of nano-sized YSZ powders is similar to that made of micron-sized YSZ powders after sintering at 1400 °C (i.e., 96–97%). In addition, the grain size of micron-sized YSZ pellet was slightly smaller than that of nano-sized YSZ pellet. Therefore, the difference of R_G of YSZ pellets is not due to the difference in their particle size. Instead, the high R_G of nano-sized YSZ pellets sintered at 1400 °C may be due to the existence of some impurity. The nano-sized YSZ sintered at 1200 °C exhibits the highest R_G, 1.56×10^5 Ωcm, most likely due to its high porosity level

(~8%), small average particle size (~0.3 μm) and thus poor contact as porosity and small average particle size will significantly increase the grain boundary resistance.

The significant difference in R_G for the YSZ electrolyte pellets is related to the microstructure of the sintered electrolyte. Both the nano-sized and micron-sized YSZ pellets sintered at 1400 °C are quite dense with some isolated pinholes. The average grain size of micron-sized YSZ pellets is ~ 1.3 μm, and it is ~ 1.7 μm for nano-sized YSZ pellets. At reduced sintering temperature of 1200 °C, the pellets prepared from nano-sized YSZ powders show sintering and densification, but with much smaller grain size (~0.3 μm) and some pinholes. The significant difference in the sintering and densification behavior of pellets made of nano- and micron-sized YSZ powders could be related to the nature of the sintering behavior of the powder. This indicates that the sintering behavior of micron-sized and nano-sized YSZ powder could be very different.

11.7 Other Properties

11.7.1 Oxygen Surface Diffusion and Exchange Coefficient Measurements

Oxygen surface coefficient (k) and oxygen diffusion coefficient (D) are the two important parameters for the investigation of oxygen transportation-adsorption/diffusion processes in the cathode reaction kinetics. Typically, two kinds of experimental methods can be used to obtain D and k: isotope exchange depth profile (IEDP) [33] and electrical conductivity relaxation (ECR). In IEDP, isotope ^{18}O is used in the experiment and the diffusion profile within the sample is determined by secondary ion mass spectroscopy (SIMS), while in ECR, an abrupt oxygen partial pressure change is applied and the total conductivity is measured until the sample reaches a new equilibrium.

The principle of both IEDP and ECR is based on Fick's second law and both require dense samples. For the IEDP method, the sample usually is first annealed in the labeled $^{16}O_2$ atmosphere for a long time. It is important that the sample is in chemical equilibrium with the desired temperature and atmosphere. Then the sample is quenched to room temperature and reheated with the $^{18}O_2$ gas and the ^{18}O penetration profile is measured by SIMS. During this process, the rate of isotope exchange across gas/solid interface is assumed to be directly proportional to the difference in isotope concentration between the gas and the solid. This leads to the boundary condition:

$$-D\left.\frac{\partial C}{\partial x}\right| = k\left(C_s - C_g\right) \tag{11.24}$$

C_s and C_g are the ^{18}O fraction in the gas phase and at the sample surface, respectively.

The k and D values can also be measured by ECR methods, based on the principle that a variation of oxygen partial pressure in the ambient atmosphere leads to a change in the oxygen vacancy concentration of the MIEC, which can be measured by a change in electronic conductivity. For the ECR method, the sample's total conductivity is assumed to be related to the oxygen stoichiometry. With a small oxygen partial pressure step change (usually $\Delta \log p_{O_2} \leq 1$), the relationship between the conductivity and stoichiometry can be written as below:

$$\frac{\sigma(t) - \sigma(0)}{\sigma(\infty) - \sigma(0)} = \frac{d(t) - d(0)}{d(\infty) - d(0)} \tag{11.25}$$

where $\sigma(t)$, $\sigma(0)$ and $\sigma(\infty)$ refer to the real time conductivity, initial conductivity, and equilibrium conductivity values, respectively, and $d(t)$, $d(0)$ and $d(\infty)$ represent the d value of $ABO_{3\text{-}d}$ for the real time, initial point, and the equilibrium point, respectively. Therefore the measured conductivity values can be used to solve the diffusion equation using the boundary conditions

$$-D\frac{\partial C}{\partial x}\bigg| = k(C_\infty - C_s) \tag{11.26}$$

C_∞ and C_s are the equilibrium and real-time surface oxygen concentration, respectively.

Figure 11.33 shows an example of electrical conductivity relaxation profiles of LSCF electrode samples under dry and humidified air at 800 °C as a function of exposure time to Cr_2O_3 [1]. Oxygen partial pressure changes from 0.05 bar to 0.21 bar. At the start of the measurement (the exposure time to Cr_2O_3 is zero), the relaxation time is approximately 600 s. With the increase of exposure time, the relaxation time increases gradually, to ~ 2000s after exposed to Cr_2O_3 under dry air for 96 h and ~ 5000 s under humidified air for 96 h (Fig. 11.33 a and b). This indicates that the rate of oxygen reduction decreases after exposed to a Cr source, clearly due to the Cr deposition and poisoning. Water vapor in the air increases the poisoning effect of Cr. The surface exchange coefficients were obtained by fitting the conductivity relaxation curve on the LSCF samples after heat treatment under different conditions. In the absence of chromium poisoning, the K_{chem} value of LSCF decreases slightly, probably due to segregated SrO and Co_3O_4 on the surface of LSCF. After heat treatment in the presence of Cr_2O_3 in dry and humidified air for 96 h, K_{chem} at 800 °C is reduced to 5.5×10^{-5} cm/s and 3.1×10^{-5} cm/s, respectively, one order of magnitude lower than the 5.8×10^{-4} cm/s measured on as-prepared LSCF sample. This indicates that Cr deposition poisons the electrochemical activity of the LSCF cathode for the O_2 reduction reaction by reducing the oxygen exchange coefficient on the LSCF surface.

Even for the LSCF with the same composition, the values for the surface exchange coefficient vary by orders of magnitude. This indicates that the results obtained by

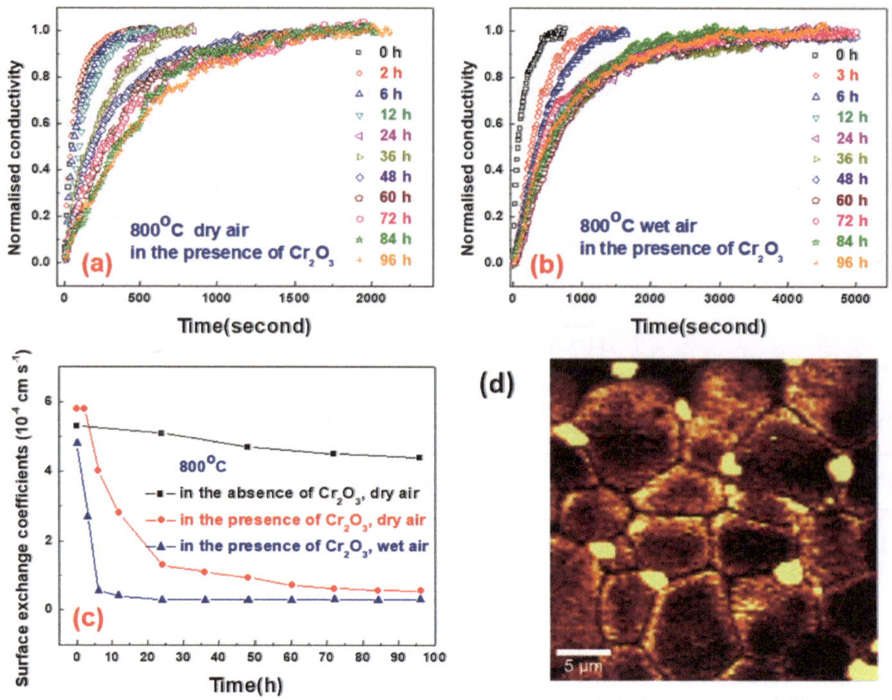

Fig. 11.33 Experimental electrical conductivity relaxation profiles measured in the presence of Cr_2O_3 at 800 °C in **a** dry air and **b** wet air as a function of exposure time, and **c** the fitted surface exchange coefficients as a function of time with and without Cr_2O_3 exposure. Raman mapping at wavelength of 860 cm^{-1} is given in **d** and the yellow regions are the $SrCrO_4$ phase formed on the LSCF surface. Modified from Ref. [1] with permission from IOP Publishing, Copyright 2014

different techniques are not conclusive and in addition the k and D values would also be affected by the surface microstructure and properties (such as composition, grain size, density etc.) and non-stoichiometric composition. Thus, it may not be practical to directly compare the k and D measured by different techniques and by different research groups. However, under the identical and controlled testing conditions, the changes in the k and D values can still be used as a good indicator for the changes of the bulk diffusion and oxygen surface diffusion properties of the electrode materials.

11.7.2 Porosity and Density Measurement

Porosity of the electrode coating is important in determining the concentration polarization. A high porosity will in general lead to a lower concentration polarization loss and may also result in a decrease in the three phase boundary areas for reaction. Therefore, the degree of porosity of the porous coating must be optimized to ensure the high transfer channels for the reactants and products and to maintain the high active sites for the reaction. For electrolyte materials in SOFCs, density is most critical for high ionic conductivity.

The basic method in determining porosity and density is Archimedes' method. In this method, the volume of a sample with an arbitrary shape is measured by either the displaced liquid after the sample is submerged or by weighing the submerged sample. The porosity obtained by Archimedes' method is generally the open porosity properties of the samples. For the theoretical density of oxide materials in SOFCs, the X-ray diffraction (XRD) technique is often used.

Taking YSZ as an example, crystallography parameters of YSZ can be obtained from XRD data:

Crystal system:	Cubic
Space group:	Fm-3 m
Space number:	225
a (Å):	5.1291
b (Å):	5.1291
c (Å):	5.1291
alpha (°):	90.000
bata (°):	90.000
gamma (°):	90.000

The theoretical density of the as-synthesized YSZ powder can be calculated using the following equation:

$$d_{\mathrm{th}} = \left(\frac{n}{N_{\mathrm{A}} V_{\mathrm{cell}}}\right)[(1-x)M_{\mathrm{Zr}} + xM_{\mathrm{Y}} + 2M_{\mathrm{O}}] = 6.16\,\mathrm{g/cm^3} \tag{11.27}$$

where $x = 0.2$, N_{A} is the Avogadro's number (6.022×10^{23} /mol), M_{Zr}, M_{Y}, and M_{O} are the atomic weight of Zr, Y, and O, V_{cell} is the volume of the unit cell and n is the number of formula units in the unit cell. In the case of YSZ, $V_{\mathrm{cell}} = 134.93$ and $n = 4$. The theoretical density of YSZ is 6.16 g/cm^3.

11.7.3 Focus Ion Beam and Scanning Transmission Electron Microscopy (FIB-STEM)

There are wide range microstructure and elemental analytical techniques for the phase and composition analysis of SOFC electrode and electrolyte materials. Among them, focused ion beam (FIB) is a popular technique in the preparation of TEM lamella for the identification of interface, microstructure and element distribution of relatively large-sized samples. Unlike inert gas ion sputtering, FIB ionizes Ga into Ga^{+}, and

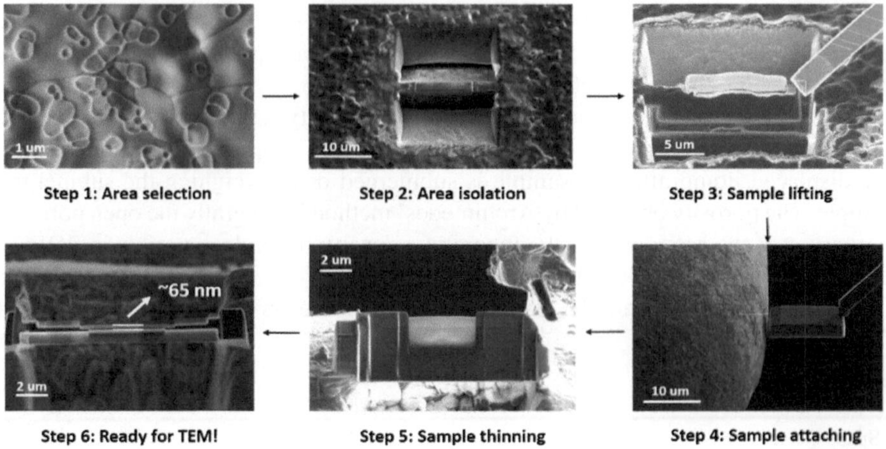

Fig. 11.34 Flowchart of FIB lamella milling procedures

then uses electric field acceleration and electrostatic lens aggregation to drive high-energy, high-speed Ga^+ ions to a specified position with micron-scale precision. The combined FIB-STEM technique offers opportunities of more precise analysis of the microstructure and electrode/electrolyte interface on a nanoscale. The FIB technique not only allows a TEM lamella to be prepared from a specific area of interest with submicron precision but also can uniformly mill heterogeneous microstructures that consist of compositionally and structurally distinct regions. Figure 11.34 shows the typical FIB milling process.

11.7.3.1 Sample Preparation and FIB Milling Procedure

In order to perform a high-quality FIB milling, sample of interest should be conductive and ideally flat. For non-conductive samples, it is necessary to coat the region of interest with carbon or platinum. Carbon coating is preferred if the crystalline structure of the sample surface is desired. In the case of samples with rough surface or the interface between electrode and electrolyte are of particular interests, resin embedding should be considered. The prepared sample needs to be free from drifting under the microscope working conditions.

Carbon or platinum deposition—First, in order to prevent the damage caused during ion milling, a Pt or C protection layer is deposited on the area of interest using electron beam, then a thick layer (~2 um) is deposited at the same area using ion beam (Step 1);

Area isolation—Trenches on both sides of the area with protection layer are milled using ion beam. Typically, large ion beam current is used to make the milling more efficient (Step 2);

Fig. 11.35 SEM images of FIB sample milling process

U-cut—The side view of the sample sheet after the trenches milled is shown in Fig. 11.35a. Then a "U-cut" is performed with ion beam. It is important to ensure that the sample is cut through, which is indicated by the appearance of a black line below the sample lamella (see Fig. 11.35b). After the U-cut, the lamella is welded to the tungsten needle using Pt deposition, see Fig. 11.35c and d. The welded lamella is then detached from the bulk (Step 3);

Lamella attachment—The thick sample lamella is then welded on a FIB sample grid using Pt deposition, see Fig. 11.36. To ensure a good attachment, one side of the lamella is welded on the grid, and contact between the needle and the lamella is cut off, then the other side of the lamella is welded on the grid (Step 4);

Sample thinning—Since the sample is still quite thick (~1 μm), further thinning process is performed by ion milling until the thickness of the lamella is less than 100 nm, see Fig. 11.37. The thinning process is critical, as it determines the imaging quality of further TEM analysis (Step 5).

Fig. 11.36 SEM image showing the FIB lamella mounting process

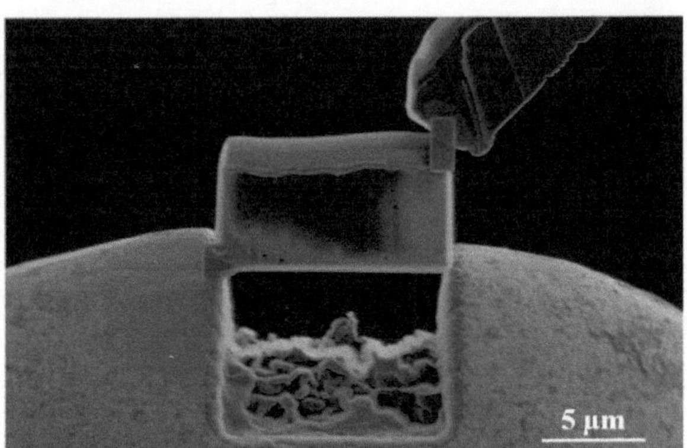

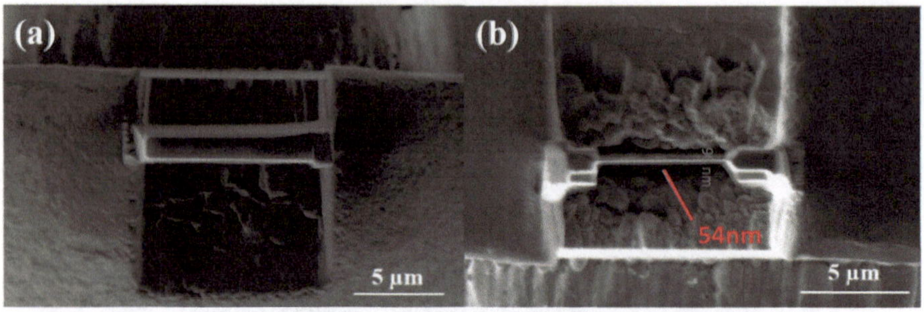

Fig. 11.37 SEM images of FIB lamella **a** before and **b** after thinning

Fig. 11.38 Typical FIB lamella for high resolution TEM

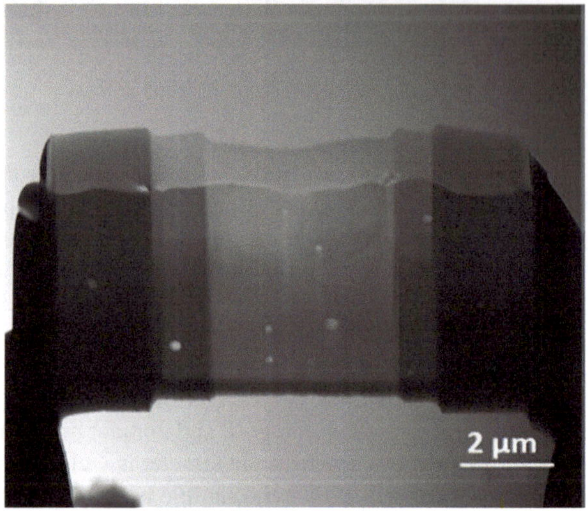

After final low voltage ion beam cleaning, the lamella is ready for TEM examination. The thickness of a FIB lamella should be less than 150 nm, and if high resolution TEM (HRTEM) imaging is required, a sample thickness of ~ 60 nm is desired. Figure 11.38 shows a typical FIB lamella for HRTEM. The FIB instrument adopted here is Dual Beam Focused Ion Beam—Scanning Electron Microscope (FIB-SEM, Helio Nanolab G3 CX, FEI company, US) with Ga^{+} source at 30 kV for the ion beam.

11.8 Summary

In this chapter, we have introduced and discussed the following topics:

- In the development of SOFC technologies, the first step is to establish the methodology and test procedures for the performance evaluation of cell components. Therefore the test station design and cell configuration and arrangement are the most important to ensure the reliability and reproducibility of the measured data. In cell configuration considerations such as the relevance of the test conditions to the practical and real operating conditions and significance of the data to the targeted reaction steps need to be taken. The comparative experimental approach can be very effective to identify the individual reaction steps or the rate determining step of the reaction. In SOFCs, devitation from ideal polarization or *i-V* curves sometimes occurs most likely due to the Joule effect particularly in the case of high cell ohmic resistance and high current densities.
- In cell configurations adopted in the R&D of SOFC, the position of reference electrode (RE) relative to the working and counter electrodes (WE and CE) is most critical particularly for the application and use of RE in thin electrolyte cells. This issue is controversial because there is no clear methodology or rule to define the minimum thickness of the electrolyte in the applicability of RE. However, as discussed in this chapter, experimentally the cell resistance can be separated by the RE placed on the electrolyte surface either on the cathode or anode side. The effective partitioning of cell resistance is a strong function of the electrolyte thickness as well as the interface contact at the cathode and anode sides.
- One unique feature of SOFC is the solid-to-solid contacts at the electrode/electrolyte interface and at the electrode/current collector interface. This will cause the distortion and shift of equipotential line as shown by numerous modeling studies. Experimentally the shift of the equipotential line due to asymmetric contacts at the cathode and anode interfaces can be measured by using the special voltage probe. The shift of the equipotential line within the electrode side of the electrode/electrolyte interface is identified by the "depressed potential decay" as shown in the GCI curves.
- EIS is the most common and popular technique in the measurement of cell performance and activity in SOFC and most common and straight forward method in the analysis of EIS data is the equivalent circuit approach. Nevertheless, we need to be cautions in the use of equivalent circuit as it is not unique to the reactions. Physical significances of the elements and subcircuit need to be verified prior to the application and adoption of specific equivalent circuit. Another issue associated with EIS in SOFC is the frequently observed high- and low-frequency inductances, which can be related to the test station design such as the heating elements of the furnace or the nature of the reaction, as shown in Fig. 11.8.
- The solid-to-solid contact can lead to the constriction effect in solid electrolyte cells like SOFC, *i.e.*, the measured cell resistance is larger than the calculated resistance based on the electrolyte thickness. It can also lead to the loss of the

effective reactive areas of the electrode, resulting in the increase of the overpotentials and decrease in the power output. Thus, the contact between the electrode coating and current collector is extremely important for the performance of SOFC.

- Finally focus ion beam (FIB) is introduced due to its specific advantage in preparation of TEM lamella from a specific area of interest with submicron precision and in uniformly milling heterogeneous microstructures that consist of compositionally and structurally distinct regions. The combined FIB-TEM technique offers extensive opportunities of precise analysis of the microstructure and electrode/electrolyte interface on a nanoscale.

11.9 Questions

16.1. Explain the following terms and materials:

Cell configuration, static arrangement, and dynamic arrangement
Comparative investigation approach
Symmetric cell configuration and reaction reversibility
Polarization curves and Joule heating effect
Performance scalability and electronic conductivity of electrode coating
Equipotential line location and reversed ohmic resistance
Constriction effect and electrolyte thickness
High frequency inductance, low frequency inductance loop and constant-phase elements
Frequency domain and time domain
Thermally activated process
Bulk resistance and grain boundary resistance
Electrical conductivity relaxation and FIB-STEM

16.2. Using the comparative approach principles, design an experimental plan for the investigation of reaction mechanism and kinetics for reactions such as the O_2 reduction reaction or the H_2 oxidation reaction on selected electrodes.

16.3. Referred to Fig. 11.16, discuss the factors which would influence the value and direction of the depressed iR curves.

16.4. Explain why the constriction effect depends on the electrochemical activities of the electrodes. Why is the constriction effect is more pronounced at low temperatures? (Clue: using electrical field distribution).

16.5. As shown in Fig. 11.27, for the H_2 oxidation reaction (HOR) on Ni/YSZ cermet electrodes the activation energy is ~ 76 kJ/mol and 170 kJ/mol, measured by GCI and EIS techniques, respectively. Explain the reason for the difference in the activation energies.

16.6. For O_2 reduction reactions (ORR) on LSM electrodes, the overall electrode behavior (e.g., the activation energy and the dependence of the reaction on partial pressure of oxygen) can be obtained by GCI or EIS techniques. Would the parameters obtained by GCI and EIS be the same for ORR? Explain your answers with evidences.

16.7. Why in general the electronic conductivity of ceramic oxides is much lower in reducing environment than that in oxidizing environment?

16.8. Given the lattice parameter, $a = 0.5425$ nm for $Ce_{0.8}Gd_{0.2}O_{1.9}$, calculate the theoretical density of $Ce_{0.8}Gd_{0.2}O_{1.9}$ (Answer: $n = 4$ and $d_{th} = 7.2327$ g/cm^3).

16.9. $LaMnO_3$ is perovskite with orthorhombic distortion. From XRD analysis, we have lattice parameters of a = 5.70 Å, b = 7.72 Å, and c = 5.54 Å. Calculate the theoretical density of $LaMnO_3$ (Answer: $n = 4$ and $d_{th} = 6.59$ g/cm^3).

References

1. Zhao L, Zhang J, Becker T, Jiang SP (2014) Raman spectroscopy study of chromium deposition on $La_{0.6}Sr_{0.4}Co_{0.2}Fe_{0.8}O_3$ cathodes of solid oxide fuel cells. J Electrochem Soc 161:F687–F693
2. Zhao L, Drennan J, Kong C, Amarasinghe S, Jiang SP (2014) Insight into surface segregation and chromium deposition on $La_{0.6}Sr_{0.4}Co_{0.2}Fe_{0.8}O_{3-\delta}$ cathodes of solid oxide fuel cells. J Mater Chem A 2(29):11114–11123
3. Chen K, Ai N, Jiang SP (2013) Chemical compatibility between boron oxides and electrolyte and cathode materials of solid oxide fuel cells. Fuel Cells 13(6):1101–1108
4. Wang CC, Chen KF, Jiang SP (2014) Sulfur Deposition and Poisoning of $La_{0.6}Sr_{0.4}Co_{0.2}Fe_{0.8}O_{3-\delta}$ cathode materials of solid oxide fuel cells. J Electrochem Soc 161(12):F1133–F1139
5. Jiang SP, Zhang JP, Foger K (2001) Deposition of chromium species at Sr-doped $LaMnO_3$ electrodes in solid oxide fuel cells—III. Effect of air flow. J Electrochem Soc 148(7):C447–C455
6. Chen K, Liu S-S, Guagliardo P, Kilburn MR, Koyama M, Jiang SP (2015) A fundamental study of boron deposition and poisoning of $La_{0.8}Sr_{0.2}MnO_3$ cathode of solid oxide fuel cells under accelerated conditions. J Electrochem Soc 162(12):F1282–F12991
7. Jiang SP, Chen XB (2014) Chromium deposition and poisoning of cathodes of solid oxide fuel cells—a review. Int J Hydrog Energy 39(1):505–531
8. Jiang SP, Badwal SPS (1997) Hydrogen oxidation at the nickel and platinum electrodes on yttria-tetragonal zirconia electrolyte. J Electrochem Soc 144(11):3777–3784
9. Jiang SP (2002) A comparison of O_2 reduction reactions on porous (La,Sr)MnO_3 and (La,Sr)(Co,Fe)O_3 electrodes. Solid State Ionics 146(1–2):1–22
10. Ostergard MJL, Clausen C, Bagger C, Mogensen M (1995) Manganite-zirconia composite cathodes for SOFC—influence of structure and composition. Electrochim Acta 40(12):1971–1981
11. Pan W, He S, Chen K, Ai N, Lü Z, Jiang SP (2020) Verification and applicability of symmetric cell configuration for mechanistic study of oxygen electrode reactions of solid oxide cells. Solid State Ionics 357:115457
12. Zhao F, Virkar AV (2005) Dependence of polarization in anode-supported solid oxide fuel cells on various cell parameters. J Power Sources 141(1):79–95
13. Leng YJ, Chan SH, Khor KA, Jiang SP (2004) Performance evaluation of anode-supported solid oxide fuel cells with thin film YSZ electrolyte. Int J Hydrog Energy 29(10):1025–1033
14. Jiang SP (2003) Issues on development of (La, Sr)MnO_3 cathode for solid oxide fuel cells. J Power Sources 124(2):390–402
15. Jiang SP, Love JG, Apateanu L (2003) Effect of contact between electrode and current collector on the performance of solid oxide fuel cells. Solid State Ionics 160(1–2):15–26

16. Polverino P, Gallo M, Pianese C (2021) Development of mathematical transfer functions correlating solid oxide fuel cell degradation to operating conditions for Accelerated Stress Test protocols design. J Power Sources 491:229–521
17. Jiang SP (2001) Resistance measurement in solid oxide fuel cells. J Electrochem Soc 148(8):A887–A897
18. Jiang SP (2017) Placement of reference electrode, electrolyte thickness and three-electrode cell configuration in solid oxide fuel cells: a brief review and update on experimental approach. J Electrochem Soc 164(7):F834–F844
19. Jiang SP, Ramprakash Y (1999) H_2 oxidation on Ni/Y-TZP cermet electrodes—a comparison of electrode behaviour by GCI and EIS techniques. Solid State Ionics 122(1–4):211–222
20. Jiang SP, Love JG, Ramprakash Y (2002) Electrode behaviour at (La, Sr)MnO_3/Y_2O_3-ZrO_2 interface by electrochemical impedance spectroscopyy. J Power Sources 110(1):201–208
21. Jiang SP (2004) Cell configuration for performance evaluation in planar solid oxide fuel cells. J Appl Electrochem 34(10):1045–1055
22. Jiang SP (2008) Dependence of cell resistivity on electrolyte thickness in solid oxide fuel cells. J Power Sources 183(2):595–599
23. Huang QA, Hui R, Wang BW, Zhang JJ (2007) A review of AC impedance modeling and validation in SOFC diagnosis. Electrochim Acta 52(28):8144–8164
24. Jiang SP, Zhang JP, Foger K (2000) Deposition of chromium species at Sr-doped $LaMnO_3$ electrodes in solid oxide fuel cells—II. Effect on O_2 reduction reaction. J Electrochem Soc 147 (9):3195–3205.
25. Jiang SP, Badwal SPS (1999) An electrode kinetics study of H_2 oxidation on Ni/Y_2O_3-ZrO_2 cermet electrode of the solid oxide fuel cell. Solid State Ionics 123(1–4):209–224
26. Vanhassel BA, Boukamp BA, Burggraaf AJ (1991) Electrode polarization at the Au, O_2 yttria stabilized zirconia interface. 2. Electrochemical measurements and analysis. Solid State Ionics 48(1–2):155–171
27. Chen KF, Ai N, Jiang SP (2016) Origin of low frequency inductive impedance loops of O_2 reduction reaction of solid oxide fuel cells. Solid State Ionics 291:33–41
28. Jiang SP (2001) Use of gaseous Cr species to diagnose surface and bulk process for O_2 reduction in solid oxide fuel cells. J Appl Electrochem 31(2):181–192
29. Teraoka Y, Nobunaga T, Okamoto K, Miura N, Yamazoe N (1991) Influence of constituent metal-cations in substituted $LaCoO_3$ on mixed conductivity and oxygen permeability. Solid State Ionics 48(3–4):207–212
30. Jiang SP, Liu L, Ong KP, Wu P, Li J, Pu J (2008) Electrical conductivity and performance of doped $LaCrO_3$ perovskite oxides for solid oxide fuel cells. J Power Sources 176:82–89
31. Guo X, Waser R (2006) Electrical properties of the grain boundaries of oxygen ion conductors: acceptor-doped zirconia and ceria. Prog Mater Sci 51(2):151–210
32. Leng YJ, Chan SH, Khor KA, Jiang SP, Cheang P (2003) Effect of characteristics of Y_2O_3/ZrO_2 powders on fabrication of anode-supported solid oxide fuel cells. J Power Sources 117(1–2):26–34
33. Chater RJ, Carter S, Kilner JA, Steele BCH (1992) Development of a novel sims technique for oxygen self-diffusion and surface exchange coefficient measurements in oxides of high diffusivity. Solid State Ionics 53–6:859–867

Chapter 12
Solid Oxide Fuel Cells: Fabrication and Microstructure

12.1 Introduction

As SOFCs mainly consist of ceramic materials, the processing techniques used in the ceramic industries have been extensively employed and adopted for the synthesis of electrolyte and electrode powders and for the fabrication of cell components of SOFCs. Due to the fact that the cost of ceramic raw materials is generally low, the overall cost of a SOFC stack is dominated by the fabrication processes, which in turn depend on the stack design (e.g., planar or tubular design) and cell structure (e.g., electrolyte supported or electrode supported). This is particularly relevant for the development of intermediate temperature SOFCs or IT-SOFCs. The reduction in operating temperature substantially increases the flexibility of material selection and durability of SOFCs, but also results in a significant increase in the electrolyte and electrode resistivity and polarization losses. To compensate for the performance losses due to the reduced operating temperature, polarization losses at both anode and cathode need to be reduced, and the thickness of the electrolyte layer has to be decreased in order to lower the ohmic resistance of the cell. This requires the selection and optimization of powder synthesis methods and thin-film fabrication technologies for SOFCs [1].

This chapter will start with a description of various synthesis processes for electrode and electrolyte powders, followed by fabrication techniques for electrode coatings and thin-film components such as electrolyte, electrode, and protective coatings in SOFCs. The interface formation and microstructrue development related to the synthesis and fabrication processes will also be discussed.

S. P. Jiang and Q. Li, *Introduction to Fuel Cells*,
https://doi.org/10.1007/978-981-10-7626-8_12

12.2 Powder Synthesis Methods

12.2.1 Solid-State Reaction and Physical Mixing Methods

The most common and conventional method to prepare electrode and electrolyte powders of SOFCs is the *solid-state reaction* (SSR) method. In this method, precursors, e.g., oxides, nitrate, or carbonate salts, are mixed by ball milling and sintered in powder or pressed disk form at selected high temperatures. Calcination of the powder is usually carried out in more than one stage. For example, to produce single-phased $La_{0.8}Sr_{0.2}MnO_3$ perovskite, stoichiometric amounts of pure La_2O_3, $SrCO_3$, and MnO_2 or $MnCO_3$ are mixed and pre-sintered at intermediate temperatures of 950–1000 °C, followed by final calcination at 1350–1450 °C for 24 h. The SSR is as follows:

$$La_2O_3 + SrCO_3 + MnCO_3 \xrightarrow{\text{calcination}} La_{0.8}Sr_{0.2}MnO_3 + CO_2 + O_2 \qquad (12.1)$$

The SSR method is simple and fast and is particularly useful in the initial stage of the material development due to its precision in the composition control. Nevertheless, the SSR method generally requires long processing time and the repetition of milling processes to break the agglomerates and to improve the particle size distribution and surface area, and to obtain the desired phase purity and homogeneity of the mixture. The types of ball mill, rotation speed, milling time, type of milling balls, and operation mode of ball milling processes are the operation parameters affecting quality of the powder prepared by SSR. The main drawbacks are the high calcination temperature, poor compositional homogeneity, low surface area, poor sintering ability of the powder and difficulty in controlling the porous structure and particle size distribution.

Physical mixing of oxide powders with desired ratios is the most common method to prepare composite electrodes such as LSM/YSZ cathodes and Ni/YSZ cermet anodes. Highly crystalline solid-state powders of each phase are first prepared through independent synthesis methods or obtained from commercial suppliers and are then mixed extensively by ball milling. Ball milling serves to decrease the particle size of each phase, increase the contacts between phases, and, thus enhance the uniformity of the composites. The particle size achieved for each component is typically in the order of one to several microns. This intimately mixed powder is then combined with an organic vehicle such as glycerol to form a slurry or ink, which can be applied as electrode or electrolyte coatings by techniques such as spraying, screen printing, or slurry painting. Physical mixing is commonly used in industry for composite cathodes and anodes of SOFCs due to its reproducibility and the scalability of the process.

12.2.2 Solution Combustion Method

Solution combustion synthesis is a common powder synthesis route in which a metal nitrate solution containing a fuel is heated in a furnace at high temperatures, e.g., 500–700 °C. To assist the combustion synthesis, urea ($CO(NH_2)_2$), glycine (NH_2CH_2COOH), citric acid (CA, $C_6H_8O_7$), and sucrose ($C_{12}H_{22}O_{11}$) are commonly used as the fuel. The combustion method is a simple method with advantages of using inexpensive precursors and leading to the formation of nano-sized, homogeneous, highly reactive powders. It is particularly useful in the production of binary and ternary oxide powders with fine particles and high porosity. Glycine is one of the most popular fuels in the combustion reaction due to its low ignition temperature. Glycine also acts as a complexing agent for metal cations as it has a carboxylic group at one end and an amino group at the other end. This zwitterionic characteristic of glycine enables the effective formation of stable complexes with metal ions of varying ionic sizes, increasing solubility, and preventing selective precipitation of metal ions during water vaporization. The one using nitrite precursor and glycine as the fuel is also called *glycine-nitrate process* (GNP).

Figure 12.1 shows an example of synthesis steps in the formation of $La_{0.6}Sr_{0.4}Co_{0.2}Fe_{0.8}O_{3-\delta}$ (LSCF) powders by GNP. $La_{0.6}Sr_{0.4}Co_{0.2}Fe_{0.8}(NO_3)_x$ solution is prepared by dissolving stoichiometric amounts of the metal nitrates $La(NO_3)_3{\cdot}6H_2O$, $Sr(NO_3)_2$, $Co(NO_3)_2{\cdot}6H_2O$, $Fe(NO_3)_3{\cdot}9H_2O$, and glycine in deionized water. The molar ratio of metal cations to glycine is 1:1.5. The solution is stirred and heated on a hotplate to evaporate water, forming a transparent gel. The gel then spontaneously ignites to flame, forming black ash. The ash is calcined at 900 °C in air for 2 h, forming LSCF powder. The synthesis process is carried out inside a well-ventilated fume hood. The process converts precursor materials into fine oxide powders. In this combustion reaction, glycine acts as a fuel and is oxidized by the nitrate ions, NO_3^-. The stoichiometric combustion reaction can be written as follows:

$$\begin{aligned} &0.6La(NO_3)_3 + 0.4Sr(NO_3)_2 + 0.2Co(NO_3)_2 + 0.8Fe(NO_3)_3 \\ &\quad + 3.0H_2NCH_2COOH \rightarrow La_{0.6}Sr_{0.4}Co_{0.2}Fe_{0.8}O_3 \\ &\quad + 6.0CO_2 + 7.5H_2O + 4.2N_2 \end{aligned} \tag{12.2}$$

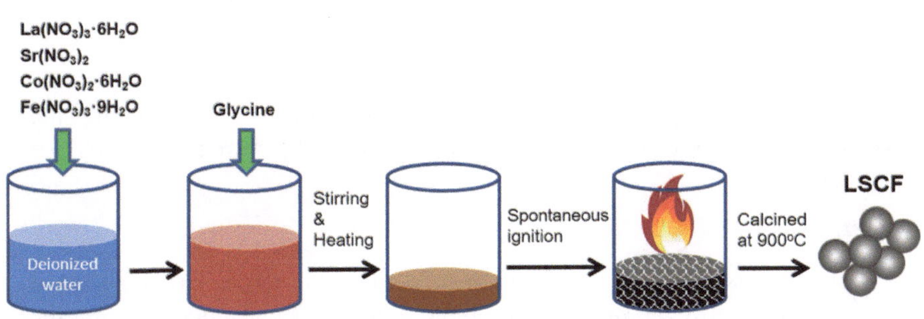

Fig. 12.1 Flowchart of the solution–combustion process for the synthesis of $La_{0.6}Sr_{0.4}Co_{0.2}Fe_{0.8}O_{3-\delta}$ powders

The hydration of metal nitrates is neglected because the reactants are dissolved in water. XRD analysis of the powder indicates that the formation of the perovskite crystalline phase, $La_{0.6}Sr_{0.4}Co_{0.2}Fe_{0.8}O_{3-\delta}$ begins at relatively low temperature of 700 °C, and single-phase perovskite occurs at 900 °C. Further increase in the calcination temperature will increase the crystallinity of the perovskite phase, but also significantly increase the particle size.

The evolution of CO_2, N_2, and H_2O gases during the combustion process helps in the formation of fine oxide powders by limiting the interparticle contact. Moreover, the combustion process occurs at such a fast rate that there would not be a sufficient time for long-distance diffusion of atoms or molecules for grain growth. Thus, the solution–combustion process can result in the formation of nano-sized powders with fine distribution. However, the thermal stability, phase formation, and particle size are strongly dependent on the precursor, nature of the fuel (urea, glycine, CA, or sucrose), and fuel-to-nitrate ratio. In general, the combustion reaction is more complete with glycine, while in the case of CA, the ignition of the resultant gel slowly propagates and the reaction is less violent, promoting the formation of materials of high crystallinity and homogeneity.

The *stoichiometric amount* of fuel or propellant like urea or glycine can be determined based on the valences of oxidizing and reducing elements, according to the propellant chemistry. In combustion reactions, C and H are considered as reducing elements with valences of +4 and +1, respectively, while N and O are treated as oxidizing elements with valences of 0 and −2, respectively. Typical valences for the metallic elements in nitrates are +2 for Sr, Co, Ni, and Cu, +3 for Sm, Fe, La, Mn and Ba and +4 for Zr and Ce. Thus, equivalent valences can be obtained for the reactants of the combustion reaction. For example, the equivalent valence for $La(NO_3)_3 = 3 + 3 \times (0 - 2 \times 3) = -15$. The equivalent valences can be used as *stoichiometric coefficients* of the reaction, and the amount of fuel can be calculated by solving the following equation:

$$\sum_i x_i n_i = 0 \tag{12.3}$$

where x_i and n_i are the equivalent valence and the number of moles of each components, respectively. Based on Eqs. (12.3 and 12.2), the stoichiometric amount of fuel or fuel-to-nitrate ratio can be obtained.

Example 12.1 Calculate the minimum amount of fuel required for the synthesis of 100 g of $La_{0.6}Sr_{0.4}Co_{0.2}Fe_{0.8}O_{3-\delta}$ (LSCF) powder using glycine as fuel.

Solution:

For the combustion synthesis of $La_{0.6}Sr_{0.4}Co_{0.2}Fe_{0.8}O_{3-\delta}$ (LSCF) using glycine as fuel, the equivalent valences and mole number of reactants based on the LSCF stoichiometric composition are as follows:

$$La(NO_3)_3 : x_1 = -15;\ n_1 = 0.6; \tag{12.4}$$

$$Sr(NO_3)_2 : x_2 = -10;\ n_2 = 0.4; \tag{12.5}$$

$$Co(NO_3)_2 : x_3 = -10;\ n_3 = 0.2; \tag{12.6}$$

$$Fe(NO_3)_3 : x_4 = -15; n_4 = 0.8. \tag{12.7}$$

$$\text{Use } NH_2CH_2COOH \text{ as fuel} : x_5 = +9;\ \text{what is } n_5?$$

$$\sum_i x_i n_i = -0.6 \times 15 - 0.4 \times 10 - 0.2 \times 10 - 0.8 \times 15 + n_5 \times 9 = 0 \tag{12.8}$$

The corresponding number of moles of glycine, n_5 is 3.0. This indicates that the stoichiometric ratio between metal nitrates and glycine corresponds to 5.4 groups of NO_3^- for every 3.0 molecules of glycine, as shown in Eq. (12.2). Thus, the stoichiometric molar ratio of nitrate-to-fuel, i.e., nitrate-to-glycine (n/g) in this case is 1:1.8, and the molar ratio of metal cations to glycine is 1:1.5. In general, excess fuel (i.e., higher n/g ratio of 1:2.5) is used to improve the combustion reaction.

In a similar way, the combustion reaction for the synthesis of $La_{0.8}Sr_{0.2}MnO_3$ can be written as follows:

$$\begin{aligned} &0.8La(NO_3)_3 + 0.2Sr(NO_3)_2 + Mn(NO_3)_2 \\ &\quad + 2.7H_2NCH_2COOH + 0.4O_2 \rightarrow La_{0.8}Sr_{0.2}MnO_3 \\ &\quad + 5.4CO_2 + 6.75H_2O + 3.75N_2 \end{aligned} \tag{12.9}$$

The solution–combustion reaction can also be used for the synthesis of composite electrodes like Ni/YSZ cermet:

$$5CO(NH_2)_2 + 3Ni(NO_3)_2 \rightarrow 3NiO + 8N_2 + 5CO_2 + 10H_2O \tag{12.10}$$

$$\begin{aligned} &1.8CO(NH_2)_2 + 0.84ZrO(NO_3)_2 + 0.16Y(NO_3)_3 \rightarrow Zr_{0.84}Y_{0.16}O_{1.92} \\ &\quad + 1.8CO_2 + 3.6H_2O + 2.88N_2 \end{aligned} \tag{12.11}$$

However, in the synthesis of more complicated oxides, impurities are frequently observed in the final products of the solution–combustion method using urea or glycine as the fuel. These impurity phases are most likely due to the low conditional stability constant of urea- or glycine-metal chelates. Using alternative fuels such as alanine ($NH_2C_2H_4COOH$, an aliphatic amino acid) may improve the crystallinity of the powder by increasing the stability of alanine-metal chelates. Ethylene glycol (EG, $C_2H_6O_2$), polyvinyl alcohol (PVA, $[C_2H_4O]_n$), and carbohydrazide ($OC(N_2H_3)_2$) can also be used as the fuel in solution–combustion processes.

To improve and sustain the combustion process, a combustion aid can be added. The most common combustion aid is ammonium nitrate (NH_4NO_3) because of its low cost and highly exothermic decomposition. Also, it generates only gaseous products without altering the proportion of other elements that are produced.

The combustion reaction can be explosive, yielding a high-temperature flame that quickly propagates through the reactant mixture. Therefore, a safety procedure must be in place prior to experiments for the solution–combustion process for the synthesis of oxide powders.

12.2.3 Co-precipitation Method

The *co-precipitation method* is a facile chemical process that is often used to prepare well-defined and less-agglomerate oxide powders. With the precipitation method, the solubility of individual hydroxide component is of prime importance, which is a function of the pH value. The solubility constant K_{sp} for the precipitation reaction of metal hydroxide, $M(OH)_n$,

$$M^{n+} + n(\mathrm{OH})^- \rightarrow M(\mathrm{OH})_n \tag{12.12}$$

is as follows:

$$K_{sp} = \left[M^{n+}\right] \times \left[\mathrm{OH}^-\right]^n \tag{12.13}$$

$$\log K_{sp} = \log\left[M^{n+}\right] + n \log\left[\mathrm{OH}^-\right] \tag{12.14}$$

where n is the valence of metal ions in the solution and $[M^{n+}]$ and $[OH^-]$ are the molar concentration of M^{n+} and OH^- ions, respectively. The ionic product of water at 25 °C under ambient pressure:

$$\left[\mathrm{H}^+\right]\left[\mathrm{OH}^-\right] = 10^{-14} \tag{12.15}$$

$$-\log\left[\mathrm{H}^+\right] - \log\left[\mathrm{OH}^-\right] = \mathrm{pH} + \mathrm{pOH} = 14 \tag{12.16}$$

Thus, Eq. (12.14) can be re-arranged as follows:

$$\log K_{sp} = \log\left[M^{n+}\right] - 14n - n \log\left[\mathrm{H}^+\right] \tag{12.17}$$

or

$$\log\left[M^{n+}\right] = \log K_{sp} + 14n - n\mathrm{pH} \tag{12.18}$$

Based on the above equations, solubility curves can be obtained for different metal hydroxides. $M(OH)_n$ starts to precipitate once the solution pH value exceeds the solubility curves. To synthesize complex ceramic oxide compounds with multiple metals, the solubility curves of metal hydroxides involved in the precipitation should be as close as possible. However, in practice, such conditions are difficult to meet.

Co-precipitation can also occur with ammonium carbonate, $(NH_4)_2CO_3$. The use of carbonates as precursors shows characteristics of being non-gelatinous and exhibits significantly weaker agglomeration after drying as compared to hydroxide co-precipitation. Figure 12.2 shows a typical TEM micrograph of $Ce_{0.8}Gd_{0.2}O_{1.9}$ (GDC) powders synthesized by the carbonate co-precipitation method [2]. To synthesize GDC, high-purity nitrate salts ($Ce(NO_3)_3 \cdot 6H_2O$ and $Gd(NO_3)_3 \cdot 6H_2O$) in stoichiometric amounts are dissolved in deionized water, forming a 0.3 M 150 mL solution. The mixed solution is then dropped very slowly into 150 mL of a 0.75 M $(NH_4)_2CO_3$ solution under vigorous stirring. In this case, the ammonium carbonate/metal cation ($Ce^{3+} + Gd^{3+}$) molar ration (R) of 2.5 is used. The resultant precipitate is filtered under vacuum and washed thoroughly with deionized water and rinsed with ethanol, respectively. After drying at 70 °C for 20 h, the powder is calcined at a high temperature (~600–800 °C) to decompose the hydroxide, forming nano-sized GDC powder. Formation of a fluoride crystalline phase starts at 600 °C. The particle size is in the range of 25–30 nm. The resultant oxides show good dispersion and good sinterability.

The carbonate co-precipitation reaction and calcination of precursor in the synthesis of GDC powders can be written as follows [2]:

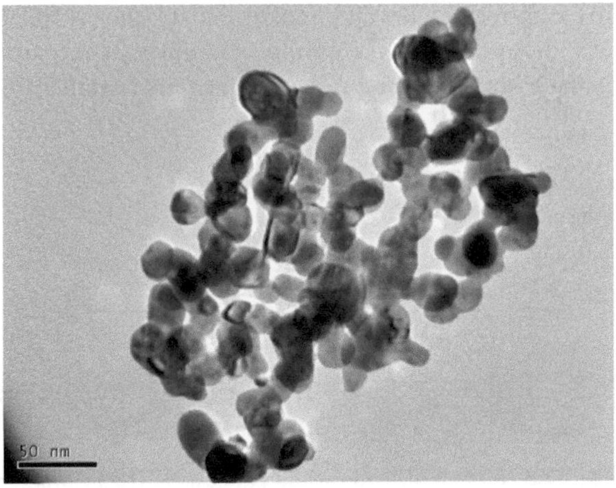

Fig. 12.2 TEM micrograph of the $Ce_{0.8}Gd_{0.2}O_{1.9}$ powder synthesized by carbonate co-precipitation method, calcined at 700 °C

$$0.8\mathrm{Ce(NO_3)_3} + \mathrm{(NH_4)_2CO_3} + 0.2\mathrm{Gd(NO_3)_3} \rightarrow \mathrm{(NH_4)_{0.25}CeGd_{0.25}(CO_3)_2 \cdot 2H_2O} \quad (12.19)$$

$$\mathrm{(NH_4)_{0.25}CeGd_{0.25}(CO_3)_2 \cdot 2H_2O} \xrightarrow{\text{calcination}} \mathrm{Ce_{0.8}Gd_{0.2}O_{1.9}} \quad (12.20)$$

The molar ratio, R, significantly affects the morphology and properties of the precursors. For compete precipitation, R should be greater than 2.

12.2.4 Pechini and Polymeric Complexing Method

The *Pechini method* [3] or polymeric complexing methods have been extensively used for the synthesis of homogeneous and fine oxide powders. The method relies on the polyesterification of chelated cations to produce metal complexation polymeric networks in which the precipitation of secondary phases is inhibited through steric entrapment of metal cations, thereby maintaining molecular-level mixing at all stages of processing. This method employs carboxylate chelating agents and glycol groups. Citric acid (CA), ethylenediaminetetraacetic acid (EDTA), polyacrylic acid, polyvinyl alcohol, and urea formaldehyde are used as chelating agents for the synthesis of perovskite powders by this route. The most common combination is the CA-driven chelating with ethylene glycol (EG)-driven polyesterification. The resulting polymeric gel is then thermally decomposed under oxidizing conditions to remove organic content and form an oxide with homogeneously dispersed cations.

Figure 12.3 shows a typical Pechini method with double chelating agents, CA and EDTA, for the synthesis of LSM powders. The process shown in the figure is also called a modified Pechini method as EG is not used. The pH of the mixed solution affects the dissociation of CA, as well as the metal complexation process. Under the pH limit for a stable and transparent sol, the complexing and combustion processes are improved with increasing pH values of the mixed solution. Thus, tuning pH is particularly important in systems with several different metal cations in order to

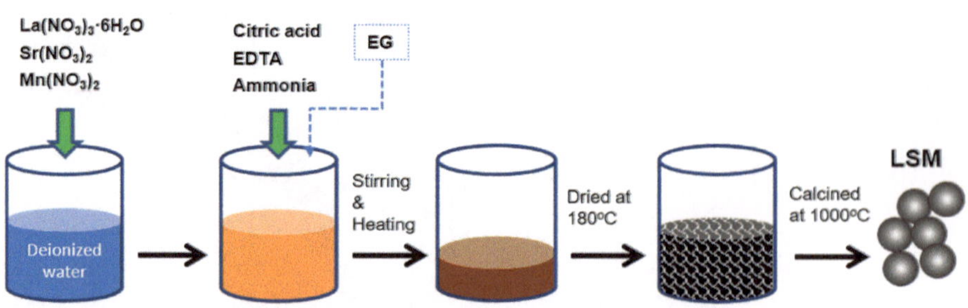

Fig. 12.3 Pechini method for the synthesis of (La,Sr)MnO_3 powder. EG can be added to assist polyesterification reaction

optimize the formation of stable metal citrate/EDTA species and prevent precipitation of individual hydroxides. Single-phase perovskite structures can be formed after calcination at relatively low temperatures, e.g., 900–1000 °C.

The polymeric complexing method involves two basic chemical reactions: (i) complexation or chelation between metal ions and chelants and (ii) polyesterification of complexes with EG. The complexation and polyesterification reactions preserve the homogeneity of the metal salts in the solution when it transforms into a gel. This polymerization reaction forms three-dimensional structures and minimizes cation segregation. The advantage of the method is the high purity that can be obtained and the excellent control that can be exercised over the resulting powder composition.

Figure 12.4 shows an example of the morphology and XRD of $La_{0.8}Sr_{0.2}MnO_3$ (LSM) synthesized by the Pechini method. In this process, LSM solution is prepared by dissolving stoichiometric amounts of metal nitrates or salts $La(NO_3)_3{\cdot}6H_2O$, $Sr(NO_3)_2$, and $Mn(NO_3)_2$, EDTA and ammonia into deionized water. The metal ions/CA/EDTA molar ratio is 1:1.5:1. Ammonia solution is added to adjust the pH of the solution to ~5.5 at which no precipitation occurs. Then the solution is stirred on a hotplate for 2 h to form a gel. The gel is thoroughly dried in an oven at 180 °C for 8 h and calcined at 900 °C in air for 2 h, forming black LSM powder. As shown in the figure, the particle size of as-synthesized LSM powder is in the range of 50–100 nm, and a single-phase perovskite LSM structure with *R-3c* space group forms after calcination at 900 °C.

The dried gel decomposes at low temperatures to form an amorphous oxyhydroxy carbonate phase, the crystallization occurs around 600–700 °C, and formation temperature of pure perovskite phase occurs around 900–1000 °C. The method can be modified to synthesize composite oxides like LSM/YSZ using a polymeric solution consisting of an emulsion of two intermediate polymeric solutions of La-Sr-Mn and Y-Zr. Direct calcination leads to the formation of LSM/YSZ composite powder.

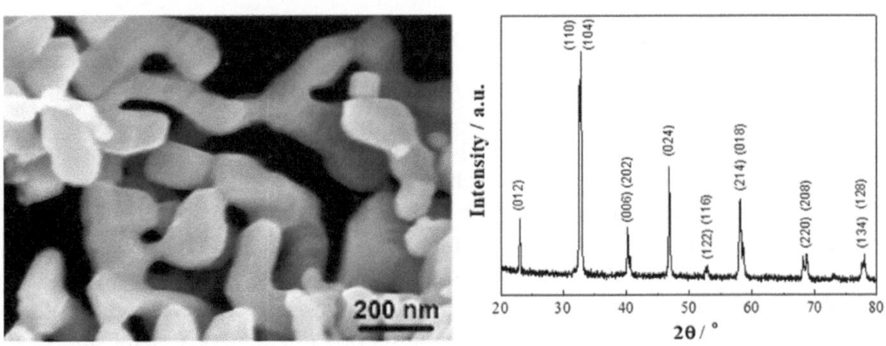

Fig. 12.4 SEM micrograph and XRD patterns of as-synthesized $La_{0.8}Sr_{0.2}MnO_3$ powder after calcination at 900 °C

12.2.5 Sol–Gel Method

The *sol–gel process* is a wet-chemical technique used for the synthesis of electrode and electrolyte powders and the fabrication of thin electrolyte coatings in SOFCs. In this process, the sol (or solution) evolves gradually toward the formation of a gel-like network containing both a liquid phase and a solid phase. The sol–gel process is an attractive route that starts from molecular precursors and forms a colloidal network via hydrolysis and condensation reactions or inorganic polymerization reactions. The basic structure of the solid phase can be in the range anywhere from discrete colloidal particles to continuous chain-like polymer networks. This process presents many advantages, including high phase purity, good composition homogeneity, and a high surface activity of the resulting powders. Typical precursors are metal alkoxides and metal chlorides. The sol–gel chemistry originating with the hydrolysis and condensation of metal alkoxides has been well illustrated with the production of silica from tetramethyl orthosilicate. In this case, the sol–gel chemistry of silica can be driven by either acid or base catalysts, but the resulting gel structure is significantly different.

The alkoxide precursor-based sol–gel process has one significant limitation, namely that many metal alkoxides do not exist or are simply not stable. Alkoxides are also generally sensitive to moisture. An alternative is to use metal nitrates instead of metal alkoxides. In this case, chelating agents are often used to modify the aqueous hydrolysis chemistry of metal ions. The main function of chelating agents is to change the hydrolysis equilibria of dissolved metal ions [4]. For example, the addition of EDTA to aqueous iron can significantly reduce the equilibrium constant of hydrolysis, K_{h}:

$$\left[\mathrm{Fe(H_2O)_6}\right]^{3+} + \mathrm{H_2O} \rightleftharpoons \left[\mathrm{Fe(OH)(H_2O)_5}\right]^{2-} + \mathrm{H_3O^+} \quad K_{\mathrm{h}} = 10^{-3} \tag{12.21}$$

$$\begin{aligned}\left[\mathrm{Fe(H_2O)_2EDTA}\right]^{-} + \mathrm{H_2O} &\rightleftharpoons [\mathrm{Fe(OH)(H_2O)EDTA}]^{2-} \\ &+ \mathrm{H_3O^+} \quad K_{\mathrm{h}} = 10^{-7.5}\end{aligned} \tag{12.22}$$

Thus, hydrolysis of $[\mathrm{Fe(H_2O)_2EDTA}]^-$ is considerably less favorable as compared to $[\mathrm{Fe(H_2O)_6}]^{3+}$. The removal of solvent from the metal-chelate solutions would result in homogeneous glassy solids or resins rather than precipitates. These can be heated to form powders or nanostructures of a wide range of binary and ternary metal oxides. CA is one of the most common and effective chelating agents used in the sol–gel method. CA is a weak triprotic acid with three carboxylic acid moieties that are able to dissociate. In a typical synthesis process, aqueous metal salts (e.g., nitrates) are mixed with CA, and the resulting solution is heated to form a viscous solution or gel. Figure 12.5 shows a typical procedure for the sol–gel synthesis of $\mathrm{La_{0.6}Sr_{0.4}CoO_{3-\delta}}$ powder. Similar to the Pechini method, ammonia or ethylene diamine can be added to modify the pH and enhance the binding of metal ions to

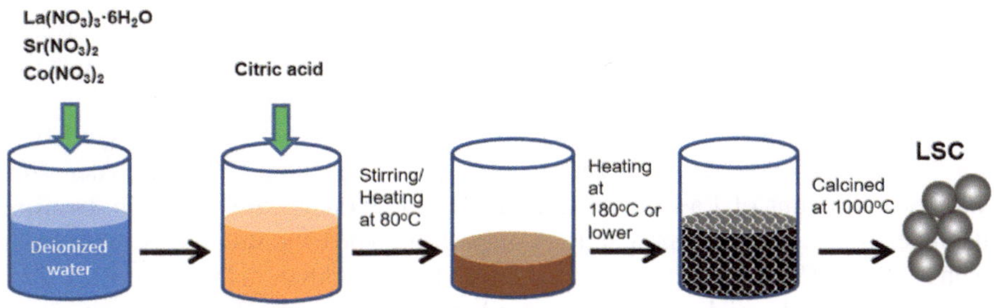

Fig. 12.5 Sol–gel synthesis of $La_{0.6}Sr_{0.4}CoO_{3-\delta}$ powder

the citrates. Other chelating agents used in sol–gel synthesis are other mono- and di-carboxylic acids including tartaric acid, glycolic acid, and oxalic acid.

There are differences in the synthesis processes of the Pechini and sol–gel methods. In the conventional sol–gel process, interconnected metal–oxygen M–O polymer networks are formed by hydrolysis and condensation of metal alkoxides or salts. In the formation of these M–O networks, segregation of cationic species and the formation of nanocrystals of binary compounds can occur owing to moisture sensitivity and/or preferential hydrolysis kinetics of individual metal precursors. In contrast, bonding in the Pechini method occurs through a polymer which is interconnected by covalent and coordinate bonding established during polyesterification between metal cations chelated with carboxylate groups and polyhydroxy alcohols. This method traps cations in a polymer gel that retains a uniform cation distribution during mixing, calcination, pyrolysis, and crystallization.

12.2.6 Gel-Casting Method

Gel casting is a low-cost and common processing technique in the ceramic industry to form objects with complex shapes and sizes in a mold using a suspension containing ceramic powder, premixed monomer and cross-linking solution. In SOFCs, gel casting has been employed for the production of electrode and electrolyte powders and thin electrolyte films for SOFCs. In this method, a slurry of high solid-state loading is obtained by dispersing oxide or carbonaceous precursors in the premixed monomer and crosslinking agents such as acrylamide (AM, $C_2H_3CONH_2$) and *N,N″*-methylenebis-acrylamide (MBAM). Gelation occurs under heating (e.g., ~80 °C for 24 h) in the presence of an initiator such as ammonium persulfate (APS) to form a three-dimensional structure. In some cases, *N,N,N′N′*-tetra methyl-ethylene diamine (TEMED) is added as a catalyst to facilitate the polymerization process. Polyacylic acid (PAA) and ammonium poly-(methacrylate) (PMAA, $C_3H_5CO_2NH_4)_n$) are commonly used as dispersants in the aqueous gel-casting process. Polymerization solidifies and immobilizes the powder into the designed shape in the case of thin electrolyte films or tapes. Gel-casting method has advantages of low cost, short-forming time, good product homogeneity, high green capacity and low-cost machining.

Figure 12.6 shows a typical aqueous gel-casting procedure to synthesize $(La_{0.75}Sr_{0.25})Cr_{0.5}Mn_{0.5}O_3$ (LSCM) powders [5]. To prepare LSCM powder, stoichiometric amount of La_2O_3, $SrCO_3$, Cr_2O_3, and $MnCO_3$ are mixed in a rolling machine for 24 h with isopropanol as dispersant. Then the mixture is dried in an oven at 80 °C. The dried mixture is transferred to a jar with the addition of PAA solution, AM, MBAM and deionized water and mixed in planetary ball milling machine at 180 rpm for 45 min. This is followed by adding APS and TEMED into the mixture. The mixture is then poured into a beaker, sealed with aluminum foil, and placed in an oven at 80 °C for ~1 h to allow the polymerization reaction to proceed. The as-prepared gel is then calcined in a high-temperature furnace at 1200 °C for 20 h, forming LSCM perovskite powder. The precursor and agents used in the synthesis of the LSCM powder are given in Table 12.1.

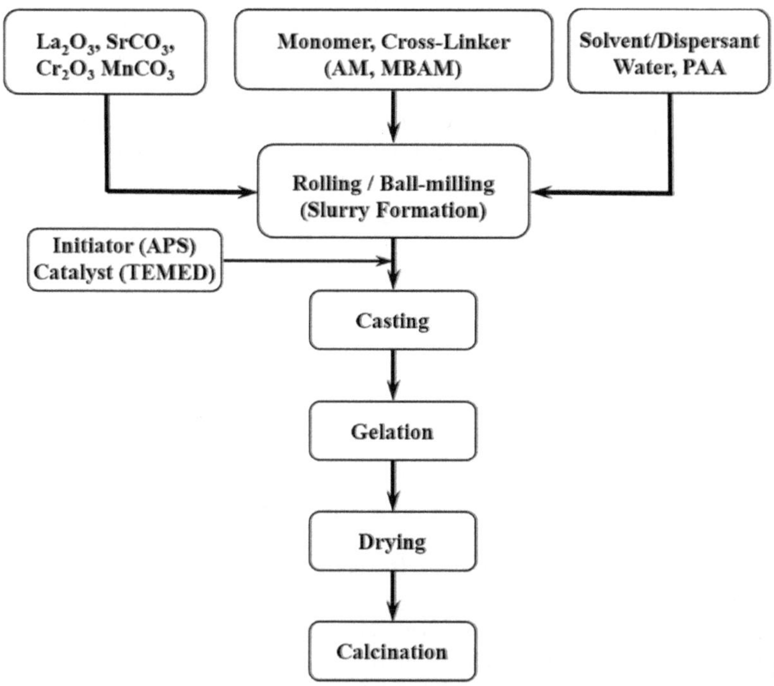

Fig. 12.6 Gel-casting method for the synthesis of LSCM powders

Table 12.1 Precursor and agents used in the synthesis of 0.5 mol $(La_{0.75}Sr_{0.25})Cr_{0.5}Mn_{0.5}O_3$ (LSCM) powders (i.e., ~114 g LSCM powder)

Name	Formula	MW g/mol	0.5 mol LSCM/g
Lanthanum oxide	La_2O_3	325.81	61.089
Strontium carbonate	$SrCO_3$	147.63	18.454
Chromium(III) oxide	Cr_2O_3	151.99	18.999
Manganese carbonate	$MnCO_3$	114.95	28.738
Agents		**Function**	
Isopropanol		Solvent	
Water, H_2O		Solvent	31.82
Acrylamide (AM, $C_2H_3CONH_2$)		Monomer	7.95
N,N″-methylenebis-acrylamide (MBAM)		Crosslinker	0.53
Ammonium persulfate (APS, 2wt% solution)		Initiator	7.95
N,N,N′N′-tetra methyl-ethylene diamide (TEMED, 2wt% solution)		Catalyst	7.95
Polyacylic acid (PAA), 30wt% solution		Dispersant	6.36

12.2.6.1 Comparison of Gel-Casting and Solid-State Reaction Methods

In the gel-casting process, chain initiation, chain propagation, and chain termination reactions and crosslinking reactions occur in the same solution. When the reactions are completed, the metal nitrates are fixed in the polymer network. The evaporation of the water of the gel causes the organic framework to shrink, while small units of metal nitrates are fixed and isolated inside the organic framework [6]. The electrode powder prepared from gel casting shows a fine particle size and uniform distribution, as compared to that prepared by SSR. The phase formation temperature of the gel-casting powder is, in general, 100–150 °C lower than that prepared by the SSR method. The significantly reduced phase formation temperature of the gel-casting powder is due to the homogeneously distributed and immobilized precursor particles in a polymeric network, which promotes the sintering and crystallization process. The gel-casting technique has been used to prepare a wide range electrode and electrolyte powders.

Figure 12.7 compares the XRD patterns of LSM powders prepared by SSR and gel-casting techniques, calcined at different temperatures [7]. In this case, A-site non-stoichiometric $(La_{0.8}Sr_{0.2})_{0.9}MnO_3$ (LSM) powders were synthesized by an aqueous gel-casting method using La_2O_3 (99.9%), $SrCO_3$ (99.9%), and $MnCO_3$ (99%) as precursors. The raw materials were weighed and ball milled in propanol for 24 h. A premix solution of monomers was prepared in deionized water with acrylamide, AM and MBAM. The AM to MBAM ratio is 15. The precursors were mixed in the premix solution with PMAA as a dispersant, followed by planetary ball milling for 30 min. APS was added as the initiator, and the slurry was poured into a container and subsequently heated in an oven at 80 °C for 24 h. After drying, the green body

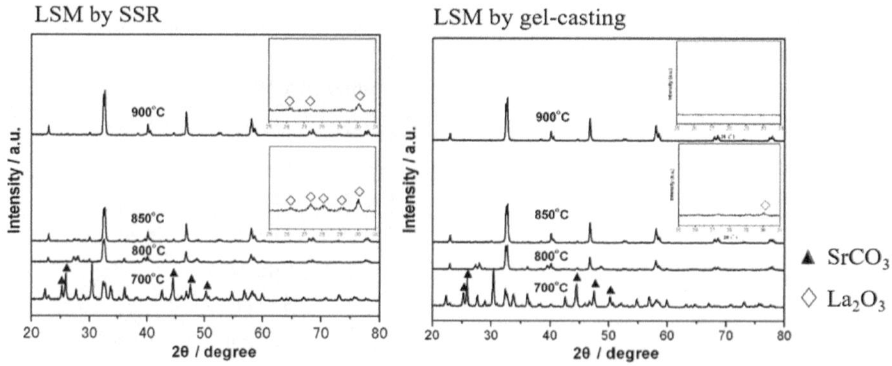

Fig. 12.7 XRD patterns of LSM powders prepared by SSR and gel-casting techniques and calcined at different temperatures. Modified from Ref. [7] with permission from John Wiley, Copyright 2007

was calcined at a temperature in the range of 700–900 °C for 20 h in air. For the LSM powders prepared by the SSR technique, La_2O_3, $SrCO_3$, and $MnCO_3$ with desired composition were mixed in propanol and ball milled for 24 h, followed by calcination at a temperature in the range of 700–900 °C for 20 h in air.

As shown from XRD patterns, for the gel-casting and SSR powders heat treated at 700 °C, the $SrCO_3$ phase is detected. For LSM prepared by the SSR technique, La_2O_3 is clearly detected in the powder after calcination at 900 °C. On the other hand, all the high intensity reflections in the gel-casting sample calcined at 850 °C are identified to the pseudo-cubic phase of LSM, indicating the formation of single perovskite phase LSM. This indicates the significantly reduced phase formation temperature in LSM powders synthesized by the gel-casting technique.

12.3 Fabrication Techniques for Electrolyte and Electrode Coatings

Fabrication methods for electrolyte and electrode coatings of SOFCs can be categorized into two groups: particulate and deposition approaches. In the particulate approach, ceramic or oxide powders are formed into a cell component and sintered at elevated temperatures to achieve the desired density or porosity. Types of fabrication methods included in this approach are die or uniaxial pressing, tape casting, screen printing, gel casting, slurry coating, slip casting, and dip dipping. The deposition approaches consist of chemical or physical processes such as spin coating, spray pyrolysis, sputtering, pulsed laser deposition, and plasma spray deposition.

12.3.1 Die Pressing Method

Die (uniaxial) pressing or dry pressing is the most simple and rapid method in the fabrication of electrolyte pellets or plates as support layers for SOFCs. This forming technology involves filling a stainless steel die cavity with a known amount of powder, compacting the powder within the die through punches under a high mechanical or hydraulic pressure, and releasing the pressure to obtain the compact or pellets. Figure 12.8 shows a stainless steel die and die pressing process to prepare pellet samples. The weighed powder is transferred to the die and evenly dispersed between two steel punches. The compaction is carried out by applying pressure of 50–500 MPa and held for 1 min. The apparent pressure is determined by the force related to the pressure gage and the area of pellets. Then, the pressure is released to take out the die. The die is placed upside down carefully to remove the base and the steel punches, and the sample pellet is collected. The density of the green pellets or disk samples can be further increased by isostatic pressing. The green disks are then sintered at high temperatures to form electrolyte or electrode substrates. The die pressing can be performed using a desktop press, and the density of green pellets can be in the range of 40–50% relative density. Depending on the raw powders, 1 wt% polyvinyl butyral (PVB) can be added as a binder to enhance the workability of the die pressing process. The thickness of the sintered disks is generally in the range of 0.5–1 mm.

The die pressing method can be modified to fabricate a thin electrolyte film such as GDC on NiO/GDC anode substrates [8]. In this method, the green NiO/GDC substrate is formed by uniaxial pressing of NiO/GDC powder under 200 MPa. A highly porous or "foam" GDC powder, synthesized by a GNP process, is carefully spread on top of the prepressed NiO/GDC substrate and then compressed at 250 MPa, forming a bilayer structure. Co-firing of the bilayer structure at 1350 °C for 5 h yields a dense GDC membrane on a porous NiO/GDC electrode substrate. Care must be taken to ensure the uniform distribution of the GDC powder on the substrate. The thickness of the GDC electrolyte is controlled by the amount of GDC powder. However, to form a uniform and thin electrolyte film requires experience and skill, particularly for powders with high filling density.

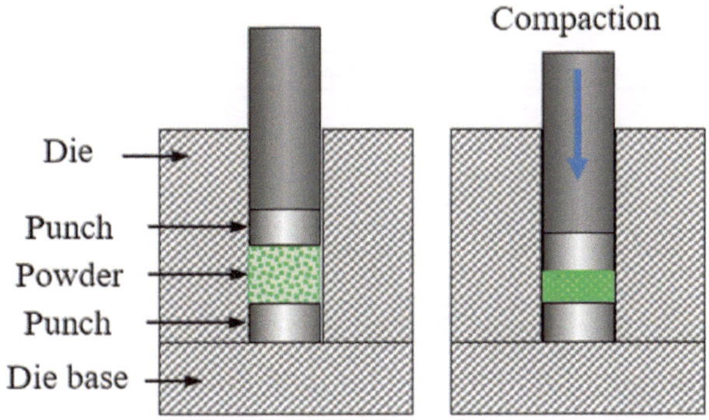

Fig. 12.8 Die and die pressing for the preparation of pellet samples

The characteristics of the electrolyte powders have a significant effect on the densification of freestanding electrolyte pellets and thin electrolyte YSZ films on anode support/substrates [9]. Pellets made of nano-sized YSZ powders (50–60 nm and specific surface area of 120–130 m^2/g) achieve a relative density of 92 and 96% at a sintering temperature of 1200 and 1400 °C, respectively. When the nano-sized YSZ powders are coated onto anode support substrates, however, thin nano-sized YSZ films are unable to sinter and densify even at 1400 °C, forming a porous film with a relative density of 62%. The reason is due to the low green density of the nano-sized YSZ film and the significant difference in the sintering kinetics of the nano-sized YSZ layer and the Ni/YSZ cermet substrate. On the other hand, with micron-sized YSZ powder (0.5 μm and specific surface area of 6.1 m^2/g), a dense and almost pore-free YSZ thin film can be achieved on the Ni/YSZ cermet substrate after sintering at 1400 °C for 2 h. This indicates the importance of the sintering behavior and the nature of the electrolyte powder in the development of an anode-supported thin electrolyte cells for IT-SOFCs.

The die pressing method is difficult to scale up and is limited to simple geometric shapes. Nevertheless, die pressing or its modified version is particularly useful for laboratory scale experiments and for the initial evaluation of electrode and electrolyte materials of SOFCs due to its simplicity and low cost.

12.3.2 Tape-Casting Method

Tape casting is a commercial processing technology which has been used extensively for the manufacturing of electronic and structural ceramics with thickness typically ranging from 25 to 1000 μm. The tape-casting technique is particularly suitable for making electrolyte and anode substrates with different thicknesses for planar SOFCs. Tape casting involves spreading of slurries of the ceramic powders and organic ingredients such as the modifier, binders, and solvents onto a flat surface, i.e., a Mylar™ or Teflon™ carrier film via a doctor blade apparatus where solvents are allowed to evaporate (see Fig. 12.9). After drying, the resulting tape develops a

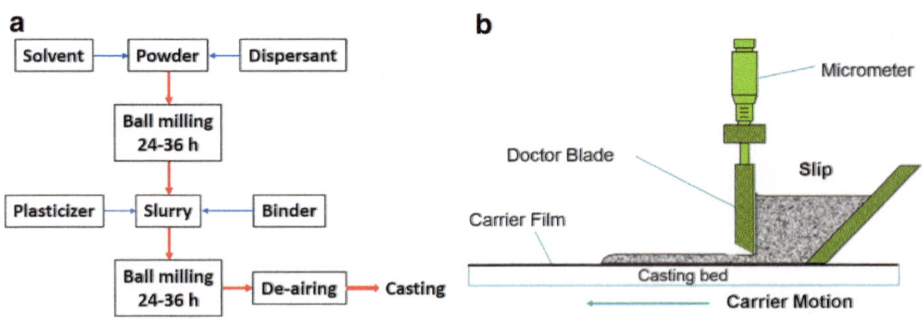

Fig. 12.9 Tape-casting procedure (**a**) and apparatus (**b**)

Table 12.2 Typical compounds and functions in the preparation of tape-casting slurry

Items	Compounds	Function
Dispersing agent	Fish oil, terpineol, phosphate ester	Disperse primary particles and hold them in homogeneous suspension, preventing reformation of powder flocs via steric hindrance and/or ionic repulsion
Solvent	Toluene, isopropanol, trichloroethylene, 50% ethyl alcohol/50% xylene	Dissolve ingredients, uniformly distributes powder particles and other additives, and can be evaporated quickly
Binder	Polyvinyl butyral, acrylics, cellulose	Attach to multiple particles in flocs and bind them together and can also act as a powerful steric hindrance dispersant but with much larger molecules
Plasticizer		Increase the flexibility, workability, or distensibility of the binder
Plasticizer type I	Butyl benzyl phthalate	T_g modifier or internal plasticizer
Plasticizer type II	Polyalkylene glycol	Lubricant or external plasticizer
Homogenizer	Cyclohexanone	Prevent the formation of skin

leather-like consistency and can be stripped off from the casting surface. The slip contains sufficient organic binders that make the tape strong and flexible enough after solvent evaporation for handling and cutting. Typically, a tape cast anode support layer of ~1–1.5 mm thick and a tape cast electrolyte layer from 40 to 100 μm thick are produced separately and are laminated. After rolling or calendaring, the green laminates are cut to size and sintered at 1300–1400 °C in a specially designed kiln furnace to ensure the parts are all flat. The final anode-supported structures are produced with an anode support thickness of 500–1000 μm and electrolyte thickness of 15–40 μm. The tape-casting technique is also used to produce a thick electrolyte of ~100 μm for electrolyte-supported cells as well as a porous anode substrate for anode-supported thin electrolyte cells. Table 12.2 lists typical chemical compounds used in the preparation of tape-casting slurry.

Example 12.2 Ni/YSZ cermet anode substrate by tape casting.

For tape-casting slurry preparation, NiO powders (1–2 μm) and YSZ powders (1–2 μm) in a 57:43 mass ratio are mixed with a mixture of 50% ethyl alcohol and 50% xylene as solvent and menhaden fish oil as dispersant and ball milled for 30 h. The mixture is further ball milled for another 30 h to obtain a homogeneous slurry by the addition of polyvinyl butyral as organic binder and polyalkylene glycol as plasticizer. De-airing is carried out by slowly stirring the slurry for 24 h. The slurry can also be degassed in a vacuum desiccator using a magnetic stirrer. After tape casting and being dried at room temperature for 24 h, a green sheet of NiO/YSZ anode-supporting layer with thickness of about 1.5 mm is obtained.

The quality of slurry for the tape casting is critical. Well-dispersed and deflocculated slurry will produce tape with higher green density. The slurry must be well deflocculated before addition of the binder. Before the casting, de-airing treatment must be performed to get rid of air bubbles in the slurry. The particle size of NiO and YSZ needs to be controlled within the designed range. Poor quality of the slurry would lead to the formation of crow's foot cracks; see Fig. 12.10.

A thin electrolyte film on a thick electrode support or substrate can be made by the progressive rolling of a green (unfired) electrolyte/anode bilayer to produce a thin electrolyte film (typically 10–20 μm) on an electrode support using a two-roll mill, so called the tape-calendering process. In this fabrication process, electrolyte and anode tapes of certain thickness ratios are laminated and rolled into a bilayer tape. The process can be repeated with different tape thickness ratios until a desired electrolyte film thickness is obtained. The bilayer is fired at elevated temperatures to remove the binders and sinter the ceramics. To form a thin-film SOFC single cell, a cathode coating is applied on the electrolyte surface of the sintered bilayer by screen printing. Cell sizes as large as 500 cm^2 can be fabricated by this method. However, mismatches in densification of such bilayer structure will cause intolerable bending and/or cracking of the structure during co-firing. Thus, a match in the sintering profile of the laminated anode and electrolyte tapes is critical in producing flat anode-supported structures.

The quality and consistency of the tape-casting tape are critically related to the rheological properties of the anode slurry, which can be monitored by a viscometer.

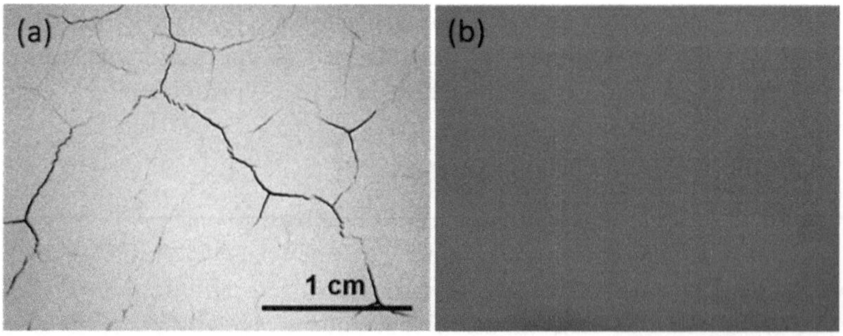

Fig. 12.10 **a** YSZ tape with crow's foot cracks and **b** Ni/YSZ anode tape with good quality

12.3.3 Screen-Printing Method

Screen printing is a common ceramic process and has been extensively adopted to fabricate porous electrode coatings as well as dense electrolyte films in the thickness range of 10–100 μm [10]. In the screen-printing process, a highly viscous ink consisting of a mixture of ceramic powder, organic binder, and plasticizer is forced through the open meshes of a screen onto a substrate using a squeegee. The screen-printed films are dried to remove solvent and then sintered at high temperatures to produce a coating on a freestanding substrate. In electrolyte and anode-supported cells, screen printing is the most common technique to produce porous electrodes.

The screen mesh count, printer settings, substrate preparation, ink rheology, sintering temperature and time, and thickness of the component layer are among the significant parameters affecting the quality of the screen-printed films or coatings. Screen mesh count is the number of wires or openings (the linear distance between one wire to the next adjacent wire) per linear inch. Rheological properties of the ink are probably the most important for producing high-quality films and coatings, which can be evaluated by determining the steady-state (e.g., viscosity, yield stress, thixotropic) and dynamic (e.g., elastic, viscous, and complex moduli) properties. The particle size and distribution of the powder, solid content, and the composition of the ink (e.g., powder, solvent, binder, and dispersant) are important parameters for controlling the rheological properties of the resulting ink.

The functionality of solvent, binder, and dispersant is similar to that in the preparation of a slurry for tape casting. Terpineol and texanol are commonly used solvents in the preparation of screen-printing inks. Ethylene cellulose (EC), polyvinyl butyral (PVB), polymethyl methacrylate (PMMA), and polyvinyl acetal are binders commonly used in the ink formulation. The content of binder has a significant effect on the rheology of inks. Too high a binder content may result in increased tackiness of the ink which in turn affects the printability and quality of the films. On the other hand, a low content of binder may cause film cracking during drying due to the reduced particle network strength within the ink. Therefore, a balance between the solid and binder contents is important. In some cases, a combination of solvent and binder such as a mixture of 94 wt% of terpineol and 6 wt% of EC is used as a vehicle in ink preparation. Dispersant agents (e.g., polyester/polyamine condensation polymer, pentastearic acid oligomer, and polyoxyethylene phosphoric acid ester) are added to improve the separation of solid particles by the formation of an electrostatic or steric barrier around the particle surface. Ink that consists of a ceramic powder with suitable solvent, dispersant, and binder is usually milled using roll mills to achieve homogeneity before screen printing.

In the formulation of an ink, the optimum solid content in the ink and the selection of screen-printing parameters are interrelated and need to be determined experimentally for each specific electrolyte and electrode coatings.

12.3.4 Chemical Vapor Deposition and Atomic Layer Deposition

Chemical vapor deposition or CVD is the method used to deposit a non-volatile solid thin film on a substrate by the reaction of vapor phase chemicals (reactants) that contain the required constituents. The basic mechanism of CVD is that the reactant substance to be deposited is introduced as a gas in a reaction chamber and is reacted and deposited on a heated substrate through high-temperature decomposition or high-temperature chemical reaction, forming thin films. CVD is a commonly used technique in the semiconductor industries to produce thin films. Therefore, the underlying processes are well understood.

To obtain a high-quality thin film by CVD, a clean vacuum environment is necessary. The deposition processes of CVD are related to the mass transfer rate and the chemical reaction that takes place at the surface. The mass transfer rate is mainly a function of concentration differences, while the surface reaction depends significantly on the reaction temperature. CVD is excellent in terms of adhesion and has been used for the deposition of dense and thin electrolyte films of SOFCs because the crystal structure and orientation of the film can be controlled by manipulating the vacuum environment and various other conditions [11]. The most significant feature of CVD is the dependence of growth rates of the film thickness on deposition temperature. The capital cost for the CVD equipment is generally high.

Atomic layer deposition (ALD) is a modified CVD technique. In ALD, the substrate surface is exposed alternately to different vaporized precursors. Because gaseous precursors are strictly separated from each other during deposition and the precursors have self-limiting chemistry, one reaction cycle may produce only one atomic layer. For this reason, ALD can be an ideal technique to grow ultra-thin oxide films because the composition of ALD films can be altered at each atomic layer with desired ratios. ALD is capable of depositing layers of significant complexity and is particularly suitable for fundamental studies in SOFCs. In the case of the deposition of thin films, the ALD process is limited by slow deposition rate and difficulty in deposition over large areas.

12.3.5 Electrochemical Vapor Deposition

Electrochemical vapor deposition or EVD is a modified CVD process, originally developed by Siemens-Westinghouse for the fabrication of a thin YSZ electrolyte layer on tubular SOFCs [12]. EVD is a two-step process. The first step involves the pore closure by a normal CVD-type reaction between steam (or oxygen), metal chloride, and hydrogen through the porous air electrode (phase I). These react to fill the air electrode pores with YSZ electrolyte according to the following reaction:

$$2\mathrm{MeCl}_y + y\mathrm{H_2O} \rightarrow 2\mathrm{MeO}_{y/2} + 2y\mathrm{HCl} \tag{12.23}$$

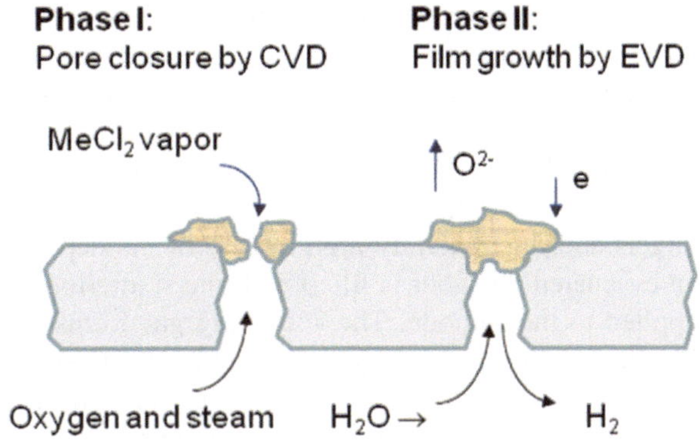

Fig. 12.11 Schematic diagram of the EVD process

$$4\text{MeCl}_y + y\text{O}_2 + y\text{H}_2 \rightarrow 4\text{MeO}_{y/2} + 4y\text{HCl} \quad (12.24)$$

where Me is the cation species (zirconium and/or yttrium) and y is the valence associated with the cation.

After pores in the air electrode are closed, film growth then proceeds due to the presence of an electrochemical potential gradient across the deposited film. In this step, oxygen ions formed on the water vapor side of the substrate (i.e., the high oxygen partial pressure side) diffuse through the thin metal oxide layer to the metal chloride side (i.e., the low oxygen partial pressure side). The oxygen ions react with the metal chloride vapor to form the metal oxide products. The solid product or electrolyte is deposited as a thin film spreading over the internal pore surface in a desired region across the air electrode or membrane substrate. This second stage of the reaction is termed the electrochemical vapor deposition or EVD (Phase II). Figure 12.11 shows the schematic diagram of the basic principles of the EVD process.

The growth of YSZ electrolyte film by the EVD process is parabolic with time, and the rate determining step has been found to be the electronic transport through the electrolyte film. The dissociated oxygen from metal oxide substrates, such as NiO, is also suggested as an oxygen source for the reaction instead of gaseous oxygen to form dense YSZ electrolyte films. The high reaction temperature, the presence of corrosive gases, and relatively low deposition rates are some of the limiting factors in the application of EVD process in SOFC.

12.3.6 Magnetron Sputtering Techniques

Magnetron sputtering is one of the most common physical deposition techniques that is widely used to grow alloy and component films in which one or more of the constituent elements are volatile. Low defect density films of high-melting point materials can be grown on unheated substrates because phase formation is mainly governed by kinetics, rather than by thermodynamics.

Radio frequency (RF) magnetron sputtering can be used when the target is an oxide or metal, and *direct current* (DC) magnetron sputtering is used when the target is metallic. RF sputtering has frequently been utilized to deposit electrolyte thin films in part because of the ability to use either metallic or electrically insulating ceramic targets and the generally high quality of the deposit. In RF-sputtering deposition, an evacuated chamber is filled with the sputtering gas. A large negative voltage is applied to the cathode. The sputtering gas forms a self-sustained glow discharge. Physical sputtering of the target occurs when positive ions from the plasma that are accelerated across the space strike the target surface. A metal oxide film is grown by sputtering a metal target in a discharge containing oxygen, usually in conjunction with a noble gas. The metal, metal oxide, and oxygen species that arrive at the substrate are adsorbed and ultimately incorporated into stable nuclei to form a continuous film. Oxygen is often included in the sputtering gas mixture as a means of controlling the metal to oxygen ratio in the target. However, the RF sputtering deposition rate decreases significantly with increased oxygen partial pressure.

DC magnetron sputtering has also been used to deposit thin electrolyte onto porous supports. The target can be a single target consisting of an alloy of zirconium and yttrium or can be multiple targets of the pure metals, where the composition of the deposit is controlled by the relative exposed surface area of the targets. Similar to RF sputtering, deposition rates are adversely affected by the oxygen partial pressure in the sputtering gas for DC reactive magnetron sputtering of thin films.

The deposition rate, microstructure, and film thickness depend on various operational parameters. The parameters for controlling the sputtering process include the substrate bias, DC or RF power, flow rate of Ar or oxygen, deposition pressure, and annealing temperatures.

The surface morphology of the substrate such as pore size and pore size distribution is critical for the deposition of a thin and dense electrolyte film on the electrode substrates as the morphology of a deposited film follows that of the substrate surface. Figure 12.12 shows the morphology of the YSZ film deposited on YSZ and Ni/YSZ cermet substrate by magnetron sputtering [13]. The thin YSZ film deposited is characterized by a columnar structure and has the same grain and grain boundary pattern as that of the YSZ substrate, indicating that the surface morphology of the YSZ thin film follows closely the morphology of the substrate. Pinholes and cracks can form due to step coverage effects and/or shadow effects. The large pores on the substrate surface result in open pores on the deposited YSZ electrolyte films. Thus, the pores diameter of the substrate surface needs to be in the order of the grain size of the deposited film to achieve a dense and uniform YSZ electrolyte film. This can be done by using an interlayer or functional layer with fine microstructure and low porosity. The functional layer can also be used to promote the electrode reaction at the electrode/electrolyte interface. Post-annealing treatment can also smooth the surface of deposited film to a certain degree.

Magnetron sputtering techniques combined with photolithography are used to produce unique, patterned electrodes such as LSM, Ni, platinum, and gold. The

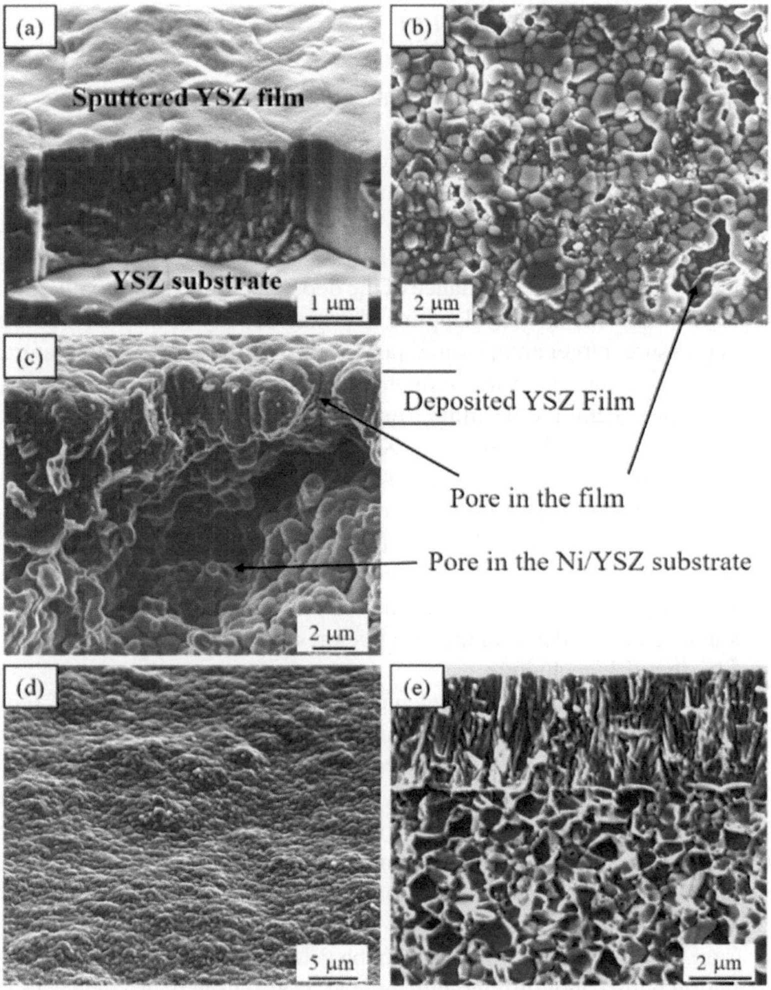

Fig. 12.12 Morphology of the YSZ thin film deposited on a YSZ and NiO/YSZ cermet anode substrate by magnetron sputtering. **a** YSZ film deposited on YSZ substrate, **b** and **c** YSZ film deposited on NiO/YSZ cermet substrate, showing that the defects in the YSZ electrolyte film are due to the large pores on the anode substrate surface, **d** and **e** dense and uniform YSZ film on a NiO/YSZ anode substrate with optimized surface morphology. Reproduced from Ref. [13] with permission from Springer Nature, Copyright 2004

patterned, thin, dense, and uniformly structured electrodes allow the fundamental aspects of the reaction mechanism to be studied, something that is not possible with a conventional, heterogeneous, porous electrode structure. The sputtering method is also used to prepare protective coatings for metallic interconnect.

12.3.7 Pulsed Laser Deposition

Pulsed laser deposition (PLD) or laser ablation is a physical method of thin-film deposition. During the PLD process, a pulsed laser beam, usually of wavelength in the UV range, is irradiated on the oxide targe in a vacuum chamber and the target is evaporated to form a plasma plume over the target, which is subsequently deposited onto a substrate in the plume. The usual range of laser wavelengths for thin-film growth by PLD lies between 200 and 400 nm for most materials. In PLD, the deposition process and the performance of the deposited film are critically influenced by many factors such as the wavelength and power density of the laser, background gas and its pressure, target compositon, target-substrate distance, temperature of the substrate, substrate bias, gas-surface interaction, etc. PLD enables the fabrication of multicomponent stoichiometric films from a single target, and with an appropriate choice of the laser (e.g., Nd:YAG, KrF, XeCl), any material can be ablated and the growth can be carried out in the presence of reactive or non-reactive gases.

The microstructure of the PLD films depends on the substrate temperature and gas pressure. One potential problem is that PLD techniques result in crystalline microstructures that usually show columnar grains and texture, resulting in unfavorable anisotropic electrical conductivity. The surface artifacts of the functional layer should also be smaller than the film thickness in order to avoid defects in films deposited by the PLD technique.

PLD can be used to deposit multielement films such as LSM and LSCF on electrolytes for SOFC. The low substrate temperature of 700 °C can avoid the interfacial reaction between YSZ and lanthanum cobaltite-based cathodes, leading to the formation of a crystalline phase on YSZ electrolyte. Due to its unique method of dislodging atoms from the target, PLD provides some distinct advantages among thin-film techniques. Due to the weak interaction of lasers with gaseous species, ambient atmospheres can be used with little contamination, a significant advantage as compared to other physical deposition techniques. Thus, stoichiometric material transfer from the target to the substrate can be achieved. A microfabrication process in combination with a photolithographic process is also used to produce well-defined three-dimensional geometries for the investigation of electrode reaction processes and reaction sites in SOFC electrode materials. Due to the dense structure of the PLD films, the PLD method is best suited for electrode applications with high mixed ionic and electronic-conducting oxides.

12.3.8 Plasma Spray Deposition

Plasma spray (PS) belongs to the category of vacuum deposition techniques[14]. The PS process is a well-established industrial line-of-sight surface-coating technique. The technique utilizes plasma created by high-voltage electrodes to partially or fully melt particles that traverse the plasma jet and deposit on a substrate. Plasma jet temperatures generally vary between 7000 and 20,000 K and anode nozzle exit velocities are typically in the order of 900–2000 m/s. The plasma spray process is based on the generation of a plasma jet consisting of argon or argon with admixture of H_2 and He, which are ionized by a high current arc discharge in a plasma torch. Powders to be sprayed are injected into the plasma where they are accelerated, melted, and finally projected onto a substrate. The coating is formed by solidification and flattening of the particles at impact on the substrate. This technique offers the possibility to deposit thin or thick layers in the mm range which is not possible with other physical methods such as magnetron sputtering and PLD.

Atmosphere plasma spray (APS), vacuum plasma spray (VPS), and plasma spray-physical vapor deposition (PS-PVD) can be used to prepare dense electrolyte layers. It is difficult to fabricate dense electrolyte layers by single-step deposition and 3–5 coating cycles of PS are generally required. The deposited layers often show the characteristics of anisotropic microstructure, or feather-like columnar structure. The microstructure of the deposited layer on the substrate can be manipulated by optimization of the input power and other operating parameters and limiting vaporization of feedstock. The films produced by VPS typically possess higher density than those produced by APS and are more applicable in fabrication of SOFC electrolytes. The addition of pore formers such as carbon is necessary for spraying a porous coating.

PS has been used to prepare the porous anode and dense electrolyte of metal-supported SOFCs (MS-SOFCs). The key issue in this application is the development of a thin, dense, and gas-tight electrolyte. The quality of the deposited coating depends strongly on the powder feed rate and carrier gas flow rate, robot speed, spraying distance, and plasma gas composition. Marcano et al. fabricated MS-SOFC by PS techniques [15]. The metallic substrate is Fe–Cr alloy fabricated by tape casting, and Ni/YSZ anode and YSZ electrolyte with thickness of ~40 μm are prepared by APS and PS-PVD techniques, respectively. A LSCF cathode is applied to the cell by screen printing. The cell achieves a good open circuit voltage of 1.033 V at 817 °C and power density of 0.89 W/cm^2 at 0.75 V in H_2/air.

The disadvantage of the method is that rapid heating and quenching induce non-stoichiometry and residual strain. For example, anisotropy developed in the deposited layer can affect the ionic conductivity of YSZ electrolytes. The cross-plane conductivities could be several times lower than in-plane conductivities. However, such anisotropy can be reduced by sintering the deposited YSZ layer at high temperatures, e.g., 1500 °C. In general, post-treatment at high temperatures (~1000 °C) has been shown to be able to recover the crystallinity and stoichiometry of plasma-sprayed electrode coatings and to increase the electrical conductivity. The processing parameters such as the heat treatment, melting temperatures of the raw materials, and particle size distribution have been shown to have a significant influence in the microstructure and conductivities of the deposited coatings.

12.3.9 Slurry-Coating, Spin-Coating and Dip-Coating Methods

Slurry coating, *spin coating*, and *dip or slip coating* are convenient wet particulate processes and are simple, fast, and cost effective for the fabrication of various components of SOFCs. In spin coating, slurry is deposited onto the center of a substrate and the substrate is spun at a high and constant speed. Centripetal acceleration causes the slurry to spread and then produces a thin and uniform film on the substrate surface. Similar to screen printing and tape casting, the slurry formulation consists of powders, dispersants (e.g., phosphate ester, fish oil), binder (e.g., PVB), and solvent (e.g., methyl ethyl ketone-ethyl alcohol, methyl ethyl ketone-isopropanol, or isopropanol), forming a suitable and stable colloidal particle suspension. Moreover, in order to avoid cracks during removal of the organic additives, careful drying and heat treatment procedures need to be adopted. The coating process in general is repeated several times to form a dense and pore-free thin electrolyte layer.

In slip casting, a thick slurry is fed into a porous mold, and the liquid is withdrawn through the porous surface, thus forming a structure of the required shape. Dip coating is used to prepare a thin coating layer on a planar or tubular substrate. Slip casting and dip coating are the techniques commonly used for the fabrication of functional ceramic components such as layered ceramics. Compared to the more conventional extrusion method, slip casting and dip-coating techniques have many advantages such as simple equipment, easy processing, the capability to fabricate products with complex shapes and homogeneous cast composition. Figure 12.13 shows the schematic diagram for spin-coating, dip-coating, and slip-casting apparatus.

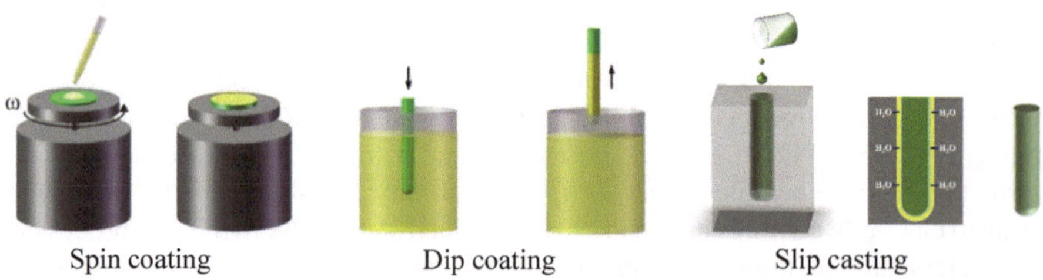

Fig. 12.13 Schematic diagrams of spin-coating, dip-coating, and slip-casting processes

Example 12.3 YSZ thin film by spin coating.

NiO/YSZ electrolyte substrates are prepared by uniaxial die pressing and pre-sintered at 1000 °C for 2 h. YSZ slurry is prepared by mixing YSZ powder with an ink vehicle (e.g., a mixture of 96 wt% of terpineol solvent and 4 wt% of EC binder) at a weight ratio of 3:7 (solid content of 30%). Then the YSZ slurry is applied on a NiO/YSZ anode substrate via a spin-coating method. The spinning rate is 6000–7000 rpm with a spinning time of 30 s, followed by heat treatment at 600 °C for 1 h. The spin-coating process is repeated, and the second layer is heat treated at 80 °C for 1 h. The heat treatment of the first spin-coated layer at a higher temperature is to burn off the ink vehicle completely, so as to avoid the formation of cracks or pores during the co-sintering process. The NiO/YSZ anode/YSZ film bilayer is then calcined at 1450 °C for 4 h. The thickness of the YSZ film formed on the anode-supported cell is ~10 μm or lower.

Aerosol spray deposition is also used to fabricate YSZ thin electrolyte films on metal supported structure from an isopropanol-based solution. Spray coating in combination with tape casting can be used for the fabrication of Ni/YSZ anode-supported thin-film YSZ cells. In this case, Ni/YSZ tape cast green tapes are cut to the desired size and annealed in air at ~1000 °C for 1 h to obtain a porous substrate with a thickness between 200 and 250 μm. A YSZ thin-film electrolyte is then deposited by spray coating a YSZ-water suspension on substrates under controlled conditions and sintered at 1400 °C for 4 h. A YSZ electrolyte film as thin as 3 μm can be obtained.

Figure 12.14 shows typical SEM micrographs of an anode-supported tubular cell with a thin YSZ film and a multilayer cathode prepared by the slip casting

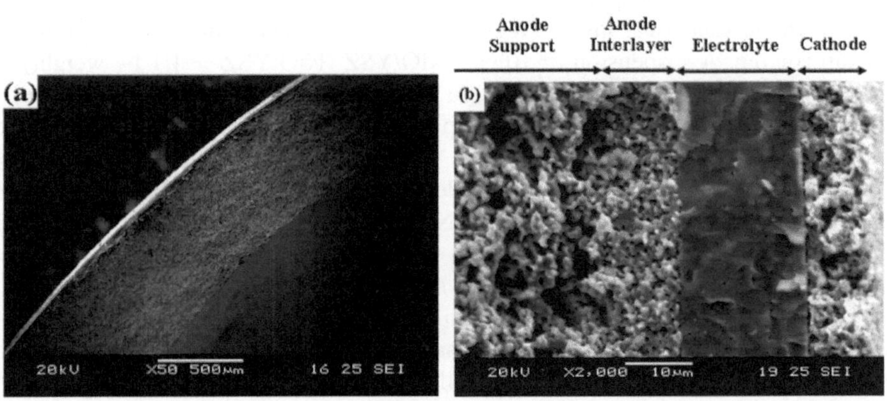

Fig. 12.14 SEM micrographs of **a** an overview of the tubular cell and **b** the cross section of the tubular cell after testing, showing the thin YSZ electrolyte film and multilayer cathode prepared by the dip-coating method. Reproduced from Ref. [16] with permission from John Wiley & Sons, Inc., Copyright 2009

and dip-coating method [16]. In this process, a NiO/YSZ slurry with NiO:YSZ = 65:35 by weight is prepared from NiO and YSZ powder, with the addition of gum Arabic powder as binder and poly((isobutylene-alt-maleic acid), ammonium salt-co(isobutylene-alt-maleic anhydride)) (PCAA) as dispersant in distilled water. NiO powder is pre-calcined at 700 °C for 2 h in air. The powders are mixed in a planetary ball mill. 10 wt% graphite is added to the NiO/YSZ slurry to increase the porosity of the anode support and PVP (5wt% of the graphite in the mixture) is added to the Ni/YSZ slurry to improve the dispersion of the graphite in the mixture. PCAA is a polyelectrolyte with one end of the chain adsorbed onto the surface of NiO and YSZ particles, and the other end of the chain extended into the solvent. The addition of PCAA dispersant increases the electrostatic and steric stabilization, reduces the agglomeration, and improves the dispersion of NiO and YSZ particles. The viscosity of the slurry decreases with an increase in the PCAA content and reaches a minimum at 0.6 wt% PCAA. On the other hand, solid loading has a significant effect on the viscosity. A low viscosity favors the easy flow of the slurry, suspension homogeneity, degassing, and easy casting, while a high solid loading is beneficial to achieve a thicker anode tube support during the slip-casting process. A NiO/YSZ slurry with good flow property and reasonable solid loading can be obtained by adjusting the solid loading (e.g., 65 wt%) and PCAA content (e.g., 0.6 wt%).

The tubular cells can be prepared by slip casting technique. The as-prepared NiO/YSZ slurry is poured into a plaster mold. Water in the slurry is absorbed by the porous plaster mold, and a layer of solid NiO/YSZ is formed on the side wall of the mold. The level of slurry in the mold is maintained during the tube formation until a required wall thickness is reached. The excess slurry is poured out and the plaster mold is dried at room temperature. The green body of the NiO/YSZ tube in the mold shrinks during drying and is separated from the mold wall. The anode tubular supports are dried thoroughly and then pre-fired at 1000 °C for 2 h to decompose the organic additives and graphite pore formers.

The anode functional layer (AFL) and YSZ electrolyte thin film are prepared by dip coating. A suspension of 10wt% NiO/YSZ (NiO:YSZ = 1:1 by weight) in *iso*propanol is dip coated onto the anode tubular substrate to form AFL. Similarly, a suspension of 10 wt% YSZ in *iso*propanol is dip coated onto the outer surface of the AFL to form an electrolyte layer, followed by drying at room temperature. The process is repeated 10 times in order to prepare a dense and pore-free YSZ thin electrolyte layer. The YSZ-coated NiO/YSZ tubular substrate tubes are then sintered at 1400 °C for 2 h to form a bilayer structure with a porous anode substrate, an AFL, and a dense electrolyte layer. A multilayered cathode of LSM, LSM/YSZ, and $Sm_{0.5}Sr_{0.5}O_3$ (SSC) is deposited onto the thin YSZ electrolyte by using the dip-coating method. With impregnation of nano-sized GDC particles into both anode supports and multilayer cathode, the cell achieved a peak power density of 1104 mW/cm^2 in H_2/air and 770 mW/cm^2 in CH_4/air at 800 °C [16].

In the spin-coating or dip-coating methods, sol–gel can be used as precursors. In this method, organo-metallic salts such as metal alkoxides (e.g., zirconium propoxide and yttrium propoxide) are deposited on a porous electrode substrate and hydrolyzed under controlled conditions, forming a colloidal sol and a condensation step with organic monomers to form a gel. The key feature in this sequence is sol–gel polymerization which can be described by a two-step reaction: initiation via the hydrolysis of alkoxy ligands and polycondensation via an oxylation reaction. The electrolyte film is deposited by spin coating or dip coating, followed by a drying and firing process. The particle concentration, viscosity, concentration, and stability of the sol–gel influence the deposition parameters and film quality and have to be controlled carefully. The nature of the porous substrate is also critical in the sol–gel processes. Large pores could lead to pore-induced defects. Thus, the substrate should have a porosity that is both submicron and uniform.

Thin films based on the sol–gel process also have several problems. For example, a large shrinkage during heat treatment and low density from inherent high organic content may cause local defects, resulting in serious gas leakage and crossover.

12.3.10 Electrophoretic Deposition

Electrophoretic deposition (EPD) is one of the colloidal processes by which ceramic films are shaped directly onto substrates from an electrostatically stabilized colloidal suspension in a DC electrical field [17]. An EPD system consists of a DC power supply, suspension, conductive substrate, and counter electrode (Fig. 12.15). A DC electrical field causes the charged particles to move forward and deposit on a conductive substrate with opposite charge. The positively charged particles form a layer on the negatively charged substrate in anodic EPD, while the negatively charged particles form a layer on the positively charged substrate in cathodic EPD. To facilitate the deposition process, the solution is stirred using magnetic stirrer.

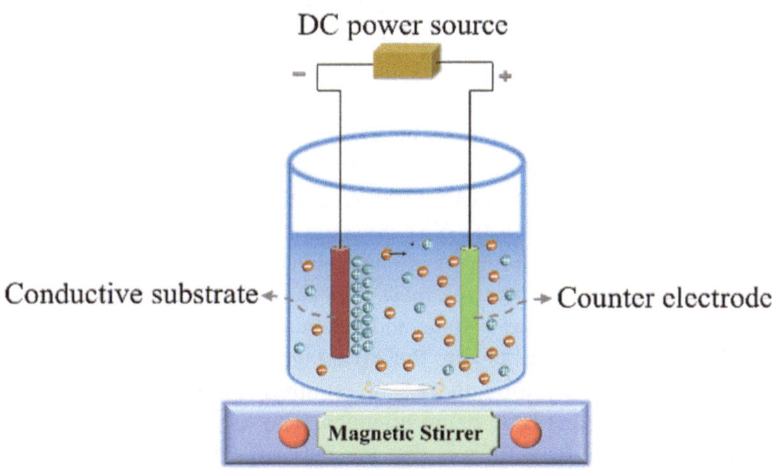

Fig. 12.15 Schematic diagram of electrophoretic deposition apparatus

The mechanism of an EPD process generally involves three stages: double layer distortion, thinning of the double layer, and coagulation. Under the application of an electric field, the electrical double layer region of ions associated with particles is distorted when the charged particles are forced to migrate toward an electrode with the opposite charge. The migration depends on the properties of the colloidal dispersion (bath conductivity, viscosity, particle concentration, size distribution, and surface charge density) and the actual field strength in the bath. The charged particles coagulate at or near the surface of the deposited electrode, forming a solid deposit layer. A suitable heat treatment is usually required in order to further densify the deposited film and to eliminate porosity.

A stable suspension without agglomeration and with a slow settling rate is important for the formation of a dense and strong adherence layer on the substrate. Agglomeration is influenced by attractive and repulsive forces between suspended particles. Electrostatic forces caused by the electrical double layer separate the particles, while van der Waals forces attract particles. Suspensions with relatively high zeta potentials are generally electrically stable as the repulsion between adjacent and similarly charged particles in the solution exceeds the attractive forces. The zeta potential can be manipulated by changing the pH of the suspensions to maintain suspension stability. In the case of the deposition of YSZ thin electrolyte layers, YSZ nanoparticles are dispersed in an organic suspension medium, such as ethanol, acetylacetone, or iodine (I_2)-dissolved acetylacetone, instead of water, to avoid the detrimental effect of water electrolysis on the quality of the deposited film. Positive charges are developed on the YSZ particles due to the presence of some residual water. Iodine-dissolved acetylacetone is an effective solvent for EPD as it has a much lower resistance than ethanol. Lower solution resistance reduces the applied voltage, depressing evolution of gases on the negative (i.e., deposited) substrate. Dispersion of YSZ in an I_2-acetone solution is effective to form charged YSZ particles as the reaction between acetone and iodine produces protons which are absorbed by the YSZ particles; consequently, the YSZ particles become positively charged as a result of the addition of I_2.

A prerequisite for EPD is that the substrate must be electrically conductive. Thus, for deposition of YSZ or GDC electrolyte thin films on non-conducting NiO/YSZ anode supports, thin-conducting layers such as graphite can be coated onto the porous NiO/YSZ composite substrates to facilitate conduction on the substrate surface. For YSZ powders deposited onto porous NiO/YSZ substrates whose reverse side is coated with conducting graphite layers, the YSZ powders are subsequently transformed into dense and continuous films after co-firing with substrates. The thickness and density of the deposited layer depend on the mass deposited during the EPD process. This is closely related to the applied voltage and deposition time.

The EPD process is simple and has the advantage of uniformity of deposition and high deposition rates. In addition, there is no requirement of binder burnout because the green coating contains little or no organics. The most significant challenge in the application of EPD in SOFCs is that the majority of the cell components are not electrically conductive at room temperature. Moreover, five or more successive repetitions of the process may be necessary to produce a dense electrolyte layer, and it is also difficult to control the quality and reproducibility of the thin film deposited by the EPD method.

12.3.11 Spray Pyrolysis and Flame-Assisted Vapor Deposition

Spray pyrolysis is a flame-assisted colloidal process. Figure 12.16 shows a typical spray pyrolysis setup. A sufficient force, e.g., using a stream of gas at high speed, applied to the surface of a metal salt precursor solution (usually aqueous or alcoholic) or colloidal suspension in the atomizer/nozzle causes the emission of droplets. The solution can also be forced through the nozzle with a syringe pump. Sprayed droplets reaching the substrate surface undergo pyrolytic decomposition. Newly deposited flat droplets, with thickness in the 10–20 nm range, pile up on the previously deposited droplets and undergo pyrolytic decomposition as well. This process continues until a desired film thickness is reached. The degree of decomposition is determined by the relationship between the substrate temperature, the boiling temperature of the solvents, and the melting point of the salts used for the precursor. A compromise has to be found between a sufficiently high deposition temperature to achieve complete decomposition if possible and the facility to deposit the droplets while still in a wet state on the substrate for a piling up of the droplets. The spray pyrolysis synthesis method has been used widely for the preparation of SOFC electrolyte thin films as well as thin porous electrodes. The spray pyrolysis method offers high film quality and low processing costs compared to other thin-film deposition techniques such as PLD and chemical or physical vapor deposition. The spray pyrolyzed thin films are usually amorphous after deposition, but can be converted to a nano-crystalline structure with further calcination.

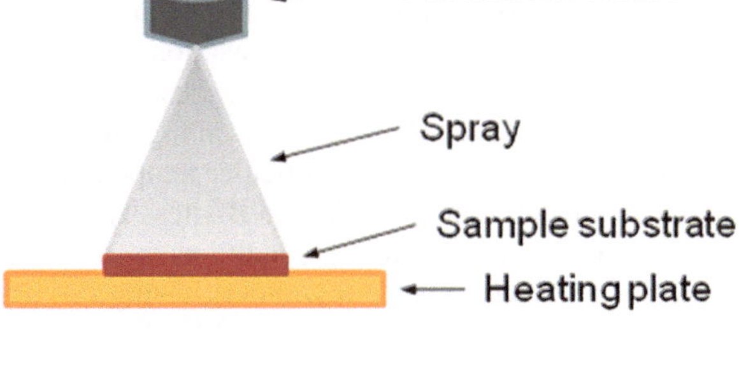

Fig. 12.16 Schematic diagram of a typical spray pyrolysis setup

The composition of the deposited film can be different from that of the precursor solutions. The pyrolysis process could also lead to thin films that contain trace amounts of carbon and hydroxyl groups from the precursors. The cracks formed during the thermal decomposition and heating treatment could be eliminated by optimizing the substrate temperature and repeating the film deposition and heating cycle. Spray pyrolysis is also used to deposit porous Ni-based composite anodes and LSM cathodes. Using a precursor sol consisting of a solution prepared from strontium acetate, lanthanum nitrate and manganese nitrate dissolved in propane-1,2-diol or LSM powder at a concentration of 0.2 ml/L, a thin porous LSM electrode can be obtained by spray pyrolysis.

Flame-assisted vapor deposition (FAVD) is a combination of spray pyrolysis and flame synthesis. In this method, an atomized solution is sprayed through a flame in an open atmosphere in which decomposition and combustion reactions occur, resulting in a stable film deposited on a heated substrate. Ethanol is added to the nitrite precursor solution to make the solution more flammable. This method requires simple apparatus and is performed at a relatively low temperature and at a high deposition rate. The morphology and porosity of the deposited coating depend strongly on the deposition temperature (i.e., the substrate temperature).

The spray pyrolysis deposition has the advantages of simple setup, inexpensive and non-toxic precursors, high deposition efficiency, and direct deposition under ambient atmosphere. However, the adhesion and uniformity of the porous coating would be difficult to control.

12.4 High-Temperature Sintering Process

In electrochemistry, the electrode/electrolyte interface is the most important region for electrode reactions, directly determining the efficiency and stability of electrochemical devices like fuel cells. In an aqueous electrolyte system, the electrode/electrolyte interface is formed almost instantaneously once the metal electrode (e.g., Pt) is in contact with aqueous electrolyte, forming a double layer with charge and compositional distribution across the two phases. In solid-state devices such as SOFCs, such an intimate interface between two ceramic phases generally forms under high-temperature sintering. Thus, high-temperature sintering is an important process in the fabrication of cell components of SOFCs. However, solid-state reaction between two components can also occur during the high-temperature sintering. The energy required to induce the reaction depends on the stability of the electrode and electrolyte phases. Perovskite oxides in particular are active, and they readily react with other fuel cell components at high temperatures. The relative reactivity of the perovskite electrode at a particular temperature critically depends on the nature of the cations present in the A- and B-sites. Thus, it is important to carefully control the high-temperature sintering conditions to form the intimate interface between the electrode and electrolyte, but without causing the detrimental solid-state reactions.

12.4.1 Anode

In SOFCs, the principal anode materials are Ni/YSZ and Ni/GDC (or Ni/SDC) cermets due to their high electrochemical performance and high stability under SOFC operating conditions. The presence of ionic-conducting electrolyte phases like YSZ and GDC in the cermets not only inhibits the sintering and growth of the Ni phase but also actively promotes the electrochemical and catalytic activity of the anodes by extending the reaction zones or the three-phase boundary (TPB). However, the wettability of Ni on YSZ and doped ceria is poor, e.g., the contact angles between liquid Ni (melting temperature 1455 °C) and YSZ are over 90° [18]. This indicates that the good interface contact between Ni/YSZ cermets and YSZ (or GDC) electrolyte can only be through the YSZ phase in the cermets. To achieve a good bonding or interface between YSZ phase in the cermets and YSZ electrolyte, sintering at 1350–1400 °C is generally required.

However, the sintering behavior of the zirconia phase in the cermets affects not only the formation of a YSZ-to-YSZ ionic network but also the formation of a Ni-to-Ni electronic network of the Ni/zirconia cermet anodes and thus the electrical contact between the Ni phase in the cermet and YSZ electrolyte. Formation of Ni-to-Ni electronic contact and YSZ-to-YSZ ionic contact networks is closely related to the sintering temperatures of the anode [19]. Figure 12.17 shows SEM micrographs of Ni/YSZ cermet electrodes sintered at 1350 and 1400 °C. For the Ni/YSZ cermet anode sintered at 1350 °C, YSZ particles exist in isolation, and there is no interconnection between the YSZ particles (Fig. 12.17a). With the increase of the sintering temperature to 1400 °C, there is interconnectivity between phases, forming YSZ-to-YSZ and Ni-to-Ni networks (see the white and yellow lines in Fig. 12.17b).

The formation of ionic and electronic network can be confirmed by the measurement of both the contact resistance (R_Ω) between the cermet anode and electrolyte and the polarization resistance (R_E) for HOR as a function of sintering temperatures. For example, at 1000 °C, R_E and R_Ω are 12.5 and 6.2 Ω cm^2 for the cermet anode sintered at 1350 °C, respectively. After sintering at 1400 °C, R_E is reduced to 0.75 Ω cm^2, while R_Ω is reduced to 1.03 Ω cm^2, substantially smaller than that measured for the anode sintered at 1350 °C (Fig. 12.17c) [19]. The significantly reduced R_Ω indicates that sintering of the YSZ phase in the cermet has a significant effect not only on the formation of the Ni-to-Ni electrical contact network in the cermet but also on the formation of electric contact between the Ni phase in the cermet and YSZ electrolyte. On the other hand, a high sintering temperature is essential to create a good bonding between the YSZ phase in the cermet and the YSZ electrolyte, leading to the formation of a rigid YSZ structure to support the Ni phase and the formation of the Ni-to-Ni electronic contact network. This is most likely the reason for the significantly reduced polarization resistance, high electrode performance, and high electronic conductivity of the anode coatings sintered at high temperatures.

For ceramic oxide anodes, the anode sintering temperature is generally lower than that of the Ni-based cermet anodes. In this case, the interface contact is formed between oxide and oxide with no metallic phase. However, the interface contact

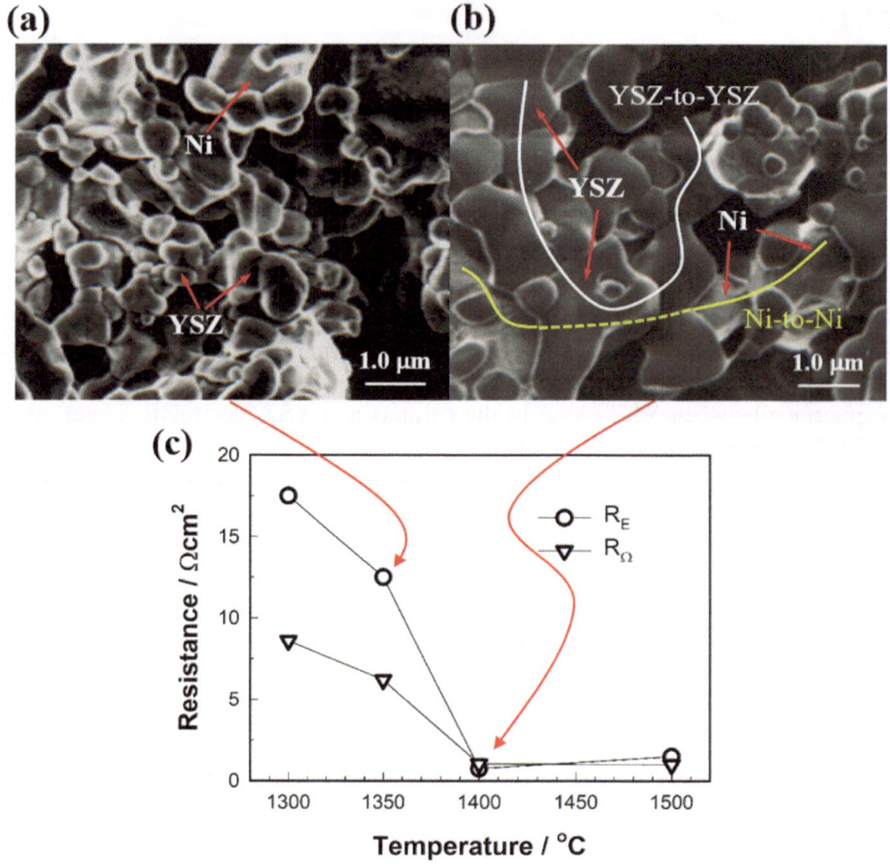

Fig. 12.17 SEM micrographs of the surface of Ni/YSZ cermet anodes sintered at **a** 1350 °C and **b** 1400 °C. The lines indicate the interconnected YSZ-to-YSZ and Ni-to-Ni networks. The change of R_Ω and R_E of Ni/YSZ cermet anodes for HOR measured at 1000 °C as a function of anode sintering temperature is shown in (**c**)

depends on the activity of the oxide powder, which is closely related to the synthesis process and the calcination temperature of the powder. Figure 12.18 shows an example of the relationship between the performance of $(La_{0.75}Sr_{0.25})Cr_{0.5}Mn_{0.5}O_3$ (LSCM)/YSZ composite anodes and the synthesis methods of LSCM powders [20]. A LSCM/YSZ composite anode prepared by the SSR method shows a much lower electrochemical activity for the H_2 oxidation reaction, as compared to that prepared by the gel-casting method. On the other hand, LSCM/YSZ composite anodes show a significant dependence on the calcination temperature of LSCM powder prepared by the gel-casting method with the best performance obtained for the powder calcined at 1100 °C. On the other hand, the activity of the LSCM/YSZ composite is much less dependent on the calcination temperature of LSCM powder prepared by the SSR method. Microstructurally, the LSCM powder prepared by gel casting is much finer

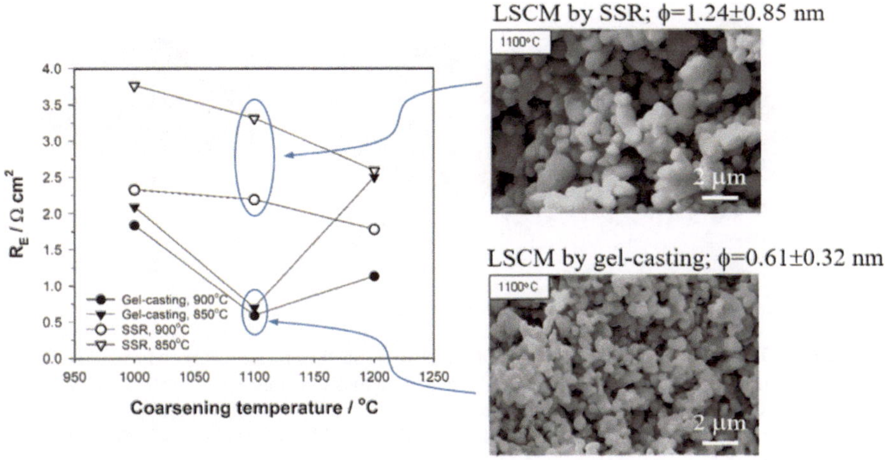

Fig. 12.18 Plots of the electrode polarization resistance for the H_2 oxidation reaction on LSCM/YSZ composite anodes as a function of calcination temperature of LSCM powders prepared by gel-casting and solid-state reaction methods. LSCM/YSZ composite anodes are sintered at 1200 °C in air for 2 h. The SEM pictures show the microstructure of LSCM powders prepared by SSR and gel-casting methods after calcination at 1100 °C. Modified from Ref. [20] with permission from Royal Society of Chemistry, Copyright 2007

and more uniformly distributed, as compared to that prepared by SSR under the same calcination temperature. This indicates that the activity, i.e., the particle size, of the oxide powder plays an important role in the formation of an intimate interface with the electrolyte, thus resulting in a very different activity of the electrode formed.

12.4.2 Cathode

High-temperature sintering is also required for the formation of an intimate contact/interface between a perovskite oxide-based cathode and an ionic-conducting electrolyte such as YSZ and GDC. Different to the Ni-based cermet anodes, the interface between the oxide cathode and electrolyte can be formed at a much lower temperature, e.g., 1150 °C or lower. Sintering at high temperatures will cause the interface formation and in the case of perovskite oxide cathodes such as LSM and LSCF electrodes, the interface formation is characterized by the formation of convex rings.

Figure 12.19 shows an example of the interface contact between $(La_{0.8}Sr_{0.2})_{0.90}MnO_3$ (LSM) and $La_{0.6}Sr_{0.4}Co_{0.2}Fe_{0.8}O_{3-\delta}$ (LSCF) electrodes with YSZ (8 mol % Y_2O_3–ZrO_2) and GDC ($Gd_{0.2}Ce_{0.8}O_2$) electrolytes after sintering at high temperatures [21]. The size of the convex contact rings varies and is related to the size of the cathode particles. In the case of the LSM/YSZ interface, significant diffusion of Mn in the form of Mn^{2+} and to a lesser extent La

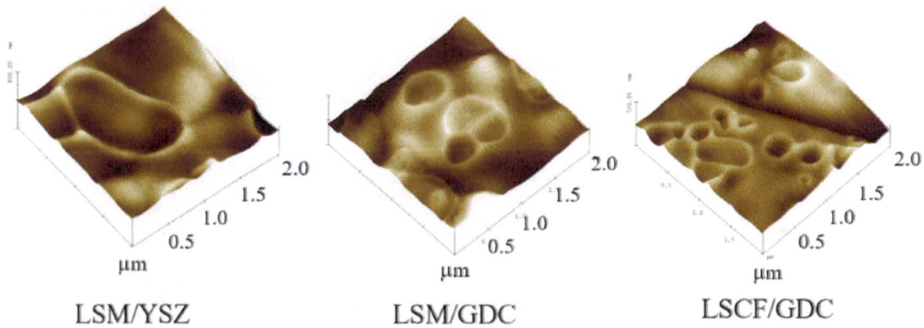

Fig. 12.19 AFM micrographs of YSZ or GDC electrolyte surface in contact with pre-sintered LSM or LSCF electrodes. LSM on YSZ and GDC electrolyte disks is sintered at 1150 °C in air for 2 h and LSCF on GDC electrolyte disk is sintered at 1100 °C in air for 2 h. LSM and LSCF electrodes were removed by HCl treatment. Modified from Ref. [21] with permission from IOP Publishing, Copyright 2015

along the YSZ electrolyte surface and the edge of the convex contact rings has been observed, replacing Y in the YSZ lattice. On the other hand, no cation interdiffusion is observed between the LSM cathode and the GDC electrolyte despite the fact that similar convex contact rings are formed on the GDC electrolyte surface. The fundamental reason for the convex ring formation is due to the diminishing the curvature of the interface as well as the reduction in surface free energy during the high-temperature sintering of the LSM or LSCF on YSZ or GDC electrolyte [22]. This is similar to the grain growth and neck formation in the high-temperature sintering of ceramic powders.

The interface stability between cathode and electrolyte is related to the sintering temperature. Increase in the sintering temperature will increase the interface stability and thus enhance the long-term stability of the electrode, but on the other hand, the increased sintering temperature will cause the grain growth and agglomeration of the electrode particles. This in turn will reduce the electrochemical activity of the electrode. Therefore, in practice, the optimum sintering temperature needs to consider the initial electrochemical performance as well as long-term stability under polarization conditions.

One commonly used strategy to avoid the interaction between LSCF and YSZ during the high-temperature sintering stage is the spatial separation of LSCF and YSZ by insertion of an interlayer, generally consisting of mixed ionic and electronic-conducting-doped ceria such as GDC and SDC. However, the doped ceria interlayer does not completely prevent the reaction between LSCF and YSZ. This is due to the fact that the surface segregated Sr species is volatile and highly mobile and can migrate through the porous-doped ceria barrier layer to reach the YSZ electrolyte, forming a resistive $SrZrO_3$ phase. Thus, the density and microstructure of the ceria interlayer are critical to the electrochemical performance of cobaltite cathodes like LSCF. It is generally considered that secondary phase formation between LSCF and YSZ could be minimized if the GDC interlayer is sintered dense or contains only a small volume of closed porosity. Very dense yet ultra-thin GDC interlayers can be produced by techniques such as magnetron sputtering and PLD techniques.

12.4.3 Sintering Profile

Another important function of high-temperature sintering or calcination is to control the *sintering profile* of the powder and cell components such as the anode support and electrolyte layers in the case of tape-casting planar cells [13, 23]. The sintering profile control is particularly important for the electrode-supported cells as the structural integrity of the electrode-supported cells depends strongly on the characteristics of the starting powders and the sintering profile of both the substrate and electrolyte layers. In this case, carefully matching the sintering profiles of both the electrode substrates (e.g., Ni-based cermets in most cases) and the thin electrolyte layer is critical; otherwise the sintering profile mismatch will cause bending and/or cracking of the structure during the co-sintering step. Large internal stress accumulated within the bilayer structure due to the mismatch of the sintering profiles can be released during operation at high temperatures, causing the formation of cracks across the YSZ film and the subsequent failure of cells [24].

The shrinkage profile of the cell components such as anode substrates and the electrolyte layer can be manipulated by using pore formers, heat treatment, or calcination of the starting powder (this will change the particle size and particle size distribution), variation of the packing density of the anode substrate, or changing the characteristics of the electrolyte slurry such as ink solids loading. The characteristics of the starting electrolyte powders not only affect the sintering profile but also affect the densification of the electrolyte film [9]. It is known that particle size, particle shape, green density, and aggregation of the precursor powders all affect the morphology and sinterability of the powders, thus greatly influencing the density of the sintered electrolyte film.

The characteristics of the starting powders can be effectively controlled by ball milling and calcination treatments. Figure 12.20 shows an example of the change in the particle size distribution of $Sm_{0.2}Ce_{0.8}O_{1.9}$ (SDC) powder prepared by the GNP method [23]. The as-prepared SDC powder (calcined at 800 °C) before ball milling treatment shows a rather symmetric and narrow distribution with a maximum at ~6 μm. The particle size and particle size distribution change significantly with the ball milling, indicated by the significant reduction in particle size and a new bimodal distribution of particle size. After ball milling for 20 h, the powder shows a strong peak at 0.2 μm and a weak one at 1 μm, and the particle distribution range is reduced to 0.06–3.0 μm. Ball milling significantly decreases the degree of agglomeration in the as-prepared SDC powders. The heat treatment is also effective in changing the particle size profile of the powder. In the case of SDC powder, as the calcination temperature increases, the peak at 1 μm decreases while the peak at 0.2 μm increases. This indicates that calcination at a higher temperature leads to a smaller secondary particle size after ball milling. This may be related to the ball

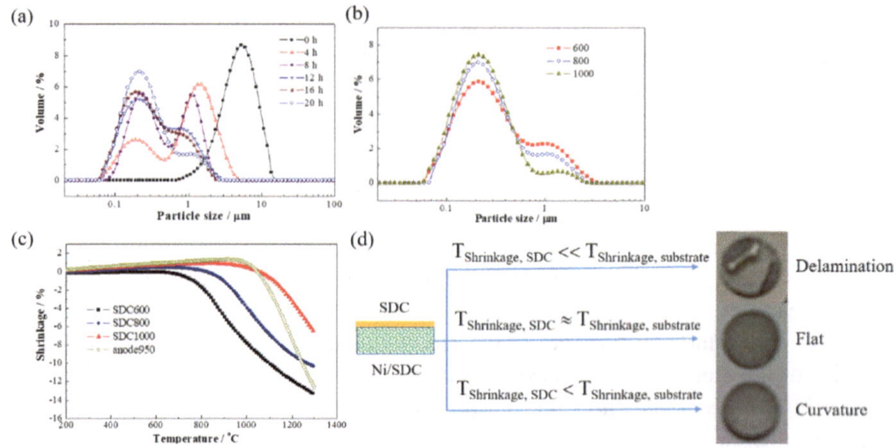

Fig. 12.20 **a** Change of the particle size distribution of $Sm_{0.2}Ce_{0.8}O_{1.9}$ (SDC) powder calcined at 800 °C as a function of ball milling time, **b** particle size distribution of SDC powder calcined at 600, 800, and 1000 °C (SDC600, SDC800, and SDC1000), **c** sintering profiles of SDC powder and Ni/SDC substrate calcined at 950 °C (anode950), and **d** relationship between the shrinkage profile of the powder and substrate and flatness of the SDC-Ni/SDC bilayer structure. Modified from Ref. [23] with permission from Elsevier, Copyright 2012

milling treatment, which is more effective in breaking the agglomerates with large crystalline particles formed at high calcination temperatures.

The characteristics of the particle size distribution have a significant effect on the sintering behavior of the powders and subsequently on the microstructure and flatness of the thin electrolyte film cells. The shrinkage onset temperature of the SDC powders changes significantly with the calcination temperature of the powder. The shrinkage onset temperature of SDC600 is 556 °C and increased to 843 °C for the SDC powder calcined at 1000 °C (SDC1000). But the overall linear shrinkage percentage decreases with the increase in the powder calcination temperature. In the case of the anode substrate pre-fired at 950 °C (anode950), the shrinkage onset temperature is 920 °C and the shrinkage is high due to the thick Ni/SDC substrates. In the case of the anode substrate, the maximum shrinkage occurred at 1200 °C, close to the SDC powder calcined at 1000 °C.

SDC powders that are ball milled for 20 h are applied to NiO/SDC anode substrates by slurry spin coating, and the anode-supported SDC bilayer structures are co-sintered at 1400 °C for 4 h at an average heating rate of ~3 °C/min. For the bilayer sample with SDC600, electrolyte films delaminated from the anode substrate. In the case of cells with SDC800 and SDC1000 powders, there is no electrolyte delamination, indicating the good adhesion between the anode substrate and the SDC layer. However, the cell with SDC calcined at 800 °C (SDC800) electrolyte thin film concaves toward the electrolyte side. A flat bilayer structure is obtained on the cells with SDC1000 electrolyte powder. As the anode substrates are the same for all the cells, the significant difference in the morphologies and flatness of the bilayer cells

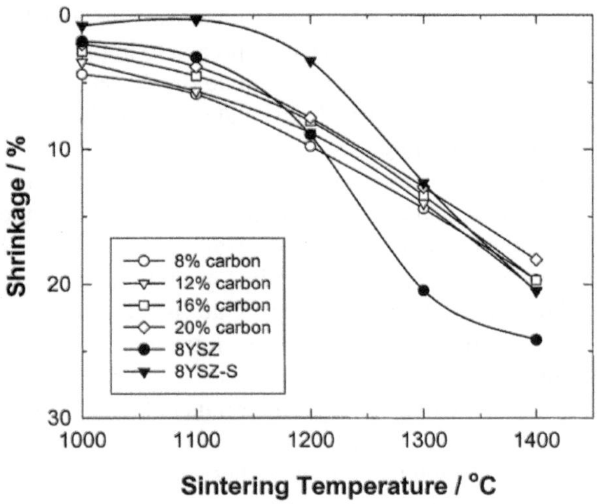

Fig. 12.21 Shrinkage profiles of Ni/YSZ cermet tape with addition of various carbon contents (empty symbols) and YSZ electrolyte tape (filled symbols) as function of sintering temperature. Reproduced from Ref. [13] with permission from Springer Nature, Copyright 2004

is evidently due to the closer matching of the sintering profile between the SDC electrolyte powder and Ni/SDC anode substrate.

Due to the significant difference in the thickness of Ni/YSZ anode substrate support and the YSZ electrolyte, it is critical to match the sintering profile of both tapes as close as possible to ensure the flatness of the sintered cell. Typically, an anode substrate layer from 1 to 2 mm thick and an electrolyte layer from 40 to 100 μm thick are produced by tape-casting technique, these two tapes are laminated and after rolling or calendering, and the green laminates are cut to size and sintered at 1400 °C. The final anode-supported structures are produced with anode thickness of 500–1000 μm and electrolyte thickness of 15–40 μm. Figure 12.21 shows a typical example of manipulating the sintering profile of the Ni/YSZ cermet tape with the addition of carbon fillers[13]. Close matching in sintering profiles is observed for the Ni/YSZ cermet tape and 8YSZ-S electrolyte tape at a high sintering temperature of 1300–1400 °C. The sintering profile of the YSZ tape can be further adjusted by using mixed 8YSZ and 8YSZ-S powders.

12.4.4 High-Temperature Sintering Aids and Solid-State Reactive Sintering

High-temperature sintering is one of the most important processes to achieve the densification of electrolyte films or layers in SOFCs. However, densification of electrolyte materials depends on a number of factors, and one of the most influential factors is related to the nature of the electrolyte materials. Some materials are more difficult to sinter due to their poor sinterability. To achieve full densification, sintering temperatures as high as 1600–1700 °C are required, for example for doped $BaZrO_3$

proton conductors, lanthanum silicate apatite, and lanthanum chromite interconnect materials. The most effective approach to improve the sinterability is to use sintering aids usually in the forms of simple oxides such as Y_2O_3, Al_2O_3, CaO, MgO, NiO, Cr_2O_3, as well as tetraethoxysilane (TEOS). The use of sintering aids is a common practice in ceramics to reduce the densification temperatures. For example, boron carbide is one of the most important structural ceramics due to its favorable physical, mechanical, and chemical properties, but boron carbide has a very poor sinterability due to the existence of strong covalent bonds in pure boron carbide and its low self-diffusion, high resistance against grain boundary slipping, and low surface energy. Adding iron oxide as a sintering aid is effective in reducing the sintering temperature and improves the mechanical properties of boron carbide via the liquid phase formation.

In the case of electrolyte materials in fuel cells, the addition of sintering aids such as NiO, Al_2O_3, ZnO, CuO, etc., generally improves the sinterability and reduces the densification temperatures, but at a cost of reduced ionic conductivity[25]. This may be related to the fact that sintering aids remain as residue in the material after the sintering process. The presence of foreign particles either in the grain boundary region or in the lattice of the ionic-conducting oxides will have an adverse effect on the ionic conductivity due to the increased grain boundary resistance contribution to the total conductivity of the oxide materials.

The use of *solid-state reactive sintering* (SSRS) could reduce the detrimental effect of sintering aids on the electrical properties of the electrolyte materials. Tong et al. used SSRS to improve the sinterability and densification of $BaZr_{0.8}Y_{0.2}O_3$ (BZY20) proton-conducting oxides, using NiO as a sintering aid [26]. In this process, raw materials of $BaCO_3$, CeO_2, and Y_2O_3 were mixed with NiO sintering aid (<2 wt%), and BCY20 ceramic pellets were fully densified at a sintering temperature of 1400 °C for 12 h. The formation of an intermediate, $BaYNi_2O_5$, was detected, and it is considered that the decomposition of $BaYNi_2O_5$ after pellet densification allows the diffusion of Ba, Y, and Ni into the BZY lattice, enhancing the sintering kinetics. Some sintering aids, like Li_2O, are not stable and tend to evaporate at high sintering temperatures, leaving the sintering oxides free of sintering aids. However, the evaporation of the sintering aids at high temperature would inevitably reduce the effect of the sintering aids in the grain growth and densification at the high-temperature sintering stage.

12.5 Direct Assembly and Polarization-Induced Interface

Direct assembly is arelatively new process for preparing cathodes on YSZ or GDC electrolyte without the high-temperature sintering step [21, 27]. In this process, the electrode slurry is directly applied to the electrolyte surface by screen printing or slurry painting, dried and tested at the electrode evaluation temperatures. Under the influence of the cathodic polarization, the electrode/electrolyte interface such as LSM and LSCF on YSZ or GDC electrolyte can be formed. Due to the bypassing

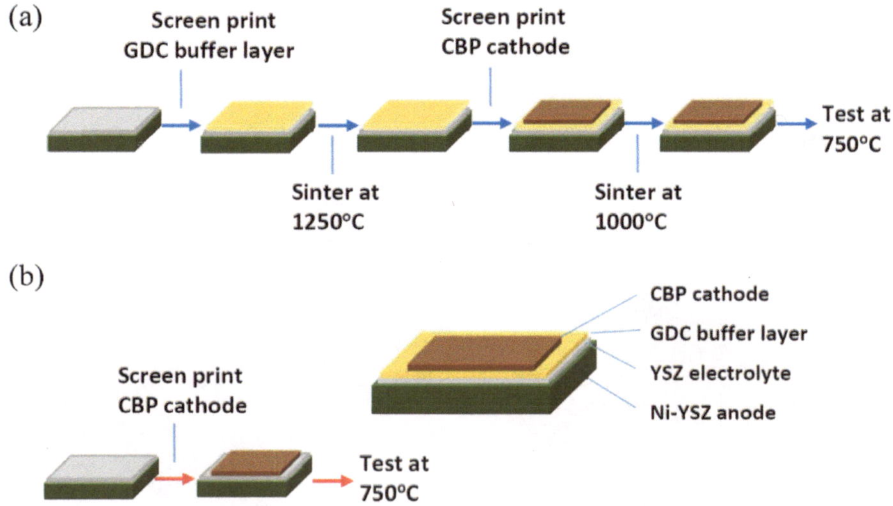

Fig. 12.22 **a** Conventional and **b** direct assembly processes for cobaltite-based perovskite (CBP) cathodes of SOFCs. Reproduced from Ref. [27] with permission from Royal Society of Chemistry, Copyright 2016

of the high-temperature sintering steps, *cobaltite-based perovskite* (CBP) cathodes such as LSCF can be directly applied on barrier layer-free YSZ electrolyte via a direct assembly approach at the SOFC operating temperature of 750 °C or lower. Figure 12.22 compares the conventional fabrication and direct assembly processes of CBP cathodes on YSZ electrolyte. The direct assembly is a single screen-printing step with no additional high-temperature sintering steps.

The principle of direct assembly is based on the phenomenon of a polarization-induced interface [21]. Take directly assembled LSM on YSZ and GDC electrolytes, LSM/YSZ and LSM/GDC, as an example. Figure 12.23 shows the change of the electrode impedance for the O_2 reduction reaction on LSM/YSZ and LSM/GDC measured at 1000 mA/cm^2 and 800 °C and the corresponding interface formed. The initial R_E for the directly assembled LSM cathode is 38 Ωcm^2, which is substantially higher than 14–17 Ωcm^2 obtained on pre-sintered LSM electrodes measured under identical conditions. Very different from the constant electrode ohmic resistance (R_Ω) of pre-sintered LSM electrodes, the initial R_Ω is 3.1 Ω cm^2 and decreases to 1.4 Ω cm^2 after polarization at 1000 mA/cm^2 for 15 min, a reduction of 1.7 Ω cm^2 ($\Delta R_\Omega = 1.7$ Ω cm^2). The reduction in R_Ω indicates the improvement in the interface contact between the assembled LSM particles and YSZ electrolyte. The significant decrease of electrode polarization resistance (R_E) under cathodic current passage is a typical activation behavior of LSM cathodes. In the case of a directly assembled LSM/GDC cell, similar reduction in electrode impedance is also observed. The initial R_Ω is 1.9 Ω cm^2 and decreases to 1.0 Ω cm^2 after polarization at 1000 mA/cm^2 for 3 h, $\Delta R_\Omega = 0.9$ Ω cm^2. R_E for the reaction also decreases with the cathodic polarization. In both electrodes, contact marks or clusters are formed on the YSZ

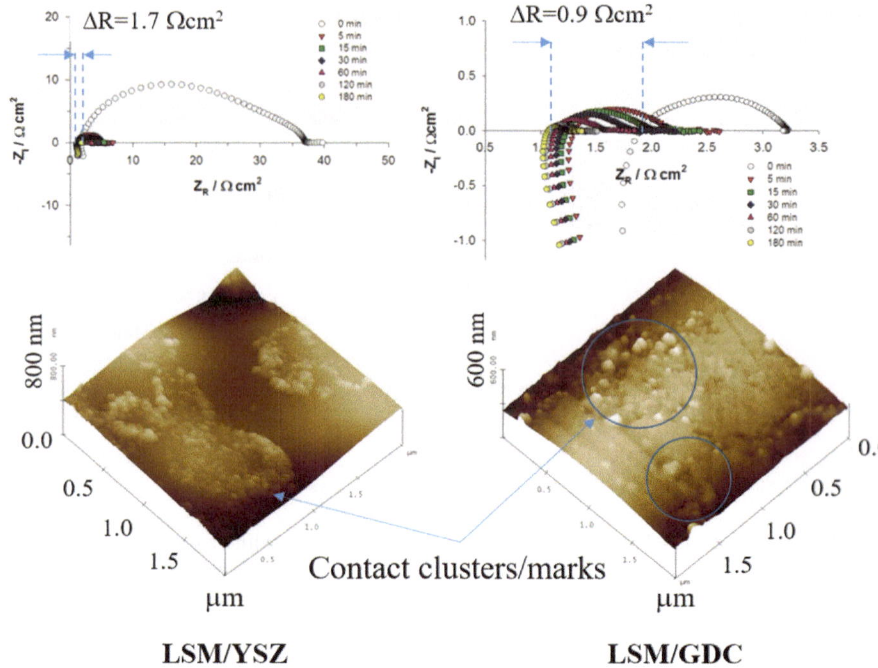

Fig. 12.23 Impedance responses for the O_2 reduction reaction on directly assembled LSM electrode on YSZ and GDC electrolyte (LSM/YSZ and LSM/GDC) as a function of polarization times at 1000 mA/cm^2 and 800 °C. AFM micrographs of YSZ and GDC electrolyte surface in contact directly assembled LSM after polarization for 3 h. The LSM electrode was removed by HCl treatment

and GDC electrolyte surface. Studies also show the number of the contact clusters increases with the polarization current density. In addition to the formation of the contact marks, there is also a formation of nano-indents on the GDC surface. Contact clusters and nano-indents appear to spread uniformly to the whole GDC electrolyte surface.

Different to the situation in the pre-sintered electrode/electrolyte interfaces, there is no pre-formed convex ring-type interface between the electrode and electrolyte before the cathodic polarization passage, and the contact resistance between the directly assembled electrode and electrolyte is high. By passing a cathodic current, the localized current density will be very high due to the limited contact points between the LSM particles and electrolyte. This will in turn generate Joule heat, causing the localized sintering at the contact points and effectively leading to the formation of the interface. The formation of such contact clusters/marks is indicated by the reduction in R_Ω. In fact, despite the significant differences in the morphology of the LSM/YSZ interface, the polarization-induced LSM/YSZ interface shows a comparable electrochemical performance as compared to that of a high-temperature sintered LSM/YSZ interface.

Anodic polarization also induces the interface formation in a porous Ni anode/YSZ electrolyte. The polarization-induced interface, e.g., at 250 mA/cm^2, probably has a strong interaction between the electrode and electrolyte, indicated by the remarkable suppression of microstructure coarsening and decrease of both electrode ohmic and polarization resistance. Similar to that formed at the cathode/electrolyte interface, one possible reason is the local sintering effect ascribed to the release of Joule heat, due to extremely high local current at the very limited contact region between the directly assembled Ni electrodes and YSZ electrolyte initially. The localized sintering process is greatly dependent on the level of current density applied to the anode and thus affects the evolution of the anode's microstructure and the interaction at the electrode/electrolyte interface.

The direct assembly method offers an excellent platform for the fundamental studies of the interface formation, surface segregation, and interphase reaction of a wide range of cathode materials in SOFCs [28], which may not be possible with the conventional high-temperature fabrication route. Examples include the LSCF/YSZ system [28], LSM/YSZ and LSM/GDC [29], and a wide range of cobaltite-based perovskites on YSZ electrolyte such as $(La_{0.6}Sr_{0.4})_{0.95}Co_{0.2}Fe_{0.7}Nb_{0.1}O_{3-\delta}$ (LSCFN), $(La_{0.6}Sr_{0.4})_{0.95}(Co_{0.2}Fe_{0.7}Nb_{0.1})_{0.95}Pd_{0.05}O_{3-\delta}$ (LSCFNPd), $Sm_{0.95}Co_{0.95}Pd_{0.05}O_{3-\delta}$ (SmCPd), and $PrBa_{0.5}Sr_{0.5}Co_{1.5}Fe_{0.5}O_{5+\delta}$ (PBSCF) [27].

12.6 Nano-Structured Electrodes

12.6.1 Wet-Infiltration Techniques

Due to the fact that high processing temperatures are often required to achieve good an interface between electrode and electrolyte, e.g., 1000–1150 °C for LSM cathodes and 1300–1400 °C for Ni/YSZ cermet anodes, the conventional approach to introduce nanoparticles in the green stage of the electrodes such as that commonly used in low-temperature PEMFCs is not very successful for the high-temperature SOFC electrodes. Thus, the key consideration in the development of nano-structured SOFC electrodes is to separate the catalytic phase formation temperature from the high electrode/electrolyte interface formation temperature. This can be achieved by *wet infiltration or impregnation* of catalytically and/or electrochemically active nanoparticles into a rigid and pre-fired electronic and/or ionic-conducting electrode or electrolyte scaffold/framework [30]. The nano-sized metal or metal oxide catalysts are often realized by in situ reduction or oxidation treatment. Figure 12.24 shows schematically a typical route for the deposition of nano-sized particles into a pre-fired electrode or electrolyte scaffold or framework via infiltration of metal salt solutions or nanoparticle suspensions. For example, wet infiltration of CeO_2 and GDC into a pre-fired LSM cathode can be carried out by simply placing a drop of a $Ce(NO_3)_3$ or mixed $Gd(NO_3)_3$ and $Ce(NO_3)_3$ nitrate solution with stoichiometric composition $Gd_{0.2}Ce_{0.8}(NO_3)_x$ on top of the coating, which infiltrates the porous

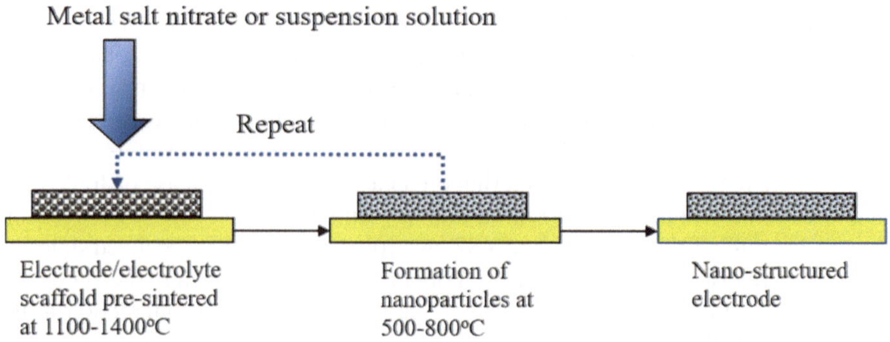

Fig. 12.24 Typical process for the infiltration of a metal salt nitrate solution or nanoparticle suspension into a pre-sintered electrode or electrolyte porous structure/scaffold

coating by capillary action. The infiltration process can be facilitated under vacuum. Then the infiltrated LSM is heated at 500–800 °C to decompose the metal salt solution, forming CeO_2 or GDC crystal phases. The phase formation temperature of the catalytic phase (i.e., CeO_2 or GDC) is much lower than the processing temperature of the standard LSM, ~1150 °C. The significantly reduced phase formation temperature reduces the grain growth, resulting in the deposition of nano-sized GDC particles on the surface of the porous of LSM scaffold. The infiltration technique is particularly useful for introducing low-melting point phases into SOFC electrodes. Due to the high surface area of the nano-sized catalysts, a small loading of typically 1–10 wt% is required to obtain high activity.

The infiltrated catalytic nanoparticles can form discrete distribution or a thin and continuous network on the surface of the porous scaffold, as schematically shown in Fig. 12.25. The porous scaffold can be an electronic-conducting electrode materials such as LSM, a mixed ionic and electronic conductor such as LSCF and BSCF, or an ionic-conducting electrolyte materials such as YSZ and GDC. The latter requires deposition of continuous and interconnected nanoparticles with high electronic conductivity for sufficient electrical conduction as well as high electrocatalytic activity for electrochemical reactions. This, in general, requires multiple infiltration steps. However, the need for multiple infiltrations to achieve the desired loading of electrocatalysts (or connectivity ofs phases) is time-consuming and costly, thus hindering the scalability and practical application of the infiltration process in SOFCs.

The distribution and phase formation of infiltrated components depend on a number of operation factors including the surface chemistries of the scaffold and the impregnated materials and the matching of the wettability of the precursor solution to the scaffold materials. Adding surfactants or complexing agents is beneficial for the uniform distribution and phase formation of infiltrated nanoparticles. The addition of urea and polymeric dispersant can facilitate the formation of perovskite phases such as LSCF and LSM at low temperatures, presumably due to the precipitation and

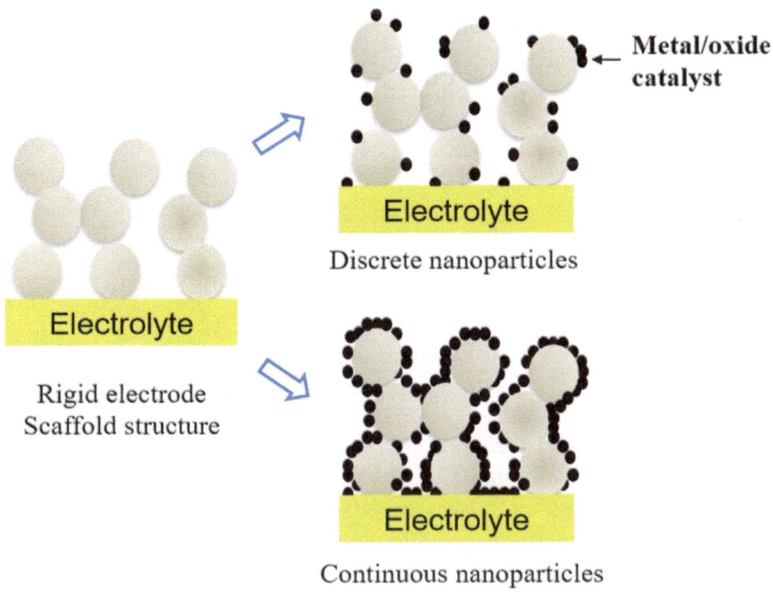

Fig. 12.25 Schematic of the infiltrated nano-structured electrodes on pre-sintered porous electrode or electrolyte scaffold/skeleton

complexing effect of the additives. Adding glycine has also been shown to be effective in promoting the uniform distribution of infiltrated nanoparticles with reduced infiltration steps and to promote the formation of the infiltrated LSM perovskite phase. It has been shown that direct decomposition of nitrate precursors at 800 °C does not produce pure LSM perovskite phase, but a pure LSM perovskite phase is formed in the presence of glycine in the nitrate solution [31]. This indicates that glycine acts as a chelating agent to form a metal ion complex in the solution; thus, the individual oxides do not segregate upon firing of the metal nitrate precursor. The uniformity of the porous scaffold morphology with large specific area is essential for the infiltration of continuous and monolayered catalytic nanoparticles, while the porous structure and pore size of the scaffold can be controlled by adding organic or polymeric pore formers.

12.6.1.1 Promotion Factor

The primary function of the wet infiltration of nanoparticles is to enhance the electrochemical performance of SOFC electrodes. The performance enhancement of nano-structured electrodes can be described by a *promotion factor*, f_P, defined as the ratio of the ASR or the overpotential, η of the nano-structured electrodes or the peak power density (PPD) of the cell with nano-structured electrodes to the performance of the baseline electrodes fabricated by conventional mixing and sintering processes, measured under identical conditions.

$$f_{\mathrm{P}} = \frac{\mathrm{ASR}_{\text{conventional electrode}}}{\mathrm{ASR}_{\text{infiltrated electrode}}} \tag{12.25}$$

or

$$f_{\mathrm{P}} = \frac{\eta_{\text{conventional electrode}}}{\eta_{\text{infiltrated electrode}}} \tag{12.26}$$

or

$$f_{\mathrm{P}} = \frac{\mathrm{PPD}_{\text{cell with infiltrated electrodes}}}{\mathrm{PPD}_{\text{cell with conventional electrodes}}} \tag{12.27}$$

Thus, f_{P} is greater than 1 for a promotion effect and for a neutral or deactivation effect, f_{P} is equal to 1 or less than 1.

In the case of electrode-based scaffolds such as LSM, LSCF, LSM/YSZ, and Ni/YSZ electrodes, the scaffolds provide the continuous electronic conductivity as well as electrocatalytic activity. A wide range of nanoparticles can be infiltrated, including ionic-conducting phases like GDC, YSZ, yttria-stabilized bismuth oxide (YSB), and catalytic active phases such as Pd, Co_3O_4, and $Sm_{0.5}Sr_{0.5}CoO_{3-\delta}$ (SSC). For example, an infiltrated Pd phase plays a significant role in the reduction of the electrode polarization resistance of LSM-based cathodes for the O_2 reduction reaction.

Use of the ionic-conducting electrolyte material-based scaffolds can increase the scaffold's structural stability and thus the stability of the infiltrated electrodes. One distinctive advantage of ionic-conducting scaffolds is that, in general, the ionic-conducting scaffold can be the same material as the electrolyte, thus, excellent bonding and perfect TEC match between the scaffold and electrolyte can be realized. The most common ionic-conducting scaffold materials are YSZ and doped ceria such as GDC and SDC. Figure 12.26 shows an example of the intrinsic relationship between the microstructure of the nano-structured electrodes and the performance [32]. The anode-supported cell with infiltrated nano-structured LSM + YSZ cathode (type II) achieves a power density of 0.83 W/cm^2 at 750 °C in H_2/air, more than 4 times higher than that in similar cells with conventional LSM-YSZ composite cathodes (type I). The promotion factor, f_{P} is ~4. With the infiltration of Pd to conventional LSM-YSZ composite cathode (type III), the cell performance is enhanced substantially to 1.42 W/cm^2, achieving $f_{\mathrm{p}} = 7$. Though the microstructure of LSM-infiltrated YSZ (LSM + YSZ) and Pd-infiltrated LSM-YSZ (Pd + LSM-YSZ) composite electrodes appears similar (Fig. 26b), the reaction paths on the electrodes II and III are in fact very different. For electrode II, the YSZ scaffold only provides the path for the oxygen ion conductivity and infiltrated LSM nanoparticles on the YSZ surface provides the TPB for the ORR. In the case of nano-structured Pd + LSM-YSZ (type III) electrode, the LSM-YSZ scaffold is both electronic and ionic conductive and the deposition of Pd nanoparticles layers on the surface of the LSM-YSZ scaffold

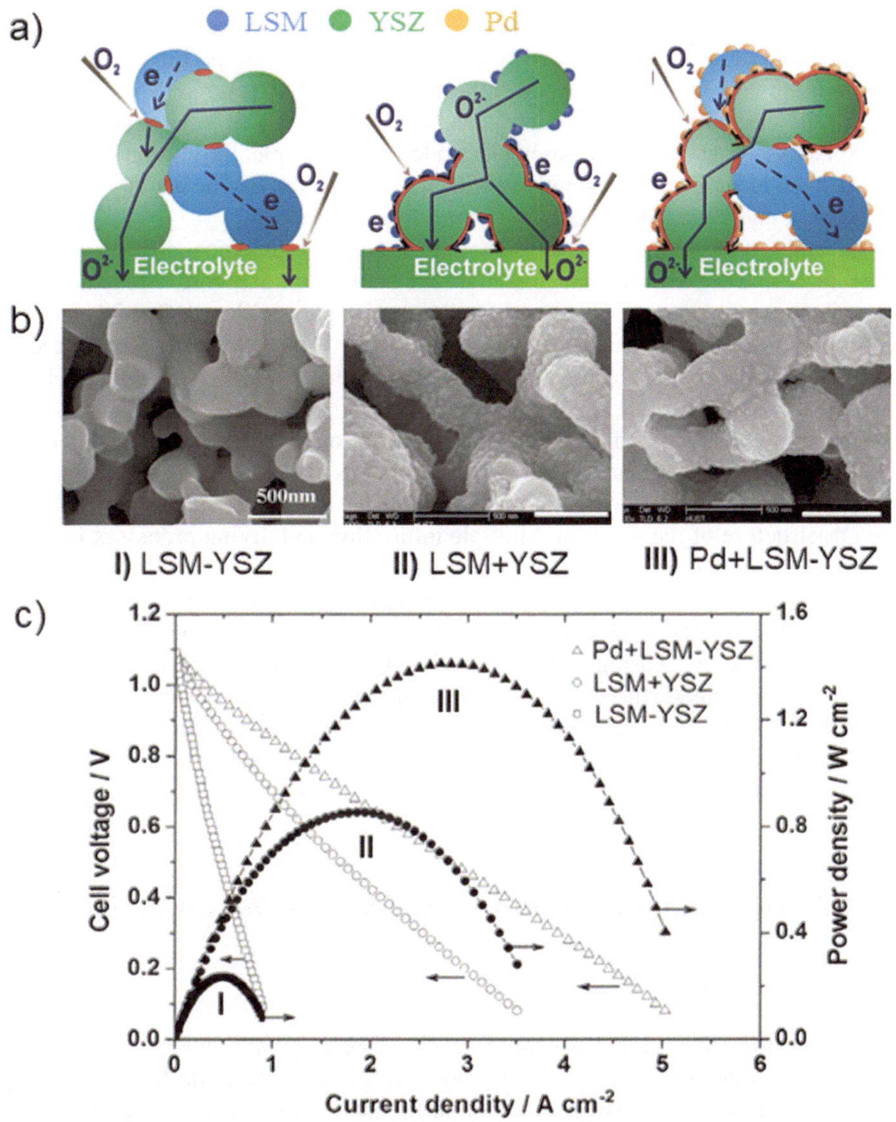

Fig. 12.26 Examples of the performance and microstructure of the nano-structured electrodes. **a** Schematic illustrations of electrode types I, II, and III, showing the paths of transport for electrons and oxygen ions and reaction sites for the ORR; **b** SEM micrographs of types I, II, and III; **c** Power output of anode-supported cells with electrode types I, II, and II, measured at 750 °C in H_2/air. Electrodes: (I) standard LSM-YSZ composite cathode by mechanical mixing of LSM and YSZ phases; (II) nano-structured LSM-infiltrated YSZ (LSM + YSZ) composite cathode; and (III) nano-structured Pd-infiltrated LSM-YSZ (Pd + LSM-YSZ) composite cathode[32]. Bar = 500 nm. Modified from Ref. [32] with permission from Elsevier, Copyright 2009

not only provides the additional reaction sites but also significantly accelerates the reaction rate of the dissociation and diffusion of oxygen species (Fig. 12.26a). This explains the very high performance of the cell with nano-structured Pd + LSM-YSZ composite cathodes. The infiltration technique has been used to alter and enhance the specific properties of the parent electrodes. For example, infiltration of catalytic active nanoparticles can substantially enhance the electrocatalytic activity as well as the carbon and sulfur tolerance of Ni-based cermet anodes.

Ceramic oxide-based materials generally have a low electrical conductivity and in some case low electrochemical activity for the fuel oxidation reaction and ORR. Introduction of ionic-conducting and catalytic active phases like GDC or Pd nanoparticles by infiltration substantially enhances the activity of oxide-based electrode such as LSCM anodes for the electrochemical oxidation reaction in H_2, CH_4, and C_2H_5OH fuels [33, 34].

There are two main limitations to the wet infiltration process in the development of nano-structured electrodes: multiple infiltration steps to reach the desired loading and the relatively low loading of the infiltrated phase due to the limited porosity and open porous structure of the scaffold. Multiple infiltration and drying processes increase the cost of fabrication. It is often the case that the narrow and small pores of the scaffold lead to the non-uniform distribution and blocked pores due to the difficulty of penetration of the infiltrated solution by capillary forces. Optimization of the infiltration solution by using surfactants and surface agents to improve the wettability between the porous surface of the host electrode and infiltrated solution can improve the efficiency and effectiveness of the infiltration processes. Nevertheless, the loading of the infiltrated phase is limited by the extent of open porosity of the scaffold and electrode coatings.

12.6.2 Decoration and Polarization-Induced Phase Migration

12.6.2.1 Decoration Technique

The direct assembly and low-temperature approach opens a new opportunity for the direct application of nano-sized phases in the host electrode by a decoration technique without the loading limitation of the incorporated phase. The *decoration technique* is a process to add the particular phase to the host electrode powders via a gelation process, thus forming a nano-structured coating on the surface of electrode powders at low calcination temperatures. In particular, the application of direct assembly of the decorated electrode allows the use of highly active but thermally unstable phases at high temperatures such as doped bismuth oxides in the composite electrodes. Figure 12.27 shows the procedure of producing erbium-doped bismuth (ESB) decorated LSM and the direct assembly of the decorated ESB/LSM electrodes [35]. Different to the infiltration technique, the decoration technique is not limited by the loading or content of the decorated phase. The decorated phase and host phase can be mixed in any ratio in the composite electrode.

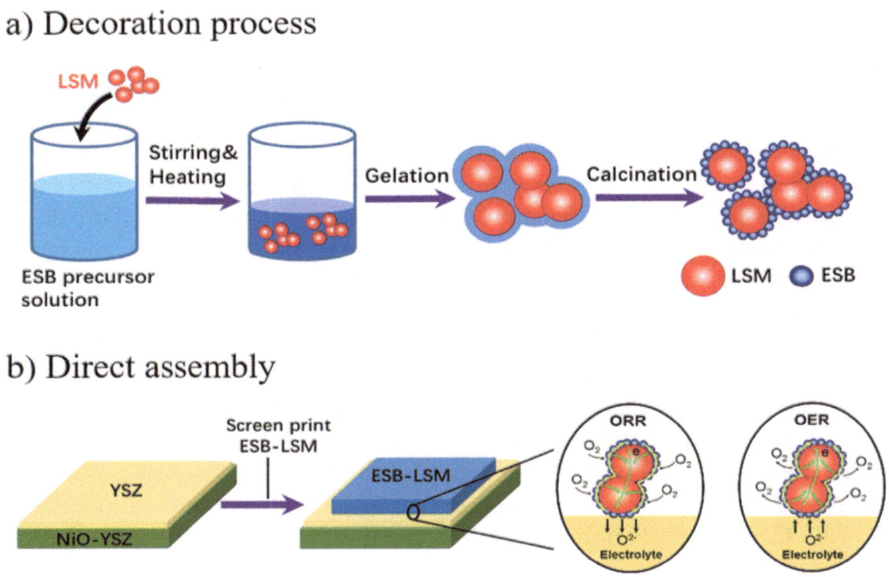

Fig. 12.27 Schematic of **a** synthesis of nanostructured ESB decorated LSM (ESB-LSM) powder via the gelation process and **b** direct assembly of the ESB-LSM electrode on a Ni-YSZ anode-supported YSZ electrolyte cell. Reproduced from Ref. [35] with permission from Royal Society of Chemistry, Copyright 2017

In this process, a $Er_{0.4}Bi_{1.6}O_3$ (ESB) aqueous precursor solution made from $Er(NiO_3)_3$, $Bi(NO_3)_3$, citric acid, EDTA, and ammonia solution is stirred on a hot plate for 2 h. Subsequently $La_{0.76}Sr_{0.19}MnO_{3+\delta}$ (LSM) powder, synthesized using a modified Pechini method and calcined at 900 °C in air for 2 h, is added to the solution and stirred continuously until the occurrence of gelation. The gel is thoroughly dried in an oven at 180 °C for 8 h and calcined at 600 °C in air for 2 h, forming ESB decorated LSM powder. ESB decorated LSM (ESB/LSM) powder with 40 wt% ESB and 60 wt% LSM is formed by a gelation process in this case. Er doping can stabilize the cubic structure of δ-Bi_2O_3 in a wider temperature region and increase the thermal stability of Bi_2O_3 [36], thus reducing the volatility of bismuth. The as-prepared ESB/LSM powder is blended with an ink vehicle at a weight ratio of 7:3 in a mortar and is screen printed on the YSZ electrolyte surface of a Ni/YSZ cermet-supported cell and dried at 100 °C for 2 h to form the directly assembled ESB/LSM electrode without further pre-sintering at high temperatures. The direct assembly process bypasses the high-temperature sintering steps and avoids the melting issues associated with the use of bismuth oxides. The Ni/YSZ cermet-supported YSZ cells with directly assembled ESB/LSM oxygen electrodes show excellent activity and stability for reversible O_2 reduction reaction (ORR) and O_2 evolution reaction (OER) in solid oxide cells [35].

12.6.2.2 Polarization-Induced Phase Migration

Polarization-induced phase migration is associated with the phase migration under the influence of polarization during SOFC operating conditions. Such phenomena have been reported on the ESB decorated electrodes[37]. The ESB phase introduced into the host composite electrode via the decoration process not only substantially increases the electrochemical activity of the composite electrodes, but is also driven to the electrode/electrolyte interface region under the cathodic polarization conditions. Such polarization-induced phase migration has a significant implication in the development of highly active and stable cathodes of IT-SOFCs. Figure 12.28 shows an example of the phase change of composite electrodes of Nb-doped $La_{0.6}Sr_{0.4}Co_{0.2}Fe_{0.7}Nb_{0.1}O_{3-\delta}$ (LSCFNb) electrode decorated with 40% $Er_{0.4}Bi_{1.6}O_3$ (ESB) composite electrodes on a barrier layer-free YSZ electrolyte cells and the corresponding stability and activation energy plots for ORR [37]. There is a redistribution and migration of the ESB phase in the ESB/LSCFNb composite toward the YSZ electrolyte under the influence of cathodic polarization at 250 mA/cm^2 and 750 °C, forming a thin ESB layer at the cathode/YSZ electrolyte interface. The

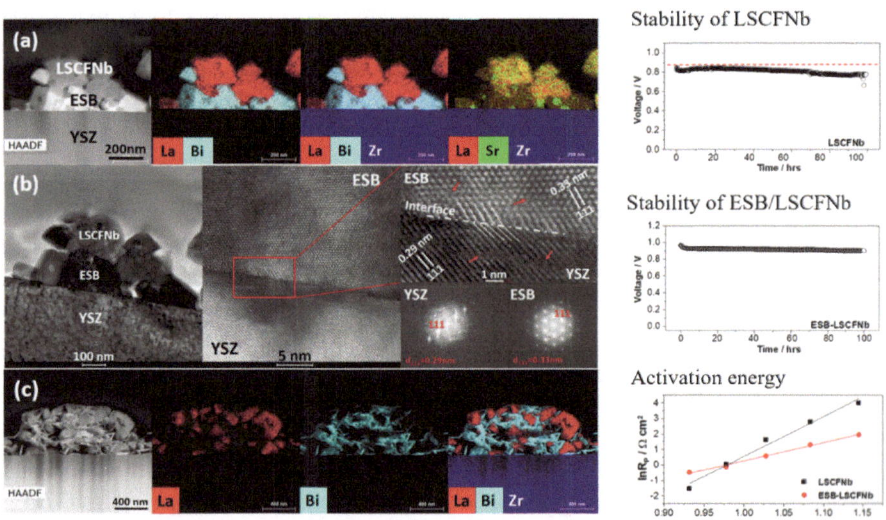

Fig. 12.28 **a** STEM-EDS element mapping and **b** HRTEM and FFT images of the ESB/LSCFNb and YSZ interface of an anode-supported YSZ electrolyte cell with a directly assembled ESB/LSCFNb electrode after polarization at 250 mA/cm^2 and 750 °C for 100 h. The red arrows indicate the lattice distortion. **c** STEM-EDS element mapping of the ESB/LSCFNb and YSZ interface of an anode-supported YSZ electrolyte cell with a directly assembled ESB-LSCFNb electrode after dwelling at 750 °C for 100 h without polarization. Stability of the anode-supported cell with directly assembled ESB/LSCFNb and LSCFNb cathodes is measured at 750 °C and 250 mA/cm^2. Activation energy plots are obtained on electrode polarization resistance for ORR on LSCFNb and ESB/LSCFNb cathodes. Modified from Ref. [27] with permission from American Chemical Society, Copyright 2018

ESB/YSZ interface is very sharp and abrupt as a result of their similar intrinsic cubic structure with space group symmetry of *Fm-3m* and lattice constants of 5.48 Å and 5.13 Å, respectively. YSZ lattice planes with a plane spacing of 0.29 nm and ESB lattice planes with a plane spacing of 0.33 nm are associated with YSZ and ESB phases, respectively, at the interface. Lattice distortion for the two phases is observed around the interface, but no additional or amorphous phases are found at the ESB/YSZ interface. The in situ formed ESB layer prevents the direct contact and subsequent reaction between segregated SrO and YSZ electrolyte, inhibiting the formation of the insulating $SrZrO_3$ phase. Such phase migration occurs under the influence of cathodic polarization. In the case of open circuit at 750 °C, no bismuth phase diffusion occurs (Fig. 12.28c).

With no decorated ESB phase, the cell performance deteriorates, due to the formation of the insulating $SrZrO_3$ phase at the interface between the segregated Sr species and the YSZ electrolyte. The presence of the ESB phase in the electrode prevents the contact of segregated Sr species with the YSZ electrolyte, resulting in a remarkable increase in power output and stability. The ESB decorated phase also promotes the oxygen migration/diffusion at the interface and significantly reduces the activation energy from 214 kJ/mol on a pristine LSCFNb electrode to 98 kJ/mol for the O_2 reduction reaction. The reduced activation energy is particularly beneficial for the intermediate and low-temperature SOFCs.

The in situ formation of a highly active and protective layer such as doped bismuth oxide has a significant and practical implication for the development of highly active electrodes, in particular cobaltite-based perovskite oxide materials on YSZ electrolyte for IT-SOFCs. However, the mechanism of polarization-induced phase migration is still not clear at this stage.

12.6.3 Nanosized Catalysts by in Situ Exsolution

In SOFCs, nanoparticles can also be introduced on the electrode surface via exsolution. *Exsolution* is a phenomena generally associated with the segregation of the B-site cations from perovskite-based electrodes under reducing conditions, forming metallic nanoparticles. In this approach, catalytically active transition metals are substituted in a host oxide lattice under oxidizing conditions, forming a solid solution, and released or exsolved on the surface as metallic nanoparticles, once the oxide lattice is reduced. Exsolution is an effective method to substantially improve the electrochemical performance and activities of SOFC electrodes. Different to the infiltration technique, the exsoluted nanoparticles are strongly bonded to the parent substrate, forming socketed nanoparticles on the substrate surface. Such formed nanoparticles are highly tolerant toward agglomeration and carbon cracking at high operation temperature.

Different from the commonly observed detrimental effect of the segregation of A-site cations such as Sr to the surface of oxygen electrodes for the ORR, the exsoluted nanoparticles from the B-site cations can significantly enhance the electrocatalytic activity and stability of the electrodes. The perovskite-based structure is probably the most effective redox exsolution host due to its unique ability to accommodate defects of different size and charge in a range of non-stoichiometry classes. Neagu et al. studied the exsoluted Ni nanoparticles on a freshly cleaved bulk surface of $La_{0.52}Sr_{0.28}Ni_{0.06}Ti_{0.94}O_3$ reduced at 900 °C in 5%H_2/Ar for 15 h[38]. Ni with its large ion size (the radius of Ni^{2+} is 0.069 nm) and high reducibility exhibits a greater tendency to segregate to the surface under a reducing environment. However, on the restructured (native) perovskite surface, the exsoluted Ni particles are diminished. The cell with in situ exsoluted Ni nanoparticles on A-site-deficient $(La_{0.6}Sr_{0.3})Cr_{0.85}Ni_{0.15}O_3$ anodes achieved a peak power density of 460 mW/cm^2 in 5000 ppm H_2S-H_2, significantly higher than 135 mW/cm^2 of fuel cells with stoichiometric $(La_{0.7}Sr_{0.3})Cr_{0.85}Ni_{0.15}O_3$ anode with limited exsoluted Ni nanoparticles [39]. The introduction of the A-site deficiency helps the formation of oxygen vacancies and remarkably enhances the reducibility of Ni nanoparticles, thus significantly increasing the electronic conductivity and catalytic activity simultaneously. This indicates the importance of surface chemistry or structure in controlling redox exsolution process. In addition to Ni, metallic nanoparticles such as Pd, Pt, Rh, Ru, Cu, Co, Fe–Ni alloy, and Ag can also be exsoluted on the surface of perovskite-based electrodes.

The redox exsolution of B-site cations from perovskites is a phase decomposition process driven by chemical/electrochemical reduction and controlled by bulk/surface defects as well as external conditions. When the lattice is reduced, surface defects such as oxygen vacancy formation become more dominant, which destabilizes the lattice stoichiometry. Meanwhile, the generation of oxygen vacancies accompanied by the reduction of B-site transition metals can further lead to metal nucleation at the surface. Nucleation is likely to occur on the lattice surface where the nucleation barrier is lowered by crystal defects. This would drive B-site cations to the surface from inside the bulk continuously and cause the growth of exsolved particles until an equilibrium is reached.

12.6.4 Microstructure and Microstructure Stability

The remarkable promotion effect of the nano-structured electrodes on the performance of anodes and cathodes of SOFCs is a direct result of the formation and uniform deposition of nano-sized and catalytically active phases on the surface of porous electrode or electrolyte scaffolds. Figure 12.29 shows typical microstructures of selected nano-structured electrodes made by infiltration techniques [30]. The GDC phase is formed after heat treatment at 800 °C, and the particle size is in the range of ~50 nm, which is much smaller than 1000–1500 nm of the LSM grains of the pre-sintered LSM electrode scaffold. LSM can also be infiltrated into a pre-sintered YSZ electrolyte scaffold to form nano-structured LSM/YSZ composite cathodes. The infiltrated LSM phase is characterized by the formation of continuous

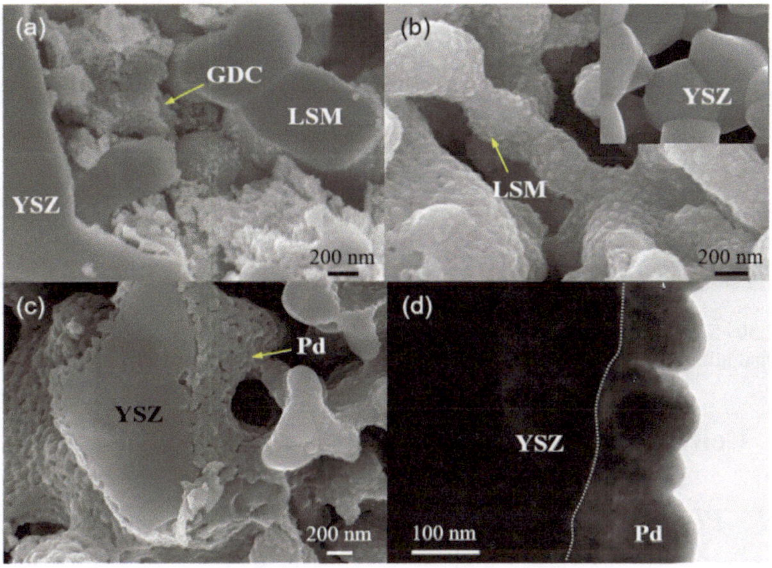

Fig. 12.29 Typical SEM micrographs of **a** GDC-infiltrated LSM, **b** LSM-infiltrated YSZ, **c** Pd-infiltrated YSZ, and **d** high resolution TEM of (**c**), showing the intimate contact between Pd nanoparticles and YSZ electrolyte. Modified from Ref. [30] with permission from Elsevier, Copyright 2012

LSM nanoparticles on the surface of YSZ grains. Continuous Pd nanoparticles on the YSZ grain surface can also be formed via a simple palladium nitrate solution infiltration process. The Pd nanoparticles are well interconnected and form an intimate contact with YSZ electrolyte scaffold. The formation of a continuous LSM or Pd nano-film on a YSZ scaffold ensures the electrical conductivity of the electrode since the YSZ scaffold is essentially a non-electrical conductor.

One of the challenges associated with the application and development of nano-structured electrodes is the long-term stability of the microstructure and activity of the infiltrated nanoparticles. Specifically, the high surface area nanoparticles (20–100 nm) are highly prone to sintering and grain growth under typical SOFC operation conditions. Figure 12.30 shows a typical example of the significant agglomeration of the infiltrated Pd nanoparticles on $La_{0.75}Sr_{0.25}Cr_{0.5}Mn_{0.5}O_3$/YSZ composite anodes after polarization at 100 mA/cm^2 and 800 °C for 15 h in CH_4 fuel. The agglomeration of infiltrated nanoparticles of nano-structured electrodes can have a significant detrimental effect on the performance stability. Microstructural and long-term stability of infiltrated electrodes are still an open issue and require more systematic and comprehensive studies to understand the effects of several factors such as the nanoparticle size and distribution, infiltration firing temperature, scaffold structure, operation temperature, and current load on the structural stability of infiltrated phases.

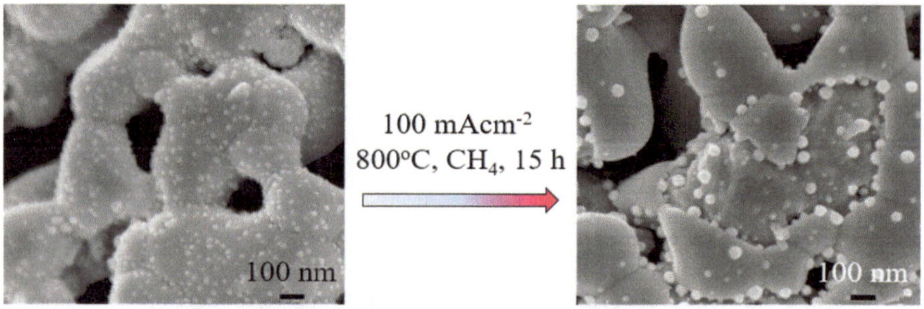

Fig. 12.30 SEM images of Pd impregnated $La_{0.75}Sr_{0.25}Cr_{0.5}Mn_{0.5}O_3$/YSZ composite anodes before and after testing under a current density of 100 mA/cm^2 in CH_4 at 800 °C for 15 h

12.7 Cell Fabrication

12.7.1 Planar Cells

Planar-type cells can be classified as electrolyte- and electrode-supported cells; see Fig. 12.31. In electrolyte-supported cells, the dense electrolyte is mechanically strong and is used as support for the deposition or screen printing of porous anode and cathode coatings. The most common techniques for the fabrication of dense electrolyte supports are die pressing and tape-casting techniques. Die pressing is typically used for the fabrication of button-sized cells, and tape casting can be used for both button cells and large planar cells.

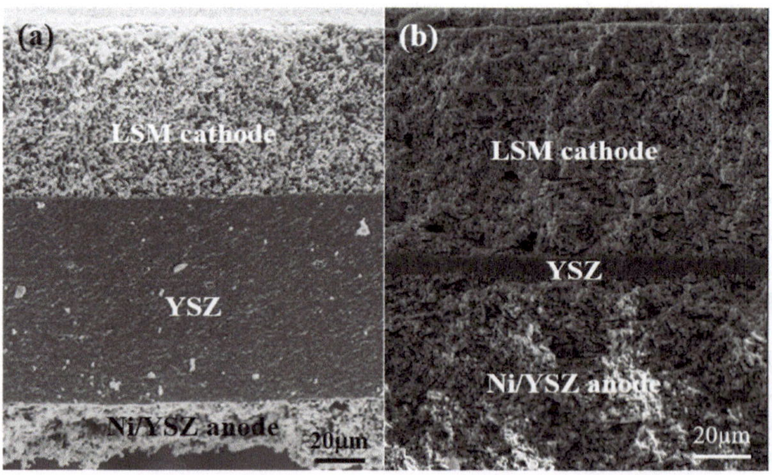

Fig. 12.31 **a** YSZ electrolyte-supported cell and **b** Ni/YSZ anode-supported cells. The thin electrolyte can be prepared by tape casting and other thin-film fabrication techniques

For electrode-supported cells, the anode-supported thin-film electrolyte design is probably the most popular one in SOFCs. For such configurations, the major technical challenge is to fabricate a dense and gas-tight YSZ (or GDC) electrolyte layer (20 μm or less in thickness) well adhered to the porous Ni/YSZ (or Ni/GDC) anode substrate, minimizing the interfacial polarization resistance. Among various fabrication techniques for the anode-supported cells, tape casting in combination with lamination, screen printing or sputtering is the most simple, cost-effective, and scalable technique that has been used to fabricate the anode supports and anode-supported thin electrolyte cells.

The formation of anode-supported YSZ electrolyte cells by tape casting, lamination, and co-sintering at high temperatures has been described in Sect. 12.2.2. The cathode functional layer and current collector layers are then applied on the YSZ surface of the half-cell by a screen-printing technique. Optimization of different processing parameters such as tape-casting slurry composition, viscosity of the slurry, and heat treatment schedule is required to achieve defect-free and flat anode-supported YSZ electrolyte half-cells. In order to have enhanced cell performance, normally two functional layers, e.g., anode functional layer and cathode functional layer, having a much finer microstructure than the electrodes, are used between the anode–electrolyte and cathode–electrolyte interfaces, respectively.

The same fabrication techniques can be used for the fabrication of cathode-supported cells. Cathode-supported cells with a thin anode layer have particular advantages when cells are operated on hydrocarbon fuels with low steam/carbon ratio. This is because most of the anode area is electrochemically active, reducing the risk of carbon deposition. In addition, the cathode-supported cells reduce the risk associated with the large volume contraction and expansion resulting from accidental thick anode support redox cycles. On the other hand, a co-firing process reduces the operation steps and can also effectively reduce the thermal stress and manufacturing cost. Thus, developing co-firing is significant to make the cathode-supported SOFCs cost effective.

12.7.2 Tubular Cells

A tubular SOFC essentially consists of two porous electrodes, separated by a dense, oxygen ion-conducting tubular electrolyte. Mechanically, tubular designs can be anode supported, cathode supported, electrolyte supported or extra porous tube supported. There are various techniques to fabricate tubular SOFC supports that include extrusion, electrophoretic deposition (EPD), tape casting, and gel casting.

Extrusion is the most common and established technology to fabricate tubular supports with a constant cross section. The process has the advantages of uniform density distribution, formation of long thin sections, high production rate, and relatively low costs in equipment and operation. In extrusion, shaping occurs by forcing a cohesive plastic material through the orifice of a rigid die. The design of the extrusion die significantly affects the paste flow and quality of extrudates. Approaches such

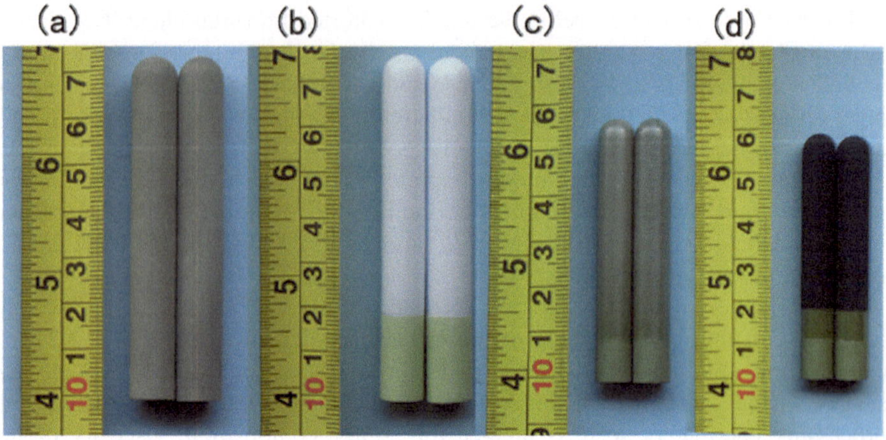

Fig. 12.32 Optical micrographs of the anode-supported tubular cells at different stages of fabrication, **a** green tubular supports after drying; **b** anode tubular supports with dip-coated YSZ electrolyte after pre-sintering at 1000 °C; **c** anode tubular supports with dip-coated YSZ electrolyte layer sintered at 1400 °C; and **d** anode-supported tubular cells with dip-coated YSZ electrolyte and a multilayer cathode after sintering at 1100 °C. Reproduced from Ref. [16] with permission from John Wiley & Sons Inc., Copyright 2009

as enlarging the paste flow pathways, tapering the flow entry and exit regions, and reducing the cross section of the die effectively reduce the static flow zones, enhance the separate flow streams joining, and reduce the amount of materials left in the die. The properties and performance of anode-supported tubular SOFCs fabricated by extrusion strongly depend on the materials, the processing and the manipulation of the microstructure.

Slip-casting and dip-coating methods are combined to fabricate microtubular cells; see Fig. 12.32 [16]. Green Ni/YSZ tubular supports were successfully fabricated by the slip-casting method. After subsequent and repeated steps of pre-sintering, dip-coating, drying, and final sintering at high temperatures, thin and dense YSZ electrolyte film and porous cathode layers can be formed on the anode tubular supports, forming tubular cells.

12.8 Summary

In this chapter, we have introduced and discussed the following topics:

- In SOFC, majority of powder and cell fabrication techniques are adopted from ceramic industry but with more precision control and modification to suit the specific requirements of SOFC components. For Ni-based cermets, physical mixing and ball milling are the techniques used in the industry due to the easy control and scale up. In the case of cathode materials, solution-based combustion and Pechini method or their derivatives are the common synthesis methods. Gel casting is more feasible for the synthesis of electrode and electrolyte materials with multiple elements and complex composition.

- Die pressing is particularly suitable for button-sized cells and very useful for the initial testing and evaluation of electrode and electrolyte materials. For large planar cells, tape casting is the most common one. To obtain good quality electrode and electrolyte tapes, the composition control and procedures of the tape casting are important.
- Wide range deposition techniques are available for the deposition of thin electrode or electrolyte coatings. Depending on the deposition technique and deposition conditions, the coating can be pinhole-free and dense or porous. Magnetron sputtering and pulsed laser deposition are popular techniques for the deposition of doped ceria barrier layer on zirconia-based electrolyte. Plasma-based spray techniques are commonly used for the one-step fabrication of metal-supported SOFC.
- Sintering is a very important step in the formation of intimate electrode/electrolyte interfaces in SOFC. In the case of Ni-based cermets, sintering leads to the establishment of good interface not only at the electrode and electrolyte but also within the electrode between electronic contact between Ni-to-Ni phase and ionic contact between YSZ-to-YSZ phase. Such intimate contact is essential for the high performance and stability of the anode. Pre-sintering or calcination is an effective way to control the shrinkage profile of electrode and electrolyte powder in order to produce the final plates or cells with acceptable flatness.
- High-temperature sintering is not the only way to for the formation of intimate electrode/electrolyte interface. By passing polarization current, an intimate electrode/electrolyte interface can also be formed most likely due to the localized Joule heating between the electrode particles and electrolyte.
- Wet impregnation or infiltration and exsolution are the effective method to introduce nano-scaled structure and nanoparticles to porous electrodes of SOFC. Infiltration and exsolution have their specific advantages and disadvantages. For the infiltration technique, wide range element and composition can be introduced either in the form of precursor solution or in the form of nanoparticle suspension solution. The main problem of the infitration technique is the poor microstructure stability due to the easy agglomeration of freestanding nanoparticles. Exsoluted nanoparticles offer a much high structural stability due to the bonding of the nanoparticles to the host supports, but the choices of the exsoluted particles are limited as they needs to be dissolved or incorporated in the oxide lattice in the first place.
- Phase migration of the electrode materials under the influence of polarization is relatively new phenomena and requires more research effort. Nevertheless, if the phase with particular function and activity can be in situ diffused and moved to specific areas such as interface under polarization conditions, this will have significant implications in the design and development of much more effective interface and electrode structure of SOFC.

12.9 Questions

12.1 Explain the following terms and materials:

Solid state reaction, glycine-nitrate process, stoichiometry coefficient, and fuel-to-nitrate ratio
Pechini method, sol–gel method, and gel-casting method
Tape casting, dispersing agent, binder, plasticizer, and homogenizer
Chemical vapor deposition and electrochemical vapor deposition
Magnetron deposition, pulsed laser deposition, and plasma spray deposition
Spin coating, dip coating, and slip casting
Ni-to-Ni electronic contact and YSZ-to-YSZ ionic contact
Solid-state reactive sintering
Thermally and polarization-induced interface
Nano-structured electrode *via* infiltration, decoration, and exsolution
Promotion factor
Polarization-induced phase migration.

12.2 Balance the following solid-state reaction for the synthesis of $La_{0.8}Sr_{0.2}MnO_3$ powder.

$$La_2O_3 + SrCO_3 + MnCO_3 \rightarrow La_{0.8}Sr_{0.2}MnO_3 + CO_2 + O_2$$

12.3 Calculate the fuel-to-nitrate ratio for the synthesis of $Ba_{0.5}Sr_{0.5}Co_{0.8}Fe_{0.2}O_{3-\delta}$, using urea ($CO(NH_2)_2$), glycine ($NH_2CH_2COOH$), citric acid ($C_6H_8O_7$) and sucrose ($C_{12}H_{22}O_{11}$) as fuels, respectively. Write the balance reactions for the solution–combustion reactions for each fuel.

12.4 List the similarities and differences between the Pechini and sol–gel methods.

12.5 Would the chelating agent affect the phase formation of a perovskite oxide phase? Please give specific examples.

12.6 As shown in Example 12.4, the phase formation temperature of $(La_{0.8}Sr_{0.2})_{0.9}MnO_3$ (LSM) powders prepared by SSR technique is significantly higher than that prepared by gel-casting technique. Explain the reasons for the differences in the phase formation temperatures of the LSM powders prepared by SSR and gel-casting techniques.

12.7 What is the morphology of the SDC-Ni/SDC bilayer cells after sintering at 1400 °C if the shrinkage of the anode substrates is greater than that of the SDC electrolyte powder? Explain the reasons behind the morphology formed.

12.8 Discuss the advantages and disadvantages of the nano-structured electrodes fabricated by the in situ exsolution method.

References

1. Shao ZP, Zhou W, Zhu ZH (2012) Advanced synthesis of materials for intermediate-temperature solid oxide fuel cells. Prog Mater Sci 57(4):804–874
2. Li JG, Ikegami T, Mori T, Wada T (2001) Reactive $Ce_{0.8}RE_{0.2}O_{1.9}$ (RE = La, Nd, Sm, Gd, Dy, Y, Ho, Er, and Yb) powders via carbonate coprecipitation. 1. Synthesis and characterization. Chem Mat 13(9):2913–2920
3. Pechini MP (1967) Method of preparing lead and alkaline-earth titanates and niobates and coating method using the same to form a capacitor. US Pat. No. 3330697, July 11, 1967
4. Danks AE, Hall SR, Schnepp Z (2016) The evolution of 'sol-gel' chemistry as a technique for materials synthesis. Mater Horiz 3(2):91–112
5. Zhang L, Jiang SP, Cheng CS, Zhang YJ (2007) Synthesis and performance of $(La_{0.75}Sr_{0.25})_{1-x}(Cr_{0.5}Mn_{0.5})O_3$ cathode powders of solid oxide fuel cells by gel-casting technique. J Electrochem Soc 154(6):B577–B582
6. Cheng CS, Zhang L, Zhang YJ, Jiang SP (2008) Synthesis of $LaCoO_3$ nano-powders by aqueous gel-casting for intermediate temperature solid oxide fuel cells. Solid State Ionics 179(7–8):282–289
7. Zhang L, Zhang YJ, Zhen YD, Jiang SP (2007) Lanthanum strontium manganite powders synthesized by gel-casting for solid oxide fuel cell cathode materials. J Am Ceram Soc 90(5):1406–1411
8. Xia CR, Liu ML (2001) A simple and cost-effective approach to fabrication of dense ceramic membranes on porous substrates. J Am Ceram Soc 84(8):1903–1905
9. Leng YJ, Chan SH, Khor KA, Jiang SP, Cheang P (2003) Effect of characteristics of Y_2O_3/ZrO_2 powders on fabrication of anode-supported solid oxide fuel cells. J Power Sources 117(1–2):26–34
10. Somalu MR, Muchtar A, Daud WRW, Brandon NP (2017) Screen-printing inks for the fabrication of solid oxide fuel cell films: a review. Renew Sust Energ Rev 75:426–439
11. Meng GY, Song HZ, Dong Q, Peng DK (2004) Application of novel aerosol-assisted chemical vapor deposition techniques for SOFC thin films. Solid State Ionics 175(1–4):29–34
12. Pal UB, Singhal SC (1990) Electrochemical vapor-deposition of yttria-stabilized zirconia films. J Electrochem Soc 137(9):2937–2941
13. Jiang SP, Chan SH (2004) A review of anode materials development in solid oxide fuel cells. J Mater Sci 39(14):4405–4439
14. Pederson LR, Singh P, Zhou XD (2006) Application of vacuum deposition methods to solid oxide fuel cells. Vacuum 80(10):1066–1083
15. Marcano D, Mauer G, Vassen R, Weber A (2017) Manufacturing of high performance solid oxide fuel cells (SOFCs) with atmospheric plasma spraying (APS) and plasma spray-physical vapor deposition (PS-PVD). Surf Coat Technol 318:170–177
16. Zhang L, He HQ, Kwek WR, Ma J, Tang EH, Jiang SP (2009) Fabrication and characterization of anode-supported tubular solid-oxide fuel cells by slip casting and dip coating techniques. J Am Ceram Soc 92(2):302–310
17. Besra L, Liu M (2007) A review on fundamentals and applications of electrophoretic deposition (EPD). Prog Mater Sci 52(1):1–61
18. Tsoga A, Naoumidis A, Nikolopoulos P (1996) Wettability and interfacial reactions in the systems Ni/YSZ and Ni/Ti-TiO_2/YSZ. Acta Mater 44(9):3679–3692
19. Jiang SP (2003) A comparative study of fabrication and performance of Ni/3 mol % Y_2O_3–ZrO_2 and Ni/8 mol% Y_2O_3–ZrO_2 cermet electrodes. J Electrochem Soc 150(11):E548–E559
20. Jiang SP, Zhang L, Zhang Y (2007) Lanthanum strontium manganese chromite cathode and anode synthesized by gel-casting for solid oxide fuel cells. J Mater Chem 17(25):2627–2635
21. Jiang SP (2015) Thermally and electrochemically induced electrode/electrolyte interfaces in solid oxide fuel cells: an AFM and EIS study. J Electrochem Soc 162(10):F1119–F1128
22. He S, Chen KF, Saunders M, Li J, Cui CQ, Jiang SP (2017) A FIB-STEM Study of $La_{0.8}Sr_{0.2}MnO_3$ Cathode and Y_2O_3–ZrO_2/Gd_2O_3–CeO_2 electrolyte interfaces of solid oxide fuel cells. J Electrochem Soc 164(13):F1437–F1447.

23. Ai N, Chen KF, Liu SM, Lu Z, Su WH, Jiang SP (2012) Effect of characteristics of (Sm, Ce)O_2 powder on the fabrication and performance of anode-supported solid oxide fuel cells. Mater Res Bull 47(1):121–129
24. Chen KF, Chen XJ, Lu Z, Ai N, Huang XQ, Su WH (2009) Performance evolution of NiO/yttria-stabilized zirconia anodes fabricated at different compaction pressures. Electrochim Acta 54(4):1355–1361
25. Li J, Wang C, Wang XF, Bi L (2020) Sintering aids for proton-conducting oxides—a double-edged sword? A mini review. Electrochem Commun 112.
26. Tong JH, Clark D, Bernau L, Subramaniyan A, O'Hayre R (2010) Proton-conducting yttrium-doped barium cerate ceramics synthesized by a cost-effective solid-state reactive sintering method. Solid State Ionics 181(33–34):1486–1498
27. Chen KF, Li N, Ai N, Li M, Cheng Y, Rickard WDA, Li J, Jiang SP (2016) Direct application of cobaltite-based perovskite cathodes on the yttria-stabilized zirconia electrolyte for intermediate temperature solid oxide fuel cells. J Mater Chem A 4(45):17678–17685
28. He S, Saunders M, Chen KF, Gao HF, Suvorova A, Rickard WDA, Quadir Z, Cui CQ, Jiang SP (2018) A FIB-STEM study of strontium segregation and interface formation of directly assembled $La_{0.6}Sr_{0.4}Co_{0.2}Fe_{0.8}O_{3-\delta}$ cathode on Y_2O_3–ZrO_2 electrolyte of solid oxide fuel cells. J Electrochem Soc 165(7):F417–F429.
29. He S, Chen KF, Saunders M, Quadir Z, Tao SW, Irvine JTS, Cui CQ, Jiang SP (2018) Interface formation and Mn segregation of directly assembled $La_{0.8}Sr_{0.2}MnO_3$ cathode on Y_2O_3–ZrO_2 and Gd_2O_3–CeO_2 electrolytes of solid oxide fuel cells. Solid State Ionics 325:176–188
30. Jiang SP (2012) Nanoscale and nano-structured electrodes of solid oxide fuel cells by infiltration: advances and challenges. Int J Hydrog Energy 37(1):449–470
31. Jiang Z, Lei Z, Ding B, Xia C, Zhao F, Chen F (2010) Electrochemical characteristics of solid oxide fuel cell cathodes prepared by infiltrating (La, Sr)MnO_3 nanoparticles into yttria-stabilized bismuth oxide backbones. Int J Hydrog Energy 35(15):8322–8330
32. Liang FL, Chen J, Jiang SP, Chi B, Pu J, Jian L (2009) High performance solid oxide fuel cells with electrocatalytically enhanced (La, Sr)MnO_3 cathodes. Electrochem Commun 11(5):1048–1051
33. Jiang SP, Chen XJ, Chan SH, Kwok JT (2006) GDC-impregnated $(La_{0.75}Sr_{0.25})(Cr_{0.5}Mn_{0.5})O_3$ anodes for direct utilization of methane in solid oxide fuel cells. J Electrochem Soc 153(5):A850–A856
34. Jiang SP, Ye YM, He TM, Ho SB (2008) Nanostructured palladium-$La_{0.75}Sr_{0.25}Cr_{0.5}Mn_{0.5}O_3/Y_2O_3$–$ZrO_2$ composite anodes for direct methane and ethanol solid oxide fuel cells. J Power Sources 185(1):179–182
35. Ai N, Li N, He SA, Cheng Y, Saunders M, Chen KF, Zhang T, Jiang SP (2017) Highly active and stable $Er_{0.4}Bi_{1.6}O_3$ decorated $La_{0.76}Sr_{0.19}MnO_{3+\delta}$ nanostructured oxygen electrodes for reversible solid oxide cells. J Mater Chem A 5 (24):12149–12157
36. Jiang N, Wachsman ED (1999) Structural stability and conductivity of phase-stabilized cubic bismuth oxides. J Am Ceram Soc 82(11):3057–3064
37. He S, Zhang Q, Maurizio G, Catellani L, Chen KF, Chang QB, Santarelli M, Jiang SP (2018) In situ formation of $Er_{0.4}Bi_{1.6}O_3$ protective layer at cobaltite cathode/Y_2O_3–ZrO_2 electrolyte interface under solid oxide fuel cell operation conditions. Acs Appl Mater Interfaces 10(47):40549–40559
38. Neagu D, Tsekouras G, Miller DN, Menard H, Irvine JTS (2013) In situ growth of nanoparticles through control of non-stoichiometry. Nat Chem 5(11):916–923
39. Sun YF, Li JH, Zeng YM, Amirkhiz BS, Wang MN, Behnamian Y, Luo JL (2015) A-site deficient perovskite: the parent for in situ exsolution of highly active, regenerable nano-particles as SOFC anodes. J Mater Chem A 3(20):11048–11056

Part IV
Other Fuel Cells

Chapter 13
Alkaline Fuel Cells

13.1 Introduction

Alkaline fuel cell (AFC) is the first fuel cell initiated in 1930s and demonstrated 20 years later by Francis Thomas Bacon [1]. The early AFC development since 1960´s was focused on the space programs such as a series of Apollo missions and the Space Shuttle.

Alkaline fuel cell (AFC) is based on the use of liquid potassium hydroxide (KOH) as electrolyte, which has a high conductivity in a wide temperature range from subfreezing points to above 200 °C. As the electrode reaction environment at the catalyst/electrolyte interface, KOH displays a much better kinetics for the oxygen reduction reaction, allowing for the use of less expensive, Pt or precious group metal (PGM)-free catalysts. The technology is therefore potentially cost effective compared to other types of low temperature fuel cells. In addition, the electrolyte has a low activation energy for electrode reactions, which means that the cell performance has a relatively small temperature coefficient. In other words, AFC exhibits the high power output at or below ambient temperature [2]. For terrestrial applications, the technology is restrained due to the demand of high-purity hydrogen and CO_2-scrubbed air.

The newest development in the field is the anion exchange membrane (AEM) fuel cells, which have achieved significant progress in the last decade. AEMs are polymer electrolytes with fixed cationic groups bounded to the polymer backbone allowing for conduction of anions such as hydroxide and carbonate. This solid membrane electrolyte eliminates the management issues of the liquid electrolyte and mitigates the carbonation, so that carbon-containing alcohols can also be used in addition to pure hydrogen.

S. P. Jiang and Q. Li, *Introduction to Fuel Cells*,
https://doi.org/10.1007/978-981-10-7626-8_13

13.2 Alkaline Fuel Cells

13.2.1 Electrolyte for AFC

Potassium and sodium hydroxide solutions (KOH and NaOH) are interchangeable electrolyte materials for many industrial applications. For fuel cell uses, KOH is more popularly used simply due to its high conductivity while NaOH is of lower cost. The same argument applies for battery applications such as nickel-cadmium, nickel-hydrogen, and manganese dioxide-zinc batteries. Figure 13.1 shows the conductivity of KOH as well as NaOH as a function of concentration at different temperatures. The maximum conductivity of 0.57 S/cm is observed at room temperature for 26 wt% KOH, corresponding to *ca.* 6 M KOH. This conductivity increases rapidly with temperature, and the maximum conductivity reaches 1.4 S/cm for 32 wt% at 80 °C and further to 2.8 S/cm for 45 wt% at 200 °C.

Potassium hydroxide is hygroscopic and often forms a series of crystalline hydrates. For example, the monohydrate $KOH{\cdot}H_2O$, corresponding to a water content of about 25%, in fact, is the composition of solid pellets of the chemical that are prepared by melt casting. The melting point of KOH is around 400 °C, at which

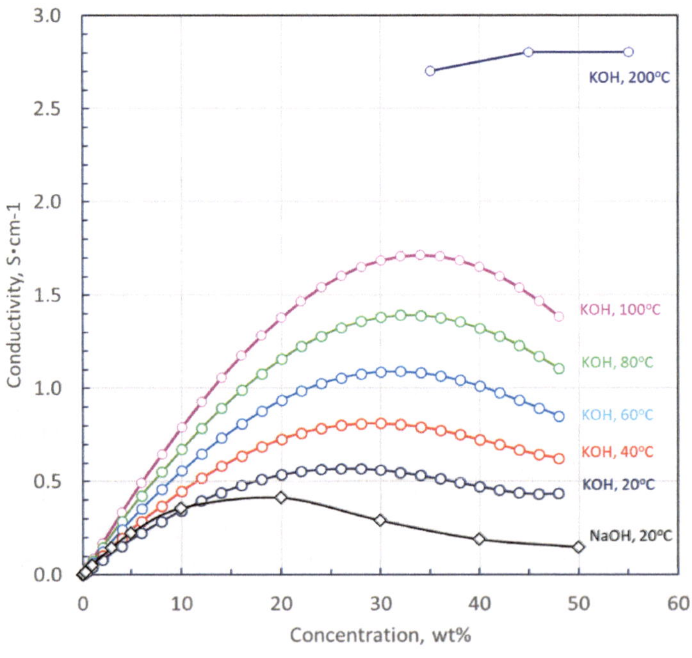

Fig. 13.1 Conductivity of sodium and potassium hydroxide. Data for KOH from 20 to 100 °C were taken from Ref. [3] and at 200 °C from Ref. [4] with permission from Elsevier, Copyright 2007 and 2012

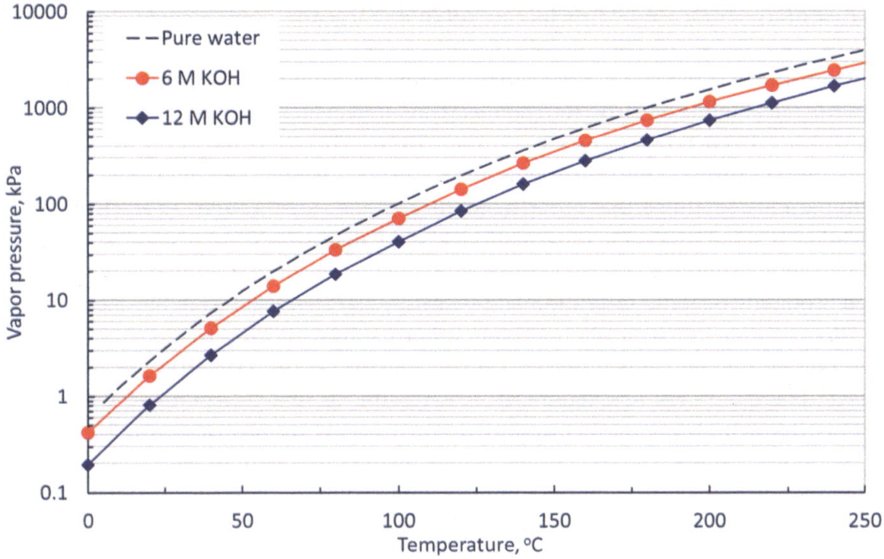

Fig. 13.2 Vapor pressure over pure water and KOH solutions as a function of temperature. Data were taken from Ref. [5] with permission from Elsevier, Copyright 1985

partial decomposition of KOH may occur with release of water and formation of potassium oxide:

$$2KOH_{(s)} \rightarrow K_2O_{(s)} + H_2O_{(g)} \quad (13.1)$$

For applications at elevated temperatures, concentrated KOH is preferred in order to handle with low water vapor pressure, which is shown in Fig. 13.2. For example, the water vapor pressure of KOH solutions over 12 M (45 wt%) is 185 mbar at 80 °C and reaches 1 bar at 125 °C, its boiling point. For water at 80°C, as a comparison, the vapor pressure is 470 mbar.

13.2.2 Catalysts

13.2.2.1 Oxygen Reduction Catalysts

The kinetics of the ORR reaction in alkaline media is faster than in acidic electrolytes. The early year study showed that the cathodic ORR potential of platinum-based electrodes at a constant current density of 100 mA/cm^2 at 70 °C is about 0.67 V in concentrated (13.9 M) H_3PO_4 and 0.89 V in 6.9 M KOH. The higher ORR performance is explained by the preferred formation and easier desorption of peroxide species in alkaline than in acids [6]. Due to the faster ORR kinetics, a wide range of catalysts have been studied including noble metals, non-noble metals, perovskites, spinels, etc.

Noble metal catalysts such as platinum, palladium, and gold are known to be super active in the alkaline electrolyte. Their uses are, however, limited to low concentration electrolytes and at temperatures lower than 70 °C. These metals, likely in their oxide forms, are slightly soluble in concentrated KOH at higher temperatures. The dissolved platinum ions are found to migrate from the cathode side through the electrolyte and are reduced at the anode. A consequent risk of the deposition of the metal at the anode side is the possible short circuit between the two electrodes. Gold does not seem to have problems of ionic migration and therefore is often used as either a catalyst or an alloying element with platinum and other catalyst metals. In the Space Shuttle Orbiter alkaline fuel cells, the cathode catalysts used consist of 10 wt% Au and 90 wt% Pt black with gold-coated nickel mesh as the current collector.

For terrestrial uses, the catalyst cost is decreased by use of non-noble metals. Silver, more precisely its oxide, shows reasonable ORR activity and is widely used in low concentration (<12 M KOH) electrolyte at low temperatures (<100 °C). In alkaline solutions, silver is first oxidized to Ag_2O which is then partly oxidized to AgO. The catalytic ORR proceeds via a 2 or 4 electron process depending on the surface and particularly on its oxidation state. Nickel oxide has a very limited ORR activity at low temperatures. Due to its good stability and conductivity, nickel is often used as support, for e.g., silver catalysts. It has long been found that nickel oxide is stable in molten KOH at temperatures higher than 200 °C where it exhibits a significant activity toward the ORR.

The ORR activity of both silver and nickel oxides can be enhanced in form of finely dispersed Raney metals. The Raney silver or nickel, more often their alloys, can be prepared by dissolving the metals in molten aluminum, preferably with addition of another promoter metal, e.g., zinc. The obtained powder is then treated with a concentrated solution of sodium hydroxide in order to selectively leaching out aluminum and zinc from the alloy phases. In addition to the silver-nickel alloys, many other metals have been alloyed with silver, nickel or both, showing increased activity or/and stability. Examples of alloying elements include alkaline earth (e.g., calcium), transitional (e.g., chromium, cobalt), rare earth (e.g., lanthanum), and refractory (e.g., molybdenum) metals. These high surface area Raney metals or their alloys are converted into the corresponding oxides when exposed to the cathode atmosphere in fuel cells. Particularly interesting is that when the sintered porous nickel oxides are soaked in a lithium hydroxide or lithium nitrate aqueous solution, following by a high-temperature treatment in air, their oxidative resistance is significantly improved at temperatures of up to 200 °C. This process is called lithiation and has also been further employed in molten carbonate fuel cells as well as high-temperature alkaline electrolyzers.

13.2.2.2 Hydrogen Oxidation Catalysts

The hydrogen oxidation reaction (HOR) in alkaline media meets more challenges than in acidic electrolytes [7]. First of all, the exchange current density at the Pt surface is about 100 times lower in alkaline than in acidic electrolytes, suggesting that the catalyst possesses active surface sites for both H- and OH-bonding. PtRu alloys have a better HOR activity in alkaline media than pure Pt, having an exchange current density of 2–6 times higher than that of Pt [8]. This is explained by the fact that Ru is oxophilic and promotes the adsorption of the OH species and hence the hydrogen oxidation.

In practice, however, nickel is preferably used as the anode catalyst, simply due to its low cost. Nickel, in its reduced form, most likely consisting of β-$Ni(OH)_2$, has a capability of absorbing hydrogen, as well demonstrated in nickel-metal hydride batteries. The material has an HOR activity of about three orders of magnitude lower than that of platinum. By means of the high surface area Raney metal and with a mass loading of 100 times higher than that of platinum, the performance of a nickel-based anode may reach that of the noble metal. In the sintered form, nickel has been used as HOR catalysts in AFCs with concentrated KOH at 200 °C as well as molten carbonate fuel cell at 650 °C.

13.2.3 Gas Diffusion Electrodes for Liquid Electrolytes

The success of the AFC ad PAFC, with liquid electrolytes, was based on development of gas diffusion electrodes (GDE), which is further extended to PEMFC. An important function of gas diffusion electrodes is to create a porous structure where both liquid electrolytes and gaseous reactants have access to the active sites of the catalysts [9].

Inside a micropore, the wetting force of a curvature liquid surface along the solid wall is determined by the surface tension, contact angle, and the pore radius ($2\pi R \cdot \gamma \cos\theta$), see Fig. 13.3. When a stable interface is established, this is balanced by the pressure difference between the liquid and gas ($\Delta P \cdot \pi R^2$),

$$2\pi R \times \gamma cos\theta = \Delta P \times \pi R^2 \tag{13.2}$$

$$\Delta P = \frac{2\gamma \times \cos\theta}{R} \tag{13.3}$$

Here, ΔP is the pressure difference, R is the pore radius, γ is the surface tension of the liquid, and θ is the contact angle of the liquid on the solid surface.

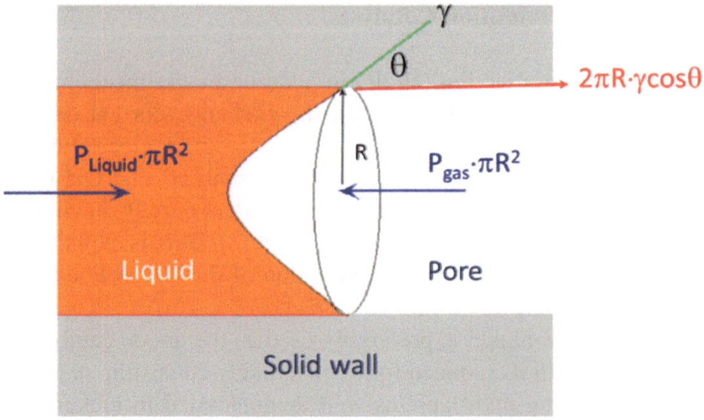

Fig. 13.3 Schematic representation of the interface balance of liquid and gas inside a micropore of the catalyst layer

Example 13.1 Estimation of the Pressure Difference of a Curvature Liquid Surface

Solution:

For 100% phosphoric acid at 200 °C, the surface tension is estimated to be $\gamma = 0.06$ N/m, and the contact angle on the carbon wall is assumed to be 60°, i.e., $\cos\theta = 0.5$. For a pore of radius $R = 5$ micron (5×10^{-6} m), the pressure difference of the curvature liquid surface within the pore is

$$\Delta P = \frac{2\gamma \times \cos\theta}{R} = \frac{2 \times 0.06 \times 0.5\,\text{N/m}}{5 \times 10^{-6}\text{m}} = 1.2 \times 10^4\,\text{pa} = 120\,\text{mbar} \tag{13.4}$$

In early years since 1950s, most electrodes were prepared from the sintered metal particles. In that case, the liquid surface tension and contact angle are constant, but the pore size of a sintered electrode is varied. The usual practice is to make electrodes in layered structures, often by several sintering steps. Each layer is made of powders of different grain sizes. The liquid electrolytes fills up the catalyst layers of fine powders where the pores are small, while the layers of coarse particles with large size pores are open for the gases to access. Such electrodes are in general thick, typically in about 2 mm. This double-layer sintered electrode as starting point of this technology has led to the first success of AFCs in the Apollo missions to the moon and later in the shuttle Orbiter. Example 13.2 collects information of early year development, application, and technical specifications of AFC [10].

Example 13.2 Early AFC development and specifications

		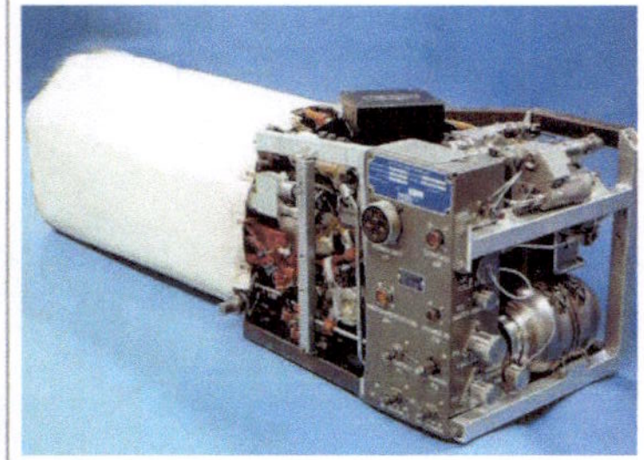

(continued)

(continued)

Example 13.2 Early AFC development and specifications

F.T. Bacon constructed the first 5 kW AFC stack in 1952. It operated on H_2 and O_2 at 200–240 °C with 30–45% KOH electrolyte. Pressure was maintained at 40–55 atm to prevent the electrolyte from boiling. At this high temperature and pressure, a good performance of 0.78 V at 800 mA/cm^2 was achieved. The anode consisted of a dual-porosity Ni electrode (two-layer structure with porous Ni of 16 μm maximum pore diameter on the electrolyte side and 30 μm pore diameter on the gas side). The cathode consisted of a porous structure of lithiated NiO. In 1960s, National Aeronautics and Space Administration (NASA, US) selected alkali fuel cells for the Space Shuttle and Apollo programs.	Pratt & Whitney Fuel Cell built the alkaline fuel cell for Apollo. The pressure was decreased from 40 to 55 atm to less than 4 atm by using concentrated KOH of 75–85%, while the temperature was further increased to 260 °C. The high temperature activates the nickel-based catalysts, which enable a cell performance of 0.87 V at 160 mA/cm^2. The Apollo fuel cell consisted of three modules connected in parallel. It weighed 95 kg and generated an average power of 1.5 kW with a peak power of 2.3 kW at 20.5 V.	United Technologies Corp. built the alkaline fuel cell for the Space Shuttle Orbiter. The Orbiter cell used 32% KOH electrolyte immobilized in an asbestos matrix. It operated with H_2 and O_2 at 4 atm and 85–95 °C. The anode was made of 10 mg/cm^2 catalysts consisting of 80% Pt–20% Pd. The cathode contained 90% Au–10% Pt catalysts at a loading of 20 mg/cm^2. The catalyst layers were made with PTFE as binder and applied on Ag-plated Ni screen substrates. The Orbiter fuel cell stack weighed 118 kg and generated an average power of 7 kW with a peak power of 12 kW at 27.5 V. The single performance reached 0.88 V at current density of 470 mA/cm^2.

Advancement was made in 1960s when polytetrafluoroethylene (PTFE) was introduced to use as the binder of the catalyst powders. Due to the high electronegativity of fluorine, the polymer exhibits diminished London dispersion forces and therefore is hydrophobic, i.e., water or water-containing liquids do not wet its surface.

Electrodes made with PTFE in varied contents may have both hydrophilic and hydrophobic characteristics. The change of hydrophobicity of the catalysts radically changes the liquid distribution in the porous layer. For a hydrophilic porous phase, e.g., carbon-supported catalysts with a low content of PTFE, the aqueous electrolyte preferably wets the small pores leaving the large pore open for gases to access. For a hydrophobic phase with a high content of PTFE, the liquid electrolyte rather stays in large pores.

Figure 13.4 illustrates the wetting of hydrophilic and hydrophobic pores of electrodes by aqueous electrolyte solutions. In the very first demonstration fuel cell with liquid KOH electrolyte, T. Bacon prepared the anode from nickel and cathode from lithiated NiO powders. The electrodes are of hydrophilic structure by sintering the catalyst particles. The particles are of varied sizes, so that the gas diffusion layer have pores of about 30 μm size, while the catalyst active layers have pores of about 16 μm size. The electrolyte preferably fills the small pores and leaves the large pores dry and open to gaseous reactants. The three-phase boundary in the porous electrodes is maintained by a differential gas pressure across the electrode.

The catalyst sites that are not in contact with liquid electrolyte are dry and inactive, while those sites that are completely flooded will only be active where the dissolved gas reactant is available. Only the catalyst sites that are in direct contact with both the liquid electrolyte and gas reactant are active for the electrode reaction. This is schematically shown in Fig. 13.5.

When the electrolyte is a liquid solution, the gas diffusion electrode should have another function to confine the electrolyte without flooding all the catalyst sites. In this way, the electrodes provide a large surface area of the catalysts with minimum mass transport resistance for the reactant access and product removal. The concept of

Fig. 13.4 Illustration of hydrophobicity, size of micropores and distribution of the aqueous electrolyte

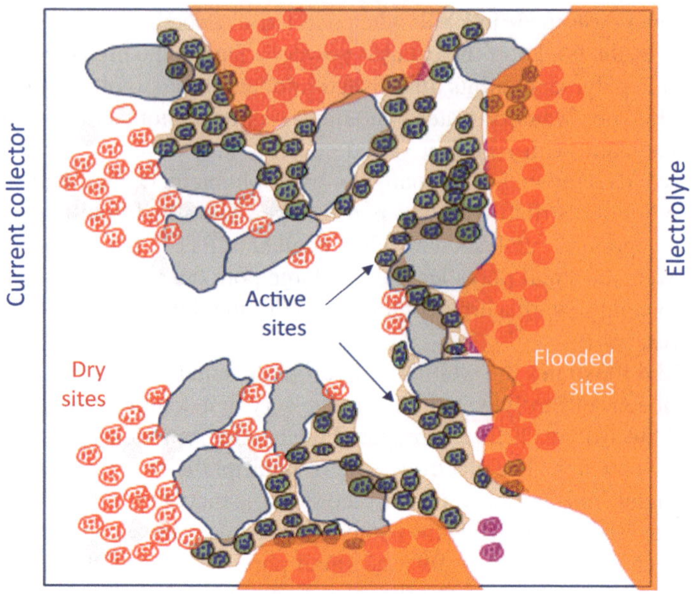

Fig. 13.5 Schematic representation of a PTFE-bonded catalyst player

the two-layer structure—a gas diffusion layer and a catalyst layer, is later developed by using PTFE when carbon-supported catalyst particles are used. PTFE is a binder to the powders and also tailors the hydrophobicity of the layers. The gas diffusion layer is often made of a high content of PTFE (20–50%), while the active layer is made on carbon-supported catalyst powders with a low content (5–15%) of PTFE.

PTFE is often supplied as an aqueous suspension containing proprietary surfactants. The electrodes with PTFE as binder should be treated at temperatures around 350 and 360 °C when the crystalline form of PTFE melts. This so-called sintering process makes powders bonded in order to give the mechanical integrity of the electrode structure. By tailoring the PTFE content as well as fabrication process, electronic conductivity, ionic transport, and gas transport properties are optimized.

In the case of carbon black-supported platinum catalysts, the platinum particles are in general in a size of 2–5 nm, the carbon powder support is in clusters of carbon particles of 30–80 nm size, while PTFE is of dispersed polymer particles in size of a few microns.

13.2.4 Electrolyte Configuration

The liquid KOH electrolyte is configured in fuel cells in two ways, immobilized or circulating electrolyte. The immobilized electrolyte fuel cells operate with pure hydrogen and oxygen with high power density, primarily used in space programs, while the circulating electrolyte fuel cells may use scrubbed air, mainly for terrestrial applications. The two types of fuel cells are compared in Table 13.1.

13.2.4.1 Circulating Electrolyte

The circulation of liquid electrolyte can be achieved by control of the electrolyte level in the storage tank underneath a stack. With circulating, the liquid electrolyte is contained between two electrode layers and separates the air at the cathode and hydrogen at the anode. The presence of the electrolyte layer between the two electrodes is essential with a constant pressure, which is slightly higher than any of the gas pressures. For the purpose of circulation of a liquid electrolyte, a large inter-electrode gap (2–3 mm) is needed, which means a large ohmic resistance of the electrolyte. To compensate this, the high conductivity KOH solution of concentrations ranging from 6 to 12 M KOH is always used.

The electrolyte is stored in a tank beside the fuel cell and pumped through. During operation, water is formed in liquid form on the anode side and removed with circulation of the electrolyte. The temperature of this type of cell configuration is hence limited to a range of below 70 °C for effective removal of liquid water. In the storage tank of electrolyte, water is evaporated and the concentration of the KOH solution can be monitored by the level. This is an easy option as there is no chance for the electrode to dry out.

The circulating electrolyte enables simple cooling and possible heat recovery with help of an external heat exchanger. The special cooling plates or loops for a cooling fluid are not required. The electrolyte flow serves also to eliminate any developed concentration gradient of the electrolyte and to remove any contamination of the electrolyte, e.g., carbon from electrodes or carbonates. The CO_2 scrubbing is therefore a less issue and sufficient by means of soda lime absorbing to achieve a lifetime of 4000 h [12].

At the same time, the electrolyte facilitates the cooling of the stack. Any concentration gradient due to the water consumption at the cathode and production at the anode will be eliminated. Of course, the expense is the need of a pump and a storage tank of electrolyte. Moreover the build-up of gas bubbles in the gap between the electrodes is also minimized. It should also be considered a plus that the cost of the KOH electrolyte is not so high and frequent replacement with fresh electrolyte, for every a few hundred operation hours, to increase overall lifetime.

Challenging problems for the circulating electrolyte option are the electrolyte leakage and parasitic losses due to the fact that each cell is linked by the KOH circulation loop (leading to shunt current), which consists of a KOH tank, a KOH pump, and a heat exchanger.

Table 13.1 Two types of electrolyte configurations of AFC with construction and operation characteristics [11]

	Circulating electrolyte	Immobilized porous matrix
	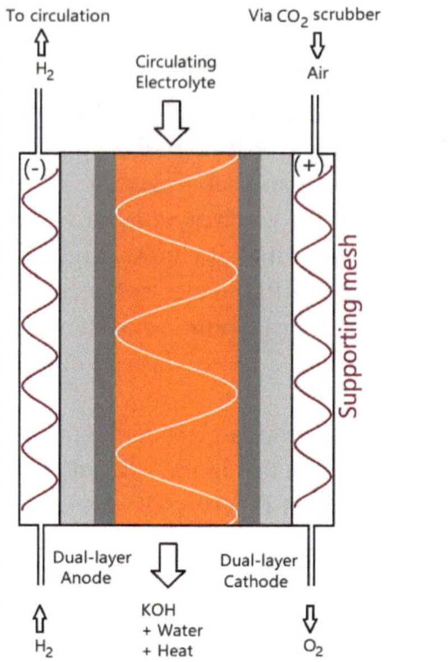	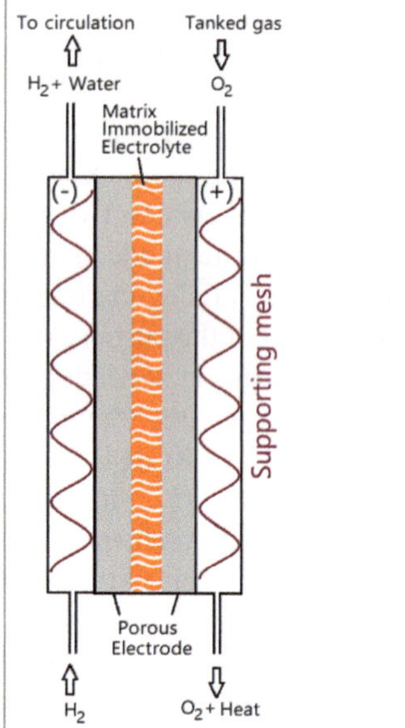
KOH solution	– Large electrolyte thickness – Periodic replacement though cost is low	+ Small electrolyte thickness + Simple construction
Cell construction	Dual–layer electrode	Single-layer and porous electrode
System issue	– KOH tank, pump, and heat exchanger – Liquid leakage issues – Shunt current due to liquid connection	+ Simple (no pumping no flowing) + Lightweight
H_2O management	+ Removal as liquid with electrolyte – Monitoring electrolyte concentration	– Removal as vapor by hydrogen loop – Varying with loads
Heat management	+ Effective via heat exchanger + Heating KOH for startup + Possible heat recovery	– Additional cooling loop at cathode
CO_2 poisoning	+ Less sensitive as carbonate removed by electrolyte + Possible use of air with CO_2 scrubber	– Sensitive to CO_2 and with no carbonate removal
Air supply and release	+ Use of air via CO_2 scrubber + Outlet air directly released	– High-pressure tanked pure H_2 and O_2
Hydrogen	– Recirculated (dead ended) + No gas humidification system	– Recirculated – Dew point control (humidified hydrogen)

(continued)

Table 13.1 (continued)

	Circulating electrolyte	Immobilized porous matrix
Gas bubbling	– Gas bubbling issue	+ No

Note plus signs (+) indicate advantages and minus signs (−) indicate disadvantages

13.2.4.2 Immobilized Electrolyte

Early development of AFC was aimed at aerospace applications using electrolyte immobilized within a matrix. Compared to the circulating electrolyte option, this is a more compact design, as there is no electrolyte loop where pumps, valves, tanks, or regulators are needed to handle circulation of the hot caustic liquid. The technology is well developed with good reliability—this is demonstrated by the fact that the Space Shuttle Orbiter was only equipped with an AFC system with no other backup electric power.

A matrix is an essential material for the immobilization of electrolyte. It soaks the liquid electrolyte and conducts the OH^- at the same time separates gases of the anode and cathode chambers. Traditionally the matrix is made of chrysotile asbestos, an important type of naturally occurring silicate minerals with a composition of $Mg_3Si_2O_5(OH)_4$. The mineral is thermally stable up to around 550 °C and then starts to dehydrate. Chemically, it is attacked by acids but resistant to strong bases like KOH at low temperatures. At above 120 °C, the silica component dissolves forming potassium silicates. Chrysotile can easily be crumbled into bundles of fibrils in length of millimeters to centimeters. The individual fibrils, in diameter of as small as 20–30 nm, are mechanically strong and can be made into thin sheet. The fibrils form very fine pores, which strongly retain the hydroxide electrolyte through capillary forces. Thus, made matrix can withstand a certain gas pressure without losing the liquid.

The matrix immobilized electrolyte can be in thin layers, and hence, the small ohmic resistance and the absence of free liquid make the cell assembling and stacking easier. This kind of electrolyte configuration makes the gas diffusion electrodes also easier — they do not need an additional barrier layer like the one with circulating liquid electrolyte. As the electrolyte is immobilized, the product water has to be removed in the vapor form with the hydrogen steam. For doing this, the operational temperature of the cells is slightly higher, 90–110 °C. And the hydrogen as purging gas should be circulated after removal of the carried water in the condenser. Another issue is that an additional cooling loop is needed in association with the cathode.

One drawback is the poor tolerance of the immobilized electrolyte to carbon dioxide contamination. This type of AFC optionally operates with pure oxygen instead of air—it was not an issue for the early year aerospace applications. The target lifetime of the Orbiter program was 2600 h for each flight and fulfilled during more than ten flights. After each flight, the AFC stack is simply replaced with a new one. Longer lifetime of more than 5000 h has also been reported. Typical performances of the two types of AFC are shown in Fig. 13.6.

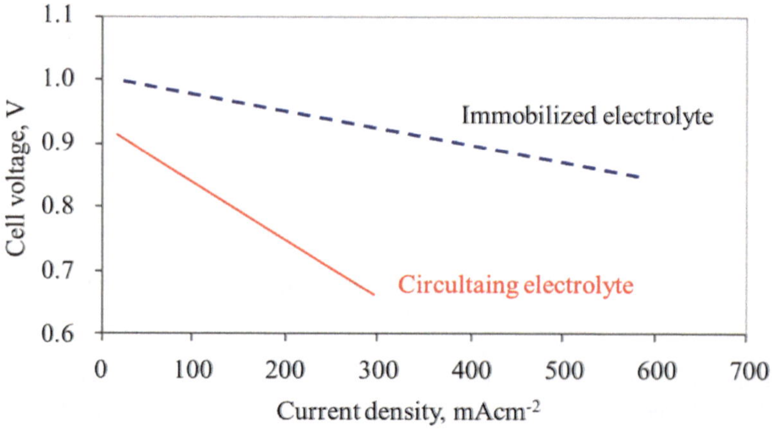

Fig. 13.6 Typical performance of two types of AFC; the immobilized electrolyte cell operated with 4 atm H_2/O_2 at 95 °C using 35% KOH and the circulating electrolyte cell operated with ambient pressure H_2/air at 75 °C with 46% KOH

13.2.5 Stack Design

There are two basic designs of AFC stacks using monopolar and bipolar plates. Using monopolar design the current is collected at the end side of electrodes which are shared by two adjacent cells and connected by external circuits (See Fig. 13.7). For each electrode, the current passes in the in-plane direction and good conductivity is required. In case of carbon electrode, metal screen or felt is used, and operation at high current densities is always an issue. To achieve a good current distribution, the electrode is limited to a certain size, e.g., smaller than 40 cm^2. An advantage is that, in case of malfunction of one cell, it can be simply disconnected without major interfering of the stack performance.

A series connection between many cells can be achieved using bipolar plates. Bipolar cells are not limited in the electrode size, because the current flows perpendicular to and distributes over the electrode surface. Electrodes made of carbon-supported catalysts can be directly used as their conductivity is not a critical issue. Of course when one cell fails, the entire stack of such a design will not function.

13.2.6 Carbonization and CO_2 Scrubbing

13.2.6.1 CO_2 Effect

Chemically KOH reacts with carbon dioxide when it is present in the fuel steam or air:

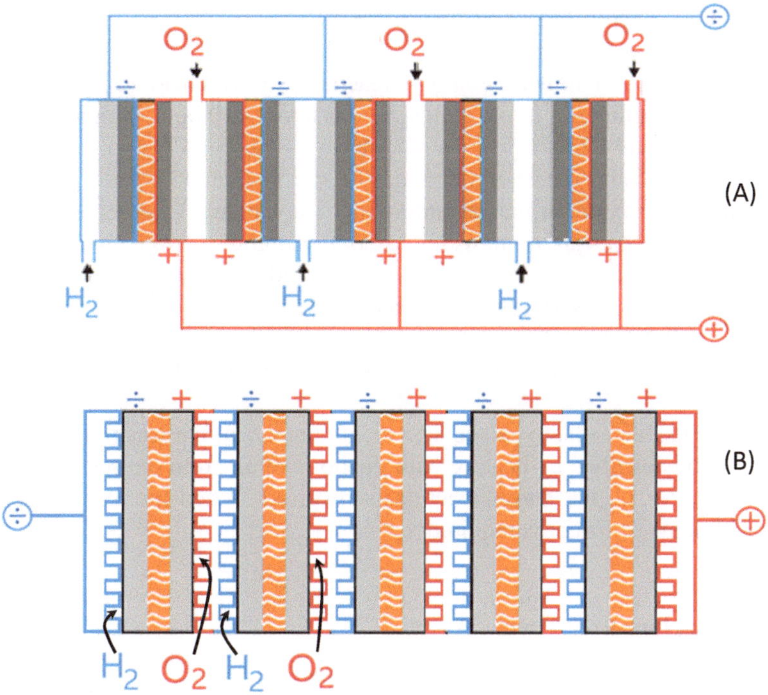

Fig. 13.7 Scheme of AFC stack design in (**a**) monopolar and (**b**) bipolar configurations

$$CO_2 + 2KOH = K_2CO_3 + H_2O \tag{13.5}$$

The formation of potassium carbonate in electrolyte has severe consequences. First of all, the carbonate has a limited solubility, though higher in KOH than in NaOH, and will precipitate as solid. At higher contents, the carbonate may lead to co-precipitation with the hydroxide, forming a solid phase of $(KOH)_2{\cdot}(K_2CO_3)_3$. The solid precipitates fill up the pores on the catalyst layer and decrease the effective active area of the electrode. The carbonization reduces the OH^- concentration and increase the viscosity of the electrolyte, both affecting the electrolyte conductivity and electrode kinetics. In addition, formation of carbonate increases the water vapor pressure. If water management is not accordingly adjusted, this may result in water loss and therefore the change of the electrolyte volume. Any drying out or wetting of the electrolyte will lead to loss of the catalytic active sites or even the crossover of gases.

These effects are more severe with immobilized electrolytes than with circulating electrolytes. Using high concentration KOH electrolyte and operation at elevated temperatures can mitigate the effects to some extent because of the increased solubility of carbonate in the electrolyte.

13.2.6.2 CO_2 Scrubbing and Carbon-Free Fuels

Careful purification of air is necessary to reach a level well below 50 ppm CO_2. The current CO_2 concentration in the atmosphere is about 410 ppm, which is rising at a rate of approximately 2 ppm/year. In order to minimize the CO_2 poisoning effect, a number of methods have been developed. The simple method is absorption by soda lime, lithium hydroxide, or magnesium oxide in a canister. Regenerative absorbers using molecular sieves are more widely used. For doing so two identical containers are filled with amine-based adsorbing materials, and one operates as an absorber of CO_2 in the inlet air. Absorption takes place below 40 °C:

$$R - \text{amine} + H_2O + CO_2 \rightarrow R - \text{amine} - H_2CO_3 \quad (13.5)$$

When the usable capacity of the adsorbent is reached, the weakly bound carbon dioxide can be desorbed by heating with the hot exit air from the stack. At temperatures above 70 °C, the above reaction is reversed. When the exit air is at low temperatures, an air heater is required.

13.2.6.3 Carbon-Free Fuels

CO_2 removal is necessary when H_2 is produced from carbon-containing fuel sources, e.g., methane, methanol, gasoline, propane, and others. Hydrogen from carbon-free fuels is therefore an option for AFC. For example, hydrogen from ammonia cracking, consisting of 25% N_2, 75% H_2 plus residual NH_3, can be fed directly to the cell. The diluting effect of nitrogen has a very small effect on the performance compared to that with pure hydrogen, of course at low to medium current densities. Other non-carbon hydrogen carriers include sodium borohydride ($NaBH_4$) and hydrazine (N_2H_4).

13.2.7 Performance and Durability

Owing to a wide choice of catalyst materials and a great variation of operational temperatures from room to 260 °C and pressures from ambient to 40 atm, the AFC performance changes a great deal. The AFC performances should be compared with literature data with a complete set of materials information and operating parameters.

AFC often operates at elevated pressures to increase the cell performance. Pressurization has only a moderate effect on the cell performance. The voltage gain at a certain current density is a semi-logarithm function of the cell pressure, by following the Nernst equation. With circulating electrolyte, the voltage gain due to increase of the pressure is actually even less because the gas solubility of hydrogen and oxygen is significantly increased, which causes higher parasitic current loss through the electrolyte.

At around 100 °C, an increase in pressure from 1 to 10 atm leads to a cell voltage improvement by 50–60 mV. This performance gain can be realized in systems where compressed gases are available such as in space crafts or submarines. In other cases where compressors are used, higher pressures than 5 atm would need a second unit. Compressors are noisy and the parasitic power consumption as well as increased weight and volume would usually outweigh the performance gain. High pressures have been often used in association with the temperature elevation which means better kinetics and lower ohmic loss with non-noble metal catalysts, e.g., Ni or oxides.

Alkaline fuel cells have a low activation energy for electrode reactions, which means that the cell performance has a relatively small temperature coefficient. In other words, AFC exhibits the high power output at or below ambient temperature. From room temperature up to about 70 °C, the voltage of an H_2/air cell with noble metal catalysts may increase at a rate of 2-3 mV/°C. Compared to the power output at room temperature, an AFC operating at 70 °C may have a power output boost by a factor of 2. At higher temperatures, this effect is much smaller, about 0.3–0.5 mV/ °C in a current density range of 50–150 mA/cm^2.

For long-term operation, catalysts degrade via oxidation of carbon, dissolution, and intercalation of metals. A higher performance degradation rate ranging from 10 to 30 μV/h is often observed for the cathode based on Pt- and Ag-catalysts, compared to a range of 5–10 μV/h for anode catalysts when pure hydrogen is used. This includes the effect of PTFE degradation after long time exposure to KOH which causes changes of hydrophobicity, leading to electrode flooding and hence reduced accessibility of reactants to reacting sites. Traces of CO_2 in the scrubbed air as well as the hydrogen stream, within a tolerance range of 5–30 ppm, may also cause changes in the conductivity, viscosity, and surface tension of the electrolyte and electrodes.

13.3 Anion Exchange Membrane Fuel Cells

The system complexity of AFC such as the electrolyte management and poor CO_2 tolerance is associated with the liquid hydroxide and can be decisively simplified by using anion exchange membrane electrolytes. An anion exchange membrane (AEM) is composed of a polymer backbone with tethered cationic sites which allow for the passage of anions by ionic exchange. The essentially solid membrane eliminates the necessity to handle the liquid and corrosive electrolyte and avoids weeping or seeping out of liquid KOH solution. In addition, the lack of mobile cations, e.g., K^+ in AEM improves the CO_2 tolerance of the electrolyte as there is no precipitation of crystalline carbonates, though the carbonation of the immobilized functional groups still leads to decreases in the membrane conductivity.

13.3.1 AEM Electrolyte

For an ideal AEM electrolyte, following properties are required for the good performance under fuel cell operating conditions

- sufficient ion-exchange capacity (IEC) with optimal water uptake in order to achieve high hydroxide conductivity above 10^{-2} S/cm;
- high thermal and chemical stability at temperatures above 80°C and high pH;
- sufficient mechanical strength and stability allowing for small thickness ($\leq 10\ \mu m$) membranes;
- available in form of ionomer solution or dispersion for catalyst layer fabrication.

The greatest challenge for AEMs is the chemical stability at temperatures above 80 °C and under high pH. Significant efforts in recent years have focused on improving the chemical stability of both functional groups and polymer backbones. The most well-studied functional cationic group is of type of quaternary ammonium (QA), e.g., benzyltrimethylammonium, which has shown reasonable stability under the fuel cell operational conditions. Other types of cationic groups include nitrogen-based cyclic cations, e.g., imidazolium, guanidinium, phosphonium, and tertiary sulfonium. Commonly used cationic head-groups are exemplified in Fig. 13.8.

The AEM stability is not always well correlated with cationic group as polymer backbones also degrade particularly via the linkages between polymer backbones and the incorporated cationic groups. A variety of polymer structures such as polyolefins (e.g., polyethylene, polypropylene, polystyrene) and polyaromatics (e.g.,

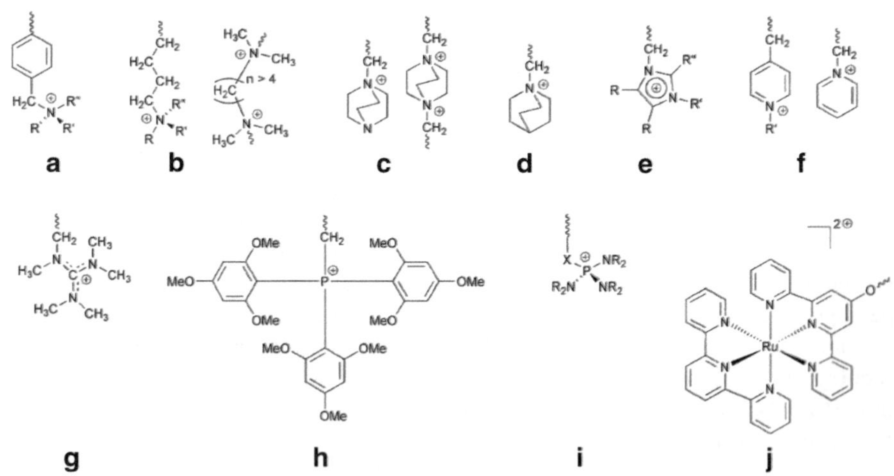

Fig. 13.8 Commonly used cationic head-groups of AEMs. **a** benzyltrialkylammonium groups; **b** alkyl-side-chain QA groups; **c** bicyclic ammonium groups; **d** quinuclidinium-based QA groups; **e** imidazolium groups; **f** pyridinium groups; **g** pentamethylguanidinium groups; **h** alkali-stabilized quaternary phosphonium groups; **i** phosphonium groups; and **j** metal-containing groups. Reproduced from Ref. [13] with permission from Royal Society of Chemistry, Copyright 2014

polysulfones, poly(arylene ether ketones), poly(phenylene oxides)) have been widely employed. These non-ionic polymers have relatively good stability against hydrolysis under alkaline conditions. The quaternary ammonium is an electron-deficient moiety, which enhances the rate of hydrolysis of the polymer backbones. In addition, the aryl ether linkages (C–O–C bonds) on the polyaromatic backbones are always the weak point for cleavage. In this context, polystyrene has been considered as a potential candidate due to its ether-free polymer backbone structure. Other polymers that are stable under high-pH conditions include perfluorinated polymers with PTFE backbones and polybenzimidazoles.

Aromatic polymers are the most widely used backbone materials, onto which QA groups can readily be introduced through chloromethylation, followed by amination with a tertiary alkyl or aryl amine to give the chloride salt of the ionomer. In the case where phenyl groups are not present in the polymer chain, vinylbenzyl chloride can be radiation grafted onto the polymer chain. Among university groups, important developments have been made from Varcoe and coworkers from University of Surrey, who prepared AEMs by radiation grafting of vinylbenzyl chloride onto completely or partially fluorinated polymers. A radiation-grafted poly(ethylene-co-tetrafluoroethylene) (PETFE)-based AEM such prepared has demonstrated good chemical stability in 1 M KOH up to 80 °C and high ionic conductivity when fully hydrated (~0.06 S/cm at 60 °C, compared to > 0.1 S/cm for Nafion PEMs) [14].

In both cases, the chloromethyl groups need to be converted into the hydroxide form by treatment with a dilute aqueous hydroxide (KOH) solution to obtain a hydroxide ion-conducting AEM as shown in Fig. 13.9.

Both protons and hydroxides conduct by breaking and forming hydrogen bonds through extensive networks. As seen from Fig. 5.3 in Chap. 5, the inherent ionic mobility of OH^- is about $2 \times 10^{-7} m^2/V{\cdot}s$ compared to $3.6 \times 10^{-7} m^2/V{\cdot}s$ for H^+ or more precisely hydrated proton, H_3O^+ in a diluted aqueous solution at room temperature. The specific conductivity of hydroxide ions is hence half of the proton conductivity. The PEM has perfluorinated backbones which enables a strong dissociation of the sulfonic acid groups as well as a hydrophobic-hydrophilic phase separation and therefore a higher degree of aqueous nano-phase agglomeration around the sulfonic

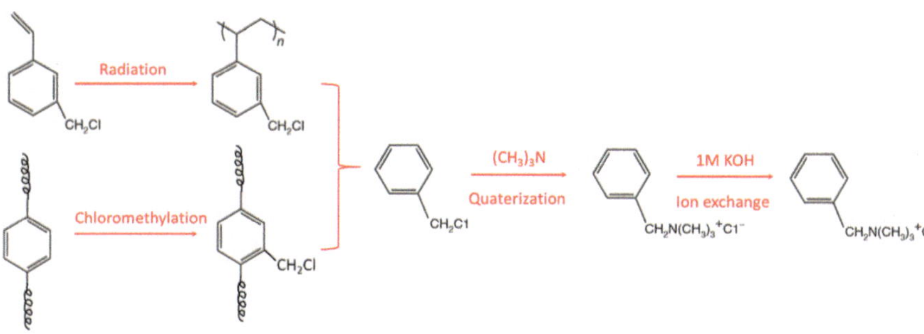

Fig. 13.9 Generic chemical routes for synthesis of AEM polymer with quaternary ammonium as cationic end groups

acid, both promoting the proton conductivity. In the case of AEM, however, the anion exchange groups are less dissociated and the formation of a sub-phase of solvated ions requires a higher water content. At a given IEC, the water uptake of PEMs and AEMs is similar while the hydroxide conductivity of AEMs is 2–8 times lower than the proton conductivity of PEMs, as seen from Fig. 13.10. Typically, the AEM membranes with an IEC of 1.5–2.0 meq/g have a conductivity of 30–40 mS/cm at room temperature under fully hydrated conditions. The largely scattered values of the reported conductivities in the figure are apparently due to a wide range of the covered materials as well as non-standardized testing conditions. This difference is larger at low relative humidities but smaller under fully hydrated state for both PEM and AEM of similar IEC. In general, an IEC > 2.0 meq/g results in excessive membrane swelling and poor mechanical strength of the membranes.

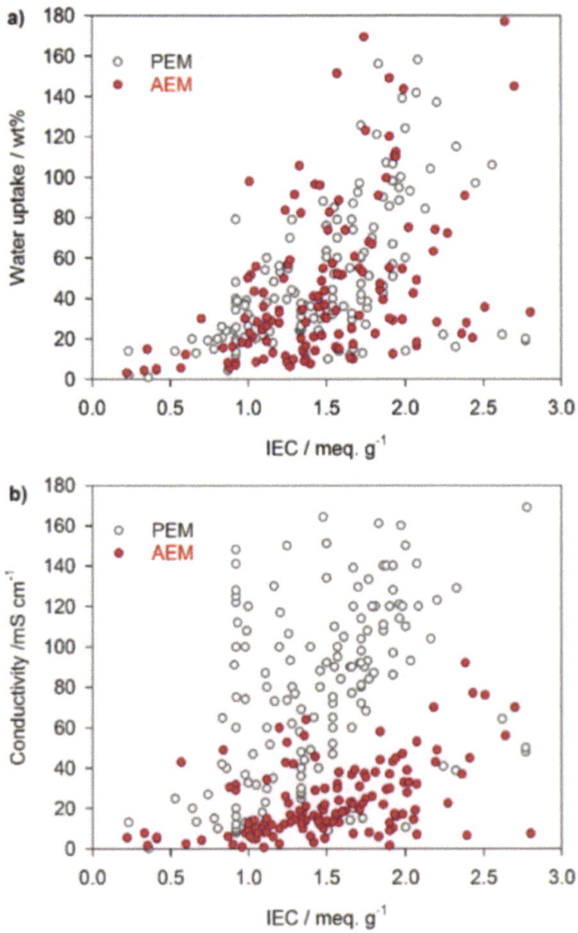

Fig. 13.10 Water uptake (**a**) and ionic conductivity (**b**) of proton exchange membrane (PEM) and anion exchange membrane (AEM) as a function of ion-exchange capacity (IEC) at 30 °C. Reproduced from Ref. [14] with permission from Elsevier, Copyright 2018

13.3.2 Catalyst Development

From an electrocatalyst point of view, fuel cell tests focusing on the AEM evaluation are conducted with Pt/C catalysts. One significant advantage of operating the fuel cell in AEM is the potential to eliminate the use of PGM catalysts such as Pt and enable the application of non-PGM materials as the catalysts. Due to the difficulties of working with liquid electrolytes and carbonate formation and precipitation, complex balance of plant such as electrolyte circulation and CO_2 scrubbers typically associated with AFCs, the development of NPGM metal or metal-free catalysts is more specifically toward AEMFCs [15].

13.3.2.1 Oxygen Reaction Catalysts

The alkaline media are less challenging for alternative ORR catalysts. Precious metal-based catalysts are the most active, while non-precious metals and their oxides have also been widely used in AEMFCs. In addition, nitrogen-coordinated transition metal catalysts as well as metal-free carbon-based catalysts are also under active development.

From an electrocatalyst point of view, fuel cell tests focusing on membrane evaluation are conducted with Pt/C catalysts. When hydrogen is used as fuel, Pt catalysts have the best performance. When liquid alcohols, e.g., methanol or ethanol, are used as fuel, the crossover of the liquid fuel poisons the platinum on the cathode side. The PtRu alloy is an ORR catalysts with enhanced tolerance to the alcohol though the alloy itself has a relatively lower activity than that of the Pt/C. Palladium and its alloy catalysts have also been shown to be less sensitive to the alcohol contamination. Ag and its alloys in form of carbon-supported nanoparticles are the most widely used catalyst for AEMFCs with a good combination of high activity, good long-term stability, and low price. In addition, transition metals in their oxide forms, e.g., Co_3O_4 and MnO_x, have also been recognized as potential ORR catalysts due to their electrical conductivity and stability in alkaline media.

Nitrogen-coordinated $M\text{-}N_x$-based catalysts have been generally recognized as the ORR catalysts in both acidic and alkaline media. The metal atoms are typically Fe, Co, and Mn, among which the $Fe\text{-}N_x$ has shown the best performance. In this connection, single atom catalysts (SACs), e.g., Fe single atoms coordinated with the doping nitrogen in the graphitic carbon nitride (C_3N_4) networks, have been prepared with a high metal loading. The catalysts showed high ORR activity, stability, and methanol tolerance in alkaline media [16].

Carbon atoms in graphene and CNTs have little catalytic activities as they are chemically sp^2 hybridized. Doping by heteroatoms such as N, B, S, F, and P can substantially enhance the activity [17]. The doping-induced charge transfer from carbon atoms to the adjacent nitrogen atoms strengthens the chemisorption of O_2 and weakens the O-O bond. In N-doped carbon, the doped nitrogen in the graphitic lattice can exist as pyridinic, pyrrolic, graphitic and quaternary nitrogen (See Sect. 5.4.2.3).

13.3.2.2 Anode Catalysts

Hydrogen oxidation reaction (HOR) on the Pt electrode surface is kinetically a sluggish process in alkaline media and is a major source of cell overpotential loss in AEMFC. The fundamental reason for the sluggish kinetics of HOR in alkaline electrolyte is the involvement of hydroxide ions in the formation of water, as compared to the simple and kinetically rapid proton formation process of HOR in acidic media. Due to the limited electrocatalytic activities of pristine Pt for HOR in alkaline media, there have been extensive studies on the bimetallic or composite Pt-based alloy catalysts such as PtRu, PtCu, PtNb, and PtNi. In general, these alloy catalysts show a much better HOR activity as compared to that of pure Pt. As the platinum-free anode catalyst, Pd-based electrocatalysts supported on either carbon or cerium oxides have been reported to have excellent performance as anode catalysts for AEMFCs. It is proposed that the presence of intimate contact between CeO_2 and Pd enhances the OH^- transfer at the electrode interface. Other HOR catalysts reported for AEMFCs include NiCoMo ternary alloys and nitrogen doped carbon nanotubes, which have an activity that is still about 20 times lower than that of Pt.

Great efforts are being made to develop direct alcohol fuel cells. For direct alcohol oxidation in alkaline electrolyte, the reaction rate in term of current density at an operating cell potential is an order of magnitude greater than in acidic electrolytes. In addition the chemisorption of intermediates, e.g., CO_{ads} species on the catalyst surface is weaker than in acidic media. As a result, in addition to the noble metal-based materials such as Pt alloys with Ru, Mo, Sn as well as their composites with oxide additives, a variety of platinum-free catalysts have been developed including metals and alloys of Pd, Ni, Co as well as perovskite-type oxides [18].

13.3.3 Performance

13.3.3.1 Performance of Hydrogen Cells

Commercial AEMs are currently available from Solvay (a cross-linked fluorinated polymer with QA groups), Fumatech (FAA, a perfluorosulfonic polymer), Tokuyama (A201 and A901, QA-containing polyolefinic(aliphatic)-type), and others. A large number of fuel cell tests have been conducted and great progress made in the last decade. Aiming at evaluation of AEM electrolyte materials, Pt is used as the cathode catalyst and Pt or PtRu as the anode catalyst, often operating with pure hydrogen and oxygen, the fuel cell performance has reached a current density of 450–620 mA/cm^2 at cell voltage of 0.8 V or a peak power density over 1 W/cm^2 [19].

More cell testing work is carried out with Pt-free cathode catalysts as well as Pt-free anode catalysts. A recent comparison study is based on using a Pd/CeO_2 anode catalyst prepared by flame spray pyrolysis and a Co_3O_4/C cathode catalyst prepared by hydrothermal synthesis. The H_2/O_2 AEMFC performance with a peak power of 309 mW/cm^2 at 70 °C, which is compared with 388 mW/cm^2 when the anode is

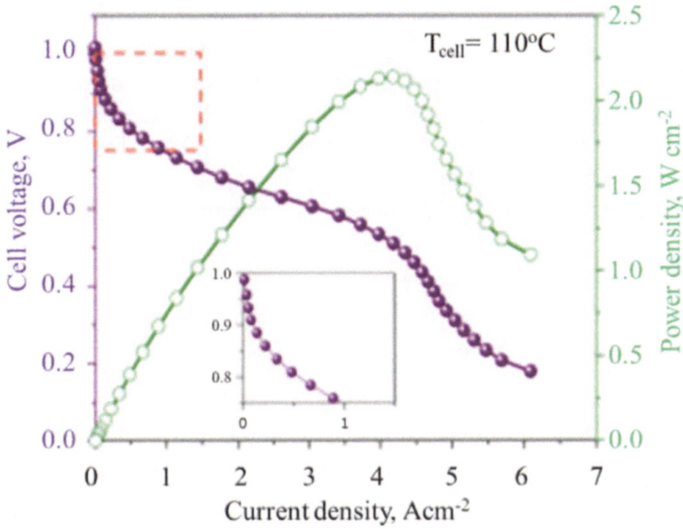

Fig. 13.11 Polarization curve of a high-temperature AEMFC operating at 110 °C under H_2/O_2 at 100% RH and 1.5 Barg back-pressurization on both anode and cathode. Reproduced from Ref. [21] under Creative Commons CC BY license

replaced by Pt/C catalysts and 590 mW/cm^2 when both electrodes are replaced with Pt/C catalysts [20].

So far most of the published performance and stability of AEMFCs have been done at temperatures lower than 60–70 °C and a few at 80 °C. A high-temperature AEMFCs operating at 110 °C with H_2 and O_2 of 1.4 Barg is recently reported using a radiation-grafted anion exchange membrane containing covalently-bonded benzyltrimethylammonium, PtRu/C anode, and Pt/C cathode. The fuel cell exhibits a current density of 574 mA/cm^2 at 0.8 V and a maximum current density of 6 A/cm^2 at 0.4 V peak power density of 2.1 W/cm^2, as shown in Fig. 13.11. A slow degradation of the performance is observed in the following 50 h at a constant current density of 0.2 A/cm^2.

13.3.3.2 Performance in Direct Alcohol Cells

An important issue related to traditional AFC is the carbonate poisoning effect by CO_2 in the feed stream. When mobile cations (K^+ or Na^+) are present in the hydroxide electrolyte, the carbonate precipitates as solid salts in the porous electrode. It is therefore that no effort has been made to develop direct alcohol fuel cells with alkaline electrolytes until recently when the AEM is available. AEMs as solid polymer electrolytes do not contain any mobile cations and hence can significantly mitigate the carbonate precipitation. It should be noted that the presence of CO_2 from the anode reaction will react with the quaternary ammonium group and convert hydroxide anions to bicarbonate or/and carbonate anions. The anodic formation of CO_2 leads to decrease in the anion conductivity of the AEM. Anyway the development of direct alcohol fuel cells using AEMs has received increasing attention. Despite the perceived advantages of the faster kinetics of the ORR and alcohol oxidation reactions in alkaline media, the power density of alkaline AEM fuel cells based on direct alcohol fuels is very low (~20 mW/cm^2) [15, 22], apparently due to the poor catalyst utilization, fuel crossover, and low anion conductivity caused by the carbonate formation.

13.4 Summary

In this chapter, we have introduced and discussed the following topics:

- The AFC technology using KOH electrolyte has been demonstrated for all possible applications. Potassium hydroxide is selected as liquid KOH electrolyte due to its high conductivity which reaches 1.4 S/cm for 32 wt% at 80 °C. For operation at above 200 °C, molten hydroxide in a concentration range of 75–85% is used in order to reduce the vapor pressure of the electrolyte.
- The activation losses of electrode reactions are lower in alkaline medium. Noble metals (Pt, Pd, Au, Ag) are the super active ORR catalysts, however, often used at low temperatures. Nickel oxide, on the other hand, has very limited ORR activity at low temperatures but are stable, conductive, and sufficiently active at above 200 °C. The performance can be enhanced in form of finely dispersed Raney and their stability improved by lithiation. More challenging are the HOR catalysts which are based on noble metal alloys or Ni Raney metal.
- Gas diffusion electrodes for AFC were constructed with varied pore sizes and bonded using PTFE to achieve desirable hydrophobic/hydrophilic characteristics.
- The liquid electrolyte is configured in immobilized or circulating modes. Advantages and disadvantages of each mode are discussed in terms of materials, performance, and system perspectives.
- One issue of AFCs is that they are very sensitive to CO_2 from either the fuel stream or ambient air, leading to precipitation of solid carbonates and hence degradation of performance. The use of pure hydrogen and oxygen or scrubbed air has limited the application of AFC for decades when an abundant hydrogen source from hydrocarbon reforming was considered for large scale power generation.
- Recent development of stable anion exchange membrane, an analog to the proton exchange membrane in thriving PEMFCs, has led to the renaissance of the technology. Advances in materials science and chemistry have enabled production of membrane and ionomer materials as well as non-noble metal catalysts. Further challenges to address are enhancement in the membrane conductivity and long-term durability.

13.5 Questions

13.1 Explain the following terms and materials:

Lithiation and lithiated nickel oxide
Raney and Raney-type metals
Monopoplar and bipolar configurations
CO_2 scrubber
Circulating electrolyte and immobilized electrolyte
Anion exchange membrane

13.2 What are the advantages and disadvantages of alkaline electrolytes over acidic electrolytes?

13.3 Describe and comment the two modes of AFC stacking: matrix electrolyte and circulating electrolyte.

13.4 A hydrazine-oxygen fuel cell with alkaline electrolyte has an overall reaction (See Example 4.1)

$N_2H_4 + O_2 \rightarrow N_2 + 2H_2O \quad E^0 = +1.56\ V$

Hydrazine is considered as a fuel because it does not release CO_2, the greenhouse gas and source of the electrolyte carbonation. The hydrazine oxidation is not involving formation of the intermediate species, *e.g.*, CO, that poison the catalysts. Dr. Karl Kordesch in 1967 built a hydrazine fuel cell motorcycle and ran it for nearly 500 km. It was found that the open circuit voltage of the hydrazine cell is far lower than the thermodynamic value (1.56 V) and often close to a value of the hydrogen-oxygen fuel cell. What can be the reason for this phenomenon? (Clue: catalytic decomposition of hydrazine)

13.5 Similar idea has been proposed to use ammonia as a direct fuel in alkaline fuel cells. Any comments?

13.6 Direct alcohol fuel cells, for example, liquid methanol fuel cell, with alkaline electrolytes are an attractive option. The anodic oxidation product of CO_2 is a critical issue as it reacts with KOH forming solid crystalline carbonate. One solution has been suggested to use alkaline electrolytes containing soluble carbonates, e.g., concentrated solutions of Cs_2CO_3 instead of alkali metal hydroxide. What do you think about this idea? (Clue: Electrode kinetics and possibility of high-temperature operation)

13.7 A 1 kW AFC stack operates with circulation of 6 M KOH solution electrolyte. The stack works at an average single cell voltage of 0.8 V using ambient air containing 400 ppm CO_2 and at stoichiometry of $\lambda_{Air} = 2$. The CO_2 leads to loss of the electrolyte alkalinity. What is the consumption rate of KOH in mol KOH/h or L/h of the 6 M KOH solution? (Answer: 0.09 mol KOH/h or 0.015 L/h of the 6 M KOH solution)

References

1. Bacon FT (1969) Fuel cells, past, present and future. Electrochim Acta 14:569–585
2. Kordesch K (1983) The choice of low-temperature hydrogen fuel-cells—acidic or alkaline. Int J Hydrog Energy 8(9):709–714
3. Gilliam RJ, Graydon JW, Kirk DW, Thorpe SJ (2007) A review of specific conductivities of potassium hydroxide solutions for various concentrations and temperatures. Int J Hydrog Energy 32(3):359–364
4. Allebrod F, Chatzichristodoulou C, Mollerup PL, Mogensen MB (2012) Electrical conductivity measurements of aqueous and immobilized potassium hydroxide. Int J Hydrog Energy 37(21):16505–16514
5. Balej J (1985) Water-vapor partial pressures and water activities in potassium and sodium-hydroxide solutions over wide concentration and temperature ranges. Int J Hydrog Energy 10(4):233–243
6. Blurton KF, McMullin E (1969) Acomparison of fuel cell performance in acid and alkaline electrolyte. Energy Conv Manag 9:141–144
7. Davydova ES, Mukerjee S, Jaouen F, Dekel DR (2018) Electrocatalysts for hydrogen oxidation reaction in alkaline electrolytes. ACS Catalysis 8(7):6665–6690
8. Strmcnik D, Uchimura M, Wang C, Subbaraman R, Danilovic N, van der Vliet D, Paulikas AP, Stamenkovic VR, Markovic NM (2013) Improving the hydrogen oxidation reaction rate by promotion of hydroxyl adsorption. Nature Chem 5(4):300–306
9. Bidault F, Brett DJL, Middleton PH, Brandon NP (2009) Review of gas diffusion cathodes for alkaline fuel cells. J Power Sources 187(1):39–48
10. Warshay M, Prokopius PR (1990) The fuel-cell in space—Yesterday, today and tomorrow. J Power Sources 29(1–2):193–200
11. Strasser K (1990) The design of alkaline fuel-cells. J Power Sources 29(1–2):149–166
12. Kordesch K, Gsellmann J, Cifrain M, Voss S, Hacker V, Aronson RR, Fabjan C, Hejze T, Daniel-Ivad J (1999) Intermittent use of a low-cost alkaline fuel cell-hybrid system for electric vehicles. J Power Sources 80(1–2):190–197
13. Varcoe JR, Atanassov P, Dekel DR, Herring AM, Hickner MA, Kohl PA, Kucernak AR, Mustain WE, Nijmeijer K, Scott K, Xu TW, Zhuang L (2014) Anion-exchange membranes in electrochemical energy systems. Energy Environ Sci 7(10):3135–3191
14. Gottesfeld S, Dekel DR, Page M, Bae C, Yan YS, Zelenay P, Kim YS (2018) Anion exchange membrane fuel cells: Current status and remaining challenges. J Power Sources 375:170–184
15. Ramaswamy N, Mukerjee S (2019) Alkaline anion-exchange membrane fuel cells: Challenges in electrocatalysis and interfacial charge transfer. Chem Rev 119(23):11945–11979
16. Cheng Y, He S, Lu SF, Veder JP, Johannessen B, Thomsen L, Saunders M, Becker T, De Marco R, Li QF, Yang SZ, Jiang SP (2019) Iron single atoms on graphene as nonprecious metal catalysts for high-temperature polymer electrolyte membrane fuel cells. Adv Sci 6(10):1802066
17. Hu CG, Dai LM (2019) Doping of carbon materials for metal-free electrocatalysis. Adv Mater 31 (7):1804672
18. Yu EH, Krewer U, Scott K (2010) Principles and materials aspects of direct alkaline alcohol fuel cells. Energies 3(8):1499–1528
19. Dekel DR (2018) Review of cell performance in anion exchange membrane fuel cells. J Power Sources 375:158–169
20. Truong VM, Tolchard JR, Svendby J, Manikandan M, Miller HA, Sunde S, Yang H, Dekel DR, Barnett AO (2020) Platinum and platinum group metal-free catalysts for anion exchange membrane fuel cells. Energies 13(3):582
21. Douglin JC, Varcoe JR, Dekel DR (2020) A high-temperature anion-exchange membrane fuel cell. J Power Sources Adv 5:100023
22. Antolini E, Gonzalez ER (2010) Alkaline direct alcohol fuel cells. J Power Sources 195(11):3431–3450

Chapter 14
Phosphoric Acid Fuel Cells

14.1 Introduction

For commercial applications particularly in the terrestrial uses, fuel cells must use the inexpensive hydrogen of an abundant supply, which is based on steam reforming of hydrocarbon, e.g., the natural gas and/or coal gasification. The reformate hydrogen typically contains about 20% carbon dioxide and a small amount of carbon monoxide, and therefore, the fuel cell electrolyte must be the CO_2-rejecting acids.

Of the inorganic acids are HCl and HNO_3 volatile, while H_2SO_4 can be reduced at the anode at temperatures above 80 °C. Phosphoric acid has a low conductivity, as seen from Fig. 14.1; however, it has excellent thermal stabilitywith extremely low vapor pressure at temperatures of up to 200 °C. This temperature range is of importance in order to enhance the kinetics of electrode reactions and suppress the poisoning effect of impurities, e.g., carbon monoxide. In addition easy removal of the product water is possible as a result of the significant vapor pressure difference of water and phosphoric acid. The temperature range increases also the value of the heat for cogeneration of heat and power (CHP), with an overall energy efficiency of well above 80%. The product water is generated in form of steam, which can be directly used in the fuel processors, i.e., steam reforming.

PAFC is the first fuel cell power unit that has been commercialized since the early 1990s. More than 250 units of the 200 kW PC-25 systems have been installed worldwide and lifetime demonstrated with an overall lifetime of over 40,000 h.

14.2 Phosphoric Acid

14.2.1 Acid Chemistry

Pure phosphoric acid (H_3PO_4) has a melting point of 42 °C and a boiling point of 256 °C. At temperatures far below its boiling point, e.g., above 160 °C, the acid

S. P. Jiang and Q. Li, *Introduction to Fuel Cells*,
https://doi.org/10.1007/978-981-10-7626-8_14

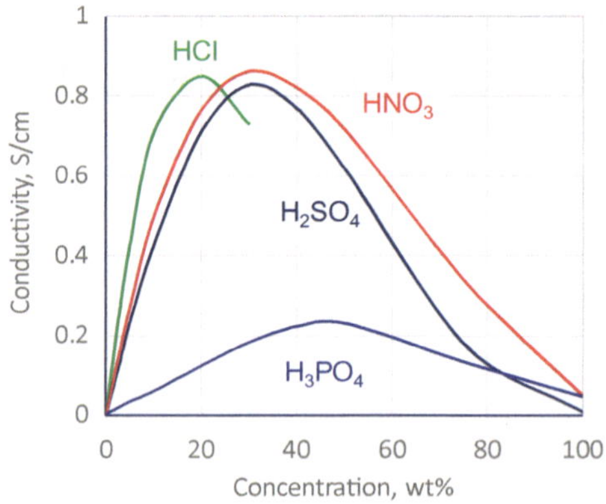

Fig. 14.1 Conductivity of various acids at 25 °C

starts to condense or dehydrate, leading to formation of biphosphoric acid (also called pyrophosphoric acid, $H_4P_2O_7$), triphosphoric acid ($H_5P_3O_{10}$), or higher polymeric acids:

$$H_3PO_4 + H_3PO_4 \rightarrow H_4P_2O_7 + H_2O \tag{14.1}$$

$$H_3PO_4 + H_4P_2O_7 \rightarrow H_5P_3O_{10} + H_2O \tag{14.2}$$

$$H_3PO_4 + H_{n+2}P_nO_{3n+1} \rightarrow H_{n+3}P_{n+1}O_{3n+4} + H_2O \tag{14.3}$$

These polymeric acids may reversibly hydrolyze in the presence of water at low temperatures, though the kinetic rate varies. The concentration of phosphoric acid is often defined for the H_3PO_4–H_2O mixtures:

$$\text{wt\%}H_3PO_4 = \frac{M_{H_3PO_4}}{M_{H_3PO_4} + M_{H_2O}} \times 100 \tag{14.4}$$

This expression is, however, confusing when the bi, tri, and higher phosphoric acids are considered where the concentrations defined on basis of H_3PO_4 are higher than 100%. It is therefore that the concentration based on definition for the P_2O_5–H_2O mixtures is also used:

$$\text{wt\%}P_2O_5 = \frac{M_{P_2O_5}}{M_{P_2O_5} + M_{H_2O}} \times 100 \tag{14.5}$$

In 85 wt% H_3PO_4, corresponds to the monohydrate $H_3PO_4{\cdot}H_2O$ and with a boiling point of 154 °C, the major component is othorphosphoric acid. In a solution above 95 wt% H_3PO_4, the pyrophosphoric acid appears. In 100 wt% H_3PO_4, the content of $H_4P_2O_7$ is about 15 wt%; this number increases to about 30 wt% in 105 wt% H_3PO_4, in which the triphosphoric acid first appears.

In the absence of free water, phosphoric acid starts to protonate itself according to reaction (14.6), the so-called auto-protolysis, leading to formation of the phosphazidium ($H_4PO_4^+$) ions.

$$2H_3PO_4 \rightarrow H_4PO_4^+ + H_2PO_4^- \tag{14.6}$$

The $H_4PO_4^+$ has a $pK_a = -3$, a value of that for the strong sulfuric acid. All the higher acids from condensation are also chemically strong acid and are able to protonate the orthophosphoric acid:

$$H_3PO_4 + H_4P_2O_7 \rightarrow H_3P_2O_7^- + H_4PO_4^+ \tag{14.7}$$

14.2.2 Conductivity and Mechanism

As discussed in Chap. 5, the proton conductivity can be of two mechanisms: the vehicle mechanism or/and the Grotthuss mechanism. The Grotthuss mechanism involves migration of protons with no parasitic movement of any proton carriers and hence also called the structural diffusion mechanism. For a proton-conducting material with conductivity of both mechanisms, the fraction of the current by the structural diffusion mechanism is expressed as $i_{\text{structural}}$ and the fraction by the vehicle mechanism as i_{vehicle}. The transference number of the structural diffusion is therefore defined as

$$t_{\text{structural}} = \frac{i_{\text{structural}}}{i_{\text{structural}} + i_{\text{vehicle}}} \tag{14.8}$$

For 100% neat H_3PO_4, this fraction is as high as 96–98%, i.e., only a few percent of the charge transport involves movement of the carrying acid molecules [1].

It is the recent finding that this is only true for 100% neat H_3PO_4 where a severe imbalance exists between proton donors and acceptors in the acid molecule, i.e., each acid molecule contains three –OH groups as proton donors and only one =O as the acceptor. This so-called the peculiar hydrogen bond frustration favors the proton conduction via the structural diffusion mechanism.

Water molecules are proton acceptors when added into the acid:

$$H_3PO_4 + H_2O \rightarrow H_2PO_4^- + H_3O^+ \quad pK_a^1 = -2.16 \tag{14.9}$$

The promoted dissociation of the acid hinders the propagation of proton through structural diffusion and the fraction of the vehicle mechanism increases. For 85 wt% phosphoric acid ($H_3PO_4 \cdot H_2O$) where the ratio of the proton donor sites to the proton acceptor sites is 3/2, the structural diffusion transference number is decreased to about 0.15 at 127 °C. In 73 wt% phosphoric acid ($H_3PO_4 \cdot 2H_2O$), where the proton

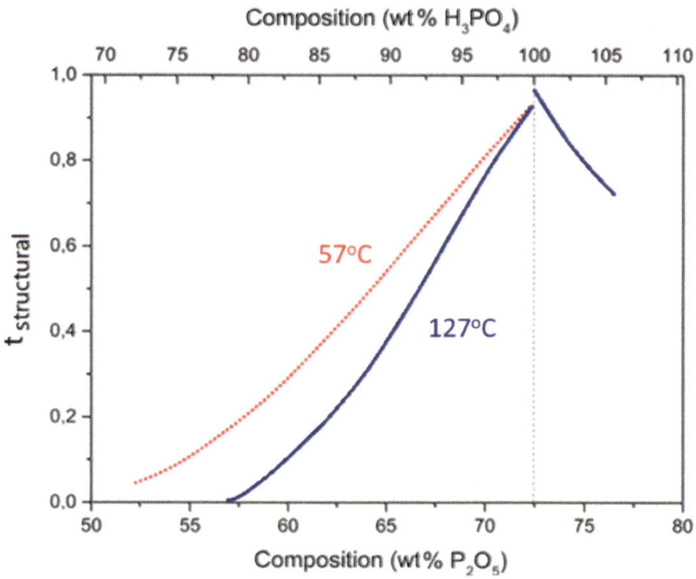

Fig. 14.2 Transference number of the structural diffusion mechanism in phosphoric acid at two temperatures as a function of concentration. Modified from Ref. [2] with permission from Royal Society of Chemistry, Copyright 2017

donor to acceptor site ratio is equal, i.e., with no proton frustration, the structural diffusion mechanism completely vanished, as seen from Fig. 14.2. In the acid of higher concentrations (above 100%), the condensation of acid molecules creates a number of highly charged phosphate anions which contribute to the vehicular mechanism [2].

In PAFC practice operating at around 200 °C, the used acid is nearly 100 wt%. The 2–4% vehicle mechanism of proton conduction through the electrolyte is not negligible and raises an issue of the acid management, as to be discussed in Sect. 14.4.3.

The conductivity of phosphoric acid of varied concentrations is shown in Fig. 14.3 at temperatures from zero to 170 °C. 100% H_3PO_4 exhibits a significant conductivity of about 0.025 S/cm at room temperature. On addition of water into the 100% acid, the conductivity increases, as water promotes the dissociation of the acid and decreases the viscosity of the solution. A maximum conductivity of about 0.22 S/cm appears at the concentration around 45 wt%. At 160 °C the maximum conductivity reaches a value of 0.64 S/cm at concentration of 90 wt%.

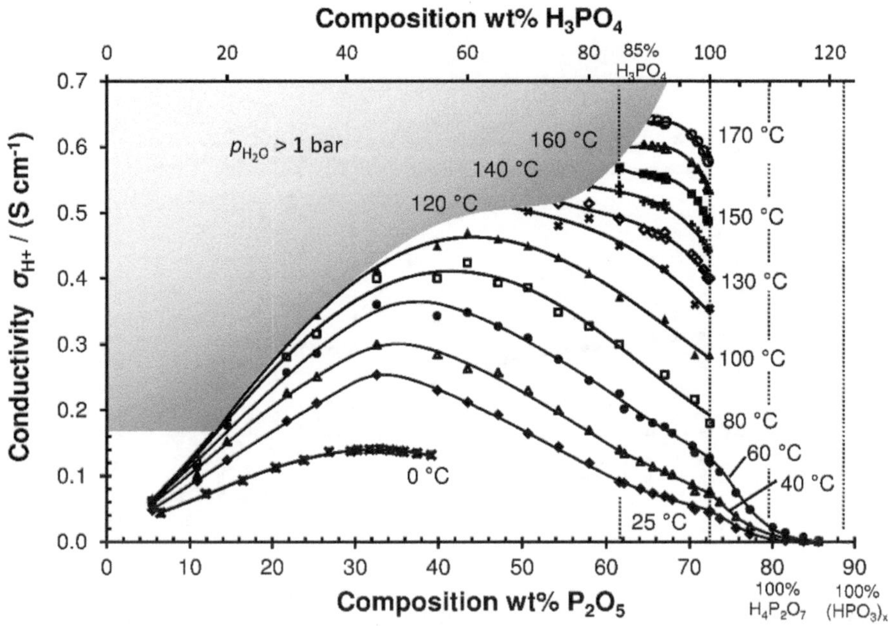

Fig. 14.3 Conductivity of phosphoric acid as a function of concentration at ambient pressure and varied temperatures. In the shaded area, the water vapor pressure reaches above 1 bar where no data are available. Reproduced from Ref. [3] with permission from Springer Nature, Copyright 2016

14.2.3 Vapor Pressure and Thermal Stability

The vapor pressure of water above phosphoric acid is compiled in Fig. 14.4 as a function of the acid concentration expressed as wt% P_2O_5 as well as H_3PO_4. The top-dashed line is for the atmospheric pressure of 1013 mbar. The bottom-dashed line presents the water vapor pressure of $p_{H_2O} = 190$ mbar. This value correlates to the typical water vapor pressure exiting in the cathode outlet gas of PAFC at any current density when the air stoichiometry (λ_{air}) is kept at a value of 2.

It can be seen from the figure that the water vapor pressure over 90 wt% H_3PO_4 may reach 1013 mbar at 170 °C, the boiling point of the acid. When exposed to an air stream containing 190 mbar water vapor at this temperature, the stable acid should have a concentration of at least 98 wt% in order to avoid further dehydration or loss of water.

A more important issue is the vapor pressure of acid in the steam above the electrolyte. The acid vapor pressure has been measured at higher temperatures, but in the temperature range of PAFC operation it is too low to measure. By extrapolating the high-temperature data to low temperatures (see Fig. 14.11), it was estimated that the acid content in an acid saturated vapor phase is about 0.8 ppm at 160 °C and 4 ppm at 190 °C. These numbers are, however, not negligible when considering the evaporation loss of the acid during fuel cell operation because the total amount of

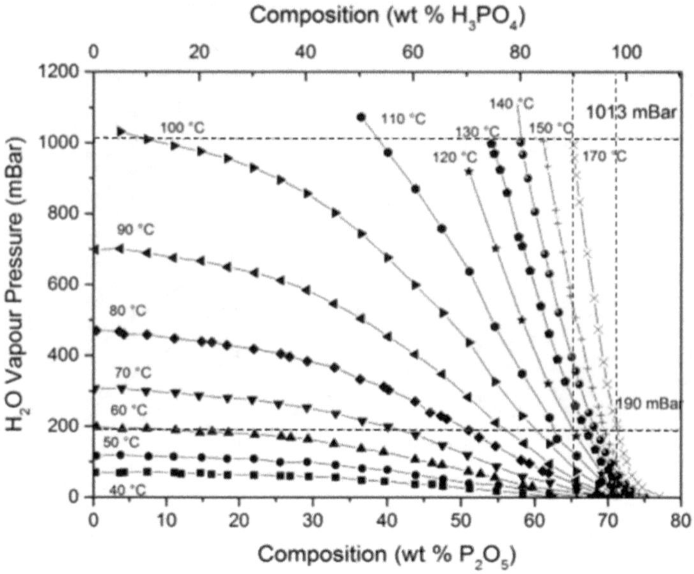

Fig. 14.4 Vapor pressure of water over H_3PO_4 of different concentrations. Reproduced from Ref. [4] with permission from Technical University of Denmark, Copyright 2017

acid retained in the cell is very small and the total volume flow of the cathode air during a period of 5–7 year lifetime is huge.

It is also noted that the difference between the water vapor and acid vapor pressures is of significance for effective rejection of the product water in order to keep the concentration and volume of the electrolyte unchanged during operation.

14.3 Cell Components

14.3.1 Catalysts

In the phosphoric acid-based electrolyte, platinum and its alloys are the only state-of-the-art catalysts, as they are (1) catalytically active for both hydrogen oxidation and oxygen reduction, (2) chemically sufficiently stable in hot phosphoric acid, and (3) readily available in resources and supply.

Electrode reactions take place on the catalyst surface and therefore depend on the total surface area of the catalysts. In early years, platinum black was employed. The metal black was prepared by, for example, reduction of chloroplatinic acid with formaldehyde in alkaline media. The metal particles are in general of around 10 nm size, displaying a surface area of 20–30 m^2/g. Electrodes made of platinum blacks have a metal loading of 5–10 mg/cm^2. This high loading is prohibitive for commercial applications of PAFC other than military or space programs. Another problem associated with this high loading of platinum blacks is that the structure of catalyst layers, upon sintering and agglomeration, may collapse.

Supported catalysts of noble metals have then been developed in connection to PAFC in the 1970s [5, 6]. The use of carbon black support materials allowed Pt to be finely dispersed to maximize active surface area. Use of high surface area carbon blacks as catalyst support was one of the major breakthroughs in the early PAFC development and led to a high degree of the metal dispersion and therefore a dramatic reduction in the Pt loading without sacrificing or even improving the electrode performance. High surface area carbon blacks, e.g., Vulcan XC-72 (Cabot Corporation) with a surface area of 250 m^2/g are the most widely used supports for PAFC catalysts. A problem associated with the supported catalysts was a significant loss of the performance, apparently due to the catalyst surface area loss through Pt sintering.

Thermodynamically carbon is unstable in acidic media:

$$C + 2H_2O = CO_2 + 4H^+ + 4e^- \quad E^0 = +0.207\,V \tag{14.10}$$

The cathode potential of a fuel cell is always significantly higher than this potential value, and therefore, the electrochemical oxidation of carbon is of concern. Figure 14.5 shows the corrosion current for carbon blacks in hot phosphoric acid. At

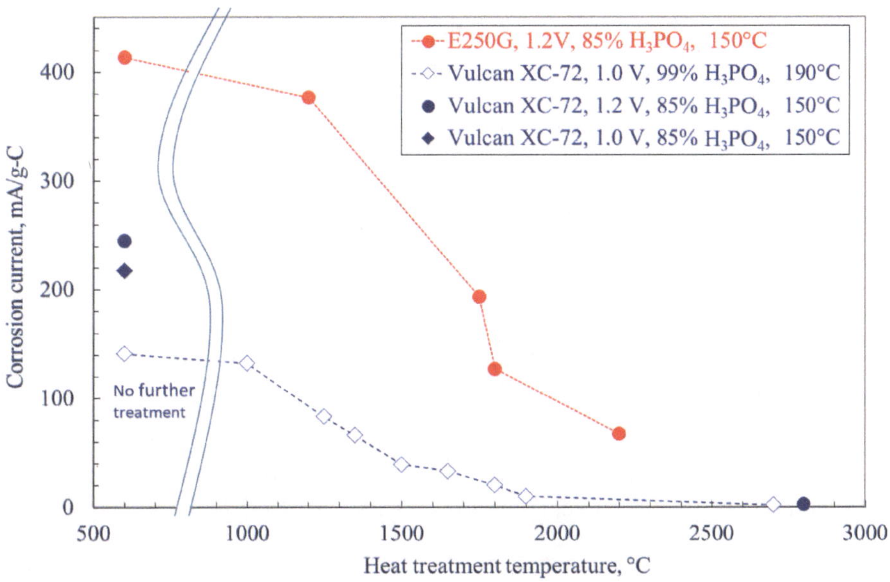

Fig. 14.5 Corrosion current for two types of carbon blacks in hot phosphoric acid as a function of further heat treatment temperatures. The measurement temperatures and potentials versus SHE are indicated in the figure. Data for Vulcan at 190 °C was taken from Ref. [7] and rest from author's group

a potential close to the open circuit value of fuel cells, the as-received Vulcan carbon black exhibits a corrosion current of about 0.1–0.2 A/g-carbon. It should be noted that the carbon support corrosion is faster in the presence of platinum nanoparticles. When the carbon black is further heat treated under an inert atmosphere at elevated temperatures, the corrosion current is significantly decreased as shown in Fig. 14.5. It is a common practice that the carbon supports for cathode catalysts are further heat treated at high temperatures in order to impart graphitic character to the carbon black.

Example 14.1 Carbon corrosion in hot phosphoric acid.

A PAFC cathode has a platinum loading on 0.5 mg/cm^2, prepared from 10% Pt/C catalysts. Assuming that the carbon black can be oxidized in hot phosphoric acid at a rate of 0.2 A/g-C under the open circuit potential, estimate how long time it will take to oxidize all the carbon black on the catalyst layer.

Solution:

The catalyst has a carbon to platinum mass ratio of 9. For an electrode with Pt loading of 0.5 mg$_{Pt}$/cm^2, the carbon black loading is 4.5 mg/cm^2 or 4.5×10^{-3} g-C/cm^2.

At a carbon corrosion rate of 0.2 A/g-C, the carbon black in the catalyst layer will generate a corrosion current density of

$$i = 0.2\frac{\text{A}}{\text{g-C}} \times 4.5 \times 10^{-3}\frac{\text{g-C}}{\text{cm}^2} = 9 \times 10^{-4}\ \text{A/cm}^2 \tag{14.11}$$

The carbon loading of 4.5×10^{-3} g-C/cm^2 corresponds to a total charge of

$$Q = 4.5 \times 10^{-3}\frac{\text{g-C}}{\text{cm}^2} \times \frac{1\ \text{mol C}}{12\ \text{g-C}} \times 4\frac{\text{mol e}^-}{\text{mol C}} \times 96{,}485\frac{\text{Coulobm}}{\text{mol e}^-} = 145\ \text{Coulobm/cm}^2 \tag{14.12}$$

And the time needed to oxidize all the carbon

$$\text{t} = \frac{Q}{i} = \frac{145\ \text{Coulobm}/\text{cm}^2}{9 \times 10^{-4}\ \text{A}/\text{cm}^2} = 1.6 \times 10^5\ \text{s} \approx 45\ \text{h} \tag{14.13}$$

It tells that, if a PAFC cathode is left at OCV, all the carbon in the catalyst layer might be oxidized to carbon dioxide within a couple of days.

Commercial Pt/C catalysts for PAFC are typically containing a noble metal content of 10 wt%, which should be compared to the PEMFC catalysts having a metal loading of as high as 60 wt% in relation to the carbon support.

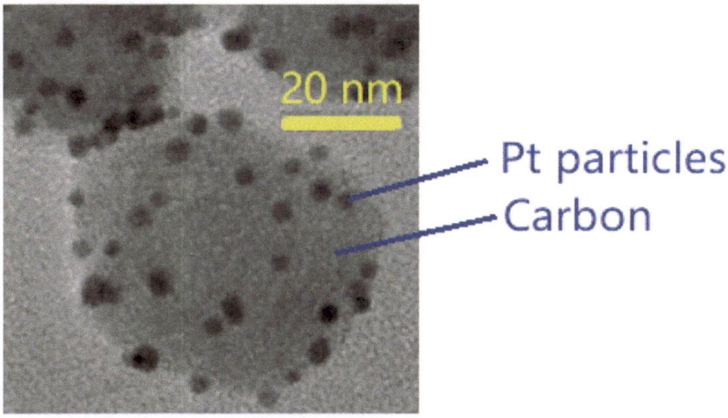

Fig. 14.6 TEM image of carbon-supported platinum nanoparticles (10 wt% Pt/C)

The size of platinum nanoparticles is typically in the range of 3–5 nm (see Fig. 14.6). Smaller platinum particles exhibit higher specific surface area, e.g., platinum particles of 1.4–1.5 nm diameter may give a surface area as high as 180–200 m^2/g. The catalytic activity is, however, found not to increase proportionally with the surface area in this range of the particle size. In other words, the area-specific activity of the catalyst, defined as the oxygen reduction current density per unit surface area of platinum particles, in e.g., A/cm^2_{Pt} at a fixed electrode potential, e.g., 0.9 V versus SHE, decreases for small platinum particles. This is called the *crystalline size effect.*

A small platinum particle of radius r possesses an excess of surface energy:

$$\Delta E = 3\gamma/\rho r \tag{14.14}$$

where γ is the surface tension and ρ the density. Platinum has a high surface energy of 2.7 J/m^2 and a density of 21.45 g/cm^3. For a platinum particle of 2 nm, its excess surface energy is as high as 189 J/g.

Under PAFC operation conditions, platinum nanoparticles are easy to agglomerate via either dissolution/precipitation or surface diffusion mechanisms. In the former case, platinum particles dissolve in hot phosphoric acid. Both dissolution rate and solubility increase with potentials in the range from 0.6 V to the open circuit potential. Figure 14.7 shows a set of platinum solubility data in hot phosphoric acid as a function of potential. As the cathode potential is increased from 0.8 to 0.95 V, the solubility changes by, *ca.*, 5–600 times. A potential cycling during the PAFC operation would cause the cathode catalyst dissolution and reprecipitation and therefore a significant growth of the catalyst particles. In addition, corrosion of the carbon support triggers further platinum particle sintering.

To further improve the activity and stability, platinum alloy catalysts have been extensively explored for PAFC. The first alloy catalysts were platinum–vanadium and platinum–chromium. A long list of binary and ternary alloys of platinum with

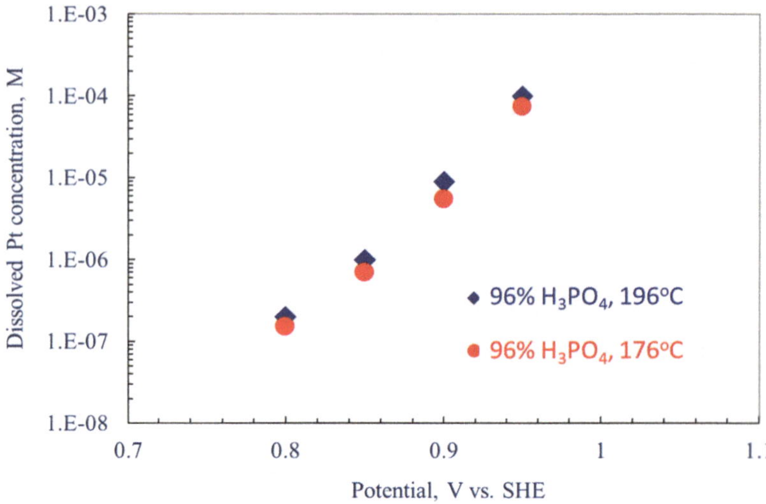

Fig. 14.7 Equilibrium concentration of dissolved platinum in phosphoric acid as a function of potential. Modified from Ref. [8] with permission from IOP Publishing, Copyright 1979

refractory and transitional metals, as the ORR catalysts, was reported in early 1980s [9]. In terms of mass activity, Pt alloys have been claimed to be up to 1.5–2 times more active than pure Pt. A voltage gain of 30–40 mV has been observed in H_2/air PAFCs in a practical range of current densities around 200 mA/cm^2. Different mechanisms have been suggested for the activity enhancement, e.g., due to the decrease in the interatomic spacing of the elements in the alloys, while the surface roughening of the left platinum surface after the leaching out of the alloying element may also play a role as a result of slow surface diffusion of platinum atoms after the leaching out. These alloying elements are insufficiently stable in phosphoric acid, and long-term durability of the alloy catalysts is always a concern.

Example 14.2 PAFC stack

The 200 kW PAFC stack is commercialized as PC-25 FC power plants. A set of technical parameters are listed below.

Specifications of PC-25 PAFC system.

- Power capacity: 0–200 kW
- Voltage and phasing 480/277 V at 60 Hz
- Thermal energy: 740,000 kJ/h at 60 °C
- Energy efficiency >40%
- Total fuel efficiency >80%
- Fuel: Natural gas 53 Nm^3/h at 1 atm
- Dimensions 3 m × 3 m × 5.5 m

- Weight 17 tons.

Assuming that the average single cell performance is 0.215 A/cm^2 at 0.60 V and the platinum load is 0.5 mg/cm^2 for the cathode and 0.25 mg/cm^2 for the anode. Calculate the total amount of platinum used in the stack.

Solutions:

The average power density of a single cell is

$$P = 0.60\,\mathrm{V} \times 0.215\,\mathrm{A/cm^2} = 0.129\,\mathrm{W/cm^2} \tag{14.15}$$

For each kW power, the cathode and anode working area is

$$A = 1000\,\mathrm{W}/0.129\,\mathrm{W/cm^2} = 7752\,\mathrm{cm^2} \tag{14.16}$$

The amount of Pt used for a kW is then

$$(0.5 + 0.25)\,\mathrm{mg_{Pt}/cm^2} \times 10^{-3}\,\mathrm{g/mg} \times 7752\,\mathrm{cm^2} = 5.81\,\mathrm{g/kW} \tag{14.17}$$

Assuming the platinum metal price of 30 US$/g the Pt cost is

$$\mathrm{Pt\ cost} = 30\,\mathrm{US\$/g} \times 5.81\,\mathrm{g/kW} = 175\,\mathrm{US\$/kW} \tag{14.18}$$

This is not significant in the current system price of about 3000 US$/kW for PAFC systems. For the final target of the system cost less than 400 US$/kW, platinum is apparently an issue.

14.3.2 Gas Diffusion Electrodes

A gas diffusion electrode is a layer of catalysts, consisting of loose powders with a low content of polymer binder, supported on a substrate material. The electrode substrate is a porous structure allowing for easy gas diffusion from the channels to the catalytic site near the electrolyte interface. Other functions of the substrate include removal of product water and heat, conduction of electrons through the cells and maintaining a uniform contact pressure at the interface between the catalyst surface and liquid electrolyte held in the matrix.

In concentrated phosphoric acid at temperatures above 150 °C, a few candidate materials resistant to corrosion are available including (1) noble metals as either bulk or coating layers, (2) refractory metals, e.g., Ta, Mo, and W with limited processibility, and (3) carbon, most preferably graphite and its composites (see Fig. 14.8).

The very first materials used as the electrode substrate were made of woven 100 mesh tantalum screen. Tantalum, when oxidized, forms a thin and very stable oxide layer that provides passivation. Due to the passivation, tantalum is resistant to most acids except hydrofluoric acid. The oxide layer is, however, not conductive, and when used on the anode side the metal forms brittle hydrides.

Carbon fibers were then used in the form of carbon paper bonded by impregnation of thermoset resins. The resins are then pyrolyzed to carbon followed by heat treatment at above 2000 °C to improve the electrical and thermal conductivity as well as corrosion resistance. This type of carbon paper substrate is the precursor of gas diffusion layers. Used carbon fibers are typically of 5–10 μm in diameter. The

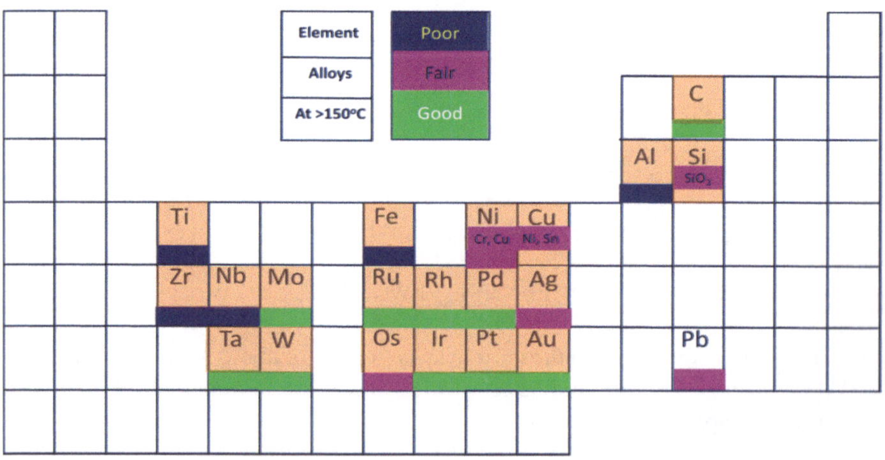

Fig. 14.8 Possible materials for PAFC construction with effective corrosion resistance in hot phosphoric acid. The three rows represent the element, alloys, and at temperatures above 150 °C while the color codes are good (green), fair (pink), and poor (blue) resistances

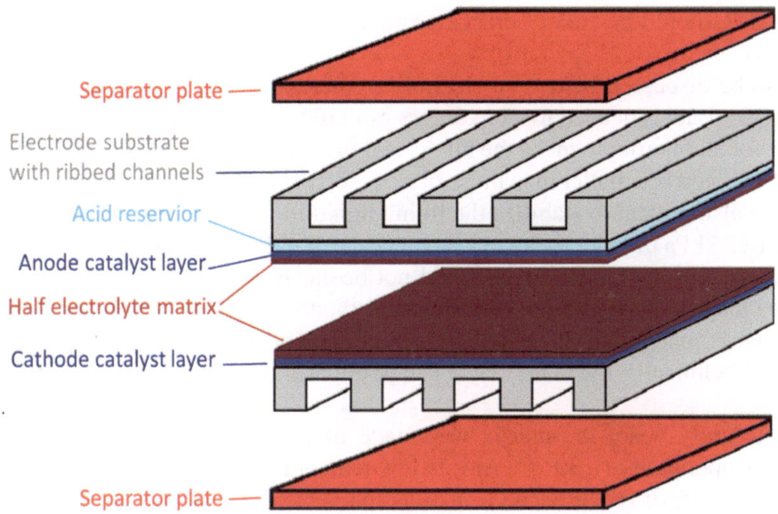

Fig. 14.9 Thick electrode substrates with ribs and the assembling in a PAFC stack

carbon paper is in thickness of 100–300 μm, having a porosity of 70–80% with a mean pore size of 20–30 μm.

Another type of the electrode substrate is used in connection to a different cell configuration where a thick electrode substrate, of about 2 mm thickness, is processed with ribbed gas channels. The assembling of such cells is schematically represented in Fig. 14.9. The ribbed substrate is made of chopped or milled carbon fibers, using the similar procedure of pyrolysis and heat treatment as described above.

Polytetrafluoroethylene (PTFE) was first introduced in later 1960s for manufacturing catalyst layers [10]. The substrate is normally also wet proofed by PTFE, which needs to be sintered in order to soften the polymer for covering the fiber surface and remove the surfactants of the PTFE dispersion. The polymer loading is 15–25 wt% to compromise the wet proofing, diffusional resistance, and electronic conductivity.

14.3.3 Matrix and Acid Reservoir

The matrix of a PAFC is a sponge material like the battery separator. The main function of the matrix is to immobilize the liquid electrolyte allowing for proton transport from the anode to the cathode. For that purpose the matrix layer should be as thin as possible in order to reduce the ohmic resistance, and at the same time the layer should be porous to accommodate the liquid electrolyte likely by the wetting or capillary force, which requires small pores with a narrow pore size distribution.

The matrix should be free of cracks or pinholes to avoid any acid draining or gas crossover.

As to be discussed below, an electrolyte reservoir layer, in association with the anode electrode, is needed to store excess of liquid electrolyte, also via the capillary driving force. In order to make the liquid acid first filling up the matrix layer, the pore size of it should be smaller than those in the reservoir layer. The small pore size of the matrix is also critical to stabilize the filling liquid for maintaining a bubble pressure of at least 35 kPa in order to prevent any leakage of either reactant gas across the cell. In addition, the matrix material should not be electronically conductive so that no shorting may occur between the anode and cathode. Of course the chemical stability of the matrix material is also essential against reaction with the acid or dissolution in the acid. Technically, the matrix layer can be either freestanding or coated on catalyst layers.

In the early years the matrix was made of phenolic fibers as a freestanding paper for operation at 130 °C. At this low temperature, the anode catalyst was an alloy/composite of Pt/Rh/Ni/WO_3 in order to tolerate the CO from steam reforming. To avoid using the anode alloy catalyst, the tolerance to CO must be enhanced by elevating the temperatures. 135 °C is enough to generate stream from the cooling or product water for the reforming, which is a major part of the heat need. The acid inventory is low, limiting the lifetime (to *ca.* 5000 h) of the stack compared to the acid loss by absorption of the graphite-polymer bipolar plates and evaporation into the reactant streams leaving the cell.

Breault [11] was the first who found that silicon carbide, a commercially available insulating powder, is very stable in concentrated phosphoric acid at up to 230 °C. Silicon carbide containing about 5% PTFE as binder was coated on the catalyst layer of each electrode. One advantage of using coatings on each electrode is that it is less critical to make crack-free layers—any defect in a coating layer on one electrode might be healed by the other. As mentioned above, the PTFE binder needs to be treated at above 330 °C.

Using SiC powders of 5 μm size, the thickness of the matrix can hardly be made thinner than 50 μm, which corresponds to a packing thickness of five–ten particles. To further decrease the thickness down to 20–30 μm each, smaller SiC powders of 0.5–1.0 μm size are used. Such a matrix layer is the state-of-the-art matrix, with a porosity of 50% and a bubble pressure of as high as 70 kPa. It has been reported that such a matrix filled with the acid exhibits an effective ionic resistivity of 6.7 Ω cm, corresponding to an ASR of 0.033 Ω cm^2 or an ionic conductivity of 0.15 S/cm.

The volume of the electrolyte expands and contracts when there are changes in the fuel cell temperature, reactant pressure, and stoichiometry. Any increase in the air stoichiometry will decrease the relative humidity or dew point of the air flow. For the air stoichiometry (λ_{air}) of 2, for example, the water vapor pressure in the exiting air at the cathode outlet is 190 mBar. Any increase of the air stoichiometry will lower the equilibrium water vapor pressure and, therefore, increase the acid concentration and reduce the acid volume. The change of the electrolyte volume may lead to flooding of the catalyst layer or dry out of the matrix. The flooding decreases the electrode performance while the dry out may lead to gas crossover. In addition the lifetime of a PAFC stack is, to a great extent, limited by the acid loss, either via diffusion into the gas diffusion layers and porous graphite platesor evaporation loss.

An additional acid inventory by means of an electrolyte reservoir layer within the cell is, therefore, necessary. This can be achieved by tailoring the hydrophilicity of the two electrodes. The anode, often with a thicker microporous and catalyst layer, as thick as 350 μm, is made more hydrophilic and therefore absorbs extra acid, functioning as an acid reservoir.

14.3.4 Bipolar or Separate Plates

A typical PAFC stack contains cells connected in series, i.e., the back side of a hydrogen feeding anode is in direct contact with the backside of the air feeding cathode of the adjacent cell. In such an arrangement, individual cells are stacked with bipolar plates between the cells.

A bipolar plate separates the individual cells in a fuel cell stack and electrically connects them in series. The bipolar plate also contains gas channels that feed the reactant gases to the porous electrodes and remove the reaction products.

Cheap metals, e.g., Austenite 316, 317 and 904 L are insufficiently resistant to corrosion in hot concentrated phosphoric acid. In early years of the PAFC development the used bipolar plates were made of single pieces of high density graphite with gas channels machined on either side to direct the flow of fuel and oxidant. Both the material and machining are very costly.

To facilitate the easy processing, bipolar plates are often made of two parts, a porous plate (a thick electrode substrate) with gas channel ribs to direct the gas flow and a thin impervious plate, called the separator plate. The separator plate has two major functions, to provide electronic interconnection between adjacent cells and to separate the fuel and oxidant of adjacent cells. Both plates are based on graphite materials. Graphite is sufficiently stable to replace the more expensive gold-plated tantalum cell hardware used in the early years. In fact it has the best combination of the necessary properties for PAFC applications—the electronic conductivity, sufficient corrosion resistance, low density, and low cost. The plates are prepared from a mixture of graphite and resin, followed by graphitization treatment at temperatures of up to 2700 °C. High density or low porosity of the separator plates is critical as it will after thousand hours of operation be filled with acid—an important mechanism of acid loss.

14.3.5 Seals and Coolers

In small cells, gasket of PTFE films is used as the edge seal of the matrix-uncoated area in order to prevent the leakage of reactants and electrolyte. Between the two

Table 14.1 Summary of material issues of PAFC

			1960s	Today
Catalysts	Anode		Pt black	10% Pt on C (Vulcan XC-72R)
	Cathode		Pt black	10% Pt on C (HT-Vulcan XC-72R)
Electrodes	Catalyst layer		PTFE bonded	PTFE bonded mg_{Pt}/cm^2
	Pt loading	Anode	9 mg_{Pt}/cm^2	0.1–0.25 mg_{Pt}/cm^2
		Cathode	9 mg_{Pt}/cm^2	0.5 mg_{Pt}/cm^2
	Gag diffusion layer		Ta mesh screen	Carbon fiber paper
Electrolyte			85% H_3PO_4	100% H_3PO_4
Electrolyte matrix			Phenot fiber	PTFE-bonded SiC
Electrolyte reservoir			None	Graphite composite plate
Bipolar and separate plates			Gold-coated tantalum	Multicomponent graphite plates
Cooler				Imbedded tubes in bipolar plates

electrodes, a middle layer gasket is extended a small distance into the matrix-coated area to prevent the crossover of the reactant at the edge.

In large size cells, the active area of the electrode is coated with the desired thickness of the matrix while at the edge outside the active electrode area, an increased thickness of the matrix is coated on both the anode and cathode. When the two electrodes are assembled and compressed during stacking, the edge matrix layer is densified with finer porosity. The following impregnation of the edge area with phosphoric acid makes the cell "wet-sealed".

Cooling, often in association with heat recovery, is often achieved through a separate cooling plate, typically every module of 5–7 cells. A liquid coolant, e.g., water is often used though air cooling can be simpler.

The above discussion on PAFC material issues is summarized in Table 14.1.

14.4 Performance and Cell Management

14.4.1 Performance and Lifetime

The phosphoric acid electrolyte suffers from slow kinetics for oxygen reduction due to the strong adsorption effect of anions onto the active sites of catalysts as well as the low solubility and diffusivity of oxygen. Figure 14.10 compares the polarization

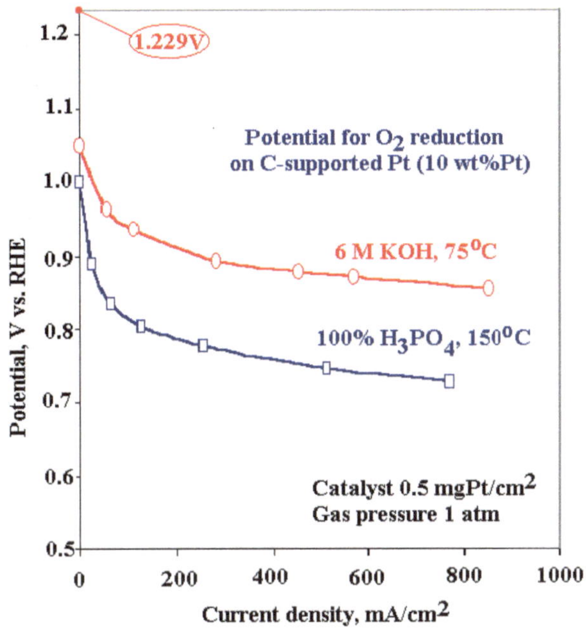

Fig. 14.10 Polarization curves for oxygen reduction in 6 M KOH at 75 °C and 100% H_3PO_4 at 150 °C measured in half-cell using gas diffusion electrodes with 10% Pt/C catalysts at a loading of 10% mg/cm^2

curves of ORR in 6 M KOH and 100% phosphoric acid. Through the entire current density range, the ORR electrode showed about 120–140 mV lower potential in the latter than in the former.

Great efforts have been made to investigate alternative acidic electrolytes. Some perfluorinated acids are found to be able to enhance the kinetics of ORR. For example, trifluoromethansulfonic acid (CF_3SO_3H), with a melting point of 33 °C and a boiling point of 162 °C, possesses a high ionic conductivity and chemical as well as electrochemical stability. As a super strong acid with pKa of about −14, this acid exhibits little anion adsorption on platinum catalyst and with a high solubility of O_2. The oxygen reduction in this acid has about 100 times faster kinetics than in phosphoric acid. Because it has a similar perfluoro structure as PTFE, this acid wets and floods the catalyst layers of gas diffusion electrodes. These types of perfluorinated compounds, in form of acids or salts, have been suggested to use as additives to the phosphoric acid electrolyte to improve the oxygen reduction performance.

Despite the extensive R&D effort in the last 30–40 years, no breakthroughs in improving PAFC performance have been achieved, compared to, e.g., PEMFC. Typical PAFCs operate in a low current density range, from 100 to 400 mA/cm^2. The single cell voltage falls in between 0.6 and 0.7 V under ambient pressure and operating with air and reformate hydrogen. Compared to pure hydrogen, the reformate contains contaminants such as carbon monoxide (CO), carbonyl sulfide (COS), and hydrogen sulfide (H_2S). The tolerance of a PAFC anode at 200 °C is about 1 vol% CO and 20 ppm H_2S or 50 ppm (COS + H_2S).

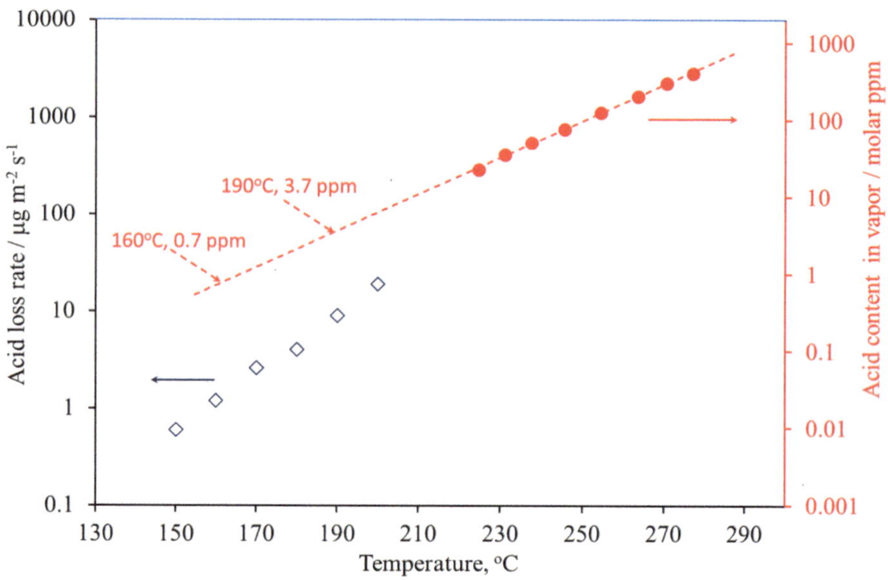

Fig. 14.11 Saturated content of phosphorus oxide in the vapor phase above the boiling acid and the acid evaporation rate in PAFC. The solid circles are measurements from Ref. [12]. The dashed trend line is extrapolated to the low-temperature range. The open diamonds are measurements of acid evaporation loss by Okae et al.[13]

It has been reported that the effect of pressure within a range of 1–10 atm the PAFC performance can be expressed as

$$\Delta V_P(\mathrm{mV}) = 146 \times \log\left(\frac{P_2}{P_1}\right) \tag{14.19}$$

When the pressure is doubled, the cell voltage may increase by about 40 mV in the operational current density range from 100 to 400 mA/cm^2. It should be noted that any increase in the pressure will lessen the condensation of the acid electrolyte and therefore increase the conductivity and decrease the ohmic loss.

Any elevation of the operating temperature will improve the activation polarization, mass transfer, and ohmic losses of PAFC. As a result, the cell voltage increases at a rate of around 1 mV/°C within the practical current density range. When the reformate hydrogen contains a significant amount of CO and sulfide, the temperature effect will be more on the anode. Any gains in performance at higher temperatures will eventually be compromised with expense of the lifetime reduction.

The target lifetime of 40,000 h for utility applications is a more critical challenge. Under normal operation, the state-of-the-art PAFC systems show a degradation rate of 2–3 mV for every 1000 h. Assuming a beginning of life (BoL) cell voltage of 700 mV, the 2 mV/1000 h rate means 80 mV or 11% performance loss after 40,000 h.

14.4.2 Acid Evaporation and Mitigation

There are several mechanisms of acid loss in PAFC stacks, for example, evaporation in the gas phase with anodic and cathodic off-gases, absorption by the bipolar plates and GDL/MPL layers. The acid vapor content of phosphoric acid in the steam phase above an acid solution is very low. Brown and Whitt [12] made an elegant measurement by collection and analysis of the P_2O_{10} content above the boiling acid of varied concentration at temperatures down to 221 °C. At lower temperatures that PAFC operates the acid content is too low to measure. By extrapolating of these high-temperature data it was estimated that the acid content in an acid saturated vapor phase is about 0.8 ppm at 160 °C and 4 ppm at 190 °C, i.e., an increase by a factor of 5 through this temperature interval (see Fig. 14.11). Considering the fact that a large volume of reactant gases pass through the stack during an operation period of more than 40,000 h, the acid loss is in fact a significant factor limiting for the lifetime of PAFC. Based on these values, the acid loss rate with the cathode off-gas is estimated to be in a range of a few $\mu g/m^2s$ at the PAFC operating temperatures. A set a data presented by Okae et al. [13] are within the range and dependence on temperature. From 160 to 190 °C, the acid evaporation loss increased from 1.2 to 9 $\mu g/m^2s$. The additional losses may be due to the aerosol formation, i.e., the acid loss in the form of mist or droplet.

The long-term durability of PAFC is limited by the acid loss which must be mitigated. In addition to the above-mentioned electrolyte reservoir plate or layer to achieve increased acid inventory, an acid condensation zone was also developed. It was Breault who developed this concept that at the exit edge of a cell the electrodes were made without catalysts. In this zone, there is no reaction and therefore no generation of any current or heat. Furthermore, additional cooling is designed into this zone. The combination of no heat and additional cooling results in the temperature of this condensation zone about 30 °C lower. This temperature difference make 4 times reduction in the acid vapor pressure and significant increase of the lifetime.

14.4.3 Migration of Acid and Management

The well-known contribution of proton conductivity from the vehicular mechanism introduces parasitic mass transport of acid molecules. The proton hopping mechanism prevails predominantly in a narrow range of the acid concentration (close to 100%) and temperature. At lower acid concentrations, the higher water content promotes the acid dissociation and results in increased conductivity and contribution from the vehicular charge transport. At concentrations higher than 100 wt% H_3PO_4, on the other hand, the acid condensation leads to formation of pyro- and higher polymeric acids, which increase the contribution of the vehicular mechanism due to the addition of charge carrier in the system (e.g., $H_2P_2O_7^{=}$ and $H_3P_3O_{10}^{=}$). As a result, a phosphoric acid concentration gradient is developed between the anode and cathode side during fuel cell operation particularly at higher current densities. For PAFC workers this phenomenon is called the acid pumping, which causes increases in the acid volume as well as pressure within the anode layer. This is the driver for and balanced by back diffusion or hydraulic flow from the anode to the cathode, and the acid distribution is in fact under a dynamic equilibrium. Under such a dynamic equilibrium, the intracell acid concentration gradient between the cathode and anode could be steadily established depending on the operational current density and hydrophobicity of the catalyst layers. In a PAFC stack intercell acid concentration gradient exists also between the anode of one cell and the adjacent cathode of the next cell. At the worst case, the cells with acid inadequate to saturate the matrix and the edge seals may fail due to gas crossover.

14.5 Summary

In this chapter we have introduced and discussed the following topics:

- A main technical feature of PAFC is rejection of CO_2 in both the fuel and air steams. This allows use of inexpensive hydrogen from, e.g., natural gas reforming and of air without scrubbing. The operating temperature around 200 °C further enhances the CO tolerance, another inevitable component of the reformat hydrogen. Another benefit from this operating temperature is the increased value of heat for CPH or other stationary applications, enabling a cogeneration of power and heat at an overall efficiency of 80–90%.
- Chemistry of phosphoric acid as electrolyte is discussed. The condensation of phosphoric acid at elevated temperatures leads to formation of polymeric acids which are chemically more acidic and thermally more stable. The acid is unique in terms of proton conductivity in the concentrated hot acid with a small fraction of the vehicular mechanism. The combination of the extremely low vapor pressure and high anhydrous conductivity allows for operation of PAFC at around 200 °C. The vapor phase over the hot acid is predominantly consisting of water with the acid content in the ppm level. This makes it easy to remove the product water with little loss of the acid during the fuel cell operation.
- The acid electrolyte is immobilized in a matrix composed of typically SiC powders bonded by PFTE. PTFE is also the bonder materials in construction of gas diffusion electrode. Retention of phosphoric acid in the electrolyte matrix and electrode layers is a limiting factor of the cell lifetime. The loss of the acid is primarily via an evaporation mechanism and additional inventory of the acid, often in the anode side, is therefore necessary.
- In acidic electrolytes Pt and its alloys are the state-of-the-art catalysts for both electrodes where the slow kinetics of the cathodic ORR is a challenge. High platinum loading and cost are a well-recognized issue, but cost of all other components and their processing should also be significantly reduced. New catalysts based on

platinum alloys with higher activity may increase the cell voltage and therefore energy efficiency, though their long-term stability is yet demonstrated.

- The Pt-based catalysts are supported on high specific surface area carbons. Materials for other construction components such as electrode substrate, bipolar plates, and cell separate plates are also carbon based. The corrosion resistance of carbon is an issue for the long-term lifetime.
- In terms of performance PAFCs operate in a low current density range from 100 to 400 mA/cm^2 at a typical cell voltage of 0.6–0.7 V under ambient pressure and operating with air and reformate hydrogen. Operation of PAFC at higher current densities and hence higher power densities is, however, not easy to achieve because of the high iR due to the thickness of the electrolyte matrix. Any further decrease in the matrix thickness is limited by the permeation of gaseous reactants.
- The fuel cell can tolerate about 1 vol% CO and 20 ppm H_2S or 50 ppm (COS + H_2S). Doubling the stack pressure may result in a voltage increase by about 40 mV. Under normal operation the state-of-the-art PAFC systems show a degradation rate of 2–3 mV for every 1000 h. The target lifetime of 40,000 h for utility applications is still a challenge.
- The technology is commercialized in 1990s with trademark PC-25™. The commercial viability of the power plants has been demonstrated as an electrical power sources, though their heavy and bulky characteristics make them unsuitable for mobile applications. The greatest obstacle for the PAFC commercialization is its high cost, probably 2–3 times higher than the required for significant market penetration at 300–400 US$/kW. The manufacturing rate of 30–40 PC-25™ power plants per year does not seem to be inadequate to drive the cost reduction and to attract further investment.

14.6 Questions

•

14.1 Explain the following terms and materials:

Pyrophosphoric acid
Transference number of structural diffusion
Crystalline size effect and platinum dissolution
Matrix and electrolyte reservoir
Carbon corrosion
Acid loss.

14.2 It was estimated that the equilibrium water vapor pressure over 100% H_3PO_4 at 200 °C is *ca.* 0.15 Bar. Justify that this acid is thermally stable during PAFC operation at this temperature using an air stoichiometry of $\lambda_{Air} = 2$. (Clue: at $\lambda_{Air} = 2$ the water content in the cathode chamber is *ca.* 0.19 bar)

14.3 The phosphoric acid loading in the SiC matrix is estimated to be *ca.* 40 mg H_3PO_4/cm^2. The platinum loading of the cathode is 0.5 mg_{Pt}/cm^2. Under OCV the platinum solubility in 100% H_3PO_4 at 200 °C is 10^{-4} M (mol/L). What is the maximum fraction of the total cathode Pt metal can go into the acid phase under OCV at 200 °C? The density of 100% PA is 1.88 g/cm^3). (Answer: 0.3%)

14.4 It was reported that the acid loss rate in the gas stream of a PAFC operating at 190 °C is 2 $\mu g/cm^2$ h. When 50% of the total acid inventory is lost, it is assumed to be the end-of–the-life of the fuel cells. Estimate the lifetime of the above PAFC having an acid loading of 40 mg H_3PO_4/cm^2. (Answer: 10,000 h)

14.5 For the above PAFC to operate for 50,000 h, how much acid is needed to store in the acid reservoir plate? (Answer: 80 mgH_3PO_4/cm^2)

14.6 A lifetime test of PAFC operating with reformate hydrogen and air at 190 °C showed that the average specific surface area (SA) of platinum nanoparticles was decreased to 61% and 36% of the initial value after 15,000 h and 60,000 h, respectively. The polarization curve of the fuel cell exhibited a Tafel slope of b = 100 mV/dec. Estimate the fuel cell voltage losses due to the Pt particle growth during the lifetime test. (Clue: The voltage loss can be estimated using $\Delta V = b \times \log(SA_{EoL}/SA_{BoL})$, where SA_{EoL} and SA_{BoL} are the Pt specific surface area at the end-of-the-life and beginning-of-the-life, respectively) (Answer: 21 and 43 mV).

References

1. Dippel T, Kreuer KD, Lassegues JC, Rodriguez D (1993) Proton conductivity in fused phosphoric-acid—a H^1P^{31} PFG-NMR and QNS study. Solid State Ionics 61(1–3):41–46
2. Melchior JP, Kreuer KD, Maier J (2017) Proton conduction mechanisms in the phosphoric acid-water system ($H_4P_2O_7$–H_3PO_4 $2H_2O$: a H^1, P^{31} and O^{17} PFG-NMR and conductivity study. Phys Chem Chem Phys 19(1):587–600
3. Korte C, Conti F, Wackerl J, Lehert W (2016) Phosphoric acid and its interactions with polybenzimidazole-type polymers. In: Li Q, Aili D, Hjuler HA, Jensen JO (eds) High temperature polymer electrolyte membrane fuel cells—Approaches, status and perspectives. Springer, pp 169–194
4. Becker H (2017) Phosphoric acid issues in high temperature polymer electrolyte membranes. Ph.D. Technical University of Denmark
5. Kunz HR, Gruver GA (1975) Catalytic activity of platinum supported on carbon for electrochemical oxygen reduction in phosphoric acid. J Electrochem Soc 122(10):1279–1287
6. Bett JAS, Kinoshita K, Stonehart P (1976) Crystallite growth of platinum dispersed on graphitized carbon black. 2. Effect of liquid environment. J Catal 41(1):124–133
7. Landsman DA (2003) Catalyst studies and coating technologies. In: Vielstich W, Lamm A, Gasteiger H (eds) Handbook of fuel cells - fundamentals, technology and applications, vol 3. Wiley, pp 811–831
8. Bindra P, Clouser SJ, Yeager E (1979) Platinum dissolution in concentrated phosphoric acid. J Electrochem Soc 126(9):1631–1632

9. Thompsett D (2003) Pt alloys as oxygen reduction catalysts. In: Vielstich W, Gasteiger H, Lamm A (eds) Handbook of fuel cells—fundamental, technology and applications, vol 3. Wiley, pp 467–480
10. Niedrach LW, Alford HR (1969) Polytetrafluoroethylene coated and bonded cell structure. US Patent 3,432,355
11. Breault RD (2003) Stack materials and stack design. In: Vielstich W, Lamm A, Gasteiger H (eds) Handbook of fuel cells—fundamentals, technology and applications, vol 4. Wiley, pp 797–810
12. Brown EH, Whitt CD (1952) Vapor pressure of phosphoric acids. Ind Eng Chem 44:615–618
13. Okae I, Kato S, Seya A, Kamonoshita T (1990) The 67th Spring Meeting of the Chemical Society of Japan, 1990, p 148

Chapter 15
Molten Carbonate Fuel Cells

15.1 Introduction

In 1896, William W. Jacques reported the first "carbon electric generator," an essential fuel cell with a solid carbon electrode. An iron container was filled with potash (potassium hydroxide) and held over a gas flame to keep the potash electrolyte in the molten state at about 500°C. A coal (carbon) electrode was submerged in the potash, and air was blown into the molten electrolyte through a nozzle. The coal acted as the anode electrode and the air (container) electrode as the cathode, as shown in Fig. 15.1. The invention was by then speculated to be a more efficient and cleaner use of coal for power generation.

The electrode reactions are assumed to be

$$\text{Cathode}: \quad O_2 + 2H_2O + 4e^- \rightarrow 4OH^- \tag{15.1}$$

$$\text{Anode}: \quad C + 4OH^- \rightarrow CO_2 + 2H_2O + 4e^- \tag{15.2}$$

$$\text{Overall}: \quad C + O_2 \rightarrow CO_2 \tag{15.3}$$

This is, however, not true as the anode product CO_2 chemically reacts with hydroxide forming carbonate. The real cell reactions are in fact involving consumption of hydroxide ions

$$\text{Cathode}: \quad O_2 + 2H_2O + 4e^- \rightarrow 4OH^- \tag{15.4}$$

$$\text{Anode}: \quad C + 6OH^- \rightarrow CO_3^= + 3H_2O + 4e^- \tag{15.5}$$

$$\text{Overall}: \quad C + O_2 + 2OH^- \rightarrow CO_3^= + H_2O \tag{15.6}$$

S. P. Jiang and Q. Li, *Introduction to Fuel Cells*,
https://doi.org/10.1007/978-981-10-7626-8_15

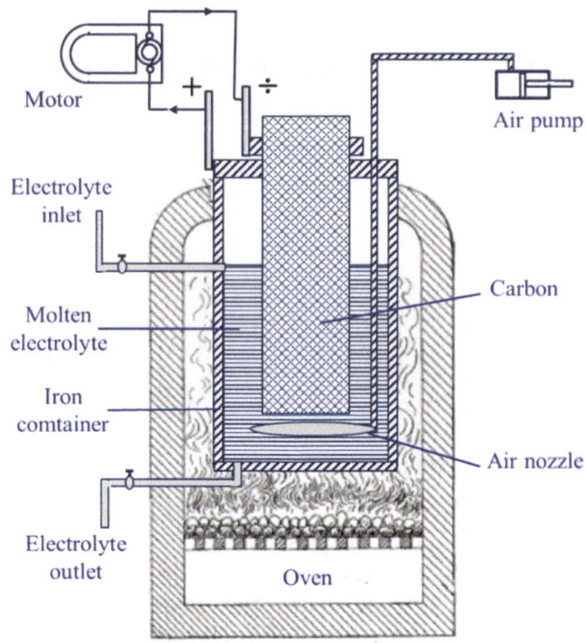

Fig. 15.1 Schematic of carbon fuel cell by William Jacques from 1896

It is seen from above reactions that for formation of every 4 mole OH^- at the cathode, 6 mole OH^- are consumed at the anode. As a result, the cell reaction is accompanied by a net consumption of the high value caustic polash (KOH) or soda (NaOH), being converted to the low value caustic ash (K_2CO_3 or Na_2CO_3). The concept of carbon fuel cells using molten hydroxide electrolytes is still a topic in literature today [1], but when consumption of a valuable chemical is involved, the electrochemical device is not a fuel cell, rather called a battery. The idea of using molten carbonates has, however, inspired the early effort of molten carbonate fuel cell (MCFC) development in 1950s [2] where, instead of a solid carbon electrode, hydrogen-rich fuels from coal gasification and natural gas reforming are used.

The fuel flexibility is an advantage of MCFC resulting from the relatively high operating temperature at around 650 °C. Another advantage is the fast electrode kinetics so that nickel-based low-cost catalysts provide sufficient catalytic activity. In addition, the waste heat is also of high value.

Challenges for MCFC originate from the mobile electrolyte, which corrodes electrode and construction materials and impacts the mechanical stability and lifetime of key materials and components. In addition a source of CO_2 is required at the cathode which can be recycled from the anode exhaust, however, needs additional balance of plant components. The high contact and cathode kinetic resistances limit the fuel cell power output which is commonly below 200 mW/cm^2 at practical operating voltages. This low power density limits the MCFC to larger stationary and marine applications, where the relatively large size and weight of MCFC and slow start-up time are not an issue. The technology has been demonstrated in a power range from 300 kW to 3

MW. The operating pressure is from ambient to 10 atm, while the temperatures are between 600 to 700 °C. International developers include Fuel Cell Energy (USA), CFC Solution (Germany), Ansaldo Fuel Cells (Italy), Ishikawajima-Harima Heavy Industries (Japan), POSCO Energy (Korea), among others.

Based on hydrogen and oxygen, the electrochemical reaction involves formation of the carbonate ion at the cathode which transports to anode where hydrogen oxidization occurs:

$$\text{Cathode}: \quad \frac{1}{2}O_2 + CO_{2(c)} + 2e^- \rightarrow CO_3^= \tag{15.7}$$

$$\text{Anode}: \quad H_2 + CO_3^= \rightarrow CO_{2(a)} + H_2O + 2e^- \tag{15.8}$$

$$\text{Cell}: \quad H_2 + \frac{1}{2}O_2 \rightarrow H_2O \qquad \Delta G^0_{650^\circ C} = -197\,\text{kJ/mol} \tag{15.9}$$

The Nernst reversible voltage for MCFC can be expressed as follows:

$$E = E^0 + \frac{RT}{2F}\ln\frac{p_{O_2}^{1/2}p_{H_2}}{p_{H_2O}} + \frac{RT}{2F}\ln\frac{p_{CO_{2(c)}}}{p_{CO_{2(a)}}} \tag{15.10}$$

where the subscript *a* and *c* indicate the carbon dioxide partial pressure at the anode and cathode side, respectively. When the partial pressures of carbon dioxide are identical for the two electrodes, the cell voltage is a function of the partial pressure of oxygen, hydrogen, and steam.

The high operating temperature of MCFC enables the use of CO as a fuel, and the reactions are as follows:

$$\text{Anode}: \quad CO + CO_3^= \rightarrow 2CO_{2(a)} + 2e^- \tag{15.11}$$

$$\text{Cathode}: \quad \frac{1}{2}O_2 + CO_{2(c)} + 2e^- \rightarrow CO_3^= \tag{15.12}$$

$$\text{Cell reaction}: \quad CO + \frac{1}{2}O_2 \rightarrow CO_2 \qquad \Delta G^0_{650^\circ C} = -210\,\text{kJ/mol} \tag{15.13}$$

It is interesting to notice that the standard free energies for reactions (15.9) and (15.13) at 650°C are nearly the same so are the reversible voltages of the cells.

With the carbonate $\left(CO_3^=\right)$ as the charge carrier through the electrolyte, there is an issue of CO_2 management, i.e., CO_2 is consumed at the air side and supplied at the hydrogen side. This sounds complicated but can be easily achieved in practice. When the fuel cell operates at around 650°C, the inlet air needs to be preheated by burning the anode exhaust gas, which is a mixture of methane, steam, and CO_2. The preheated air containing CO_2 is fed into the cathode, as shown in Fig. 15.2.

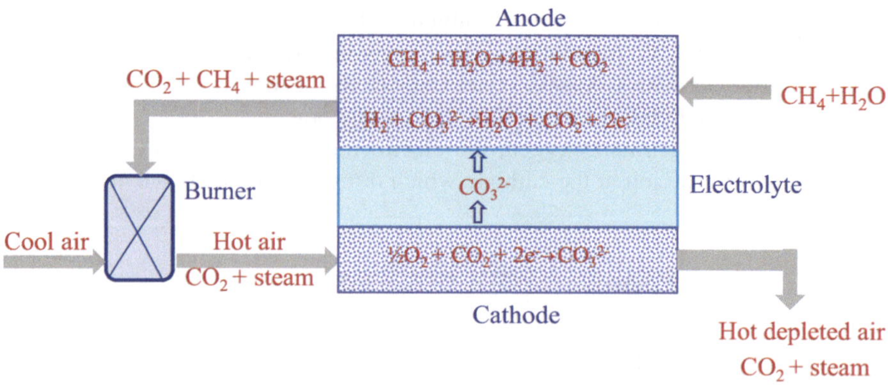

Fig. 15.2 Principle of the molten carbonate fuel cell operation

15.2 Electrolytes

15.2.1 Alkali Metal Carbonates

The MCFC electrolyte is based on a mixture of molten alkali metal carbonates. The melting point and ionic conductivity above the melting point of each salt are listed in Table 15.1. Of these lithium carbonate has the lowest melting point but the highest conductivity.

The Li_2CO_3–Na_2CO_3 mixture forms a simple eutectic system with a eutectic point of 53.3 mol% Li_2CO_3 at 501° C, as seen from Fig. 15.3a.The Li_2CO_3-K_2CO_3 binary system exhibits two eutectic points at 42.7 and 62.0 mole % Li_2CO_3, with melting points of 498 and 488° C, respectively. They form an equal molar compound $LiKCO_3$ with a congruent melting point at 505 °C, see Fig. 15.3b. K_2CO_3 and Na_2CO_3 are completely miscible, exhibiting a minimum melting point of 710 °C at 56–59 mol% Na_2CO_3.

The Li_2CO_3-Na_2CO_3-K_2CO_3 ternary mixture has a eutectic point of 397 °C at composition of 43.5-31.5-25.0 mol%. This composition nearly corresponds to a simple molar ratio of 1.75:1.25:1.0.

Table 15.1 Melting points and conductivities of alkali metal carbonates

Electrolyte (mol%)	Melting point (°C)	Conductivity (S/cm)
Li_2CO_3	730	5.66 (800°C)
Na_2CO_3	858	3.03 (900°C)
K_2CO_3	901	2.18 (950°C)
$(Li_{0.53}Na_{0.47})_2CO_3$	501	0.85 (600°C)
$(Li_{0.435}Na_{0.315}K_{0.25})_2CO_3$	397	0.75 (500°C)

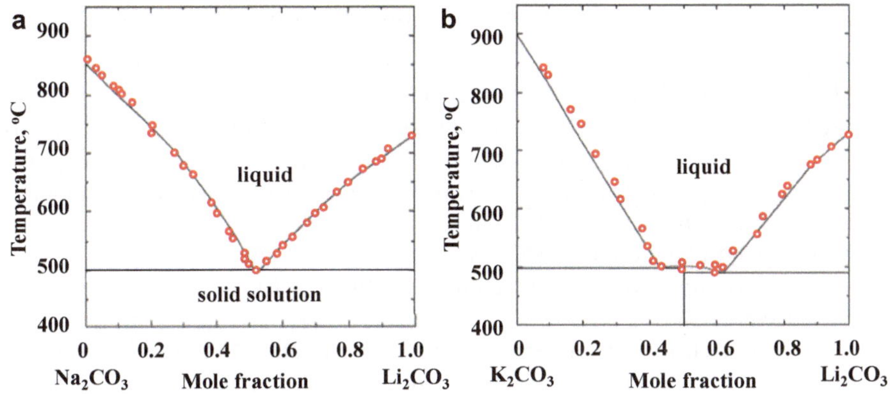

Fig. 15.3 Na_2CO_3—Li_2CO_3 **a** and K_2CO_3—Li_2CO_3 **b** binary phase diagrams

15.2.2 Acid—Base Chemistry

Upon melting, alkali metal carbonates dissociate according to the following equilibria:

$$M_2CO_3 \rightleftharpoons 2M^+ + CO_3^= \tag{15.14}$$

$$CO_3^= \rightleftharpoons O^= + CO_2 \tag{15.15}$$

Chemically the oxide ion, $O^=$, is considered to be a basic component and the gaseous CO_2 an acidic component in the carbonate melt. The basicity of the melt is defined by $-\log p_{O^=}$ and the acidity by $-\log p_{CO_2}$, both being influenced by the partial pressure of CO_2 and the content of alkali metals. Mixtures rich of Li_2CO_3 or containing alkaline earth metals, e.g., Ca and Ba are considered to be of higher basicity.

Water, as always present as the fuel cell reaction product or in humidified gas, is involved in the electrolyte chemistry *via* the following hydrolysis

$$CO_3^= + H_2O = CO_2 + 2OH^- \tag{15.16}$$

This affects the melt equilibrium (15.15) by

$$O^= + H_2O \rightarrow 2OH^- \tag{15.17}$$

$$2OH^- + CO_2 \rightarrow CO_3^= + H_2O \tag{15.18}$$

These reactions may provide alternative reaction paths for the oxygen reduction reaction.

15.2.3 Properties of Molten Carbonate Systems

For MCFC the electrolyte issues include conductivity or ohmic loss, evaporation loss, NiO solubility, material corrosion, and the oxygen reduction kinetics. Frequently used compositions of MCFC electrolyte materials are binary $(Li_{0.52}Na_{0.48})_2CO_3$, $(Li_{0.62}K_{0.38})_2CO_3$, or eutectic ternary $(Li_{0.435}Na_{0.315}K_{0.25})_2CO_3$ melt. A large fraction of Li carbonate is often used due to its higher ionic conductivity. The content of Li in these melts is, however, limited to 62% since too high Li contents lower the solubility and ionic diffusivity of oxygen. In addition, the high Li_2CO_3 content electrolyte is very corrosive.

The Li/Na electrolyte has been extensively investigated. Compared to the Li/K melt, the Li/Na-based electrolyte has shown a higher conductivity and therefore a better fuel cell performance and improved stability of the cathode NiO. Another advantage of Li/Na melt over Li/K is a lower vapor loss of the molten electrolyte, which is essential for enhancement of the stack durability. The Li/K system, on the other hand, exhibits a better electrode kinetics due to its high oxygen solubility and better electrolyte wetting in the electrode.

The oxygen solubility and electrolyte wettability of the Li/Na melt can be improved by using electrolyte additives. The Li/Na electrolyte doped with a small amount of alkaline earth metals (Ca and Ba) offers a high performance at low temperatures resulting from the increased oxygen solubility. The addition of rare earth metal oxides particularly La_2O_3 to the Li/Na electrolyte, in addition to suppressing NiO dissolution, has been found to improve the oxygen solubility.

The electrolyte conductivity is essential for the fuel cell performance. The density (or molar volume) of the electrolyte is related to the optimal loading of the electrolyte in the matrix and electrode. The surface tension determines the wetting of the electrolyte on the porous electrodes and therefore the electrochemical processes. The surface tension of the electrolyte is also associated to the retention of the electrolyte in the matrix *via* a capillary mechanism. In addition the electrolyte creeping on cell construction materials is critical for the long-term stability of fuel cells. For the eutectic ternary electrolyte, the expressions and value ranges of these properties are given in Table 15.2.

Moreover, the vapor pressure of the electrolyte is connected to the electrolyte losses during operation which is a lifetime-limiting issue for the MCFC systems. The most volatile species is potassium, which is known to evaporate as the volatile

Table 15.2 Density, surface tension, and electrical conductivity of molten $(Li_{0.435}Na_{0.315}K_{0.25})_2CO_3$ ternary eutectic from reference [3]

Properties	Temperature range (K)	Value at 1073 K
Density $\rho = 2.39 - 4.45 \times 10^{-4}$ T (g/cm^3)	770–1163	1.91 g/cm^3
Surface tension $\gamma = 27.0 - 0.0063$ T (N/cm)	770–1163	20.2 N/cm
Conductivity $\sigma = 2.80 + 4.61 \times 10^{-3}$ T$- 2.91 \times 10^{-4}$ T^2 (S/cm)	675–1134	2.11 S/cm

potassium hydroxide. The vapor pressure of the corresponding sodium and lithium hydroxides is somewhat smaller. Therefore, the electrolyte loss due to evaporation is critical in fuel cells with Li/K or Li/Na/K melts. Use of Li/Na electrolytes is, therefore, expected to extend the lifetime.

15.2.4 Comparison of Li/K and Li/Na Systems

Table 15.3 summarizes the discussion made above. In brief, Li/K-based electrolytes show high ionic conductivity and better kinetics for the oxygen reduction due to high oxygen solubility and better wetting characteristics in the electrode; however, they suffer from higher NiO solubility and vapor pressure. For operation under ambient pressure, the Li/K electrolyte is estimated to be adequate for a lifetime of up to 5 years. For high-pressure operation, on the other hand, the Li/Na-based electrolyte is needed in order to achieve the long lifetime. Additives of alkaline earth metal carbonates have been extensively investigated in order to increase the melt basicity primarily for retarding the NiO cathode dissolution. No significant effects of the additives have been reported on the electrode polarization and construction material corrosion. The negative effect of the additives is the decrease in the electrolyte conductivity.

Table 15.3 Qualitative comparison of electrolyte compositions [4]

Characteristics	Electrolyte systems
Cathodic oxygen reduction kinetics	(Li/K) > (Li/Na)
NiO dissolution	(Li/K) > (Li/Na) > (Li/K(Na)/Additive)
Ionic conductivity	(Li/Na) > (Li/K) > (Li/K/Additive)
Vaporization loss	(Li/K) > (Li/Na)
310S corrosion in oxidizing gas and wet seal area	(Li/K) > (Li/Na)
Cell performance under atmospheric pressure	(Li/K) > (Li/Na)
Cell performance under pressurized conditions	(Li/Na) > (Li/K)

15.3 Cathode Reactions and Materials

15.3.1 Oxygen Reduction Reaction in Molten Carbonates

The cathode exhibits the major polarization of the MCFC, which is one of the reasons for the lower power density of the technology. On the porous nickel oxide, the oxygen reduction reaction has an exchange current density of a few mA/cm^2 in molten carbonates at the MCFC temperatures, which is about two orders of magnitude smaller than that of the hydrogen oxidation on the nickel metal anode [5].

Extensive efforts have been made to elucidate the reaction mechanism of the oxygen reduction in molten carbonates, in which the oxide ($O^{=}$) and carbonate $CO_3^{=}$ exist under the equilibrium (15.15). In the presence of oxygen as in the vicinity of the cathode, peroxide ($O_2^{=}$ or $O^- \cdot O^-$) and superoxide (O_2^- or $O \cdot O^-$) species can be formed:

$$O_2 + CO_3^{=} = 2O_2^{=} + 2CO_2 \tag{15.19}$$

$$3O_2 + 2CO_3^{=} = 4O_2^{-} + 2CO_2 \tag{15.20}$$

As seen from these reactions, the predominance of the peroxide and superoxide depends on the partial pressure of oxygen and carbon dioxide. It should be of note that, in a practical MCFC system operated under atmospheric pressure, the cathode gas is recycled and air is used to provide oxygen. As a result, the partial pressures of oxygen and carbon dioxide are always lower than 0.2 atm and frequently lower than 0.1 atm.

These species are of significance for the electrode reaction mechanism, based on which two reaction mechanisms, the superoxide mechanism and peroxide mechanism, have been proposed as schematically represented in Fig. 15.4. According to

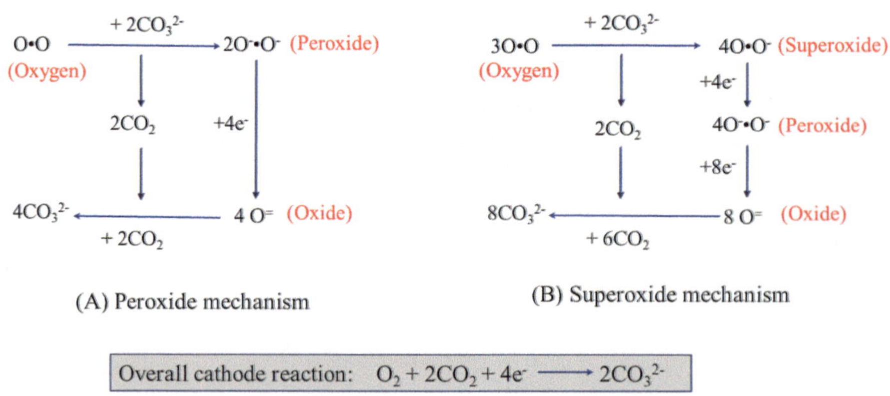

Fig. 15.4 Schematic representation of the oxygen reduction mechanisms in molten carbonates

Appleby et al. [6], the peroxide mechanism is proposed to prevail in pure lithium and lithium-rich melts, while the superoxide mechanism seems to be involved in sodium and potassium-rich melts for the oxygen reduction reaction.

15.3.2 Cathode Materials

Porous nickel oxide (NiO), in its lithiated form, is the almost exclusive cathode material for MCFC. The oxide is often *in situ* formed by oxidation of porous metallic nickel during cell conditioning. NiO is a *p*-type semiconductor. Its conductivity is significantly enhanced by lithiation, reaching a value about 15 S/cm. The lithiation involves the oxidation of Ni^{2+} to Ni^{3+} and is achieved *in situ* with lithium provided by the Li_2CO_3 component of the molten electrolyte:

$$(1-x)\mathrm{NiO} + x\mathrm{Li} + x\mathrm{CO}_3^{=} = \mathrm{Li}_x\mathrm{Ni}_{1-x}\mathrm{O} + x\mathrm{CO}_2 + xe^- \tag{15.21}$$

15.3.3 NiO Dissolution and Precipitation

The NiO cathode has a solubility of the low ppm level in molten carbonates. The dissolution is believed to be controlled by the acid–base chemistry of the oxyanion-containing molten carbonates, where the oxide anion, $O^{=}$, is considered to be the basic component of the melt in the same sense as the hydrogen ion defines the acidity in aqueous solutions.

In melts with a low oxide ion activity, corresponding to a high partial pressure of CO_2, nickel oxide is found to dissolve in an acidic mode:

$$\mathrm{NiO} = \mathrm{Ni}^{2+} + \mathrm{O}^{=} \tag{15.22}$$

In melts of high oxide ion activity or under low CO_2 partial pressures, on the other hand, the basic mode of dissolution occurs with the formation of complex anions:

$$\mathrm{NiO} + \mathrm{O}^{=} = \mathrm{NiO}_2^{=} \tag{15.23}$$

or

$$\mathrm{NiO} + \mathrm{CO}_3^{=} = \mathrm{NiO}_2^{=} + \mathrm{CO}_2 \tag{15.24}$$

For single alkali metal carbonates, the NiO solubility in both acidic and basic regimes increases in the order from Li_2CO_3, Na_2CO_3 to K_2CO_3. In binary mixtures of Li-K carbonates, both acidic and basic dissolution of NiO have been observed. Figure 15.5 shows the NiO solubility as a function of log p_{CO_2} in Li/K carbonates

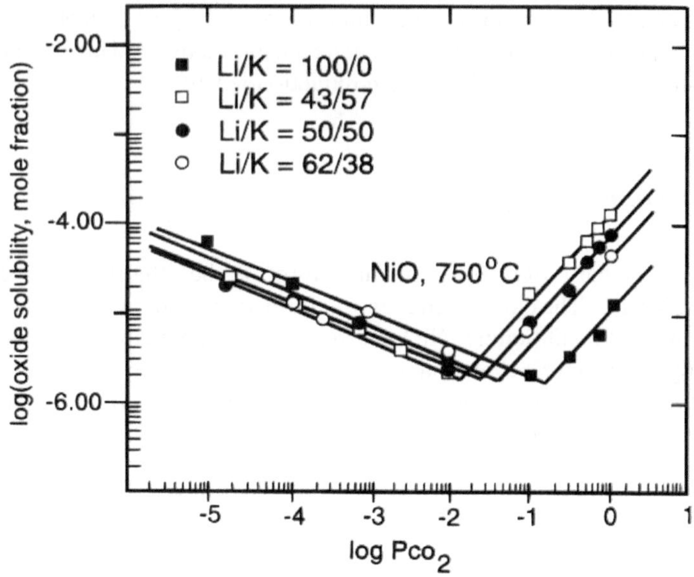

Fig. 15.5 Solubility of nickel oxides as a function of carbon dioxide partial pressure in Li/K carbonates. Modified from Ref. [7] with permission from IOP Publishing, Copyright 1992

[7]. As seen from the figure the solubility decreases with the content of Li_2CO_3 in the high CO_2 partial pressure region.

During the fuel cell operation the dissolved nickel ions diffuse toward the anode and are subsequently reduced when encounter the anode H_2. This leads to reprecipitation of nickel which may cause an internal short circuit between the two electrodes. For fuel cell operation under atmospheric pressure, i.e., with lower CO_2 partial pressure, the nickel shorting is fairly manageable as already demonstrated by the long-term field operation of MCFC for over 40,000 hours. More basic Li/Na electrolytes can further extend the cell lifetime. For high-pressure operation corresponding to higher CO_2 partial pressures, however, the nickel shorting is a critical concern.

15.3.4 Electrolyte Additives and Alternative Materials

Several factors affect the dissolution of the cathode material. Since the early beginning of the carbonate fuel cell research, it has been known that the acidity and basicity of the electrolyte have a strong influence on the dissolution. The extent of dissolution is controlled mainly by electrolyte composition, applied gas atmosphere, operation pressure (and therefore CO_2 partial pressure), and temperature.

Electrolyte additives are widely used to modify the acid–base properties of the carbonates. In addition, applied current, thermal convection, porosity of the cathode

Table 15.4 Solubility of NiO and $LiCoO_2$ in Li/Na and Li/K carbonates at 650°C under 30%CO_2–70% atmosphere [8]

Oxide and pressure	Solubility (molar ppm)		Deposition rate in $(Li_{0.62}K_{0.38})_2CO_3$ (μg/cm^2·h)
	$(Li_{0.52}Na_{0.48})_2CO_3$	$(Li_{0.62}K_{0.38})_2CO_3$	
NiO, 1 atm	4	12	1.2–1.5
NiO, 7 atm	25	45	–
$LiCoO_2$, 1 atm	3	4	0.01–0.04
$LiCoO_2$, 7 atm	20	40	–

and the matrix, and particle characteristics are also important for the transport of the dissolved nickel oxide species.

A set of solubility and the deposition rate data of NiO is listed in Table 15.4 under rich gas conditions (air:CO_2 = 70:30) at 650°C. The solubility of NiO in the Li/Na melt is significantly lower under atmospheric conditions than in the Li/K melt. Under pressurized conditions the solubility increases.

Alternative materials to nickel oxide have been explored, and the most promising is lithium cobaltite which shows a better stability in molten carbonates. The solubility and deposition rate of the lithium cobaltite are also listed in Table 15.4. Similar results have been reported for the cobalt oxide powder as well as cerium oxide and magnesium oxide. Efforts have also been made to prepare cobalt or lithium cobaltite coatings on either external surface or internal surface inside the porous NiO structure. Solution gel and solution impregnation techniques have been developed for this purpose, and single cell tests showed significant reduction of the nickel precipitation on the coating layer.

15.4 Anode Reactions and Materials

The working temperature of MCFC at around 650°C allows for the integration of the electrochemical cell with a fuel processing reactor, the internal steam reformer, for production of hydrogen:

$$CH_4 + H_2O = CO + 3H_2 \tag{15.25}$$

in association with the water–gas shift reaction

$$CO + H_2O = CO_2 + H_2 \tag{15.26}$$

The hydrogen is immediately consumed by electrochemical oxidation at the fuel cell anode:

$$H_2 + CO_3^{=} \rightarrow H_2O + CO_2 + 2e^- \tag{15.27}$$

Active metals commonly used as reforming catalysts are Ni and noble metals. Nickel is the most widely employed because of its low-cost and sufficiently high activity. Decomposition of methane on the nickel surface is believed to be the first step of the steam reforming, followed by reactions of the carbon species formed on the surface with steam or surface oxygen species. Challenges for the reforming catalysts include the coke formation and growth of Ni particles. Under MCFC conditions the poisoning of alkali carbonate electrolytes, in form of alkali metal hydroxide vapor, is an issue. The effect of the alkali vapor is most pronounced for K and less pronounced for Li and Na and can be mitigated by supporting the nickel catalysts on magnesia, alumina or silica. The evaporation of alkali metals leads to poisoning of the internal reforming catalyst. A porous membrane has been suggested to be placed between the cell internal reforming catalyst and the electrolyte-containing cells in order to trap the volatile alkali species, while the hydrogen-rich fuel generated from the internal reforming can pass through the membrane. Significant extension of the fuel cell lifetime has been reported [9].

The requirements for the MCFC anode are good electrical conductivity and structural stability with, of course, a suitable catalytic activity. In molten carbonates, the number of possible electrode materials is very limited. For the anode, a very few metals or alloys are stable enough which are in general noble metals, being too expensive to be of practical use. Since the early stage of the MCFC development, the pure Ni-based anodes have demonstrated a reasonable electrochemical activity with a polarization loss of less than 30 mV at current densities of 150–200 mA/cm^2.

These anodes are commonly manufactured by tape casting or hot pressing of metallic powders. A critical issue with the pure Ni anode is the performance degradation after a period of operation. The high creep strains of up to 40% have been observed for pure Ni anodes. At the same time the decrease in the surface area and anode shrinkage occurs due to sintering of the catalyst particles. To overcome these problems, alloying of Ni anode with metals such as Cr, Al or both is a common practice and the Ni alloys show improved creep resistance and stability of its electrochemical activity under MCFC operating conditions. The creep resistance is an ability of a material to resist any kind of distortion under the influence of stresses over an extended period of time. The Ni–Al anode shows higher creep resistance, while the Ni–Cr anode can be easily lithiated by the electrolyte to produce $LiCrO_2$ during cell operation. For this purpose a layer of aluminum or Ni_3Al alloy is applied by the sol–gel precipitation technique on the surface of the nickel or Ni-Cr anode. This layer is expected to increase the electrolyte wettability of the internal surface of the pores and strengthen the pore structure of the anode.

15.5 Matrix and Immobilization of Molten Carbonate Electrolyte

15.5.1 *Matrix Materials*

A molten carbonate mixture is stored in a matrix, which functions as electrolyte to facilitate the ionic transport and isolate the fuel from the oxidant. The matrix consists of ultra-fine lithium aluminate ($LiAlO_2$) powders, often with incorporated fibers or coarse particulates for strengthening, crack attenuation, and thermal cycleability enhancement.

Two allotropic phases of $LiAlO_2$ are present in the MCFC temperature range, the α phase with a hexagonal structure and stable at temperatures lower than 650 °C, while the γ phase with a tetragonal structure and stable at temperatures above 700 °C. The former has a density of 3.4 g/cm^3 while the latter of 2.6 g/cm^3. Though stable and preferred, the α-$LiAlO_2$ is more expensive as it is produced in a batch method with long process time and low yield. Instead the high surface area submicron γ-LiAlO powder (*ca.* 10 m^2/g) is the baseline material for practical use. It has been reported that during the long-term fuel cell operation, significant transformation of the γ-phase to the α-phase occurs. The transformation may result in changes in the pore volume as well as the mean pore size. As a result, this will cause the electrolyte loss and hence the reactant crossover, ionic voltage loss, and the performance degradation of the fuel cell.

15.5.2 *Immobilization and Distribution of Electrolyte*

The immobilization of the molten electrolyte in the microporous ceramic structure is by the capillary force. From Chap. 14 it is learnt that PTFE is used as a binder and wet-proofing agent to the SiC matrix for retention of the liquid acid electrolyte. Such a material like PFTE is not available for MCFC at the high temperatures. To establish a stable electrolyte/gas interface in porous MCFC electrodes, the capillary pressures should be in equilibria as follows:

$$\frac{\gamma_c \cos\theta_c}{D_c} = \frac{\gamma_a \cos\theta_a}{D_a} = \frac{\gamma_e \cos\theta_e}{D_e} \tag{15.28}$$

where γ is the interfacial surface tension of the electrolyte, θ is the wetting angle of the electrolyte, D is the pore diameter, and the subscripts a, c, and e refer to the anode, cathode, and electrolyte matrix, respectively. Fixing the electrolyte surface tension and wetting characteristics, the filling up of the electrolyte into each of the porous components is determined by the equilibrium pore size. Small pores are filled with the molten electrolyte, while large pores remain open for gases.

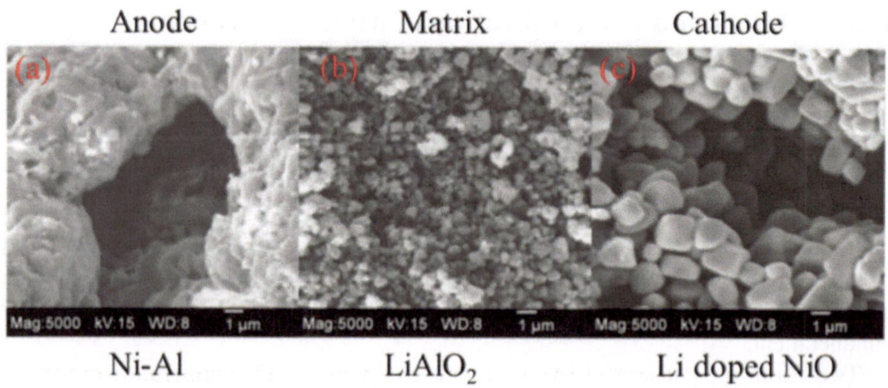

Fig. 15.6 SEM images of MCFC active components. **a)** anode of the larger pore size and low surface area, **b)** matrix of fine pore size and **c)** cathode of the dual-porosity agglomerated structure. Reproduced from Ref. [10] with permission from IOP Publishing, Copyright 2009

By adjusting the particle size and therefore the pore diameter in the three components, the electrolyte matrix with smallest pores is to be completely filled with the molten electrolyte while the two porous electrodes are partially filled. Figure 15.6 shows examples of particles as well as pore size of the anode, cathode, and electrolyte matrix. Specifications of these materials in terms of the specific surface area are also listed in Table 15.5. By formulation into the electrode and electrolyte matrix layers, these powder specifications correlate to the pore size and porosity of the fuel cell components.

15.5.3 Matrix Stability and Fabrication

To achieve long-term operation, it is essential that the matrix is made of α-$LiAlO_2$ powders with stable and small size pores. A uniform particle size distribution is also essential in order to alleviate the Ostwald-ripening coarsening process. Growth of lithium aluminate particles and the phase transformation are believed to take place *via* a dissolution–precipitation mechanism. The solubility of α-$LiAlO_2$ in molten carbonates is, therefore, crucial for the long-term durability of the electrolyte matrix. The $LiAlO_2$ solubility (S) in the Li/Na melt increases with temperature and has an inverse relation with the partial pressure of CO_2, as expressed by

$$\text{Log}\, S = -\frac{18.4 \times 10^3}{T} - \log p_{CO_2} + 10.5 \qquad (15.29)$$

In $Li_{52}Na_{48}$ melt at 650 °C and under ambient air (410 ppm CO_2) atmosphere, the $LiAlO_2$ powder has a solubility around 10 mol ppm. This value becomes about 3 times lower in $Li_{62}K_{38}$ melt [11]. In addition, the solubility of the α phase is lower

Table 15.5 Summary of anode, cathode, and electrolyte matrix material and component characteristics

Components	Materials and specifications	Issues	Solutions
Anode	• Porous Ni, Ni-Cr or Ni-Al • Pore size: 3–6 μm • Specific area: 0.1–1 m^2/g • Thickness: 0.2–0.5 mm • Initial porosity: 50–70%	• Susceptible to creep • Particle sintering • Poor wetting by carbonate	• Alloying
Cathode	• In situ oxidized and lithiated NiO • Pore size: 7–12 μm • Specific area: 0.5 m^2/g • Thickness: 0.5–1.0 mm • Initial porosity: 70–80% • Porosity after lithiation: 60–65%	• Dissolution in electrolyte • Ni reprecipitation and short circuit between electrodes	• Oxide additives • Coatings • Modification of electrolyte
Matrix	• Porous γ-$LiAlO_2$ • Pore size 0.2–0.5 μm • Porosity 50–60% • Specific area: 0.1–12 m^2/g • Thickness: 0.5–1.0 mm	• Cracking due to thermal cycling • Loss of capillary retention of carbonates	• Reinforcement
Bipolar plates / current collector	• Stainless steel, e.g., S316	• Corrosion	• Ni coating or cladding

than that of the γ phase in the Li/Na melt, verifying the better stability of the α phase. MCFC stacks typically operate in a temperature range from 550 to 700 °C. Control of the operational temperature in a narrow range can avoid the undesirable $LiAlO_2$ phase transformation and therefore the particle growth.

In early years the matrix was fabricated by hot press of a mixture of $LiAlO_2$ and carbonates at temperatures slightly below the melting point of the carbonates, for example, 490 °C for the $(Li_{0.62}K_{0.38})_2CO_3$-based electrolyte. This structure, called the electrolyte tile, was about 1 to 2 mm thick and therefore had high ohmic resistance. Thinner electrolytes, also more uniform structures, are currently manufactured by tape casting. Issues for cost-effective fabrication of full-area thin matrices include faster milling, casting, and drying as well as a yield higher than 95%. Environmentally friendly slurry using an aqueous solvent is highly desired for mass production; however, a fast drying process is needed.

15.6 Construction Materials

Construction of MCFC stacks is schematically shown in Fig. 15.7. The anode and cathode catalyst layers are applied on perforated plates which are in direct contact with corrugated current collector plates. Through the current collector plates are fuel and oxidant gas channels in crossing flow directions. The separate plate, also called bipolar plates, is a dense plate facing the fuel channel on one side and oxidant channel on the other.

The temperature range around 650 °C allows use of high-temperature-resistant metal alloys as construction materials. Hot corrosion in the presence of molten carbonates presents a challenge to the material which causes formation of an oxide surface layer and hence an increase in the electrical contact resistance. In the corrosive carbonate environment at high temperatures, the basic requirements of these construction materials are chemical stability, good electrical conductivity, and low contact resistance at the interfaces with electrodes. The bipolar plates are exposed to different atmospheres on the anodic side, cathodic side, and wet seal area of MCFC, and hence the ideal materials for bipolar plate current collectors should have acceptable corrosion resistance for all these atmospheres. Metallic Ni meets these requirements but pure Ni is of high cost and lacking of desired mechanical flexibility. The low-cost austenitic stainless steels like 310S (24–26 wt% Cr) and 316 L (16–18 wt% Cr) have been the materials of choice.

The mechanism of corrosion in these steels involves formation of a layer of iron oxides on the chromium-rich oxide layer which undergo the lithiation. In a typical MCFC, the anodic atmosphere is more problematic than the cathodic atmosphere due to the continuous reduction of the protective oxide layer by hydrogen. It is a common practice that the anode-side surface of the stainless steel plates is coated with a nickel layer of typically 50 μm thick. This can be achieved by either electrolytic plating or electroless coating. In general the electroless nickel coating is dense, uniform but expensive compared with the electrolytic plating.

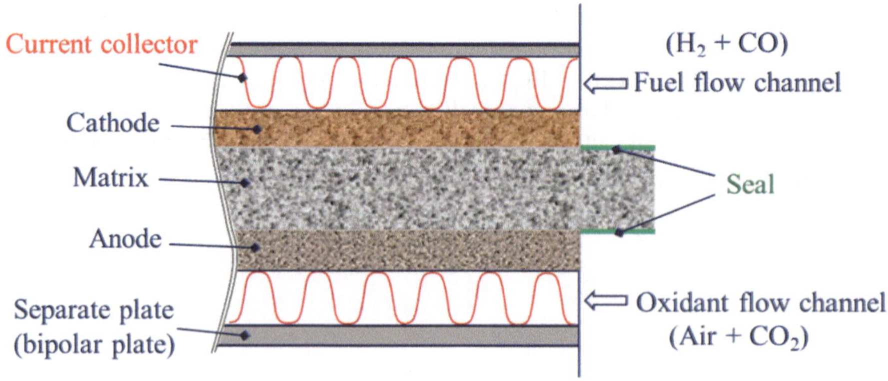

Fig. 15.7 Schematic of MCFC stack construction

The gas seal of the MCFC stack is achieved by compressing the contact area between the electrolyte structure and the extended bipolar plate so that the molten carbonate at operating temperature prevents the gases to permeate through. This technique is called the wet seal. The wet seal simultaneously experiences reducing and oxidizing environments. The wet seal area is the place where there is no need for electronic conductivity. The local bipolar plate area is, therefore, coated with a thin aluminum layer for corrosion protection. The aluminum layer is oxidized to form oxide which is then in situ converted to an insulating $LiAlO_2$ layer after the reaction of aluminum with lithium carbonate. Such a protective layer is only useful for the wet seal area but not for providing corrosion resistance to the electrode area, where the bipolar plate must be electrically conductive.

15.7 Performance and Durability

The thickness of the catalyst layer of MCFC is about 0.2–0.5 mm for the anode and 0.5–1.0 mm for the cathode. The matrix-immobilized electrolyte layer is about 0.5–1.2 mm. These values are significantly larger than those in, e.g., PEMFC. A thick cathode layer is necessary in order to tolerate the dissolution of NiO cathode, while a thick electrolyte layer is to achieve mechanical strength for handling and to prevent the nickel shorting by the precipitated Ni dendrites between the electrodes. It is fortunate that molten carbonates have a high conductivity, and therefore, the cell area-specific resistance remains still acceptable. For example, an electrolyte layer of 1 mm thick with a conductivity of 0.3 S/cm displays an area-specific resistance of 0.33 Ωcm^2. At a current density of 0.2 A/cm^2 this gives an ohmic voltage loss of 66 mV. It is therefore that one of the operational characteristics of MCFC is the low current density, typically in the range of 0.1 to 0.2 A/cm^2 at a single cell voltage of 0.75–0.80 V under ambient operation and of 0.8–0.9 V under a pressure of 10 bar.

The target for the MCFC durability is 40,000 h, which can be translated into an average cell voltage decay rate of 2 mV/1000 hours assuming a lifetime by 10% performance degradation from a beginning-of-the-life voltage of 0.8 V. The MCFC stack technology has demonstrated a performance decay rate of less than 2 mV/1000 h over a period of 10,000-15,000 h. During this so-called linear decay period, the major degradation mechanisms are (1) sintering and growth of anode and cathode catalysts and hence increase in the polarization; (2) increase in the cell ohmic resistance due to the material corrosion and electrolyte loss; and (3) Ni dissolution from the cathode. In the extended period the performance degradation occurs in an accelerated decay mode when the main causes include (1) short circuiting of the reprecipitated Ni from the cathode dissolution and (2) significant electrolyte loss from the matrix due to the particle growth and phase transformation of $LiAlO_2$.

This is illustrated in Fig. 15.8 with the recorded open circuit voltage, cell voltage at 150 mA/cm^2 and the internal ohmic resistance presented by Tanimoto et al. [12]. The test was based on single cells of 256 cm^2 area with detailed parameters given in the figure caption. Table 15.6 lists the postmortem analysis of material changes of

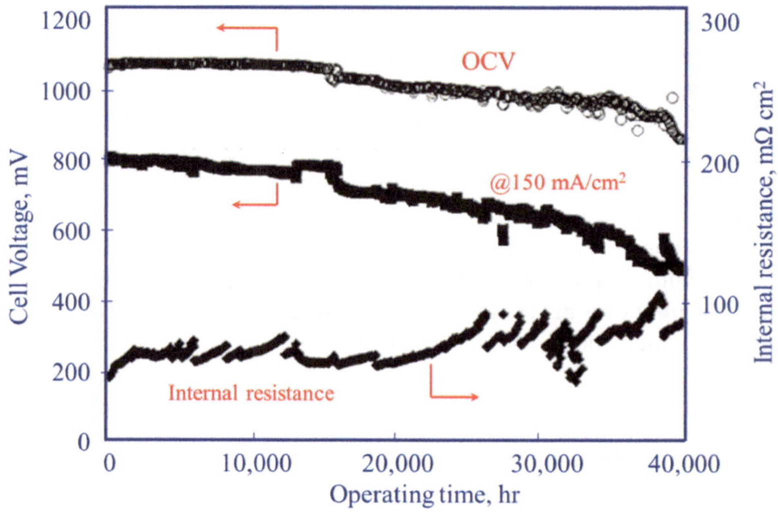

Fig. 15.8 Performance degradation of MCFC in term of open circuit voltage, cell voltage at 150 mA/cm^2, and internal resistance. The cell of 256 cm^2 size was operating at 650 °C under atmospheric pressure. Anode was made of porous Ni–Al operated with $H_2/CO_2 = 80/20$ (dry base) humidified with fuel utilization of 40% at 150 mA/cm^2. Cathode was made of lithiated porous NiO and operated with air/$CO_2 = 70/30$ for first 16,000 h and Air/$CO_2 = 78/22$ from 16,000 to 40,000 h with oxygen utilization of 40% at 150 mA/cm^2. The electrolyte matrix was made of 1.2 mm-thick $LiAlO_2$ filled with $(Li_{0.62}K_{0.38})_2CO_3$. Modified from Ref. [12] with permission from Elsevier, Copyright 1998

Table 15.6 Postmortem analysis results of anode, cathode, and electrolyte matrix changes during the long-term durability test of MCFC shown in Fig. 15.8

Operating time (h)	Anode	Cathode			Matrix	
	Thickness change %	Thickness change %	Particle size (μm)	Dissolved Ni in matrix (mg/cm^2)	Particle size	Content of α phase (%)
0	–	–	1	0	0.1–0.5	0
15,000	1.4	0.5–5.0	2.5	26	0.4–4.0	31
40,000	2.5–4.2	13.6–20.5	2.4	45.9	0.5–8.0	80

the anode, cathode, and matrix during the two long-term tests for 15,000 and 40,000 hours.

The sintering and growth occur for $LiAlO_2$ powders of the electrolyte matrix, NiO powders of the porous cathode, and Ni powders of the metallic anode. These changes lead to decrease in the active area and hence increase in the electrode polarization. At the same time, it causes the changes of the micropore structure of both the electrode and electrolyte matrix, leading to electrolyte loss and poor contact between components. More critical is the significant growth and phase transformation of the $LiAlO_2$ matrix, as seen from Table 15.6. It should be noted that this study was performed with refilling of the molten carbonate during the test. After 15,000 hours the fuel cell voltage and performance cannot be recovered by the electrolyte refilling, indicating the permanent deterioration of the matrix, which is the cause for the fast degradation after the prolonged operation. The dissolution of NiO results in a steady decrease in the cathode thickness, an issue that has to be mitigated. As a comparison the Ni alloy anode showed acceptable creep strength and long-term stability. The used submicron γ-$LiAlO_2$ powder in the matrix is subject to significant particle growth and γ to α phase transformation after the long-term operation. This leads to a pore structure change and loss of the electrolyte.

15.8 Summary

In this chapter we have introduced and discussed the following topics:

- Advantages of MCFC include fuel flexibility, i.e., hydrogen-containing CO from internal reforming of natural gas by using waste heat. Fast electrode kinetics makes it possible to use low-cost catalysts. Challenges originate from the mobile and corrosive electrolyte, which corrodes electrode and construction materials and degrades the mechanical stability and lifetime. The fuel cell power output is commonly below 200 mW/cm^2, limiting the technology for stationary applications in a power range from 300 kW to 3 MW.
- The electrolyte is based on binary (Li/K, Li/Na) or ternary (Li/Na/K) molten alkali metal carbonates. The melt chemistry is influenced by CO_2, H_2O, $O^=$, and metal ions, which in turn impacts ion conductivity, vapor pressure, NiO solubility, wettability, and corrosiveness of the melt. To achieve a long lifetime, the Li/K melt is preferred for ambient pressure operation while the Li/Na is used for high-pressure operation.
- The molten electrolyte is immobilized in a ceramic matrix consisting of fine lithium aluminate ($LiAlO_2$) powders, often with incorporated fibers or coarse particulates for strengthening. The particle growth and the phase transition of the oxide powders are a key mechanism of the electrolyte loss and long-term performance degradation.
- Nickel, alloyed with Cr or/and Al for improving stability, is the most widely employed anode catalyst for natural gas reforming and anode oxidation. Porous nickel oxide is the exclusive cathode material, often *in situ* oxidized from the metal. Lithiation of the oxide in the Li_2CO_3-containing melt enhances its conductivity and stability. NiO has a ppm-level solubility in the electrolyte, leading to reprecipitation of the metal on the anode causing short circuit and fuel cell failure.
- The thickness of the electrolyte matrix limits the ohmic loss. As a result, MCFC is typically operating in the low current density of 0.1 to 0.2 A/cm^2 at a single cell voltage of 0.75 to 0.80 V under ambient operation and of 0.8-0.9 V under a pressure of 10 bar. The technology has been demonstrated on a pilot system level from a kilowatt to megawatt scale. Lifetime operation of pilot plants has shown

promising durability for up to 30,000 h with an electric efficiencies of up to 50 %.

- Significant development work is still required to improve the power density of MCFCs and tackle the materials issues such as cathode dissolution, corrosion of bipolar plates and current collectors, and electrolyte loss due to creep in the matrix and electrodes.

15.9 Questions

15.1. Explain the following terms and materials:

Binary and ternary alkaline metal system
Eutectic points
Li/K and Li/Na molten carbonate systems
α-$LiAlO_2$
Capillary pressure and equilibrium

15.2. What are the technique issues for molten carbonates compared to other hot liquid electrolytes such as potassium hydroxides and phosphoric acid?

15.3. Why is the Li/K melt selected for the ambient MCFC while the Na/Li melt for high-pressure MCFC systems?

15.4. What is the poisoning effect of alkali metals on the internal reforming catalysts? Give examples of techniques to mitigate the poisoning-

15.5. Explain the degradation mechanism of the $LiAlO_2$-based matrix during the MCFC operation.

15.6. MCFC typically operates in a low current density range primarily due to the thick electrolyte layer, which is typically of 0.5-1.2 mm. What are the main considerations preventing further decrease in the electrolyte layer thickness?

References

1. Kacprzak A, Kobylecki R, Wlodarczyk R, Bis Z (2016) Efficiency of non-optimized direct carbon fuel cell with molten alkaline electrolyte fueled by carbonized biomass. J Power Sources 321:233–240
2. Roers GHJ, Ketelaar JAA (1960) High temperature fuel cells. Ind Eng Chem Fundam 52:303–306
3. Kojima T, Miyazaki Y, Nomura K, Tanimoto K (2008) Density, surface tension, and electrical conductivity of ternary molten carbonate system Li_2CO_3-Na_2CO_3-K_2CO_3 and methods for their estimation. J Electrochem Soc 155(7):F150–F156
4. Yuh CY, Huang CM, Farooque M (1997) Advances in carbonate fuel cell matrix and electrolyte. In: Electrochemical Society Proceedings 97(4), 66–78
5. Yuh CY, Selman JR (1992) Porous-electrode modeling of the molten-carbonate fuel-cell electrodes. J Electrochem Soc 139(5):1373–1379

6. Appleby AJ, Nicholson SB (1980) Reduction of oxygen in lithium-potassium carbonate melt. J Electroanal Chem 112(1):71–76
7. Ota K, Mitsushima S, Kato S, Asano S, Yoshitake H, Kamiya N (1992) Solubilities of nickel-oxide in molten carbonate. J Electrochem Soc 139(3):667–671
8. Fukui T, Okawa H, Tsunooka T (1998) Solubility and deposition of $LiCoO_2$ in a molten carbonate. J Power Sources 71(1–2):239–243
9. Passalacqua E, Freni S, Barone F (1998) Alkali resistance of tape-cast SiC porous ceramic membranes. Mater Lett 34(3–6):257–262
10. Yuh C, Hilmi A, Farooque M, Leo T, Xu G (2009) Direct fuel cell materials experience. ECS Trans 17(1):637–654
11. Terada S, Higaki K, Nagashima I, Ito Y (1999) Stability and solubility of electrolyte matrix support material for molten carbonate fuel cells. J Power Sources 83(1–2):227–230
12. Tanimoto K, Yanagida M, Kojima T, Tamiya Y, Matsumoto H, Miyazaki Y (1998) Long-term operation of small-sized single molten carbonate fuel cells. J Power Sources 72(1):77–82

Chapter 16
Protonic Ceramic Oxide Fuel Cells, Microbial Fuel Cells, and Biofuel Cells

In this chapter, some important and emerging fuel cells are briefly introduced. This includes protonic ceramic oxide fuel cells (PCFCs), microbial fuel cells (MFCs), and biofuel cells (BFCs).

16.1 Protonic Ceramic Oxide Fuel Cells

In addition to oxygen ion conducting oxides, certain ceramic oxides such as acceptor-doped $Ba(Zr,Ce)O_3$ can conduct protons in the presence of a proton source (such as steam) and can be used as the proton-conducting electrolyte for fuel cells. Since the first report of proton conductive behavior of cerate based perovskites like $SrCeO_3$ in the presence of water at high temperatures in 1981 by Iwahara et al. [1], there have been significant developments in the fundamental understanding of proton-conducting oxides [2, 3] and protonic ceramic oxide fuel cell (PCFC) technologies [4–6]. Proton conductors are attractive for low and intermediate temperature range fuel cells (300–600 °C) due to the high mobility of hydrogen ions which are small in the ionic size and have a lower activation energy for conduction, as compared to the oxide conduction. The development of low and intermediate temperature protonic ceramic fuel cells (PCFCs) would significantly simplify the sealing and probably reduce the cell degradation. Another advantage of PCFCs is that the steam produced at the cathode side does not dilute the fuel and thus benefits the operation with high fuel utilization. Proton-conducting oxides offer the promising potential and alternatives in the energy conversion and storage including fuel cells for electricity generation, electrolysis for hydrogen production, electrochemical reactor for fuel production, and ammonia synthesis.

S. P. Jiang and Q. Li, *Introduction to Fuel Cells*,
https://doi.org/10.1007/978-981-10-7626-8_16

16.1.1 Operation Principle of Protonic Ceramic Oxide Fuel Cells

In PCFC, hydrogen is oxidized at the anode, forming protons and releasing electrons:

$$H_2 \rightarrow 2H^+ + 2e^- \tag{16.1}$$

The formed protons will transfer through the electrolyte to the cathode where oxygen reduction reaction (ORR), and water formation reaction (WFR) occur simultaneously, producing water at the cathode:

$$2H^+ + \frac{1}{2}O_2 + 2e^- \rightarrow H_2O \tag{16.2}$$

The electrons flow from the anode to the cathode to do the electrical work. The overall reaction for a hydrogen fuel cell based on a proton-conducting electrolyte is the same as that of oxygen ion conducting solid oxide fuel cells, except that water is produced at the cathode:

$$H_{2(\text{anode})} + \frac{1}{2}O_{2(\text{cathode})} \rightarrow H_2O_{(\text{cathode})} \tag{16.3}$$

Figure 16.1 shows the operation principle of protonic ceramic fuel cells.

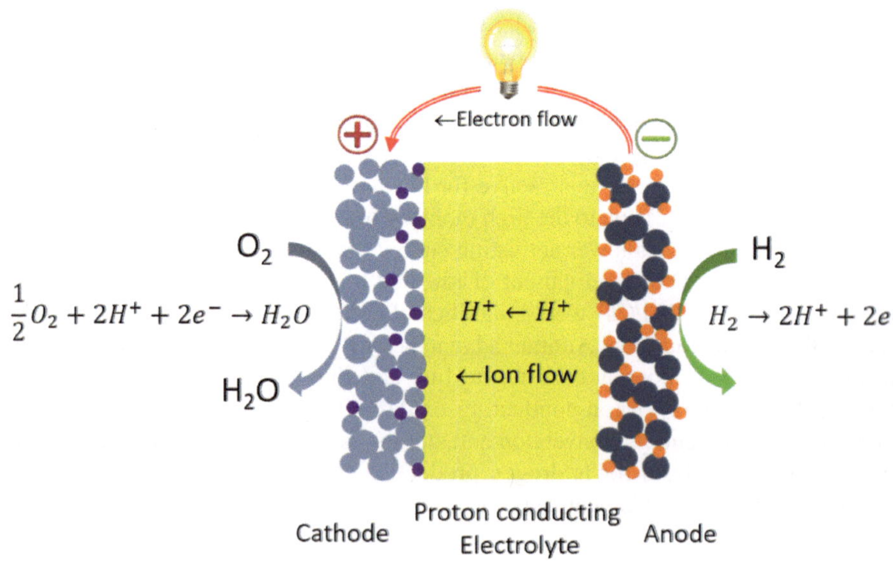

Fig. 16.1 Operating principle of a PCFC

These reactions are the same as PAFC and PEMFC that are also based on proton-conducting electrolytes. However, the proton carrier in PCFCs is different from that in low-temperature PEMFCs. In PEMFCs, the charge carrier is a proton associated with water molecules, i.e., hydrated proton, H_3O^+, while in PCFCs, the charge carrier is a proton associated with an oxide ion as positively charged hydroxide defects (acid–base reaction), i.e., protons are incorporated into the electrolyte through the hydration of oxygen vacancies as shown below:

$$H_2O + V_O^{\bullet\bullet} + O_O^x \rightleftharpoons 2OH_O^{\bullet} \tag{16.4}$$

where $OH_O^{\bullet}$ represents a hydroxide defect or protonic defect at a lattice oxygen site. This is due to the fact that protons are not native to the lattice of ceramic oxides. A water molecule dissociates into a hydroxide ion and a proton; the former fills an oxygen vacancy while the latter can form a covalent bond with lattice oxygen. However, it has been known that protons do not occupy regular interstitial position in the lattice since they generally interact with the electronic cloud of neighboring oxygen ions, forming as hydroxide defects. Thus, the hydration of oxygen vacancies in Eq. (16.4) is critical in determining the proton concentration of the material, which is a function of temperature and in the case of incomplete hydration, a function of water vapor partial pressure. The hydration equilibrium constant and resulting proton concentration can be expressed as:

$$k_{\text{hydration}} = \frac{\left[OH_O^{\bullet}\right]^2}{\left[V_O^{\bullet\bullet}\right] \times \left[O_O^x\right] \times p_{H_2O}} \tag{16.5}$$

$$\left[OH_O^{\bullet}\right] = k_{\text{hydration}}^{1/2} \times \left[V_O^{\bullet\bullet}\right]^{1/2} \times \left[O_O^x\right]^{1/2} \times p_{H_2O}^{1/2} \tag{16.6}$$

$k_{\text{hydration}}$ is realted to the Gibbs free energy of hydration. As the lattice oxygen concentration is a bulk property, the concentration of O_O^x can be taken as unity, $[O_O^x] = 1$. The above equation can then be written as:

$$\left[OH_O^{\bullet}\right] = k_{\text{hydration}}^{1/2} \times \left[V_O^{\bullet\bullet}\right]^{1/2} \times p_{H_2O}^{1/2} \tag{16.7}$$

This indicates that the hydroxide defect concentration increases with the partial pressure of water vapor and concentration of oxygen vacancies. This also implies that a reasonable concentration of oxygen vacancies is required for measurable proton concentration or proton conductivity. Similar to the oxide ion conductivity of oxide ionic conductors, proton conduction is also a temperature activated process and the relationship of proton conductivity with temperature obeys the Arrhenius equation in water-containing atmosphere:

$$\sigma_{OH_O^{\bullet}} = \sigma_{OH_O^{\bullet}}^0 \exp\left(-\frac{E_a}{RT}\right) \tag{16.8}$$

where $\sigma_{OH^{\bullet}}$ is the pre-exponential factor and E is the activation energy. The activation energy of various doped $BaZrO_3$ is in the range of 0.4-0.6 eV. At high temperatures, the incorporation of molecular oxygen into oxygen vacancies with the concomitant formation of electron holes also occurs, which reduces the concentration of oxygen vacancies and thus the proton conductity. Another consideration is that water adsorption is an exothermic reaction, favorable at relatively low temperatures. Thus, the proton uptake or concentration in perovskite oxides increases with decreasing temperatures and follows a decreasing trend with increasing temperature. This explains why the proton-conduction mechanism is dominant at low temperatures and oxygen ionic conduction prevails at high temperatures. Proton concentration is generally measured through thermogravimetric analysis (TGA) by correlating the weight increase from water uptake to the proton concentration.

The formed hydroxide defects connect with the lattice oxygen via hydrogen bonding and create O–H•••O chains in the perovskite structure. The proton conduction occurs by proton hopping along the O–H•••O chains via a Grotthus-type mechanism [2]. A quantum molecular dynamics simulation on the dynamics of $BaCeO_3$ lattice structure has shown that the proton transport involves hopping of protons from one oxygen ion to another by breaking the weak O–H bonds (i.e., the Grotthuss mechanism) as well as rotational motion of the hydroxyl unit [7]. Proton diffusion involves the orientation or motion of a proton towards a neighboring oxygen atom and the jump of the proton towards the neighboring oxygen atom. It is generally considered that the rate-determining step in the proton conduction is the proton transfer, i.e., the proton jump, and the proton rotational diffusion or motion is a relatively fast process with lower activation energy.

16.1.2 Proton-Conducting Oxides

The most common proton-conducting oxides are ABO_3 cubic perovskites due to the minimum distortion and large crystallographic unit cell volume, which is beneficial to proton mobility and proton formation through hydration. Examples include $BaCeO_3$, $SrCeO_3$ and $BaZrO_3$ and can be doped with various oxides including samarium, neodymium and ytterbium to increase the proton conductivity and stability. For example, in the $BaCeO_3$ system, substituting Ce^{4+} with M^{3+} cations (e.g., Y^{3+}) is electrically compensated by the formation of oxygen vacancies.

$$Y_2O_3 \xrightarrow{BaCeO_3} 2Y'_{Ce} + V_O^{\bullet\bullet} + 3O_O^x \quad (16.9)$$

Doped $BaCeO_3$ and $SrCeO_3$ show a very good proton conductivity among the high-temperature proton conductors; for example, the conductivity of $BaCe_{0.95}Yb_{0.05}O_{3-\delta}$ in wet air reached 10^{-2} S/cm at 800 °C. However, they are not stable in the CO_2 and H_2O environment. $BaCeO_3$ can form hydroxides in the presence of water vapor and carbonates in the presence of CO_2.

$$BaCeO_3 + H_2O \rightarrow Ba(OH)_2 + CeO_2 \quad (16.10)$$

$$BaCeO_3 + CO_2 \rightarrow BaCO_3 + CeO_2 \quad (16.11)$$

The chemical stability of barium cerates can be improved by doping. Matsumoto et al. studied the dependence of the conductivity and chemical stability of $BaCeO_3$-based proton conductors on the size of trivalent dopants and found that the doping at the ceria site in $BaCeO_3$ with large ionic size dopants increase the stability [8]. On the other hand, proton-conducting perovskite zirconates such as $BaZrO_3$ and $SrZrO_3$ exhibit a better chemical stability in water vapor and CO_2 atmosphere. The total conductivity of pure $BaZr_{0.8}Y_{0.2}O_3$ (BZY20) prepared from fine BZY powder with large grain size of ~ 1 μm is as high as 0.01 S/cm at 450 °C [9]. However, the sinterability of BZY is very poor, requiring sintering temperatures of 1600–2200 °C [10].

One strategy to form a proton conductor with high conductivity and chemical stability is to manipulate the solid solutions of $BaZrO_3$ and $BaCeO_3$ as they are mutually soluble. Increasing Ce in $Ba(Zr,Ce)O_3$ would improve the proton conductivity and sinterability, but decrease the chemical stability in CO_2 and water vapor. On the other hand, adding Zr would increase the chemical stability at the cost of proton conductivity and sinterability. Thus, a solid solution of $BaZrO_3$ and $BaCeO_3$ would result in a tradeoff between two B-site cations, Ce and Zr, in terms of the conductivity, sinterability and chemical stability. The tradeoff is closely related to the nature of Ce and Zr cations, and one of the key parameters is the basicity of B-site cations in the determining proton uptake [11]. Ce is more basic than Zr. Thus, increasing Ce (proton-accepting) cation will result in basic oxygen sites and facilitate proton incorporation. This increases the solubility and stability of protonic defects, $OH_O^{\bullet}$, in the lattice. On the other hand, the increase in the basic oxide would increase the activity with acidic species such as CO_2, leading to the reduced chemical stability.

By further manipulation and doping of B-site cations, significant improvement in the proton conductivity, sinterability and chemical stability of $Ba(Zr,Ce)O_3$-based electrolyte materials have been reported. In both barium cerates and barium zirconates, Y has been found to be the best B-site acceptor dopant among a host of candidates such as Sc, Yb, Gd, Nd, etc. Yang et al. [12] studied in details the conductivity behavior of $Ba(Ce_{0.7}Zr_{0.1}Y_{0.2})O_{3-\delta}$ (BZCY) and $Ba(Ce_{0.7}Zr_{0.1}Y_{0.1}Yb_{0.1})O_{3-\delta}$ (BZCYYb7111), and the results indicate that partial replacing Y with Yb increased the conductivity at the cost of reduced sinterability. The conductivity of BZCYYb7111 is 1.41 × 10^{-2} S/cm, 1.63 × 10^{-2} S/cm and 2.14 × 10^{-2} S/cm in air, H_2 and wet H_2, respectively, at 600 °C, higher than that measured on BZCY. The results indicate that the BZCY and BZCYYb are excellent low-temperature proton and oxygen ion mixed conductors. Further development shows that the increase in the Zr content would enhance the stability and the cell with $Ba(Ce_{0.4}Zr_{0.4}Y_{0.1}Yb_{0.1})O_{3-\delta}$ (BZCYYb4411) electrolyte and $PrBa_{0.5}Sr_{0.5}Co_{1.5}Fe_{0.5}O_{5+\delta}$ (PBSCF) cathode achieved peak power densities of 500 mW/cm^2 at 500 °C with very good stability in CO_2 [13].

One of the fabrication challenges of zirconate-based protonic ceramics is the poor sinterability to form dense electrolyte membrane, requiring sintering at temperatures as high as 1700 °C. Sintering at high temperatures would lead to barium evaporation which increases the grain boundary impedance, causing a decrease in the proton conductivity. One of the most effective approaches to improve the sinterability is the use of sintering aids. Tao and Irvine used ZnO as the sintering aid and effectively reduced the sintering temperature of $BaZr_{0.8}Y_{0.2}O_{3-\delta}$ (BZY20) from 1700 to 1325 °C [14]. Similar results were reported on $BaCe_{0.5}Zr_{0.3}Y_{0.16}Zn_{0.04}O_{3-\delta}$ (BCZYZn) which can be sintered to dense at 1250 °C when it is prepared by a soft-chemical method [15]. High ionic conductivity has been reported for $BaCe_{0.5}Zr_{0.3}Y_{0.16}Zn_{0.04}O_{3-\delta}$ and $BaZr_{0.1}Ce_{0.7}Y_{0.1}Yb_{0.1}O_{3-\delta}$, exceeding 10^{-2} S/cm at 600 °C. Tong et al. studied effect of adding NiO as the sintering aid on the conductivity and electrochemical performance of BZY20, using a solid-state reactive sintering (SSRS) process [16]. In this process, raw materials of $BaCO_3$, CeO_2, and Y_2O_3 were mixed with less than 2 wt.% NiO as the sintering aid, and fully densified BCY20 ceramic pellets were obtained at a sintering temperature of 1400 °C for 12 h. Microstrcutre analysis indicates accumulated nanosized nickel particles in grain boundary regions of the NiO-modified BCY20 pellets. NiO-modified BZY electrolyte cells showed significantly improved conductivity and outstanding cell performance, reaching 455 mW/cm^2 at 500 °C [17].

However, as shown by Li et al. [18], a majority of studies show that sintering aids decrease the ionic conductivity and ionic transfer number, thus reducing the electrochemical performance of proton-conducting oxides. In most cases, sintering aids remain as a residue in the material, either in the grain boundary region or entering into proton-conducting perovskite structure, altering the proton conductivity properties of the proton-conducting oxide materials. The presence of sintering aids in the form of oxide nanoparticles in the grain boundaries decreases the hydration ability of BZY and reduces the proton uptake. One way to reduce the detrimental effect of sintering aids is to exsolve the sintering aids from the lattice of the proton-conducting oxides under fuel-reducing environment and to recover the conductivity. Nevertheless, the best option is to densify the proton-conducting electrolytes without the addition of sintering aids if technologically possible.

In addition to cerates and zirconates, some other oxides such as $Ba_2SnYO_{5.5}$ and niobate- and tanalate-based proton-conducting oxides (e.g., $La_{0.99}Ca_{0.01}NbO_4$, $La_{0.99}Ca_{0.01}TaO_4$) have also been investigated as proton-conducting electrolyte materials for PCFCs.

Proton-conducting oxides are also found to be active promotors to increase the activity and tolerance of anodes toward sulfur and carbon deposition in hydrocarbon fuels (see Chap. 10).

16.1.3 Electrode Materials for PCFCs

In principle, electrode materials used for SOFCs can be adopted for PCFCs. However, there are distinct differences between the electrode materials used for SOFCs and PCFCs due to the involvement of transportation and reaction of protons at the electrode and electrolyte interfaces.

16.1.3.1 Cathode

In addition to the general criteria of cathode materials in high-temperature SOFCs, the water formation at the cathode side imposes additional requirements on cathode materials of PCFCs. Cathode materials need to be chemically stable and thermally compatible with electrolyte under high water partial pressures. Due to the simultaneous ORR and WFR at the cathode, oxide materials for the PCFC cathode also require the oxide ion and electronic conductivity as well as proton conductivity, i.e., the *triple conducting oxide* (TCO). Without proton conductivity or with poor proton conductivity, the reaction will be limited to the interface between cathode-and proton-conducting electrolyte. The water formed at the interface could cause delamination if the formed water cannot be removed quickly. The TCO is important to extend the electrochemical reaction sites to the bulk of the electrode effectively.

Different to the ORR in SOFCs, the reduced oxygen ions are combined with proton defects, forming water. Thus, the reaction can occur at the electrode surface (i.e., the surface pathway) and/or the three phase boundary (i.e., TPB pathway), depending on the conductivity properties of the oxide electrodes [5]. Similar to that derived for ORR in SOFCs, the reaction can be written as follows.

Surface pathway:

Step 1. Adsorption of an oxygen molecule at the active sites or adsorption sites and dissociation into adsorbed oxygen atoms:

$$O_2 + 2\text{ad} \xrightarrow{\text{adsorption and dissociation}} 2O_{\text{ad, surface}} \tag{16.12}$$

Step 2. Transport of protons through electrode bulk to the electrode surface:

$$OH^{\bullet}_{O,\,\text{electrode bulk}} \rightarrow OH^{\bullet}_{O,\,\text{surface}} \tag{16.13}$$

Step 3. Charge transfer and water formation at the electrode surface (charge transfer step):

$$O_{\text{ad, surface}} + 2OH^{\bullet}_{O,\,\text{surface}} + 2e^- \rightarrow H_2O + 2O^{x}_{O} \tag{16.14}$$

TPB pathway:

Step 1. Adsorption of an oxygen molecule at the active sites or adsorption sites and dissociation into adsorbed oxygen atoms:

$$O_2 + 2ad \xrightarrow{\text{adsorption and dissociation}} 2O_{ad,\ surface} \quad (16.15)$$

Step 2. Migration of adsorbed oxygen from surface to TPB:

$$O_{ad} \rightarrow O_{ad,\ TPB} \quad (16.16)$$

Step 3. Transport of protons from electrolyte to TPB:

$$OH^{\bullet}_{O,\ electrolyte} \rightarrow OH^{\bullet}_{O,\ TPB} \quad (16.17)$$

Step 4. Charge transfer and water formation at TPB (charge transfer step):

$$O_{ad,\ TPB} + 2OH^{\bullet}_{O,\ TPB} + 2e^- \rightarrow H_2O + 2O^x_O \quad (16.18)$$

Figure 16.2 shows the reaction pathways of ORR on different cathode materials. For electronic conducting and MIEC electrodes, the reaction is limited to the TPB; however, the bulk diffusion of oxygen species is much faster in the case of MIEC cathode as compared to the surface diffusion-dominant electronic-conducting oxides. The presence of proton conductivity of the oxide materials will extend the reactive sites to the surface of the electrode, significantly enlarging the active areas for the reaction. However, in practice, protonic-conducting electrolytes are generally mixed oxygen ion, proton and *p*-type electronic conductors, especially at temperatures higher than 500–600 °C. In this case, the reaction sites could also extend to the surface of MIEC cathode, similar to ORR in the case of SOFCs.

As shown by Eq. (16.4), the oxygen vacancy plays an important role in the proton uptake reaction from water. MIEC materials such as $La_{0.6}Sr_{0.4}Co_{0.2}Fe_{0.8}O_{3-\delta}$ (LSCF), $Ba_{0.5}Sr_{0.5}Co_{0.8}Fe_{0.2}O_{3-\delta}$ (BSCF), $La_{0.6}Sr_{0.4}CoO_{3-\delta}$ (LSC), $PrBaCo_2O_{5+\delta}$ (PBCO) or $PrBa_{0.5}Sr_{0.5}Co_{1.5}Fe_{0.5}O_{5+\delta}$ show protonic properties in certain degree and have been explored as cathodes of PCFCs. In general, the highly oxygen-deficient oxides like BSCF, LSC, and PBCO endow the high hydration properties and more active for the ORR and water formation reaction of PCFCs [6].

One strategy to develop the single-phased TCO is to increase the oxygen ion and electrical conductivity of proton conductors such as $BaCeO_3$, $SrCeO_3$, and $BaZrO_3$ by heavy doping of transition metal ions such as Co and Fe. Shang et al.

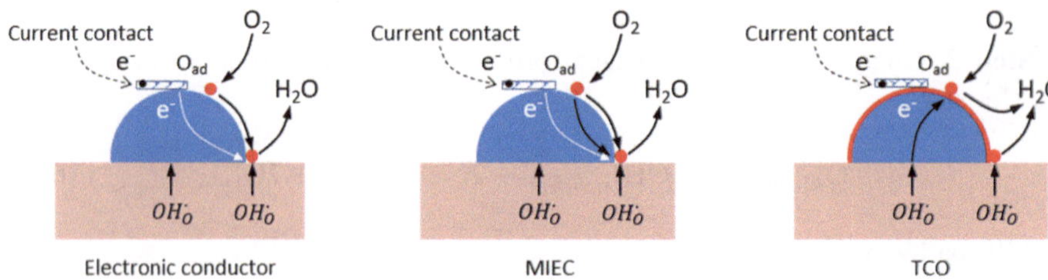

Fig. 16.2 Reaction pathways of ORR on different cathode materials

doped $BaZrO_3$ with Co and Fe, $BaCo_{0.4}Fe_{0.4}Zr_{0.2}O_{3-\delta}$ (BCFZ), as cathode materials for PCFCs and observed promising electrochemical performance and activity on $BaZr_{0.1}Ce_{0.7}Y_{0.1}Yb_{0.1}O_{3-\delta}$ (BZCYYb) electrolyte [19]. The activation energy of BCFZ cathode for ORR is 76 kJ/mol, which is significantly lower than that of the LSCF cathode (138 kJ/mol). This indicates that the proton conductivity of the TCO cathode can significantly enhance the kinetics of the ORR and WFR of PCFCs. The basicity of the oxide ions in particular the B-site cations is a key factor in proton concentration and uptake and in barium-based perovskites, highest proton concentration of 10% at 250 °C was identified for $Ba_{0.95}La_{0.05}Fe_{0.8}Zn_{0.2}O_{3-\delta}$ [11].

Similar to the composite cathodes in SOFCs, the TCO can also be made by forming composite cathodes of MIEC oxides and proton-conducting electrolyte generally with a weight ratio of 50/50. Examples of composite cathodes include $Sm_{0.5}Sr_{0.5}CoO_{3-\delta}/BaZr_{0.1}Ce_{0.7}Y_{0.2}O_{3-\delta}$ (SSC/BZCY), LSCF/$BaCe_{0.9}Yb_{0.1}O_{3-\delta}$ (LSCF/BCYb), LSCF/$BaZr_{0.7}Pr_{0.1}Y_{0.2}O_{3-\delta}$ (LSCF/BZPY), etc. Composite cathodes create additional TPB for the reactions, similar to that in the case of SOFCs.

16.1.3.2 Anode

The requirement of PCFCs for anode materials is similar to that of SOFCs. For the same reasons as described in SOFCs, Ni is the most common and active anode material for the hydrogen oxidation reaction of PCFCs. To increase the microstructure stability and thermal compatibility between the Ni anode and proton-conducting electrolyte, Ni/proton conductor cermets are commonly used as anodes of PCFCs, similar to the situation of Ni/YSZ cermet anodes in SOFCs. For example, Ni/BZCYYb with a weight ratio of 50/50 is commonly used as anode in BZCYYb proton-conducting oxide electrolyte cells. However, the use of proton-conducting oxide in the cermet anode makes the relatively basic surface of the proton-conducting oxide phase in the anode. This explains the high carbon deposition tolerance and sulfur resistance of Ni/BZY cermet anodes tested in a wide range of hydrocarbon fuels [17].

16.1.4 Other Applications of Proton-Conducting Oxides

In addition to power generation, PCFCs can also be applied in other areas including high-temperature water electrolysis, ammonia synthesis, etc.

16.1.4.1 Protonic Ceramic Electrolysis Cells

PCFCs can be used for water electrolysis by reversal operation, protonic ceramic electrolysis cells (PCECs). With proton-conducting oxide electrolyte, the principle the water electrolysis reaction can be written by the following reactions.

Air electrode reaction (anode reaction):

$$H_2O \rightarrow \frac{1}{2}O_2 + 2H^+ + 2e^- \quad (16.19)$$

Hydrogen electrode reaction (cathode reaction):

$$2H^+ + 2e^- \rightarrow H_2 \quad (16.20)$$

The overall reaction is the water electrolysis reaction:

$$H_2O \rightarrow H_2 + \frac{1}{2}O_2 \quad (16.21)$$

The low to intermediate temperature operation (400–600 °C) of PCECs offers the advantages of fast kinetics of the reaction as compared to low-temperature PEMFCs and of substantially reduced thermal integration and degradation issues associated with high-temperature SOECs. As oxygen evolution reactions occur at the surface of oxygen electrode in PCFCs, the common problems of electrode delamination in the case of SOECs can be avoided. Unlike SOECs, PCFCs produce high-purity hydrogen. Due to the fact that majority of proton-conducting materials are mixed ion conductor, PCECs suffer from low faraday efficiency, which is attributed to the electronic leaking across the electrolyte and parasitic transport losses.

16.1.4.2 Ammonia Synthesis

Both oxygen ion conductor and proton conductor have been employed for electrochemical ammonia synthesis. In the case of PCECs, high-purity hydrogen or protons are supplied to the fuel electrode, thus avoiding the interference of water with the electrocatalytic synthesis reaction. The electrochemical nitrogen reduction to ammonia is based on the assumption that electrochemically produced protons would be active and shift the reaction toward the ammonia production. Nitrogen hydrogenation is exothermic and tends to be thermodynamically favorable at low temperatures; however, experimentally, the ammonia production is kinetically slow at low temperature and atmosphere pressure and peaks at temperatures of 300–400 °C. Thus, PCECs have the thermodynamic as well as kinetic advantages for electrochemical ammonia synthesis.

$$N_2 + 3H_2 \rightarrow 2NH_3 \quad (16.22)$$

The production rate of ammonia by PCFCs is still low, e.g., ~ 10^{-9} mol/s.cm^2 at ~ 450–500 °C [6]. The challenge is the fundamental understanding and role of the electrochemical produced proton in the ammonia synthesis via Faradaic or non-Faradaic effects.

16.1.5 Proton Conductivity Measurement

16.1.5.1 Cell Configuration and EMF Measurement

The proton conduction or conductivity of protonic conducting materials is generally studied through the measurement of electromotive force (EMF) by control of the gas composition and partial pressure of hydrogen or water vapor at cell compartments, I and II, see Fig. 16.3. Ceramic specimens can be prepared by conventional solid state reaction (SSR) method or other powder synthesis methods (see Chap. 12) and densified at high temperatures. Total conductivity is measured by EMF of hydrogen concentration cells to estimate the proton conductivity. Air, oxygen, hydrogen or mixture gas of H_2/Ar and water vapor at 1 atm or selected partial pressures can be fed to both compartments. Gases can be humidified by passing through a water bottle at room temperature, i.e., wet gases. Water vapor content can also be introduced via a precision water micron pump. Dry gas can be obtained by passing the gas through phosphorus pentoxide powder. Water vapor partial pressure and the partial pressure of oxygen or hydrogen can be adjusted by changing the mixing ratio of the gases, e.g., O_2, air, Ar and H_2. Hydrogen permeation flux through the ceramic specimen can be detected by a gas chromatograph. Ionic and electronic conductivities are generally measured by electrochemical impedance spectroscopy technique.

To distinct the protonic and oxide ionic conductivity, EMF measurements can be conducted in four different types of concentration cells as shown in Fig. 16.3 (in the figure subscript I and II denotes the cell compartment I and II) [20].

- Case 1) - H_2 concentration cell with $p_{H_2(I)} \neq p_{H_2(II)}$ and $p_{H_2O(I)} = p_{H_2O(II)}$. $p_{O_2(I)} \neq p_{O_2(II)}$ is established by thermodynamics

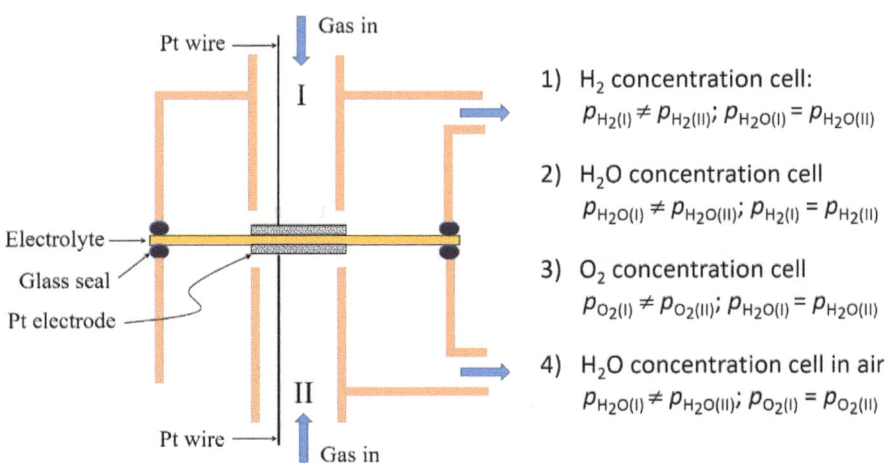

Fig. 16.3 Schematic diagram of experimental setup for the hydrogen permeation and proton conductivity measurements, showing four types of concentration cells

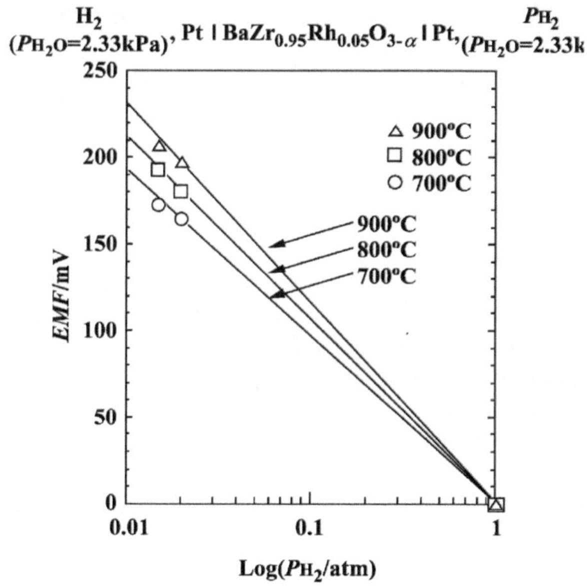

Fig. 16.4 Plots of EMF of hydrogen concentration cell as a function of the logarithm of partial pressure of hydrogen on a $BaZr_{0.95}Rh_{0.05}O_{3-\delta}$ electrolyte disk sample at different temperatures. Reproduced from Ref. [20] with permission from Elsevier, Copyright 2002

- Case 2) - H_2O vapour concentration cell with $p_{H_2O(I)} \neq p_{H_2O(II)}$ and $p_{H_2(I)} = p_{H_2(II)}$. $p_{O_2(I)} \neq p_{O_2(II)}$ is established by the thermodynamics
- Case 3) - O_2 concentration cell with $p_{O_2(I)} \neq p_{O_2(II)}$ and $p_{H_2O(I)} = p_{H_2O(II)}$. $p_{O_2(I)} \neq p_{O_2(II)}$ is established by the thermodynamics
- Case 4) - H_2O vapor concentration cell in air with $p_{H_2O(I)} \neq p_{H_2O(II)}$ and $p_{O_2(I)} = p_{O_2(II)}$. $p_{H_2(I)} \neq p_{H_2(II)}$ is established by the thermodynamics

In the case of H_2-containing atmosphere (case 1, Fig. 16.3), the EMF of the concentration cell using protonic conducting electrolyte materials can be written by the Nernst equation (see Chap. 9):

$$E_o = \frac{RT}{2F} \ln \frac{p_{H_2(II)}}{p_{H_2(I)}} \tag{16.23}$$

If partial pressure of water vapor in both cell compartments is the same, $p_{H_2O(I)} = p_{H_2O(II)}$, the EMF of the cell becomes identical to that measured across an oxide ionic conductor:

$$E_o = \frac{RT}{4F} \ln \frac{p_{O_2(I)}}{p_{O_2(II)}} = \frac{RT}{4F} \ln \frac{(p_{H_2O(I)})^2 (p_{H_2(II)})^2}{(p_{H_2O(II)})^2 (p_{H_2(I)})^2} = \frac{RT}{2F} \ln \frac{p_{H_2(II)}}{p_{H_2(I)}} \tag{16.24}$$

This indicates that the EMF of a hydrogen concentration cell includes contribution of the protonic and oxide ionic conductivities. On the other hand, for water concentration cell (case 2, Fig. 16.3), the EMF of a protonic conductor is zero as there is no hydrogen partial pressure gradient. The EMF would be due to the differences of the partial pressure of oxygen due to $p_{H_2O(I)} \neq p_{H_2O(II)}$:

$$E_o = \frac{RT}{2F} \ln \frac{p_{H_2O(I)}}{p_{H_2O(II)}} \quad (16.25)$$

Figure 16.4 shows the results of EMF of a hydrogen concentration cell measured on $BaZr_{0.95}Rh_{0.05}O_{3\text{-}\delta}$ electrolyte disk sample at difference temperature [20]. In this hydrogen concentration cell, $p_{H_2O(I)} = p_{H_2O(II)} = 2.33$ kPa (wet H_2), $p_{H_2} = 1\ atm$ at reference compartment II and gas mixtures of H_2 and Ar are fed to compartment I to control partial pressure of hydrogen. The measured EMF (E) matches the theoretical EMF calculated for the Nernst equation (Eo). The ratio of measured EMF to the theoretical calculated one, E/E_o, is the ionic transport number of the ceramic material, t_i.

16.1.5.2 Hydrogen Permeation and Protonic Conductivity

EMF gives the estimation of total conductivity, σ_T, and for the measurement of protonic conductivity, we need to conduct hydrogen permeation measurements. The hydrogen permeation cell is similar to the hydrogen concentration cell. For example, H_2/Ar mixtures with given partial pressures of hydrogen are fed to the cell compartment (I) as the H_2 supply with $p_{(H_2,supp)}$, and argon gas is fed to cell compartment II to sweep the hydrogen evolved in the compartment, producing $p_{(H_2,perm)}$. Both supply and sweeping gases are humidified at room temperature with constant partial pressure of water vapor and fed at a constant flow rate. The amount of permeated hydrogen in the Ar sweep gas is analyzed by gas chromatography.

Experimentally the change in the water vapor partial pressure can be minimized and thus the change of partial pressure of water vapor on the hydrogen permeation flux is minor. Thus, the hydrogen flux, J_{H_2}, can be expressed as [21]:

$$J_{H_2} = \frac{RT}{4F^2\delta} \times \frac{(\sigma_{H^+} + \sigma_{O^=})(\sigma_{e^-} + \sigma_{h^\bullet})}{\sigma_T} \ln \frac{(p_{H_2,supp})}{(p_{H_2,perm})} \quad (16.26)$$

where δ is the thickness of the ceramic oxide specimen. Hydrogen flux is proportional to the logarithm of the hydrogen partial pressure ratio. The linear relation between the hydrogen flux and the logarithm of the hydrogen partial pressure ratio indicates that the hydrogen permeation occurs via the ionic-electronic mixed conduction in the ceramic oxide specimens. Due to the difference in the H_2/H_2O ratios at cell compartments, different partial pressure of oxygen at cell compartments can occur, causing the potential profile of oxygen across the specimen. Such oxygen potential difference can cause oxide ion flux from the hydrogen permeation side to the hydrogen supply side. However, the fluxes of proton and oxide ions can be quantitatively measured by the conductivity measurement. The contribution of proton and oxide ion conductivity can also be distinguished by the measurement of the water vapor formation at the anode and cathode side of the ceramic oxide electrolyte specimen under fuel cell operation conditions [22].

16.2 Microbial Fuel Cells

16.2.1 Microbial Fuel Cell Configuration

Microbial fuel cells (MFCs) are fuel cells that utilize the biocatalytic capabilities of various bacteria and microorganisms for converting the chemical energy stored in range of organic fuel sources such as glucose and organic matters in wastewaters into electricity [23]. The MFC concept was first demonstrated by Potter in 1912 using living cultures of *Escherichia coli* and *Saccharomyces* and Pt electrodes [24]. Similar to any fuel cells, a typical MFC consists of anode and cathode chambers separated by a proton exchange membrane. In anodic compartment, the microorganisms oxidize the substrates to produce electrons and protons, of which the electrons are transferred to an external electric load and then to the cathode to complete the electric circuit. At the same time, protons are transported to the cathode compartment through the membrane. In the cathode chamber, oxygen is primarily used as oxidant due to its abundance and high reduction potential. The reaction at the cathode is the oxygen reduction reaction (ORR), forming water, similar to that in PEMFCs. The main difference between an MFC and conventional fuel cells is that in an MFC the anode uses microbes rather than noble metals or oxides as catalysts to convert chemical energy into electricity. The microbes work as combinations of multiple enzymes to catalyze the oxidation of substrate such as glucose and pass electrons to the electrode. Substrates used in MFCs also include sugars, organic acids such as acetate and complex polymers such as starch and cellulose. Domestic, industrial, and animal waste streams can be used as feedstock for generating electricity in MFCs. The microbes can self-regenerate, and thus, the cost of this kind of catalyst is much lower than precious platinum. Marine sediment, soil, wastewater, freshwater sediment, and activated sludge are all rich sources for the microorganisms in MFCs.

There are different designs of MFCs for different applications. Figure 16.5 shows typical dual-chamber and single-chamber configurations of MFCs. The dual-chamber model is commonly used for fundamental study and initial evaluations in laboratories. The dual-chamber test cell is easy to build by simply connecting two chambers separated by a proton exchange membrane such as Nafion. The two chambers can be directly pressed onto either side of the membrane and clamped together to decrease the electrode distance and thus lowering the internal resistance. In some cases, ceramic membrane and salt bridges, which consists of a U-shaped glass tube filled with agar and salt, can also be used to replace the proton exchange membrane and connect the two chambers. However, the power of a salt-bridged MFC is very low

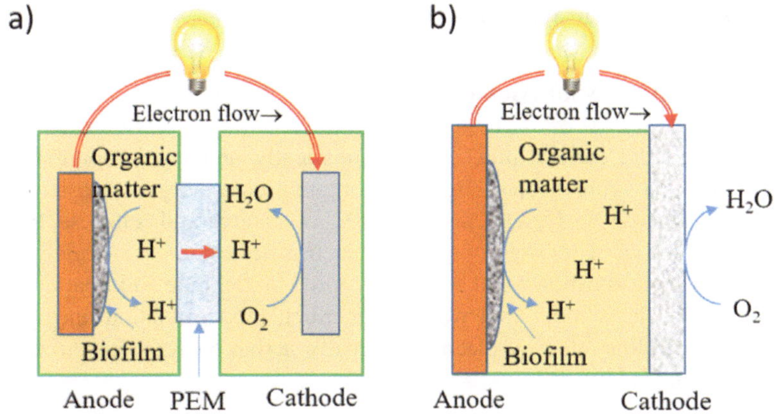

Fig. 16.5 Configurations of MFCs. **a** two-chamber MFC and **b** single-chamber MFC

due to their high internal resistance. There are various designs for the two-chamber MFCs, including conventional rectangular-shaped MFCs, cylindrical-shaped MFCs, miniature MFCs, up-flow mode configured MFCs and flat plate MFCs. Two-chamber MFCs are difficult to scale up due to the complex architectural designs.

MFC can also be in a single-chamber configuration [25]. Single-chamber MFCs are simple in design and eliminate the aqueous-cathode chamber by directly exposure of the cathode to the air (Fig. 16.5b). The protons produced at the anode are directly transferred from the electrolyte to the porous air–cathode, forming water outside of the cell. The single-chamber MFC significantly reduces the internal resistance and improves the power density. The single-chamber MFC has been used to generate electricity from wastewater and degrade or remove specific pollutants such as azo dye, Congo red, ceftriaxone sodium in the wastewater.

Many factors may affect the overall performance of MFC including the type of the utilized microorganism and substrates to be oxidized, materials of anode and cathode, the nature of the proton exchange membrane, and configurations of MFCs. The main application of current MFC technologies is to recovering energy from the biodegradation of organic-rich waster by utilizing wastewater as a fuel source. MFCs offer a significant advantage over other renewable energy sources (i.e., solar and wind) as they can be part of the wastewater treatment but without restriction of geographical location and seasonal change. However, both the cell voltage and the power density of MFCs are far below that microorganisms are theoretically capable of producing. The maximum power densities on MFCs reported in the literature vary from 2.34 mW/m^2 to 3300 mW/m^2 (2.34 $\times$ 10^{-4} mW/cm^2 to 0.33 mW/cm^2) [23].

16.2.2 Catalysis in Microbial Fuel Cell Anode

16.2.2.1 Biocatalysts for Anode

In the MFC anodic compartment, electrochemically active microorganisms can oxidize organic substrates such as sugars and organic acids/inorganic waster (e.g., the fuel), such as glucose ($C_6H_{12}O_6$), acetate ($C_2H_4O_2$), starch and cellulose, to release electrons to the anode. Thus, microorganisms such as bacteria can generate electricity by utilizing organic substrates and at the same time biodegrade and treat biodegradable products such as municipal wastewater. Such biodegradation and oxidation reactions can be illustrated by electrochemically active bacteria for the oxidation reaction of acetate [26]:

$$CH_3COO^- + 4H_2O \rightarrow 2HCO_3^- + 9H^+ + 8e^- \quad (16.27)$$

Microorganisms capable of donating electrons are termed as *exoelectrogens* and those of accepting electrons as *exoelectrontrophs*. The performance of the anode mainly depends on the biocatalysts—the microbes. Unlike other chemical catalysts, the reaction rate of substrate oxidation is not the criterion to evaluate the microbes. The ability to utilize the electrons generated by the reaction from cells to the electrode is more important. Generally, because the cell wall and plasma membrane of microbes are non-conductive, the electrons produced in cells need carriers to move them out to reach the electrode surface. One solution is via the electron mediators. Artificial electron mediators are small and biocompatible molecules that can easily access both cells and electrodes, working as external electron acceptors of the electron transfer chain and rapidly oxidizing mediators on the electrode surface to enhance the kinetics of anodic reactions. Examples of mediators are neutral red (NR) or anthraquinone-2,6-disulfonate (AQDS). Employing mediators could lower the anode overpotential and increase the current density; therefore, they were widely used in early stage MFCs. This cyclic process or *mediated electron transfer* (MET) facilitates the transfer of electrons and thus increases the power output. Microorganisms that do not require an exogenous redox mediator to transfer electrons to an electrode are important for the development of mediator-less MFCs. One of them is Fe(III)-reducing bacterium *Rhodoferax ferrireducens*, which had been isolated from anoxic subsurface sediment on Oyster Bay [27]. *Rhodoferax ferrireducens* can oxidize glucose to CO_2 and quantitatively transfer electrons to graphite electrodes without a redox mediator.

Based on the electron transfer pattern between microorganism cells and electrode, there are two main types of anodes. In type I, the anode receives electrons through artificial electron mediators, while in type II, electroactive microorganisms directly transfer electrons from electron donors to the anode. The electrocatalytic performance of the anode is mainly dependent on the electron transfer process between microorganisms and the anode.

In MFCs, the fuel sources, e.g., biodegradable organic maters, are replenished either intermittently or continuously. MFCs are generally operated under anaerobic conditions in order to facilitate the growth of obligatory anaerobic bacteria with electron transfer properties, e.g., *Geobacter sulfurreducens.*

Model bacteria currently used in MFCs are iron-reducing species such as *Shewanella* spp and *Geobacter* spp [28]. These bacterial species are capable of degrading organic matter for nutrient cycling.

16.2.2.2 Anode Materials

Reduction of the activation energy is the key to improve the reaction kinetics of the anode. The anode materials in MFCs must also be conductive, biocompatible and chemically stable in the reactor solution. Metal anodes consisting of non-corrosive stainless steel mesh can be utilized, but copper is not useful due to the toxicity of even trace copper ions to bacteria.

Carbon materials are the most versatile electrode materials in MFCs due to their high conductivity, good biocompatibility, low cost, and easy handling. For non-modified carbon-based anodes, electrocatalytic capability depends on the oxidation of microorganisms on the anode. Increasing the surface area of anode is an efficient and widely utilized approach, and various strategies have been used to make carbon materials with a large electrochemically active surface area. The most favorable materials are carbon cloth and carbon paper, both of which are carbon-fiber-based porous materials. Although these porous carbon materials can enhance the power and current density, the clogging of the material can be a problem, especially in saturated flow systems. To resolve this problem, Logan et al. [29] developed a new type of brush electrode that bundled large-surface-area graphite fibers into a central core. When they examined this brush anode in cube and bottle-air–cathode MFCs inoculated with wastewater, they found that the cube MFCs achieved a maximum power density of 2400 mW/m^2 or 73 W/m^3 based on liquid volume. This brush architecture could be scaled up for use in larger reactors.

However, because the electrocatalytic activity of carbon is very low, carbon-based materials simply serve as current collectors accepting the electrons from the biofilm. In addition, they cannot provide a greater number of active reaction centers because their active surface area is too small. Carbon nanomaterials such as graphene and carbon nanotubes (CNTs) possess many advantages such as flexible microstructure, high specific surface areas, high electromobility and conductivity, high physical strength at room temperatures. CNTs have attracted considerable attention for potential applications as supporting materials for immobilizing enzymes or bacterial species owing to their unique electrical conductivity, high chemical stability, as well as biocompatibility; however, a key issue in such systems is the efficient electron transfer between the active centers and the supporting substrate or electrode.

16.2.3 Catalysis in Microbial Fuel Cell Cathode

Cathodic catalysis, including the cathodic electron acceptors and the catalytic activity of the cathodic materials, greatly affects MFC performance. The kinetics of the reduction reaction and the standard redox potential of the oxidants determine the performance of the cathode.

16.2.3.1 Abiotic or Aerobic Cathode

The reaction at the cathode compartment can be classified into aerobic or anaerobic reactions depending on the source of the final electron acceptor available. In aerobic cathode, oxygen is the most suitable electron acceptor. Two processes could occur during the cathodic oxygen reduction: a complete reduction in a four-electron pathway or an incomplete two-electron reduction reaction via the formation of hydrogen peroxide intermediates that needs further reduction to H_2O.

$$O_2 + 4H^+ + 4e^- \rightarrow 2H_2O \quad E^0 = +1.229\,V \tag{16.28}$$

$$O_2 + 2H^+ + 2e^- \rightarrow H_2O_2 \quad E^0 = +0.695\,V \tag{16.29}$$

The sluggish kinetics of ORR is the main limitation of the power output of MFCs. To increase the rate of oxygen reduction, highly active catalysts such as Pt catalysts are usually used. However, for practical applications, the high cost of Pt would make the MFCs very expensive. Non-Pt materials such as pyrolyzed iron(II) phthalocyanine have been used as MFC cathodes. Alternative cathodes to replace oxygen are explored in MFCs. For example, the $Fe(CN)_6^{3-}$ ion is a good electron acceptor under anaerobic conditions and has been used in two-chamber MFCs [30].

$$Fe(CN)_6^{3-} + e^- \rightarrow Fe(CN)_6^{4-} \quad E^0 = +0.361\,V \tag{16.30}$$

The reduction kinetics of $Fe(CN)_6^{3-}$ ion are much more rapid than ORR, obviating the need to use catalysts in ferricyanide cathode. The peak power density of MFC using a ferricyanide cathode is increased compared to that obtained using an oxygen-dissolved cathode. The reason for this increase is that mediators that undergo reversible redox reactions (e.g., ferricyanide) can also reduce the cathodic overpotential in MFCs. However, the the use of ferricyanide cathode is not sustainable in practice since $Fe(CN)_6^{3-}$ ion must be chemically regenerated and is potentially toxic. The cathode compartment can also be maintained under anaerobic conditions. In this case, microorganisms transfer the electrons from the cathode to the final electron accepter, e.g., nitrate. Nevertheless, the performance of biocatalysts for the cathodic reaction is constrained by high activation overpotentials.

To further increase the MFC cell voltage (normally 0.5 ~ 0.8 V), permanganate can be used as the cathodic electron acceptor, which has a higher standard redox potential than oxygen. In acidic conditions, permanganate accepts three electrons and is thus reduced to manganese dioxide as follows:

$$MnO_4^- + 4H^+ + 3e^- \rightarrow MnO_2 + 2H_2O \quad E^0 = +1.70\,V \tag{16.31}$$

The permanganate cathode (pH 3.6) increases the MFC open cell voltage to 1.382 V and thus can deliver a higher power density than that produced by the ferricyanide cathode or the oxygen cathode.

Carbon-based materials such as graphene and carbon nanotubes have been extensively explored as electrode materials and catalyst supports in low temperature fuel cells. Thus, similar to the development of non-precious metal catalysts for cathodes, iron and nitrogen functionalized grapheme-based electrocatalysts have also been used as cathodes in MFCs with enhanced cell performance.

16.2.3.2 Biocatalysts for Cathode

The use of catalysts and electron acceptors in MFCs can be expensive due to the need to constantly replenish exhausted materials. Some microorganisms can also be used as catalysts in the cathode to perform functions similar to those that they perform in the anode. They are also known as biocathodes. Both aerobic and anaerobic biocathodes have been explored. The first research conducted on biocathodes was by Rhoads et al. [31], who employed a cycle of Mn(IV) reduction on a cathode catalyzed by manganese-oxidizing bacteria (MOB). As the oxygen was required for the biocatalyzed re-oxidation of Mn(II), manganese dioxide actually served as the mediator for the bacteria-catalyzed oxygen reduction.

$$MnO_{2(s)} + 4H^+ + 2e^- \rightarrow Mn^{2+} + 2H_2O \tag{16.32}$$

$$Mn^{2+} + O_2 \xrightarrow{MOB} MnO_2 \tag{16.33}$$

Similarly, iron compounds can be used as mediators in biocatalyzed oxygen reduction on MFC cathodes. Heijne et al. [32] designed an MFC with a ferric-iron-mediated biocathode catalyzed by *Acidithiobacillus ferrooxidans*. In mediator-less biocathodes, bacteria take electrons directly from the cathode and catalyze the reduction reaction of oxidants. *Geobacter metallireducens* not only exhibited superior electrocatalytic performance in the anode but also catalyzed the reduction of nitrate and fumarate in the cathode [33]. Despite the pure strains, mixed microorganism communities, such as those found in sewage or waste water, have also been used as biocatalysts in MFC cathodes [34]. The oxidant often used in the cathode of these mixed-cultures-catalyzed MFCs is nitrate, which has relatively high redox potential. Hence, their denitrification function is the major advantage of using these biocathodes [35, 36].

$$2NO_3^- + 10e^- + 12H^+ \rightarrow N_2 + 6H_2O \quad E^0 = +0.74\,V \tag{16.34}$$

16.2.4 Electron Transfer Mechanism

Exoelectrogenic bacterial species have the ability to facilitate electron transfer via two mechanisms; direct electron transfer (DET) and indirect electron transfer (IET). The DET requires a physical connection between the bacterial cell and the electrode surface, i.e., nanowires and/or redox active proteins. The presence of redox-active proteins allows for short-range electron transfer to take place. On the other hand, the IET relies on electron shuttling molecules or mediators. Thus, the IET is also called mediated electron transfer or MET.

Isolated bacterial colonies in the anode chamber of MFCs are not capable of direct transferring electrons to the electrode. However, some special bacterial species, i.e., anodophiles, have the ability to use electrons as their end electron acceptor, resulting in the generation of an electrical charge. For example, that the surface of specifically isolated *Shewanella* spp. and *Geobacter* spp. bacterial have micrometer long proteinaceous filaments that extend from their outer surface into the extracellular matrix. These proteinaceous filaments are involved in the extracellular electron transfer processes and are also termed microbial nanowires. Nanowires have the ability to partake in DET and can be either flagella or pili with distinct properties. One of the best examples is *G. sulfurreducens* with abundant micro-nanowires (i.e., pili). This bacterial has the ability to reduce metals such as iron oxide and donates electrons to the anode via electrically conductive pili, generating electrical charge in a MFC setup.

Bacteria species can generate electricity due to the production of secondary metabolites, which are able to act as endogenous redox mediators, i.e., electron shuttles or mediators. Mediators are organic molecules with a low molecular weight with ability to catalyze both oxidation and reduction reactions. Examples of effective mediators include phenazines and quinones and compounds like thionine, methylene blue, humic acid, methyl viologen, anthraquinone-2,6-disulfonic acid, ferrocene monocarboxylic acid and 2-hydroxy-1,4-naphtoquinone. Different to electrically conductive pili, mediators do not require the direct contact between the bacterial cell and the electrode (i.e., the electron acceptor). However, in MET using electroactive molecules or mediators to shuttle electrons between the bacterial species or enzyme and electrode, the maximum cell voltage of MFCs is determined by the thermodynamic redox potential of mediators. A more positive redox potential is required in order to provide the driving force for the electron transfer between enzyme active center and electrode for the oxidation biocatalysts, which contributes to cell voltage loss.

The electron transfer process is also affected by the properties of the support materials such as CNTs. The direct electrochemistry of redox enzymes/proteins such as glucose oxidase (GO_X) plays an important role in the glucose oxidation reactions in enzymes immobilized electrodes in MFC and and enzymatic fuel cells [37]. In the case of GO_X, DET with the bare electrode is difficult due to the fact that the redox active flavin adenine dinucleotide (FAD) cofactors, i.e., small nonproteinaceous electroactive species, of GO_X are deeply buried within electrically well-insulated prosthetic shells. Liu et al. showed that the number of walls or inner tubes of CNTs plays an important role in promoting the electron transfer of the electrochemical redox reaction on the surface of CNTs, and such promotion effect is beneficial to the DET of the redox reactions of enzymes/proteins [38]. The GO_X immobilized by the CNTs maintains its biocatalytic activity toward glucose. However, the direct electrochemistry and electrocatalytic activity of GO_X depend strongly on the nature of CNTs supports, showing distinct volcano curves as a function of number of walls. The results indicate that triple-walled CNTs (TWCNTs) facilitate fast DET of GO_X and enhance the electrocatalytic activity for glucose oxidation reaction, as compared to that on single-walled and multi-walled CNTs. The excellent promotion effect of TWCNTs is ascribed to the fast DET via the efficient electron tunneling between the outer wall and inner tubes of CNTs. The electron transfer process for the glucose oxidation reaction via the immobilized $GO_X$is shown schematically in Fig. 16.6.

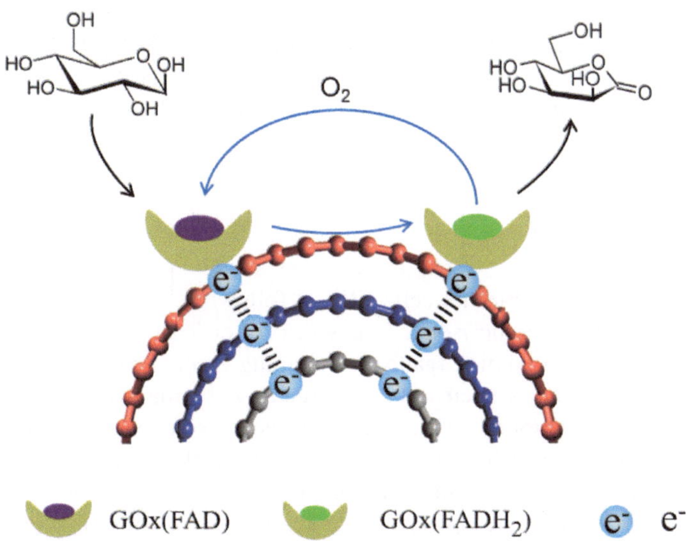

Fig. 16.6 Scheme of the DET and electrocatalytic activity of GO_X immobilized on TWCNTs via electron tunneling mechanism between the outer wall and inner tubes of TWCNTs. Reproduced from ref [38] under Creative Commons CC BY license

16.2.5 Challenges in MFCs

Power generation of MFCs is still very low (e.g., in the range of several hundred $\mu W/cm^2$ level). There are significant energy barriers to the anodic oxidation of substrates and cathodic reduction of oxygen reactions in MFCs, in addition to the significantly internal ohmic resistance losses and concentration losses. The key parameter is the rate of electron transfer to the anode and the electrochemical activities of the materials. Other factors that limit the cell performance and stability include biofouling and excess growth of biofilms, which in turn lead to electrode surface blockage, production of non-conductive debris and eventually a reduction in power output. The challenges in MFC technologies are also related to the high material cost associated with the electrode materials and the use of proton exchange membranes and the limited life spans.

MFCs are particularly suitable for powering small telemetry systems and wireless sensors that require only low power to transmit signals (e.g., temperature) to receivers in remote areas. In combination with wastewater treatment process, the power produced by MFCs could potentially halve the electricity needed in a conventional treatment process. However, in addition to the challenges of the need in the improving efficiency and long-term stability of the MFC stack and systems, the success of specific MFC application in wastewater treatment will depend on the concentration of and biodegradability of the organic matter in the influent, the wastewater temperature, and the absence of toxic chemicals.

16.3 Biofuel Cells

Biofuel cells or enzyme biofuel cells (BFCs) follow the same operating principle of MFCs. BFCs use specific isolated enzymes for at least part of their operation, while MFCs utilize the whole organism containing complete enzyme pathways. In BFCs, enzymes are used as electrocatalysts to catalyze biomass or biofuel substrate to produce electricity, or biological fluids are used as fuel sources for the electrical activation of implantable electronic medical devices or prosthetic aids. The development of enzyme-based cells (or enzymatic biofuel cells) began in the 1960s and received extensive attention due to the high turnover rates associated with enzymes that lead to a high biocatalysis rate. Enzymes such as bilirubin oxidase can have higher catalytic activity and lower overpotential than the Pt catalysts [39]. Enzymes allow for a wide variety of fuels to be utilized without the need for expensive and time-consuming purification because enzymes typically do not react with or get passivated by impurities in the fuel. The selectivity of the enzymes also allows for the elimination of the PEMs used to separate cathodic and anodic solutions by simultaneous use of a biocathode and a bioanode.

Similar to MFCs, the spatial separation of the biocatalytic redox sites from the electrode prevents the electrical contacting of the enzyme with the electrode (the dimension of redox proteins are in the range of 7–20 nm). Thus, the electrical communication between the redox centers of enzymes and electrodes needs to be established. Mediators have also been employed in BFC systems through polymerization on the electrode surface prior to enzyme immobilization, co-immobilization of enzyme and mediator simultaneously or simply allowing the mediator to be free in solution. Many low molecular weight redox-active compounds and polymers have been incorporated to mediate the electron transfer, including organic dyes such as methylene green, phenazines and azure dyes along with other redox active components such as ferrocene, ferrocene derivatives and conducting salts. The problems associated with the mediated system are that mediators have short lifetimes or are often not biocompatible. However, some enzymes are capable of DET via the active site of the enzyme. DET occurs through the ability of enzymes to act as a "molecular transducer" that converts the chemical signal to an electrical one through the transfer of charge to a stable redox species which is in turn capable of transferring this charge to another molecule or electrode surface. Many of these enzymes contain redox-active metal centers that perform the catalytic electron transfer. Enzymes such as laccase can catalyze the four-electron reduction of O_2 to water through electron transfer from the electrode surface directly to the active sites and through to the substrate.

However, DET is correlated directly to the enzyme proximity and orientation to the electrode surface in order for electron tunneling to occur, allowing only the biocatalytic reaction to be the limiting process. These issues can be addressed by implementing immobilization techniques such as crosslinking, wiring, sandwich, and encapsulation. The term "wiring" refers to chemically binding or attracting the enzyme of interest to the anode or cathode surface through covalent or ionic interactions in such a way that there is an electron pathway to the electrode. The structure of carbon electrodes has significant effect on the cell performance. When laccase was immobilized on superdispersed colloid graphite and acetylene black electrodes, the rate of ORR per enzyme molecule at the graphite electrode was found to be five times higher than that at the acetylene black electrode [40].

The development of BFCs for practical applications is a field which is still in its infancy, and there is much potential for further improvement. Limitations to the system efficiency include incomplete oxidation of fuels, short lifetimes and reduced performance due to slow DET and/or problems associated with the stability or thermodynamics of redox mediator. Maintaining the integrity and performance of the three-dimensional protein structure of both the enzyme active sites and the macromolecules as a whole requires accurate control of temperature, pH, and chemical components of the solvent environment. Moehlenbrock and Minteer [41] summarized the advances made in the immobilization and stabilization of enzymes at biocathodes and bioanodes of BFCs, and Willner et al. [42] gave an excellent review on the methods of constructing the integrated anode and cathode units in BFCs. The major challenge is to increase the power output of BFCs.

16.4 Summary

In this chapter, we have introduced and discussed the following topics:

- PCFC, MFC, and BFC are the emerging fuel cell technologies. PCFC has some inherent advantages of potentially high proton conductivity and low activation energy due to the small size of protons and high fuel utilization due to the steam production at the cathode side. The main attraction of PCFC is its low and intermediate temperature operation capability.
- Proton conductivity is crucial for the performance of PCFC, similar to that of low-temperature PEMFC. However, there is a significant difference between the proton carrier of PEMFC and PCFC. In PCFC, the charge carrier is a proton associated with an oxide ion as positively charged hydroxide defects. This is due to the fact that protons are not native to the lattice of oxide materials. Thus, mixed oxide and proton conductivity is a common phenomena of protonic ceramic electrolyte; nevertheless, the proton conductivity prevails at lower temperatures.
- $BaCeO_3$- and $BaZrO_3$-based cubic perovskites are the dominant ceramics of proton-conducting ceramic electrolyte materials, and the proton conductivity and stability can be significantly enhanced by doping with different oxides such as samarium, neodymium, and ytterbium. The typical examples are $Ba(Ce_{0.7}Zr_{0.1}Y_{0.2})O_{3-\delta}$ (BZCY) and $Ba(Ce_{0.7}Zr_{0.1}Y_{0.1}Yb_{0.1})O_{3-\delta}$. One challenge in the fabrication of $BaCeO_3$- and $BaZrO_3$-based protonic ceramic electrolytes is the high sintering temperature and poor densification process.
- The cathode reaction in PCFC involves simultaneously the ORR and water formation reaction. Therefore, the ideal cathode electrode for PCFC should be oxide ionic, protonic, and electronic conductive, i.e., triple conducting oxide (TCO). This is very different from the MIEC requirement of SOFC cathode. Composite approach by combination of different functional components appears to be the best option for the development of TCO.
- The most distinguish advantage of MFC is the utilization of the biocatalytic capabilities of various bacteria and microorganisms to convert the chemical energy stored in range of organic fuel sources such as glucose and organic matters in wastewaters into electricity. Here, microbes rather than precious metals are used as catalysts to convert chemical energy to electricity. Thus, the early driving force for the development of MFC is to combining MFC with wastewater treatment plants to treat wastewater and at the same time to produce electricity for the plant use. However, due to the inherently slow and high-energy barrier for the electron donating and accepting properties of microorganism, the charge transfer rate of microbes is extremely slow, resulting in very low-power performance of MFC. BFC has similar issue of low-power output (e.g., in the range of several hundred $\mu W/cm^2$). The grand challenges in both MFC and BFC are to substantially increase the power density for practical applications.

16.5 Questions

16.1. Explain the following terms and materials:

Protonic carrier
$BaCeO_3$, $SrCeO_3$ and $BaZrO_3$
Sintering aids
Triple conducting oxides
Exoelectrogens and exoelectrontrophs
Mediated electron transfer
Direct electron transfer and indirect electron transfer
Glucose oxidase.

16.2. What are the similarities and differences between the charge carriers and proton conductivity behavior between low temperature PEMFCs and PCFCs?

16.3. What are the similarities and differences of oxygen reduction reaction in SOFC and PCFC? Why the TCO is important for the ORR in PCFC?

16.4. Compare the advantages and disadvantages of SOFCs and PCFCs.

16.5. Give examples of direct electron transfer and indirect electron transfer of the oxidation reactions in MFCs or BFCs.

16.6. Discuss the prospective of utilization of MFC in wastewater treatment plants.

References

1. Iwahara H, Esaka T, Uchida H, Maeda N (1981) Proton conduction in sintered oxides and its application to steam electrolysis for hydrogen-production. Solid State Ionics 3–4:359–363
2. Norby T (1999) Solid-state protonic conductors: principles, properties, progress and prospects. Solid State Ionics 125(1–4):1–11
3. Kreuer KD (2003) Proton-conducting oxides. Ann Rev Mater Res 33:333–359
4. Fabbri E, Pergolesi D, Traversa E (2010) Materials challenges toward proton-conducting oxide fuel cells: a critical review. Chem Soc Rev 39(11):4355–4369
5. Duan CC, Huang JK, Sullivan N, O'Hayre R (2020) Proton-conducting oxides for energy conversion and storage. Appl Phys Rev 7(1):011314
6. Kim J, Sengodan S, Kim S, Kwon O, Bu Y, Kim G (2019) Proton conducting oxides: a review of materials and applications for renewable energy conversion and storage. Renew Sust Energ Rev 109:606–618
7. Munch W, Seifert G, Kreuer KD, Maier J (1996) A quantum molecular dynamics study of proton conduction phenomena in $BaCeO_3$. Solid State Ionics 86–8:647–652
8. Matsumoto H, Kawasaki Y, Ito N, Enoki M, Ishihara T (2007) Relation between electrical conductivity and chemical stability of $BaCeO_3$-based proton conductors with different trivalent dopants. Electrochem Solid State Lett 10(4):B77–B80
9. Yamazaki Y, Hernandez-Sanchez R, Haile SM (2009) High total proton conductivity in large-grained yttrium-doped barium zirconate. Chem Mat 21(13):2755–2762

10. Duval SBC, Holtappels P, Vogt UF, Pomjakushina E, Conder K, Stimming U, Graule T (2007) Electrical conductivity of the proton conductor $BaZr_{0.9}Y_{0.1}O_{3-\delta}$ obtained by high temperature annealing. Solid State Ionics 178(25–26):1437–1441
11. Zohourian R, Merkle R, Raimondi G, Maier J (2018) Mixed-conducting perovskites as cathode materials for protonic ceramic fuel cells: understanding the trends in proton uptake. Adv Funct Mater 28(35):1801241
12. Yang K, Wang JX, Xue YJ, Sen Wang M, He CR, Wang Q, Miao H, Wang WG (2014) Synthesis, sintering behavior and electrical properties of $Ba(Zr_{0.1}Ce_{0.7}Y_{0.2})O_{3-\delta}$ and $Ba(Zr_{0.1}Ce_{0.7}Y_{0.1}Yb_{0.1})O_{3-\delta}$ proton conductors. Ceram Int 40 (9):15073–15081
13. Choi S, Kucharczyk CJ, Liang YG, Zhang XH, Takeuchi I, Ji HI, Haile SM (2018) Exceptional power density and stability at intermediate temperatures in protonic ceramic fuel cells. Nat Energy 3(3):202–210
14. Tao SW, Irvine JTS (2006) A stable, easily sintered proton-conducting oxide electrolyte for moderate-temperature fuel cells and electrolyzers. Adv Mater 18(12):1581
15. Zhang XL, Jin MF, Sheng JM (2010) Layered $GdBa_{0.5}Sr_{0.5}Co_2O_{5+\delta}$ delta as a cathode for proton-conducting solid oxide fuel cells with stable $BaCe_{0.5}Zr_{0.3}Y_{0.16}Zn_{0.04}O_{3-\delta}$ delta electrolyte. J Alloy Compd 496(1–2):241–243
16. Tong JH, Clark D, Bernau L, Subramaniyan A, O'Hayre R (2010) Proton-conducting yttrium-doped barium cerate ceramics synthesized by a cost-effective solid-state reactive sintering method. Solid State Ionics 181(33–34):1486–1498
17. Duan CC, Kee RJ, Zhu HY, Karakaya C, Chen YC, Ricote S, Jarry A, Crumlin EJ, Hook D, Braun R, Sullivan NP, O'Hayre R (2018) Highly durable, coking and sulfur tolerant, fuel-flexible protonic ceramic fuel cells. Nature 557(7704):217
18. Li J, Wang C, Wang XF, Bi L (2020) Sintering aids for proton-conducting oxides—a double-edged sword? a mini review. Electrochem Commun 112:106672
19. Shang M, Tong JH, O'Hayre R (2013) A promising cathode for intermediate temperature protonic ceramic fuel cells: $BaCo_{0.4}Fe_{0.4}Zr_{0.2}O_{3-\delta}$. Rsc Advances 3(36):15769–15775
20. Shimura T, Esaka K, Matsumoto H, Iwahara H (2002) Protonic conduction in Rh-doped $AZrO_3$ (A=Ba, Sr and Ca). Solid State Ionics 149(3–4):237–246
21. Matsumoto H, Shimura T, Higuchi O, Tanaka H, Katahira K, Otake T, Kudo T, Yashiro K, Kaimai A, Kawada T, Mizusaki J (2005) Protonic-electronic mixed conduction and hydrogen permeation in $BaCe_{0.9-x}Y_{0.1}Ru_xO_{3-\alpha}$. J Electrochem Soc 152(3):A488–A492
22. Iwahara H, Mori T, Hibino T (1995) Electrochemical studies on ionic-conduction in Ca-doped $BaCeO_3$. Solid State Ionics 79:177–182
23. Slate AJ, Whitehead KA, Brownson DAC, Banks CE (2019) Microbial fuel cells: an overview of current technology. Renew Sust Energ Rev 101:60–81
24. Potter MC (1912) Electrical effects accompanying the decomposition of organic compounds. Proc R Soc Lond B 84:260–276
25. Liu H, Ramnarayanan R, Logan BE (2004) Production of electricity during wastewater treatment using a single chamber microbial fuel cell. Environ Sci Technol 38(7):2281–2285
26. Logan BE, Hamelers B, Rozendal RA, Schrorder U, Keller J, Freguia S, Aelterman P, Verstraete W, Rabaey K (2006) Microbial fuel cells: methodology and technology. Environ Sci Technol 40(17):5181–5192
27. Finneran KT, Johnsen CV, Lovley DR (2003) Rhodoferax ferrireducens sp. nov., a psychrotolerant, facultatively anaerobic bacterium that oxidizes acetate with the reduction of Fe(III). Int J Syst Evol Microbiol 53 (Pt 3):669–673.
28. Logan BE, Regan JM (2006) Electricity-producing bacterial communities in microbial fuel cells. Trends Microbiol 14(12):512–518
29. Logan B, Cheng S, Watson V, Estadt G (2007) Graphite fiber brush anodes for increased power production in air-cathode microbial fuel cells. Environ Sci Technol 41(9):3341–3346
30. Bard AJ, Faulkner LR (2001) Electrochemical methods-fundamentals and applications, 2nd edn. Wiley, New York
31. Rhoads A, Beyenal H, Lewandowski Z (2005) Microbial fuel cell using anaerobic respiration as an anodic reaction and biomineralized manganese as a cathodic reactant. Environ Sci Technol 39(12):4666–4671

32. Ter Heijne A, Hamelers HVM, De Wilde V, Rozendal RA, Buisman CJN (2006) A bipolar membrane combined with ferric iron reduction as an efficient cathode system in microbial fuel cells. Environ Sci Technol 40(17):5200–5205
33. Gregory KB, Bond DR, Lovley DR (2004) Graphite electrodes as electron donors for anaerobic respiration. Environ Microbiol 6(6):596–604
34. Clauwaert P, Rabaey K, Aelterman P, De Schamphelaire L, Ham TH, Boeckx P, Boon N, Verstraete W (2007) Biological denitrification in microbial fuel cells. Environ Sci Technol 41(9):3354–3360
35. Lefebvre O, Al-Mamun A, Ng HY (2008) A microbial fuel cell equipped with a biocathode for organic removal and denitrification. Water Sci Technol 58(4):881–885
36. Virdis B, Rabaey K, Yuan ZG, Rozendal RA, Keller J (2009) Electron fluxes in a microbial fuel cell performing carbon and nitrogen removal. Environ Sci Technol 43(13):5144–5149
37. Liu Y, Du Y, Li CM (2013) Direct electrochemistry based biosensors and biofuel cells enabled with nanostructured materials. Electroanalysis 25(4):815–831
38. Liu YX, Zhang J, Cheng Y, Jiang SP (2018) Effect of carbon nanotubes on direct electron transfer and electrocatalytic activity of immobilized glucose oxidase. ACS Omega 3(1):667–676
39. Mano N, Fernandez JL, Kim Y, Shin W, Bard AJ, Heller A (2003) Oxygen is electroreduced to water on a "wired" enzyme electrode at a lesser overpotential than on platinum. J Am Chem Soc 125(50):15290–15291
40. Tarasevich MR, Bogdanovskaya VA, Kapustin AV (2003) Nanocomposite material laccase/dispersed carbon carrier for oxygen electrode. Electrochem Commun 5(6):491–496
41. Moehlenbrock MJ, Minteer SD (2008) Extended lifetime biofuel cells. Chem Soc Rev 37(6):1188–1196
42. Willner I, Yan YM, Willner B, Tel-Vered R (2009) Integrated enzyme-based biofuel cells-a review. Fuel Cells 9(1):7–24